T0376689

Modeling in Fluid Mechanics

Mechanics

Instabilities and Turbulence

Modeling in Fluid Mechanics

Instabilities and Turbulence

Igor Gaissinski
Vladimir Rovenski

CRC Press
Taylor & Francis Group
Boca Raton London New York

CRC Press is an imprint of the
Taylor & Francis Group, an **informa** business

A CHAPMAN & HALL BOOK

CRC Press
Taylor & Francis Group
6000 Broken Sound Parkway NW, Suite 300
Boca Raton, FL 33487-2742

Printed on acid-free paper
Version Date: 20180517

International Standard Book Number-13: 978-1-138-50683-1 (Hardback)

Library of Congress Cataloging-in-Publication Data

Names: Gaissinski, Igor, author. | Rovenskii, Vladimir Y., 1953- author.
Title: Modeling in fluid mechanics : instabilities and turbulence / Igor Gaissinski and Vladimir Rovenski.
Description: Boca Raton : CRC Press, Taylor & Francis Group, 2018. | Includes bibliographical references and index.
Identifiers: LCCN 2018000547| ISBN 9781138506831 (hardback : alk. paper) | ISBN 9781351029063 (ebook)
Subjects: LCSH: Fluid dynamics--Mathematical models.
Classification: LCC QA901 .G3485 2018 | DDC 532.001/5118--dc23
LC record available at https://lccn.loc.gov/2018000547

Visit the Taylor & Francis Web site at
http://www.taylorandfrancis.com

and the CRC Press Web site at
http://www.crcpress.com

Dedicated to the bright memory of my wife, Mrs. Olga Gaissinski. I am very grateful for her help and support throughout our entire life together.
Igor Gaissinski

Dedicated to my family.
Vladimir Rovenski

Contents

Foreword

The turbulence and instability of motion is an important topic in fluid mechanics. The book is devoted to theoretical study and modeling of such flows. The results about stability/instability of jets are used in many practical areas (aircraft, rockets and vehicle) and in the process industry.

The authors of the book have many years of experience in fluid mechanics and applied mathematics.

Igor Gaissinksi graduated from the Moscow Institute of Chemical Engineering in 1971, receiving an MSc in technical sciences, and Moscow State University in 1975, receiving an MSc in mathematics. He was awarded a PhD in mathematics and physics in 1980 from the Institute of High Temperature, Academy of Sciences of the USSR, and a DSc in mathematics and physics from the Russian Academy of Sciences in 1989. In 1982, Dr. Gaissinski was senior researcher in the Department of Theoretical Physics of the All-Union Center for Surface and Vacuum Research, Academy of Sciences of the USSR. From 1980-1988, Dr. Gaissinski was a senior lecturer in the Department of High Mathematics at Moscow Institute of Chemical Engineering. Since 2000, he has been a senior scientist in the faculty of Aerospace Engineering at Technion—Israel Institute of Technology, Haifa. His fields of research are differential and integral equations, the Fokker-Planck equation, shell models, turbulence, spectroscopy and physical kinetics. He is a consultant to various researchers in fluid mechanics. From 2011-2015, Dr. Gaissinski was awarded Grant ISF, No: 661/11. He has published more than 100 scientific papers and 4 books.

Vladimir Rovenski graduated from Novosibirsk State University, Russia in 1976, received his PhD in mathematics and physics from the Mathematical Institute at Novosibirsk in 1985, and was awarded a DSc in mathematics and physics from the Russian Academy of Sciences in 1994. In 1996, Dr. Rovenski was a full professor of mathematics and the geometry chair at Krasnoyarsk Pedagogical University, Russia. Since 1998, he has been a senior scientist in the faculty of Aerospace Engineering at Technion—Israel Institute of Technology, Haifa, and from 2004 until the present he has been a professor and a senior scientist in the Department of Mathematics at the University of Haifa. His fields of research are Riemannian and computational geometry and applied mathematics (modeling in anisotropic elasticity, piezoelectricity and fluid mechanics). From 2008-2010, Dr. Rovenski was awarded Grant P-IEFF of Marie-Curie action No: 219696, and became a visiting professor at the University of Lodz; and from 2011-2014, he was awarded Grant P-ERG No: 276919 of Marie-Curie action. Dr. Rovenski has published more than 90 scientific papers and 8 books.

Together, Drs. Rovenski and Gaissinski have published several research papers and two monographs. The book *Modeling in Fluid Mechanics: Instabilities and Turbulence* is devoted to the theory and analytical modeling of the instability of motion and turbulence in fluid mechanics; this is based on geometrical methods of ordinary differential equations and analytical aspects of perturbation theory. The authors present in this book original results:

– The linear perturbation model for the Kelvin-Helmholtz instabilities is developed and applied to a jet of low viscosity; the dispersion relation based on continuity and momentum conservation equations for a viscid compressible jet is deduced.

– The nonlinear approach to the Rayleigh-Taylor instability is first performed and developed for hollow jets and thin liquid films; the obtained models accept precise analytical solutions.

– The Shell model for fully developed turbulence and phenomenological relation/test based on the main conservation laws (energy, enstrophy or helicity) is developed, the Entropy Principle Maximum for Hierarchical Dynamic Systems and its application for shell models is deduced.

– A fairly complete and, at the same time, compact review devoted to hypersonic flow over blunted bodies is presented.

The book contains theoretical and practical materials, examples and exercises, and can be recommended to a wide audience of researchers and engineers, students and lecturers, specialists in mathematic, physics, chemistry, biology and engineering, who are interested in mathematical methods and models, applied mathematics and theoretical physics.

Irina Albinsky
Doctor of Technical Sciences (Ph.D.)
Computer Sciences and Modeling Specialist

Preface

Nonlinear phenomena arise in all fields of physics, chemistry, biology and engineering, and are responsible for a variety of effects such as jumping between modes, sudden onset or vanishing of periodic oscillations, loss or gain of stability, buckling of frames and shells, ignition and combustion. Their analysis requires, on one hand, tools that provide quantitative results and, on the other hand, the theoretical knowledge of nonlinear behavior that allows one to interpret the quantitative results.

The volume is dedicated to modeling in fluid mechanics and consists of four chapters, which contain a significant number of useful exercises with solutions. We provide relatively complete references on relevant topics in the bibliography at the end of each chapter. The book is designed for graduate and post-graduate students, engineers and researchers (mathematicians and theoretical physicists) working in the field of fluid mechanics. It requires the reader to have basic knowledge in functional analysis, differential equations and statistical mechanics.

The book is written with the view that mathematical methods and models are branches of applied mathematics and theoretical physics.

Chapter 1 contains mathematical background, and is aimed to make the text as self-contained as possible and also to provide a logically ordered context for the subject matter and to motivate later development. We present mainly those basic mathematical methods and topics that are directly related to Chapters 2–4 and not often found in textbooks. Dynamical systems deal with the long-time qualitative behavior of evolving systems. The theory of dynamical systems originated at the end of the 19th century with fundamental questions concerning the stability and evolution of the solar system. By analogy with celestial mechanics, the evolution of a particular state of a dynamical system is referred to as an orbit. A number of themes appear repeatedly in the study of dynamical systems: properties of individual orbits, periodic orbits, typical behavior of orbits, bifurcations, statistical properties of orbits, randomness determinism, chaotic behavior and stability of individual orbits and patterns. The detailed derivation of the Navier-Stokes equation is presented; we bring also a short review about the current state of the problem of existence and uniqueness of solutions to the Navier–Stokes equation for nonrelativistic and relativistic fluid mechanics. Some of these themes are discussed through the examples.

Chapter 2 is devoted to hydrodynamic instabilities. About 20 hydrodynamic instability types are known to be named after people. The most well-known are Rayleigh–Taylor instability (occurring in density-stratified flows, astrophysics), Kelvin–Helmholtz instability (Shear flow), Richtmyer–Meshkov instability (plasma physics, astrophysics), Taylor–Couette instability (flow in rotating cylinder), Bénard instability (natural convection, Rayleigh–Bénard convection), Buneman instability (plasma physics), Darrieus–Landau instability (flame propagation), Kruskal–Shafranov stability (line-tied partial-toroidal plasmas), Velikhov instability (nonequilibrium MHD), etc. In this chapter we study two types of them (because of restrictions on the volume of the book): Rayleigh–Taylor (Section 2.2, which presents pioneer studies performed and developed by the authors in the field of the nonlinear approach to the Rayleigh-Taylor instability of a thin liquid film), and Kelvin–Helmholtz

(Section 2.3, where the linear perturbation model for the Kelvin–Helmholtz instabilities is developed and applied to a jet of low viscosity; the dispersion relation based on continuity and momentum conservation equations for a viscous compressible jet is deduced).

Chapter 3 represents the basic knowledge in the turbulence theory (symmetries, conservation laws, Euler's and Navier–Stokes equations). It introduces the Richardson–Kolmogorov concept, and then develops multifractal and hierarchical (shell) models of turbulence. Three sets of anomalous scaling exponents (scaling structure functions, dissipative scaling and fusion rules) are considered. Sections 3.4–3.5 are based on original authors' results. The main goal is deducing of phenomenological relation based on the main conservation laws (energy, enstrophy and helicity). This relation plays the key role in the choice of shell model for the energy cascade in two- and three-dimensional fully developed turbulence. In Chapter 3 we assume that the readers have basic knowledge in functional analysis, differential equations, classical and statistical mechanics in the scope of the basic courses of theoretical physics and the Fokker-Planck equation (derivation and solutions).

Chapter 4 is devoted to mathematical methods and models to study supersonic flow over blunted bodies. In Sections 4.1–4.3, we study the main laws of the multicomponent gas dynamics. Sections 4.4–4.6 include boundary value problem for the Navier–Stokes equation, for multicomponent chemical nonequilibrium gas mixture. This part also contains a matched asymptotic expansion method, which belongs to perturbation methods. Sections 4.7–4.8 are dedicated mainly to the flow regimes of hypersonic viscous gas flow over blunted bodies with a permeable surface. In Chapter 4 we assume that the readers have basic knowledge in differential equations, tensor analysis and perturbation theory.

December 2017   Igor Gaissinski and Vladimir Rovenski

Symbol Description

NSE	Navier–Stokes equation	ν	kinematic viscosity, and
$\mathbb{R}$	real numbers	ε	dissipation energy
$\mathbb{C}$	complex numbers	$\lambda_{K_r} = (\nu^3/\xi)^{1/6}$	Kraichnan length
$\mathbb{N}$	natural numbers: $1, 2, \dots$	$\xi = \nu L^{-2}\langle\|\nabla\omega\|_2^2\rangle$	enstrophy dissipation rate
Re	real part		
Im	imaginary part	$O(n)$	orthogonal group
$L_p(\mathbb{R}^n)\ p \geq 1$	the p-norm denoted by	$SO(n)$	special orthogonal group
	$\|\psi\|_p = (\int_{\mathbb{R}^n}|\psi(\mathbf{r})|^p d\mathbf{r})^{\frac{1}{p}}$	g_{ij}	metric tensor
$L_\infty(\mathbb{R}^n)$	the norm denoted by	$ds^2 = g_{ij}dx^i dx^j$	metric of the space
	$\|\psi\|_\infty = \lim\limits_{p\to\infty}\|\psi\|_p$	R_{ijkl}	curvature tensor
$C^k(\mathbb{R}^n)$	functions f such that $\nabla^\alpha f$ is continuous for $|\alpha| \leq k$	τ_{w}	local shear stress at the wall
$C^\infty(\mathbb{R}^n)$	smooth functions: $\cup_{k=0}^\infty C^k(\mathbb{R}^n)$	$c_f = \tau_{\mathrm{w}}/(\rho_\varepsilon u_\varepsilon^2/2)$	skin-friction coefficient (index "ε" means boundary layer)
$C_0^\infty(\mathbb{R}^n)$	compactly supported smooth functions	$\Gamma(z)$	Gamma function
$H^p(\mathbb{R}^n)$	Sobolev space of functions $\{f \in L_2(\mathbb{R}^n):$	$G^{\alpha\beta}(\mathbf{r}_0|X_1, x_2)$	Green's function
		$P_l(\theta)$	Legendre polynomials
	$(1+|\omega|^2)^{p/2}\hat{f}(\omega)\in L_2(\mathbb{R}^n)\}$	$J_n(z)$	Bessel functions
$\hat{f}(\omega)$	Fourier image of $f(\mathbf{r})$	$I_n(z), K_n(z)$	modified Bessel functions
$\operatorname{curl}\mathbf{u} = \nabla \times \mathbf{u}$	swirl of the vector field $\mathbf{u}$	$H_n^{(1)}(z)$	Hankel functions
$\partial\Omega$	the boundary of domain Ω	$Y_n(z)$	Neumann functions
		$\tilde{B}_n(z)$	Heun functions
$Re = uL\rho/\mu$	Reynolds number	**KHI**	Kelvin-Helmholtz instability
$We = u^2 L\rho/\sigma$	Weber number		
Kn	Knudsen number	**RTI**	Rayleigh-Taylor instability
k_B	Boltzmann constant		
$M = u/a$	Mach number	∇	Hamilton operator (gradient)
$a = \sqrt{\gamma RT/m}$	sound velocity		
$\gamma = c_p/c_V$	specific heats ratio	$\Delta = \nabla^2$	Laplace operator
c_p	heat capacity under constant pressure	$[\mathbf{p}(t), \mathbf{p}(t)]$	Poisson bracket
		$\langle\,\cdot\,\rangle$	averaging
c_V	heat capacity under constant volume	$a \cdot b$	scalar product
		$a \times b$	vector product
$R \approx 8.314\,\mathrm{J/K\text{-}mol}$	universal gas constant	A^T	transpose of matrix
$C \approx 0.5772$	Euler constant	$\|A\|$	norm of operator
$\mathrm{Lf} = \mathrm{Eu/We}$	Lefebvre number	$\operatorname{Tr} A$	trace of matrix
$\mathrm{Gr} = g\beta(T_{\mathrm{w}} - T_\infty)$	Grashof number, where	$\Lambda(t) \circ Q(t)$	binary operation
$\times L^3/\nu^2$	β – the thermal expansion	sup	supremum
		e_i	covariant vector
	T_{w} – the surface temperature	e^i	contravariant vector
		Γ_{ij}^k	Christoffel symbol
	T_∞ – the bulk temperature	$\otimes$	tensor product
$\lambda_K = (\nu^3/\varepsilon)^{1/4}$	Kolmogorov length scale, where	$g_{ij} = e_i \cdot e_j$	the product of basis vectors
		$F(\mathbf{r}_0\,|\,X_1, \dots X_n)$	correlation function

Chapter 1

Mathematical Background

Chapter 1 contains mathematical background to help readers to navigate better while reading Chapters 2–4. We present mainly basic mathematical methods, which are not often found in textbooks.

Dynamical systems deal with the long-time qualitative behavior of evolving systems. The theory of dynamical systems originated at the end of the 19th century with fundamental questions concerning the stability and evolution of the solar system. Attempts to answer those questions led to the development of a rich and powerful field with applications to physics, biology, meteorology, astronomy, economics and other areas of science. By analogy with celestial mechanics, the evolution of a particular state of a dynamical system is referred to as an orbit. A number of themes appear repeatedly in the study of dynamical systems: properties of individual orbits, periodic orbits, typical behavior of orbits, bifurcations, statistical properties of orbits, randomness determinism, chaotic behavior, and stability of individual orbits and patterns. We discuss some of these themes through the examples in this chapter. The geometrical properties of chaotic motion, fractals and attractors are investigated in much the same way as one might study the geometry of classical figures such as circles, spirals, or ellipses: locally a circle may be approximated by a line segment, the projection or "shadow" of the circle is generally an ellipse, a circle typically intersects a straight line segment in two points (if at all), and so on. There are fractal analogues of such properties, with dimension playing a key role. For example, the local forms of fractals, projections and intersections of fractals will be considered. This chapter includes examples from symbolic dynamics, chaotic dynamical systems, attracting sets and attractors, random fractals and some physical applications including Navier–Stokes equations.

1.1 Dynamical systems

In this section we define the important concept of a *vector field* on an open set $U \subset \mathbb{R}^n$, which allows us to think of a first-order system of ordinary differential equations (ODEs) in a geometric way.

1.1.1 Vector fields and dynamical systems

Autonomous systems. Denote by $\mathbb{R}^n$ the set of all n-tuples $y = (y_1, \ldots, y_n)$ of real numbers $y_i \in \mathbb{R}$. We'll view $\mathbb{R}^n$ either as the n-dimensional *Euclidian space*, whose elements are points in this space, or as an n-dimensional *vector space*, whose elements $\mathbf{y} = (y_1, \ldots, y_n)$ are (position) vectors. Denote by $|\mathbf{y}|$ the length of the vector and by $\mathbf{y} \cdot \mathbf{z}$ the scalar product of two vectors. When $n = 2$, or $n = 3$, we often use coordinates (x, y) or (θ, v) instead of (y_1, y_2) for a vector or a point in $\mathbb{R}^2$; similarly (x, y, z) or other variants are often used instead of (y_1, y_2, y_3) for a point or a vector in $\mathbb{R}^3$.

Definition 1 A *vector field* on $U \subset \mathbb{R}^n$ is a map $X : U \to \mathbb{R}^n$, in component form given by

$$X(y) = (X^1(y), \ldots, X^n(y)),$$

where $y \in U$ and $X^i : U \to \mathbb{R}$ $(i = 1, \ldots, n)$ are the component functions of X. The terminology of differentiability, $C^1, \ldots, C^\infty$, is applicable to vector fields, so we speak of C^1 vector fields, and so on.

Vector fields have an important geometric interpretation – for a point $y \in U$, one interprets $X(y)$ as a vector attached to the point y. Namely, we take its initial point to be y instead of plotting the vector $X(y)$ with its initial point at the origin; see Fig. 1.1(a). Doing plots like this at a number of different points $x, y, z, \ldots$ in U, as shown, gives a geometric picture of a *field of vectors*. This geometric picture also explains the origin of the name *vector field*. The notion of a t-independent vector field is synonymous with an autonomous system. In what follows $'$ or $\dot{}$ denotes derivation.

Definition 2 (Autonomous system) (a) A system of differential equations

$$\dot{y} = X(y), \tag{1.1}$$

where X is a vector field on some open subset $U \subseteq \mathbb{R}^n$ and the independent variable t does not occur explicitly, is called *autonomous*. A simple property of such systems is that if $y = y(t)$, $\alpha < t < \beta$ is the solution of (1.1), then $y = y(t + t_0)$ is also a solution for $t \in (\alpha - t_0, \beta - t_0)$, for any constant t_0.

(b) Recall that a *curve* in $\mathbb{R}^n$ is a map $\alpha : J \to \mathbb{R}^n$ from some interval $J \subset \mathbb{R}$ into $\mathbb{R}^n$. If α is differentiable, then it is called a *differentiable curve*. A *solution* of (1.1) is a differentiable curve $\alpha : J \to \mathbb{R}^n$ with the properties (a) $\alpha(t) \in U$ for all $t \in J$; (b) $\dot{\alpha}(t) = X(\alpha(t))$ for all $t \in J$. Such a curve $\alpha(t)$ is also called an *integral curve* of the vector field X (or of the system of ODEs).

Property (a) says that the curve lies in the open set U, which is necessary in order for the expression $X(\alpha(t))$ in property (b) to make sense. Property (b) says that $\alpha(t)$ satisfies the system of ODEs:

$$\dot{\alpha}_i(t) = X^i(\alpha_1(t), \ldots, \alpha_n(t)), \quad i = 1, \ldots n,$$

for all $t \in J$; this means that $\alpha(t)$ is a curve, whose tangent vector at the point $\alpha(t)$ coincides with vector $X(\alpha(t))$ at the same point; see Fig. 1.1(b). There are many realizations of this in physics.

Example 1 i) If X is an electrostatic force field on $U \subset \mathbb{R}^3$, then the integral curves of X are the force field lines. If $X = -\nabla \phi = -(\partial \phi / \partial y_1, \ldots, \partial \phi / \partial y_n)$ (∇ is the gradient) is derived from a potential ϕ, then a positively charged particle will move along a force field line toward region of lower potential.

ii) Let U be a tank in which a fluid is circulating with a steady flow, and $X(y)$ represents the velocity of the fluid flowing through the point $y \in U$ at any time. If we follow motion of any particle in the fluid over time, then it will describe a trajectory that is an integral curve of the vector field $X(y)$.

iii) The heat flow is yet another situation, where the vector field and its integral curves have physical meaning. There U is the unit cube, X is the heat flux vector and the integral curves of $X(y)$ are the lines along which the heat must flow in order to maintain the distribution of temperatures $T(y)$.

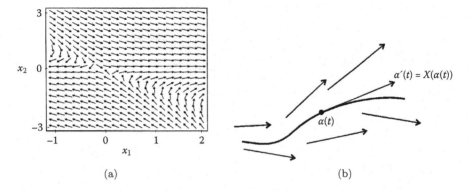

FIGURE 1.1: (a) Vector field $X = (\frac{1}{2}(y_1 + y_2), -\frac{1}{2}y_2)$. (b) An integral curve $\alpha(t)$ of X.

The geometrical idea of viewing solutions of (1.1) as integral curves of a vector field is of great importance to the theory of ODEs. In fact, an accurate plot (by computer!) of the vector field at a large number of points in U will almost delineate the picture of what all the integral curves look like.

Prior to era of computers, all plots of vector fields were limited to simple examples. Most software actually plots what is known as the *direction field* for X. This is the plot of $kX(y)/|X(y)|$ at all points of y on a specified grid (here $|X(y)|$ denotes the length of the vector $X(y)$). All these vectors have the same length, k, so the resulting plot only indicates the direction of $X(y)$ at each point on the grid. The scale factor k depends on the software package and the grid size. A plot of the direction field gives an overall view of what to expect, gives the direction of flow for the system, and helps locate fixed points, if any. Since each integral curve of $X(y)$ traces out a path in $\mathbb{R}^2$ that is tangent to the direction field element at each point, it is not hard to roughly discern from the figure what some of the curves look like.

Example 2 Let a vector field $X : \mathbb{R}^2 \to \mathbb{R}^2$ be given by $X(y_1, y_2) = (\frac{1}{2}(y_1 + y_2), -\frac{1}{2}y_2)$. A direction field plot is shown in Fig. 1.1(a) on the rectangle $[-1, 2] \times [-3, 3]$ with a grid of 20×20 points. The two components of X are $X^1(y_1, y_2) = \frac{1}{2}(y_1 + y_2)$ and $X^2(y_1, y_2) = -\frac{1}{2}y_2$; the associated system is

$$\dot{y}_1 = (y_1 + y_2)/2, \quad \dot{y}_2 = -y_2/2.$$

The simplicity of this example allows us to explicitly exhibit formulas for the solutions. Thus, $y_1 = b_1 e^{t/2} - \frac{1}{2}b_2 e^{-t/2}$, $y_2 = b_2 e^{-t/2}$ is the general solution, involving arbitrary constants b_1, b_2. The resulting integral curve is then $\alpha(t) = (b_1 e^{t/2} - 0.5 b_2 e^{-t/2}, b_2 e^{-t/2})$, where b_1, b_2 can be chosen so that $\alpha(0) = c$, where $c = (c_1, c_2)$ is any specified point. This gives the system $b_1 - b_2/2 = c_1$, $b_2 = c_2$, which is solvable for b_1, b_2. For example, if $c = (1, 2)$, then $b_1 = 2$, $b_2 = 2$ and thus the curve $\alpha(t) = (2e^{t/2} - e^{-t/2}, 2e^{-t/2})$ is the integral curve that satisfies $\alpha(0) = (1, 2)$. Observe that each integral curve must lie on a branch of a hyperbola or on a straight line. The straight line case occurs for $b_1 = 0$ and then the last equation gives a pair of lines $2y_1 + y_2 = 0$, $y_2 = 0$.

It is convenient to study the behavior of an autonomous equation of order n when it is in the form of a system of n coupled first-order equations.

Example 3 (Phase plane) The equation $\ddot{y}(t) + y(t) = 0$ describes a particle undergoing harmonic motion (sinusoidal oscillation). To represent this motion in the phase plane we convert this equation into a first-order system: let $y_1(t) = y(t)$ be the position of a particle,

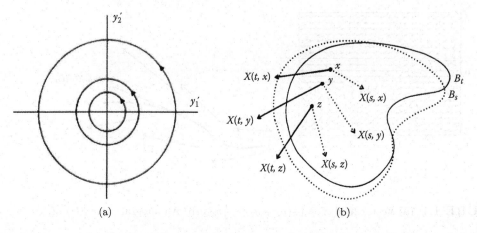

(a)   (b)

FIGURE 1.2: (a) Trajectories for $\dot{y}_1(t) = y_2(t)$, $\dot{y}_2(t) = -y_1(t)$. (b) Direction fields for X_t and X_s.

and $y_2(t) = \dot{y}(t)$ its velocity. Then, $\dot{y}_1(t) = y_2(t)$, $\dot{y}_2(t) = -y_1(t)$ is an equivalent first-order system. In phase space trajectories are concentric circles; see Fig. 1.2(a).

By convention we will think of the independent variable of the system as time t and the independent variables $y_1, \ldots, y_n$ as position coordinates. The general form of such system is

$$\dot{y}_i = f_i(y_1, \ldots, y_n), \quad i = 1, \ldots n,$$

where the dots indicate differentiation with respect t. The solution of this system is a curve of *trajectory* in an n-dimensional space called *phase space*. The trajectory is parameterized in terms of t: $y_i = y_i(t)$ $(1 \leq i \leq n)$. We'll assume that $f_1, \ldots, f_n$ are continuously differentiable with respect to each of their arguments. Thus, by the existence and uniqueness theorem of ODEs, any initial condition $y_i(0) = y_{i0}$ $(1 \leq i \leq n)$ gives rise to a *unique* trajectory through the point $(y_{10}, y_{20}, \ldots, y_{n0})$. To understand this uniqueness property geometrically, note that at every point on the trajectory $\{\dot{y}_1(t), \dot{y}_2(t), \ldots \dot{y}_n(t)\}$, which is tangent to the trajectory at the point. It immediately follows that *two trajectories $y_i(t)$ and $y_j(t)$ $(i \neq j)$ cannot cross*; otherwise, the tangent vector at the crossing point would not be unique.

Nonautonomous systems. A *time-dependent vector field* is a map $X : B \to \mathbb{R}^n$ defined on an open subset $B = \{(t, y) \in \mathbb{R}^{n+1} : y \in \mathbb{R}^n, \ t \in \mathbb{R}\} \subseteq \mathbb{R} \times \mathbb{R}^n$. This gives a component form for X:

$$X(t, y(t)) = (X^1(t, y(t)), \ldots, X^n(t, y(t))).$$

For each $t \in \mathbb{R}$ we let $B_t = \{y \in \mathbb{R}^n : (t, y) \in B\}$. This is a *slice* through B at time t (and may be empty). At this instant t in time we get a vector field on B_t in the previous sense (when B_t is open). It is the map $X_t(y) \equiv X(t, y)$. Plotting the direction field for X_t gives us a snapshot of X at time t. As t varies so do the plots of the direction fields; see Fig. 1.2(b). Often B is a product $B = J \times U$, of an open interval J and an open set $U \subseteq \mathbb{R}^n$, which greatly simplifies things.

Definition 3 (Nonautonomous systems) Let $X : B \to \mathbb{R}^n$ be a time-dependent vector field. A *solution*, or *integral curve*, of the corresponding nonautonomous system

$$\dot{y}p = X(t, y)$$

is a curve $\alpha : J \to \mathbb{R}^n$, in $\mathbb{R}^n$, with the following properties:

(a) $(t, \alpha(t)) \in B$ for every $t \in J$; (b) $\dot{\alpha}(t) = X(t, \alpha(t))$ for every $t \in J$.

An *initial value problem* consists of finding a solution of a system of ODEs which passes through a given point at a given time. More specifically, suppose $(t_0, y_0) \in B$ is given. The corresponding initial value problem is written symbolically as $\dot{y}(t) = X(t, y)$, $y(t_0) = y_0$.

A *solution of the initial value problem* is a curve $\alpha : J \rightarrow \mathbb{R}^n$ that is a solution of the system above and which also satisfies $\alpha(t_0) = y_0$ (note this implies that the interval J contains t_0). This condition is known as an *initial condition*; y_0 is called an *initial point* for the integral curve in the problem.

Example 4 One of the oldest examples of a dynamical system is the *two-body system*: two bodies with masses m_1, m_2, attract each other mutually with the force of attraction along the line joining the bodies and of magnitude reciprocally as the square of the distance. In terms of Newton's second law of dynamics, the model for this system is

$$m_1 \ddot{\mathbf{r}}_1 = g m_1 m_2 (\mathbf{r}_2 - \mathbf{r}_1)/r_{12}^3, \quad m_2 \ddot{\mathbf{r}}_2 = g m_1 m_2 (\mathbf{r}_1 - \mathbf{r}_2)/r_{12}^3. \tag{1.2}$$

This is actually a second-order system of ODEs written in vector form. Here

$$\mathbf{r}_1(t) = (x_1(t), y_1(t), z_1(t)), \quad \mathbf{r}_2(t) = (x_2(t), y_2(t), z_2(t))$$

are the position vectors for the two bodies, and we have used the notation $r_{12}(t) \equiv |\mathbf{r}_1(t) - \mathbf{r}_2(t)|$ for the distance between bodies at time t. For the sake of comparison, write the two-body system as

$$\begin{cases} \dot{x}_1 = u_1, & \dot{y}_1 = v_1, & \dot{z}_1 = w_1, \\ \dot{x}_2 = u_2, & \dot{y}_2 = v_2, & \dot{z}_2 = w_2, \\ \dot{u}_1 = g m_2 (x_2 - x_1)/r_{12}^3, & \dot{v}_1 = g m_2 (y_2 - y_1)/r_{12}^3, & \dot{w}_1 = g m_2 (z_2 - z_1)/r_{12}^3, \\ \dot{u}_2 = g m_1 (x_1 - x_2)/r_{12}^3, & \dot{v}_2 = g m_1 (y_1 - y_2)/r_{12}^3, & \dot{w}_2 = g m_1 (z_1 - z_2)/r_{12}^3. \end{cases} \tag{1.3}$$

In addition to the six unknown functions x_i, y_i, z_i ($i = 1, 2$), the above system involves the six functions u_i, v_i, w_i, $i = 1, 2$, which we recognize (via the first six equations in the system) as the components of the velocity vectors for the two bodies: $\mathbf{v}_i \equiv (u_i, y_i, w_i) = \dot{\mathbf{r}}_i$. The system (1.3) of first-order scalar ODEs is not as convenient as the second-order vector form (1.2), but exhibits a general technique for reducing higher-order systems to first-order systems (introduce extra unknown functions). A solution of (1.3) is a curve in the 12-dimensional space $\mathbb{R}^{12} \simeq \mathbb{R}^6 \times \mathbb{R}^6$, of positions and velocities. Thus, α has the form $\alpha(t) = (\mathbf{r}_1(t), \mathbf{r}_2(t), \mathbf{v}_1(t), \mathbf{v}_2(t))$ for t in some interval. More precisely, since r_{12} is nonzero on the right-hand side (RHS) of the two-body system, each solution curve lies in the 12-dimensional domain $M \subset \mathbb{R}^{12}$ defined by $M \equiv U \times \mathbb{R}^6$, where $U \equiv \{(\mathbf{r}_1, \mathbf{r}_2) : \mathbf{r}_1 \neq \mathbf{r}_2\}$. While we cannot visualize (except possibly mentally) the graph of solution curve in this M, it is nevertheless a useful theoretical notion, or in this special case, center of mass coordinates, we get around this visualization limitation and plot the orbits of the two bodies relative to the center of mass.

By analogy with the two-body system, the space $M = \mathbb{R}^2 \backslash O$ for the row of vortices example and the space $M \equiv U$ for the heat flow in a cube example are also called the phase spaces for those dynamical systems, even though historically the term phase space referred to the spaces of positions and velocities (or positions and moments). In the two-body problem, the system in vector form is

$$\begin{cases} \dot{\mathbf{r}}_1(t) = \mathbf{v}_1, & \dot{\mathbf{r}}_2(t) = \mathbf{v}_2, \\ \dot{\mathbf{v}}_1(t) = G m_2 (\mathbf{r}_2 - \mathbf{r}_1)/r_{12}^3, & \dot{\mathbf{v}}_2(t) = G m_1 (\mathbf{r}_1 - \mathbf{r}_2)/r_{12}^3 \end{cases}$$

and the initial condition just specifies the initial position and initial velocities of the two bodies

$$\mathbf{r}_1(0) = \mathbf{a}_1, \quad \mathbf{r}_2(0) = \mathbf{a}_2, \quad \mathbf{v}_1(0) = \mathbf{b}_1, \quad \mathbf{v}_2(0) = \mathbf{b}_2.$$

Here $\mathbf{a}_1 \neq \mathbf{a}_2$, $\mathbf{b}_1$, $\mathbf{b}_2$ are given points (vectors) in $\mathbb{R}^3$. With some effort we actually exhibit a solution of this problem with explicit dependence on the initial condition data that gives uniqueness of solutions. For a larger number of bodies, 3-bodies (like the earth, moon, sun), 4-bodies, ..., or in general N-bodies, the possibility of exhibiting an exact solution of the initial value problem (except in highly special cases) is too much to hope for. Proving existence and uniqueness by other means is more tractable. The existence and uniqueness question here can be interpreted as saying that Newtonian mechanics is deterministic, i.e., knowing the initial positions and velocities of all the bodies determines uniquely their evolution in time thereafter.

1.1.2   Critical points in phase space

If there are any solutions to the set of simultaneous algebraic equations

$$f_i(y_1, \ldots, y_n) = 0, \quad i = 1, \ldots n,$$

then there are special degenerate trajectories in phase space which are just points. The velocity at these points is zero so the position vector does not move. Such points are called *critical points*. While a trajectory can approach a critical point as $t \to \infty$, it cannot reach such point at a finite time. The proof is simple. Suppose it were possible for a trajectory to reach a critical point in time T. Then the time-reversed system of equations (under the equations obtained to replace f_i with $-f_i$) would exhibit an impossible behavior: the position vector $(y_1(t), \ldots, y_n(t))$ would rest motionless at the critical point and then suddenly begin to move at time T. Recall that in an autonomous system, the components of the velocity vector depend only on the position of the particle and not on the time.

One-dimensional phase space. One-dimensional phase space, which may be called *phase line*, is used to study solutions to the first-order autonomous system $\dot{y} = f(y)$. There are only three possibilities for the global behavior of a trajectory on a phase line:

1. The trajectory $y = y(t)$ may approach a critical point as $t \to \infty$.
2. The trajectory $y = y(t)$ may approach $\pm\infty$ as $t \to \infty$.
3. The trajectory $y = y(t)$ may remain motionless at a critical point for all t.

In the neighborhood of a critical point there are three positions for the local behavior:

1. All trajectories may approach the critical point as $t \to \infty$. Call such a critical point a *stable node*; see Fig. 1.3(a). The point $y = 0$ is a stable node for the equation $\dot{y} = -y$.

2. All trajectories may move from the critical point as $t \to \infty$. Call such critical point an *unstable node*; see Fig. 1.3(a). The point $y = 0$ is an unstable node for the equation $\dot{y} = y^3$.

3. Trajectories on one side of the critical point may move toward it while trajectories on the other side of the critical point move away from it as $t \to \infty$. Call such a point as a *saddle point*; see Fig. 1.3(a). The point $y = 0$ is a saddle point for the equation $\dot{y} = y^2$.

Example 5 (Critical-phase analysis of one-dimensional system) The equation $\dot{y} = y^2(t) - y(t)$ represents a population $y(t)$ having a quadratic birthrate (y^2) and a linear deathrate ($-y$). It is easy to find the exact time dependence of $y(t)$ by solving the ODE because the equation is separable. It is even easier to obtain a rough picture of the global behavior using critical-point analysis. The critical points for $\dot{y} = y^2(t) - y(t)$ solve the algebraic equation $y^2(t) - y(t) = 0$. Hence, there are critical points at $y_1 = 0$ and at $y_2 = 1$. Next, we classify these critical points by finding the local behavior of $y(t)$ near y_1 and y_2. To do this we *linearize* the ODE near these points, i.e., we let $y(t) = \varepsilon(t)$, when $y = 0$, and

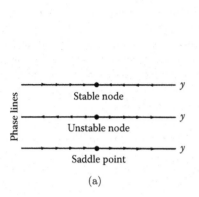
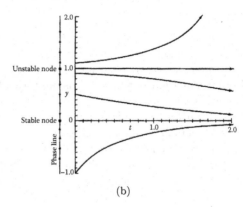

FIGURE 1.3: (a) Three kinds of critical points. (b) Phase-line analysis of equation $\dot{y} = y^2 - y$.

approximate the exact ODE by $\dot{\varepsilon}(t) \simeq -\varepsilon(t)$ $(\varepsilon \to 0)$. The solutions of this equation decay exponentially, so $y = 0$ is a stable node. Similarly, when y is near 1 we let $y(t) = 1 + \varepsilon(t)$ and approximate the exact ODE by $\dot{\varepsilon}(t) \simeq \varepsilon(t)$ $(\varepsilon \to 0)$. Since the solutions of this equation grow exponentially with t, we identify $y = 1$ as an unstable node.

We can infer behavior of $y(t)$ for any initial value $y(0)$. Any trajectory on the phase line to the left of $y = 0$ must move to the right and approach $y(t) = 0$ as $t \to \infty$ because, by continuity, trajectories sufficiently near $y = 0$ move toward $y = 0$. If some trajectories were to move leftward toward $y(t) = -\infty$ as $t \to \infty$, then dividing point between left-moving and right-moving trajectories would be the critical point. But there is no critical point to the left of $y(t) = 0$. Similarly, all trajectories to the right of $y = 1$ move rightward toward $y(t) = \infty$ as $t \to \infty$ and all trajectories between $y = 0$ and $y = 1$ move leftward toward $y(t) = 0$ as $t \to \infty$. This completes our critical-point analysis of the ODE and our conclusions about the structure of the phase line are presented in Fig. 1.3(b). From this diagram we can predict the behavior of $y(t)$ as $t \to \infty$ for any $y(0)$; see Fig. 1.3(b):

if $y(0) < 0$, then $y(t) \to 0$ as $t \to \infty$,   if $y(0) = 0$, then $y(t) = 0$ for all t,
if $0 < y(0) < 1$, then $y(t) \to 0$ as $t \to \infty$,   if $y(0) = 1$, then $y(t) = 1$ for all t,
if $y(0) > 1$, then $y(t) \to +\infty$ as $t \to \infty$.

Critical-point analysis does not predict how fast $y(t)$ approaches 0 or ∞; that requires further *local* analysis. Rather, it answers the *global* question of how $y(0)$ affects the behavior of $y(t)$ as $t \to \infty$.

Two-dimensional phase space. The phase plane is used to study a system of two coupled first-order equations. It is more complicated than the phase line, but it is still possible to make elegant global analysis of systems of two coupled ODEs. First, we estimate the possible global behavior of a trajectory in a two-dimensional system. Trajectory may:

1. approach a critical point as $t \to \infty$.
2. approach ∞ as $t \to \infty$.
3. remain motionless at a critical point for all t.
4. describe a closed orbit or circle; see Fig. 1.2(a).
5. approach a closed orbit (by spiraling inward/outward toward the orbit) as $t \to \infty$.

The first three possibilities also occur in one-dimensional systems, but the fourth and fifth ones are new configurations which cannot occur in a phase space of less than two dimensions.

Next, we enumerate the possible local behaviors for trajectories near a critical point:

1. All trajectories may approach the critical point along curves which are asymptotically straight lines as $t \to \infty$. Call such critical point a *stable node*.

2. All trajectories may approach the critical point along spiral curves as $t \to \infty$. Such critical point is called a *stable spiral point*. It is also possible for trajectories to approach the critical point along curves which are neither spiral nor asymptotic to straight lines.

3. All time-reversed trajectories (that is, $y(t)$ with t decreasing) may move toward the critical point along paths which are asymptotically straight lines as $t \to -\infty$. Such critical point is an *unstable node*. As t *increases*, all trajectories that start near an unstable node move *away* from the node along paths that are approximately straight lines at least until the trajectory gets far from the node.

4. All time-reversed trajectories may move toward the critical point along spiral curves as $t \to -\infty$. Such critical point is called an *unstable spiral point*. As t *increases*, all trajectories move *away* from an unstable spiral point along trajectories that are, at least initially, spiral shaped.

5. Some trajectories may approach the critical point while others move away from the critical point as $t \to \infty$. Such a critical point is called a *saddle point*.

6. All trajectories may form closed orbits about the critical point. Such critical point is called a *center*. In Fig. 1.2(a) we see an example of a center.

While nodes and saddle points occur in one-dimensional phase space, spiral points and centers cannot exist in less than two dimensions.

Linear autonomous systems. Two-dimensional linear autonomous systems can exhibit any of the critical point behaviors that have been described above. With this in mind we introduce an easy method (based on elementary matrix algebra) to solve linear autonomous systems. A two-dimensional linear autonomous system $\dot{y}_1 = ay_1 + by_2$, $\dot{y}_2 = cy_1 + dy_2$ may be written in matrix form

$$\dot{\mathbf{Y}}(t) = \mathbf{A}\mathbf{Y}(t), \quad \mathbf{Y}(t) = \begin{pmatrix} y_1(t) \\ y_2(t) \end{pmatrix}, \quad \mathbf{A} = \begin{pmatrix} a & b \\ c & d \end{pmatrix}. \qquad (1.4)$$

If the eigenvalues λ_1 and λ_2 of the matrix $\mathbf{A}$ are distinct and $\mathbf{V}_1$ and $\mathbf{V}_2$ are eigenvectors of $\mathbf{A}$ associated with the eigenvalues λ_1 and λ_2, then the general solution of this system has the form $\mathbf{Y}(t) = c_1\mathbf{V}_1 e^{\lambda_1 t} + c_2\mathbf{V}_2 e^{\lambda_2 t}$, where c_1 and c_2 are constants of integration which are determined by the initial position $\mathbf{Y}(0)$. The linear system (1.4) has a critical point at the origin $(0,0)$, and λ_1 and λ_2 satisfy the equation

$$\det(\mathbf{A} - \lambda\mathbf{I}_2) = \det\begin{pmatrix} a - \lambda & b \\ c & d - \lambda \end{pmatrix} = \lambda^2 - \lambda(a + d) + (ad - bc) = 0. \qquad (1.5)$$

If λ_1 and λ_2 are real and negative, then all trajectories approach the origin as $t \to \infty$ and $(0,0)$ is a stable node. Conversely, if λ_1 and λ_2 are real and positive, then all trajectories move away from $(0,0)$ as $t \to \infty$ and $(0,0)$ is an unstable node. Also, if λ_1 and λ_2 are real but $\lambda_1\lambda_2 < 0$ then $(0,0)$ is a saddle point. Trajectories approach the origin in the direction $\mathbf{V}_2$ and move away from the origin in the direction $\mathbf{V}_1$. Solution λ_1 and λ_2 of (1.5) may be complex. When the matrix $\mathbf{A}$ is real, then λ_1 and λ_2 must be a complex conjugate pair. If λ_1 and λ_2 are pure imaginary, then the vector $\mathbf{Y}(t)$ represents a closed orbit for any c_1 and c_2, and the critical point at $(0,0)$ is a center. If λ_1 and λ_2 are complex with nonzero real part, then the critical point at $(0,0)$ is a spiral point. If $\mathrm{Re}\,\lambda_{1,2} < 0$ then $\mathbf{Y}(t) \to 0$ as $t \to \infty$ and $(0,0)$ is a stable spiral point; conversely, if $\mathrm{Re}\,\lambda_{1,2} > 0$ then $(0,0)$ is an unstable spiral point.

To determine whether the rotation (of the trajectories in the vicinity of a spiral point or a center) is counterclockwise we simply let $y_2 = 0$, $y_1 > 0$, and see whether $\dot{y}_2$ is positive.

Example 6 (Identification of critical points for linear systems)

(a) The system $\dot{y}_1 = -y_1$, $\dot{y}_2 = -3y_2$ has a *stable node* at the origin because the eigenvalues of $\mathbf{A} = \begin{pmatrix} -1 & 0 \\ 0 & -3 \end{pmatrix}$ are $\lambda_1 = -1$ and $\lambda_2 = -3$ which are both negative.

(b) The system $\dot{y}_1 = y_2$, $\dot{y}_2 = -2y_1 - 2y_2$ has a *stable spiral point* at $(0,0)$ because the eigenvalues of $\mathbf{A} = \begin{pmatrix} 0 & 1 \\ -2 & -2 \end{pmatrix}$ are $\lambda_{1,2} = -1 \pm i$ which have negative real parts. All orbits approach $(0,0)$ following spirals in a clockwise direction.

(c) The system $\dot{y}_1 = y_1$, $\dot{y}_2 = 2y_2$ has an *unstable node* at $(0,0)$.

(d) The system $\dot{y}_1 = y_1 + 2y_2$, $\dot{y}_2 = -y_1 + y_2$ has an *unstable spiral point* at the origin.

(e) The system $\dot{y}_1 = y_1 + 2y_2$, $\dot{y}_2 = y_1 + y_2$ has a *saddle point* at $(0,0)$.

(f) The system $\dot{y}_1 = y_2$, $\dot{y}_2 = -y_1$ has a *center* at the origin because the eigenvalues of the matrix $\mathbf{A} = \begin{pmatrix} 0 & 1 \\ -1 & 0 \end{pmatrix}$ are $\lambda_{1,2} = \pm i$ which are pure imaginary.

Two-dimensional nonlinear systems. To illustrate the power of critical-point analysis, we use it to deduce the global features of some nonlinear systems. The approach we take in the following examples is the same as in Example 5. First, identify the critical points. Then, we perform a local analysis of the system very near these critical points. As in Example 5, the exact system can be approximated by linear autonomous system near a critical point. Using matrix methods we identify the nature of critical points of linear systems. Finally, we assemble the results of our local analysis and synthesize a qualitative global picture of solutions for the nonlinear system.

Example 7 (System with two critical points) The nonlinear system

$$\dot{y}_1 = y_1 - y_1 y_2, \qquad \dot{y}_2 = -y_2 + y_1 y_2,$$

known as *Volterra equations*, is a simple model of a predator-prey relation between two populations like that of rabbits and foxes: $y_1(t)$ (rabbit population) will grow out of bounds if $y_2(t)$ (the fox population) is 0; but if y_1 is 0 then y_2 will decay to 0 because of starvation. There are two critical points $(0,0)$ and $(1,1)$. Near $(0,0)$ we let $(y_1, y_2) = (\varepsilon_1, \varepsilon_2)$ and approximate the exact ODE by $\dot{\varepsilon}_1 \simeq \varepsilon_1$, $\dot{\varepsilon}_2 \simeq -\varepsilon_2$ ($\varepsilon_1, \varepsilon_2 \to 0$). The solutions of this system exhibit saddle-point behavior; trajectories near $(0,0)$ approach the origin vertically and move away from the origin horizontally as $t \to \infty$. Near $(1,1)$ we let $(y_1, y_2) = (1 + \varepsilon_1, 1 + \varepsilon_2)$ and approximate the exact equation by $\dot{\varepsilon}_1 \simeq -\varepsilon_2$, $\dot{\varepsilon}_2 \simeq \varepsilon_1$ ($\varepsilon_1, \varepsilon_2 \to 0$). The eigenvalues of the matrix $\mathbf{A}$ for this linear system are $\pm i$. Therefore, the critical point at $(1,1)$ is a center having closed orbits going counterclockwise as $t \to \infty$. The counterclockwise rotation about $(1,1)$ is consistent with the direction of incoming and outgoing trajectories near the saddle point $(0,0)$.

Suppose the initial condition $y_1(0) = y_2(0) = a$ ($0 < a < 1$). Then as t increases from 0, the vector $[y_1(t), y_2(t)]$ would move counterclockwise around the point $(1,1)$. As t increases, the vector must *continue* to rotate around $(1,1)$. It cannot cross the y_1 or y_2 axes because they are themselves trajectories. A deeper analysis shows that this trajectory cannot approach ∞. Therefore for some t, the vector must encircle the point $(1,1)$ and eventually recross the line connecting $(0,0)$ and $(1,1)$. Moreover, it must cross at the initial point (a, a). In summary, all trajectories with $y_1(0) > 0$, $y_1(0) > 0$ are closed and encircle the point $(1,1)$ regardless of the initial condition at $t = 0$. Thus, the populations y_1 and y_2 oscillate with time; see Fig. 1.4(a). The conclusion that the trajectories are exactly closed cannot be justified by local analysis alone. Although it is possible to infer global behavior from local analysis of stable and unstable critical points, in this example local analysis gives an incomplete description of the nature of phase-space trajectories.

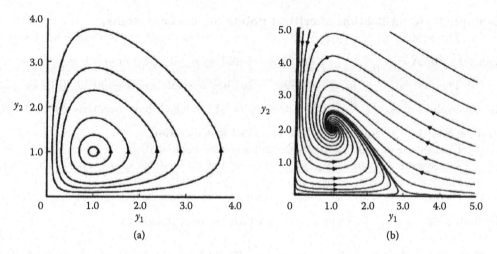

FIGURE 1.4: Solutions of nonlinear systems.
(a) $\dot{y}_1 = y_1 - y_1 y_2$, $\dot{y}_2 = -y_2 + y_1 y_2$. (b) $\dot{y}_1 = y_1(3 - y_1 - y_2)$, $\dot{y}_2 = y_2(y_1 - 1)$.

Fig. 1.4(b) illustrates a solution to the system with three critical points, one of them a stable spiral point with counterclockwise rotation.

Example 8 (System with four critical points) The system

$$\dot{y}_1 = y_1^2 - y_1 y_2 - y_1, \quad \dot{y}_2 = y_2^2 + y_1 y_2 - 2y_2,$$

is a more complicated version of the predator-prey equations in Example 7, because it contains linear deathrate and quadratic birthrate terms. This system does not predict oscillating populations of rabbits y_1 and foxes y_2. Rather it predicts an unstable situation in which either the rabbits or the foxes grow out of bounds or else both species become extinct. There are four critical points. It is good practice to verify that $(0,0)$ is a stable node; $(1.5, 0.5)$ is an unstable spiral point having a counterclockwise rotation; $(1,0)$ is a saddle point in whose neighborhood the general behavior of a solution has the form $\left(\begin{array}{c} y_1 \\ y_2 \end{array} \right) \simeq c_1 \left(\begin{array}{c} 1 \\ 0 \end{array} \right) e^t + c_2 \left(\begin{array}{c} 1 \\ 2 \end{array} \right) e^{-t}$, so that trajectories of slope 2 move inward and trajectories of slope 0 move outward; $(0,2)$ is a saddle point in whose neighborhood the general solution has the form $\left(\begin{array}{c} y_1 \\ y_2 \end{array} \right) \simeq c_1 \left(\begin{array}{c} 5 \\ -2 \end{array} \right) e^{-3t} + c_2 \left(\begin{array}{c} 0 \\ 1 \end{array} \right) e^{2t}$, so that trajectories of slope $-2/5$ move inward and vertical trajectories move outward. By our local analysis, there is an approximate trapezoid of initial values bounded by the four critical points at $(0,0)$, $(1,0)$, $(1.5, 0.5)$ and $(0,2)$ for which trajectories all approach the origin and both populations $y_1(t)$ and $y_2(t)$ die out as $t \to \infty$. Outside this trapezoid the initial conditions are unstable with either the fox or the rabbit populations becoming infinite as $t \to \infty$. These predictions are consistent with the exact solution in Fig. 1.5(a). It is possible for both $y_1(t)$ and $y_2(t)$ to become infinite as $t \to \infty$ or one of $y_1(t)$ and $y_2(t)$ must always vanish.

The analysis in Examples 7 and 8 is called *linear* critical-point analysis because there is a linear approximation to the nonlinear system in the neighborhood of each critical point. Unfortunately, it is not always possible to find a linear approximation to a nonlinear system. For example, the system

$$\dot{y}_1 = y_1^2 + y_2^2 + y_1^3, \quad \dot{y}_2 = \sin(y_1^4 + y_2^4)$$

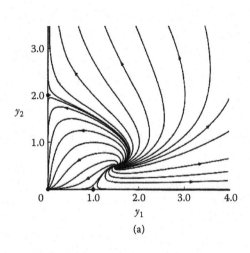

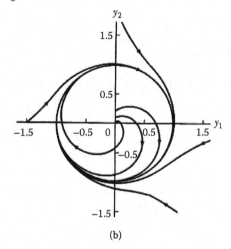

(a)   (b)

FIGURE 1.5: Solutions of nonlinear systems.
(a) $\dot{y}_1 = y_1^2 - y_1 y_2 - y_1$, $\dot{y}_2 = y_2^2 + y_1 y_2 - 2y_2$. (b) $\dot{y}_1 = y_1 + y_2 - y_1(y_1^2 + y_2^2)$, $\dot{y}_2 = -y_1 + y_2 - y_2(y_1^2 + y_2^2)$.

has a critical point at $(0,0)$, but in the vicinity of this critical point the equations have the appropriate form $\dot{\varepsilon}_1 \simeq \varepsilon_1^2 + \varepsilon_2^2$, $\varepsilon_2 \simeq \varepsilon_1^4 + \varepsilon_2^4$, $\varepsilon_1, \varepsilon_2 \to 0$, which is still nonlinear.

Example 9 (System with a limit circle) The nonlinear system

$$\dot{y}_1 = y_1 + y_2 - y_1(y_1^2 + y_2^2), \quad \dot{y}_2 = -y_1 + y_2 - y_2(y_1^2 + y_2^2)$$

has a critical point at $(0,0)$. An analysis of the approximate linear system near this point $\dot{\varepsilon}_1 \simeq \varepsilon_1 + \varepsilon_2$, $\dot{\varepsilon}_2 \simeq -\varepsilon_1 + \varepsilon_2$ $(\varepsilon_1, \varepsilon_2 \to 0)$ shows that $(0,0)$ is an unstable spiral point having clockwise rotation. Next, we examine the system above at ∞. If y_1 and y_2 are large, then $\dot{y}_1 \simeq -y_1 z$, $\dot{y}_2 \simeq -y_2 z$ $(y_1, y_2 \to \infty)$, where $z = y_1^2 + y_2^2$, which implies that distant trajectories move toward the origin as $t \to \infty$.

By continuity, there must be at least one trajectory at a moderate distance from the origin that neither moves inward nor outward. Such a trajectory must be a closed orbit which encircles the origin. Trajectories outside this orbit may approach but not cross it. We've thus inferred the existence of a limit cycle from pure local analysis. By local analysis alone it is impossible to say how many limit cycles there are. There might be several concentric orbits. It is easy to solve the system exactly and show that there is just one. We multiply the first and second equations by y_1 and y_2, and add the results. Setting $z = y_1^2 + y_2^2$ we obtain $\dot{z} = 2z - 2z^2$. This equation is similar to that analyzed in Example 5. There is a circular limit cycle having a radius of $z = y_1^2 + y_2^2 = 1$. The exact solution is plotted in Fig. 1.5(b). It exhibits all that we have deduced from local analysis $\dot{y}_1 \simeq -y_1 z$, $\dot{y}_2 \simeq -y_2 z$ $(y_1, y_2 \to \infty)$.

The structure of nonlinear critical points can be much more complicated than that of linear critical points. There can be saddle points having many in and out directions and nodes for which the trajectories are not asymptotic to straight lines as $t \to \infty$. No simple matrix methods exist for identifying the structure of such nonlinear critical points. Extemporaneous analysis is often required. There is a more subtle difficulty with linear critical-point analysis when the critical point is a center. If a linear approximation to a nonlinear system has a center (i.e., the eigenvalues are imaginary), it is still *not correct* to conclude that the nonlinear system also has a center. A center is a specific kind of critical point for which the orbits *exactly* close. Any distortion or perturbation of a closed orbit can give an open orbit. Small distortions of nodes, spiral points, and saddle points do not change the qualitative features of these critical points. Therefore, even though a linear approximation to a nonlinear system may have a center, the nonlinear system may actually have a spiral point.

Example 10 (System with a spurious center) The system

$$\dot{y}_1 = -y_2 + y_1(y_1^2 + y_2^2), \quad \dot{y}_2 = y_1 + y_2(y_1^2 + y_2^2)$$

has a critical point $(0,0)$. A linear approximation in the vicinity of $(0,0)$ is $\dot{\varepsilon}_1 \simeq -\varepsilon_2$, $\dot{\varepsilon}_2 \simeq \varepsilon_1 (\varepsilon_1, \varepsilon_2 \to 0)$. This linear system has a center. The exact nonlinear system has a *spiral point* and not a center. To see this we multiply the first and second equations by y_1 and y_2, and add the resulting equations. Hence, we obtain $y_1\dot{y}_1 + y_2\dot{y}_2 = (y_1^2 + y_2^2)^2$. In polar coordinates this becomes $\frac{1}{2}\frac{d}{dt}(r^2) = r^4$. Thus, the radius r *increases* with increasing t and we have an unstable spiral point. In fact, since the exact solution is $r(t) = r(0)\sqrt{1 - 2r^2(0)t}$, we see that $r(t)$ reaches ∞ in the finite time $t = 1/2r^2(0)$. To prove that a critical point is really a center one must demonstrate the existence of closed orbits. No approximate methods may be used in the proof. The usual technique consists of integrating the system once to construct a time-independent quantity and is often called energy integral. An energy integral for the system will be very difficult to find when the appropriate integrating factors are not obvious.

1.1.3 Higher-order autonomous systems

While the behavior of first- and second- order autonomous systems of ODEs is simple and easy to analyze in the phase plane, solutions to autonomous systems of order three or more can be very complicated. The properties of solutions to high-order systems are the subject of much current research interest and our discussion of them here is limited to a brief nontechnical survey.

A critical point is *stable* if the eigenvalues of the system of equations obtained by linearizing the nonlinear system in the neighborhood of the critical point have negative real parts. This is by far the simplest case. It can be proved that all trajectories of the full nonlinear equations that originate sufficiently close to a stable point always decay toward that critical point as $t \to \infty$. The nonlinear effects do not change the qualitative (or even quantitative) behavior of a system near a stable critical point.

Example 11 (Behavior of third-order system near a stable critical point) A point $(0,0,0)$ is a stable critical point of the system

$$\dot{x} = -x + x^3y^2 - x^2y^3z, \quad \dot{y} = -y + z^3, \quad \dot{z} = -z + x^4 - z^4$$

because the linear system has three negative eigenvalues. The linearized system is just $\dot{x} = -x$, $\dot{y} = -y$, $\dot{z} = -z$, so the linearized behavior is just $x(t) = e^{-t}x(0)$, $y(t) = e^{-t}y(0)$, $z(t) = e^{-t}z(0)$. This behavior persists in the nonlinear system. There, it can be shown that

$$x(t) \simeq \sum_{n=1}^{\infty} a_n e^{-nt}, \quad y(t) \simeq \sum_{n=1}^{\infty} b_n e^{-nt}, \quad z(t) \simeq \sum_{n=1}^{\infty} c_n e^{-nt}, \quad t \to \infty$$

for sufficiently small $|x(0)|$, $|y(0)|$, $|z(0)|$. In fact, $a_2 = b_2 = c_2 = a_3 = c_3 = 0$, $b_3 = -c_1^2/2$, and so on. To test these conclusions, the ratios $x(t)/z(t)$ and $y(t)/z(t)$ are presented in Fig. 1.6(a) for initial conditions $x(0) = y(0) = z(0) = 1$. Equations predict that these ratios should approach the constants a_1/c_1 and b_1/c_1, respectively, as $t \to \infty$. Figure 1.6(a) verifies this prediction.

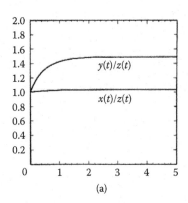 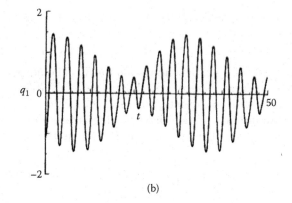

(a)   (b)

FIGURE 1.6: (a) The ratios y/z and x/z of $\dot{x} = -x + x^3 y^2 - x^2 y^3 z$, $\dot{y} = -y + z^3$, $\dot{z} = -z + x^4 - z^4$. (b) $q_1(t)$ with $m = 3$ and the initial conditions $p_{10} = -1$, $p_{20} = 0.6$, $p_{30} = 0.3$, $q_{10} = -1$, $q_{20} = 0$ and $q_{30} = 1$.

A *simple center* is a critical point at which all the eigenvalues of the linearized system are pure imaginary and distinct. This case is perhaps the most difficult. To begin with, the solution to the linearized system need not be periodic because the eigenfrequencies may not be commensurate. For example, a real fourth-order linear system could have eigenvalues $\pm i$, $\pm i\sqrt{2}$. Can we see that there are solutions to be not periodic? All solutions of a linear system at a simple center are *almost periodic* in the sense that there exist arbitrary large time periods T over which the solution repeats itself to any prespecified tolerance. Mathematically speaking, for any $\varepsilon > 0$, there exists an unbounded sequence of time periods T_1, T_2, T_3, such that, for each T_i $|y(t + T_i) - y(t)| < \varepsilon$ for all t. The behavior of a nonlinear system in the vicinity of a simple center can be even more complicated. The orbits may no longer be almost periodic; they may exhibit very complicated random behavior.

Example 12 (Toda lattice: a simple center with almost periodic orbits) The system

$$\dot{q}_j = p_j, \quad \dot{p}_j = e^{q_{j-1} - q_j} - e^{q_j - q_{j+1}},$$

where $j = 1, 2, \ldots, m$, $m > 1$, and $q_0 = q_m$, $q_{m+1} = q_1$, is known as equations of the Toda lattice. The point $p_1 = p_2 = \cdots = p_m = 0$, $q_1 = q_2 = \cdots = q_m = 0$ is a simple center. Despite the nonlinearity of this system, it can be proved that the solutions to these equations are almost periodic for all m. In Fig. 1.6(b) there is plotted $q_1(t)$ versus t; the solution is almost periodic.

Hamiltonian systems. The system considered in Example 12 is a *Hamiltonian system*.

Definition 4 Let a function $H \in C^{r+1}(\mathcal{D})$, $r \geq 1$, be defined on an open subset $\mathcal{D}$ of $\mathbb{R}^{2n}$ (or of a $2n$-dimensional differentiable manifold). The integer n is called the number of degrees of freedom, and H is the *Hamilton function* or *Hamiltonian*. Denote elements of $\mathcal{D}$ by $(\mathbf{q}, \mathbf{p}) = (q_1, \ldots, q_n, p_1, \ldots, p_n)$. The *canonical equations* associated with H define a *Hamiltonian flow* on $\mathcal{D}$:

$$\dot{q}_i = \partial H / \partial p_i, \quad \dot{p}_i = \partial H / \partial q_i \quad (i = 1, \ldots, n). \tag{1.6}$$

The "skew-symmetric" structure of (1.6) is called *symplectic structure*.

The simplest example of a Hamiltonian is $H = \frac{1}{2}|p|^2 + V(q)$ $(n = 1)$, whose canonical equations

$$\dot{q} = p, \quad \dot{p} = -\dot{V}$$

are those of a particle in a potential V. The advantage of (1.6) is that they are invariant under a large class of coordinate transformations, and thus are adapted to mechanical systems with constraints. For the Toda lattice case, $H = \frac{1}{2}\sum_j p_j^2 + \sum_j e^{q_j-1-q_j}$.

Hamiltonian systems have two important properties. First, the Hamiltonian is an energy of the motion. This means that $H(\mathbf{p}(t), \mathbf{q}(t)) = H(\mathbf{p}(0), \mathbf{q}(0))$, hence

$$\frac{d}{dt} H(\mathbf{p}(t), \mathbf{q}(t)) = \sum_{j=1}^{n} \left(\frac{\partial H}{\partial p_j} \dot{p}_j + \frac{\partial H}{\partial q_j} \dot{q}_j \right) = 0.$$

Second, Hamiltonian systems conserve volume in phase space $\mathcal{D}$. This means that the end points of trajectories, which ordinate from all points inside a region of volume V in $\mathcal{D}$ at any t, at time t fill a region with the same volume V for all t. In other words, the Jacobian $J(t) = \partial(\mathbf{p}(t), \mathbf{q}(t))/\partial(\mathbf{p}(0), \mathbf{q}(0))$ satisfies $J(t) = 1$. Since Hamiltonian systems preserve volumes in phase space, stable critical points must be centers (e.g., a Hamiltonian cannot exhibit the kind of stable critical point in Example 11).

Proposition 1 H *is a constant of the motion. The Hamiltonian flow preserves the volume element* $\mathbf{dq}\,\mathbf{dp} = dq_1 \ldots dq_n\, dp_1 \ldots dp_n$. *A curve* $\gamma = \{(\mathbf{q}(t), \mathbf{p}(t)) : t_1 \leq t \leq t_2\}$ *is a trajectory of (1.6) if and only if the action integral*

$$S(\gamma) = \int_{\gamma} (\mathbf{p} \cdot \mathbf{dq} - H\,dt) = \int_{t_1}^{t_2} [\mathbf{p}(t) \cdot \dot{\mathbf{q}}(t) - H(\mathbf{q}(t), \mathbf{p}(t))]\,dt$$

is stationary with respect to all variations of γ with fixed end points.

Proof. In the *extended phase space* $\mathcal{D} \times \mathbb{R}$, the canonical equations define the vector field

$$f(\mathbf{q}, \mathbf{p}, t) = \left(\frac{\partial H}{\partial \mathbf{p}}, -\frac{\partial H}{\partial \mathbf{q}}, 1 \right).$$

Let $M(t)$ be the image of a compact set $M(0)$ in $\mathcal{D}$ under the flow $\varphi(t)$. Define a cylinder $\mathcal{C} = \{(M(s), s) : 0 \leq s \leq t\}$ in $\mathcal{D} \times \mathbb{R}$. Then, by Gauss's theorem,

$$\int_{M(t)} \mathbf{dq}\,\mathbf{dp} - \int_{M(0)} \mathbf{dq}\,\mathbf{dp} = \int_{\partial \mathcal{C}} f \cdot \mathbf{dn} = \int_{\mathcal{C}} (\text{div}\, f)\, \mathbf{dq}\,\mathbf{dp}\,dt = 0,$$

where $\mathbf{dn}$ is a normal vector to $\partial \mathcal{C}$, and we used the fact that $f \cdot \mathbf{dn} = 0$ on the sides of $\mathcal{C}$.

Let $\delta = \{\xi(t), \eta(t) : t_1 \leq t \leq t_2\}$ be a curve vanishing for $t = t_1$ and $t = t_2$. The first variation of the action in the direction δ is defined as

$$\frac{d}{d\varepsilon} S(\gamma + \varepsilon\delta)|_{\varepsilon=0} = \int_{t_1}^{t_2} \left(\mathbf{p}\dot{\xi} + \dot{\mathbf{q}}\eta - \frac{\partial H}{\partial \mathbf{p}}\eta - \frac{\partial H}{\partial \mathbf{q}}\xi \right) dt = \int_{t_1}^{t_2} \left(\xi(-\dot{\mathbf{p}} - \frac{\partial H}{\partial \mathbf{q}}) + \eta(\dot{\mathbf{q}} - \frac{\partial H}{\partial \mathbf{p}}) \right) dt,$$

where we integrated the term $\mathbf{p}\dot{\xi}$ by parts. Now, this expression vanishes for all perturbations δ if and only if $(\mathbf{q}(t), \mathbf{p}(t))$ satisfied the canonical equations (1.6). $\qquad\square$

Definition 5 The *Poisson bracket* of two differentiable functions $f, g : \mathcal{D} \to \mathbb{R}$ is given as

$$[f, g] = \sum_i \frac{\partial f}{\partial q_i} \frac{\partial g}{\partial p_i} - \frac{\partial f}{\partial p_i} \frac{\partial g}{\partial q_i}.$$

In particular, $\dot{f} = [f, H]$. Since $[\cdot, \cdot]$ is an antisymmetric bilinear form, we find that $\dot{H} = [H, H] = 0$. A function $J : \mathcal{D} \to \mathbb{R}$ is a *constant of motion* if and only if $[J, H] = 0$. If J_1 and J_2 have this property then $[J_1, J_2]$ is also a constant of motion; and J_1 and J_2 are *in involution* if $[J_1, J_2] = 0$.

For one degree of freedom ($n = 1$), orbits of the Hamiltonian flow are the level curves of H. With more degrees of freedom, the manifolds $\{H = \text{const}\}$ are invariant, but the motion within each manifold may be complicated. If there are additional constants of motion, there will be invariant manifolds of smaller dimension, in which the motion may be easier to analyze. To do this, we would like to choose coordinates which exploit these constants of motion, but keeping the symplectic structure of the differential equation. Transformations which achieve this are called *canonical transformations*.

Definition 6 A twice continuously differentiable function $S(\mathbf{q}, \mathbf{Q})$ in an open set of $\mathbb{R}^{2n}$, such that

$$\det \frac{\partial^2 S}{\partial \mathbf{q} \, \partial \mathbf{Q}} \neq 0 \quad \forall \, \mathbf{q}, \mathbf{Q}, \tag{1.7}$$

is called a *generating function*. Introduce the following functions:

$$\mathbf{p}(\mathbf{q}, \mathbf{Q}) = \frac{\partial S}{\partial \mathbf{q}}(\mathbf{q}, \mathbf{Q}), \quad \mathbf{P}(\mathbf{q}, \mathbf{Q}) = -\frac{\partial S}{\partial \mathbf{Q}}(\mathbf{q}, \mathbf{Q}). \tag{1.8}$$

By (1.7), we can invert the relation $\mathbf{p} = \mathbf{p}(\mathbf{q}, \mathbf{Q})$ with respect to $\mathbf{Q}$, and thus the relations (1.8) define a transformation $(\mathbf{q}, \mathbf{p}) \to (\mathbf{Q}, \mathbf{P})$ in some open domain of $\mathbb{R}^{2n}$.

Proposition 2 *The transformation* $(\mathbf{q}, \mathbf{p}) \to (\mathbf{Q}, \mathbf{P})$ *is volume-preserving and canonical. The new equations of motion are given by*

$$\dot{\mathbf{Q}}_i = \partial K / \partial P_i, \quad \dot{\mathbf{P}}_i = -\partial K / \partial Q_i \quad (i = 1, \ldots, n),$$

where $K(\mathbf{Q}, \mathbf{P}) = H(\mathbf{q}, \mathbf{p})$ *is the old Hamiltonian expressed in the new variables.*

Proof. Note that $dS = \sum_i p_i dq_i - P_i dQ_i$. For any i and a given $M \subset \mathbb{R}^2$, we get by Green's theorem

$$\int_M (dq_i dp_i - dQ_i dP_i) = \int_{\partial M} (p_i dq_i - P_i dQ_i) = \int_{\partial M} \left(\frac{\partial S}{\partial q_i} dq_i - \frac{\partial S}{\partial Q_i} dQ_i \right) = \int_{\partial M} dS,$$

but this last integral vanishes since ∂M consists of one or several closed curves. This proves that $dq_i dp_i = dQ_i dP_i$ for every i. For any solution of (1.6) we have

$$\int_{t_1}^{t_2} (\mathbf{P} \cdot d\mathbf{Q} - H \, dt) = \int_{t_1}^{t_2} (\mathbf{p} \cdot d\mathbf{q} - H \, dt) - \int_{t_1}^{t_2} dS.$$

The first integral on the right-hand side is stationary, and the second depends only on the end points. Thus the integral of $\mathbf{P} \, d\mathbf{Q} - H \, dt$ is stationary with respect to variations of the curve, and by Proposition 1, $(\mathbf{Q}, \mathbf{P})$ satisfies the canonical equations (1.8). $\qquad \square$

Consider for the moment the case $n = 1$. Then, the orbits are level curves of H. It would thus be useful to construct a canonical transformation such that one of the new variables is constant on each invariant curve. Such variables are called *action-angle variables*.

Definition 7 Let the level curve $H^{-1}(h) = \{(\mathbf{q}, \mathbf{p}) \in \mathcal{D} : H(\mathbf{q}, \mathbf{p}) = h\}$ be bounded. Then its *action* is

$$I(h) = \frac{1}{2\pi} \int_{H^{-1}(h)} \mathbf{p} \, d\mathbf{q}. \tag{1.9}$$

Assume further that $\dot{I}(h) \neq 0$ on some interval, so that the map $I = I(h)$ can be inverted to give $h = h(I)$. If $\mathbf{p}(h, \mathbf{q})$ is the solution of $H(\mathbf{q}, \mathbf{p}) = h$ with respect $\mathbf{p}$, define a generating function

$$S(I, \mathbf{q}) = \int_{\mathbf{q}_0}^{\mathbf{q}} \mathbf{p}(h(I), s) \, ds. \tag{1.10}$$

The function $\mathbf{p}(h, \mathbf{q})$ is not uniquely defined in general, but the definition (1.10) makes sense for any parameterization of the curve $H(\mathbf{q}, \mathbf{p}) = h$. The transformation from $(\mathbf{q}, \mathbf{p})$ to action-angle variables (I, φ) is defined implicitly by

$$\mathbf{p}(I, \mathbf{q}) = \partial S/\partial \mathbf{q}, \quad \varphi(I, \mathbf{q}) = \partial S/\partial I.$$

The normalization in (1.9) has been chosen in such a way that $\varphi \to 2\pi - 0$ as $\mathbf{q} \to \mathbf{q}_0 - 0$, and thus φ in indeed an angle. The Hamiltonian in action angle takes the form $K(\varphi, I) = h(I)$, and the associated canonical equations are simply

$$\dot{\varphi} = \partial K/\partial I = h'(I), \quad \dot{I} = -\partial K/\partial \varphi = 0.$$

The first relation means that the angle φ is an "equal-time" parameterization of the level curve. Indeed,

$$\varphi(t) = \int_{\mathbf{q}_0}^{\mathbf{q}(t)} \frac{\partial \mathbf{p}}{\partial H}(h(I), s) \frac{dh}{dI} \, ds = \int_{\mathbf{q}_0}^{\mathbf{q}(t)} \frac{ds}{\dot{\mathbf{q}}} \frac{dh}{dI} = (t - t_0) \frac{dh}{dI}.$$

In the case $n > 1$, there is no reason for constants of motion other than H (and functions of H). There are many cases where additional constants of motion exist, for instance when H is invariant under a (continuous) symmetry group (Noether's theorem, see [6]).

Definition 8 A Hamiltonian with n degrees of freedom is called *integrable* if it has n constants of motion $J_1, \ldots, J_n$ such that

  a) the J_i are the involution, i.e., $[J_i, J_k] = 0$, $\forall i, k$;

  b) the gradients of the J_i are everywhere linearly independent in $\mathbb{R}^{2n}$.

It has been shown by V.I. Arnold that if a given level set of $J_1, \ldots, J_n$ is compact, then it typically has the topology of an n-dimensional torus. One can then introduce action-angle variables in a similar way as for $n = 1$, and the Hamiltonian takes the form $K(\varphi, I) = H_0(I)$.

Systems that exhibit random behavior. One class of equations known as C-systems exhibits random behavior. Roughly speaking, a C-system is one for which:

  1. *There exists an energy integral which confines the trajectory to a finite volume in the phase space.*

  2. *There are no critical points.*

  3. *For each trajectory Γ in phase space there exist nearby trajectories, some of which move away from and some of which approach the trajectory Γ with increasing t.*

There are no C-systems in less than three dimensions. There are systems of ODEs which are not C-systems (because they have critical points) which also exhibit random behavior.

Example 13 (Lorentz model) The Lorentz model is the third-order system with parameter $r \in \mathbb{R}$,

$$dx/dt = -3(x - y), \quad dy/dt = -xz + rx - y, \quad dx/dt = xy - z.$$

If $r < 1$ then the only critical point is $(0, 0, 0)$ and it is stable; thus the system cannot

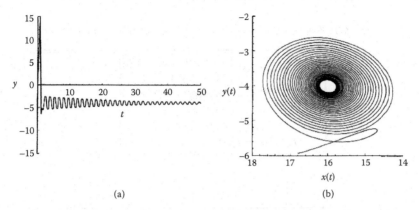

(a)   (b)

FIGURE 1.7: The Lorentz model for $r = 17$ with initial conditions $x(0) = z(0) = 0$, $y(0) = 1$. (a) $y(t)$. (b) $y(t)$ versus $z(t)$ ($1 \leq t \leq 50$). The slow spiral approaches to the point $(-4, -4, 16)$.

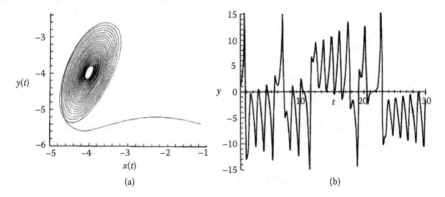

(a)   (b)

FIGURE 1.8: The Lorentz model for $r = 17$ and the same initial conditions as in Fig. 1.7. (a) $y(t)$ versus $x(t)$ for $1 \leq t \leq 50$. (b) $y(t)$.

exhibit random behavior when $x(0)$, $y(0)$, $z(0)$ are small. If $r > 1$ then the origin is an unstable critical point; there also appear critical points at $x = y = \pm\sqrt{r} - 1$, $z = r - 1$, which are stable for $1 < r < 21$.

To illustrate the behavior of trajectories near these stable critical points, take $r = 17$ and $x(0) = y(0) = z(0) = 1$, and plot $y(t)$; see Fig. 1.7(a). Observe that $y(t) \to -4$ as $t \to \infty$. In Fig. 1.7(b) we plot $y(t)$ versus $z(t)$, and in Fig. 1.8(a) we plot $x(t)$ versus $y(t)$. There is a lovely oscillatory approach to the stable critical point $(-4, -4, 16)$. If $r > 21$, the critical points at $x = y = \pm\sqrt{r - 1}$, $z = r - 1$ and at $x = y = z = 0$ are all unstable. In Figure 1.8(b) we plot $y(t)$ versus $x(t)$ for the trajectory starting at $x(0) = y(0) = 0$, $z(0) = 1$ with $r = 26$. The randomness of the trajectory consists of haphazard jumping back and forth from neighborhoods of the critical points $(\pm 5, \pm 5, 25)$.

1.1.4   Dirac delta function

Dirac delta function is the most important *generalized function*. We will show shortly that the concept of Green's functions is tried to this most "unusual" function. Despite the delta function's fundamental role in electrical engineering and quantum mechanics, there existed several schools of thought concerning its exact nature because Dirac's definition,

$$\delta(t) = \begin{cases} \infty, & t = 0, \\ 0, & t \neq 0 \end{cases}$$

TABLE 1.1: *Useful relationships involving the Dirac delta function.*

$\delta(t) = \begin{cases} \infty, & t = 0 \\ 0, & t \neq 0 \end{cases}$	$\int_{-\infty}^{\infty} \delta(t)dt = 1$
$\int_{-\infty}^{\infty} \delta(\tau - t)f(\tau)\mathrm{d}\tau = f(t)$	
$\delta(ct) = \delta(t)/|c|$	$\delta(-t) = \delta(t)$
$\delta(t) = -t\delta'(t)$	
$f(\tau)\delta(\tau - t) = f(t)\delta(\tau - t)$	
$\delta(t^2 - \tau^2) = \frac{\delta(t+\tau)\delta(t-\tau)}{2|\tau|}$	
$x^n \delta^{(m)}(x) = 0$	if $0 \leq m < n$
$x^n \delta^m(x) = (-1)^n \frac{m!}{(m-n)!}\delta^{(m-n)}(x)$	if $0 \leq n < m$

such that $\int_{-\infty}^{\infty} \delta(t)dt = 1$, was unsatisfactory; no conventional function could be found to be satisfactory. One approach, especially popular with physicists intuition of a point mass or charge, sought to view the delta function as a limit of a sequence of peaked functions $\delta_n(t)$: $\delta(t) = \lim_{n\to\infty} \delta_n(t)$. Candidates include

$$\delta_n(t) = \frac{n}{\pi} \cdot \frac{1}{1 + n^2 t^2}, \quad \delta_n(t) = \frac{n}{\sqrt{\pi}} e^{-n^2 t^2}, \quad \delta_n(t) = \frac{1}{n\pi} \cdot \frac{\sin^2(nt)}{t^2}. \tag{1.11}$$

The difficulty with this approach is that the limit of the sequences may not exist.

Another approach, favored by electrical engineers, involves the *Heaviside step function*:

$$H(t) = \begin{cases} 1, & t > 0, \\ 0, & t < 0. \end{cases}$$

Then $\delta(t) = H'(t)$ – the derivative of $H(t)$. The difficulty here is that the derivative does not exist at $t = 0$. One can define the delta function on the basis of its *sifting property*, see Table 1.1,

$$\int_{-\infty}^{\infty} \delta(\tau - t)f(\tau)\mathrm{d}\tau = f(t). \tag{1.12}$$

Example 14 (i) Using the periodic set of pulsed with period $2L$ and defined over the interval $[-L, L]$ by $\delta_n(x) = \begin{cases} n/2, & |x| < 1/n, \\ 0, & |x| > 1/n, \end{cases}$ the Fourier series representation of $\delta_n(x)$ is

$$\delta_n(x) = \frac{1}{2L} + \sum_{m=1}^{\infty} \frac{n}{m\pi} \sin\left(\frac{m\pi}{nL}\right) \cos\left(\frac{m\pi x}{L}\right).$$

The above representation for the delta function follows by letting $n \to \infty$ and we find that $\delta(x) = \frac{1}{2L} + \frac{1}{L}\sum_{n=1}^{\infty} \cos(n\pi x/L)$. This series is divergent; otherwise, $\delta(x)$ will be a conventional function.

(ii) Let us show that our Fourier series representation of $\delta(x)$ possesses the sifting property (1.12). Consider integral $\int_{-\infty}^{\infty} f(x)\delta(x)\,\mathrm{d}x$, where $f(x)$ is a conventional function. Thus,

$$\int_{-L}^{L} f(x)\delta(x)\mathrm{d}\tau = \frac{1}{2L}\int_{-L}^{L} f(x)\,\mathrm{d}x + \frac{1}{L}\sum_{n=1}^{\infty}\int_{-L}^{L} f(x)\cos(\frac{n\pi x}{L})\,\mathrm{d}x = \frac{A_0}{2} + \sum_{n=1}^{\infty} A_n = f(0),$$

where A_0 and A_n are the Fourier coefficients of $f(x)$.

(iii) Let us use the delta sequence technique to compute the Fourier transform of the delta function. As in the case of Fourier series, we first find the Fourier transform of $\delta_n(x)$ with $L \to \infty$ and then take the limit as $n \to \infty$. Therefore,

$$\Delta_n(\omega) = \int_{-\infty}^{\infty} \delta_n(t)e^{-i\omega t}\,dt = \int_{-1/n}^{1/n} \frac{n}{2}e^{-i\omega t}\,dt = \frac{\sin(\omega/n)}{\omega/n}.$$

Finally, the Fourier transform of delta function follows by taking $n \to \infty$, $F[\delta(t)] = \lim_{n\to\infty} \Delta_n(\omega) = 1$.

Although our use of delta sequences has netted several useful results, they are suspect because the limits of these sequences do not exist according to common definitions of convergence. It is the purpose of what is known as the *theory of distributions of generalized functions* to put $\delta(x)$ on a firm mathematical basis and to unify the many *ad hoc* mathematical approaches used by engineers and scientists. The theory of distributions is concerned with the problem of extending the definition of a function that we may include expressions such as $\delta(x)$. This process might be thought of as akin to the manner in which natural numbers were extended to include integers, integers and rational numbers, and rational to real numbers. Although there are several methods open to us, we will use a sequential approach. Consider integrals of sequences of functions of the type $\int_{-\infty}^{\infty} g_n(x)\varphi(x)\,dx$, where $n \in \mathbb{N}$. A sequence of functions $g_n(x)$, such as delta sequences (1.11), leads to a new concept, such as the delta function, provided such sequence of integrals converges for any suitable function $\varphi(x)$. What does one mean by "suitable" for $\varphi(x)$? Because we wand to define concepts such as $\delta'(x)$, $\delta''(x)$, and so forth, then $\varphi(x)$ should be infinitely differentiable (possess derivatives of all orders). Furthermore, $\varphi(x)$ should vanish identically outside of a bounded region so that it behaves properly at infinity. A function $\varphi(x)$ that satisfies these requirements is called a *test function*. An example of a test function is

$$\varphi(x) = \begin{cases} e^{\frac{-a^2}{a^2-x^2}}, & |x| < a, \\ 0, & |x| \geq a, \end{cases} \qquad a > 0.$$

Having introduced test functions, we proceed to define the class of *admissible functions* from which the functions $g_n(x)$ will be selected. Although there is some choice in this matter, we require that these admissive functions be infinitely differentiable over the entire range $(-\infty, \infty)$, with their behavior at infinity left arbitrary. It is these functions that we are extending to encompass other (not infinitely differentiable) functions as well as distributions such as the delta function. Having introduced test and admissible functions, we define a *weakly convergent sequence* as one where the limit $\lim_{n\to\infty} \int_{-\infty}^{\infty} g_n(x)\varphi(x)\,dx$ exists for all test functions $\varphi(x)$. A weakly convergent sequence may or not be convergent in any of the conventional definitions such as pointwise convergent, uniformly convergent, convergent in the mean, and so forth. Although we could have extended the admissible functions by means of other types of convergence, the extension by weak convergence turns out to be very powerful.

Definition 9 A *distribution* $g(x)$ is a "function" associated with a weakly convergent sequence of admissible functions for which the symbolic integral $\int_{-\infty}^{\infty} g(x)\varphi(x)\,dx$ means

$$\int_{-\infty}^{\infty} g(x)\varphi(x)\,dx = \lim_{n\to\infty} \int_{-\infty}^{\infty} g_n(x)\varphi(x)\,dx.$$

For example, the sequences in (1.11) are equivalent because $\lim_{n\to\infty} \int_{-\infty}^{\infty} g_n(x)\varphi(x)\,dx = \varphi(0)$, for all $\varphi(x)$. The distribution $\delta(x)$ defined by any of these sequences is defined as delta function.

TABLE 1.2: Dirac delta function in various coordinate systems.

Coordinate/Dim-s	Three	Two	One
Cartesian	$\delta(x-\xi)\delta(y-\eta)\delta(z-\zeta)$	$\delta(x-\xi)\delta(y-\eta)$	$\delta(x-\xi)$
Cylindrical	$\dfrac{\delta(r-\rho)\delta(\varphi-\varphi')\delta(z-\zeta)}{r}$	$\dfrac{\delta(r-\rho)\delta(z-\zeta)}{2\pi r}$	$\dfrac{\delta(r-\rho)}{2\pi r}$
Spherical	$\dfrac{\delta(r-\rho)\delta(\theta-\theta')\delta(\varphi-\varphi')}{r^2\sin\theta}$	$\dfrac{\delta(r-\rho)\delta(\theta-\theta')}{2\pi r^2\sin\theta}$	$\dfrac{\delta(r-\rho)}{4\pi r^2}$

Example 15 We solve ODEs that, in their simplest form, are similar to

$$\dot{y} + ay(t) = \delta(t-\tau). \tag{1.13}$$

The left-hand side (LHS) is a conventional differential operator; on the RHS we have the peculiar delta function. Clearly $y(t)$ must be very strange in its own right. How would the theory of distributions handle? Because (1.13) involves a delta function, we should multiply it by a test function $\varphi(t)$ and integrate from $t=-\infty$ to $t=\infty$. From the definition of the delta function,

$$\int_{-\infty}^{\infty}\left(\dot{y}+ay(t)\right)\varphi(t)dt = \int_{-\infty}^{\infty}\delta(t-\tau)\varphi(t)dt = \varphi(\tau)$$

for every test function $\varphi(t)$. Integrating by parts, we obtain

$$\int_{-\infty}^{\infty} y(t)[a\varphi(t)-\dot{\varphi}(t)]dt = \varphi(\tau) \tag{1.14}$$

for all functions $\varphi(t)$ because $y(t)\varphi(t)$ vanishes at infinity. Therefore, $y(t)$ is a distribution or generalized function such that (1.14) is satisfied for all sufficiently good test functions $\varphi(t)$. Although (1.14) is formally correct, it still does not give us a method to find $y(t)$. So far we deal with the delta function only as it applies in one dimension. When it comes to PDEs, we need a corresponding definition of the multidimensional delta function $\delta(\mathbf{r})$, $\mathbf{r}\in\mathbb{R}^d$,

$$\int_V f(\mathbf{r})\delta(\mathbf{r}-\mathbf{r}_0)dV = \begin{cases} f(\mathbf{r}_0), & \mathbf{r}_0\in V, \\ 0, & \mathbf{r}_0\notin V, \end{cases}$$

where $\mathbf{r}$ is the vector from the origin to some point $P(x,y,z)$ and $\mathbf{r}_0$ is the vector from the origin to another point $P(\xi,\eta,\zeta)$. Although there is no restriction on the number of dimensions involved and $f(\mathbf{r})$ can be a scalar or vector function, $f(\mathbf{r})$ must be defined at the point $P(\xi,\eta,\zeta)$. If $f(\mathbf{r})$ equals unity, then the delta function is normalized and is of unit magnitude in the sense that the integral of the delta function over the coordinates involved equals one. Following this convention, we list in Table 1.2 the representation for $\delta(\mathbf{r}-\mathbf{r}_0)$ in the three most commonly used coordinate systems.

For a three-dimensional orthogonal coordinate system with elements of length h_iu_i, where h_i are the scale factors and u_i are the curvilinear coordinates, the definition of delta function can be written as

$$\delta(\mathbf{r}-\mathbf{r}_0) = \frac{\delta(u_1-\xi_1)}{h_1}\cdot\frac{\delta(u_2-\xi_2)}{h_2}\cdot\frac{\delta(u_3-\xi_3)}{h_3}. \tag{1.15}$$

For two dimensions, one of the terms on the RHS of (1.15) vanishes. This does not allow us to arbitrarily omit one of the terms because the integral of the delta function must still equal unity. The proper procedure replaces the denominator of the RHS of (1.15) by the integral of the three scale factors over the coordinate that is ignored. For example, in the case of spherical coordinates with no φ-dependence, the denominator becomes $\int_0^{2\pi} r^2 \sin(\theta) d\varphi = 2\pi r^2 \sin\theta$. If the problem involves spherical coordinates, but with no dependence on either φ, or θ, the denominator becomes $\int_0^{\pi} \int_0^{2\pi} r^2 \sin(\theta) \, d\varphi \, d\theta = 4\pi r^2$.

1.1.5 Special functions

Special functions are particular mathematical functions which have established names and notations due to their importance in analysis, physics or other applications. Consider a second-order ODE

$$y'' + P(x)y' + Q(x)y(x) = 0, \tag{1.16}$$

where $P(x)$, $Q(x)$ are continuous functions on R. The general solution to (1.16) has a form $y(x) = C_1 y_1(x) + C_2 y_2(x)$, where $y_1(x)$ and $y_2(x)$ are two linearly independent particular solutions.

Definition 10 If $P(x)$ or $Q(x)$ are analytic functions at $x = x_0$ then x_0 is called a *regular point*. If either $P(x)$ or $Q(x)$ are not defined at $x = x_0$, then x_0 is called a *singular point*. If x_0 is a singular point but $(x - x_0)P(x)$ and $(x - x_0)^2 Q(x)$ remain to be analytic functions at $x = x_0$ and they have finite limits at $x \to x_0$, then x_0 is called a *regular singular point*.

Example 16 The point $x = 0$ is a regular singular point of Euler equation

$$x^2 y'' + axy' + by(x) = 0 \quad (a, b \neq 0).$$

Theorem 1 (Frobenius) *If $x = x_0$ is a regular point of (1.16), then we have the first solution to (1.16) in the form $y_1(x) = \sum_{n=0}^{\infty} A_n(x - x_0)^{n+\lambda}$, where λ is a root of the indicial equation*

$$\lambda(\lambda - 1) + p_1\lambda + p_2 = 0, \quad p_1 = \lim_{x \to x_0} (x - x_0)P(x) < \infty, \quad p_2 = \lim_{x \to x_0} (x - x_0)^2 Q(x) < \infty.$$

There are three possibilities:

1. If $\lambda_1 \neq \lambda_2$ and $\lambda_1 - \lambda_2 \notin \mathbb{Z}$ then there are two independent solutions to (1.16) of the form

$$y_1(x) = \sum_{n=0}^{\infty} A_n(x - x_0)^{n+\lambda_1}, \quad A_0 \neq 0,$$
$$y_2(x) = \sum_{n=0}^{\infty} B_n(x - x_0)^{n+\lambda_2}, \quad B_0 \neq 0.$$

2. If $\lambda_1 \neq \lambda_2$ and $\lambda_1 - \lambda_2 \in \mathbb{Z}$ then there are two independent solutions to (1.16) of the form

$$y_1(x) = \sum_{n=0}^{\infty} A_n(x - x_0)^{n+\lambda_1}, \quad A_0 \neq 0,$$
$$y_2(x) = ay_1(x) \ln x + \sum_{n=0}^{\infty} B_n(x - x_0)^{n+\lambda_2}, \quad B_0 \neq 0, \quad a = \text{const}.$$

3. If $\lambda_1 = \lambda_2$ then there are two independent solutions to (1.16) of the form

$$y_1(x) = \sum_{n=0}^{\infty} A_n(x - x_0)^{n+\lambda_1}, \quad A_0 \neq 0,$$
$$y_2(x) = y_1(x) \ln x + \sum_{n=0}^{\infty} B_n(x - x_0)^{n+\lambda_2}, \quad B_0 \neq 0.$$

Bessel functions, first defined by D. Bernoulli, then generalized by F. Bessel, are canonical solutions $y(x)$ of Bessel's ODE for an arbitrary real or complex α (the order of Bessel function):

$$x^2 y'' + xy' + (x^2 - \alpha^2)y = 0. \tag{1.17}$$

A point $x = 0$ is regular singular. The solutions to (1.17) are *Bessel functions* of order α.

The most common and important special case is when α is an integer. Although α and $-\alpha$ produce the same ODE, it is conventional to define different Bessel functions for these two cases (so that the Bessel functions are mostly smooth functions of α). Bessel functions are also known as *cylinder functions* or *cylindrical harmonics* because they are found in the solution to Laplace's equation in cylindrical coordinates.

Although the order α can be any real number, the book is limited to nonnegative integers, n, unless specified otherwise. Since the Bessel equation is a second-order ODE, two sets of functions, the Bessel function of the first kind $J_n(x)$ and the Bessel function of the second kind (also known as the *Neumann function*) $X_n(x)$, are needed to form the general solution: $y(x) = C_1 J_n(x) + C_2 X_n(x)$. For integer order n, J_n is often defined via a *Laurent series* for a generating function

$$\exp\left[\frac{1}{2}x(t - 1/t)\right] = \sum_{n \in \mathbb{Z}} J_n(x) t^n.$$

An important relation for integer orders is the *Jacobi-Anger expansion*

$$e^{iz \cos \phi} = \sum_{n \in \mathbb{Z}} i^n J_n(x) e^{in\phi},$$

which is used to expand a plane wave as a sum of cylindrical waves. Another way to define Bessel functions is *Poisson representation formula*

$$J_k(z) = \frac{(z/2)^k}{\Gamma(k + 1/2)\Gamma(1/2)} \int_{-1}^{1} e^{izs}(1 - s^2)^{k - \frac{1}{2}} ds,$$

where $k > -\frac{1}{2}$ and z is a complex number. This formula is useful when working with Fourier transformation. Bessel functions have asymptotic forms for $\alpha \geq 0$. For $0 < x \ll \sqrt{\alpha + 1}$, see [1, Chapter 3]:

$$J_\alpha(x) \to \frac{1}{\Gamma(\alpha + 1)}\left(\frac{x}{2}\right)^\alpha, \quad X_\alpha(x) \to \begin{cases} \frac{2}{\pi}\left[\ln\left(\frac{1}{2}x\right) + \gamma\right], & \alpha = 0, \\ -\frac{\Gamma(\alpha)}{\pi}\left(\frac{x}{2}\right)^{-\alpha}, & \alpha > 0, \end{cases}$$

where $\gamma = e^C \simeq 0.5772$ is *Euler-Mascheroni constant* and Γ denotes *Gamma function*. For large arguments $x \gg \sqrt{\alpha + 1}$, they become

$$J_\alpha(x) \to \sqrt{\frac{2}{\pi x}} \cos\left(x - \frac{1}{2}\pi\alpha - \frac{1}{4}\pi\right), \quad X_\alpha(x) \to \sqrt{\frac{2}{\pi x}} \sin\left(x - \frac{1}{2}\pi\alpha - \frac{1}{4}\pi\right).$$

Definition 11 Two important linearly independent solutions to Bessel's equation are *Hankel functions* $H_\alpha^{(1)}$ and $H_\alpha^{(2)}$, defined by $H_\alpha^{(1,2)} = J_\alpha(x) \pm i X_\alpha(x)$. Thus, we find

$$H_\alpha^{(1,2)} = \frac{J_{-\alpha}(x) - e^{\mp i\pi\alpha} J_\alpha(x)}{\pm i \sin(\pi\alpha)}.$$

Functions J_α, X_α, $H_\alpha^{(1)}$ and $H_\alpha^{(2)}$ all satisfy the recurrent relations, where Z is J_α, X_α, $H_\alpha^{(1)}$, or $H_\alpha^{(2)}$:

$$\frac{2\alpha}{x} Z_\alpha(x) = Z_{\alpha-1}(x) + Z_{\alpha+1}(x), \quad 2\frac{dZ_\alpha}{dx} = Z_{\alpha-1}(x) - Z_{\alpha+1}(x).$$

These identities are often combined (added or subtracted) to yield various other relations. For example, one can compute Bessel functions of higher orders (or higher derivatives) given the values at lower orders:

$$\left(\frac{d}{x\,dx}\right)^m [x^\alpha Z_\alpha(x)] = x^{\alpha-m} Z_{\alpha-m}(x), \quad \left(\frac{d}{x\,dx}\right)^m [Z_\alpha(x) x^{-\alpha}] = (-1)^m x^{-(\alpha+m)} Z_{\alpha+m}(x).$$

Approximate formulas for the Bessel functions roots. For approximate calculation of the Bessel functions $J_\nu(x)$ positive roots $\mu_n^{(\nu)}$ one may apply the McMahon formulas [14],

$$\mu_n^{(\nu)} = \alpha_n - (4\nu^2 - 1)(8\alpha_n)^{-1} - (4/3)(4\nu^2 - 1)(28\nu^2 - 41)(8\alpha_n)^{-3}, \qquad (1.18)$$

where $\alpha_n = \frac{1}{2}\pi[\nu + (4n-1)/2]$. For $\nu = 0$ we have

$$\mu_n^{(0)} = \frac{1}{2}\theta_n \cdot \left\{1 + \frac{1}{2}\theta_n^{-2} - \frac{31}{24}\theta_n^{-4}\right\}, \qquad \theta_n = \frac{1}{2}\pi(4n-1). \qquad (1.19)$$

For approximate calculation of the Bessel functions $J_\nu(x)$ positive roots (high numbers) they use asymptotic representation

$$J_\nu(x) = \sqrt{2/\pi}\, x^{-1/2} \cos\left(x - (\pi/2)(\nu + 1/2)\right) + O(x^{-3/2}) \quad \text{for } x \to \infty,$$

and hence, $\mu_n^{(\nu)} \approx \frac{3}{4}\pi + \frac{1}{2}\pi\nu + \pi n$ for large $n \in \mathbb{N}$. Again, $\mu_n^{(0)} \approx \pi(n + \frac{3}{4})$ for $\nu = 0$ and large n.

Heun functions. The representative form of the Bessel-Heun equation (BHE) of order $2m$ is

$$z'' + \frac{1}{r}z' + \sum_{p=0}^{2m} A_p r^{p-2} z = 0 \qquad (1.20)$$

under condition $A_{2m} + \frac{1}{4}m^2 = 0$; see [37, p. 196]. For $m = 2$, equation (1.20) becomes

$$z'' + \frac{1}{r}z' + \left(\frac{A_0}{r^2} + \frac{A_1}{r} + A_2 + A_3 r + A_4 r^2\right)z = 0 \qquad (1.21)$$

with $A_4 = -1$. Using transformation

$$z(r) = r^{B_0/2} e^{-\frac{1}{2}(B_1 r + r^2)} y(r), \qquad A_0 = -\frac{1}{4}B_0^2, \qquad A_3 = -B_1$$

and its derivations

$$z'(r) = e^{-\frac{1}{2}(B_1 r + r^2)}\left\{\left[\frac{1}{2}B_0 r^{\frac{1}{2}B_0 - 1} - r^{\frac{1}{2}B_0}\left(\frac{1}{2}B_1 + r\right)\right]y(r) + r^{\frac{1}{2}B_0}y'\right\},$$
$$z'' = r^{\frac{1}{2}B_0 - 2} e^{-\frac{1}{2}(B_1 r + r^2)}\left\{r^2 y'' + r\left[B_0 - 2r\left(\frac{1}{2}B_1 + r\right)\right]y'(r)\right.$$
$$\left. + \frac{1}{2}B_0\left[(r^2 - 1)\left(\frac{1}{2}B_1 + r\right) + \left(\frac{1}{2}(B_0 - B_1) - 1\right)\right]y(r)\right\},$$

we find a canonical form of (1.21) with $A_1 = -\frac{1}{2}B_3$, $A_2 = B_2 - \frac{1}{4}B_1^2$; see [37, pp. 194–195]:

$$y'' + \left(r^{-1}(1 + B_0) - B_1 - 2r\right)y' + \left\{(B_2 - B_0 - 2) - \frac{1}{2}r^{-1}(B_3 + (1 + B_0)B_1)\right\}y(r) = 0. \qquad (1.22)$$

Denote by $N(B_0, B_1, B_2, B_3; r)$ the solution of (1.22) with initial conditions:

$$N(B_0, B_1, B_2, B_3; 0) = 1, \qquad N'(B_0, B_1, B_2, B_3; 0) = \frac{1}{2}[B_3 + B_1(1 + B_0)].$$

Since $r = 0$ is a regular point for (1.22) then due to the Frobenius Theorem we find a particular solution ([37, p. 203]) in the form $y(r) = \sum_{n=0}^{\infty} b_n r^{n+\rho}$. Substituting $y(r) = \sum_{n=0}^{\infty} b_n r^{n+\rho}$ in (1.22) yields

$$\begin{aligned}
&\rho(\rho + B_0) = 0, \\
&(\rho + 1)(\rho + 1 + B_0)b_1 - \left\{B_1\rho + \frac{1}{2}[B_3 + B_1(1 + B_0)]\right\} = 0, \\
&(\rho + 2 + n)(\rho + 2 + n + B_0)b_{n+2} - \left\{B_1(\rho + 1 + n) + \frac{1}{2}[B_3 + B_1(1 + B_0)]\right\} b_{n+1} \\
&- [(B_2 - B_0) - 2(\rho + n + 1)] = 0.
\end{aligned} \qquad (1.23)$$

For $B_0 \in \mathbb{N}$, let $N(B_0, B_1, B_2, B_3; r)$ be such a solution to (1.22) with $\rho = 0$ that $b_0 = 1$. We write it as

$$N(B_0, B_1, B_2, B_3; r) = \sum_{n=0}^{\infty} \frac{a_n}{(1 + B_0)_n} \frac{r^n}{n!}, \qquad (1.24)$$

where $(B_0)_n = \Gamma(B_0 + n)/\Gamma(B_0)$ for $n \geq 0$, and

$a_0 = 1$, $a_1 = \frac{1}{2}[B_3 + B_1(1 + B_0)]$,
$a_{n+2} = [(j + 1)B_1 + a_1]a_{n+1} - (n + 1)(n + 1 + B_0)[(B_2 - B_0) - 2(n + 1)]a_n$.

The convergence radius for expansion (1.24) may be found following the D'Alembert test,

$$R_n = \left| \frac{(1 + B_0)_{n+1}}{(1 + B_0)_n} \frac{(n + 1)!}{n!} \right|.$$

The convergence radius $R_n = |(n + 1)(B_0 + n)|$ tends to infinity when $n \to \infty$ because $\Gamma(n) = (n - 1)!$

Remark 1 The *Gamma function* $\Gamma(n)$ is an extension of the factorial to complex and real number arguments. It is defined as a definite integral for $\operatorname{Re} z = 0$ (Euler's integral form)

$$\Gamma(z) = \int_0^{\infty} t^{z-1} e^{-t} dt. \qquad (1.25)$$

It is analytic everywhere except at $z = -1, -2, \ldots$, and $\operatorname{Res}_{z=-k} \Gamma(z) = \frac{(-1)^k}{k!}$ is the *residue* at $z = -k$. There are no solutions of equation $\Gamma(z) = 0$. Integrating (1.25) by parts for a real argument gives

$$\Gamma(z) = \int_0^{\infty} t^{z-1} e^{-t} dt = [-t^{z-1} e^{-t}]\big|_0^{\infty} + (z - 1) \int_0^{\infty} t^{z-2} e^{-t} dt = (z - 1)\Gamma(z - 1).$$

For $n \in \mathbb{N}$ we have $\Gamma(n) = (n - 1)\Gamma(n - 1)$, thus the function reduces to factorial in this case.

Using the Frobenius Theorem one can prove the following proposition.

Proposition 3 *Canonical equation (1.22) has the following fundamental solution (depending on B_0):*
   (a) *If $B_0 \in \mathbb{N}$ then*

$$\begin{aligned} y_1(r) &= N(B_0, B_1, B_2, B_3; r), \\ y_2(r) &= A(B_0, B_1, B_2, B_3)N(B_0, B_1, B_2, B_3; r)\ln r + \sum_{n=0}^{\infty} d_n r^{n - B_0} \end{aligned}$$

with $d_0 = 1$, d_{B_0} to be arbitrary, and

$A(B_0, B_1, B_2, B_3) = \frac{1}{B_0} \left\{ \frac{1}{2} d_{B_0 - 1}[B_3 + B_1(B_0 - 1)] - d_{B_0 - 1}(B_2 - B_0 + 2) \right\}$;
$(1 + B_0)d_{B_0 + 1} - [B_3 + B_1(B_0 - 1)]d_{B_0} + (B_2 - B_0)d_{B_0 - 1} = -A[(B_0 + 2)b_1 - B_0]$,
$d_1 = \frac{1}{2}[B_3 + B_1(B_0 - 1)]/(1 - B_0)$, $B_0 \geq 2$;
$(n + 2)(n + 2 - B_0)d_{n+2} - \frac{1}{2}[B_3 + B_1(2n + 3 - B_0)]d_{n+1} + [B_2 - 2(n + 1) + B_0]d_n = 0$,
$0 \leq n \leq B_0 - 3$, $B_0 \geq 3$;
$(n + 2)(n + 2 - B_0)d_{n + B_0 + 2} - \frac{1}{2}[B_3 + B_1(2n + 3 - B_0)]d_{n + B_0 + 1}$
$+[B_2 - 2(n + 1) + B_0]d_{n + B_0} + A[(2n + 4 + B_0)b_{n+2} - B_1 b_{n+1} - 2b_n] = 0$, $n \geq 0$,

where the sequence $\{b_n\}_{n \geq 0}$ is given by (1.23).

(b) *If $B_0 = 0$, then*

$$y_1(r) = N(0, B_1, B_2, B_3; r),$$

$$y_2(r) = A(0, B_1, B_2, B_3)N(0, B_1, B_2, B_3; r)\ln r + \sum_{n=1}^{\infty} d_n r^{n-B_0},$$

where

$$d_1 = -B_3, \quad d_2 = \frac{1}{4}\left(B_2 - 2B_1B_3 - \frac{3}{4}B_3^2 - \frac{1}{4}B_1^2\right),$$

$$(n+3)^2 d_{n+3} - \left[(n+2)B_1 + \frac{1}{2}(B_1 + B_3)\right]d_{n+2} + (B_2 - 2n - 4)d_{n+1}$$

$$= -2(n+3)b_{n+3} + B_1 b_{n+2} + 2b_{n+1}, \quad n \geq 0.$$

1.1.6 Green's function

An important aspect of mathematical physics is the solution of linear nonhomogeneous ODEs. We express this problem as $Ly = f$, where L is an ordinary linear differential operator involving the variable t such that $f(t)$ is a known function and $y(t)$ the desired solution. We solve the problem using expansions of eigenfunctions given by the eigenvalue problem $Ly = \lambda y$, where $\lambda \in \mathbb{R}$. The method consists of finding solutions to $Lg(t, \tau) = \delta(t - \tau)$, where τ is an arbitrary point of excitation. The solution $y(t)$ is given by an integral involving the *Green's function* $g(t, \tau)$ (more commonly written as $g(t|\tau)$) and $f(\tau)$.

Example 17 Let us find the displacement $u(x)$ of a string of length L that is connected at both ends to supports and is subjected to a load (external force per unit length) of $f(x)$. If the load $f(x)$ acts downward (in negative direction), the displacement $u(x)$ of the string is given by ODE

$$T\ddot{u} = f(x), \tag{1.26}$$

where T is the uniform tensile force of the string. Because the string is stationary at both ends, the displacement $u(x)$ satisfies the boundary conditions $u(0) = u(L) = 0$. For the load $\delta(x - \xi)$ concentrated at $x = \xi$, (1.26) becomes

$$T\ddot{g} = \delta(x - \xi), \quad g(0|\xi) = g(L|\xi) = 0. \tag{1.27}$$

The Green's function $g(x|\xi)$ corresponds to the displacement of the string when it is subjected to an impulse load $x = \xi$. In line with our circuit theory example, it gives the *Green's function* for the static problem. The displacement $u(x)$ of the string subject to arbitrary load $f(x)$ can be found by convolving the load $f(x)$ with the Green's function $g(x|\xi)$ as follows: $u(x) = \int_0^L g(x|\xi)f(\xi)\,d\xi$.

Let us find this Green's function. At a point $x \neq \xi$, we get ODE $g_x''(x|\xi) = 0$, whose solution is

$$g(x|\xi) = \begin{cases} ax + b, & 0 \leq x < \xi, \\ cx + d, & \xi < x \leq L, \end{cases}$$

where a and c are constants. Applying the boundary conditions $g(0|\xi) = g(L|\xi) = 0$, we find $g(0|\xi) = 0$, and $g(L|\xi) = 0$, or $d = -cL$. Therefore,

$$g(x|\xi) = \begin{cases} ax, & 0 \leq x < \xi, \\ c(x - L), & \xi < x \leq L, \end{cases} \tag{1.28}$$

At $x = \xi$, the displacement $u(x)$ must be continuous; otherwise, the string would be broken. Therefore, $g(x|\xi)$ given by (1.28) must be continuous. Hence, $a\xi = c(\xi - L)$ or $c = \frac{a\xi}{\xi - L}$. By

(1.27) the second derivative of $g(x|\xi)$ is equal to the impulse function. Therefore, the first derivative of $g(x|\xi)$, obtained by integrating (1.27), is discontinuous by the amount $1/T$ or $\lim\limits_{\varepsilon \to 0} \left[\frac{dg(\xi+\varepsilon|\xi)}{dx} - \frac{dg(\xi-\varepsilon|\xi)}{dx} \right] = \frac{1}{T}$, in which case

$$\frac{dg(\xi^+|\xi)}{dx} - \frac{dg(\xi^-|\xi)}{dx} = \frac{1}{T}, \tag{1.29}$$

where ξ^+ and ξ^- are the points lying above or below ξ, respectively. Using (1.28), we find $\frac{dg(\xi^-|\xi)}{dx} = a$ and $\frac{dg(\xi^+|\xi)}{dx} = \frac{a\xi}{\xi-L}$. Hence (1.29) leads to $\frac{a\xi}{\xi-L} - a = \frac{aL}{\xi-L} = \frac{1}{T}$ or $a = \frac{\xi-L}{LT}$, and

$$g(x|\xi) = \frac{1}{TL}(x_> - L)x_< = \frac{1}{TL} \cdot \begin{cases} x(\xi - L), & 0 \le x < \xi \\ \xi(x - L), & \xi < x \le L \end{cases}.$$

To find the displacement $u(x)$ subject to the load $f(x)$, we proceed as we did in the above:

$$u(x) = \int_0^L f(\xi)g(x|\xi)\,d\xi = \frac{x-L}{TL}\int_0^x f(\xi)\xi\,d\xi + \frac{x}{TL}\int_x^L f(\xi)(\xi-L)\,d\xi,$$

since $\xi < x$ in the first integral and $x < \xi$ in the second one.

Maxwell's reciprocity. By the theory of ODEs, every differential equation of the form

$$Ly = a_0(x)u^{(n)} + a_1(x)u^{(n-1)} + \ldots + a_{n-1}(x)\dot{u} + a_n(x)u(x) = 0$$

has associated with it an *adjoint equation*

$$L^*y = (-1)^n(a_0(x)v)^{(n)} + (-1)^{n-1}(a_1(x)v)^{(n-1)} + \ldots - (a_{n-1}(x)v)' + a_n(x)v = 0.$$

The concept of adjoint is useful because

$$vL(u) - uL^*(v) = \frac{d}{dx}P(u,v), \tag{1.30}$$

where $P(u,v)$ is known as the *bilinear concomitant*[1]. Applying these considerations to Green's functions, we discover the important property of *reciprocity*. We will prove for second-order ODEs (the results can be generalized to higher orders):

$$a_0(\xi)\frac{d^2g(\xi|x)}{d\xi^2} + [2a_0'(\xi) - a_1(\xi)]\frac{dg(\xi|x)}{d\xi} + [a_0''(\xi) - a_1'(\xi) + a_2(\xi)]g(\xi|x) = \delta(\xi - x). \tag{1.31}$$

Let $g(\xi|x)$ obey homogeneous boundary conditions at $\xi = a$ and $\xi = b$. The associated adjoint problem is

$$a_0(\xi)\frac{d^2g^*(\xi|x')}{d\xi^2} + a_1(\xi)\frac{dg^*(\xi|x')}{d\xi} + a_2(\xi)g^*(\xi|x') = \delta(\xi - x') \tag{1.32}$$

plus suitable homogeneous boundary conditions. Here we add a prime to x because x and x' are not necessarily the same. Let us compute (1.30) and find

$$g(\xi|x)\big[a_0(\xi)\tfrac{d^2g^*(\xi|x')}{d\xi^2} + a_1(\xi)\tfrac{dg^*(\xi|x')}{d\xi} + a_2(\xi)g^*(\xi|x')\big]$$
$$-g^*(\xi|x')\big\{a_0(\xi)\tfrac{d^2g(\xi|x)}{d\xi^2} + [2a_0'(\xi) - a_1(\xi)]\tfrac{dg(\xi|x)}{d\xi} + [a_0''(\xi) - a_1'(\xi) + a_2(\xi)]g(\xi|x)\big\}$$
$$= \tfrac{d}{d\xi}\big\{a_0(\xi)g(\xi|x)\tfrac{dg^*(\xi|x')}{d\xi} + [a_1(\xi) - a_0'(\xi)]g(\xi|x)g^*(\xi|x') - a_0(\xi)g^*(\xi|x')\tfrac{dg(\xi|x)}{d\xi}\big\}.$$

---

[1]Ince E. L. Ordinary Differential Equations. Dover Publ., Inc., 1956, Section 5.3.

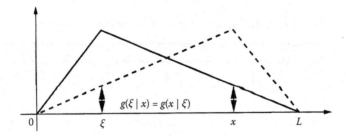

FIGURE 1.9: Maxwell's reciprocity.

Integrating from a to b, we get

$$\int\limits_a^b g(\xi|x)[a_0(\xi)\frac{d^2g^*(\xi|x')}{d\xi^2} + a_1(\xi)\frac{dg^*(\xi|x')}{d\xi} + a_2(\xi)g^*(\xi|x')]d\xi$$

$$= \int\limits_a^b g^*(\xi|x')\{a_0(\xi)\frac{d^2g(\xi|x)}{d\xi^2} + [2a_0'(\xi) - a_1(\xi)]\frac{dg(\xi|x)}{d\xi} + [a_0''(\xi) - a_1'(\xi) + a_2(\xi)]g\}d\xi,$$

because both $g(\xi|x)$ and $g^*(\xi|x')$ satisfy homogeneous boundary conditions. Substituting (1.31) and (1.32), we find $\int_a^b \delta(\xi - x')g(\xi|x)d\xi = \int_a^b g^*(\xi|x')\delta(\xi - x)d\xi$, or $g(x'|x) = g^*(x|x')$. If the differential operators in (1.31) and (1.32) are *self-adjoint*, i.e., $L = L^*$, then $g(x|\xi) = g(\xi|x)$.

Example 18 In Section 1.1.6 it was shown that the Green's function for the ODE

$$g''(x|\xi) = \delta(x - \xi), \quad g(0|\xi) = g(L|\xi) = 0,$$

is $g(x|\xi) = (L - x_>)x_<$, and it is symmetric, $g(x|\xi) = g(\xi|x)$, because the differential operator $L = d^2/dx^2$ is a self-adjoint one. This result is illustrated in Figs. 1.11(b) and 1.9.

Laplace transform. The *Laplace transform* is an integral transform perhaps next to the Fourier transform in its utility in solving physical problems. The Laplace transform is particularly useful in solving linear ODEs such as those arising in the analysis of electronic circuits. The Laplace transform is defined by

$$F[f](s) = \int_0^\infty f(t)e^{-st}dt, \quad t > 0, \quad s = \sigma + i\omega \in \mathbb{C}.$$

The inverse Laplace transform is known as the Bromwich integral, sometimes known as the Fourier-Mellin integral (see also the Duhamel's convolution principle). Several important one-sided Laplace transforms are given in Table 1.3. Here, $J_0(t)$ is the zero-order Bessel function of the first kind, $\delta(t)$ is the delta function, and $H_c(t)$ is the Heaviside step function

$$H_c(t) = \begin{cases} 1, & t > c, \\ 0, & t < c. \end{cases}$$

The Laplace transform existence theorem states that if $f(t)$ is piecewise continuous on every finite interval in $[0, \infty)$ satisfying $|f(t)| \le Me^{at}$ for all $t \in [0, \infty)$, then $F[f](s)$ exists for all $s > a$. The Laplace transform is also unique, in the sense that given two functions $f_1(t)$ and $f_1(t)$ with the same transform so that $F[f_1](s) = F[f_2](s) \equiv F(s)$, then *Lerch's theorem* guarantees that the integral $\int_0^a N(t)dt = 0$ vanishes for all $a > 0$ for a null function

TABLE 1.3: *Important one-side Laplace transforms.*

f	$F[f]$	range
1	$1/s$	$s > 0$
t	$1/s^2$	$s > 0$
t^n	$n!/s^{n+1}$	$n \in \mathbb{N}$
t^a	$\Gamma(a+1)/s^{a+1}$	$a > 0$
e^{at}	$1/(s-a)$	$s > a$
$\cos(at)$	$s/(s^2 + a^2)$	$s > 0$
$\sin(at)$	$a/(s^2 + a^2)$	$s > 0$
$\cosh(at)$	$s/(s^2 - a^2)$	$s > |a|$
$\sinh(at)$	$a/(s^2 - a^2)$	$s > |a|$
$e^{at}\sin(bt)$	$b/[(s-a)^2 - b^2]$	$s > a$
$e^{at}\cos(bt)$	$(s-a)/[(s-a)^2 - b^2]$	$s > a$
$\delta(t-\tau)$	$e^{-\tau s}$	$\tau > 0$
$H_c(t)$	e^{-cs}/s	$s > 0$
$J_0(t)$	$1/\sqrt{s^2 + 1}$	
$J_n(at)$	$(\sqrt{s^2 + a^2} - s)^n/(a^n \sqrt{s^2 + a^2})$	$s > 0, \ n > -1$

defined by $N(t) \equiv f_1(t) - f_2(t)$. The Laplace transform is linear since

$$F[af + bg] = \int_0^\infty [af(t) + bg(t)]e^{-st}dt = aF[f] + bF[g].$$

The Laplace transform of a convolution $f(t) * g(t) := \int_0^t f(t-s)g(s)ds$ is given by

$$F[f * g] = F[f]F[g], \quad F^{-1}[F(s)G(s)] = F^{-1}[F(s)] * F^{-1}[G(s)].$$

Now consider differentiation. If $f(t) \in C^{n-1}([0,\infty))$ and $|f(t)| \le Me^{at}$ then

$$F[f^{(n)}](s) = s^n F[f](s) - s^{n-1}f(0) - s^{n-2}f'(0) - \ldots - f^{(n-1)}(0).$$

This can be proved by integration by parts,

$$F[f'] = \lim_{a\to\infty} \int_0^a e^{-st}f'(t)dt = \lim_{a\to\infty} \left\{ \left[e^{-st}f(t)\right]\big|_0^a + s\int_0^a e^{-st}f(t)\,dt \right\} = sF[f](s) - f(0).$$

Continuing for higher-order derivatives then gives $F[f''](s) = s^2 F[f](s) - sf(0) - f'(0)$ etc. The Laplace transform has a number of useful properties, e.g., if $F[f](s) = F(s)$ for $s > a$ (i.e., $F(s)$ is the inverse Laplace transform of $f(t)$) then $F[e^{\alpha t}f](s) = F(s-\alpha)$ for $s > a + \alpha$. This follows from

$$F(s-\alpha) = \int_0^\infty f(t)e^{-(s-\alpha)t}dt = \int_0^\infty [f(t)e^{\alpha t}]e^{-st}dt = F[e^{\alpha t}f(t)](s).$$

The Laplace transform also has nice properties when applied to integrals of functions. If $f(t)$ is piecewise continuous and $|f(t)| \le Me^{at}$, then $F\left[\int_0^t f(\tau)d\tau\right] = \frac{1}{s}F[f(t)](s)$.

Example 19 Consider a problem for analytical description of populations of highly excited levels in rarified plasma. Let $N(n)$ be the population of atoms at sublevel with a set of quantum numbers n. The balance equation can be transformed to the form; see [3],

$$N(\varepsilon)\ln\left(\frac{\varepsilon}{\varepsilon_0}\right) - \int_0^\varepsilon \frac{N(\xi) - N(\varepsilon)}{\varepsilon - \xi}d\xi = \varphi(\varepsilon),$$

where $\varphi(\varepsilon)$ is the density of the rate of upper level population by external processes to the levels of nondimensional ionization energy $\varepsilon = 1/n^2$ of the level n; the value ε_0 refers to the lowest level for a radiative transition. After integration by part we obtain

$$N(\varepsilon)\ln(\varepsilon_0 - \varepsilon) - N(0)\ln\varepsilon = \int_0^{\varepsilon - \delta} \ln(\varepsilon - \xi)N'(\xi)d\xi + \varphi(\varepsilon).$$

Assuming $\varepsilon \ll \varepsilon_0$ for $n \gg 1$, we may write

$$N(\varepsilon)\ln(\varepsilon_0) = N(0)\ln\varepsilon + \int_0^{\varepsilon - \delta} \ln(\varepsilon - \xi)N'(\xi)d\xi + \varphi(\varepsilon).$$

Using Laplace transform we obtain $N(\varepsilon) = \int_0^{\varepsilon} g(\varepsilon|\xi)\varphi(\xi)d\xi$, with $g(\varepsilon|\xi) = g(\varepsilon - \xi) = \frac{1}{\varepsilon_0 e^C}\nu'\left(\frac{\varepsilon - \xi}{\varepsilon_0 e^C}\right)$, where $\nu'(x) = \int_0^{\infty} \frac{x^{s-1}s}{\Gamma(s+1)}ds$ is a tabulated function, and $C = 0.5772$ is the Euler constant.

1.1.7 Boundary and initial value problems

The simplest application of Green's functions to solving ODEs arises with the initial-value problem

$$y^{(n)} + a_1 y^{(n-1)} + \cdots + a_{n-1}y' + a_n y = f(t), \quad t > 0, \tag{1.33}$$

where $a_1, a_2, \ldots$ are constants and all of the valued $y(t), y'(t), \ldots, y^{(n-1)}(t)$ are zero at $t = 0$. Although there are several methods for solving (1.33), let us employ Laplace transform. So,

$$(s^n + a_1 s^{n-1} + \ldots + a_{n-1}s + a_n)X(s) = F(s), \tag{1.34}$$

where $X = F[y]$ and $F = F[f]$ denote the Laplace transforms of $y(t)$ and $f(t)$, respectively. Consequently, once $F(s)$ is known, (1.34) is inverted to give $y(t)$. The most of obvious difficulties is the dependence of solutions on the forcing function $f(t)$. For each new $f(t)$ the inversion process must be repeated. Is there any way to avoid this problem? Rewrite (1.34) as

$$X(s) = \frac{F(s)}{s^n + a_1 s^{n-1} + \ldots + a_{n-1}s + a_n} = G(s)F(s),$$

where $G(s) = (s^n + a_1 s^{n-1} + \cdots + a_{n-1}s + a_n)^{-1}$. By the convolution theorem for Laplace transform,

$$y(t) = g(t) * f(t) = \int_0^t g(t - s)f(s)ds, \tag{1.35}$$

where $g(t)$ is the inverse Laplace transform of $G(s)$. Once we have $g(t)$, we can compute the response $y(t)$ to the force $f(t)$ by an integration. So far, we assumed homogeneous initial conditions.

For nonhomogeneous initial conditions, we would find a homogeneous solution that satisfies the initial conditions and add that solution to our nonhomogeneous solution (1.35). Having shown the usefulness of Green's function, define it for the initial-value problem (1.33). Given a linear ODE, the function $g(t|\tau)$ is the solution of the equation

$$g^{(n)}(t|\tau) + a_1 g^{(n-1)}(t|\tau) + \ldots + a_n g(t|\tau) = \delta(t - \tau), \quad t, \tau > 0. \tag{1.36}$$

The forcing occurs at time $t = \tau$, which is later than the initial time $t = 0$. This avoids the problem of the Green's function *not* satisfying all of the initial conditions.

Remark 2 Although we will use (1.36) as the fundamental definition of the Green's function as it applies to the ODE, we can also find it by solving the initial-value problem

$$u^{(n)} + a_1 u^{(n-1)} + \ldots + a_n u = 0, \quad t > \tau, \tag{1.37}$$

with initial conditions

$$u(\tau) = u'(\tau) = \ldots = u^{(n-2)}(\tau) = 0, \quad u^{(n-1)}(\tau) = 1.$$

Equation (1.37) is an example of a *stationary* system, one where the ODE and initial conditions are invariant under translations in time. The Green's function is related to $u(t)$ via $g(t|\tau) = u(t-\tau)H(t-\tau)$. Indeed, introducing a new time variable $\tilde{t} = t - \tau$, one may show that the Laplace transforms of (1.36) and (1.37) are identical.

Our final task is to prove that $y(t) = \int_0^t g(t|\tau)f(\tau)\,d\tau$. We have

$$\begin{cases} y'(t) = g(t|t)f(t) + \int_0^t g'(t|\tau)f(\tau)\,d\tau, \\ \ldots\ldots\ldots\ldots\ldots\ldots\ldots\ldots\ldots\ldots\ldots\ldots\ldots \\ y^{(n)}(t) = g^{(n-1)}(t|t)f(t) + \int_0^t g^{(n)}(t|\tau)f(\tau)\,d\tau. \end{cases} \tag{1.38}$$

Since the Green's function possesses the properties $g(t|t) = g'(t|t) = \ldots = g^{(n-2)}(t|t) = 0$ and $g^{(n-1)}(t|t) = 1$, then, upon substituting (1.38) into (1.33), we find that

$$y^{(n)} + a_1 y^{(n-1)} + \ldots + a_n y = f(t) + \int_0^t \left[g^{(n)} + a_1 g^{(n-1)} + \ldots + a_n g \right] f(\tau)\,d\tau = f(t),$$

since the bracketed term vanishes except at $\tau = t$. The contribution from this point is infinitesimally small.

Example 20 Let us find the transfer and Green's functions for $y'' - 3y' + 2y = f(t)$ with $y(0) = y'(0) = 0$. Replace the equation by $y'' - 3y' + 2y = \delta(t-\tau)$, with $g(0|\tau) = g'(0|\tau) = 0$. Using the Laplace transform of the above equation, we find $G(s|\tau) = e^{-s\tau}/(s^2 - 3s + 2)$, which is the transfer function for this system when $\tau = 0$. The Green's function equals the inverse of $G(s|\tau)$,

$$g(t|\tau) = \left[e^{2(t-\tau)} - e^{t-\tau} \right] H(t-\tau).$$

Regular boundary value problems. Sturm-Liouville systems. One of our purposes is the solution of a wide class of nonhomogeneous ODEs

$$[p(x)y']' + s(x)y(x) = -f(x), \quad x \in [a, b], \tag{1.39a}$$

with

$$\alpha_1 y(a) + \alpha_2 y'(b) = 0, \quad \beta_1 y(a) + \beta_2 y'(b) = 0. \tag{1.39b}$$

Although this may appear to be a fairly restrictive class of equations, any second-order linear nonhomogeneous ODE can be written as (1.39a). From the countless equations of the form (1.39a), the most commonly encountered are those of the Sturm-Liouville type,

$$[p(x)y']' + [q(x) + \lambda r(x)]y(x) = -f(x), \quad x \in [a, b], \tag{1.40}$$

where λ is a parameter. This is devoted to *regular Sturm-Liouville problems*. From the theory of ODEs, a regular Sturm-Liouville problem has three distinct elements: a finite interval $[a, b]$, continuous $p(x)$, $p'(x)$, $q(x)$ and $r(x)$, and strictly positive $p(x)$ and $r(x)$ on the interval $a \le x \le b$. If any of these conditions is absent then the problem is *singular*.

Example 21 Consider the nonhomogeneous boundary value problem (BVP)

$$y'' + k^2 y = -f(x), \quad y(0) = y(L) = 0.$$

We can rewrite the equation above as $(1 \cdot y')' + (0 + k^2 \cdot 1)y = -f(x)$, and it can be brought into Sturm-Liouville form by choosing $p(x) = 1$, $q(x) = 0$, $r(x) = 1$ and $\lambda = k^2$.

Let us determine the Green's function for the equation

$$[p(x)g']' + s(x)g(x) = -\delta(x - \xi), \tag{1.41}$$

subject to yet undetermined boundary conditions. We know that such a function exists for the special case $p(x) = 1$ and $s(x) = 0$, and we now show that this is *almost always* true in general case. We construct Green's function by requiring that they satisfy the following conditions:

- $g(t|\tau)$ satisfies the *homogeneous* equation $f(x) = 0$ *except* at $x = \xi$,
- $g(t|\tau)$ satisfies the *homogeneous* conditions,
- $g(t|\tau)$ is continuous at $x = \xi$.

These homogeneous boundary conditions for a finite interval (a, b) will be

$$\alpha_1 g(a|\xi) + \alpha_2 g'(b|\xi) = 0, \quad \beta_1 g(a|\xi) + \beta_2 g'(b|\xi) = 0,$$

where g' is the x derivative of $g(x|\xi)$ and neither a nor b equals ξ. The coefficients α_1 and α_2 cannot both be zero; this also holds for β_1 and β_2. These conditions include the commonly encountered Dirichlet, Neumann and Robin boundary conditions.

Because $g(x|\xi)$ is a continuous function of x, (1.41) dictates that there must be discontinuity in $g'(x|\xi)$ at $x = \xi$. Indeed, integrating (1.41) from $\xi - \varepsilon$ to $\xi + \varepsilon$ yields

$$p(x)g'(x|\xi) \big|_{\xi-\varepsilon}^{\xi+\varepsilon} + \int_{\xi-\varepsilon}^{\xi+\varepsilon} s(x)g(x|\xi)\, \mathrm{d}x = -1.$$

Because $g(x|\xi)$ and $s(x)$ are both continuous at $x = \xi$, $\lim_{\varepsilon \to 0} \int_{\xi-\varepsilon}^{\xi+\varepsilon} s(x)g(x|\xi)\, \mathrm{d}x = 0$. Hence, applying the limit $\varepsilon \to 0$ we have $p(x)[g'(\xi^+|\xi) - g'(\xi^-|\xi)] = -1$, where ξ^+ and ξ^- denote points just above and below $x = \xi$, respectively. Consequently, our last requirement on $g(x|\xi)$ is that dg/dx must have a jump discontinuity of magnitude $-1/p(\xi)$ at $x = \xi$. Similar conditions hold for higher-order ODEs.

Let $y_1(x)$ be a nontrivial solution of the *homogeneous* ODE satisfying the boundary condition at $x = a$; then $\alpha_1 y_1(a) + \alpha_2 y_1'(a) = 0$. Because the function $g(x|\xi)$ must satisfy the same boundary condition, we obtain $\alpha_1 g(a|\xi) + \alpha_2 g'(a|\xi) = 0$. Since the set α_1, α_2 is nontrivial, then the Wronskian of $y_1(x)$ and $g(x|\xi)$ vanishes at $x = a$ or $y_1(a)g'(a|\xi) + y_1'(a)g(a|\xi) = 0$. For $a \leq x < \xi$, both $y_1(x)$ and $g(x|\xi)$ satisfy the same ODE, the homogeneous one. Therefore, their Wronskian is zero at all points and $g(x|\xi) = c_1 y_1(x)$ for $a \leq x < \xi$ (here c_1 is an arbitrary constant). Similarly, if the nontrivial function $y_2(x)$ satisfies the homogeneous equation and the boundary condition at $x = b$, then $g(x|\xi) = c_2 y_2(x)$ for $\xi < x \leq b$. The continuity condition of $g(x|\xi)$ and the jump discontinuity of $g'(x|\xi)$ at $x = \xi$ imply

$$c_1 y_1(\xi) - c_2 y_2(\xi) = 0, \quad c_1 y_1'(\xi) - c_2 y_2'(\xi) = 1/p(\xi). \tag{1.42}$$

This system can be solved for c_1 and c_2 provided the Wronskian of y_1 and y_2 does not vanish at $x = \xi$ or $y_1(\xi)y_2'(\xi) - y_2(\xi)y_1'(\xi) \neq 0$. In other words, $y_1(x)$ must *not* be a multiple of $y_2(x)$ that is "generally" true. If the homogeneous equation admits no trivial solutions satisfying both boundary conditions at the same time, then $y_1(x)$ and $y_2(x)$ must be linearly independent. On the other hand, if the homogeneous equation possesses a single solution, say $y_0(x)$, which also satisfies $\alpha_1 y_0(a) + \alpha_2 y_0'(a) = 0$ and $\beta_1 y_0(a) + \beta_2 y_0'(a) = 0$, then $y_1(x)$ will be multiple of $y_0(x)$ and so is $y_2(x)$. Then they are multiples of each other and their Wronskian $W(y_1, y_2)$ vanishes. This would occur, for example, if the ODE is a Sturm-Liouville equation, λ equals the eigenvalue, and $y_0(x)$ is the corresponding eigenfunction. No Green's function exists in this case!

Example 22 Consider the problem of finding the Green's function for

$$g''(x|\xi) + k^2 g(x|\xi) = -\delta(x - \xi), \quad 0 < x < L,$$

subject to the boundary conditions $g(0|\xi) = g(L|\xi) = 0$ with $k \neq 0$. Consequently,

$$g(0|\xi) = c_1 y_1(x) = c_1 \sin kx, \quad 0 \leq x \leq \xi,$$

while $g(x|\xi) = c_2 y_2(x) = c_2 \sin[k(l - x)]$ for $\xi \leq x \leq L$. Let us compute the Wronskian:

$$W(y_1, y_2)(x) = y_1(x)y_2'(x) - y_1'(x)y_2(x) = -k \sin(kx) \cos[k(L - x)]$$
$$-k \cos(kx) \sin[k(L - x)] = -k \sin[k(x + L - x)] = -k \sin kL,$$

and $W(y_1, y_2)(\xi) = -k \sin kL$. Therefore, the Green's function will exist as long as $kL \neq n\pi$. If $kL = n\pi$, $y_1(x)$ and $y_2(x)$ are linearly *dependent* with the solution $y_0(x) = c_3 \sin(n\pi x/L)$ of the regular Sturm-Liouville problem $y'' + \lambda y = 0$ and $y(0) = y(L) = 0$. The system (1.42) has a unique solution

$$c_1 = -\frac{y_2(\xi)}{p(\xi)W(y_1, y_2)(\xi)}, \quad c_2 = -\frac{y_1(\xi)}{p(\xi)W(y_1, y_2)(\xi)}.$$

Therefore,

$$g(x|\xi) = -\frac{1}{p(\xi)W(y_1, y_2)(\xi)} \times \begin{cases} y_1(x)y_2(\xi), & 0 \leq x \leq \xi \\ y_1(\xi)y_2(x), & \xi \leq x \leq L \end{cases} = -\frac{y_1(x_<)y_2(x_>)}{p(\xi)W(y_1, y_2)(\xi)}. \quad (1.43)$$

Clearly $g(x|\xi)$ is symmetric in x and ξ. It is also unique. The proof of the uniqueness is as follows. Choose a different $y_1(x)$, but it will be a multiple of "old" $y_1(x)$, and the Wronskian will be multiplied by the same factor, keeping $g(x|\xi)$. This is also true if we modify $y_2(x)$ in a similar manner.

Example 23 Let us find the Green's function for $g'' + k^2 g = -\delta(k - \xi), 0 < x < L$, subject to the boundary conditions $g(0|\xi) = g(L|\xi) = 0$. As we showed in Example 22,

$$y_1(x) = c_1 \sin kx, \quad y_2(x) = c_2 \sin[k(L - x)], \quad W(y_1, y_2)(\xi) = -k \sin kL.$$

Substituting into (1.43), we obtain

$$g(x|\xi) = \frac{1}{k \sin kL} \times \begin{cases} \sin(kx) \sin[k(L - \xi)], & 0 \leq x \leq \xi \\ \sin(k\xi) \sin[k(L - x)], & \xi \leq x \leq L \end{cases} = \frac{\sin(kx_<) \sin[k(L - x_>)]}{k \sin kL}.$$

Eigenfunction expansion for regular boundary value problems. We present the most common method *series expansion*.

Example 24 Consider the nonhomogeneous problem

$$y'' = -f(x), \quad y(0) = y(L) = 0. \quad (1.44)$$

Then $g(x|\xi)$ must satisfy

$$g''(x|\xi) = -\delta(x-\xi), \quad g(0|\xi) = g(L|\xi) = 0. \tag{1.45}$$

Because $g(x|\xi)$ vanishes at the ends of the interval $(0, L)$, it can be expanded in the series of suitably chosen orthogonal function such as, for instance, the Fourier sine series $g(x|\xi) = \sum_{n=1}^{\infty} G_n(\xi)\sin(\frac{n\pi x}{L})$, where the coefficients G_n are dependent on ξ. Although we have chosen to use the orthogonal functions $\{\sin(n\pi x/L)\}$, we could use other orthogonal functions as long as they vanish at the end points. Because

$$g''(x|\xi) = \sum_{n=1}^{\infty} (-n^2\pi^2/L^2)G_n(\xi)\sin(n\pi x/L) \tag{1.46}$$

and

$$\delta(x-\xi) = \sum_{n=1}^{\infty} A_n(\xi)\sin(\frac{n\pi x}{L}), \quad A_n(\xi) = \frac{2}{L}\int_0^L \delta(x-\xi)\sin(\frac{n\pi x}{L})\,dx = \frac{2}{L}\sin(\frac{n\pi\xi}{L}), \tag{1.47}$$

after substituting (1.46) and (1.47) into (1.45) we obtain

$$-\sum_{n=1}^{\infty} \frac{n^2\pi^2}{L^2} G_n(\xi)\sin(\frac{n\pi x}{L}) = -\frac{2}{L}\sum_{n=1}^{\infty} \sin(\frac{n\pi\xi}{L})\sin(\frac{n\pi x}{L}). \tag{1.48}$$

Since (1.48) must hold for any x, we get $\frac{n^2\pi^2}{L^2}G_n(\xi) = \frac{2}{L}\sin(\frac{n\pi\xi}{L})$. Hence,

$$g(x|\xi) = \frac{2L}{\pi^2}\sum_{n=1}^{\infty} \frac{1}{n^2}\sin(\frac{n\pi\xi}{L})\sin(\frac{n\pi x}{L}).$$

We use this series to construct the solution of (1.44) via the formula $y(x) = \int_0^L g(x|\xi)f(\xi)d\xi$,

$$y(x) = (2L/\pi^2)\sum_{n=1}^{\infty} \frac{1}{n^2}\sin(n\pi x/L)\int_0^L f(\xi)\sin(n\pi\xi/L)d\xi,$$

or $y(x) = \frac{L}{\pi^2}\sum_{n=1}^{\infty} \frac{a_n}{n^2}\sin(\frac{n\pi x}{L})$, where a_n are the Fourier sine coefficients of $f(x)$.

Example 25 Consider the more complicated boundary value problem

$$y'' + k^2 y = -f(x), \quad y(0) = y(L) = 0. \tag{1.49}$$

Then $g(x|\xi)$ must satisfy $g''(x|\xi) = -\delta(x-\xi)$ and $g(0|\xi) = g(L|\xi) = 0$. Again, we use the Fourier sine expansion $g(x|\xi) = \sum_{n=1}^{\infty} G_n(\xi)\sin(n\pi x/L)$. Substitution of this expression and (1.47) into ODE for Green's function and grouping by corresponding harmonics yields

$$-\frac{n^2\pi^2}{L^2} G_n(\xi) + k^2 G_n(\xi) = -\frac{2}{L}\sin\left(n\pi\xi/L\right),$$

or $G_n(\xi) = \frac{2}{L}\frac{\sin(n\pi\xi/L)}{n^2\pi^2/L^2 - k^2}$. Then

$$g(x|\xi) = (2/L)\sum_{n=1}^{\infty} \frac{\sin(n\pi\xi/L)\sin(n\pi x/L)}{n^2\pi^2/L^2 - k^2}.$$

By the last expression, the function enjoys the symmetry property: $g(x|\xi) = g(\xi|x)$.

Summarize the expansion technique as follows. Suppose that we want to solve the ODE $Ly(x) = -f(x)$, with some condition $By(x) = 0$ along the boundary, where L now denotes the Sturm-Liouville differential operator $L = \frac{d}{dx}\left[p(x)\frac{d}{dx}\right] + [q(x) + \lambda r(x)]$, and B is the boundary condition operator

$$B = \begin{cases} \alpha_1 + \alpha_2 \frac{d}{dx} & \text{at} \quad x = a, \\ \beta_1 + \beta_2 \frac{d}{dx} & \text{at} \quad x = b. \end{cases}$$

We begin by seeking $g(x|\xi)$, which satisfies $Lg = -\delta(x - \xi)$, $Bg = 0$. We utilize the set of eigenfunctions $\varphi_n(x)$ associated with the regular Sturm-Liouville problem

$$[p(x)\varphi_n{}']' + [q(x) + \lambda_n r(x)]\varphi_n(x) = 0,$$

where $\varphi_n(x)$ satisfies the same boundary conditions as $y(x)$. If $g(x|\xi)$ exists and if the set $\{\varphi_n(x)\}$ is complete, then $g(x|\xi)$ can be represented by the series $g(x|\xi) = \sum_{n=1}^{\infty} G_n(\xi)\varphi_n(x)$. Applying operator L to this $g(x|\xi)$, we obtain

$$Lg(x|\xi) = \sum_{n=1}^{\infty} G_n(\xi)L\varphi_n(x) = \sum_{n=1}^{\infty} G_n(\xi)(\lambda - \lambda_n)r(x)\varphi_n(x) = -\delta(x - \xi),$$

provided that λ does not equal any of the eigenvalues λ_n. Multiplying both sides of the last expression by $\varphi_m(x)$ and integrating over x,

$$\sum_{n=1}^{\infty} G_n(\xi)(\lambda - \lambda_n) \int_a^b r(x)\varphi_n(x)\varphi_m(x)\,dx = -\varphi_m(\xi).$$

If the eigenfunctions are orthogonal,

$$\int_a^b r(x)\varphi_n(x)\varphi_m(x)\,dx = \begin{cases} 1, & n = m, \\ 1, & n \neq m, \end{cases} \quad \text{and} \quad G_n(\xi) = \frac{\varphi_n(\xi)}{\lambda_n - \lambda}.$$

This leads directly to the *bilinear formula*:

$$g(x|\xi) = \sum_{n=1}^{\infty} \frac{\varphi_n(\xi)\varphi_n(x)}{\lambda_n - \lambda}, \tag{1.50}$$

which permits us to write the Green's function at once if the eigensystem of L is known.

An intriguing aspect of the bilinear formula is that the Green's function possesses poles, called the *point spectrum* at λ_n. For regular second-order ODEs, there are an infinite number of simple poles that all lie along the real λ-axis according to Sturm's oscillation theorem. This leads to the following.

Theorem 2 *The integral*

$$\frac{1}{2\pi i} \oint_C g(x|\xi)d\lambda = -\frac{\delta(x - \xi)}{r(\xi)}, \tag{1.51}$$

where C is a closed contour in the complex λ-plane that includes all of the singularities of $g(x|\xi)$.

Proof. Starting with the bilinear formula (1.50), we apply Cauchy's integral formula and find

$$\frac{1}{2\pi i} \oint_C g(x|\xi)d\lambda = -\frac{1}{2\pi i} \sum_{n=1}^{\infty} \varphi_n(\xi)\varphi_n(x) \oint_C \frac{d\lambda}{\lambda - \lambda_n} = -\sum_{n=1}^{\infty} \varphi_n(\xi)\varphi_n(x).$$

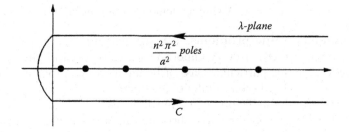

FIGURE 1.10: Closed contour C enclosing the poles of the Green's function (1.52).

From Sturm-Liouville theory, $\delta(x - \xi) = \sum_{n=1}^{\infty} c_n \varphi_n(x)$, where

$$c_n = \frac{\int_a^b r(x) \delta(x - \xi) \varphi_n(x) \, dx}{\int_a^b r(x) \varphi_n^2(x) \, dx} = r(\xi) \varphi_n(\xi),$$

since $\varphi_n(x)$ are orthonormal eigenfunctions. Using $\frac{\delta(x-\xi)}{r(\xi)} = \sum_{n=1}^{\infty} \varphi_n(\xi) \varphi_n(x)$ completes the proof. $\qquad \square$

Example 26 By Example 23, the Green's function for the problem

$$g''(x|\xi) + \lambda g(x|\xi) = -\delta(x - \xi) \quad (0 < x < L)$$

with $g(0|\xi) = g(L|\xi) = 0$ is

$$g(x|\xi) = \frac{\sin(\sqrt{\lambda} x_<) \sin[\sqrt{\lambda}(L - x_>)]}{\sqrt{\lambda} \sin \sqrt{\lambda} L}. \tag{1.52}$$

Let us examine our theorem using the solution over the interval $[0, \xi]$. We must evaluate

$$\oint_C \frac{\sin(\sqrt{\lambda} x_<) \sin[\sqrt{\lambda}(L - x_>)]}{\sqrt{\lambda} \sin \sqrt{\lambda} L} \, d\lambda,$$

where the closure of contour C is shown in Fig. 1.10.

Our first concern is the presence of square roots in the integrand; we might have a multivalued function. Then the Taylor expansion for sine is substituted into the integral, the square roots disappear and we have a single-valued integrand. Next, we must find the location and nature of singularities. Clearly $\lambda = 0$ is one, while $\sqrt{\lambda_n} L = n\pi$, or $\lambda_n = n^2 \pi^2 / L^2$, $n = 1, 2, 3, \ldots$, is another. To discover the nature of the singularities, we use the infinite product formula for sine and find

$$\frac{\sin(\sqrt{\lambda} x) \sin[\sqrt{\lambda}(L - \xi)]}{\sqrt{\lambda} \sin \sqrt{\lambda} L} = \frac{\sqrt{\lambda} x \left(1 - \frac{\lambda x^2}{\pi^2}\right)\left(1 - \frac{\lambda x^2}{9\pi^2}\right) \cdots \left[1 - \frac{\lambda(L-\xi)^2}{\pi^2}\right]\left[1 - \frac{\lambda(L-\xi)^2}{9\pi^2}\right] \cdots}{\sqrt{\lambda}\left(1 - \frac{\lambda L^2}{\pi^2}\right)\left(1 - \frac{\lambda L^2}{9\pi^2}\right) \cdots}.$$

Consequently, $\lambda = 0$ is a removable singularity; $\lambda_n = n^2 \pi^2 / L^2$ are simple poles. Therefore,

$$\frac{1}{2\pi i} \oint_C g(x|\xi) d\lambda = \sum_{n=1}^{\infty} \operatorname{Res}\left\{ \frac{\sin(\sqrt{\lambda} x) \sin[\sqrt{\lambda}(L-\xi)]}{\sqrt{\lambda} \sin(\sqrt{\lambda} L)}; \frac{n^2 \pi^2}{L^2} \right\}$$

$$= \sum_{n=1}^{\infty} \lim_{\lambda \to \lambda_n} \frac{\lambda - \lambda_n}{\sin(\sqrt{\lambda} L)} \lim_{\lambda \to \lambda_n} \frac{\sin(\sqrt{\lambda} x) \sin[\sqrt{\lambda}(L-\xi)]}{\sqrt{\lambda}}$$

$$= \sum_{n=1}^{\infty} \frac{2\sqrt{\lambda_n} \sin(\sqrt{\lambda_n} x) \sin[\sqrt{\lambda_n}(L-\xi)]}{L \cos(\sqrt{\lambda_n} x) \sqrt{\lambda_n}} = \frac{2}{L} \sum_{n=1}^{\infty} (-1)^n \sin[\frac{n\pi}{L}(L - \xi)] \sin(\frac{n\pi x}{L})$$

$$= -\frac{2}{L} \sum_{n=1}^{\infty} (-1)^n \sin(\frac{n\pi \zeta}{L}) \sin(\frac{n\pi x}{L}) = -\delta(x - \xi)$$

and (1.51) holds because $r(\xi) = 1$. A similar demonstration exists for $\xi \leq x \leq L$.

Singular boundary value problems. The *singular Sturm-Liouville problem* (that is not regular) and singular ODE are very important for mathematical physics. Here we seek the Green's function for this particular class of boundary value problems. For example, if $p(x) = 0$ or $p(x)$, see (1.39a), becomes infinite at a point $x = c$ on the interval (a, b), then we have a singular Sturm-Liouville problem. We will focus on two cases: 1) (1.40) is defined on a finite interval $[a, b]$ while $p(x)$ is zero or becomes infinite at some point on this interval, 2) one or both end points lie at $x = \pm\infty$.

The difficulty with the singular Sturm-Liouville problem becomes clear when we construct the Green's function. If we use homogeneous solutions to the second-order ODE, say $y_1(x)$ and $y_2(x)$, as we did in Section 1.1.7, then one or both of these solutions may diverge to infinity at the singular boundary, and it is hopeless to satisfy any boundary condition there. Similarly, a linear combination of these solutions may refuse to satisfy the boundary conditions. On the other hand, both solutions may satisfy the boundary condition and it becomes ambiguous which one to choose. Since most singular Sturm-Liouville problems involve a singular end point, their analysis begins by classifying these points along the lines. As such, there are two types of singular end points:

• *Limit-circle.* In this case, all solutions $y(x)$ from the singular Sturm-Liouville differential equation are *square-integrable*, i.e., $\int |y(x)|^2 r(x)\, dx$ converges at $x = b$. Then, we impose the boundary condition

$$\lim_{x \to b}[p(x)y'(x)f(x) - p(x)f'(x)y(x)] = 0$$

for a suitable $f(x)$. This is an extension of the regular boundary condition to the singular boundary point.

• *Limit-point.* In this case, there is just one solution $y(x)$ (up to scalar factor) for which $\int |y(x)|^2 r(x)\, dx$ converges on $[a, b]$. In this case, no boundary condition is required at $x = b$.

Example 27 A popular method for converting (1.49) into a singular BVP is to consider the limiting process of $L \to \infty$. In that case, the eigenfunction is still $y = \sin kx$. Because that is the only solution valid at $x = \infty$, we have a limit-point. To find the Green's function for this singular Sturm-Liouville problem, we use results from Example 25, and write

$$g(x|\xi)'' + k^2 g(x|\xi) = -\delta(x - \xi), \quad g(0|\xi) = g(L|\xi) = 0.$$

It was shown that

$$g(x|\xi) = \frac{2}{L} \sum_{n=1}^{\infty} \frac{\sin(n\pi\xi/L)\sin(n\pi x/L)}{n^2\pi^2/L^2 - k^2}.$$

As before, this problem is converted into a singular Sturm-Liouville problem by taking the limit of $L \to \infty$. One of the properties of bilinear expansions for regular Sturm-Liouville problems is that the Green's function possesses poles at $\lambda = \lambda_n$. What happens for singular Sturm-Liouville problems? From the example we see that the poles move closer together. When $L = \infty$, the poles coalesce into a continuous branch out in the λ-plane, called the *continuous spectrum*. Similarly to a regular case, the continuous spectrum for the singular Sturm-Liouville problem lies on the λ-axis. Not all ODEs will possess a continuous spectrum. Some contain only the points spectrum, and some have both.

In Theorem 2 the completeness relationship (1.51) was defined, which we formally extended in the case of singular Sturm-Liouville problems. For example, if we only have a continuous spectrum and no point spectrum (a common situation), the completeness relationship becomes

$$\frac{1}{2\pi i} \oint_{C_\infty} g(x|\xi) d\lambda = -\frac{\delta(x-\xi)}{r(\xi)}, \tag{1.53}$$

where the contour C_∞ denotes a circle taken in the positive direction with infinite radius so that it encloses every pole or branch cut in the complex λ-plane. In this way, formula (1.53) holds both for regular as well as for singular problems. For the singular problem, it is generally easier to find the Green's function by solving the ODE as it was done in Section 1.1.7 rather than calculate the equivalent bilinear expansion. Generalizing the completeness relation (1.51), we can derive the equivalent "eigenfunction" expansion for the Green's function that contains a branch cut. We do this by substituting (1.43) into (1.51) and then performing the contour integration over C. Simplifying the final integral, the completeness relation becomes $\delta(x-\xi) = \int \varphi_\omega(x)\tilde{\varphi}_\omega(\xi) d\omega$, where $\varphi_\omega(x)$ and $\tilde{\varphi}_\omega(\xi)$ are any two "eigenfunctions" of the ODE (1.40). The limit of integration over ω will be from 0 to ∞ or from $-\infty$ to ∞, depending how the contour integral was simplified. An application of this completeness relationship is the definition of the integral transform pair:

$$G(\omega|\xi) = \int_0^\infty \tilde{\varphi}_\omega(x) g(x|\xi) \, dx \quad \text{and} \quad g(x|\xi) = \oint_C \varphi_\omega(x) G(\omega|\xi) d\omega. \tag{1.54}$$

In this example, $\varphi_\omega(x) = 2\sin(\omega x)/\pi$ and $\tilde{\varphi}_\omega(x) = \sin(\omega x)$. If the integral over ω in (1.54) occurs between 0 and ∞, we have Fourier sine transform.

Example 28 (Fokker-Planck equation) The *Fokker-Planck equation* is important in stochastic differential equations and plasma physics PDE. In finding its Green's function, we must solve the BVP

$$(x^2 g')' - a\,(xg)' - bg + \lambda g = -\delta(x-\xi) \quad (x, \, \xi > 0). \tag{1.55}$$

This equation is singular for two reasons:
  1) the boundary point $x = \infty$ is singular by definition.
  2) the point $x = 0$ is singular because the coefficients of the derivative terms of (1.55) vanish.
Two independent solutions of the homogeneous form of (1.55) are $u_1(x) = x^{-1+\delta_+}$ and $u_2(x) = x^{-1+\delta_-}$, where $\delta_\pm = \frac{a+1}{2} \pm \mu$, $\mu = \sqrt{\lambda_0 - \lambda}$ and $\lambda_0 = \left(\frac{a+1}{2}\right)^2 + b$. From the classification of singular boundary conditions, $x = 0$ is a limit point and $u_1(x)$ is the only possible solution for $x < \xi$. Similarly, $x = \infty$ is also a limit point and $u_2(x)$ is the only possible solution for $x > \xi$. Using the techniques outlined in Section 1.1.7, we find

$$g(x|\xi) = \frac{1}{2\xi\mu} \begin{cases} (x/\xi)^{-1+\delta_+}, & x < \xi, \\ (x/\xi)^{-1+\delta_-}, & x > \xi. \end{cases} \tag{1.56}$$

By (1.56), this problem has continuous spectrum but no discrete one. Substituting (1.56) into (1.53), we find $\delta(x-\xi) = \frac{1}{2\pi\xi}(x/\xi)^{-(a+1)/2} \int_{-\infty}^\infty (x/\xi)^{i\omega} d\omega$. Thus, the eigenfunctions in this case are $\varphi_\omega(x) = \frac{1}{2\pi} x^{(a-1)/2-i\omega}$ and $\tilde{\varphi}_\omega(\xi) = \xi^{(a+1)/2+i\omega}$, thus, from (1.54) we get the *Mellon transform* pair

$$G(\omega|\xi) = \int_0^\infty x^{-(a+1)/2+i\omega} g(x|\xi) \, dx, \quad g(x|\xi) = \frac{1}{2\pi} \int_{-\infty}^\infty x^{(a-1)/2-i\omega} G(\omega|\xi) d\omega.$$

Example 29 (Bessel's equation) Let us find the Green's function for Bessel's equation

$$xg'' + g' + (k^2 x - m^2/x)g = -\delta(x-\xi), \quad 0 < x < L,$$

where $m \geq 0$ and is an integer. The boundary conditions are $\lim_{x \to 0} |g(x|\xi)| < \infty$ and $g(L|\xi) = 0$. The homogeneous solutions that satisfy the boundary conditions are

$$y_1(x) = J_m(kx), \quad y_2(x) = J_m(kx)X_m(kL) - J_m(kL)X_m(kx). \tag{1.57}$$

Compute the Wronskian using a result from [48]:

$$
\begin{aligned}
W(x) &= y_1(x)y_2'(x) - y_1'(x)y_2(x) = kJ_m(kx)[J_m'(kx)X_m(kL) - J_m(kL)X_m'(kx)] \\
&\quad -kJ_m'(kx)[J_m(kx)X_m(kL) - J_m(kL)X_m(kx)] \\
&= kJ_m(kL)[J_m'(kx)X_m(kx) - J_m(kx)X_m'(kx)] = -\tfrac{2}{\pi x}J_m(kL).
\end{aligned}
$$

Substituting (1.57) and $W(x)$ from the last formula into (1.43) with $p(x) = x$ yields

$$g(x|\xi) = \frac{\pi J_m(kx_<)[J_m(kx_>)X_m(kL) - J_m(kL)X_m(kx_>)]}{2J_m(kL)}, \tag{1.58}$$

which is illustrated in Fig. 1.11(a) when $kL = 7$, and $m = 0$. There are two goals in an expression like (1.58) over an eigenfunction expansion: first, there is no summation of a series outside of evaluating J_m and X_m, and second, we do not have to find the zeros of Bessel functions.

Example 30 (Whittaker equation) Find the free-space Green's function governed by *Whittaker's equation*

$$\left(\frac{d^2}{dz^2} - \frac{1}{4} + \frac{k}{z} + \frac{1/4 - \mu^2}{z^2}\right)g(z|\zeta) = \delta(z - \zeta) \quad (z, \zeta > 0). \tag{1.59}$$

Motivated by (1.43), assume that $g(z|\zeta) = CM_{k,\mu}(z_<)W_{k,\mu}(z_>)$, where $M_{k,\mu}(z)$ and $W_{k,\mu}(z)$ are the homogeneous solutions, known as Whittaker functions (Whittaker-Watson [50]). To compute C, we integrate (1.59) over small interval $[\zeta^-, \zeta^+]$, where ζ^- and ζ^+ denote points just above and below $z = \zeta$, respectively. Thus, $\frac{dg(z|\zeta)}{dz}\big|_{z=\zeta^+} - \frac{dg(z|\zeta)}{dz}\big|_{z=\zeta^-} = 1$. Thus, $C = -\frac{\Gamma(\mu-k+1/2)}{\Gamma(2\mu+1)}$, and

$$g(z|\zeta) = -\frac{\Gamma(\mu - k + 1/2)}{\Gamma(2\mu + 1)}M_{k,\mu}(z_<)W_{k,\mu}(z_>).$$

Finally, using integral formulas for the product of two Whittaker functions, we obtain the integral representation when $\mathrm{Re}(k - \mu + 1/2) > 0$:

$$g(z|\zeta) = -\sqrt{z\zeta}\int_1^\infty \left(\frac{\xi+1}{\xi-1}\right)^k e^{(-\frac{z+\zeta}{2}\xi)}I_{2\mu}[\sqrt{z\zeta(\xi^2-1)}]\frac{d\xi}{\sqrt{\xi^2-1}}.$$

This Green's function is symmetric in z and ζ since Whittaker's equation is self-adjoint.

1.2 Asymptotic behavior and stability

1.2.1 Asymptotic expansions

Preliminaries. The use of infinite series goes back to the time of Newton, though the mathematicians of the eighteenth century were often more interested in the formal

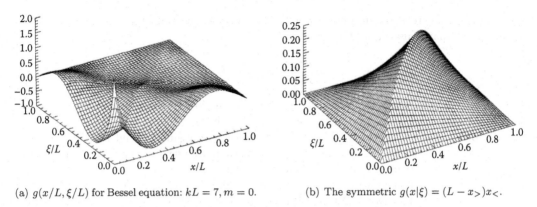

(a) $g(x/L, \xi/L)$ for Bessel equation: $kL = 7, m = 0$. (b) The symmetric $g(x|\xi) = (L - x_>)x_<$.

FIGURE 1.11: The Green's function g.

manipulation of these series than in rigorous proofs. Stirling gave an infinite series for limit $\ln m!$,

$$\ln(m!) = x \ln x - x + \frac{1}{2}\ln(2\pi) + \sum_{k=1}^{\infty} \frac{B_{2k}(1/2)}{(2k-1)2kx^{2k-1}}, \quad x = m + \frac{1}{2},$$

where $B_n(x)$ is Bernoulli's polynomial defined implicitly by $\frac{ze^{xz}}{e^z-1} = \sum_{k=0}^{\infty} B_k(x)\frac{z^k}{k!}$. De Moivre gave a similar formula in terms of Bernoulli's numbers $B_{2n} := B_{2n}(0)$,

$$\ln(m!) = \left(m + \frac{1}{2}\right)\ln m - m + \frac{1}{2}\ln(2\pi) + \sum_{k=1}^{\infty} \frac{B_{2k}}{(2k-1)2km^{2k-1}}.$$

For large n we have $B_{2n} \simeq (-1)^{n-1}4n\sqrt{\frac{2n-1}{2\pi}}\left(\frac{2n-1}{2\pi e}\right)^{2n-1}$, so the second of the above series does not converge for any m; also the Stirling's series does not converge. Nevertheless, Stirling was able to calculate $\log(1000!)$ to ten decimate places using only the first few terms of his series; De Moivre's approximation provides the same accuracy. The theory which put Stirling's work on a rigorous basis was developed by Poincaré, and this is the foundation of a very important tool widely used in applied mathematics. One example comes from *Laplace's Théorie des Probabilités*, published in 1812. Here Poincaré showed that the error function $\text{erf}(x) = \frac{2}{\sqrt{\pi}}\int_0^x e^{-t^2}dt$ can be represented by the uniformly convergent Taylor's series

$$\text{erf}(x) = \frac{2}{\sqrt{\pi}}\sum_{k=0}^{\infty} \frac{(-1)^k x^{2k+1}}{(2k+1)k!} = \frac{2}{\sqrt{\pi}}\left(x - \frac{1}{3}x^3 + \frac{1}{10}x^5 - \frac{1}{42}x^7 + \dots\right),$$

obtained by expanding the integrand and integrating term by term. This series converges uniformly for all x. At $x = 1$ eight terms of this series gives a relative accuracy of 10^{-6}; for $x = 2$, 17 terms are required, etc. to obtain the same degree of accuracy. The large magnitude of individual terms means that in practice, rounding errors limit the magnitude of x for which this type of series can be used. Laplace also derived the approximation

$$\text{erf}(x) = 1 - \frac{e^{-x^2}}{x\sqrt{\pi}}\left(1 - \frac{1}{2x^2} + \frac{1\cdot 3}{(2x^2)^2} - \frac{1\cdot 3\cdot 5}{(2x^2)^3} + \dots\right), \quad x > 0,$$

which diverges for all finite x. At $x = 2$ the first three terms in the curly bracket provides a relative accuracy of less than 10^{-4}, and for $x > 3$ the first two terms give a relative accuracy of less than 10^{-6}. At this stage it is natural to ask why such series exist and why they are so important. There is no simple answer to this type of question, but the main source of asymptotic expansions is the ODE:

$$\varepsilon^2 \frac{d^2 u}{dx^2} + g(x)u = 0,$$

where ε is a small parameter and $g(x)$ is some known function. Such equations arise naturally in the description of wave motion – such as sound, light or water, and in the quantum description of matter, in which the wavelength is relatively small – and also in fluid flow where viscous effects are significant only close to a boundary. We expect some kind of small ε expansion to be useful; but when $\varepsilon = 0$ the equation reduces to $g(x)u(x) = 0$, suggesting that the behavior of solutions as $\varepsilon \to 0$ is far from simple. Integrals of the form $\int_0^\infty e^{\frac{1}{\varepsilon} ih(t)} dt$, $|\varepsilon| \ll 1$, where $h(t)$ is a real function, also give rise to asymptotic expansions, because in many cases the dominant contribution comes from a particular value of t and the expansion about this point estimates the exponentially small contributions from other values of t incorrectly. Such approximations are called *stationary phase approximations*. We start with the example of the function $g(x)$ defined by integral $g(x) = \int_0^\infty \frac{e^{-t} dt}{x+t}$ $(x > 0)$, which is related to the *exponential integral*

$$E_1(x) = \int_x^\infty \frac{e^{-t} dt}{t},$$

defined by the equality $g(x) = e^x E_1(x)$. We expect $g(x) \to 0$ as $x \to \infty$ and wish to determine exactly how it approaches zero. Integrating by parts gives $g(x) = \frac{1}{x} - \int_0^\infty \frac{e^{-t} dt}{(x+t)^2}$. This integral is smaller than the original one if $x > 1$, which suggests repeating this operation. Using $\int_0^\infty \frac{e^{-t} dt}{(x+t)^n} = \frac{1}{x^n} - n \int_0^\infty \frac{e^{-t} dt}{(x+t)^{n+1}}$, obtained by integrating by parts, we find

$$g(x) = \frac{1}{x} - \frac{1}{x^2} + \ldots + \frac{(-1)^{n-1}(n-1)!}{x^n} + (-1)^n n! \int_0^\infty \frac{e^{-t} dt}{(x+t)^{n+1}}.$$

Example 31 1. Often one can obtain approximate values of integrals by expanding the integrand in a power series. For instance, the following integral converges for all x:

$$\int_0^x e^{-t^2} dt = \int_0^x \left(1 - t^2 + \frac{t^4}{2} - \ldots\right) dt = x - \frac{x^3}{3} + \frac{x^5}{10} - \ldots.$$

To estimate the integral, restrict ourselves to the first few terms of the series; the result will of course be appropriate only for $x \leq 1$. To estimate this integral for large x, integrate repeatedly by parts,

$$\int_0^x e^{-t^2} dt = \frac{\sqrt{\pi}}{2} - e^{-x^2} \left[\frac{1}{2x} - \frac{1}{4x^3} + \ldots + (-1)^n \frac{(2n-1)!!}{2^{n+1} x^{2n+1}} + \ldots\right].$$

The series is divergent: the factorial $(2n+1)!!$ increases faster than x^{2n+1} at $n \to \infty$. This series is an example of so-called asymptotic series. Since it diverges, it does not pay to take a very large number of terms when one estimates the integral; this actually makes it less accurate. So we should try to find the optimum number of terms to keep. For large x the terms of the series first decrease in absolute magnitude and then subsequently begin to increase. The optimum number of terms is evidently defined by the requirement that the remainder of the series should be a minimum. The remainder is of the order of the $(n+1)$-th term of the series. The correct prescription is to sum as far as the smallest term of the series. The condition for the minimum to be reached at the n-th term can be approximated by setting the n-th and $(n+1)$-th terms equal: $\frac{(2n-1)!!}{2^{n+1} x^{2n+1}} \simeq \frac{(2n+1)!!}{2^{n+2} x^{2n+3}}$. Hence, $n \simeq x^2$.

2. Many integrals can be estimated by separating out the most important part of the integrand. For example, if $x \ll 1$ then the exponential in the integrand of $I(x) = \int_0^x \frac{e^{t^2} dt}{\sqrt{x^2 - t^2}}$ is approximately 1. Consequently,

$$I(x) \simeq \int_0^x \frac{dt}{\sqrt{x^2 - t^2}} = \int_0^1 \frac{dz}{\sqrt{1 - z^2}} = \frac{\pi}{2}.$$

If $x \gg 1$, then, in view of the exponential increase of e^{t^2}, the main contribution to the integral comes from the region near $t = x$. If we write $\xi = x - t$ then

$$I(x) = \int_0^x \frac{e^{(x^2 - 2\xi x + \xi^2)} d\xi}{\sqrt{2\xi x - \xi^2}}.$$

The region of ξ for which the integrand is large is concentrated near the lower limit, an its width is of order $1/(2x)$. In this region we have $\xi^2 \sim 1/(4x^2) \ll 1$, hence $e^{\xi^2} \simeq 1$ and so

$$I(x) = e^{x^2} \int_0^x \frac{e^{-2\xi x} d\xi}{\sqrt{2\xi x}} \simeq \frac{e^{x^2}}{2x} \int_0^\infty \frac{e^{-z} dz}{\sqrt{z}} = \frac{\sqrt{\pi}}{2x} e^{x^2}.$$

If $x \sim 1$ the two expressions for $I(x)$ are of the same order of magnitude (namely, unity), as of course they must be. Thus the estimates given form a good description of $I(x)$ over the whole range of x.

3. The integral $I(\alpha, \beta) = \int_0^\infty e^{-\alpha x^2} \sin^2(\beta x)\, \mathrm{d}x$ $(\alpha > 0)$ may be rewritten as

$$I(\alpha, \beta) = \frac{1}{\sqrt{\alpha}} \int_0^\infty e^{-z^2} \sin^2\left(\frac{\beta}{\sqrt{\alpha}} z\right) dz.$$

For $z > 1$ the integrand decreases quickly so that the important region of the integration is $0 < z < 1$. If $\beta \gg \sqrt{\alpha}$ then $\sin(\beta z / \sqrt{\alpha})$ oscillates many times in the region of z. Therefore, we may replace $\sin^2(\beta z / \sqrt{\alpha})$ by $1/2$, and the integral is approximately given by

$$I(\alpha, \beta) = \frac{1}{2\sqrt{\alpha}} \int_0^\infty e^{-z^2} dz = \frac{1}{4}\sqrt{\pi/\alpha}.$$

If $\beta \ll \sqrt{\alpha}$, then in the region of z we have $\sin(\beta z / \sqrt{\alpha}) \simeq \beta z / \sqrt{\alpha}$ and, consequently,

$$I(\alpha, \beta) = \frac{\beta^2}{\alpha^{3/2}} \int_0^\infty e^{-z^2} z^2 dz = \sqrt{\pi} \frac{\beta^2}{4\,\alpha^{3/2}}.$$

For $\beta \sim \sqrt{\alpha}$ the two expressions coincide and are equal to $\frac{1}{4}(\pi/\alpha)^{1/2}$. The exact value of the integral we have estimated to be $I(\alpha, \beta) = \frac{1}{4}\sqrt{\frac{\pi}{\alpha}}(1 - e^{-\beta^2/\alpha})$. It is easy to verify that in the limits $\beta \gg \sqrt{\alpha}$ and $\beta \ll \sqrt{\alpha}$ this expression reduces to the appropriate forms given above. For $\beta = \sqrt{\alpha}$ we have $I(\alpha, \beta) = \frac{1}{4}\sqrt{\frac{\pi}{\alpha}}(1 - 1/e)$, which is of the same order of magnitude with our estimated value.

Method of steepest descents. Consider the integral $I = \int_0^\infty g(t)e^{f(t)}dt$, where $f(t)$ is a function which has a strong maximum for some $t = t_0 > 0$. Suppose that the function $g(t)$ is strongly varying near t_0. Then we can replace the function $g(t)e^{f(t)}$ near the maximum by a simpler function; to do this we expand $f(t)$ in a Taylor series around t_0: $f(t) = f(t_0) + \frac{1}{2}(t - t_0^2)f'(t_0) + \dots$ Assume that $|f''(t_0)| \gg t_0^{-2}$ (since $f(t)$ has a strong maximum). In fact, the values of $(t - t_0)^2$ which are important in the integral are of order $1/f''(t_0)$, as

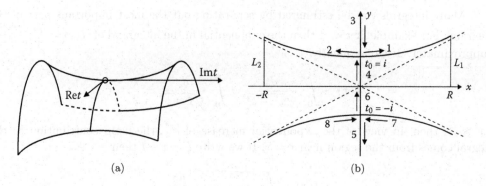

(a)   (b)

FIGURE 1.12: (a) The point t_0 is a saddle point in the complex t plane and the direction of integration is the direction of steepest descent from the saddle point. (b) Example 33.

we will see below. Thus, $(t - t_0)^2/t_0^2 \ll 1$. This condition makes it legitimate to omit the higher-order terms in the above Taylor series for $f(t)$. Then we have

$$I \simeq g(t_0) \int_{-\infty}^{\infty} \exp\left[f(t_0) - \frac{1}{2}(t - t_0)^2 |f''(t_0)|\right] dt = \left(\frac{2\pi}{|f''(t_0)|}\right)^{1/2} g(t_0)\, e^{f(t_0)}.$$

Here we replaced the limits of integration by $(-\infty, \infty)$, since the integrand decreases exponentially in the region $\delta t > \frac{1}{\sqrt{|f''(t_0)|}} \ll t_0$. Let us estimate the correction given by the subsequent terms in the Taylor series. If we keep only the cubic term and expand the exponential of it in a series, then the first term of the expansion gives no contribution, since the integrand is odd. Therefore, we consider the fourth-order term in the Taylor expansion of $f(t)$, namely, $f^{(4)}(t_0)(t - t_0)^4/4!$. Expanding the exponential of this quantity in a series, we find that the correction is of order $f^{(4)}/(f'')^2$ relative to the last expression for our integral. If the function f is characterized by a single parameter, then estimating the order of magnitude of the derivatives of f, we find $f^{(4)}/(f'')^2 \sim 1/f(t_0)$.

The method of steepest descents is applicable under condition $f(t_0) \gg 1$, that is equivalent to the assumption $|f''(t_0)| \gg t_0^{-2}$ made above. If in the last expression of the integral we were to make the substitution $t - t_0 = i\xi$, the integrand would become an increasing exponential; see Fig. 1.12(a).

Hence there is the name, saddle-point method or method of steepest descents. We actually considered a special case, when the direction of the steepest descent coincides with the real axis; one can also consider the general case, where the direction of steepest descent makes an arbitrary angle with the real axis.

Let us apply the method of steepest descent to obtain an asymptotic expression for large x for the Gamma function $\Gamma(x + 1) = \int_0^{\infty} \exp(-t + x \ln t) dt$. Write $-t + x \ln t = f(t)$. Then the condition $f'(t) = 0$ gives us the saddle point t_0: $f'(t_0) = -1 + \frac{x}{t_0} = 0$, whence $t_0 = x$. Since $f(t_0) = x \ln x - x$, the condition of applicability of the method of steepest descents, viz. $f(x) \gg 1$, means that $x \gg 1$. We then have $f''(t_0) = -x/t_0^2 = -1/x$. Using the expression above for $f'(t_0)$, we obtain $\Gamma(x + 1) \overset{x \gg 1}{\simeq} \sqrt{2\pi x}(x/c)^x$. This asymptotic formula is called the *Stirling formula*. To estimate its accuracy we use the relation $\Gamma(x + 1) = x\Gamma(x)$, and write the (unknown) exact expression for $\Gamma(x + 1)$ in the form $\Gamma(x + 1) = \sqrt{2\pi x}(\frac{x}{e})^x[1 + \phi(x + 1)]$, where ϕ is a so far unknown function. Using the above recurrence relation we obtain $\phi(x + 1) - \phi(x) \simeq -1/(12x^2)$ for large x. For $x \gg 1$ the difference $\phi(x + 1) - \phi(x)$ is approximately equal to $\phi'(x)$, and we accordingly find $\phi(x) = 1/(12x)$. Thus, for $x \gg 1$, we have $\Gamma(x + 1) \simeq \sqrt{2\pi x}(\frac{x}{e})^x[1 + \frac{1}{12x} + O(\frac{1}{x^2})]$. This formula is very accurate even for small values of x. One may check its accuracy using

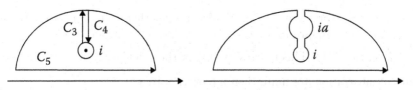

FIGURE 1.13: Example 32(1,2).

that for integral values of x the Gamma function $\Gamma(x+1)$ is just $x!$ For instance, we get $\sqrt{2\pi}\frac{1}{e}(1+\frac{1}{12}) = 0.9990 \simeq 1!$ even for $x = 1$, and $\sqrt{2\pi}\frac{1}{e^2}(1+\frac{1}{24}) = 1.9990 \simeq 2!$ for $x = 2$.

The next examples illustrate the properties of integrals of oscillating functions.

Example 32

1. $I = \int\limits_{-\infty}^{\infty} \frac{e^{i\omega t}dt}{\sqrt{1+t^2}}$ $(\omega \gg 1)$. The singularities of the integrand occur on the imaginary axis: $t = \pm i$. Let us calculate the integral by deforming the contour into upper half-plane; see Fig. 1.13(a). The contributions from C_1 and C_2 vanish when the contour is moved away to infinity, so that the integral is just the contributions from parts $C_3 + C_4 + C_5$, which around the branch point $t = i$. The denominator of the integrand $\sqrt{1+t^2}$ changes sign when we go around the branch point. Therefore, the contributions from C_3 and C_4 are equal. The contribution from C_5 tends to zero as the radius of the circle is decreased; this is easy to see if we make the substitution $t = i + re^{i\phi}$, and let $r \to 0$. Then we get $\int_{C_5} \sim \int_0^{2\pi} \frac{re^{i\phi}d\phi}{\sqrt{re^{i\phi}}} \sim \sqrt{r} \to 0$. Introducing the variable of integration y by $t = i(1+y)$ and calculating the integral over C_3, we find $I \simeq 2e^{-\omega}\int_{-\infty}^{\infty} \frac{e^{-\omega y}dy}{\sqrt{2y}} = \sqrt{\frac{2\pi}{\omega}}e^{-\omega}$. Thus, the integral is exponentially small for large ω.

2. $I = \int\limits_{-\infty}^{\infty} \frac{e^{i\omega t}dt}{\sqrt{(a^2+t^2)+(b^2+t^2)}}$, $\omega \to \infty$. In this case there are two branch points in the upper half-plane, $t = ia$ and $t = ib$; we assume that $a > b$. First of all we can reduce the number of independent parameters in I; measuring t, a, ω and I in units of the appropriate powers of b, we get $I = \int_{-\infty}^{\infty} \frac{e^{i\omega t}dt}{\sqrt{(a^2+t^2)+(1^2+t^2)}}$.

We shift our contour of integration into the upper half-plane (Fig. 1.13(b), where there is a cut from the point i to the point ia). By considerations analogous to those used in the previous example, we see that the required integral is just twice the integral along the cut. By changing the variable of integration $t = i(1+y)$, we get $I \simeq 2e^{-\omega}\int_0^{a-1} \frac{e^{\omega y}dy}{\sqrt{2y(a^2-1-2y)}}$. In calculating this integral there are two possibilities. If $a-1 \gg \omega^{-1}$, then the important region of integration is cut off by the exponential factor $e^{-\omega y}$ and is of order ω^{-1}. Consequently we have $I \simeq 2e^{-\omega}\int_0^{a-1} \frac{e^{-\omega y}dy}{\sqrt{2y(a^2-1)}} = (\frac{2\pi}{\omega(a^2-1)})^{1/2}e^{-\omega}$. If on the other hand $a-1 \ll \omega^{-1}$, then within the interval of integration we have $e^{-\omega y} \simeq 1$ and so

$$I \simeq 2e^{-\omega}\int_0^{a-1} \frac{e^{\omega y}dy}{\sqrt{2y(a^2-1-2y)}} = 2e^{-\omega}\arcsin\left(\sqrt{\frac{2}{a+1}}\right).$$

In the first case the two singularities of the integrand are distant from one another, in the second case they are close together. In both cases the exponentially small factor is determined by the singularity closest to the real axis, whereas the pre-exponential factor depends appreciably on the position of both singularities. Notice that in the limit $a \to 1$

we get from the last expression $I \sim \pi e^{-\omega}$. We can get this result immediately by noticing that for $a \to 1$ the two square-root singularities coalesce to give a simple pole.

3. $I = \int\limits_{-\infty}^{\infty} f(x)e^{i\omega x}\,dx$. This integral occurs in the calculation of scattering amplitudes. If the limits of integration are finite then integration by parts gives

$$I = f(x)\frac{e^{i\omega x}}{i\omega}\Big|_{-1}^{1} - \frac{1}{i\omega}\int_{-1}^{1} f'(x)e^{i\omega x}\,dx = \frac{f(1)e^{i\omega} - f(-1)e^{-i\omega}}{i\omega} + O(\omega^{-2}).$$

Thus, the high Fourier components decrease, and, generally, according to a power law, not exponentially as in the case of infinite limits. An exception is in the case in which, on repeatedly integrating by parts, we find that the contribution of $f(x)$ and all its derivatives vanishes at the limits of integration.

Debye's first-order approximation. There are first-order approximations to the integrals $I = e^{pg(z_0)}\int_0^\infty \frac{f(w)}{dw/dz}e^{-pw}\,dw$, where $z_0 = (x_0, y_0)$ is a saddle point. Expanding $\frac{f(w)}{dw/dz}$ in a Taylor series about $w = 0$, $\frac{f(w)}{dw/dz} = \sum_{n=0}^{\infty} F_n w^{n+\nu}$, where ν is a noninteger constant, resulting from the derivative dw/dz. The slope dw/dz has a different value on C_1 and C_2, which are the two paths of the original path C (before the saddle point z_0 and past it). Substituting the expression above into integrand and using *Watson's formula*

$$\int_0^\infty t^\nu e^{-pt}\,dt = \frac{\Gamma(\nu+1)}{p^{\nu+1}},$$

the integral becomes $I = e^{pg(z_0)}\sum_{n=0}^{\infty} F_n \frac{\Gamma(n+\nu+1)}{p^{n+\nu+1}}$. Principally, these approximations assume that the major contribution to the integral comes from the section of the path near the saddle point, especially when p is very large. This means that the first term in the formula above would suffice if p is sufficiently large. To obtain the first-order approximation, one can neglect higher-order terms in $f(w)$ and dw/dz in such a way that a closed form expression can be obtained for the first-order term. Thus an approximate value for w can be obtained by neglecting high-order terms in w $w = g(z_0) - g(z) = -(z - z_0)^{m+1}\frac{f^{(m+1)}(z_0)}{(m+1)!}$. Thus, for z near z_0, the conformal transformation between w and z can be obtained explicitly in a closed form by approximation

$$(z - z_0) = \left[-\frac{(m+1)!}{f^{(m+1)}(z_0)}\right]^{1/(m+1)} w^{\frac{1}{m+1}} = (cw)^{\frac{1}{m+1}}.$$

The $m + 1$ roots have different values along the different path C'_m. Differentiating this approximation for z with respect to w results in $\frac{dz}{dw} = \frac{c^{\frac{1}{m+1}}w^{-\frac{m}{m+1}}}{m+1}$. Similarly, the function $f(x)$ can be approximated by its value at z_0: $f(z) \simeq f(z_0)$. Thus, the integral $I_{C'_1,C'_2}$ becomes

$$I_{C'_1,C'_2} \sim \frac{c^{\frac{1}{m+1}}f(z_0)e^{pg(z_0)}}{m+1}\int_0^\infty e^{-pw}w^{-\frac{m}{m+1}}\,dw = \frac{c^{\frac{1}{m+1}}\Gamma(1/(m+1))}{m+1}\cdot\frac{f(z_0)e^{pg(z_0)}}{p^{1/(m+1)}}.$$

The first-order approximation to the integrals $I_C = \int_C e^{pg(z)}f(z)\,dz$, where C is a path of integration in the complex plane, $z = x + iy$, is thus given by

$$I_C = I_{C'_1} - I_{C'_2} = \frac{\Gamma(1/(m+1))f(z_0)e^{pg(z_0)}}{(m+1)p^{1/(m+1)}}\left\{c^{\frac{1}{m+1}}\Big|_{\text{on}C'_1} - c^{\frac{1}{m+1}}\Big|_{\text{on}C'_2}\right\},$$

where the residues on the poles were neglected. The last expression represents the leading term in the approximation of the asymptotic series. For $m = 1$, the two roots of c are opposite in signs and hence the expression in the bracket is double the first term in the bracket, i.e.,

$$I_C = f(z_0)\frac{e^{pg(z_0)}}{p^{1/2}}\sqrt{2\pi/[-f''(z_0)]}.$$

Example 33 Find the first-order approximation for *Airy's function*

$$\text{Ai}(z) = \frac{1}{\pi}\int_0^\infty \cos(s^3/3 + sz)\,ds = \frac{1}{2\pi}\int_{-\infty}^\infty \exp[i(s^3/3 + sz)]\,ds.$$

For large z, the first exponential term is also not a slowly varying function. To merge the first exponential with the second, put $s = \sqrt{z}\,t$, $\text{Ai}(z) = \frac{\sqrt{z}}{2\pi}\int_{-\infty}^\infty \exp[i\,z^{3/2}(t^3/3 + t)]\,dt$. Letting $x = z^{3/2}$ one can write out the integral as $\text{Ai}(z) = \frac{x^{1/3}}{2\pi}\int_{-\infty}^\infty \exp[i\,x(t^3/3 + t)]\,dt$. Then we can evaluate the first-order approximation for large x. In the integral $f(t) = 1$ and $g(t) = i(\frac{1}{3}t^3 + t)$. The saddle points are given by $g'(t_0) = i(t_0^2 + 1) = 0$ resulting in two saddle points $t_0 = \pm i$. To map the *Steepest Descent Path*,

$$g(\pm i) = \mp\frac{2}{3}, \quad g''(\pm i) = \mp 2 = 2\left\{\begin{matrix} e^{i\pi} \\ e^{i0} \end{matrix}\right..$$

Here $b = 2$ and $\theta = \pi$ for $t_0 = +i$, and $\theta = 0$ for $t_0 = -i$. Since $g''(t_0) \neq 0$, $m = 1$ for both saddle points. Letting $t = \xi + i\eta$, then in the Steepest Descent Path equations both saddle points are given by

$$v(\xi, \eta) = \text{Im}\,g(t) = \xi^3/3 - \xi\eta + \xi = u_0(\xi_0, \eta_0) = u_0(0, \pm 1) = 0.$$

The path of steepest ascent or descent is plotted in Fig. 1.12(b) for $t_0 = +i$ (path 1–4) and for $t_0 = -i$ (path 5–8).

For SP at $t_0 = +i$, the path "3" extends from i to $i \cdot \infty$ and the path "4" extends from i to $-i \cdot \infty$. For SP at $t_0 = -i$, the path "5" extends from $-i$ to $-i \cdot \infty$ and "6" extends from $-i$ to $i \cdot \infty$. The path "4" partially overlaps "5" and "3" partially overlaps "6". For SP at $t_0 = +i$, $g''(+i) = 2e^{ix}$, so that the steepest descent paths near $t_0 = +i$ make tangent angles given by $\theta = (n + 1/2)\pi - \pi/2 = n\pi$. Thus, the Steepest Descent Paths for $t_0 = +i$ are path "1" and "2" having tangent angles 0 and π, while the paths "3" and "4" are steepest ascent paths. For $t_0 = -i$, $g''(-i) = +2$, so that the Steepest Descent Path makes tangent angles $\pi/2$ and $3\pi/2$ near the saddle point $t_0 = -i$.

Since there are two saddle points, one can connect the original path $(-\infty, \infty)$ to either path "1" and "2" through $t_0 = +i$ or "5" and "6" through $t_0 = -i$. Considering the second choice, the closure with the original path with "6" and "5" through $t_0 = -i$ requires going through $t_0 = +i$ along paths "3" and "4", which were steepest ascent paths for $t_0 = +i$. Thus, this will result in the integrals becoming unbounded. Thus, the only choice left is to close that original path $(-\infty, \infty)$ through $t_0 = +i$ by connecting to the paths "1" and "2" by line segments L_1 and L_2. To obtain a first-order approximation,

$$\text{Ai}(x) = \frac{x^{1/3}}{2\pi}\cdot 1\cdot e^{x(-2/3)}\left(\frac{2\pi}{-(-2x)}\right)^{1/2} = \frac{x^{-1/6}}{4\sqrt{\pi}}e^{-2x/3},$$

so that $\text{Ai}(z) = \frac{z^{-1/4}}{2\sqrt{\pi}}\exp\left(-\frac{2}{3}z^{2/3}\right)$.

Method of stationary phase. The *stationary phase* method is analogous to the *steepest descent* method, although the approach and reasoning for the approximation are different. Performing the integration in the complex plane, resulting in the two methods having identical outcomes, consider integral $I(p) = \int_C f(z)e^{i\,pg(z)}dz$, where $f(z)$ is a slowly varying function, and $g(z)$ is an analytic function. Thus, as p becomes larger, the exponential term oscillates in increasing frequency. Since the exponential can be written in terms of circular functions, then p increases, and the frequency of the circular functions increases, so much that these circular functions oscillate rapidly between $+1$ and -1. This then tends to cancel out the integral of $f(z)$ when p becomes very large for sufficiently large z. The major contribution to the integral occurs when $g(z)$ has minimum so the exponential function oscillates the least. This occurs at the *Stationary Phase point* $z_0 = (x_0, y_0)$, where $g'(z_0) = 0$. Let $g(z) = u(z) + iv(z)$, then $e^{i\,pg(z)} = e^{-pv(z)}e^{i\,pu(z)}$. If $f(z)$ is a slowly varying function, then most of the contribution to the integral comes from near the Stationary Phase point z_0, where the exponential oscillates the least. Expanding the function $g(z)$ about Stationary Phase point z_0: $g(z) = g(z_0) + \frac{1}{2}g''(z_0)(z - z_0)^2 + \ldots$, and defining $w = g(z_0) - g(z) = -(1/2)g''(z_0)(z - z_0)^2 - \ldots$, then the integral becomes $I(p) = e^{-pg(z_0)} \int_{C'} \frac{f[z(w)]}{dw/dz} e^{-i\,pw} dw$, where C' is the *stationary phase path* defined by $v = \text{const} = u_0$ and $u_0 = v(x_0, y_0)$. This is the same part defined for the *steepest descent path*. For an equivalent Debye's first-order approximation for $m = 1$, let

$$w = -\frac{1}{2}g''(z_0)(z - z_0)^2, \quad \frac{dw}{dz} = -g''(z_0)(z - z_0) \simeq \sqrt{2w}, \quad f(z_0) = f(z_{|w=0}).$$

Then the integral above becomes:

$$I(p) = e^{-pg(z_0)}f(z_0) \int_C \frac{e^{-i\,pw}}{\sqrt{2w}} dw = e^{-pg(z_0)}f(z_0)\sqrt{\frac{2\pi}{pg''(z_0)}} e^{i\pi/4}.$$

1.2.2   Asymptotic behavior of autonomous systems

Autonomous systems and their phase portraits. An *orbit* will mean a set of points y on a solution $y = y(t)$ of $y'(t) = f(y)$, see (1.1), without reference to a parameterization. Any system of differential equations $y' = f(t, y)$ can be considered autonomous if the dependent variable $y(t)$ is replaced by the $(d + 1)$-vector (t, y) and the system is replaced by $t' = 1$, $y' = f(t, y)$, where the prime denotes differentiation with respect to a new independent variable. For most purposes, this remark is not useful. A point y_0 is called a *stationary (equilibrium)* or singular point of (1.1) if $f(y_0) = 0$ and a *regular* point if $f(y_0) \neq 0$. The stationary points are characterized by the property that the constant $y(t) \equiv y_0$ is a solution of (1.1). When solutions of (1.1) are uniquely determined by initial conditions, $f(y_0) = 0$ and $y(t_0) = y_0$ for some t_0 imply $y(t) = y_0$. This need not be the case in general.

If (1.1) has a solution $C^+ : y = y(t)$ on a half-line $t \geq t_0$ then its set $\Omega(C^+)$ of ω-limit points is the (possible empty) set of points y_0 for which there is a sequence $t_0 < t_1 < \ldots$ such that $t_n \to \infty$ and $y(t_n) \to y_0$ as $n \to \infty$. Similarly, if $C^- : y = y(t)$ is considered for $t \leq t_0$, we define the set $\Upsilon(C^-)$ of v-limit points, and if $C : y = y(t)$ is defined for $-\infty < t < \infty$, its set of limit points is $\Upsilon(C) \cup \Omega(C)$.

Remark 3 $\Omega(C^+)$ is contained in the closure of the set of points $C^+ : y = y(t)$, $t \geq t_0$.

Theorem 3 *Assume that $f(y)$ is continuous on an open y-set E and that $C^+ : y = y_+(t)$ is a solution of (1.1) for $t \geq t_0$. Then $\Omega(C^+)$ is closed. If C^+ has a compact closure in E, then $\Omega(C^+)$ is connected.*

Proof. Clearly, $\Omega(C^+)$ is a closed set. By Remark 3, $\Omega(C^+)$ is a compact set. Assume the contrary, that is, $\Omega(C^+)$ is the union of two closed (hence, compact) sets C_1, C_2 such that $\text{dist}(C_1, C_2) = \delta > 0$. There is a sequence $0 < t_1 < t_2 < \ldots$ of t-values satisfying $\text{dist}(y_n(t_{2n+1}), C_1) \to 0$, $\text{dist}(y_n(t_{2n}), C_2) \to 0$, as $n \to \infty$. Hence, for large n, there is a point $t = t_n^*$ such that $t_n < t_n^* < t_{n+1}$, $\text{dist}(y_+(t_n^*), C_i) \geq \delta/4$ for $i = 1, 2$. The sequence $y_+(t_1^*), y_+(t_2^*), \ldots$ has a cluster point y_0, since C^+ has a compact closure. Clearly, $y_0 \in \Omega(C^+)$ and $\text{dist}(y_n, C_i) \geq \delta/4$ for $i = 1, 2$. This contradiction completes the proof. $\square$

Definition 12 An orbit $C_0 : y = y_0(t)$, $\omega_- < t < \omega_+$, which is contained in some $\Omega(C^+)$, $C^+ \not\subset C_0$, is called an (ω_-)-*limit orbit*. If, in addition, $y = y_0(t)$ is periodic, $y_0(t + p) = y_0(t)$ for all t and some $p > 0$, the orbit $C_0 : y = y_0(t)$ is called an (ω_-)-*limit cycle*. The condition $C^+ \not\subset C_0$ assures that not every periodic solution $C_0 : y = y_0(t)$ is a limit cycle (cf. the case of a family of closed orbits).

Corollary 1 *Let* f, C^+ *be as in Theorem 3 and* $y_0 \in E \cap \Omega(C^+)$. *Then*

$$y' = f(y), \quad y(0) = y_0 \tag{1.60}$$

has at least one solution $y = y_0(t)$ *on a maximal interval* (ω_-, ω_+) *such that* $y(t_0) \in \Omega(C^+)$ *for* $\omega_- < t < \omega_+$. *In particular, when* C^+ *has a compact closure in* E, *then* $C_0 : y = y_0(t)$ *exists on* $(-\infty, \infty)$ *and* $C_0 \cup \Upsilon(C_0) \cup \Omega(C_0) \subset \Omega(C^+)$.

Remark 4 If solutions of an initial value problem associated with (1.1) are unique, then "selection" in the proof of the theorem is unnecessary; thus $y_n = y_+(t_n) \to y_0$ as $n \to \infty$ implies that $y_0 = \lim_{n \to \infty} y_+(t + t_n)$ holds uniformly on every closed, bounded interval in (ω_-, ω_+).

Corollary 2 *If* $\Omega(C^+)$ *consists of a single point* $y_0 \in E$, *then* y_0 *is a stationary point and* $y_+(t) \to y_0$ *as* $t \to \infty$.

Definition 13 A *Jordan curve* J is a topological image of a curve; in other words, J is a y-set of points $y = y(t)$, $a \leq t \leq b$, where $y(t)$ is continuous, $y(a) = y(b)$, and $y(s) \neq y(t)$ for $a \leq s < t < b$.

The Jordan curve theorem will be stated here for reference; see, e.g., [34, p. 115].

Theorem 4 (Jordan curve theorem) *If* J *is a plane Jordan curve, then its complement in the plane is the union of two disjoint connected open sets,* E_1 *and* E_2, *each having* J *as its boundary,* $\partial E_1 = \partial E_2 = J$. *One of the sets* E_1 *or* E_2 *is bounded (the interior of* J) *and is simply connected.*

Consider a continuous arc $J : y = y(t)$, $a \leq t \leq b$, in the $y = (y^1, y^2)$ plane. Let $\eta = \eta(t) \neq 0$, $a \leq t \leq b$ be a continuous two-dimensional vector attached to the point $y(t)$, i.e., $\eta \neq 0$ is a vector field on J. Consider an angle $\varphi = \varphi(t)$ from a positive y^1-direction $(1, 0)$ to $\eta(t)$, so that $\cos \varphi = \eta^1/\|\eta\|$, $\sin \varphi = \eta^2/\|\eta\|$, where $\|\eta\|^2 = (\eta_1)^2 + (\eta_2)^2$. These formulas determine $\varphi(t)$ up to an integral multiple of 2π, but if $\varphi(t)$ is fixed at some point, say $t = a$, then $\varphi(t)$ is uniquely determined as a continuous function. The $\varphi(t)$ is always meant such a continuous determination. Define $j_\eta(J)$ by $2\pi j_\eta(J) = \varphi(b) - \varphi(a)$. For example, if $\eta(t)$ is continuously differentiable, then

$$2\pi j_\eta(J) = \int_a^b \frac{\eta^1 d\eta^2 - \eta^2 d\eta^1}{\|\eta\|^2}.$$

If $J = J_1 + J_2$ in the sense that $J : y = y(t)$, $a \leq t \leq b$, and $a < c < b$, $J_1 : y = y(t)$, $a \leq t \leq c$ and $J_2 : y = y(t)$, $c \leq t \leq b$, then $j_\eta(J_1 + J_2) = j_\eta(J_1) + j_\eta(J_2)$. Actually, if $\eta(t)$ is given, $j_\eta(J)$ has nothing to do with J but, in applications, $\eta(t)$ will be a "vector beginning at the point $y = y(t)$" of J.

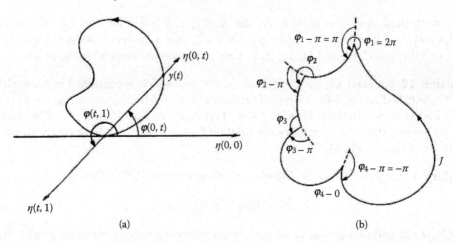

FIGURE 1.14: (a) Oppositely oriented vectors $\eta(0,t)$ and $\eta(1,t)$. (b) The angle $\varphi_k \in [0, 2\pi]$ is an angle from $\eta(t_k - 0)$ to $\eta(t_k + 0)$.

The main interest below will be in the case that J is a Jordan curve, in which case it will always be assumed that J is positively orient and $\eta(t) \neq 0$ is a continuous vector on J so that $y(a) = y(b)$ and $\eta(a) = \eta(b)$. Only Jordan curves $J : y = y(t)$ which are piecewise of class C^1 will occur, so that the positive orientation means that the normal vector $(-dy^2/dt, dy^1/dt) \neq 0$, defined at corners of J, points into the interior of J. Thus $j_\eta(J)$ is an integer. It is called the *index of η with respect to J*.

Theorem 5 *Let $J : y = y(t)$ $(0 \leq t \leq 1)$ be a positively oriented Jordan curve of class C^1 and $\eta(t) = dy/dt$ $(\neq 0)$ the tangent vector field on J. Then $j_\eta(J) = 1$.*

Proof. On the triangle $\Delta : 0 \leq s \leq t \leq 1$, define $\eta(s,t)$ as follows:

$$\eta(s,t) = [y(t) - y(s)]/(t - s) \text{ if } s \neq t \text{ or } (s,t) \neq (0,1),$$
$$\eta(s,t) = y'(t) \text{ and } \eta(0,1) = -\eta(0,0).$$

Thus $\eta(s,t)$ is continuous and $\eta(s,t) \neq 0$ on Δ. Note that $\eta(0,t)$ and $\eta(t,1)$ are oppositely orient vectors; see Fig. 1.14(a). Suppose that the point $y = y(0)$ on J is chosen so that the tangent line through $y(0)$ is parallel to the y^1-axis and no part of J lies below this tangent line. Since Δ is simply connected, it is possible to define (uniquely) a continuous function $\varphi(s,t)$ such that $\varphi(0,0) = 0$ and $\varphi(s,t)$ is an angle from the positive y^1-direction to $\eta(s,t)$. By considering $\varphi(t,t)$ we then get $2\pi j_\eta(J) = \varphi(1,1) - \varphi(0,0)$.

The position of J implies that $0 \leq \varphi(0,t) \leq \pi$ and that $\varphi(0,1)$ is an odd multiple of π, hence $\varphi(0,1) = \pi$. Similarly, a consideration of $\varphi(s,1) - \varphi(0,1) = \varphi(s,1) - \pi$ for $0 \leq s \leq 1$ shows that $\varphi(1,1) - \pi = \pi$. Consequently, $\varphi(1,1) = 2\pi$, since $2\pi j_\eta(J) = \varphi(1,1) - \varphi(0,0) = \varphi(1,1)$. $\square$

The "rounding off" of corners in the case of a J, which is piecewise C^1, gives the following:

Corollary 3 *Let $J : y = y(t)$ $(0 \leq t \leq 1)$ be a positively oriented Jordan curve which is piecewise of class C^1 with corners at the t-values $(0 <) t_1 < \cdots < t_n (< 1)$ and $\eta(t) = dy/dt$ for $t \neq t_k$. Let $J_k : y = y(t)$, $t_{k-1} < t < t_k$ for $k = 1, \ldots, n+1$ with $t_0 = 0$ and $t_{n+1} = 1$. Then $2\pi \sum_{k=1}^{n+1} j_\eta(J_k) + \sum_{k=1}^{n} (\varphi_k - \pi) = 2\pi$, where φ_k is the exterior angle $0 \leq \varphi_k \leq 2\pi$ at $y(t_k)$.*

Note that $\varphi_k - \pi$ is an angle, $-\pi \leq \varphi_k - \pi \leq \pi$, from $\eta(t_k - 0)$ to $\eta(t_k + 0)$; see Fig. 1.14(b). The essential idea in the proof of Theorem 5 is contained in the next lemma.

Lemma 1 *Let $J : y = y(t)$ $(a \leq t \leq b)$ be a Jordan curve, $\xi(t)$ and $\eta(t)$ two vector fields on J which can be deformed into one another without vanishing. Then $j_\xi(J) = j_\eta(J)$.*

The possibility of a deformation without vanishing means the existence of a continuous vector $\eta(t, 0) = \xi(t)$, $\eta(t, 1) = \eta(t)$, $\eta(a, s) = \eta(b, s)$, and that $\eta(t, s) \neq 0$. For example, $\eta(t, s) = (1 - s)\xi(t) + s\eta(t)$ is such a deformation if $\xi(t), \eta(t)$ are not in opposite directions for any t.

Proof. Let $j(s)$ be the index of $\eta(t, s)$ for a fixed s. Thus $j(s)$ is a continuous function of s. Since $j(s)$ is an integer, it is constant. In particular, $j(0) = j(1)$. $\square$

Index of stationary point. In what follows, $f(y) = f(y^1, y^2)$ is continuous on an open plane set E. As before, a point where $f = 0$ is called a stationary point and a point where $f \neq 0$ is to be called regular. Let $J : y = y(t)$ $(a \leq t \leq b)$ be an *arc* in E on which $f(y) \neq 0$. Define $j_f(J)$ to be $j_\eta(J)$, where $\eta(t) = f(y(t))$. For example, if $f(y), y(t)$ are of class C^1, then $j_f(J)$ is given by the line integral $2\pi j_f(J) = \int_J \frac{f^1 df^2 - f^2 df^1}{\|f\|^2}$, where $f = (f^1, f^2)$. When J is a positively oriented Jordan curve in E on which $f(y) \neq 0$, the integer $j_f(J)$ is called the *index of f* with respect to J.

Lemma 2 *Let J_0 and J_1 be two Jordan curves in E which can be deformed into one another in E without passing through a stationary point. Then $j_f(J_0) = j_f(J_1)$.*

The assumption here means the existence of a continuous $y(t, s)$, $a \leq t \leq b$, $0 \leq s \leq 1$, such that
(i) for a fixed s, $J(s) : y = y(t, s)$ is a Jordan curve in E;
(ii) $J(0) = J_0$, $J(1) = J_1$;
(iii) $f(y(t, s)) \neq 0$.
The proof is the same as that of Lemma 1.

Corollary 4 *Let J be a positively oriented Jordan curve in E such that the interior of J is in E and that $f(y) \neq 0$ on and inside J. Then $j_f(J) = 0$.*

Proof. Since the interior of a Jordan curve is simply connected, J can be deformed (in its interior) to a small circle J_1 around a point y_0 of its interior. Since $f(y_0) \neq 0$, then if the circle J_1 is sufficiently small, the change of the angle between $f(y)$ and the y^1-direction around J_1 is small. Since $j_f(J_1)$ is an integer, $j_f(J_1) = 0$. By Lemma 1, $j_f(J) = 0$. Let $y_0 \in E$. By Lemma 1, the integer $j_f(J)$ is independent of the Jordan curve J in the class of curves $J \subset E$ with interiors in E containing no stationary point except possibly y_0. This integer $j_f(J)$ is called the *index $j_f(y_0)$* of y_0 with respect to f. By Corollary 4, $j_f(y_0) = 0$ for y_0 to be a regular point. For this reason, only the *indices of isolated stationary points* y_0 are considered. $\square$

Corollary 5 *Let J be a positively oriented Jordan curve in E on which $f(y) \neq 0$ and let the interior of J be in E and contain only a finite number of stationary points $y_1, \ldots, y_n$. Then $j_f(J) = j_f(y_1) + \ldots + j_f(y_n)$.*

Theorem 6 *Let $f(y)$ be continuous in open set E and let $y = y_p(t)$ be a solution of $y' = f(y)$, of period p, $y_p(t + p) = y_p(t)$ for $-\infty < t < \infty$. Let $y = y_p(t)$ $(0 \leq t \leq p)$ be a Jordan curve with an interior I contained in E and $f(y_p(t)) \neq 0$. Then I contains a stationary point.*

Proof. Let J be the Jordan curve, $y = y_p(t)$ $(0 \leq t \leq p)$, with a positive orientation. By Theorem 5, $j_f(J) = 1 \neq 0$. Thus the theorem follows from Corollary 4. $\square$

Poincaré-Bendixson theory. Continue discussion of ODE $y' = f(y)$ for the plane case $(d = 2)$ with the aid of the Jordan's curve theorem.

Theorem 7 (Poincaré-Bendixson) *Let $f(y)$, $y = (y^1, y^2)$, be continuous on an open plane set E and let $C^+ : y = y_+(t)$ be a solution of*

$$y' = f(y) \qquad (1.61)$$

for $t \geq 0$ with a compact closure in E. In addition, suppose that $y_+(t_1) \neq y_+(t_2)$ for $0 \leq t_1 < t_2 < \infty$ and that $\Omega(C^+)$ contains no stationary points. Then $\Omega(C^+)$ is the set of points y on a periodic solution $C_p : y = y_p(t)$ of (1.61). If $p > 0$ is the smallest period of $y_p(t)$, then $y_p(t_1) \neq y_p(t_2)$, for $0 \leq t_1 < t_2 < p$; i.e., $J : y = y_p(t)$, $0 \leq t \leq p$ is a Jordan curve.

In the case that initial value problems associated with (1.61) have a unique solution either $y_+(t_1) \neq y_+(t_2)$ for $0 \leq t_1 < t_2 < \infty$ or $y_+(t)$ is periodic, i.e., $y_+(t + p) = y_+(t)$ for all t for some fixed positive number p. In the latter case (excluded this), $\Omega(C^+)$ coincides with the set of points C^+. Note that C^+ is a spiral, which tends to the closed curve $\Omega(C^+) : y = y_p(t)$. It will have the following consequence.

Corollary 6 *In conditions of Theorem 7, let $p > 0$ be a period of $y = y_p(t)$. Then there exists a sequence $0 \leq t_1 < t_2 < \ldots$ such that*

$$y_+(t + t_n) \to y_p(t) \qquad as \qquad n \to \infty \qquad (1.62)$$

uniformly for $0 \leq t \leq p$ and

$$t_{n+1} - t_n \to p \qquad as \qquad n \to \infty. \qquad (1.63)$$

Proof. Let L_0 be a transversal through $y^0 = y_p(0)$. Let the successive crossing of L_0 by $y = y_+(t)$ occur at $t_2 < \ldots$. Then $y_+(t_n)$ tends monotonically along L_0 to y^0. Since $y = y_p(t)$ is the unique solution of the problem $y' = f(t)$, $y(0) = y^0$ in $\Omega(C^+)$, Theorem 1 shows that (1.62) holds uniformly for bounded t-intervals, in particular for $0 \leq t \leq p$.

Note that $y_+(t_n + p) \to y_p(p) = y_0$, $n \to \infty$. Thus $\varepsilon > 0$ and n is large, $y_+(t)$ crosses L^0 in the interval $[t_n + p - \varepsilon, t_n + p + \varepsilon]$. Hence $t_{n+1} \leq t_n + p + \varepsilon$. Also $\|y_+(t_n + t) - y_p(t)\|$ is small for large n, $0 < \varepsilon \leq t \leq p - \varepsilon < p$, which implies that there is $\delta > 0$ such that $\|y_+(t_n + t) - y_0(t)\| \geq \delta$ for $0 < \varepsilon \leq t \leq p - \varepsilon$. Hence $t_{n+1} \geq t_n + p - \varepsilon$ for large n. $\square$

Theorem 8 *Let f and C^+ be as in Theorem 7 except that $\Omega(C^+)$ contains a finite number n of stationary points of (1.61). If $n = 0$, Theorem 7 applies. If $n = 1$ and $\Omega(C^+)$ is a point, Corollary 2 applies. If $1 \leq n < \infty$ and $\Omega(C^+)$ is not a point, then $\Omega(C^+)$ consists of stationary points $y_1, \ldots, y_n$ and a finite or nonfinite sequence of orbits $C_0 : y = y_0(t)$, $-\infty \leq \alpha_- < t < \alpha_+ \leq \infty$, which do not pass through a stationary point, but $y_0(\alpha_\pm) = \lim y(t)$, as $t \to \alpha_\pm$, exist and are among the set $y_1, \ldots, y_n$. It is possible that $y_0(\alpha_+) = y_0(\alpha_-)$. It is not claimed that (α_-, α_+) is the maximal interval of existence of $y = y_0(t)$. But when initial conditions uniquely determine solutions of (1.61), so that the only solution of (1.61) through a stationary point y_k is $y(t) \equiv y_k$, then it follows that $\alpha_- = -\infty$ and $\alpha_+ = \infty$.*

In the next theorem, the assumption $y_+(t_1) \neq y_+(t_2)$ for $t_1 \neq t_2$ is omitted.

Theorem 9 *Let $f(y) = f(y^1, y^2)$ be continuous on an open plane set E and $C^+ : y = y_+(t)$ a solution of (1.61) for $t \geq 0$ with compact closure in E. Then $\Omega(C^+)$ contains a closed periodic orbit $C_p : y = y_p(t)$ of (1.61) which can reduce to a stationary point $y_p(t) \equiv y_0$.*

Proof. Suppose that $\Omega(C^+)$ does not contain a stationary point. Let $y_0 \in \Omega(C^+)$ and $C_0^+ : y = y_0(t)$, $0 \leq t < \infty$, be a solution supplied by Theorem 7, so that $C_0^+ \subset \Omega(C^+)$. Since $\Omega(C^+)$ is closed, $\Omega(C_0^+) \subset \Omega(C^+)$. If $y_0(t_1) \neq y_0(t_2)$ for $0 \leq t_1 < t_2 < \infty$, then $\Omega(C_0^+)$ is closed orbit $y = y_p(t)$ by Theorem 7. If $y_0(t_1) = y_0(t_2)$ for certain t_1, t_2 with $0 \leq t_1 < t_2 < \infty$, then the orbit C_0^+ contains the periodic solution path $y = y_p(t)$ of period $p = t_2 - t_1$ which coincides with $y_0(t)$ for $t_1 \leq t \leq t_2$. In either case $\Omega(C^+)$ contains a closed orbit $y = y_p(t)$. □

Theorem 10 *Let $f(y)$ be continuous on open, simply connected plane set E where $f(y) \neq 0$ and $y = y(t)$ is a solution of (1.61) on its maximal interval of existence (ω_-, ω_+). Then $y = y(t)$ does not remain in any compact subset $E_0 \subset E$ as $t \to \omega_+$ (or $t \to \omega_-$).*

Proof. If, e.g., $C_0^+ : y = y_0(t)$, $t_0 \leq t < \omega_+$, is in a compact set $E_0 \subset E$ for some t_0, then $\omega_+ = \infty$ and $\Omega(C^+)$ contains a periodic solution $y = y_p(t)$ by the last theorem. Since $f(y) \neq 0$ on E, $y_p(t)$ does not reduce to a constant on any t-interval. Let t_1 be the first $t > t_0$, where $y_p(t_1) = y_p(t_0)$. Then $y = y_p(t)$, $t_0 \leq t \leq t_1$, is a Jordan curve J. Thus (1.61) has a periodic solution $y = y_0(t)$ of period $t_1 - t_0$ such that $y_0(t) \equiv y_p(t)$ for $t_0 \leq t \leq t_1$. Since E is simply connected, the interior of $J : y = y_0(t)$, $t_0 \leq t \leq t_1$, is contained in E. By Theorem 6, this interior contains a stationary point. This contradicts $f(y) \neq 0$ and proves the theorem. □

Rotation points, foci, nodes and saddle points. If the problem $y' = f(y)$, $y(0) = y_0$ then the following holds: if $0 < \|y_0\| < b$ and $C_0^+ : y = y_0(t)$ is the solution of (1.62) on the maximal interval $[0, \omega_+)$, then only the following (not mutually exclusive) cases can occur:
   (i)   there is at least one t_0, $0 < t_0 < \omega_+$, such that $\|y_0(t_0)\| = b$;
   (ii) $\omega_+ = \infty$ and the solution path C_0^+ is a Jordan curve with $y = 0$ in its interior;
   (iii) $\omega_+ = \infty$ and C_0^+ is a spiral approaching such a closed orbit;
   (iv) $\omega_+ = \infty$ and $y_0(t) \to 0$ as $t \to \infty$;
   (v) $\omega_+ = \infty$, C_0^+ is a spiral around $y = 0$ and $\Omega(C_0^+)$ consists of a finite or infinite sequence of orbits $y(t)$, $-\infty < t < \infty$, such that $y(t) \to 0$ as $t \to \pm\infty$.

By a *spiral* $y = y_0(t)$, $0 \leq t < \omega_+$ around $y = 0$ is meant an *arc* $y_0(t) \neq 0$ such that a continuous determination of $\arctan[y_0^2(t)/y_0^1(t)]$ tends to either ∞ or $-\infty$ as $t \to \omega_+$. If every neighborhood of $y = 0$ contains closed orbits surrounding $y = 0$, the stationary point $y = 0$ is called a *rotation point*.

When $y = 0$ is a rotation point and solutions of an arbitrary initial value problems (1.62) are unique, the set of solutions to $y' = f(y)$ in a neighborhood of $y = 0$ can be described as follows: there is a neighborhood E_0 of $y = 0$ such that the solution $C_0 : y = y_0(t)$ to (1.62) for every $y_0 \in E_0$ is either a closed orbit surrounding $y = 0$ or is a spiral such that $\Omega(C_0)$, $A(C_0)$ are closed orbits surrounding $y = 0$. This is illustrated by the following. Consider the differential equations

$$y_1' = y_1(t)u(r) - y_2(t)v(r), \quad y_2' = y_1(t)v(r) + y_2(t)u(r),$$

where $u(r), v(r)$ are continuous, real-valued functions of $r = \|y\|$ for small $r \geq 0$. In polar coordinates, these equations become

$$r' = ru(r), \quad \theta' = v(r). \tag{1.64}$$

A rotation point $y = 0$ such that all orbits, except $y \equiv 0$, in a vicinity of $y = 0$ are closed curves is called a *center*.

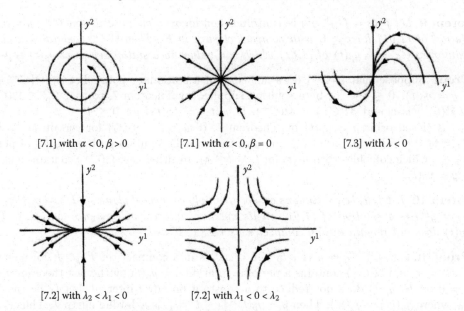

FIGURE 1.15: The cases of attractors.

Example 34 Let $u(r) = 0$, $v(r) \equiv \beta \neq 0$. Then (1.63) is $y_1' = -\beta y_2(t)$, $y_2' = \beta y_1(t)$, and (1.64) is $r' = 0$, $\theta' = 1$, so that all orbits $(y \neq 0)$ are *circles*.

Example 35 Suppose that $u(r) = r \sin(1/r)$, $v(r) = 1$. Then (1.64) is $r' = r^2 \sin(1/r)$, $\theta' = 1$. Thus, besides the trivial solution $y \equiv 0$, there are closed orbits $r = 1/\pi n$, $n = 1, 2, \ldots$. Between the orbits, $r = 1/\pi n$ and $r = 1/\pi(n+1)$, $(-1)^n r' > 0$, and so the corresponding orbits are *spirals* which tend to the circles $r = 1/\pi n$, $r = 1/\pi(n+1)$ as $t \to \infty$ or $t \to -\infty$ depending on party of n.

Example 36 In the previous example, $u(r)$ can be redefined to be 0, say, between $r = 1/\pi n$ and $r = 1/\pi(n+1)$ for a finite sequence of values $n \in \mathbb{N}$. Correspondingly, the spiral orbits between $r = 1/\pi n$ and $r = 1/\pi(n+1)$ are replaced by circular orbits. Assume that the only solution of the problem

$$y' = f(y), \quad y(0) = 0 \tag{1.65}$$

is $y(t) \equiv 0$ so that no solution $y(t) \neq 0$ can tend to 0 as t tends to a finite value. The simplest nonrotation points are called *attractors*. The isolated stationary point $y = 0$ is called an attractor for $t = \infty$ (or $t = -\infty$) if all solutions $y = y_0(t)$ of (1.65) for small $\|y_0\|$ exist for $t \geq 0$ (or for $t \leq 0$) and $y_0(t) \to 0$ as $t \to \infty$ (or $t \to -\infty$). If, in addition, all orbits $y_0(t) \neq 0$ are spirals, then the attractor $y = 0$ is called a *focus*. If all orbits $y_0(t) \neq 0$ have a tangent at $y = 0$; i.e., if a continuous determination of $\theta(t) = \arctan[y_0^2(t)/y_0^1(t)]$ tends to a limit θ_0, $-\infty < \theta_0 < \infty$, then the attractor $y = 0$ is called *node*. A node is called a *proper node* if for every $\theta_0 \mod 2\pi$, there is a unique solution $y = y_0(t)$ such that $\theta(t) \to \theta_0$; otherwise it is called an *improper node*.

Illustrations for these cases of attractors are presented by real linear equations (Fig. 1.15). The system $y_1' = \alpha y_1(t) - \beta y_2(t)$, $y_2' = \beta y_1(t) + \alpha y_2(t)$ has attractor for $t = \infty$ (or for $t = -\infty$) if $\alpha < 0$ (or $\alpha > 0$). It is a *focus* if $\alpha, \beta \neq 0$ and a *proper none* if $\alpha \neq 0$, $\beta = 0$. In the case

$$y_1' = \lambda_1 y_1(t), \quad y_2' = \lambda_2 y_2(t), \tag{1.66}$$

$y = 0$ has an *improved node* if $\lambda_1, \lambda_2 > 0$, $\lambda_1 \neq \lambda_2$. The system $y_1' = \lambda y_1(t)$, $y_2' = y_1(t) + \lambda y_2(t)$, where $\lambda \neq 0$, has also an improved node. There are nonrotation points which are not attractors and attractors which are neither foci nor nodes. The simplest example of a stationary point which is not an attractor is a *saddle point*. This is a stationary point $y = 0$ with the property that only a finite number of orbits tend to 0 as $t \to \infty$ or $t \to -\infty$. This is illustrated by (1.66) where λ_1, λ_2 are real and $\lambda_1, \lambda_2 < 0$.

Example 37 In order to analyze the system $\dot{x} = x - y - 2x(x^2 + y^2)$, $\dot{y} = x + y - y(x^2 + y^2)$ it is convenient to write it in terms of polar coordinates (r, θ). From the definitions, $r = \sqrt{x^2 + y^2}$ and $x = r\cos\theta$, $y = r\sin\theta$, the chain rule gives $\dot{r} = (x\dot{x} + y\dot{y})(x^2 + y^2)^{-1/2} = \dot{x}\cos\theta + \dot{y}\sin\theta$. Thus,

$$\dot{r} = r - r^3(1 + \cos^2\theta).$$

Since $0 \leq \cos^2\theta \leq 1$, we conclude that $r(1 - 2r^2) \leq \dot{r} \leq r(1 - r^2)$. Therefore $\dot{r} > 0$ for $0 < r < 1/\sqrt{2}$ and $\dot{r} < 0$ for $r > 1$. Remembering that if $\dot{r} > 0$, $r(t)$ is an increasing function of t, and therefore integral paths are directed away from the origin, we can define a closed, bounded, annual region $E = \{(r, \theta) : r_0 \leq r \leq r_1\}$ with $0 < r_0 < 1/\sqrt{2}$ and $r_1 > 1$, such that all integral paths enter the region I. If there are no equilibrium points in I, the Poincaré-Bendixson theorem shows that at least one limit cycle exists in I. From the definition of $\theta = \tan^{-1}(y/x)$, $\dot{\theta} = \frac{1}{r}(\dot{y}\cos\theta - \dot{x}\sin\theta)$. Hence,

$$\dot{\theta} = 1 - \frac{1}{2}r^2\sin 2\theta.$$

Since $-1 \leq \sin 2\theta \leq 1$, we conclude that $\dot{\theta} > 0$ provided $r < \sqrt{2}$, and hence that there are no equilibrium points for $0 < r < \sqrt{2}$. Therefore, provided that $1 < r_1 < \sqrt{2}$, there are no equilibrium points in I, and hence, by Poincaré-Bendixson theorem, there is at least one limit cycle. A limit cycle must enclose at least one node, focus or center. In this example, there is an unstable focus at $r = 0$, which is enclosed by any limit cycles, but which does not lie within I.

1.2.3 Stability of autonomous systems

In this section, we consider systems of autonomous ODEs having an isolated equilibrium (stationary) point at the origin.

An equilibrium point, $y = y_{eq}$ ($y \in \mathbb{R}^d$), is *Lyapunov stable* if for all $\varepsilon > 0$ there exists a $\delta > 0$ such that for all $|y(0) - y_{eq}| < \delta$, $|y(t) - y_{eq}| < \varepsilon$ for $t > 0$. In other words, integral paths that start close to a Lyapunov stable equilibrium point remain close for all time, as shown in Fig. 1.16(a).

An equilibrium point, $y = y_{eq}$, is *asymptotically stable* if there exists a $\delta > 0$ such that

$$\forall y(0) : |y(0) - y_{eq}| < \delta, \quad |y(t) - y_{eq}| \underset{t \to \infty}{\longrightarrow} 0 \quad \text{as} \quad t \to \infty.$$

This is a stronger definition of stability than Lyapunov stability, and states that integral paths that start sufficiently close to an asymptotically stable equilibrium point are attracted into it, as illustrated in Fig. 1.16(b). Stable nodes and spirals are both Lyapunov and asymptotically stable. Clearly, asymptotically stable equilibrium points are also Lyapunov stable. Lyapunov stable equilibrium points, for example centers, are not necessarily asymptotically stable.

Since we have formalized our notions of stability, we need one more new concept. Let $V : \Omega \to \mathbb{R}$, with $\Omega \subset \mathbb{R}^d$, $0 \in \Omega$. We say that the function V is *positive definite* on Ω if and only if $V(0) = 0$ and $V(y) > 0$, $\forall y \neq 0$, $y \in \Omega$. If $-V(y)$ is a positive definite function, we say that $V(y)$ is a *negative definite* function. In the following, assume that $V(y)$ is continuous and has well-defined partial derivatives with respect to each of its arguments, so that $V \in C^1(\Omega)$.

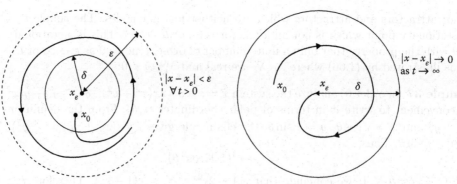

(a) A Lyapunov stable equilibrium point.   (b) An asymptotically stable equilibrium point.

FIGURE 1.16: Equilibrium points of a second-order system.

We introduce the idea of the derivative of a function $V(y)$ with respect to the system $y' = f(y) = (f_1, \ldots, f_d)$, which is defined to be scalar product $V^*(y) = \nabla V \cdot f(y) = \frac{\partial V}{\partial y_1} f_1(y) + \ldots + \frac{\partial V}{\partial y_d} f_d(y)$. This derivative can be calculated for given $V(y)$ and $f(y)$, without knowing the solutions of the differential equation. In particular,

$$\frac{dV}{dt} = \sum_i \frac{\partial V}{\partial y_i} \dot{y}_i = \sum_i \frac{\partial V}{\partial y_i} f_i(y) = V^*(y),$$

so that the total derivative of V with respect to the solution of the equations coincides with our definition of the derivative with respect to the system. This allows us to prove three important theorems.

Theorem 11 *If, in some region $\Omega \in \mathbb{R}^d$ that contains the origin, there exists a scalar function $V(y)$ that is positive definite and for which $V^*(y) \leq 0$, then the origin is Lyapunov stable. The function $V(y)$ is known as a Lyapunov function.*

Proof. Since $V(y)$ is positive definite in Ω, there exists a sphere of radius $r > 0$ contained within Ω such that $V(y) > 0$ for $y \neq 0$ and $|y| < r$, and $V^*(y) \leq 0$, $|y| \leq r$. Let $y = y(t)$ be the solution of the differential equation $y' = f(y)$ with $y(0) = y_0$. By the local existence theorem, the solution exists for $0 \leq t < t^*$ with $t^* > 0$. This solution can then be continued for $t \geq t^*$, and we denote by t_1 the largest value of t for which the solution exists. There are two possibilities, either $t_1 = \infty$ or $t_1 < \infty$. We show that for $|y_0|$ sufficiently small, $t_1 = \infty$. From the definition of the derivative with respect to a system,

$$\frac{d}{dt} V(y(t)) = V^*(y(t)) \quad \text{for} \quad 0 \leq t < t_1.$$

Integrate this equation and obtain $V(y(t)) - V(y_0) = \int_0^t V^*(y(s)) ds \leq 0$, since V^* is negative definite. This means that $0 \leq V(y(t)) \leq V(y_0)$, for $0 \leq t < t_1$. Now let ε satisfy $0 < \varepsilon \leq r$, and let S be the closed spherical shell with inner and outer radii ε and r. By continuity of V, and since S is closed, $\mu = \min_{y \in S} V(y)$ exists and is strictly positive. Since $V(y) \to 0$ as $|y| \to 0$, we can choose δ with $0 < \delta < \mu$ such that for $|y_0| \leq \delta$, $V(y_0) < \mu$, so that $0 < V(y(t)) \leq V(y_0) < \mu$ for $0 \leq t < t_1$. Since μ is the minimum value of V in S, this gives $|y(t)| < \varepsilon$ for $0 \leq t < t_1$. If there exists t_2 such that $|y(t_2)| = \varepsilon$, then, when $t = t_2$, we also have, from the definition of μ, $\mu \leq V(y(t_2)) \leq V(y_0) < \mu$, which cannot hold. We conclude that $t_1 = \infty$, and that, for a given $\varepsilon > 0$, there exists a $\delta > 0$ such that when $|y_0| < \delta$, $|y(t)| < \varepsilon$ for $t \geq 0$, and hence that the origin is Lyapunov stable. $\qquad\square$

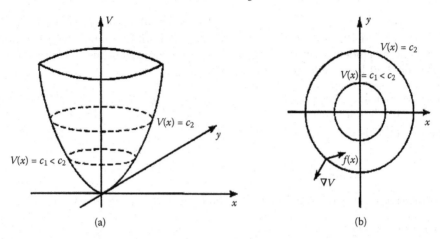

FIGURE 1.17: (a) The local behavior. (b) A contour of a Lyapunov function near the origin.

The proofs of the following two theorems are similar and so they will not be presented here.

Theorem 12 *If in some region $\Omega \in \mathbb{R}^d$ that contains the origin, there exists a scalar function $V(y)$ that is positive definite and for which $V^*(y)$ is negative definite, then the origin is asymptotically stable.*

Theorem 13 *If in some region $\Omega \in \mathbb{R}^d$ that contains the origin, there exists a scalar function $V(y)$ such that $V(0) = 0$ and $V^*(y)$ is either positive definite or negative definite, and if, in every neighborhood O of the origin with $O \subset \Omega$, there exists at least one point a such that $V(a)$ has the same sign as $V^*(a)$, then the origin is unstable.*

The last three theorems are known as *Lyapunov's theorems*, and have a geometrical interpretation that is particularly attractive for two-dimensional systems. The equation $V(y) = c$ then represents a surface in the (x, y, V)-space. By varying c through positive values only, since V is positive definite, we can obtain a series of contour lines, with $V = 0$ at the origin being a local minimum, as shown in Fig. 1.17. Since $y' = f(y)$, the vector field f represents the direction taken by an integral path at any point. The vector normal to the surface $V(y) = c$ is ∇V, so that, if $V^* = \nabla V \cdot f \leq 0$, integral paths cannot point into exterior of the region $V(y) < c$. We conclude with $V(y_0) < c_1$, cannot leave the region bounded by $V(y) = c_1$, and hence that the origin is Lyapunov stable. Similarly, if $V^* < 0$, the integral paths must actually cross from the exterior to the interior of the region. Hence $V(y)$ will decrease monotonically to zero from its initial value when the integral path enters the region Ω, and we conclude that the origin is asymptotically stable.

Although we can see why a Lyapunov function is useful, it can take considerable ingenuity to actually construct one for a given system.

Example 38 Consider the system $\dot{y}_1 = -y_1 - 2(y_2)^2$, $\dot{y}_2 = y_1 y_2 - (y_2)^3$. The origin is the only equilibrium point, and the linearized system is $\dot{y}_1 = -y_1$, $\dot{y}_2 = 0$. The eigenvalues are therefore $\lambda_1 = 0$ and $\lambda_2 = -1$, so this is a nonhyperbolic equilibrium point. Let us construct a Lyapunov function. We start by trying $V = y_1^2 + \alpha y_2^2$. This is positive definite for $\alpha > 0$, and $V(0, 0) = 0$. In addition,

$$V^* = \frac{dV}{dt} = -2y_1^2 + 2(\alpha - 2)y_1 y_2^2 - 2\alpha y_2^4.$$

If we choose $\alpha = 2$, then $dV/dt = -2y_1^2 - 4y_2^4 < 0$ for all x and y excluding the origin. By Theorems 11 and 12, we conclude that the origin is both Lyapunov and asymptotically stable. A general guideline, it is worth looking for a homogeneous, algebraic Lyapunov function when $f(y)$ has ample algebraic form.

Example 39 Consider the problem $\dot{y}_1 = y_2$, $\dot{y}_2 = -cy_2 - \sin y_1$. A suitable Lyapunov function may be chosen as $V(y_1, y_2) = \frac{1}{2}(\sqrt{c}y_1 + \frac{1}{\sqrt{c}}y_2)^2 + \frac{1}{c}(y_1^2 + \frac{1}{2}y_2^2)$. One can show that a function $\phi(|y|)$ exists such that $V \geq \phi(|y|)$. Hence,

$$\frac{dV}{dt} = -(y_1^2 + y_2^2) - (y_1 + \frac{2}{c}y_2)(\sin y_1 - y_1),$$

and this function is negative definite. Therefore, the solution $(0, 0)$ is asymptotically stable.

We'll discuss a general method for finding Lyapunov functions for asymptotically stable systems with constant coefficients. This will show that for stable linear systems with constant coefficients, Lyapunov functions do exist. This method also applies to almost linear systems like in the previous example. It is known that every matrix $\mathbf{A}$ with distinct eigenvalues can be diagonalized, i.e., there exists a nonsingular matrix $\mathbf{T}$ such that $\mathbf{T}^{-1}\mathbf{A}\mathbf{T} = \mathbf{D}$. If $\mathbf{A}$ has complex eigenvalues, $\mathbf{T}$ is a complex matrix. In many applications to differential equations, $\mathbf{A}$ is real, so that its eigenvalues are real or occur in complex conjugate pairs. In such cases one can find a slightly different normal form, operating only with real terms.

Let $\mathbf{A}$ be a real matrix with distinct eigenvalues, $\lambda_1, \ldots, \lambda_k$ the real eigenvalues, and $u_1, \ldots, u_m$ the complex eigenvalues. Suppose also that they are so indexed that $u_1 = \bar{u}_2$, $\bar{u}_3 = u_4$, $\ldots$, $\bar{u}_{m-1} = u_m$, and define a_j and b_j as follows: $u_1 = a_1 + ib_1$, $u_2 = a_1 - ib_1$, etc. For each real eigenvalue we can find real eigenvector φ_j such that $\mathbf{A}\varphi_j = \lambda_j\varphi_j$. For each complex eigenvalue we find complex eigenvector ψ_j such that, for example, $\mathbf{A}\psi_1 = u_1\psi_1$. If we decompose ψ_j into real and imaginary components, we obtain, for example, $\psi_1 = \eta_1 + i\omega_1$ so that

$$\mathbf{A}\psi_1 = \mathbf{A}\eta_1 + i\mathbf{A}\omega_1 = (a_1 + ib_1)(\eta_1 + i\omega_1) = (a_1\eta_1 - b_1\omega_1) + i(b_1\eta_1 + a_1\omega_1).$$

By equating real and imaginary terms, we have $\mathbf{A}\eta_1 = a_1\eta_1 - b_1\omega_1$, $\mathbf{A}\omega_1 = b_1\eta_1 + a_1\omega_1$, and similarly for all other complex eigenvalues. We build a matrix $\mathbf{T}$ whose columns are all real and are found with all column vectors $\varphi_i, \eta_i, \omega_i$, and this matrix is also nonsingular. The matrix $\mathbf{T}^{-1}\mathbf{A}\mathbf{T}$ is no longer diagonal, but is almost so. Its normal form consists of real diagonal elements λ_i corresponding to the real eigenvalues and of the 2×2 blocks occurring of the form $\begin{pmatrix} a_i & -b_i \\ b_i & a_i \end{pmatrix}$, corresponding to the complex eigenvalue; and all other elements vanish.

Example 40 Consider the system $Y' = \mathbf{A}Y$, $\mathbf{A} = \begin{pmatrix} -1 & 2 & 3 \\ 0 & -2 & -4 \\ 0 & 1 & -2 \end{pmatrix}$. So, $|\mathbf{A} - \lambda\mathbf{I}| =$

$-(1 + \lambda)(\lambda^2 + 4\lambda + 8)$, $\lambda_1 = -1$, $u_1 = -2 + 2i$, $u_2 = -2 - 2i$, and eigenvectors are $\varphi_1 = \langle 1, 0, 0 \rangle$, $\psi_1 = \langle 1 - 2i, 2i, 1 \rangle$, $\psi_2 = \langle 1 + 2i, -2i, 1 \rangle$. We find $\eta_1 = \langle 1, 0, 1 \rangle^*$, $\omega_1 = \langle -2, 2, 0 \rangle^*$ so that

$$\mathbf{T} = \begin{pmatrix} 1 & 1 & -2 \\ 0 & 0 & 2 \\ 0 & 1 & 0 \end{pmatrix}, \quad \mathbf{T}^{-1} = \begin{pmatrix} 1 & 1 & -1 \\ 0 & 0 & 1 \\ 0 & 1/2 & 0 \end{pmatrix}.$$

Then $\mathbf{T}^{-1}\mathbf{A}\mathbf{T} = \begin{pmatrix} -1 & 0 & 0 \\ 0 & -2 & 2 \\ 0 & -2 & -2 \end{pmatrix}$. Now we come back to equation $Y' = \mathbf{A}Y$, which will by hypothesis be assumed to be asymptotically stable. First, find a real nonsingular matrix $\mathbf{T}$ such that $\mathbf{T}^{-1}\mathbf{AT}$ is in the normal form. Then $Y = \mathbf{T}Z$, so that $Z' = \mathbf{T}^{-1}\mathbf{AT}Z = \mathbf{D}Z$, where the matrix $\mathbf{D}$ is of the normal form. For stability we require that all diagonal elements of $\mathbf{D}$ be negative. Next, we will seek a Lyapunov function in the form $V = (Z, \mathbf{B}Z)$, using inner product notation, and $\mathbf{B}$ will be a suitable matrix, which is supposed to be symmetric.

$$V' = (Z', \mathbf{B}Z) + (Z, \mathbf{B}Z') = (\mathbf{D}Z, \mathbf{B}Z) + (Z, \mathbf{B}\mathbf{D}Z) = (Z, (\mathbf{D}^T\mathbf{B} + \mathbf{B}\mathbf{D})Z),$$

where $\mathbf{D}^T$ denotes the transpose of $\mathbf{D}$. In order to be sure that V' is negative definite, we will require that $V' = -(Z, Z) = -\sum_{i=1}^{d} z_i^2$, where z_i are the components of vector function Z. In order for the normal form and V' of the form above to agree, we impose the condition that $\mathbf{D}^T\mathbf{B} + \mathbf{B}\mathbf{D} = -\mathbf{I}$, which is an equation for $\mathbf{B}$. The solution for $\mathbf{B}$ is given by diagonal matrix with elements

$$-\frac{1}{2\lambda_1}, \ \cdots, \ -\frac{1}{2\lambda_k}, \ -\frac{1}{2a_1}, \ -\frac{1}{2a_1}, \ \cdots, \ -\frac{1}{2a_m}, \ -\frac{1}{2a_m}.$$

Since by hypothesis all λ_i and a_i are negative, all diagonal elements in $\mathbf{B}$ are positive, and therefore $V = (Z, \mathbf{B}Z)$ consists of sums of positive terms so that V is positive definite:

$$V = -\frac{z_1^2}{2\lambda_1} - \cdots - \frac{z_k^2}{2\lambda_k} - \frac{1}{2a_1}(z_{k+1}^2 + z_{k+2}^2) - \cdots - \frac{1}{2a_m}(z_{k+m-1}^2 + z_{k+m}^2).$$

Next, we construct a Lyapunov function for the system

$$Y' = \begin{pmatrix} -1 & 2 & 3 \\ 0 & -2 & -4 \\ 0 & 1 & -2 \end{pmatrix} Y.$$

Set $Y = \mathbf{T}Z$ and find $Z' = \begin{pmatrix} -1 & 0 & 0 \\ 0 & -2 & 2 \\ 0 & 2 & -2 \end{pmatrix} Z$. Since a matrix $\mathbf{B}$ obeys $\mathbf{D}^T\mathbf{B} + \mathbf{B}\mathbf{D} = -\mathbf{I}$,

we get

$$\mathbf{B} = \begin{pmatrix} 1/2 & 0 & 0 \\ 0 & 1/4 & 0 \\ 0 & 0 & 1/4 \end{pmatrix}.$$

The Lyapunov function is given by $V = (Z, \mathbf{B}Z) = \frac{1}{2}z_1^2 + \frac{1}{4}z_2^2 + \frac{1}{4}z_3^2$ and $V' = -z_1^2 - z_2^2 - z_3^2$. Transforming back to Y as $V(Z) = V(\mathbf{T}^{-1}Y)$ yields a Lyapunov function for the original equation.

It is more complicated to verify that a function L is a *strong Lyapunov function*. In essence, one must show that both L and $\nabla_Y L$ have local minimum at c.

1.2.4 More on stability

Orbital stability. We will discuss a concept of stability (respectively, asymptotic stability) in a sense different from Section 1.2.3. Let $C = \{y = f(t), (y, f) \in \mathbb{R}^2\}$ be a periodic solution (limit cycle) to the system $y' = X(y)$. Consider the possible configuration of solutions near C. Some typical formations are indicated in Fig. 1.18. One can deduce all the positive configurations, e.g., by the proof of the Poincaré-Bendixson theorem (Theorem 7). On the inside of C there may be curves spiraling against C, either for increasing or for decreasing t, or else there is a sequence of limit cycles approaching C. All the curves sufficiently close to C may be limit cycles, or there may be spirals between limit cycles; see Fig. 1.18(e). A similar description applies to the outside of C.

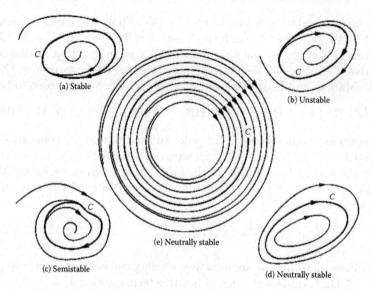

(a) Stable

(b) Unstable

(e) Neutrally stable

(c) Semistable

(d) Neutrally stable

FIGURE 1.18: Stability question for periodic solutions.

The configuration in Fig. 1.18(a) is called *stable*. Here all curves sufficiently close to C approach C in the direction of increasing t. We term C *orbitally stable*. Orbital stability does not imply that for each nearby solution $C_1 = \{y = f_1(t),\ (y, f_1) \in \mathbb{R}^2$ we have $f(t) - f_1(t) \to$ 0, as $t \to \infty$. Indeed, for each constant k, $y = f_1(t + k)$ is also a solution whose graph is the curve C_1; even if for some k we have $f(t) - f_1(t + k) \to 0$, as $t \to \infty$, this will not hold for all k. We will show in examples that even though C_1 spirals toward C, the condition $f(t) - f_1(t + k) \to 0$ need not hold for any choice of k.

This phenomenon is justified by example, where we think on the Earth's orbit as orbitally stable; the same path is continually retraced. Perturbations lead to slight differences in the length of a year, and there is no mechanism which will compensate for these differences. We call the configuration of Fig. 1.18(b) unstable and call C *orbitally unstable*. The configuration of Fig. 1.18(c) provides stability on one side, instability on the other; C is called *orbitally semistable*. In Figs. 1.18(d,e), the nearby paths do not approach C, but their maximum deviation from C can be made as small as desired; we call C *neutrally stable* in these cases.

Conditional stability. Manifold's stability. We start with some definitions.

Definition 14 The *local stable manifold* in some neighborhood of the origin is given by $\omega^s = \{(x, y) \in U : x = 0\}$.

This is the set of points that are, locally, attracted to the origin in the direction of the eigenvector with negative eigenvalue.

Definition 15 A *homeomorphism is a mapping* $f : L \to L'$ (here, $L, L' \subset \mathbb{R}^d$) that is one-to-one, onto and continuous and has a continuous inverse. A C^k *diffeomorphism* is mapping $f : L \to L'$ that is one-to-one, onto and k-times differentiable with k-times differentiable inverse. A *smooth diffeomorphism* is a C^∞ diffeomorphism, i.e., C^k under all $k \in \mathbb{N}$, and a homeomorphism is a C^0 diffeomorphism. A set $M \subset \mathbb{R}^d$ is an m-*dimensional manifold* where $m \leq d$, if each $x \in M$ has a neighborhood $U_x \subset M$ in which there exists a homeomorphism $\phi : U \to \mathbb{R}^m$. A manifold M is *differentiable* if ϕ is a diffeomorphism.

For example, a smooth curve in $\mathbb{R}^3$ is a one-dimensional differentiable manifold, and a surface, e.g., a sphere, is a two-dimensional differentiable manifold. Consider the behavior of solutions to

$$y' = X(y), \tag{1.67}$$

where $y, X \in \mathbb{R}^d$. In the neighborhood of an equilibrium point y_0, there exist three invariant manifolds.

(i) The *local stable manifold*, ω_{loc}^s, of dimension d, is spanned by the eigenvectors of A, a linear part of the Taylor series for the function $X(y)$, whose eigenvalues have real parts less than zero.

(ii) The *local unstable manifold*, ω_{loc}^u, of dimension u, is spanned by the eigenvectors of A, whose eigenvalues have real parts greater than zero.

(iii) The *local center manifold*, ω_{loc}^c, of dimension c, is spanned by the eigenvectors of A, whose eigenvalues have zero real parts.

Note that $s+u+c = n$. Solutions lying in ω_{loc}^s are characterized by exponential decay and those in ω_{loc}^u have exponential growth. The behavior on the center manifold is determined by nonlinear terms in the normal form. For a linear system these manifolds exist globally, while for a nonlinear system they exist in some neighborhood of the equilibrium point. Define a *local center manifold* for the system

$$\dot{x} = P(x, y), \quad \dot{y} = -y + Q(x, y), \tag{1.68}$$

where P and Q are nonlinear functions with $P(0,0) = Q(0,0) = 0$. The linear approximation to (1.68) is $\dot{x} = 0$, $\dot{y} = -y$, which has the solution $x = 0$, $y = y_0 e^{-t}$ that passes through the origin. This shows that points on the y-axis close to the origin approach the origin as $t \to \infty$. This local center manifold is a C^k curve ω^s in a neighborhood U of the origin such that

(i) ω^s is an invariant set within U; so that if $\xi(0) \in \omega^c$, then $\xi(t) \in \omega^c$ when $\xi(t) \in U$, where $\xi(t) \equiv (x(t), y(t))$. In other words, a solution that initially lies on the center manifold remains there when it lies in U.

(ii) ω^s is the graph of a C^k function that is tangent to the x-axis at the origin, so that

$$\omega^s = \{(x, y) : \; y = h(x), \; h(0) = \frac{dh}{dx}(0) = 0, \; (x, y) \in U\}.$$

The picture of the local stable and center manifolds is shown in Fig. 1.19. The main idea is that the qualitative behavior of solutions in the neighborhood of the origin, excluding the stable manifold, is determined by the behavior of solutions on the center manifold.

Definition 16 Let X in (1.67) be continuous in domain E. Then solution $y(t)$ is *conditionally stable in the Lyapunov's sense* if there exists $t_0 > \tau$ such that, if $y(t_0) = y_0$ and $y(t)$ is denoted by $y(t, t_0, y_0)$ and the following conditions are satisfied:

(a) There exists a real $b > 0$ such that if $|y_1 - y_0| < b$ then $y(t, t_0, y_1)$ is defined for all $t > t_0$;

(b) Given $\varepsilon > 0$, then there exists $\delta = \delta(\varepsilon, f, t_0, y_0) > 0$, such that $\delta \leq b$ and if $|y_1 - y_0| < \delta$, then for all $t \geq t_0$, $|y(t, t_0, y_1) - y(t, t_0, y_0)| < \varepsilon$;

(c) There exists $\delta^*(f, t_0, y_0) \in (0, b]$ such that $\lim_{t \to \infty} |y(t, t_0, y_1) - y(t, t_0, y_0)| = 0$ for all $t \geq t_0$ when $|y_1 - y_0| < \delta^*$.

Theorem 14 (The center manifold theorem) *The equilibrium point at the origin of the system (1.68) is stable/unstable if and only if the equilibrium point at $x = 0$ of the first-order ODE*

$$\dot{x} = P(x, h(x)), \tag{1.69}$$

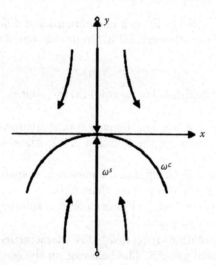

FIGURE 1.19: Local stable and center manifolds.

where $y = h(x)$ is the local central manifold, is stable/unstable. Integral paths in the neighborhood of the local central manifold are attracted onto the central manifold exponentially fast.

Proof. This may be found in [51].   □

Although it is known that a local central manifold exists, it is necessary to determine its equation $y = h(x)$. Since $\dot{y} = \dot{x}h'$, (1.67) shows that

$$h'\, P(x, h(x)) = -h(x) + Q(x, h(x)), \tag{1.70}$$

subject to

$$h(0) = h'(0) = 0. \tag{1.71}$$

We will not usually be able to solve (1.70) analytically, but we can determine the *local* form of the center manifold by assuming that $h(x)$ can be represented as a power series, which satisfies (1.71),

$$h(x) = a_2 x^2 + a_3 x^3 + \ldots \tag{1.72}$$

Example 41 Consider the system of the form (1.69), $\dot{x} = ax^3 + xy$, $\dot{y} = -y + y^2 + x^2 y + x^3$ with $a > 0$. Substituting (1.72) into our system shows that

$$(2a_2 x + 3a_3 x^2 + \ldots)(ax^3 + a_2 x^3 + a_3 x^4 + \ldots)$$
$$= -a_2 x^2 - a_3 x^3 + (a_2 x^2 + a_3 x^3 + \ldots) + x^2(a_2 x^2 + a_3 x^3 + \ldots) + x^3.$$

Equating power of x gives $a_2 = 0$ at $O(x^2)$ and $a_3 = 1$ at $O(x^3)$, so that the center manifold is given by $y = h(x) = x^3 + \ldots$. On the center manifold, we therefore have $\dot{x} = ax^3 - x^4 + \ldots$. For $|x| \ll 1$ we can ignore the quadratic term and just consider $\dot{x} \approx ax^3$, so that $\dot{x} > 0$ for $x > 0$ and $\dot{x} < 0$ for $x < 0$. Integral paths that begin on the local central manifold therefore asymptote to the origin as $t \to -\infty$, and we conclude that the local phase portrait is of a nonlinear saddle, as shown in Fig. 1.20(b). Interpret the vector X (in (1.67)) as a *velocity vector* of the point $y(t)$ as it traces the solution path. We can think of particles moving simultaneously along all solution paths. This is a *fluid motion*, and since the velocity vector does not change with time at each point $y(t)$, it is a *stationary fluid motion*. If $x = x_0$ on the central manifold when $t = 0$, $x \approx x_0 \sqrt{1 - 2ax_0^2 t}$, so that $x \sim x_0 / \sqrt{-2ax_0^2 t}$ as $t \to -\infty$. This algebraic behavior on the center manifold is in contrast to the exponential behavior that occurs on the unstable separatrix of a linear saddle point.

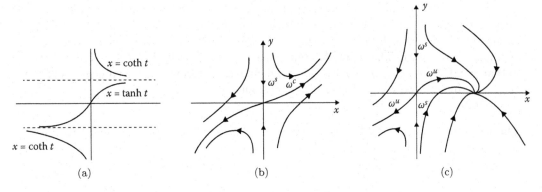

FIGURE 1.20: Integral paths starting on the local central manifold: (a) local phase portraits of asymptotes, (b) nonlinear saddle, and (c) pole.

Example 42 Consider a system $\dot{x} = x(2 - x)$, $\dot{y} = -y + x$, and try to determine the equations of the local manifolds that pass through the equilibrium point at the origin. The Jacobian at the origin is $J = \begin{pmatrix} 2 & 0 \\ 1 & -1 \end{pmatrix}$, which has eigenvalues 2 and -1, and associated eigenvectors of $(3, 1)$ and $(0, 1)$, respectively, so that $\omega_{\text{loc}}^{\text{u}}$ is the line $y = x/3$ and $\omega_{\text{loc}}^{\text{c}}$ is the y-axis. Solve the nonlinear system directly, since the equation for x is independent of y, and the equation for y is linear. The solution is

$$x = \frac{2A}{A + 2e^{-t}}, \quad y = e^{-t}\{B + 2e^{t} - \frac{4}{A}\log(Ae^{t} + 2)\},$$

where $A \neq 0$ and B are constants. There is also the obvious solution $x = 0$, $y = Be^{-t}$, the y-axis, which gives the local stable manifold, $\omega_{\text{loc}}^{\text{s}}$, and also the *global stable manifold*, $\omega^{\text{s}}(\xi_0)$, where $\xi_0 = (x_0, y_0)$ is the equilibrium point. Points that lie in the local unstable manifold, $\omega_{\text{loc}}^{\text{u}}$, have $y \to 0$ as $t \to -\infty$. Since $y \sim e^{-t}(B - 4\log 2/A)$ as $t \to -\infty$, we must have $B = A + 4/A \log 2$, so that the *global unstable manifold*, $\omega^{\text{u}}(\xi_0)$, is given in parametric form by

$$[2A/(A + 2e^{-t}), \quad A + (4/A)\log 2 + 2e^{t} - (4/A)\log(Ae^{t} + 2)].$$

The phase portrait is sketched in Fig. 1.20(c). If some (but not all) of the characteristic roots of $\mathbf{A}$ have negative real parts then some solutions $y(t)$ of the problem

$$y' = \mathbf{A}y + X(t, y), \quad |y(0)| \overset{weak}{\underset{t \to \infty}{\longrightarrow}} 0, \tag{1.73}$$

providing X is suitably restricted. Assume that X is continuous in (t, y) for small $|y|$ and $t \geq 0$; moreover, given any $\varepsilon > 0$, there exists a value $\delta > 0$ and T such that for $t \geq T$

$$|X(t, \tilde{y}) - X(t, y)| \leq \varepsilon|\tilde{y} - y| \tag{1.74}$$

for $|y| \leq \delta$, $|\tilde{y}| \leq \delta$. A sufficient condition (1.74) is that the matrix X_y' exists and that as $|y| \to 0$, $X_y' = o(1)$ uniformly in $t \geq 0$. It will be assumed that $X(t, 0) = 0$. It will also be assumed that $\mathbf{A}$ and $X(t, y)$ are real-valued but this requirement can be omitted if X is, for example, analytic.

Theorem 15 *Let for d-dimensional system the assumption hold, and let k characteristic roots of $\mathbf{A}$ have negative real parts and $d - k$ have positive real parts. Then for any large t_0 there exists in y-space a real k-dimensional manifold ω containing the origin such that any solution y of (1.73) with $y(t_0)$ on the manifold ω satisfies $y(t) \to 0$ as $t \to \infty$. Moreover, there exists an η such that any solution y near the origin but not on ω at $t = t_0$ cannot satisfy $|y(t)| \leq \eta$, $t \geq t_0$.*

More precisely, it can be shown that there exists a real nonsingular constant matrix $\mathbf{P}$ such that if $z = \mathbf{P}y$ then there are $d - k$ real continuous functions $\psi_j(z_1, \ldots, z_k)$ defined for small $|z_i|$, $i \leq k$, such that $z_j = \psi_j(z_1, \ldots, z_k)$, $j = k+1, \ldots, d$ define a k-dimensional manifold $\tilde{\omega}$ in z-space. The manifold ω in y-space is obtained from $\tilde{\omega}$ by applying $\mathbf{P}^{-1}$ to z so that $y = \mathbf{P}^{-1}[z_1, \ldots z_k, \psi_{k+1}, \ldots \psi_d]$ defines ω in terms of k curvilinear coordinates $z_1, \ldots, z_k$. In the case where $X(t, y)$ has continuous first derivatives with respect to the y_j, the manifold $\omega \in C^1$, as the following theorem shows.

Theorem 16 *The manifold ω from Theorem 15 is differentiable if $\partial f / \partial y_i$ exists and is continuous for $i = 1, \ldots, d$, and t_0 is sufficiently large. To be more precise, the functions ψ_j, $j = k+1, \ldots, d$, are of class C^1 for $|z_l|$ sufficiently small, $l \leq k$. Moreover, $\partial \psi_j / \partial z_l = 0$ at $z_1 = \cdots = z_k = 0$.*

Example 43 Consider the system

$$y' = -y + g(y), \quad g(y) = (y_1^2 + y_2^2)\begin{pmatrix} -y_2 \\ y_1 \end{pmatrix}, \quad y = (y_1, y_2) \in \mathbb{R}^2.$$

It is known that $(0,0)$ is a stable proper node of the linear system $y' = -y$. To find the general solution we should transform our system to $z' = e^{-2t}g(z)$ by using $y = e^{-t}z$. Next, introduce a new independent variable $s = e^{-2t}/2$. Then the system becomes $\frac{dz}{ds} = g(z)$. Observe that $\frac{d}{ds}(z_1^2 + z_2^2) = 2(z_1 \frac{dz_1}{ds} + z_2 \frac{dz_2}{ds}) = 0$. If a solution $y(t)$ of our system satisfies an initial condition $y(0) = \eta = \begin{pmatrix} \eta_1 \\ \eta_2 \end{pmatrix}$, the corresponding solution $z(s) = e^t y(t)$ is the solution of the initial value problem

$$\frac{dz}{ds} = r(\eta)^2 \begin{pmatrix} z_2 \\ -z_1 \end{pmatrix}, \quad z(1/2) = \eta,$$

where $r(\eta) = \sqrt{\eta_1^2 + \eta_2^2}$. Hence,

$$z_1(s) = r(\eta)\sin[r(\eta)^2 s + \theta(\eta)], \quad z_2(s) = r(\eta)\cos[r(\eta)^2 s + \theta(\eta)],$$

where (see Fig. 1.20(c), left)

$$\begin{cases} z_1(0) = r(\eta)\sin(\theta(\eta)), \\ z_2(0) = r(\eta)\cos(\theta(\eta)), \end{cases} \quad \begin{cases} \eta_1 = r(\eta)\sin[\frac{r(\eta)^2}{2} + \theta(\eta)], \\ \eta_2 = r(\eta)\cos[\frac{r(\eta)^2}{2} + \theta(\eta)]. \end{cases}$$

The general solution is

$$y_1(t) = e^{-t}r(\eta)\sin(\frac{r(\eta)^2}{2}e^{-2t} + \theta(\eta)), \quad y_2(t) = e^{-t}r(\eta)\cos(\frac{r(\eta)^2}{2}e^{-2t} + \theta(\eta)),$$

where $y(0) = (\eta_1, \eta_2)$ and $r(\eta) = \sqrt{\eta_1^2 + \eta_2^2}$. Every orbit goes around the point $(0,0)$ only a finite number of times. Fig. 1.21 (right) shows that $(0,0)$ is a stable proper node as $t \to \infty$.

1.3 Bifurcations

Bifurcation theory deals with a systematic classification of the sudden changes in the qualitative behavior of dynamical systems. The study of how the character of fixed points

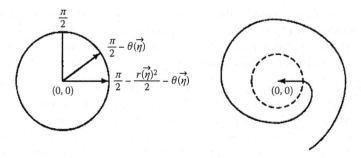

FIGURE 1.21: Stability question for periodic solutions.

(and other types of state space attractors) changes as parameters of the system change is called a *bifurcation theory*. The effort is divided into two parts. The first part of the theory focuses attention on bifurcations that can be linked to the change stability of either fixed points or limit cycles (which can be treated as fixed points in Poincaré-Bendixson theory). Call these bifurcations *local* because they can be analyzed in terms of the local behavior of the system near the relevant fixed point or limit cycle. The other part of the theory, the part which is much less well developed, deals with bifurcation events that involve larger-scale behavior in state space and hence are called *global* bifurcations. These global events involve larger-scale structures such as *basins of attraction* and *homoclinic* and *heteroclinic orbits* for saddle points. For both types of bifurcation, bifurcation theory attempts to classify the kinds of bifurcations that can occur for dynamical systems as function of the dimensionality of the state space (or, importantly, of the *effective* dimensionality associated with trajectories near the parameter dependent of the bifurcation). The number of parameters that must change to "cause" a bifurcation is called *codimension* of bifurcation.

1.3.1 Instability and bifurcations

Consider the first-order ODE $y' = f(y)$. For an equilibrium point at $y = y_0$ a simple linearization about $y = y_0$ determines the local behavior and stability. If $f'(y_0) = 0$, $y = y_0$ is a nonhyperbolic equilibrium point, and we need to retain more terms in the Taylor expansion of $f(y)$. For example, if $f(y_0) = f'(y_0) = 0$ and $f''(y_0) \neq 0$, then $\frac{d\bar{y}}{dt} \approx \frac{1}{2} f''(y_0)\bar{y}^2$, for $|\bar{y}| \ll 1$, and hence $\bar{y} = -\frac{2}{f'(y_0)(t-t_0)}$ for some constant t_0. Focusing on this case, we can see that $\bar{y} \to 0$ $(y \to y_0)$, as $t \to \infty$ for $t_0 < 0$, while $\bar{y} \to -\infty$ as $t \to \infty$ for $t > t_0$. The fact that nonhyperbolic equilibrium point therefore attracts solutions from $y \geq y_1$ and repels them from $y > y_1$ is algebraic, in contrast to the factor, exponential approach associated with stable hyperbolic equilibrium points.

Definition 17 A system with one or more nonhyperbolic equilibrium points is called *structurally unstable*.

This means that a small perturbation, not to the solution but to the model itself, for example the addition of a small extra term to $f(y)$, can lead to a qualitative difference in the structure of the set of solutions, for example, a change in the number of equilibrium points in their stability [36]. In our example the addition of a small positive constant to $f(y)$ shifts the graph upward, and a small negative constant shifts the graph downward and leads to the absence of any equilibrium solutions.

Example 44 Consider ODE $y' = \mu - y^2$. For $\mu > 0$ there are two hyperbolic equilibrium points at $y_{\pm} = \pm\sqrt{\mu}$, and for $\mu < 0$ there are no equilibrium points. If $\mu = 0$, then $y = 0$ is a nonhyperbolic equilibrium point, called a *bifurcation point*, because the qualitative nature of the phase line changes. The bifurcation associated with $y' = \mu - y^2$ is called a *saddle-node* bifurcation. A bifurcation takes place at $\mu = 0$ and we have a saddle point, which changes into a repeller-node pair as μ becomes positive.

Local bifurcations are those in which fixed points or limit cycles appear, disappear, or change their stability. This change in stability is signaled by a change in the real part of the characteristic exponents (roots) associated with that fixed point. At a local bifurcation, the real part becomes equal to 0 as parameters of the system are changed. As the real part of the characteristic exponent changes from negative to positive, the motion associated with that characteristic direction goes from stable (attracted toward the fixed point) to unstable (being repelled by the fixed point). For a Poincaré map fixed point, this criterion means that the absolute value of the characteristic multiplier equals to unity [40].

The *Center-Manifold Theorem* (Theorem 14) tells us that at a local bifurcation, we can focus our attention on just those degrees of freedom associated with the characteristic exponents whose real parts go to 0. It is the number of characteristic exponents with real parts equal to 0 that determines the number of effective dimensions for the bifurcation. The *Center-Manifold* is that "subspace" associated with the characteristic exponents whose real parts are 0. The real parts for the stable manifold are negative, and the real parts for the unstable manifold are positive.

In order to classify the types of bifurcation, it is traditional to reduce the dynamical equations to a standard form, the *normal form* in which the bifurcation event occurs when a parameter value, e.g., μ, reaches 0 and a fixed point, located at $y = 0$, has a characteristic exponent with real part equal to 0. More general situations are reduced to these normal forms by appropriate coordinate and parameter transformations; therefore, there is no less in generality in using the simplest normal forms. If the center manifold dynamics is one-dimensional and described by a differential equation then the time evolution equation can be written as

$$y' = A_0 + B_0 y(t) + C_0 y^2(t) + \cdots + \mu[A_1 + B_1 y(t) + C_1 y^2(t) + \ldots]$$
$$+\mu^2[A_2 + B_2 y(t) + C_2 y^2(t) + \ldots] + \ldots,$$

where the subscript on the constants A, B, C, and so on, tells us with which power of the control parameter μ they are to be associated. By choosing the values of the constants, a systematic classification of the bifurcations can be developed. For systems (such as Poincaré mapping of limit cycles) we can write a similar expression for the iterated map dynamics near the bifurcation point:

$$y_{n+1} = A_0 + B_0 y_n(t) + C_0 y_n^2(t) + \cdots + \mu[A_1 + B_1 y_n(t) + C_1 y_n^2(t) + \ldots]$$
$$+\mu^2[A_2 + B_2 y_n(t) + C_2 y_n^2(t) + \ldots] + \ldots$$

For a system described by differential equations, if we set $A_1 = 1$, $C_0 = -1$ and all the other constants to be zero, we arrive at the equation $y' = \mu - y^2$ which we recognize as the equation describing the behavior at a repeller-node (or, in higher dimension, saddle node) bifurcation (see below). If we use the iterated map form with $B_0 = -1$, $B_1 = -1$ and $D_0 = 1$, the equation $y_{n+1} = -(1 + \mu)y_n + y_n^3$ describes a period-doubling bifurcation. By adding a second state space dimension with its own normal form equation, we can model Hopf bifurcation. Bifurcations are also classified as *subtle* (or *supercritical*) and *catastrophic* (or equivalently, *subcritical*). In a subtle bifurcation, the stable fixed point changes smoothly with parameter value near the bifurcation point. The Hopf bifurcation is an example of a subtle bifurcation. For a catastrophic bifurcation, the stable fixed point appears (as in a saddle-node bifurcation) or disappears or jumps discontinuously to a new location. As an example of a subcritical bifurcation, consider the normal form equation $y' = \mu y(t) + y^3(t) -$

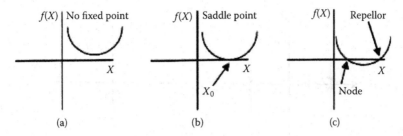

FIGURE 1.22: The *bifurcation diagram* for the repeller-node bifurcation. At the repeller-node bifurcation point, the fixed point of the system is structurally unstable. Structurally unstable points are important because their existence indicates a possible bifurcation.

$y^5(t)$, with the control parameter μ. For $\mu < 1$, the steady-state solution $y = 0$ is stable. For $\mu > 0$, there are two stable states: $y_{\pm} = \sqrt{(1 \pm \sqrt{1 + 4\mu})/2}$. A bifurcation occurs at $\mu = 0$ and the stable states jump to a new location.

Global bifurcations are bifurcation events that involve changes in basins of attraction, homoclinic or heteroclinic orbits, or their structures that extend over significant regions of state space. Such bifurcations include intermittency and crises. Since we need to take into account behavior over a wide range of state space, a different means of classifying and studying such bifurcations is obviously needed. Unfortunately, the theory of global bifurcations is both more difficult and less articulated than is the theory of local bifurcations. Specific cases, such as homoclinic tangencies and crises, have been studied in some detail, but a general classification scheme is yet to be devised.

Consider a local bifurcation $y' = (\mu - 1)(y - a)$ with the fixed point $y = a$. The characteristic value is $\mu - 1$. For $\mu < 1$ the fixed point is a node, and for $\mu > 1$ a repeller. We say that if at $\mu = 1$ there is a bifurcation: the node (a stable fixed point) changes to a repeller (an unstable fixed point). It is traditional to define a new parameter $\delta = \mu - 1$ and a new variable $z = y - a$. Then the equation becomes $z' = \delta z(t)$. We plot the characteristic value as a function of the parameter, see Fig. 1.22; if we shift the function so that the fixed point occurs at $y = 0$, then the function $f(y)$ changes as we pass through the bifurcation, and the function in question becomes tangent to be y-axis at the bifurcation point. The bifurcation just described is called a *saddle-node bifurcation*, or a *fold bifurcation*.

To visualize the trajectories, it is customary to add another dimension to the state space. Trajectories moving along the added y_2 direction are assumed to be attracted toward the y_1 axis (the original y-axis). The one-dimensional repeller then becomes a two-dimensional saddle point. Thus in this *lifted* or *suspended state space*, the bifurcation involves the intersection of a saddle point and a node. Hence, this event is called a saddle-point bifurcation. Figure 1.23 gives a sketch of the system's (lifted) phase portrait above and below the bifurcation value $\mu = 0$. For $\mu > 0$, there is a saddle point at $y_1 = -\sqrt{\mu}$ and a node at $y_1 = +\sqrt{\mu}$. There is no fixed point for $\mu < 0$.

The basic idea is that nonhyperbolic equilibrium leads to local bifurcations; the number of limit sets near such equilibrium can change under a small perturbation. Also, nonhyperbolic periodic orbits may disappear or duplicate when they are perturbed, and saddle connections can be broken by a small transversal perturbation. Finally, to show that structurally stable vector fields are dense, one proceeds as follows: Let f be structurally unstable and $\varepsilon > 0$ a small number. Then one builds a structurally stable vector field g such that $\|f - g\| < \varepsilon$. The construction proceeds by successive small deformations of f, which remove the violated conditions one by one.

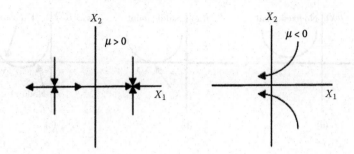

FIGURE 1.23: Phase portraits above and below a saddle-point bifurcation.

1.3.2   Saddle-node bifurcation

A first type of a singular plane vector field occurs when a nonhyperbolic equilibrium point is present, such that the linearization around this point has 0 as a simple eigenvalue. In appropriate coordinates, the vector field can be written as

$$y'_1 = f_1(y_1, y_2), \quad y'_2 = ay_2 + f_2(y_1, y_2), \quad a \neq 0, \tag{1.75}$$

where f_1, f_2 and their first-order derivatives vanish at the origin. Assume that f_1 and f_2 are of class C^2. The stable manifold theorem (Theorem 14) admits the following generalization.

Theorem 17 (Center manifold) *If $a < 0$ then (1.75) has a unique invariant curve ω^s tangent to the y_2-axis at the origin such that $\omega(y) = 0$ for all $y \in \omega^s$; ω^s is called the stable manifold of 0. If $a > 0$ then (1.75) has a unique invariant curve ω^u tangent to the y_2-axis at the origin such that $\alpha(y) = 0$ for all $y \in \omega^u$; ω^u is called the unstable manifold of 0.*

There exists, but not necessarily unique, invariant curve ω^c, tangent to the y_1-axis at the origin, called a *center manifold*. Any center manifold can be described, for sufficiently small y_1, by equation of the form $y_2 = h(y_1)$. Inserting this relation into (1.75), we obtain the equation

$$ah(y_1) + f_2(y_1, h(y_1)) = h'(y_1)f_1(y_1, h(y_1)). \tag{1.76}$$

All solutions h of this equation admit the same Taylor expansion around $y_1 = 0$. For our purposes, it is sufficient to know that $h(y_1) = O(y_2)$ as $y_1 \to 0$. Let $z = y_2 - h(y_1)$ describe the distance of a general solution to the center manifold. By (1.75) and (1.76), z satisfies an equation of the form $z' = F(z, y_1)$, $F(z, y_1) = a + O(y_1)$ for small y_1. Thus ω^c attracts nearby orbits if $a < 0$ and repels them if $a > 0$, so that there are no equilibrium points outside ω^c. The qualitative dynamics near the origin is described by the dynamics on the central manifold, which is governed by the equation

$$y'_1 = f_1(y_1, h(y_1)) = cy_1^2 + O(y_1^2), \quad c = \frac{1}{2}\frac{\partial^2 f_1}{\partial y_1^2}(0,0). \tag{1.77}$$

If $c \neq 0$ (the typical case) then the orbits on ω^c will be attracted by the origin from one side, and repelled from the other side; see Fig. 1.24(a).

Definition 18 Assume that the linearization of f at a nonhyperbolic equilibrium point y^* has the eigenvalues 0 and $a \neq 0$, and (1.77) on its center manifold satisfies $c \neq 0$. Then y^* is called an *elementary saddle node*.

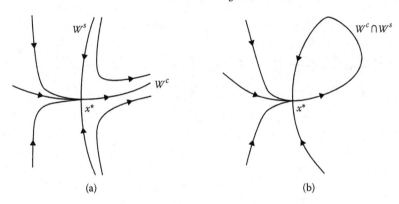

FIGURE 1.24: (a) Local phase portrait near an elementary saddle-node x^*, in the case where $a > 0$; (b) if the center and stable manifolds intersect.

In order to examine the effect of a small perturbation of f, consider a one-parameter family $f(y, \lambda)$ such that $f(\cdot, 0) = f$, and write the system in the form

$$\begin{cases} y_1' = g_1(y_1, y_2, \lambda), & g_1(y_1, y_2, 0) = f_1(y_1, y_2), \\ y_2' = g_2(y_1, y_2, \lambda), & g_2(y_1, y_2, 0) = ay_2 + f_2(y_1, y_2), \\ \lambda' = 0. \end{cases}$$

The point $(0, 0, 0)$ is a nonhyperbolic equilibrium point of this extended system, with eigenvalues $(0, a, 0)$. Theorem 17 shows the existence of an invariant manifold, locally given by $y_2 = h(y_1, \lambda)$, which attracts nearby orbits if $a < 0$ and repels them if $a > 0$. The dynamics on the center manifold is given by

$$y_1' = f_1(y_1, h(y_1, \lambda), \lambda) = F(y_1, \lambda), \tag{1.78}$$

where $G \in C^2$ satisfies the relations

$$F(0, 0) = 0, \quad \frac{\partial F}{\partial y_1}(0, 0) = 0, \quad \frac{\partial^2 F}{\partial y_1^2}(0, 0) = 2c \neq 0. \tag{1.79}$$

The graph of $F(y_1, 0)$ has a quadratic tangency with the y_1-axis at the origin. For small λ, the graph of $F(y_1, \lambda)$ can thus have zero, one or two intersections with the y_1-axis.

Theorem 18 *For small λ, there exists a differentiable function $H(\lambda)$ such that*
  a) if $H(\lambda) > 0$ then $F(y_1, \lambda)$ has no equilibrium near $y_1 = 0$;
  b) if $H(\lambda) = 0$ then $F(y_1, \lambda)$ has isolated nonhyperbolic equilibrium point near $y_1 = 0$;
  c) if $H(\lambda) < 0$ then $F(y_1, \lambda)$ has two hyperbolic equilibrium points of opposite stability near $y_1 = 0$;

Proof. Consider the function $G(y_1, \lambda) = \frac{\partial F}{\partial y_1}(y_1, \lambda)$. Then $G(0, 0) = 0$ and $\frac{\partial G}{\partial y_1}(0, 0) = 2c \neq 0$ by (1.79). The implicit function theorem shows the existence of a unique differentiable function φ such that $\varphi(0) = 0$ and $G(\varphi(\lambda), \lambda) = 0$ for all sufficiently small λ. Taylor's formula for the function $K(y, \lambda) = F(\varphi(\lambda) + z, \lambda)$ shows that

$$K(z, \lambda) = F(\varphi(\lambda), \lambda) + z^2[c + R_1(z, \lambda)], \quad \frac{\partial K}{\partial z}(z, \lambda) = z[2c + R_2(z, \lambda)],$$

for some continuous functions R_1, R_2 which vanish at the origin. Set $H(\lambda) = F(\varphi(\lambda), \lambda)/c$. Then $H(0) = 0$ and if λ and z are sufficiently small that $|R_1(z, \lambda)| < c/2$, then

$$H(\lambda) + \frac{1}{2} z^2 \leq \frac{K(z, \lambda)}{c} \leq H(\lambda) + \frac{3}{2} z^2.$$

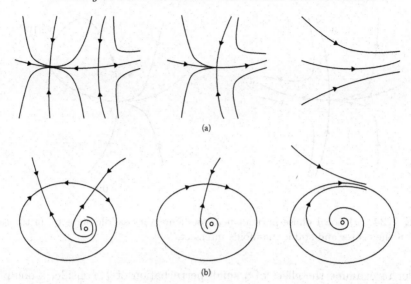

FIGURE 1.25: (a) Unfolding of a saddle node when the center manifold does not form a loop: the system has a saddle and a node if $H(\lambda) < 0$, an elementary saddle node if $H(\lambda) = 0$, and there is no equilibrium if $H(\lambda) > 0$. (b) If the center manifold does form a loop then a periodic orbit appears for $H(\lambda) > 0$.

Hence $K(z, \lambda)$ vanishes twice, once or never, depending on the sign of $H(\lambda)$. The expression for $\frac{\partial K}{\partial z}$ shows that K is monotonous in z for small positive or negative z, thus there is no other equilibrium near the origin. The existence of $H(\lambda)$, which determines entirely the topology of orbits near the equilibrium point, shows that f is indeed a singularity of codimension one, in a sufficiently small neighborhood of the equilibrium. It admits the local unfolding. $y_1' = \lambda + y_1^2$, $y_2' = ay_2$. For $H > 0$ there are no equilibrium points, while for $H < 0$ the system admits a saddle and a node (Fig. 1.25a).   □

The global behavior of the perturbed vector field $f(y, \lambda)$ depends on the global behavior of the center manifold of the saddle node at $\lambda = 0$.

Example 45 Consider the system $\dot{y}_1 = -y_1 + cy_2^2$, $\dot{y}_2 = y_1 y_2 - y_2^3$, where $c \in \mathbb{R}$. We will determine the stability of the origin. Naively, one might think that since the first equation suggests that $y_1(t)$ converges to zero as $t \to \infty$, the dynamics can be approximated by projecting on the line $y_1 = 0$. The conclusion is that the origin is asymptotically stable, because $\dot{y}_2 = -y_2^3$ when $y_1 = 0$. We will compute the center manifold in order to find the correct answer to the question of stability. For functions $\phi(y_2)$ which are continuously differentiable in a neighborhood of the origin, define operator $\hat{L}\phi \equiv \phi'(y_2)[y_2\phi(z_2) - y_2^3] + \phi(y_2) - cy_2^2$. We will find the solution $h(y_2)$ of $\hat{L}\phi(y_2) = 0$ in the form $h(y_2) = h_2 y_2^2 + h_3 y_2^3 + h_4 y_2^4 + O(y_2^5)$. Thus, $\hat{L}h(y_2) = 0$ becomes

$$\hat{L}h(y_2) = (h_2 - c)y_2^2 + h_3 y_2^3 + [h_4 + 2h_2(h_2 - 1)]y_2^4 + O(y_2^5) = 0,$$

which requires $h_2 = c$, $h_3 = 0$ and $h_4 = -2c(c-1)$. Hence the center manifold has a Taylor expansion of the form $h(y_2) = cy_2^2 - 2c(c-1)y_2^4 + O(y_2^5)$, and its motion is governed by

$$\dot{u} = uh(u) - u^3 = (c-1)u^3 - 2c(c-1)u^5 + O(u^6).$$

For instance, u^2 as a Lyapunov function that the equilibrium point $u(t) = 0$ is asymptotically stable if $c < 1$ and unstable if $c > 1$. Thus, the origin is asymptotically stable if $c < 1$ and

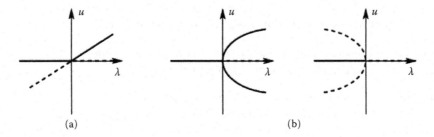

(a)   (b)

FIGURE 1.26: Pitchfork bifurcation. (a) Supercritical bifurcation and (b) subcritical bifurcation.

unstable if $c > 1$, which contradicts the approach when $c > 1$. In the case $c = 1$, the function $h(y_2) = y_2^2$ is an exact solution of $\hat{L}h(y_2) = 0$, and the curve $y_1 = y_2^2$ is the unique center manifold of our system, which has the particularity to consist only of equilibrium points. The origin is thus stable.

1.3.3 Transcritical and pitchfork bifurcations

If $F(y_1, \lambda) = \lambda^2 - y_1^2$, then $H(\lambda) = -\lambda^2$. Thus, there are two equilibrium points for all $\lambda \neq 0$, because the family $F(y_1, \lambda)$ is tangent to the manifold Ω. This case is called *transcritical bifurcation*. Particularly, if $F(y_1, \lambda) = \lambda y_1 - y_2^2$, then $H(\lambda) = -\lambda^2/4$ and we have a transcritical bifurcation.

Theorem 19 (Transitional bifurcation) *Consider the first-order ODE $\dot{y} = f(y, \mu)$ with $f(0,0) = f_y(0,0) = f_\mu(0,0) = 0$. Provided that $f_{yy}(0,0) \neq 0$ and $f_{\mu y}^2(0,0) - f_{yy}(0,0) f_{\mu\mu}(0,0) > 0$, there exist two continuous curves of equilibrium points in the neighborhood of $(0,0)$. These curves intersect transversely at $(0,0)$. For each $\mu \neq 0$ there are two hyperbolic equilibrium points in the neighborhood at $x = 0$. If $f_{yy}(0,0) < 0$ then the equilibrium point with the larger value of y is stable, while the other is unstable, and vice versa for $f_{yy}(0,0) > 0$.*

The normal form $\dot{y} = \mu y - y^3$ has a single equilibrium point $y = 0$ for $\mu < 0$, and three equilibrium points $y = 0$ and $y = \pm\sqrt{\mu}$ for $\mu > 0$; see the bifurcation diagram in Fig. 1.26(a). The bifurcation at $y = 0$, $\mu = 0$ is called *supercritical pitchfork bifurcation*. A similar situation, in which the two new equilibrium points created at the bifurcation point to be unstable, is the *subcritical pitchfork bifurcation*, with normal form $\dot{y} = \mu y + y^3$, whose bifurcation diagram is shown in Fig. 1.26(b).

Let $f \in C^3$ have a nonhyperbolic equilibrium point y^*, such that the Jacobian $f'(y^*)$ admits the eigenvalues 0 and $a \neq 0$. Unlike in the case of the saddle-node bifurcation, assume that the dynamics on the center manifold is governed by the equation $\dot{y}_1 F(y_1)$ with

$$F(0) = \frac{\partial F}{\partial y}(0) = \frac{\partial^2 F}{\partial y^2}(0) = 0, \qquad \frac{\partial^3 F}{\partial y^3}(0) = 6c \neq 0. \qquad (1.80)$$

Then $\dot{y}_1 = cy_1^3 + o(y_1^3)$, and thus y^* will attract nearby solutions on the center manifold from both sides if $c < 0$, and repel them if $c > 0$, but at a much slower rate than in the hyperbolic case.

A perturbation $f(y, \lambda)$ of f will admit a center manifold on which the dynamics is governed by $\dot{y}_1 = F(y_1, \lambda)$. Then for sufficiently small λ and y_1, this system is topologically equivalent to a member of the two-parameter family

$$\dot{y}_1 = F_{\lambda_1 \lambda_2}(y_1) = \lambda_1 + \lambda_2 y_1 \pm y_1^3, \qquad (1.81)$$

where the signs $\pm$ correspond to the cases $c > 0$ or $c < 0$. We focus on the $c < 0$ case, since the other case can be obtained by inverting the direction of time.

Theorem 20 *The two-parameter family (1.81) is a local unfolding of the singular vector field, whose restriction to the center manifold satisfies (1.80).*

Proof. The function $G(y_1, \lambda) = \frac{\partial F}{\partial y_1}(y_1, \lambda)$ satisfies the hypotheses of $F(y_1, \lambda)$ in Theorem 18, that we used in the case of the saddle-node bifurcation. There we showed the existence of a function $H(\lambda)$ such that, in a neighborhood of $y_1 = 0$, G vanishes twice if $H(\lambda) < 0$, once if $H(\lambda) = 0$ and never if $H(\lambda) > 0$. It follows that F has two extrema if $H(\lambda) < 0$, is monotonous with an inflection point if $H(\lambda) = 0$, and is strictly monotonous if $H(\lambda) > 0$. This shows us that H plays the role of λ_2 in the unfolding (1.81). Now let $\varphi(\lambda)$ be the extremum of G, i.e., the intersection point of F. Then the Taylor expansion of F around $\varphi(\lambda)$ can be written as

$$\frac{1}{c}F(\varphi(\lambda) + z, \lambda) = \frac{1}{c}F(\varphi(\lambda), \lambda) + \frac{1}{c}\frac{\partial F}{\partial y_1}(\varphi(\lambda), \lambda)z + z^3[1 + R(z, \lambda)],$$

where $R(y, \lambda)$ is a continuous function with $R(0, 0)$. The coefficient of z is $H(\lambda) = \lambda_2$. Since $F(\varphi(\lambda), \lambda)/c$ controls the vertical position of the graph of F, it plays the same role as λ_1 (since F has the same behavior as $F_{\lambda_1 \lambda_2}$ for a λ_1 of the form $c^1 F(\varphi(\lambda), \lambda)[1+\rho(\lambda)]$, with $\rho(\lambda) \to 0$ as $\lambda \to 0$).

Theorem 21 (Pitchfork bifurcation) *Consider the first-order ODE*

$$\dot{y} = f(y, \mu), \quad f(0, 0) = f_y(0, 0) = f_\mu(0, 0) = f_{yy}(0, 0) = 0.$$

Provided that $f_{\mu y}(0, 0) \neq 0$ and $f_{yyy}(0, 0) \neq 0$, there exist two continuous curves of equilibrium points in the neighborhood of $(0, 0)$. One curve passes through $(0, 0)$ transverse to the line $\mu = 0$, while the other is tangent to $\mu = 0$. In addition;

(i) if $f_{\mu y}(0, 0) \cdot f_{yyy}(0, 0) < 0$ then, close to $(0, 0)$, there is a single equilibrium point for $\mu < 0$ and three equilibrium points for $\mu > 0$;

(ii) if $f_{\mu y}(0, 0) \cdot f_{yyy}(0, 0) > 0$ then, close to $(0, 0)$, there is a single equilibrium point for $\mu > 0$ and three equilibrium points for $\mu < 0$;

(iii) if $f_{yyy}(0, 0) < 0$ then the single equilibrium point and the outer two of the three equilibrium points are stable, while the middle of three equilibrium points is unstable, and vice versa for $f_{yyy}(0, 0) > 0$.

1.3.4 Hopf bifurcation

Consider another type of nonhyperbolic equilibrium point, namely, assume that $f \in C^3$ vanishes at y^* and that the linearization of f at y^* has imaginary eigenvalues $\pm i\omega_0$ with $\omega_0 \neq 0$. In appropriate coordinates, we can write

$$\dot{y}_1 = -\omega_0 y_2 + f_1(y_1, y_2), \quad \dot{y}_2 = \omega_0 y_1 + f_2(y_1, y_2), \tag{1.82}$$

where f_1, f_2 are nonlinear terms, that is, f_1, f_2 and their derivatives vanish for $y_1 = y_2 = 0$. The easiest way to describe the dynamics is to introduce the complex variable $z = y_1 + iy_2$ that carries out the change of variables $y_1 = \frac{z+\bar{z}}{2} = \text{Re}\, z$, $y_2 = \frac{z-\bar{z}}{2i} = \text{Im}\, z$. The system (1.82) becomes

$$\dot{z} = i\omega_0 z + G(z, \bar{z}),$$

where G admits a Taylor expansion $G(z, \bar{z}) = \sum_{2 \leq n+m \leq 3} G_{nm} z^n \bar{z}^m + o(|z|^3)$. In order to simplify the nonlinear term G, using the theory of normal forms, observe that if $h(z, \bar{z})$ is a homogeneous polynomial of degree 2, hence $w = z + h(z, \bar{z})$ satisfies the equation

$$\dot{w} = i\omega_0 z + i\omega_0 \left(z \frac{\partial h}{\partial z} - \bar{z} \frac{\partial h}{\partial \bar{z}} \right) + G(z, \bar{z}) + \frac{\partial h}{\partial z} G(z, \bar{z}) + \frac{\partial h}{\partial \bar{z}} \bar{G}(z, \bar{z}).$$

Let h satisfy the equation

$$i\omega_0 \left(h - z \frac{\partial h}{\partial z} + \bar{z} \frac{\partial h}{\partial \bar{z}} \right) = G_2(z, \bar{z}) \equiv \sum_{n+m=2} G_{nm} z^n \bar{z}^m. \tag{1.83}$$

Then $\dot{w} = i\omega_0 w + G(z, \bar{z}) - G_2(z, \bar{z}) + \frac{\partial h}{\partial z} G(z, \bar{z}) + \frac{\partial h}{\partial \bar{z}} \bar{G}(z, \bar{z})$. All terms of order 2 have disappeared from this equation, while new terms of order 3 have appeared. The RHS can be expressed as a function of w and $\bar{w}$ without generating any new terms of order 2. The same procedure can be used to eliminate terms of order 3 as well, by solving an equation analogous to (1.83).

Consider this equation in more detail. If h is a sum of terms $h_{nm} z^n \bar{z}^m$ then for each (n, m) we get $i\omega_0(1 - n + m)h_{nm} = G_{nm}$. This equation can be solved if $m \neq n - 1$. Hence the only term which cannot be eliminated is $(n, m) = (2, 1)$ – the term proportional to $|z|^2 z$. Reduce $\dot{z} = i\omega_0 z + G(z, \bar{z})$ to the form

$$\dot{w} = i\omega_0 w + c_{21} |w|^2 w + o(|w|^3), \tag{1.84}$$

where c_{21} can be expressed as a function of G_{nm}. In polar coordinates $w = re^{i\varphi}$, substitute in (1.84) and take the real and imaginary parts, then we obtain the system

$$\dot{r} = \operatorname{Re} c_{21} r^3 + R_1(r, \varphi), \quad \dot{\varphi} = \omega_0 + \operatorname{Im} c_{21} r^2 + R_2(r, \varphi), \tag{1.85}$$

where $R_1 = o(r^3)$ and $R_2 = o(r^2)$. Thus, if $\operatorname{Re} c_{21} < 0$ then solutions starting sufficiently close to the origin spiral slowly towards $(0, 0)$ (r decreases like $t^{-1/2}$), and if $\operatorname{Re} c_{21} > 0$ then solutions spiral outward.

Definition 19 Let y^* be an equilibrium point such that the linearization of f at y^* admits imaginary eigenvalues $\pm i\omega_0$, $\omega_0 \neq 0$, and such that the normal form (1.85) satisfies $\operatorname{Re} c_{21} \neq 0$. Then y^* is called an *elementary composed focus*.

Consider perturbations of the form $\dot{y} = f(y, \lambda)$ of (1.82). Observe that since $f(0, 0) = 0$ and $\frac{\partial f}{\partial y}(0, 0)$ has no vanishing eigenvalue, the implicit function theorem can be applied to show the existence of a unique equilibrium $y^*(\lambda)$ near 0 for small λ. The linearization at $y^*(\lambda)$ has eigenvalues $a(\lambda) \pm i\omega(\lambda)$, where $a(0) = 0$ and $\omega(0) = \omega_0$. In appropriate coordinates, we thus have $\dot{y}_1 = a(\lambda)y_1 - \omega(\lambda)y_2 + f_1(y_1, y_2, \lambda)$ and $\dot{y}_2 = \omega(\lambda)y_1 + a(\lambda)y_2 + f_2(y_1, y_2, \lambda)$. The normal form can be computed in the same way, and becomes in polar coordinates

$$\dot{r} = a(\lambda)r + \operatorname{Re} c_{21}(\lambda)r^3 + R_1(r, \varphi, \lambda), \quad \dot{\varphi} = \omega(\lambda) + \operatorname{Im} c_{21}(\lambda)r^2 + R_2(r, \varphi, \lambda).$$

Thus, the following result holds.

Theorem 22 (Hopf bifurcation) *Let $\operatorname{Re} c_{21}(0) \neq 0$ and $H(\lambda) = a(\lambda)/\operatorname{Re} c_{21}(0)$. Then for small λ,*
- *if $H(\lambda) > 0$ then f has an isolated hyperbolic equilibrium near 0;*
- *if $H(\lambda) = 0$ then f has an elementary composed focus near 0;*
- *if $H(\lambda) > 0$ then f has a hyperbolic equilibrium and a hyperbolic periodic orbit near 0, with opposite stability.*

The proof is clear if R_1 is identically zero. If not, this result can be proved by computing $dr/d\varphi$ and examining the properties of the Poincaré map associated with the section $\varphi = 0$.

Thus, if f admits an elementary composed focus, then f is singular of codimension 1 (provided all other limit sets are hyperbolic and there are no saddle connections). The family of equations $\dot{z} = (\lambda + i\omega_0)z + c_{21}|z|^2 z$ is a local unfolding of the singularity.

1.3.5 Saddle-node bifurcation of a periodic orbit

A third class of structurally unstable vector fields $f(y)$ consists of vector fields which admit a nonhyperbolic periodic orbit Γ. If Π is the Poincaré map for a transverse section Σ through $y^* \in \Gamma$, then Γ is nonhyperbolic if $\Pi'(y^*) = 1$, hence the graph of $\Pi(y^*)$ is tangent to the diagonal at y^*.

Definition 20 A periodic *orbit* Γ is called *quasi-hyperbolic* if its Poincaré map satisfies $\Pi''(y^*) = 1$ and $\Pi'(y^*) \neq 0$.

The condition $\Pi''(y^*) \neq 0$ means that the graph of $\Pi(y)$ has a quadratic tangency with the diagonal. Hence, the orbit Γ attracts other orbits from one side, and repels them from the other side; see Fig. 1.27(a).

Consider a one-parameter family of perturbations $f(y, \lambda)$ of the vector field. For sufficiently small λ, $f(y, \lambda)$ will still be transverse to Σ, and it admits a Poincaré map $\Pi(y, \lambda)$ defined in some neighborhood of $(y^*, 0)$. Periodic orbits correspond to fixed points of Π, and are thus solutions of $\Pi(y, \lambda) - y = 0$.

Note that $\Pi(y, \lambda) - y$ is tangent to the y-axis at $y = y^*$. This situation is very reminiscent of the saddle-node bifurcation of equilibrium points; see (1.78) and (1.79). Indeed, Theorem 18 applied to $\Pi(y, \lambda) - y = 0$ yields the existence of a function $H(\lambda)$ which controls the number of fixed points of Π.

Theorem 23 *Let $f(y, 0)$ admit a quasi-hyperbolic orbit Γ. Then there is a function $H(\lambda)$ such that, for sufficiently small λ,*
- *if $H(\lambda) > 0$ then $f(y, \lambda)$ admits no periodic orbit near Γ;*
- *if $H(\lambda) = 0$ then $f(y, \lambda)$ admits a quasi-hyperbolic orbit near Γ;*
- *if $H(\lambda) < 0$ then $f(y, \lambda)$ admits two hyperbolic periodic orbits of opposite stability near Γ.*

Moreover, $f(y, 0)$ is singular of codimension 1 if and only if there do not exist two invariant manifolds of saddles which approach Γ, one for $t \to \infty$, the other for $t \to -\infty$.

The reason for this last condition is that if such manifolds exist, then one can build perturbations which admit a saddle connection and are neither structurally stable nor topologically equivalent to f; Fig. 1.27(b).

1.3.6 Global bifurcation

The remaining singular plane vector fields are those which admit a saddle connection. Such saddle connections are indeed very sensitive to small perturbations.

Consider first the case of a heteroclinic saddle connection between two saddle points y_1^* and y_2^*. To simplify the notations, choose coordinates in such a way that $y_1^* = (0, 0)$ and $y_2^* = (1, 0)$, and that the saddle connection belongs to the y_1-axis. Write the perturbed system as

$$\dot{y}_1 = f_1(y_1, y_2) + \lambda g_1(y_1, y_2, \lambda), \quad \dot{y}_2 = f_2(y_1, y_2) + \lambda g_2(y_1, y_2, \lambda), \tag{1.86}$$

where, in particular, $f_2(y_1, 0) = 0$ for $y_1 \in [0, 1]$. We will find a condition for this system

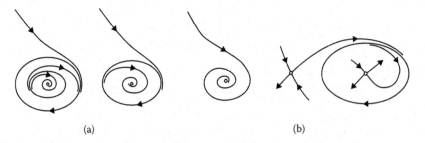

(a)   (b)

FIGURE 1.27: (a) The unfolding of a quasi-hyperbolic periodic orbit Γ is a saddle-node bifurcation of periodic orbits: A small perturbation will have either two hyperbolic periodic orbits, or one quasi-hyperbolic periodic orbit, or no periodic orbit near Γ. The focus in the center of the pictures does not take part in the bifurcation, and can be replaced by a more complicated limit set. (b) Partial phase portrait near a quasi-hyperbolic periodic orbit which is not a codimension 1 singularity; a small perturbation may admit a saddle connection.

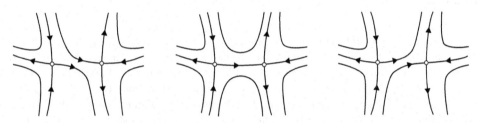

FIGURE 1.28: Breaking a heteroclinic saddle connection.

to admit a saddle connection for $\lambda \neq 0$. Note first that since y_1^* and y_2^* are hyperbolic, the perturbed system (1.86) still admits two saddles near y_1^* and y_2^* for small λ. Carrying out an affine coordinate transformation, we may assume that they are still $(0,0)$ and $(1,0)$.

Assume that the saddle connection is given by $y_2 = \lambda h_1(y_1) + O(\lambda^2)$, with $h_1(0) = h_2(0) = 0$ (such an expansion holds for sufficiently small λ). By Taylor's formula, we find that on this curve,

$$\dot{y}_1 = f_1(y_1, 0) + O(\lambda), \quad \dot{y}_2 = \lambda \frac{\partial f_2}{\partial y_2}(y_1, 0) h_1(y_1) + \lambda g_2(y_1, 0, 0) + O(\lambda^2).$$

Thus h_1 must satisfy the linear equation

$$h_1'(y_1) = \lim_{\lambda \to 0} \frac{\dot{y}_2}{\lambda \dot{y}_1} = \frac{1}{f_1(y_1, 0)} \left[\frac{\partial f_2}{\partial y_2}(y_1, 0) h_1(y_1) + g_2(y_1, 0, 0) \right],$$

which admits the solution

$$h_1(y_1) = \int_0^{y_1} \exp \left\{ \int_z^{y_1} \frac{1}{f_1(w, 0)} \frac{\partial f_2}{\partial y_2}(w, 0) dw \right\} \frac{g_2(z, 0, 0)}{f_1(z, 0)} dz$$

(using $h_1(0) = 0$ as initial value). This relation can also be written in the symmetric form

$$h_1(y_1) = \frac{1}{f_1(x_1, 0)} \int_0^{y_1} \exp \left\{ \int_z^{y_1} \frac{1}{f_1(w, 0)} \left(\frac{\partial f_1}{\partial y_1} + \frac{\partial f_2}{\partial y_2} \right)(w, 0) dw \right\} g_2(z, 0, 0) dz.$$

Only the last factor in the integrand depends on the perturbation term g. Here we have only computed the first-order term of $h(y_1, \lambda) = \lambda h_1(y_1) + O(\lambda^2)$. Including higher-order

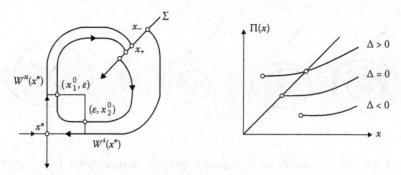

FIGURE 1.29: (a) The first return map Π is associated with a section Σ transverse to the stable and unstable manifolds of y^*. (b) Behavior of the first return map in a case where $|a_1| > a_2$ for different signs of the splitting $\Delta = y_+ - y_-$. A map Π admits a fixed point when the splitting is positive, which corresponds to a periodic orbit. If $|a_1| < a_2$, then Π admits a fixed point when the splitting is negative.

terms, the requirement $h(1, \lambda) = 0$ provides us with a codimension 1 condition of the form $H(\lambda) = 0$. This kind of integral condition is closely related to the method of *Melnikov functions*, which allows us to estimate the splitting between manifolds. If the condition $h(1, \lambda) = 0$ is not satisfied, then the unstable manifold of y_1^* and the stable manifold of y_2^* will avoid each other, and the sign of h will determine which manifold lies above the other; see Fig. 1.28. In other words, $\omega^u(y_1^*)$ may not be the graph of a function when y_1 approaches 1, and one would rather compute estimations of $\omega^u(y_1^*)$ for $0 < y_1 < 1/2$ and of $\omega^s(y_2^*)$ for $1/2 < y < 1$, and determine their splitting at $1/2$.

Finally, consider the other kind of saddle connection, called a *homoclinic loop* Γ. In order to describe the flow near Γ, introduce a transverse arc Σ through a point $y_0 \in \Gamma$. We order the points on Σ in the inward direction; see Fig. 1.29(a). Orbits starting slightly "above" y_0 with respect to this ordering will return to Σ, defining the first return map Π for small $y > y_0$. For small $\lambda \neq 0$, the stable and unstable manifold of y^* intersects Σ at points $y_-(\lambda)$ and $y_+(\lambda)$, and their distance $\Delta(\lambda) = y_+(\lambda) - y_-(\lambda)$ may be estimated as in the case of a heteroclinic saddle connection. The first return map $\Pi(y, \lambda)$ can be defined for sufficiently small $y > y_-$; see Fig. 1.29(a). We wish to determine the behavior of Π in the limit $y \to y_-$. In this limit, the first return map depends essentially on the behavior of the orbits as they pass close to y^*. Assume that the linearization of f at y^* has eigenvalues $a_1 < 0 < a_2$, and choose a coordinate system where this linearization is diagonal. Consider an orbit connecting the points (ε, y_2^0) and (y_1^0, ε) with $0 < y_2^0 < \varepsilon \ll 1$, and determine y_1^0 as a function of y_2^0. A linear approximation yields

$$\begin{cases} \dot{y}_1 = -|a_1|y_1 \implies y_1(t) = e^{-|a_1|t}\varepsilon, \\ \dot{y}_2 = |a_2|y_2 \implies y_2(t) = e^{-|a_2|t}y_2^0. \end{cases}$$

Requiring that $y_2(t) = \varepsilon$ and eliminating t, we get

$$y_1^0/y_2^0 = (\varepsilon/y_2^0)^{-|a_1/a_2| \log(\varepsilon/y_2^0)} = (\varepsilon/y_2^0)^{1-|a_1/a_2|}.$$

Thus, $\lim\limits_{y_2^0 \to 0} \frac{y_1^0}{y_2^0} = \begin{cases} 0 & \text{if } |a_1| > a_2, \\ \infty & \text{if } |a_1| < a_2, \end{cases}$ By controlling the error terms more carefully, one can show that $\lim_{vy \to y_-} \Pi'(y)$ has the same behavior, as long as $|a_1/a_2|$ is bounded above by 1.

Definition 21 Let y^* be a saddle point with eigenvalues $a_1 < 0 < a_2$ satisfying $a_1 + a_2 \neq 0$. An orbit Γ such that $\omega(z) = \alpha(z) = y^*$ for all $z \in \Gamma$ is called an *elementary homoclinic loop*.

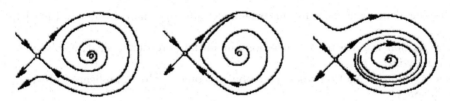

FIGURE 1.30: Unfolding an elementary homoclinic loop Γ: any small perturbation will either admit a hyperbolic periodic orbit close to Γ, or an elementary homoclinic loop, or no invariant set near Γ.

In summary, as $y \to y_-$, the first return map of an elementary homoclinic loop satisfies

$$\lim_{y \to y_-} \Pi(y) = y_- + \Delta(\lambda), \quad \lim_{y \to y_-} \Pi'(y) = \left\{ \begin{array}{ll} 0 & \text{if } |a_1| > a_2, \\ \infty & \text{if } |a_1| < a_2. \end{array} \right.$$

The shape of Π for different values of Δ is sketched in Fig. 1.29(b), in a case where $|a_1| > a_2$.

If $\Delta > 0$ then Π admits a fixed point, which corresponds to a periodic orbit (Fig. 1.30). If $|a_1| < a_2$ then there is a periodic orbit for $\Delta < 0$.

Theorem 24 *Let f admit an elementary homoclinic loop Γ. Given a one-parameter family of perturbations $f(y, \lambda)$, there exists a function $H(\lambda) = \Delta(\lambda) \log(|a_1|/a_2)$ (where $\Delta(\lambda)$ describes the splitting of stable and unstable manifolds) such that*
  · *if $H(\lambda) < 0$ then there are no invariant sets near Γ;*
  · *if $H(\lambda) = 0$ then $f(y, \lambda)$ admits an elementary homoclinic loop near Γ;*
  · *if $H(\lambda) > 0$ then $f(y, \lambda)$ admits a hyperbolic periodic orbit near Γ. Moreover, if all other equilibrium and periodic orbits of f are hyperbolic, there are no saddle connections, and no stable or unstable manifold if another saddle point tends to Γ then f is a singularity of codimension 1.*

This completes the list of all structurally unstable vector fields of codimension 1: $f \in S_1$ if and only if it satisfies the hypotheses of Peixoto's theorem (see [7], except for one limit set, chosen from the list:
  • an elementary saddle node;
  • an elementary composed focus;
  • a quasi-hyperbolic periodic orbit (satisfying hypotheses of Theorem 23);
  • a heteroclinic saddle connection;
  • an elementary homoclinic loop (satisfying hypotheses of Theorem 24).

1.4 Attractors

The characterization of chaotic dynamics is a large subject with many recent developments and ramifications in several domains of mathematics. We discuss only a few selected topics, the major objective being to provide an idea of what we call chaotic motion.

1.4.1 Chaotic motion and symbolic dynamics

In various definitions of chaos, at least one of the following elements is present:

  – the time dependence of solutions is more complicated than stationary, (quasi-)periodic behavior;

  – the motion is sensitive to variations in initial conditions: nearby solutions diverge exponentially fast;

  – the asymptotic motion takes place on a geometrically complicated object, called a *strange attractor*;

  – chaotic orbits coexist with a countable infinity of unstable periodic orbits; the number of orbits of periods less than or equal to T grows exponentially fast with T;

  – as time goes by the images under the flow of any two subjects of phase space get entangled in a complicated way.

For instance, strange attractors cannot occur in conservative systems, but are quite typical in chaotic dissipative systems, e.g., turbulence. There are subtle relations between the above properties, many of which can be characterized quantitatively by Lyapunov's exponents, topological entropy, etc.

Symbolic dynamics is a useful technique for characterizing dynamical systems. It consists of associating with every orbit a sequence of symbols, related to a partition of phase space, and describing in which order the orbit of x visits elements of the partition. In this way, one can reduce the problem to a combinatorial one, by transposing the dynamics to the space of allowed symbolic sequences.

We illustrate the method by a simple one-dimensional map called the *tent map*, and then discuss a two-dimensional map, which is useful in proving existence of chaotic orbits for a general class of systems.

One may check that $y = h(x) = \frac{1}{2}(1 - \cos \pi x)$ transforms the logistic map (see Example 46) $f_\lambda(y) = \lambda y(1 - y)$ with $\lambda = 4$ into the map (called the *tent map* because of its triangular shape)

$$g_2(x) = h^{-1} \circ f_4 \circ h(x) = \left\{ \begin{array}{ll} 2x, & 0 \leq x \leq 1/2 \\ 2 - 2x, & 1/2 \leq x \leq 1. \end{array} \right. \tag{1.87}$$

In (1.87), we have lost differentiability at $x = 1/2$, but the piecewise linearity is useful to classify the orbits. The map (1.87) has two unstable fixed points, at 0 and 2/3. Note that 1 is mapped to 0 and 1/2 is mapped to 1, so there exist "transient" orbits, ending at 0 after finitely many iterations. In aim to understand the other orbits, it turns out to be a good idea to write x in *binary expansion*. We write

$$x = \sum_{i=0}^{\infty} b_i 2^{-i}, \quad b_i \in \{0, 1\} \;\Rightarrow\; x = b_0 b_1 b_2 \ldots \tag{1.88}$$

This decomposition is not unique. Indeed, $\sum_{i=1}^{\infty} 2^{-i} = 1 \;\Rightarrow\; 0.1^\infty = 1.0^\infty$, where the subscript ∞ means that the symbol is repeated infinitely. This is the only kind of degeneracy. Let B be the set of symbolic (binary) sequences $b = 0.b_1 b_2 \ldots$, which are not terminated by 0^∞, except that we include the sequence 0.0^∞. Then (1.88) defines a bijection $b : [0, 1] \to B$. In fact, B is a metric space with the distance

$$d(b, b') = \sum_{i=1}^{\infty} (1 - \delta_{b_i b_i'}) 2^{-i}, \tag{1.89}$$

where δ_{ab} is the Kronecker delta function, and b is continuous in the resulting topology. Note that $x \leq 1/2$ if and only if $b_1 = 0$. In this case

$$g_2(0.0 b_2 b_3 \ldots) = 2 \sum_{i=2}^{\infty} b_i 2^{-i} = \sum_{i=1}^{\infty} b_{i+1} 2^{-i} = 0.b_2 b_3 \ldots \tag{1.90}$$

Otherwise, we have $x > \frac{1}{2}$, $b_1 = 1$ and

$$
\begin{aligned}
g_2(0.0b_2b_3 \ldots) &= 2 - 2\left(\frac{1}{2} + \sum_{i=2}^{\infty} b_i 2^{-i}\right) = 1 - \sum_{i=1}^{\infty} b_{i+1} 2^{-i} \\
&= \sum_{i=1}^{\infty} (1 - b_{i+1}) 2^{-i} = 0.(1 - b_2)(1 - b_3) \ldots
\end{aligned}
\tag{1.91}
$$

Hence g_2 includes map τ from B to itself, defined by the following rules:
- shift all digits of b one unit to the left, discarding b_0;
- if the first digit is 1, reverse all digits;
- replace the sequence 10^{∞}, if present, by 0.1^{∞}.

One may analyze the dynamics in B instead of $[0, 1]$, which is easier due to the relatively simple form of τ. In fact, g_2 has an even simpler representation. With every $b \in B$, we associate a sequence $\varepsilon = (\varepsilon_1, \varepsilon_2, \ldots)$ defined by

$$
\varepsilon_j = (-1)^{b_{j-1} + b_j}, \quad j = 1, 2, \ldots
\tag{1.92}
$$

Elements of ε are -1 if adjacent digits of b are different, and $+1$ if they are equal. The set Σ of sequences ε constructed in this way consists of $\{-1, +1\}^N$, from which we exclude sequences ending with $(+1)^{\infty}$ and containing an even number of -1. The correspondence (1.92) admits an inverse defined by

$$
b_0 = 0, \quad (-1)^{b_j} = \varepsilon_j (-1)^{b_{j-1}} = \prod_{i=1}^{j} \varepsilon_i, \quad b_j \in \{0, 1\}, \quad j \geq 1.
$$

Σ can be endowed with a similar distance as (1.89), and then (1.92) defines a homeomorphism between B and Σ, and, by composition with b, a homeomorphism between $[0, 1]$ and Σ.

Treating separately the cases $b_1 = 0$ and $b_1 = 1$, one may see that the tent map induces a dynamics in Σ given by $\sigma : (\varepsilon_1, \varepsilon_2, \ldots) \mapsto (\varepsilon_2, \varepsilon_3, \ldots)$ which is called the *shift map*. The sequence ε has a very simple interpretation. Indeed, $x \in [0, 1/2] \Leftrightarrow b_1 = 0 \Leftrightarrow \varepsilon_1 = +1$, so that

$$
\varepsilon_j = +1 \Leftrightarrow \sigma^j(\varepsilon)_1 = +1 \Leftrightarrow g_2^j(x) \in [0, 1/2].
$$

We have shown that the sequence $\{\varepsilon_j\}$ indicates whether the j-th iterate of x is to the left or to the right of $1/2$, and that information is encoded in the binary expansion of $b(x)$. The sequence ε associated with x is called *itinerary*.

Let us examine the different possible orbits.

Periodic orbits. The itinerary of periodic orbit must be periodic. Conversely, if an itinerary ε is periodic with period $p \geq 1$, then $\sigma^p(\varepsilon) = \varepsilon$, and by bijectivity the corresponding x is a fixed point of g_2^p. Thus x is a point of period p if and only if its itinerary is of the form $\varepsilon = A^{\infty}$, with A a finite sequence of length p. There are exactly 2^p such sequences, and thus $2^p/p$ orbits of period p (the number of orbits of least period p can be a bit smaller). For instance, we obtain against the fixed points of g_2

$$
\begin{aligned}
\varepsilon = (+1)^{\infty} &\Leftrightarrow b = 0.0^{\infty} \Leftrightarrow x = 0; \\
\varepsilon = (-1)^{\infty} &\Leftrightarrow b = 0.(10)^{\infty} \Leftrightarrow x = \frac{1}{2} + \frac{1}{8} + \frac{1}{32} + \cdots = \frac{2}{3}.
\end{aligned}
$$

Transient orbits. Itineraries of the form $\varepsilon = BA^{\infty}$ with B a finite sequence of the length q correspond to orbits that reach a periodic orbit after a finite number of steps.

Call *eventually periodic* orbits which are either periodic, or reach a periodic orbit after a finite number of iterations. Their itineraries are of the form $\varepsilon = A^{\infty}$ or $\varepsilon = BA^{\infty}$. It is easy to see that the orbit of $x \in [0, 1]$ is eventually periodic if and only if x is rational. If

x has an eventually periodic orbit (which implies that $2^{p+q}x - 2^q x \in \mathbb{Z}$, for some $p, q \geq 1$), then $x \in \mathbb{Q}$. The map $x \mapsto \{2x\}$, where $\{\cdot\}$ denotes the fractional part, shifts the bits of $b(x)$ one unit to the left. It also maps the set $\{0, 1/m, 2/m, \ldots, (m-1)/m\}$ into itself. Thus the orbit of x under this map is eventually periodic, and so are its binary expansion and its itinerary. The set $\mathbb{Q}$ of rational numbers being dense in $\mathbb{R}$, the union of all eventually periodic orbits is dense in $[0, 1]$. This set is countable and has zero Lebesgue measure.

Chaotic orbits. All irrational initial conditions, by contrast, admit itineraries, which are not periodic. The corresponding orbits will typically look quite random.

The following properties are direct consequences of the symbolic representation:

– for every sequence $\varepsilon \in \Sigma$, there exists $x \in [0, 1]$ with itinerary ε. Choose the initial condition in such a way that the orbit passes left and right of $1/2$ in any prescribed order;
– the dynamic is sensitive to initial conditions. If we only know x with finite precision δ, we are incapable of making any prediction on its orbit after n iterations, whenever $2^n \geq 1/\delta$;
– simply by looking at whether successive iterates of x lie to the left or right of $1/2$, we are able to determine the binary expansion of x (even though g_2 is not injective);
– there exists a dense orbit. Indeed, choose x in such a way that its itinerary

$$\varepsilon = (\underbrace{+1}_{\text{period 1}}, \underbrace{-1, +1}_{+,+}, \underbrace{+1, +1}_{+,-}, \underbrace{-1, -1}_{-,+}, \underbrace{+1, -1}_{-,-}, \underbrace{-1, +1, +1, +1}_{+,+,+}, \ldots, \ldots)$$

contains all possible finite sequences, ordered by increasing length. Given any $\delta > 0$ and $y \in [0, 1]$, one can find $n \in \mathbb{N}$ such that $|g_2^n(x) - y| < \delta$. It suffices to take n in such a way that the shifted sequence $\sigma^n \varepsilon$ and the itinerary of y agree for the first $[|\log \delta|/\log 2]$ bits.

Remark 5 The tent map g_2 has other properties, interesting from the point of view of measure theory. The key property is that its unique invariant measure, which is absolutely continuous with respect to the Lebesgue measure, is the uniform measure. This *measure* is *ergodic*, meaning that

$$\lim_{n \to \infty} \frac{1}{n} \sum_{i=1}^{n} \varphi(g_2^n(x)) = \int_0^1 \varphi(x) \, dx$$

for every continuous test function φ and for Lebesgue almost all $x \in [0, 1]$. Thus from a probabilistic perspective, almost all orbits will be uniformly distributed over the interval.

Itineraries can be defined for more general maps, just by choosing some partition of phase space. In general, there are no simple relations between a point and its itinerary, and the correspondence need not be one-to-one. Some symbolic sequences may never occur, and several initial conditions may have the same itinerary, for instance if they are in the basin of attraction of a stable equilibrium.

Consider the following version of the tent map: $g_3(x) = \begin{cases} 3x & \text{for } x \leq \frac{1}{2}, \\ 3 - 3x & \text{for } x \geq \frac{1}{2}. \end{cases}$ Since the interval $[0, 1]$ is not invariant, we define g_3 on all $\mathbb{R}$. First, observe that
– if $x_0 < 0$ then the orbit of x_0 converges to $-\infty$;
– if $x_0 > 1$ then $x_1 = g_3(x_0) < 0$;
– if $x_0 \in (1/3, 2/3)$ then $x_1 = g_3(x_0) > 1$ and thus $x_2 < 0$.
Hence all orbits which leave $[0, 1]$ eventually converge to $-\infty$. One could suspect that all orbits leave the interval after a certain number of iterations. This is not the case since, for instance, orbits starting in multiples of 3^{-n} reach the fixed point 0 after finitely many

FIGURE 1.31: The first steps of the construction of the Cantor set Λ.

iterations. But there exists a more subtle nontrivial invariant set. In order to describe it, we use a ternary (base 3) representation of $x \in [0,1]$:

$$x = \sum_{i=0}^{\infty} b_i 3^{-i}, \quad b_i \in \{0,1,2\} \Rightarrow x = b_0.b_1 b_2 b_3.$$

Again, this representation is not unique. We can make it unique by replacing 10^{∞} with 02^{∞} and 12^{∞} with 20^{∞} if applicable. Next, observe that

$$g_3(0.0b_2 b_3 \ldots) = 0.b_2 b_3 \ldots \quad g_3(0.2b_2 b_3 \ldots) = 0.(2-b_2)(2-b_3)\ldots, \tag{1.93}$$

Points of the form $0.1b_2 b_3 \ldots$ belong to $(1/3, 2/3)$, and we already know that these leave the interval $[0,1]$. It is thus immediate that the orbit of x never leaves the interval $[0,1]$ if and only if its ternary expansion $b(x)$ does not contain the symbol 1. The largest invariant subset of $[0,1]$ is thus

$$\Lambda = \{x \in [0,1] : \ b_i(x) \neq 1, \ \forall i \geq 1\}. \tag{1.94}$$

Definition 22 The set Λ obtaining by removing from $[0,1]$ the open intervals $(1/3, 2/3)$, $(1/9, 2/9)$, $(1/27, 2/27)$ and so on is called a *Cantor set* (Fig. 1.31).

A set Λ is a Cantor set if and only if it is closed, its interior is empty, and all its points are accumulation points. The Cantor set Λ is an example of a *fractal*.

Definition 23 Let M be a subset of $\mathbb{R}^d$. Assume that for any $\varepsilon > 0$, M can be covered by a finite number of hypercubes of side length ε. Let $N(\varepsilon)$ be the smallest possible number of such cubes. Then the *box-counting dimension* of M is defined by $D_0(M) = \lim_{\varepsilon \to 0} \frac{\log N(\varepsilon)}{\log(1/\varepsilon)}$.

One shows that $D_0 = d$ for "usual" sets $M \subset \mathbb{R}^d$, such as a d-dimensional hypercube. For the Cantor set (1.94), we find that $N(\varepsilon) = 2^n$ if $3^{-n} \leq \varepsilon < 3^{-n+1}$, and thus $D_0(\Lambda) = \log 2/\log 3$. The map g_3 restricted to Λ is conjugate to the tent map g_2. Indeed, with any point $x = 0.b_1 b_2 \ldots$ in Λ, we can associate $h(x) = 0.(b_1/2)(b_2/2)\ldots$ in $[0,1]$, and the relations (1.93) and (1.90), (1.91) show that $h \circ g_3|_\Lambda \circ h^{-1} = g_2$. Thus we can compute itineraries of points $x \in \Lambda$, and the conclusions on the behavior of orbits of g_2 can be transposed to $g_3|_\Lambda$. The set Λ is the simplest example of what is called a *hyperbolic invariant set*. Similar properties hold for nonlinear perturbations of the tent map. Indeed, the ternary representation is a useful tool, but not essential for the existence of orbits with all possible symbolic representations. It is, in fact, sufficient that f maps two disjoint intervals $I_1, I_2 \subset [0,1]$ onto $[0,1]$, and maps all other points outside $[0,1]$. Then the symbolic representation corresponds to the sequence of intervals I_1 or I_2 visited by the orbit.

Example 46 (Logistic map) *Consider a population of animals that reproduces once a year. Let P_n be the number of individuals in the year number n. The offspring being proportional to the number of adults, the simplest model for the evolution of the population from one*

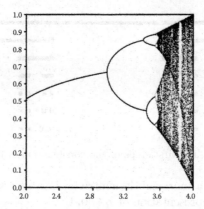

FIGURE 1.32: Bifurcation diagram of the logistic map. For each λ on the abscissa, the points x_{1001} through x_{1100} are represented on the ordinate, for an initial value $x_0 = 1/2$.

year to the next is the linear equation $P_{n+1} = \lambda P_n$, where λ is the natality rate (minus the mortality rate). This law leads to an exponential growth of the form $P_n = \lambda^n P_0 = e^{n \ln \lambda} P_0$ (the Malthus law). This model becomes unrealistic when the number of individuals is so large that the limitation of resources becomes apparent. The simplest possibility to limit the growth is to introduce a quadratic term $-\beta P_n^2$, leading to the law $P_{n+1} = \lambda P_n - \beta P_n^2$. The rescaled variable $x = \beta P$ then obeys the equation

$$x_{n+1} = f_\lambda(x_n) := \lambda x_n (1 - x_n).$$

The map f_λ is called the logistic map. For $0 \le \lambda \le 4$, f_λ maps the interval $[0, 1]$ into itself. The dynamics of the sequence x_n depends drastically on the value of λ. For $0 \le \lambda \le 1$, all orbits converge to 0, which means that the population becomes extinct. For $1 < \lambda \le 3$, all orbits starting at $x_0 > 0$ converge to $1 - 1/\lambda$, and thus the population reaches a stable equilibrium. For $3 < \lambda \le 1 + \sqrt{6}$, the orbits converge to a cycle of period 2, so that the population asymptotically jumps back and forth between two values. For $\lambda > 1 + \sqrt{6}$, the system goes through a whole sequence of period doubling. Similarly as in RB convection, the values λ_n of the parameter for which the n-th period doubling occurs obey the law

$$\lim_{n \to \infty} \frac{\lambda_n - \lambda_{n-1}}{\lambda_{n+1} - \lambda_n} = \delta \approx 4.669,$$

where δ is called the Feigenbaum constant. In 1978, Feigenbaum as well as Coulet and Tresser independently outlined an argument showing that such period doubling cascades should be observable for a large class of systems, and that the constant δ is universal. For instance, it also appears in two-dimensional Hénon map $x_{n+1} = 1 - \lambda x_n^2 + y_n$ and $y_{n+1} = b x_n$, where $0 < b < 1$. Rigorous proofs of these properties were later worked out by different authors (see [21]).

*　　For the logistic map, the period doubling accumulates at a value $\lambda_\infty \approx 3.56$ beyond which the orbits become chaotic. For larger values of λ, there is a complicated interplay of regular and chaotic motion (Fig. 1.32), containing other period doubling cascades. Finally, for $\lambda = 4$, one can prove that the dynamics is as random as coin tossing. Mappings which have been used for the biological population models and which exhibit similar features include the following:*

$$f_\lambda(x) = \lambda \sin(\pi x), \quad f_\lambda(x) = x \exp[\lambda(1 - x)],$$
$$f_\lambda(x) = x[1 + \lambda(1 - x)], \quad f_\lambda(x) = \lambda x / (1 + ax)^5.$$

1.4.2   Homoclinic tangles and Smale's horseshoe map

One-dimensional maps are rather special cases of dynamical systems, but some of their properties can often be transposed to more "realistic" systems. As a motivating example, consider the equations

$$\dot{x}_1 = x_2, \quad \dot{x}_2 = x_1 - x_1^3, \tag{1.95}$$

called the (undumped) unforced *Duffing oscillator*. This is a Hamiltonian system, with $H(x_1, x_2) = x_2^2/2 + x_1^4/4 - x_1^2/2$. The Hamiltonian is an invariant (a constant) of the motion, and thus orbits of (1.95) belong to level set of H; see Fig. 1.33(a). The point $(0,0)$ is a hyperbolic equilibrium, with the particularity that its unstable and stable manifolds W^u and W^s are identical: they form the curve $H = 0$, and are called *homoclinic loops*. We perturb (1.95) by periodic forcing: $\dot{x}_1 = x_2$, $\dot{x}_2 = x_1 - x_1^3 + \varepsilon \sin(2\pi t)$. Introduce $x_3 = t$ as additional periodic variable to obtain the autonomous system

$$\dot{x}_1 = x_2, \quad \dot{x}_2 = x_1 - x_1^3 + \varepsilon \sin(2\pi x_3), \quad \dot{x}_3 = 1. \tag{1.96}$$

We can thus use the surface $x_3 = 0$ as a Poincarè section. The Poincarè map; see Fig. 1.33(b), has a complicated structure; it seems to fill a two-dimensional region, in which its dynamics is quite random.

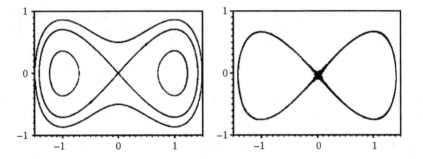

FIGURE 1.33: (a) Orbits of Duffing's equation for $\varepsilon = 0$. (b) Orbit of the Poincarè map for $\varepsilon = 0.25$.

This phenomenon can be explained by analyzing the properties of the Poincarè map P_ε. First observe that the system (1.96) is conservative. On the section perpendicular to the flow, the Poincarè map is also conservative. When $\varepsilon = 0$, $P_0 = \varphi_1$ is the time-one flow of (1.95), thus P_0 admits the origin as a hyperbolic fixed point. The implicit function theorem implies that for small ε, P_ε admits an isolated hyperbolic fixed point $x^*(\varepsilon)$ near $(0,0)$. It is unlikely that the stable and unstable manifolds of $x^*(\varepsilon)$ still form a loop when $\varepsilon > 0$, and if not, they must intersect transversally at some point y_0 because of area conservation. The point y_0 is called a *homoclinic point*; a trajectory of a homoclinic point is called a *homoclinic orbit*. A method due to Melnikov allows to prove that such a transverse intersection exists.

Consider the successive images $x_n = P_\varepsilon^n(x_0)$. If x_0 belongs to the stable manifold W^s of $x^*(\varepsilon)$, all x_n also belong to W^s, and they must accumulate at x^* for $n \to \infty$. Similarly, if x_0 belongs to the unstable manifold W^u, all x_n belong to W^u and accumulate at x^* for $n \to -\infty$. Thus W^s and W^u must intersect infinitely many times (often). Because the map is area preserving, the area between W^s, W^u, and any consecutive intersection points must be the same, and thus W^u has to oscillate with increasing amplitude when approaching W^* (Fig. 1.34). A stable manifold W^s has a similar behavior. This geometrical structure, which was first described by Poincarè, is called the *homoclinic tangle*.

The dynamics near the homoclinic tangle can be described as follows. Let Q be a small

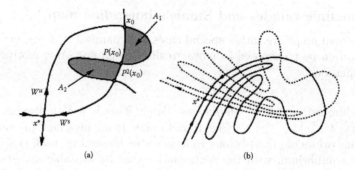

FIGURE 1.34: If the stable and unstable manifolds of a conservation map intersect transversally, they must intersect infinitely often, forming a homoclinic tangle.

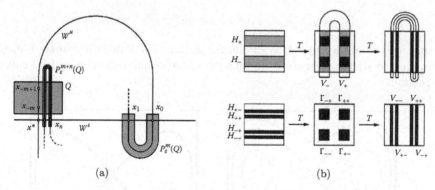

FIGURE 1.35: (a) Schematic representation of the homoclinic tangle. The homoclinic orbit $\{x_n\}_{n\in\mathbb{Z}}$ is asymptotic to x^* for $n \to \pm\infty$. One can choose m and n in such a way that the image of rectangle Q after $m+n$ iterations intersects Q twice. (b) The horseshoe map T maps the rectangles $H_\pm$ of the unit squares Q to the rectangles $V_\pm$. If $\Gamma_{\varepsilon_0\varepsilon_1} = V_{\varepsilon_0} \cap H_{\varepsilon_1}$, one also has $T(H_{\varepsilon_0\varepsilon_1}) = \Gamma_{\varepsilon_0\varepsilon_1}$ and $T(\Gamma_{\varepsilon_0\varepsilon_1}) = V_{\varepsilon_0\varepsilon_1}$.

rectangle, containing a piece of W^u and two points x_{-m}, x_{-m+1} of the *homoclinic orbit* (Fig. 1.35(a)). At least during the first iterations of P_ε, Q will be stretched in the unstable direction, and contracted in the stable one. After m iterations, the image $P_\varepsilon^m(Q)$ contains x_0 and x_1. Let n be sufficiently large that the points x_n and x_{n+1} of the homoclinic orbit are close to x^*. Due to area conservation, one can arrange that the piece of W^u between x_n and x_{n+1} crosses Q at least twice, and so does the image $P_\varepsilon^{m+n}(Q)$.

This behavior is reproduced qualitatively by *Smale's horseshoe map*; see Fig. 1.35(b). This map T takes a square $[0, 1] \times [0, 1]$, stretches it vertically by a factor $\mu > 2$, and contracts it horizontally by factor $\lambda < 1/2$. Then it bends the resulting rectangle in the shape of a horseshoe, and superimposes it with the initial square. Two horizontal strips H_+ and H_-, of size $1 \times \mu^{-1}$, are mapped, respectively, to two vertical strips V_+ and V_-, of size $\lambda \times 1$. For simplicity, the map restricted to $V_+ \cup V_-$ is assumed to be linear, but its qualitative features can be shown to remain unchanged by small nonlinear perturbations.

In order to construct an invariant set Λ of T, observe that $T(x) \in Q$ if and only if $x \in H_- \cup H_+$. Since $T(x) \in V_- \cup V_+$, $T^2(x) \in Q$ if and only if $T(x)$ belongs to one of the four rectangles $\Gamma_{\varepsilon_0\varepsilon_1} = V_{\varepsilon_0} \cap H_{\varepsilon_1}$, where $\varepsilon_0, \varepsilon_1 \in \{-, +\}$. These rectangles have size $\lambda \times \mu^{-1}$. The preimage of each $\Gamma_{\varepsilon_0\varepsilon_1}$ is a rectangle $H_{\varepsilon_0\varepsilon_1} \subset H_{\varepsilon_0}$ of size $1 \times \mu^{-2}$. The image $\Gamma_{\varepsilon_0\varepsilon_1}$ is a rectangle $V_{\varepsilon_0\varepsilon_1} \subset V_{\varepsilon_1}$, of size $\lambda^2 \times 1$. Thus T^2 maps $H_{\varepsilon_0\varepsilon_1}$ to $V_{\varepsilon_0\varepsilon_1}$. More generally, we can define

$$H_{\varepsilon_0\varepsilon_1\ldots\varepsilon_n} = \{x \in Q : x \in H_{\varepsilon_0}, T(x) \in H_{\varepsilon_1}, \ldots, T^n(x) \in H_{\varepsilon_n}\}.$$

Note that $H_{\varepsilon_0\varepsilon_1...\varepsilon_n} \subset H_{\varepsilon_0\varepsilon_1...\varepsilon_{n-1}}$, and that $H_{\varepsilon_0\varepsilon_1...\varepsilon_n} = \{x \in H_{\varepsilon_0} : T(x) \in H_{\varepsilon_1...\varepsilon_n}\}$. By induction, one may show that $H_{\varepsilon_0\varepsilon_1...\varepsilon_n}$ is a rectangle of size $1 \times \mu^{-n}$. Similarly, we introduce

$$
\begin{aligned}
V_{\varepsilon_{-m}...\varepsilon_{-1}} &= \{x \in Q : T^{-1}(x) \in H_{\varepsilon_{-1}}, \ldots, T^{-m}(x) \in H_{\varepsilon_{-m}}\} \\
&= \{x \in V_{\varepsilon_{-1}} : T^{-1}(x) \in V_{\varepsilon_{-m}...\varepsilon_{-2}}\},
\end{aligned}
$$

which is a rectangle of size $\lambda^m \times 1$ contained in $V_{\varepsilon_{-m+1}...\varepsilon_{-1}}$. Thus we have a rectangle

$$
\Gamma_{\varepsilon_{-m}...\varepsilon_{-1},\varepsilon_0\varepsilon_1...\varepsilon_n} = V_{\varepsilon_{-m}...\varepsilon_{-1}} \cap H_{\varepsilon_0\varepsilon_1...\varepsilon_n} = \{x \in Q : T^j(x) \in H_{\varepsilon_j}, \ -m \leq j \leq n\} \quad (1.97)
$$

of size $\lambda^m \times \mu^{-n}$. By comparison, $T^j(x) \in Q$ for $-m \leq j \leq n+1$ if and only if $x \in \Gamma_{\varepsilon_{-m}...,\varepsilon_0...\varepsilon_n}$ for some sequence $\varepsilon_{-m}...,\varepsilon_0...\varepsilon_n$ of symbols -1 and $+1$. Let $n, m \to \infty$ and Σ be the set of bi-infinite sequences $\varepsilon = ...\varepsilon_{-1},\varepsilon_0\varepsilon_1...$ of symbols -1 and $+1$. It is a metric space with the distance

$$
d(\varepsilon, \varepsilon') = \sum_{i=-\infty}^{\infty} (1 - \delta_{\varepsilon_i\varepsilon_i'})2^{-|i|}. \quad (1.98)
$$

By (1.97), the largest subset of Q invariant under T is $\Lambda = \bigcap_{n=-\infty}^{\infty} T^n(Q) = \{\Gamma_\varepsilon : \varepsilon \in \Sigma\}$, where Γ_ε, defined as the limit of $\Gamma_{\varepsilon_{-m}...\varepsilon_{-1},\varepsilon_0\varepsilon_1...\varepsilon_n}$ (a single point), where the finite sequence $\varepsilon_{-m}...,\varepsilon_0...\varepsilon_n$ converges to ε. Then Λ is a Cantor set, obtained by taking the product of two one-dimensional Cantor sets. The map $\phi : \varepsilon \to \Gamma_\varepsilon$ is a bijection from Σ to Λ, which is continuous in the topology defined by (1.98). By (1.97), T is conjugated to the shift map

$$
\sigma = \phi^{-1} \circ T|_\Lambda \circ \phi : \ ...\varepsilon_{-2}\varepsilon_{-1}\varepsilon_0\varepsilon_1... \ \mapsto \ ...\varepsilon_{-1}\varepsilon_0\varepsilon_1\varepsilon_2...
$$

The symbol ($\circ$) means the function composition, that is, the pointwise application of one function to the result of another to produce a third function. This conjugacy can be used to describe various orbits of $T|_\Lambda$ in a similar way as for the tent map. Thus, periodic itineraries ε correspond to periodic orbits of T, aperiodic itineraries correspond to chaotic orbits. Smale has shown that these qualitative properties are robust under small perturbations of T.

Theorem 25 (Smale) *The horseshoe map T has an invariant Cantor set Λ such that*
- *Λ contains a countable set of periodic orbits of arbitrary long periods;*
- *Λ contains an uncountable set of bounded nonperiodic orbits;*
- *Λ contains a dense orbit.*

Moreover, any map $\tilde{T}$ sufficiently close to T in the C^1 topology has an invariant Cantor set $\tilde{\Lambda}$ with $\tilde{T}|_{\tilde{\Lambda}}$ topologically equivalent to $T|_\Lambda$.

The horseshoe map example can be generalized to a class of so-called *axiom Λ* systems with similar chaotic properties. These systems are used to show the existence of chaotic orbits in a wide class of dynamical systems, including those with a transverse topologically homoclinic intersection. Examples of hyperbolic invariant sets have zero measure and are not attracting. Showing the existence of invariant sets containing chaotic orbits and attracting nearby orbits is a much more difficult task.

1.4.3 Poincaré return map

We will demonstrate how the solutions of the ODE system can be related to the solutions of a map, the Poincaré map that is easier to work with than the original system. Consider the autonomous system

$$
\dot{x} = f(x), \quad (1.99)
$$

where $f : \mathbb{R}^d \to \mathbb{R}^d$ and $x \in \mathbb{R}^d$. Introduce the *flow evolution operator* or *flow operator*, ϕ^{t_1,t_0}, which maps the point $x_0 \in \mathbb{R}^d$ and an interval $(t_0, t_1) \in \mathbb{R}$ to the point $x_1 \in \mathbb{R}^d$, where $x_1 = x(t_1)$ and $x(t)$ is the solution of (1.99) subject to $x = x_0$ when $t = t_0$. That is, $x(t_0) = x_0$ evolves to $x(t_1) = x_1$ in the d-dimensional phase space during the time interval (t_0, t_1). Note that $\phi^{t_1,t_0}(x_0) = x_0$ and $\phi^{t_2,t_1} \cdot \phi^{t_1,t_0} \equiv \phi^{t_2,t_0}$ for all $t_1 \in (t_0, t_2)$. Since (1.99) is an autonomous system, we can express the flow evolution operator in terms of the length of the time interval alone, and we write in a short form $\phi^{t_1,t_0}(x_0) = \phi^{t_1-t_0}(x_0)$. Consequently we have that φ^0 is the identity transformation, $\phi^t \cdot \phi^s \equiv \phi^{t+s}$ and $(\phi^t)^{-1} \equiv \phi^{-t}$, so that ϕ is a *group operator*.

Example 47 Construct the flow operator for the *logistic equation* $\dot{x} = x(1-x)$ with $d = 1$. This is the continuous version of the logistic map. The logistic equation is separable, and we can therefore integrate it subject to the initial condition that $x = x_0$ when $t = 0$ to obtain the solution, and hence the evolution operator is $x(t) = \phi'(x_0) = \frac{x_0 e^t}{x_0 e^t - x_0 + 1}$. Solutions of the logistic equation all have $x \to 1$ as $t \to \infty$, in complete contrast to the solutions of the logistic map. This shows that choosing to use a continuous rather than a discrete system can have dramatic consequences for the behavior of the solutions. So, we can verify that $\phi^0(x_0) = x_0$ and

$$\phi^t \cdot \phi^s(x_0) = \phi^t(\phi^s(x_0)) = \phi^s(x_0)e^t \cdot [\phi^s(x_0)e^t - \phi^s(x_0) + 1]^{-1},$$

and, since $\phi^s(x_0) = x_0 e^s/(x_0 e^s - x_0 + 1)$,

$$\phi^t \cdot \phi^s(x_0) = \phi^t(\phi^s(x_0)) = \frac{x_0 e^s e^t}{x_0 e^s e^t - x_0 e^s + (x_0 e^s - x_0 + 1)} = \frac{x_0 e^{s+t}}{x_0 e^{s+t} - x_0 + 1},$$

which is equal to $\phi^{s+t}(x_0)$, as expected.

Equilibrium points of (1.99) satisfy $\phi^t(x^*) = x^*$ for all $t \in \mathbb{R}$ or equivalently $f(x^*) = 0$. Periodic solutions of (1.99) with period T satisfy $x^*(t) = x^*(t + T)$ for all $t \in \mathbb{R}$, and can be written in terms of the flow operator as $\phi^{t+T}(x^*) = \phi^t(x^*)$ for all $t \in \mathbb{R}$. One way to obtain map from the system (1.99) is to sample the solution with period τ, so that $\phi^n(x_0) = x(t_0 + n\tau)$ for $n \in \mathbb{Z}$. This allows us to track the trajectory of particles in a stroboscopic way.

An important map, which we can use to analyze the behavior of integral paths close to a period solution, is the *Poincaré return map*. Let γ be a trajectory of (1.99), and consider the intersections of γ with $\Sigma \subset \mathbb{R}^d$ such that

- Σ has dimension $d - 1$,
- Σ is transverse (nowhere parallel) to the integral paths,
- all solutions in the neighborhood of γ pass through Σ.

If γ intersects Σ at $x = p$, and its next intersection with Σ is at $x = q$, then the Poincaré return map, $P : \Sigma \to \Sigma$, maps the point p to q. This is illustrated in Fig. 1.36(a) for $d = 3$ where Σ is a plane. An equilibrium point of the Poincaré return map corresponds to a limit cycle that intersects Σ once, and a periodic solution of period k corresponds to a limit cycle that k times intersects Σ.

Example 48 We find the Poincaré return map for (1.99) when Σ is the positive x-axis and $f(x) = \begin{pmatrix} 4x + 4y - x(x^2 + y^2) \\ 4y - 4x - y(x^2 + y^2) \end{pmatrix}$. Write down this system in terms of plane polar coordinates, and, using

$$\dot{r} = \frac{x\dot{x} + y\dot{y}}{(x^2 + y^2)^{1/2}} = \dot{x}\cos\theta + \dot{y}\sin\theta, \quad \dot{\theta} = \frac{1}{1 + (y/x)^2} \frac{x\dot{y} - y\dot{x}}{x^2} = \frac{1}{r}(\dot{y}\cos\theta - \dot{x}\sin\theta),$$

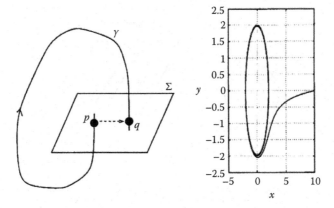

FIGURE 1.36: (a) The Poincaré return map. (b) Solution for $x_0 = 10$ and the Poincaré return map.

(see Example 37) find that $\dot{r} = r(4 - r^2)$ and $\dot{\theta} = -4$. These expressions can be integrated with initial conditions in Σ, the positive x-axis, so that $r(0) = x_0$ and $\theta(0) = 0$, to obtain $r = \sqrt{4x_0^2 e^{8t}/(4 - x_0^2 + x_0^2 e^{8t})}$, $\theta = -4t$. The orbit next returns to the axis when θ is an integer multiple of 2π, that is, when $t = n\pi/2$, so that $x_1 = P(x_0) = \sqrt{4x_0^2 e^{4\pi}/(4 - x_0^2 + x_0^2 e^{4\pi})}$. The solution for $x_0 = 10$ along with the Poincaré return map; see Fig. 1.36(b), rapidly approaches to $x = 2$.

1.4.4 Lyapunov's exponents and entropy

The concept of invariant measures is fundamental in dynamical systems theory. A measure $\mu(D)$ is *invariant* for a mapping $f : D \to D$ if for every measurable subset $A \in D$ we have

$$\mu(f^{-1}(A)) = \mu(A). \tag{1.100}$$

Assume that μ has been normalized so that $\mu(D) = 1$. Any attractor F supports at least one invariant measure; for fixed x in the basin of attraction of F and A being a Borel set, write

$$\mu(A) = \lim_{m \to \infty} \frac{1}{m} \times \{k : 1 \le k \le m, \ f^k(x) \in A\} \tag{1.101}$$

for the proportion of iterates in A. It may be shown using ergodic theory that this limit exists and is the same for almost all (w.r.t. μ) points in the basin of attraction under general circumstances. Clearly, $\mu(A \cup B) = \mu(A) + \mu(B)$ if A and B are disjoint, and $f^k(A) \in A$ if and only if $f^{k-1}(A) \in f^{-1}(A)$, giving (1.100). The measure (1.101) is concentrated on the set of points to which $f^k(x)$ comes arbitrarily close infinitely often; thus μ is supported by an attractor of f. The measure $\mu(A)$ reflects the proportion of iterates that lie in A, and may be thought of as the distribution when the large number of iterates $f^k(x)$ are plotted on a computer screen. The intensity of the measure can vary widely across attractor A; this variation is often analyzed using multifractal analysis. Concerning the size of an attractor, it is often the dimension of the set occupied by the invariant measure μ that is relevant, rather than the entire attractor. With this in mind, define the *Hausdorff dimension of a measure* μ for which $\mu(D) = 1$ as

$$\dim_H \mu = \inf\{\dim_H E : \ E - \text{Borel set with } \mu(E) > 0\}.$$

If μ is supported by F then clearly $\dim_H \mu \le \dim_H F$, but we may have strict inequality. If there are numbers $s > 0$ and $c > 0$ such that for every set U

$$\mu(U) \le c|u|^s \tag{1.102}$$

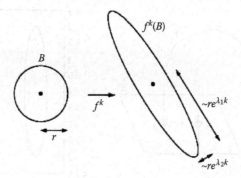

FIGURE 1.37: The definition of the Lyapunov's exponents λ_1 and λ_2.

then the *mass distribution principle* implies that for each set E with $0 < \mu(E)$ we have $\mathcal{H}^s(E) \geq \mu(E)/c > 0$, so that $\dim_H(E) \geq s$. Hence if (1.102) holds then

$$\dim_H E \geq s. \tag{1.103}$$

Once f is equipped with an invariant measure μ several other dynamical constants may be defined. Assume that $f : D \to D$ is differentiable.

Write $a_{ik}(x)$ for the lengths of the semi-axes of the ellipse $(f^k)'(B)$ where B is a unit ball. Define the *Lyapunov's exponents* as the average logarithmic rate of growth with k of these semi-axes:

$$\lambda_i(x) = \lim_{k \to \infty} \frac{1}{k} \log a_{ik}(x). \tag{1.104}$$

For simplicity, assume that the domain $D \subset \mathbb{R}^2$. The derivative $(f^k)'(x)$ is a linear mapping on $\mathbb{R}^2$. Write $a_k(x)$ and $b_k(x)$ for the lengths of the semi-axes of the ellipse $(f^k)'(B)$ where B is a unit ball. Thus the image under f^k of a small ball of radius r and center x approximates to an ellipse with semi-axes of lengths $ra_k(x)$ and $rb_k(x)$. Then the Lyapunov's exponents are $\lambda_1(x) = \lim_{k \to \infty} \frac{1}{k} \log a_k(x)$ and $\lambda_2(x) = \lim_{k \to \infty} \frac{1}{k} \log b_k(x)$. Techniques from ergodic theory show that if μ is invariant for f, these exponents exist and have the same values λ_1, λ_2 for almost all x (w.r.t. μ). Hence in a system with an invariant measure, we refer to λ_1 and λ_2 as the Lyapunov's exponents of the system. The Lyapunov's exponents represent the *average* rates of expansion of f. If B is a disk of small radius r, then $f^k(B)$ will *typically* be close to an ellipse with semi-axes of lengths $re^{\lambda_1 k}$ and $re^{\lambda_2 k}$; see Fig. 1.37. A related dynamical idea is the entropy of a mapping $f : D \to D$. Write

$$V(x, \varepsilon, k) = \{y \in D : |f^i(x) - f^i(y)| < \varepsilon, \ 0 \leq i \leq k\} \tag{1.105}$$

for the set of points with their first k iterates within ε of those of x. If μ is an invariant measure of f, we define the *μ-entropy* of f as

$$h_\mu = \lim_{\varepsilon \to 0} \lim_{k \to \infty} \{ -\frac{1}{k} \log \mu[V(x, \varepsilon, k)] \}.$$

Under reasonable conditions, this limit exists and has a constant value for almost all x (w.r.t. μ). The entropy $h_\mu(F)$ reflects the rate at which nearby points spread out under iteration by f, or alternatively the amount of extra information about an orbit $\{f^k(x)\}_{k=1}^\infty$ that is gained knowing the position of an additional point on the orbit.

The Baker's transformation (1.126) provides a simple illustration of these ideas (the line of discontinuity makes little difference). There is a natural invariant mass distribution $\mu(F)$

such that each of the 2^k strips of E_k has the mass 2^{-k}, with this mass spread uniformly across the width of the E. Just as we get that $\mu(U) \le c|u|^s$ with $s = 1 + \log 2/(-\log \lambda)$, so by (1.102) and (1.103) $s \le \dim_H \mu \le \dim_H F = s$. The Lyapunov's exponents can be found. The derivative (1.126) is $f'(x,y) = \begin{pmatrix} 2 & 0 \\ 0 & \lambda \end{pmatrix}$ (provided $x \ne \frac{1}{2}$), and so $f^{k\prime}(x,y) = \begin{pmatrix} 2^k & 0 \\ 0 & \lambda^k \end{pmatrix}$ except where $x = p/2^k$ for nonnegative integers p and k. Hence $a_k(x,y) = 2^k$ and $b_k(x,y) = \lambda^k$. By (1.104) $\lambda_1(x,y) = \log 2$, $\lambda_2(x,y) = \log \lambda$ for almost all (x,y) (w.r.t. μ), and the Lyapunov's exponents of the system are $\lambda_1 = \log 2$ and $\lambda_2 = \log \lambda$. Since f^k stretches by a factor 2^k horizontally and contracts by a factor λ^k vertically, we get, using (1.105) and ignoring the *cutting* effect of f, that $V((x,y), \varepsilon, k)$ is approximately a rectangle with sides $2^{-k}\varepsilon$ and ε, which has μ-measure approximately $\varepsilon^s 2^{-k}$, if $(x,y) \in F$. Thus $h_\mu = \lim_{\varepsilon \to 0} \varliminf_{k \to \infty} [-\frac{1}{k} \log(\varepsilon^s 2^{-k})] = \log 2$. The dimensions, Lyapunov's exponents and entropies of an invariant measure of a dynamical system can be estimated computationally or experimentally and are often useful when comparing different systems. These quantities may not be completely independent of each other. One relationship that has been derived rigorously applied to a smooth bijective transformation f on a two-dimensional surface. If μ is an invariant measure for f with Lyapunov's exponents $\lambda_1 > 0 > \lambda_2$ then $\dim_h \mu = h_\mu(F)(\lambda_1^{-1} - \lambda_2^{-1})$. The exponents calculated for the Baker's transformation satisfy this formula. The following relationship holds in many cases: if f is a plane transformation with attractor F and Lyapunov's exponents $\lambda_1 > 0 > \lambda_2$, then $\dim_B F \le 1 - (\lambda_1/\lambda_2)$. An argument to support this runs as follows. Let $N_\delta(F)$ be the least number of discs of radius δ that can cover F. If $\{U_i\}$ are $N_\delta(F)$ such discs, then $f^k(F)$ is covered by the $N_\delta(F)$ sets $f^k(U_i)$ which are approximately elliptical with semi-axis lengths $\delta \exp(\lambda_1 k)$ and $\delta \exp(\lambda_2 k)$. These ellipses may be covered by about $\delta \exp((\lambda_1 - \lambda_2)k)$ discs of radii $\delta \exp(\lambda_2 k)$. Hence $N_{\delta \exp(\lambda_2 k)}(F) \le \exp((\lambda_1 - \lambda_2)k) N_\delta(F)$, so

$$\frac{\log N_{\delta \exp(\lambda_2 k)}(F)}{-\log(\delta \exp(\lambda_2 k))} \le \frac{\log[\exp((\lambda_1 - \lambda_2)k) N_\delta(F)]}{-\log(\delta \exp(\lambda_2 k))} = \frac{(\lambda_1 - \lambda_2)k + \log N_\delta(F)}{-\lambda_2 k - \log \delta}.$$

Letting $k \to \infty$ gives us $\overline{\dim}_B F \le 1 - (\lambda_1/\lambda_2)$. This argument can often be justified, but it assumes that the Lyapunov's exponents cross the domain D, which need not be the case.

The relationships between these and other dynamical parameters are complicated, being closely interrelated with the chaotic properties of f and the fractal nature of the attractor.

1.4.5 Attracting sets and attractors

Consider a dynamical system on $D \subset \mathbb{R}^d$, of a flow φ_t. One can include the case of integrated maps by restricting t to integer and setting $\varphi_t = \phi^t$. Various subsets of D can be associated with the flow.

Definition 24

   – A *subset* $S \subset D$ is called *invariant* if $\varphi_t(S) = S$ for all t.

   – $A \subset D$ is called an *attracting set* if there exists a neighborhood $O \subset A$ such that $\varphi_t(x) \in O$ for all $t \ge 0$ and all $x \in O$, and $\text{dist}(\varphi_t(x), A) \to 0$ as $t \to \infty$.

   – A *point* x is called *nonwandering* for φ_t if for any neighborhood $O(x)$ and $T > 0$ there exists $t > T$ such that $\varphi_t(O(x)) \cap O(x) \ne \emptyset$. The set of all nonwandering points is called the *nonwandering set* Ω.

These sets can be partly determined in terms of the asymptotic behavior of orbits of the flow.

Definition 25 The *ω-limit set* of x for φ_t is the set of $y \in D$ such that there exists a sequence $t_n \to \infty$ with $\varphi_{t_n}(x) \to y$. The *α-limit set* of x for φ_t is the set of $y \in D$ such that there exists a sequence $t_n \to -\infty$ with the property $\varphi_{t_n}(x) \to y$.

The α- and ω-limit sets of any x are invariant sets, and the ω-limit set is included in the nonwandering set. One can show that asymptotically stable equilibrium points, periodic orbits and invariant tori are all ω-limit sets of the orbits in their basin of attraction, nonwandering sets and attracting sets. There is a problem with the definition of attracting set, as is shown in the following example.

Example 49 The ω-limit set of the system $\dot{x}_1 = x_1 - x_1^3$, $\dot{x}_2 = -x_2$ is $(-1, 0)$ if $x_1 < 0$; $(0, 0)$ – if $x_1 = 0$ and $(1, 0)$ if $x_1 > 0$. The origin $(0, 0)$ is also the α-limit set of all points in $(-1, 1) \times \{0\}$. The nonwandering set is composed of the three equilibrium points of the flow, while $(\pm 1, 0)$ are attracting sets. The segment $[-1, 1] \times \{0\}$ is also an attracting set.

One would like to exclude attracting set such as the segment $[-1, 1] \times \{0\}$ in the example, which contains wandering points. This is generally solved in the following way.

Definition 26 A closed invariant set Λ is called *topologically transitive* if φ_t has an orbit which is dense in Λ. An attractor is a *topologically transitive attracting set*.

Remark 6 One may use a notation of indecomposability weaker than topological transitivity, which is based on the notion of *chain recurrence*. In this case, one requires that for any points x, y and any $\varepsilon > 0$, there exist points $x_0 = x, x_1, \ldots, x_n = y$ and times $t_1, \ldots, t_n$ such that $\|\varphi_{t_i}(x_{j-1}) - x_j\| \le \varepsilon$, $\forall j, i$. Then a weaker definition of attractor is proposed. An *attractor* is an indecomposable closed invariant set Λ with the property that, given $\varepsilon > 0$ there is a set U of positive Lebesgue measure in the ε-neighborhood of Λ such that, if $x \in U$ then the forward orbit of x belongs to U and the ω-limit set of x belongs to Λ.

Strange attractor. Its dynamics is chaotic in a sense. A strange attractor is defined as an attractor containing a transversal homoclinic orbit. The existence of such an orbit implies various chaotic properties; it is not necessary to assume that the system is conservative. Modern definitions are less specific; they require that the dynamics should be sensitive to initial conditions.

Definition 27 Let Λ be a compact set such that $\varphi_t(\Lambda) \subset \Lambda$ for all $t \ge 0$. The flow φ_t *has sensitive dependence on initial conditions* on Λ if there exists $\varepsilon > 0$ with the property: for any $x \in \Lambda$ and any neighborhood $O(x)$ there exist $y \in O(x)$ and $t > 0$ such that $|\varphi_t(x) - \varphi_t(y)| \ge \varepsilon$. Λ is a *strange attractor* if it is an attractor and φ_t has sensitive dependence on initial conditions on Λ.

Sensitive dependence on initial conditions means that for any point x, one can find an arbitrarily close point y such that the orbits of x and y diverge from each other; the quantity ε should not depend on $\|x - y\|$, though t may depend on it. This is a rather weak property, and one often requires that divergence occur at an exponential rate.

Definition 28

(i) Let φ_t be a flow of ODE $\dot{x} = f(x)$ ($x \in \mathbb{R}^d$). Let $x \in D$ and $U(t)$ be a general solution of the equation linearized around the orbit of x, $\dot{y} = \frac{\partial f}{\partial x}(\varphi_t(x))y$. Under certain assumptions on f, there exists $\mathbf{L} = \lim\limits_{t \to \infty} \frac{1}{2t} \log(U(t)^T U(t))$; see [35]. The eigenvalues of $\mathbf{L}$ are called *Lyapunov's exponents of the orbit of* x.

(ii) Let F be an iterated map and $\{x_k\}$ a given orbit of F. Consider the linear equation

$$y_{k+1} = F'(x_k)y_k \ \Rightarrow \ y_k = U_k y_0, \quad U_k = F'(x_{k-1}) \cdot \ldots \cdot F'(x_0).$$

The *Lyapunov exponents of the orbit* $\{x_k\}$ are the eigenvalues of the matrix $\mathbf{L} = \lim_{k\to\infty} \frac{1}{2k} \log(U_k^T U_k)$.

Note that $U(t)^T U(t)$ is a symmetric positive definite matrix, hence it is always diagonalizable and has real eigenvalues. Thus the solution $y(t)$ of the equation linearized around $\varphi_t(x)$ obeys

$$\|y(t)\|^2 = \|U(t)y(0)\|^2 = y(0) \cdot U(t)^T U(t)y(0) \simeq y(0) \cdot e^{2t\mathbf{L}}y(0).$$

Let λ_1 be the largest eigenvalue of $\mathbf{L}$. Unless the projection of $y(0)$ on the eigenspace of $\mathbf{L}$ associated with λ_1 is zero, $\|y(t)\|$ will grow asymptotically like $e^{\lambda_1 t}$. We say that the flow has *exponentially sensitive dependence on initial conditions* in Λ if the largest Lyapunov exponent $\lambda_{\max}$ of all orbits in Λ is positive.

Definition 29 The map F is called *conservative* if $|\det(F'(x))| = 1$ for all $x \in D$. The map F is called *dissipative* if $|\det(F'(x))| < 1$ for all $x \in D$.

Theorem 26 *If φ_t is conservative, then the sum of all Lyapunov exponents of any orbit is zero. If the flow is dissipative, then this sum is negative. Assume $\{\varphi_t\}$ is an orbit of the ODE $\dot{x} = f(x)$, such that $\|f\|$ is bounded from below and above by strictly positive constants on $\{\varphi_t(x)\}$. Then this orbit has at least one Lyapunov exponent equal to zero.*

Lemma 3 *Let f be continuously differentiable and $V(t) = \mathrm{Vol}(M(t))$. Then $V'(t) = \int_{M(t)} (\mathrm{div}\, f)\,\mathrm{d}x$.*

Proof of Lemma 3. Let $y(t) = \varphi_t(x)$ for x in domain M, where $M(t) = \varphi_t(M)$. We have

$$V(t) = \int_{M(t)} \mathrm{d}y = \int_M \left|\det \frac{\partial}{\partial x}\varphi_t(x)\right| \mathrm{d}x.$$

Set $J := \frac{\partial}{\partial x}\varphi_t(x)$, $\Lambda(t) := \frac{\partial f}{\partial x}(y(t))$. By definition of φ_t, $J(0) = E$ is the identity matrix. Now we compute $J'(t) = \frac{\partial}{\partial x}\dot{y}(t) = \frac{\partial}{\partial x}f(\varphi_t(x)) = \frac{\partial f}{\partial x}(\varphi_t(x))\frac{\partial}{\partial x}\varphi_t(x)$, and thus

$$J'(t) = A(t)J(t), \quad J(0) = E.$$

This is a linear, time-dependent differential equation for $J(t)$, which is known to admit a unique global solution. This implies in particular that $\det J(t) \neq 0$, $\forall t$, since otherwise $J(t)$ would not be surjective, contradicting uniqueness. Since $\det J(0) = 1$ and $\det J(t)$ continuously vary, then $\det J(t) > 0$ for all t. Determine the evolution of $\det J(t)$. By Taylor's formula, there exists $\theta \in [0,1]$ such that

$$J(t+\varepsilon) = J(t) + \varepsilon J'(t+\theta\varepsilon) = J(t)[1 + \varepsilon J(t)^{-1} A(t+\theta\varepsilon)J(t+\theta\varepsilon)].$$

Recall from linear algebra that for any $d \times d$ matrix B, $\det(1 + \varepsilon B) = 1 + \varepsilon\,\mathrm{Tr}\, B + r(\varepsilon)$ with $\lim_{\varepsilon\to 0} r(\varepsilon)/\varepsilon = 0$ – this is a consequence of the definition of the determinant as a sum over permutations. Using identity $\mathrm{Tr}\,(AB) = \mathrm{Tr}\,(BA)$, this leads to $\det J(t+\varepsilon) = \det J(t)[1 + \varepsilon\,\mathrm{Tr}\,(A(t+\theta\varepsilon)J(t+\theta\varepsilon)J(t)^{-1}) + r(\varepsilon)]$, thus

$$(\det J)'(t) = \lim_{\varepsilon\to 0} \frac{\det J(t+\varepsilon) - \det J(t)}{\varepsilon} = \mathrm{Tr}\,(A(t))\det J(t).$$

Taking the derivative of $V(t)$ we get

$$\begin{aligned}
V'(t) &= \int_M \frac{d}{dt}\det(J(t))\,\mathrm{d}x \\
&= \int_M \mathrm{Tr}\left(\frac{\partial f}{\partial x}(y(t))\right)\det(J(t))\,\mathrm{d}x = \int_{M(t)} \mathrm{Tr}\left(\frac{\partial f}{\partial x}(y(t))\right)\mathrm{d}y,
\end{aligned}$$

and the conclusion follows from the fact that $\mathrm{Tr}\,\frac{\partial f}{\partial x} = \mathrm{div}\, f$. $\qquad\square$

Definition 30 The *vector field f* is called *conservative* if div $f(x) = 0$, $\forall x \in D$. The vector field f is called *dissipative* if div $f(x) < 0$, $\forall x \in D$.

Lemma 3 implies that $V(t)$ is constant if f is conservative, and monotonously decreases when f is dissipative. More generally, if div $f(x) \leq c$, $\forall x \in D$, then $V(t) \leq V(0)e^{ct}$.

Proof. (of Theorem 26). The determinant of $U(t)$ is constant if the system is conservative, and decreasing if the system is dissipative, which follows from Lemma 3. By definition, $\det U(0) = 1$ and by uniqueness of solutions, $\det U(t) > 0$ for all t. Thus for $t > 0$, the product of all eigenvalues of $U(t)$ is equal to 1 in the conservative case, and belongs to $(0,1)$ in the dissipative case. The same is true for $\det U(t)^T U(t)$. But for any matrix B, $\det e^B = e^{\operatorname{Tr} B}$ because the eigenvalues of e^B are exponentials of the eigenvalues of B. Assume that $x(t) = \varphi_t(x)$ is a solution of $\dot{x} = f(x)$. Then $\ddot{x}(t) = \frac{d}{dt}f(\varphi_t(x)) = \frac{\partial f}{\partial x}(\varphi_t(x))\dot{x}(t)$, and thus $\dot{x}(t) = U(t)\dot{x}(0)$ by definition of $U(t)$. Hence,

$$\dot{x}(0) \cdot U(t)^T U(t)\dot{x}(0) = \|U(t)\dot{x}(0)\|^2 = \|\dot{x}(t)\|^2 = \|f(\varphi_t(x))\|^2.$$

By the assumption on $\|f\|$, it follows that as $t \to \infty$, the function $\dot{x}(0)e^{2t\mathbf{L}}\dot{x}(0)$ is bounded from above and below by strictly positive constants. Being symmetric, $\mathbf{L}$ admits an orthonormal set of eigenvectors $u_1, \ldots, u_d$. If $c_j = u_j\dot{x}(0)$, then $\dot{x}(0) = \sum_j c_j u_j$ and thus $\dot{x}(0) \cdot e^{2t\mathbf{L}}\dot{x}(0) = \sum_{j=1}^{d} c_j^2 e^{2\lambda_j t}$. Since $\dot{x}(0) \neq 0$, at least one of the coefficients, say c_k, is different from zero, and thus λ_k must be zero. $\square$

Definition 31 Let $\gamma(t)$ be a periodic solution of period T, $x(0) = \gamma(0)$, and let Σ be a hyperplane transverse to the orbit at x_0 (Fig. 1.38(a)). By continuity of the flow, there is a neighborhood $O(x_0) \subset \Sigma$ such that for all $x = x_0 + y \in O(x_0)$, we can define a continuous map $\tau(y)$, $\tau(0) = T$, such that $\varphi_t(x)$ returns for the first time to Σ in a vicinity of x_0 at $t = \tau(y)$. The *Poincaré map* Π *associated with the period orbit* is defined by

$$x_0 + \Pi(y) := \varphi_{\tau(y)}(x_0 + y). \tag{1.106}$$

It is easy to find systems with positive Lyapunov exponents. Consider for instance the linear system $\dot{x} = Ax$. In this case, $U(t) = e^{At}$ and one may show (using the Jordan canonical form of A) that the Lyapunov exponents are exactly the real parts of the eigenvalues of A. Thus, if A has an eigenvalue with positive real part, all orbits have exponentially sensitive dependence on initial conditions.

More generally, if x^* is a linearly unstable equilibrium point, its largest Lyapunov exponent is positive. Similarly, if Γ is a periodic orbit then $U(t) = P(t)e^{Bt}$, where $P(t)$ is periodic. The Lyapunov exponents are the real parts of the eigenvalues of B, i.e., real parts of the characteristic exponents.

Theorem 27 *Let $\gamma(t)$ be a periodic solution of period T. Then the Poincaré map Π is as smooth as the vector field in a neighborhood of the origin. The characteristic multipliers of the periodic orbit are given by 1 and the $d - 1$ eigenvalues of the Jacobian matrix $\frac{\partial \Pi}{\partial y}(0)$.*

Proof. The smoothness of Π follows directly from the flow smoothness and the implicit function theorem. Observe that $\frac{d}{dt}\dot{\gamma}(t) = \frac{d}{dt}f(\gamma(t)) = \frac{d}{dx}f(\gamma(t))\dot{\gamma}(t) = A(t)\dot{\gamma}(t)$. Thus by Floquet's theorem, one can write $\dot{\gamma}(t) = P(t)e^{Bt}\dot{\gamma}(0)$, and, in particular $\dot{\gamma}(0) = \dot{\gamma}(t) = P(T)e^{BT}\dot{\gamma}(0) = e^{BT}\dot{\gamma}(0)$. This shows that $\dot{\gamma}(0)$ is an eigenvector of e^{BT} with eigenvalue 1. Let $(e_1, \ldots, e_{d-1}, \dot{\gamma}(0))$, $e_1, \ldots, e_{d-1} \in \Sigma$ be a basis of $\mathbb{R}^d$. In this basis, $e^{BT} = \begin{pmatrix} e^{B_\Sigma T} & 0 \\ \cdots & 1 \end{pmatrix}$, where B_Σ is the restriction of B to Σ, and the dots denote arbitrary entries.

TABLE 1.4: Examples of attractors of a three-dimensional flow.

Attractor	Sign of Lyapunov exponents	Asymptotic dynamics
Stable equilibrium	$(-,-,-)$	Stationary
Stable periodic orbit	$(0,-,-)$	Periodic
Attracting torus	$(0,0,-)$	Quaziperiodic
Strange attractor	$(+,0,-)$	Chaotic

If we consider momentarily y as a vector in $\mathbb{R}^d$ instead of Σ, linearization of (1.106) gives

$$\frac{\partial \Pi}{\partial y}(0) = \frac{\partial}{\partial y}(\varphi_{\tau(y)}(x_0 + y))\mid_{y=0} = \frac{\partial \varphi_T}{\partial t}(x_0)\frac{\partial \tau}{\partial y}(0) + \frac{\partial \varphi_T}{\partial t}(x_0) = \dot{\gamma}(0)\frac{\partial \tau}{\partial y}(0) + e^{BT}.$$

The first term is a matrix with zero entries except on the last line, so that $\partial_y \Pi(0)$ has the same representation as e^{BT}, save for the entries marked by dots. In particular, when y is restricted to Σ, $\frac{\partial \Pi}{\partial y}(0)$ has the same eigenvalues as $e^{\mathbf{B}_\Sigma T}$.

So, the exponent matrix e^{Bt} has at least one eigenvalue equal to 1, corresponding to translations along the periodic orbit. Thus one of the Lyapunov exponents is equal to zero. All above invariant sets (equilibrium, periodic orbit) must be unstable in order to have sensitive dependence on initial conditions, and thus they have zero measure and are not attractors. Orbits starting near these sets will not necessarily have positive Lyapunov exponents. Since orbits on the attractor are bounded and not attracted by an equilibrium point, one Lyapunov exponent is equal to zero. Hence a two-dimensional flow cannot admit a strange attractor. A theorem due to Poincaré-Bendixson states that a nonempty compact ω- or α-limit set of a planar flow is either a periodic orbit, or contains equilibrium points, which also rules out the existence of strange attractors for two-dimensional flows. If a three-dimensional flow has a strange attractor, its Lyapunov exponents must satisfy $\lambda_1 > \lambda_2 = 0 > \lambda_3$ and $|\lambda_3| > \lambda_1$; see Table 1.4. Because of volume contraction, the attractor must have zero volume, but it cannot be a surface and have positive Lyapunov exponents. This accounts for the fractal nature of many observed attractors. □

Hénon and Lorentz attractors. The first discrete dynamical system for which the existence of a strange attractor was proved is the *Hénon map* (which does not describe a physical system):

$$x_{k+1} = 1 - \lambda x_k^2 + y_k \qquad y_{k+1} = b x_k. \tag{1.107}$$

It was introduced as a two-dimensional generalization of the one-dimensional map $x_{k+1} = 1 - \lambda x_k^2$, see the logistic map. When $b = 0$, (1.107) is reduced to this one-dimensional map. If $b \neq 0$, the Hénon map is invertible, and it is dissipative if $|b| < 1$. Numerical simulations indicate that for some parameter values, the Hénon map has indeed a strange attractor with a self-similar structure; see Fig. 1.38(b).

Theorem 28 (Benedicks, Carleson) *Let $z^* = (x^*, y^*)$ be a fixed point of (1.107) with $x^*, y^* > 0$ and let W^u be its unstable manifold. For all $c < \log 2$, there is a set of positive Lebesgue measure of parameters (b, λ) for which*
   1. There is an open set $U \subset \mathbb{R}^2$ depending on λ and b such that $T^k(z) \to \bar{W}^u$ as $k \to \infty$ for all $z \in U$;
   2. There is a point $z_0 \in W^u$ such that: (a) the positive orbit of z_0 is dense in W^u; (b) the largest Lyapunov exponent of the orbit of z_0 is greater than c.

Hence the closure of the unstable manifold W^u is a strange attractor. Locally, the strange attractor is smooth in the unstable direction with positive Lyapunov exponent, and has

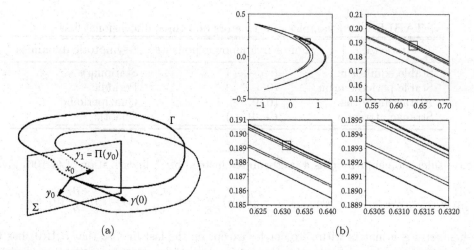

(a)   (b)

FIGURE 1.38: (a) The Poincaré map associated with periodic orbit $\gamma(t)$. (b) Hénon attractor for $\lambda = 1.4$, $b = 0.3$. Successive magnifications of details, by factor 10, show its self-similar structure.

the structure of a Cantor set in the transverse, stable direction with negative Lyapunov exponent.

Finally, we come back to the *Lorentz equations*

$$\dot{x}_1 = \sigma(x_2 - x_1), \quad \dot{x}_2 = -rx_1 - x_2 - x_1x_3, \quad \dot{x}_3 = -bx_3 + x_1x_2.$$

The existence of a strange attractor for this system has not been proved to our knowledge, although there is strong numerical evidence that such an attractor exists for certain parameter values, including in particular $\sigma = 10$, $b = 8/3$ and $r = 28$. The qualitative properties of dynamics are nonetheless quite well understood. The strange attractor seems to appear after a rather subtle sequence of bifurcations. Consider the case $\sigma = 10$, $b = 8/3$, and take r as a bifurcation parameter.

1. For $0 \leq r \leq 1$, the origin is globally asymptotically stable, that is, all orbits converge to the origin.

2. At $r = 1$, the origin undergoes a pitchfork bifurcation, and two new stable equilibria $C_\pm = (\pm\sqrt{b(r-1)}, \pm\sqrt{b(r-1)}, r-1)$, appear. These points become unstable in a Hopf bifurcation at $r = \frac{470}{19} \simeq 24.74$. The Hopf bifurcation is subcritical, and thus corresponds to the destruction of an unstable periodic orbit that must have been created somehow for smaller r.

The dynamics for $r > 1$ can be described by taking a Poincaré section on the surface $\Sigma : x_3 = r - 1$ (we only look at intersections with $\dot{x}_3 < 0$). Consider a rectangle containing segment C_-C_+ and intersection S of the two-dimensional stable manifold of the origin with Σ (Fig. 1.39). S is attracted by the origin, which has the effect to pinch the rectangle and map it to two pieces of triangular shape. One vertex of each triangle belongs to a piece of the one-dimensional unstable manifold W^u of the origin. Due to the dissipation, the angle at this vertex is quite small. In first approximation, the dynamics can thus be described by a one-dimensional map for a coordinate transverse to S, parameterizing the long side of the triangles. For r sufficiently small, this map is increasing, discontinuous at 0, and admits two stable fixed points $c_\pm$ corresponding to $C_\pm$; see Fig. 1.39(a).

3. At $r = r_1 \simeq 13.296$, a homoclinic bifurcation occurs; see Fig. 1.39(b): the unstable manifold W^u belongs to W^s, and thus the sharpest vertices of both triangles belong to S. The one-dimensional approximation of the Poincaré map is continuous, but still monotonously increasing.

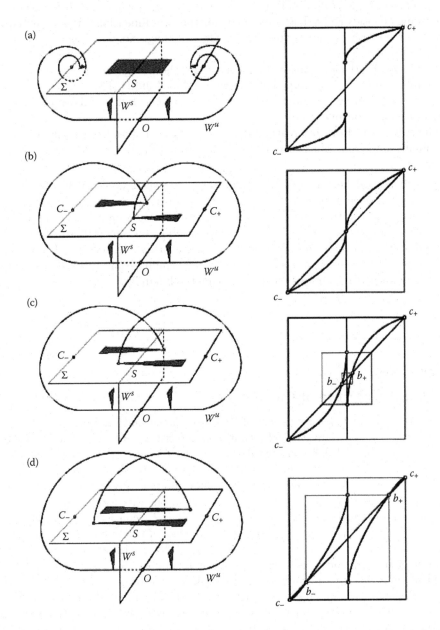

FIGURE 1.39: Schematic representation of the stable and unstable manifolds of the Lorentz equation (left), and one-dimensional approximation of the Poincaré map for the coordinate perpendicular to W^s (right): (a) $1 < r < r_1$; (b) homoclinic bifurcation at $r = r_1$; (c) existence of a nonwandering repeller for $r_1 < r < r_2$; (d) bifurcation to a strange attractor at $r = r_2$.

4. For r slightly larger than r_1, each piece of W^u hits Σ on the opposite side of S; see Fig. 1.39(c). The one-dimensional map becomes noninvertible, and has two new unstable equilibria b_- and b_+. The map does not leave the interval $[b_-, b_+]$ invariant, but maps two of its subintervals onto $[b_-, b_+]$. Similarly to the tent map g_3, there exists an invariant Cantor set containing chaotic orbits. The same holds for the two-dimensional Poincaré map, which resembles the horseshoe map. Thus there exists a strange nonwandering set Λ, but it is repelling and has zero measure.

5. As r increases, the fixed points $b_\pm$ move toward the fixed points $c_\pm$. Then, for $r \geq r_2 \simeq 24.06$ the interval $[b_-, b_+]$ is mapped into itself; see Fig. 1.39(d). The strange nonwandering set Λ becomes attracting, although it does not yet attract a full neighborhood, since orbits starting in $b_\pm$ may converge to $c_\pm$.

6. The subcritical Hopf bifurcation at $r = r_3 \simeq 24.74$ makes fixed points $C_\pm$ unstable, so that finally all orbits can converge to the nonwandering set Λ, which has become an attractor.

The method used here to describe the dynamics near a homoclinic bifurcation by a Poincaré map has been applied to other systems. Transitions to chaotic behavior are quite frequently associated with such homoclinic bifurcations. The Poincaré map in a vicinity of the unstable manifold is described as the composition of an almost linear map, reflecting the motion near the equilibrium point, and a nonlinear usually rather simple one, reflecting the motion near W^u away from the equilibrium. The composition of two maps contains all the ingredients necessary for the existence of chaotic dynamics.

1.5   Fractals

1.5.1   Local structure of fractals

The middle third Cantor set is one of the best known and easily constructed fractals; nevertheless it displays many typical fractal characteristics. It is constructed from a unit interval by a sequence of deletion operations. Let E_0 be the interval $[0, 1]$. Let E_1 be the set obtained by deleting the middle third of E_0, so that E_1 consists of the two intervals $[0, \frac{1}{3}]$ and $[\frac{2}{3}, 1]$. Deleting the middle third of these intervals gives E_2; thus E_2 comprises the four intervals $[0, \frac{1}{9}]$, $[\frac{2}{9}, \frac{1}{3}]$, $[\frac{2}{3}, \frac{7}{9}]$, $[\frac{8}{9}, 1]$. Continue in this way, with E_k obtained by deleting the middle third of each interval in E_{k-1}. Thus E_k consists of 2^k intervals each of the length 3^{-k}. The *middle third Cantor set* F consists of the numbers that are in E_k for all k; mathematically, F is the intersection $\bigcap_{k=0}^{\infty} E_k$. The Cantor set F may be thought of as the limit of the sequence of sets E_k as k tends to infinity. It is obviously impossible to draw the set F itself, with its infinitesimal detail, so "pictures F" tend to be pictures of one of the E_k, which are good approximation to F when k is reasonably large (Fig. 1.31).

At first glance it might appear that we have removed so much of the interval $[0, 1]$ during the construction of F that nothing remains. In fact, F is an infinite set, which contains infinitely many numbers in every neighborhood of each of its points. The middle third Cantor set F consists precisely of those numbers in $[0, 1]$ whose base-3 expansion does not contain the digit 1, i.e., all numbers $a_1 3^{-1} + a_2 3^{-2} + a_3 3^{-3} + \ldots$ with $a_i = 0$ or 2 for each i. Indeed, to get E_1 from E_0 we remove those numbers with $a_1 = 1$, to get E_2 from E_1 we remove those numbers with $a_2 = 1$, and so on. We list some of the features of the middle third Cantor set F; similar features are found in many fractals.

(i) F is self-similar. The part of F in the interval $[0, \frac{1}{3}]$ and the part of F in $[\frac{2}{3}, 1]$ are both geometrically similar to F, scaled by a factor $\frac{1}{3}$. Again, the parts of F in each of the four intervals of E_2 are similar to F but scaled by a factor, and so on. The Cantor set contains copies of itself at many different scales.

(ii) The set F has a "fine structure"; that is, it contains detail at arbitrarily small scales. Moreover, we enlarge the picture of the Cantor set; the more gaps become apparent to the eye.

(iii) Although F has an intricate detailed structure, the actual definition of F is very straightforward.

(iv) F is obtained by a recursive procedure. Our construction consisted of repeatedly removing the middle thirds of intervals. Successive steps give increasingly good approximations E_k to the set F.

(v) The geometry of F is not easily described in classical terms: it is not the locus of the points that satisfy some simple geometric condition, nor is it the set of solutions of any simple equation.

(vi) It is awkward to describe the local geometry of F – near each of its points are a large number of other points, separated by gaps of varying lengths.

(vi) Although F is in some ways quite a large set to be uncountably infinite, its size is not quantified by the usual measures such as length – but any reasonable F has length zero.

The second example is the von Koch curve (Fig. 1.40). We let E_0 be a line segment of unit length. The set E_1 consists of the four segments by removing the middle third of E_0 and replacing it by the other two sides of the equilateral triangle based on the removed segment. We construct E_2 by applying the same procedure to each of the segments in E_1, and so on. Thus E_1 comes from replacing the middle third of each straight line segment of E_{k-1} by other two sides of an equilateral triangle. When k is large, the curves E_{k-1} and E_k approaches a limiting curve F, called the *von Koch curve*.

The von Koch curve has features in many ways similar to those listed for the middle third Cantor set. It is made up of four *quarters* each similar to the whole, but scaled by a factor $\frac{1}{3}$. The fine structure is reflected in the irregularities at all scales; nevertheless, this intricate structure stems from a basically simple construction. While it is reasonable to call F a curve, it is much too irregular to have tangents in the classical sense. A simple calculation shows that E_k is of length $(4/3)^k$; letting k tend to infinity implies that F has infinity length. On the other hand, F occupies zero area in the plane, so neither length nor area provides a very useful description of the size of F.

Many other sets may be constructed using such recursive procedures. For example, the *Sierpinski triangle or gasket* with $\dim_H F = \dim_B F = \log 3/\log 2$ is obtained by repeatedly removing (inverted) equilateral triangles from an initial equilateral triangle of unit side-length. For many purposes, it is better to think of this procedure as repeatedly replacing an equilateral triangle by three triangles of half the length. A plane analogue of the Cantor set, a *Cantor dust*, [6, 15], is illustrated in Fig. 1.41. At each stage each remaining square is divided into 16 smaller squares of which four are kept and the rest discarded. Such examples are similar to those mentioned in connection with the Cantor set and the von Koch curve. The example depicted in Fig. 1.42 is constructed using two different similarity ratios.

The highly intricate structure of the Julia set illustrated in Fig. 1.43(a) stems from the single quadratic function $f(z) = z^2 + c$ for a suitable constant c. Although the set is not strictly self-similar (as the Cantor set and von Koch curve are), it is *quasi-self-similar* in that arbitrarily small portions of the set can be magnified and then distorted smoothly to coincide with a large part of the set. Figure 1.43(b) shows the graph of the function $f(t) = \sum_{k=0}^{\infty} (3/2)^{-k/2} \sin((3/2)^k t)$; the infinite summation leads to the graph having a fine structure, rather than being a smooth curve to which classical calculus is applicable.

The main tool of fractal geometry is dimension in its many forms.

Definition 32 In general, a set made up of m copies of itself scaled by a factor r might

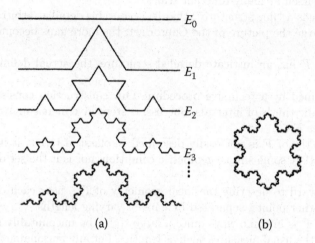

FIGURE 1.40: (a) The von Koch curve F. At each stage, the middle third of each interval is replaced by the other two sides of an equivalent triangle. (b) Three von Koch curves fitted together to form a snowflake curve.

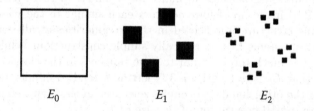

FIGURE 1.41: A *Cantor dust*: $\dim_h F = \dim_B F = 1$.

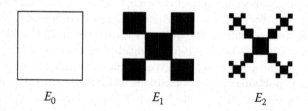

FIGURE 1.42: A self-similar fractal with two different similarity ratios.

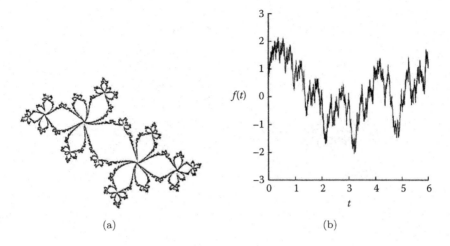

(a)   (b)

FIGURE 1.43: (a) A Julia set. (b) Graph of $f(t) = \sum_{k=0}^{\infty} (3/2)^{-k/2} \sin((3/2)^k t)$.

be thought of as having dimension $D = -\log m / \log r$. The number obtained in this way is referred to as the *self-similarity dimension* of the set.

We are familiar with the idea that a smooth curve is a one-dimensional object and a surface is two-dimensional. It is less clear that, for many purposes, the Cantor set should be regarded as having dimension $D = \log 2 / \log 3 \approx 0.631$ and the von Koch curve as having dimension $D = \log 4 / \log 3 \approx 1.262$. This latter number is, at least, consistent with the von Koch curve being larger than one-dimensional (having infinite length) and smaller than two-dimensional (having zero area).

The following argument gives one rather crude interpretation of the meaning of these "dimensions" indicating how they reflect scaling properties and self-similarity. A line segment is made up of four copies of itself, scaled by a factor $1/4$. The segment has dimension $D = -\log 4 / \log(1/4) = 1$. A square is made up of four copies of itself scaled by a factor $1/2$, i.e., with half the side length, and has dimension $D = -\log 4 / \log(1/2) = 2$. In the same way, the von Koch curve is made up of four copies of itself scaled by a factor $1/3$, and has dimension $D = -\log 4 / \log(1/3) = \log 4 / \log 3$, and the Cantor set may be regarded as comprising four copies of itself scaled by a factor $1/9$ and having dimension $D = -\log 4 / \log(1/9) = \log 2 / \log 3$.

Unfortunately, similarity dimension is meaningful only for a relatively small class of strictly self-similar sets. Nevertheless, there are other definitions of dimension that are much more widely applicable. For example, Hausdorff dimension and the box-counting dimension may be defined for any sets, and, in these four examples, may be shown to equal the similarity dimension. Very roughly, a dimension provides a description of how much space a set fills. It is a measure of the prominence of the irregularities of a set when viewed at very small scales. A dimension contains much information about the geometrical properties of a set. It is possible to define the "dimension" of a set in many ways, some satisfactory and others less so. The different definitions may give different values of dimension for the same set, and may also have very different properties. Inconsistent usage has sometimes led to considerable confusion. In particular, warning lights flash whenever the term *fractal dimension* is seen. In his original essay, Mandelbrot defined a fractal to be a set with Hausdorff dimension strictly greater than its topological dimension. The *topological dimension* of a set is always an integer and is 0 if it is totally disconnected, 1 if each point has arbitrarily

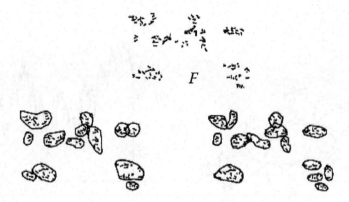

FIGURE 1.44: A set F and two possible δ-covers for F. The infimum of $\sum |U_i|^s$ over all such δ-covers $\{U_i\}$ gives $\mathcal{H}^s_\delta(F)$.

small neighborhoods with boundary of dimension 0, and so on. This definition proved to be unsatisfactory in that it excluded a number of sets that clearly ought to be regarded as fractals. Various definitions have been proposed, but they all seem to have this same drawback.

It is difficult to avoid developing properties of dimension other than in a way that applies to *fractal* and *nonfractal* sets alike. For nonfractals, such properties are of little interest – they are generally almost obvious and could be obtained more easily by other methods. When we refer to a set F as a fractal, therefore, we typically have the following in mind.

(i) F has a fine structure, i.e., detail on arbitrarily small scales.

(ii) F is too irregular to be described in traditional geometric language, both locally and globally.

(iii) Often F has some form of self-similarity, perhaps approximate or statistical.

(iv) Usually, the fractal dimension of F is greater than its topological dimension.

(v) In most cases of interest F is defined in a very simple way, e.g., recursively.

Recall that if U is any nonempty subset of d-dimensional Euclidian space $\mathbb{R}^d$, the *diameter* of U is defined as $|u| = \sup\{|x - y| : x, y \in U\}$, i.e., the greatest distance apart of any pair of points in U. If $\{U_i\}$ is a countable (or finite) collection of sets of diameter at most δ that covers F, i.e., $F \subset \bigcup_{i=i}^{\infty} U_i$ with $0 \leq |U_i| \leq \delta$ for each i, we say that $\{U_i\}$ is a δ-*cover* of F. Let F be a subset of $\mathbb{R}^d$ and $s \geq 0$. For any $\delta > 0$ we define

$$\mathcal{H}^s_\delta(F) = \inf\left\{\sum_{i=1}^{\infty} |U_i|^s : \{U_i\} \text{ is a } \delta\text{-cover of } F\right\}. \qquad (1.108)$$

Thus we look at all covers of F by sets of diameter at most δ and seek to minimize the sum of the s-th powers of the diameters (Fig. 1.44). As δ decreases, the class of permissible covers of F in (1.108) is reduced. Therefore, the infimum $\mathcal{H}^s_\delta(F)$ increases, and so approaches as $\delta \to 0$. Write

$$\mathcal{H}^s(F) = \lim_{\delta \to 0} \mathcal{H}^s_\delta(F). \qquad (1.109)$$

This limit exists for any subset $F \subset \mathbb{R}^d$, though the limiting value can be 0 or ∞. Call $\mathcal{H}^s(F)$ the *s-dimensional Hausdorff measure* of F.

Hausdorff measures generalize the familiar ideas of length, area, volume, etc. It may be shown that, for subsets of $\mathbb{R}^d$, d-dimensional Hausdorff measure is, to within a constant multiple, just d-dimensional Lebesgue measure, i.e., the usual d-dimensional volume. More precisely, if F is a Borel subset of $\mathbb{R}^d$, then $\mathcal{H}^d(F) = c_d^{-1}\text{vol}^d(F)$, where c_d is the volume of

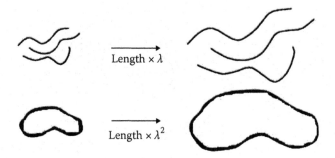

FIGURE 1.45: Scaling sets by a factor λ, area by a factor λ^2, and so on.

a d-dimensional ball of diameter 1, so that

$$c_d = \begin{cases} \frac{\pi^{d/2}}{2^d(d/2)!}, & d - \text{even}, \\ \pi^{(d-1)/2}\frac{[(n-1)/2]!}{n!}, & d - \text{odd}. \end{cases}$$

Similarly, for "nice" lower-dimensional subsets of $\mathbb{R}^d$, we have that $\mathcal{H}^0(F)$ is the number of points in F; $\mathcal{H}^1(F)$ gives the length of a smooth curve F; $\mathcal{H}^2(F) = (4/\pi) \times \text{area}(F)$ if F is a smooth surface; $\mathcal{H}^3(F) = (6/\pi) \times \text{vol}(F)$; and $\mathcal{H}^m(F) = c_m^{-1} \times \text{vol}^m(F)$ if F is a smooth submanifold of $\mathbb{R}^d$, i.e., an m-dimensional surface in the classical sense. The scaling properties of length, area and volume are well known. On magnification by a factor λ^1, the length of a curve is multiplied by λ, the area of a plane region is multiplied by λ^2 and the volume of a three-dimensional object is multiplied by λ^3. As might be anticipated, s-dimensional Hausdorff measure scales with a factor λ^s (Fig. 1.45). Such scaling properties are fundamental to the theory of fractals.

Let F be a subset of a plane. The *density* of F at x is

$$\lim_{r \to 0} \frac{\text{area}(F \cap B(x,r))}{\text{area}(B(x,r))} = \lim_{r \to 0} \frac{\text{area}(F \cap B(x,r))}{\pi r^2}, \tag{1.110}$$

where $B(x,r)$ is the closed disc of radius r and center x. The classical Lebesgue density theorem tells us that, for Borel set F, this limit exists and equal 1 when $x \in F$ and 0 when $x \notin F$, except for a set of x which is of area 0. In other words, for a typical point $x \in F$, small discs centered at x are almost entirely filled by F, but if x is outside F then small discs centered at x contain very little of F; see Fig. 1.46.

Similarly, if F is a smooth curve in the plane and x is a point of F (other than an end point), then $F \cap B(x,r)$ is close to a diameter of $B(x,r)$ for small r and $\lim_{r \to 0} \frac{\text{length}(F \cap B(x,r))}{2r} = 1$. If $x \notin F$, then this limit is 0. Density theorems such as these tell us how much of the set F, in the sense of area or length, is concentrated near x. In the same way it is natural to investigate densities of fractals – if F has dimension s, how does the s-dimensional Hausdorff measure of $F \cap B(x,r)$ behave as $r \to 0$. We look at this question when F is an s-set in $\mathbb{R}^2$ with $0 < s < 2$ (0-sets are just finite sets of points, and there is little to say, and $\mathcal{H}^2$ is essentially area, so if $s = 2$ we are in the Lebesgue density situation (1.110).

Define the *lower* and *upper densities* of an s-set F at a point $x \in \mathbb{R}^d$ as $D^s(F,x) = \underline{\lim}_{r \to 0} \frac{\mathcal{H}^s(F \cap B(x,r))}{(2r)^s}$ and $\bar{D}^s(F,x) = \overline{\lim}_{r \to 0} \frac{\mathcal{H}^s(F \cap B(x,r))}{(2r)^s}$, respectively (note that $|B(x,r)| = 2r$). If $D^s(F,x) = \bar{D}^s(F,x)$ we say that the *density* of F at x exists and we write $D^s(F,x)$ for the common value.

Definition 33 A point x at which $D^s(F,x) = \bar{D}^s(F,x) = 1$ is called a *regular* point of F; otherwise x is an *irregular* point.

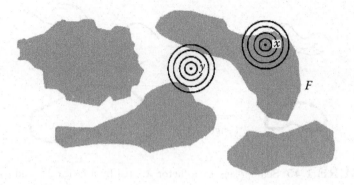

FIGURE 1.46: The Lebesgue density theorem.
The point $x \in F$, and area$(F \cap B(x,r))/$area$(B(x,r))$ is close to 1 if r is small.
The point $y \notin F$, and area$(F \cap B(x,r))/$area$(B(x,r))$ is close to 0 if r is small.

An s-set is termed *regular* if $\mathcal{H}^s$ almost all of its points, i.e., all of its points except for a set of $\mathcal{H}^s$ measure 0, are regular, and irregular if $\mathcal{H}^s$ almost all of its points are irregular. Here "irregular" does not mean "not regular"! A fundamental result is that an s-set F must be irregular unless s is an integer. If s is integral an s-set decomposes into a regular and an irregular part. Roughly speaking, a regular 1-set consists of portions of rectifiable curves of finite length, whereas an irregular 1-set is totally disconnected and dust-like, and typically of fractal form.

Lemma 4 (Covering) *Let C be a family of balls contained in some bounded region of $\mathbb{R}^d$. Then there is a finite or countable disjoint subcollection $\{B_i\}$ such that*

$$\bigcup_{B \in C} B = \bigcup_i \tilde{B}_i, \tag{1.111}$$

where $\tilde{B}_i$ is the closed ball concentric with B_i and of four times the radius.

Proof. For simplicity, let C be a finite family; the basic idea is the same in the general case. We select the $\{B_i\}$ inductively. Let B_1 be a ball in C of maximum radius. Suppose that $B_1, \ldots, B_{k-1}$ have been chosen. Take B_k to be the largest ball in C (or one of the largest) that does not intersect $B_1, \ldots, B_{k-1}$. The process terminates when no such ball remains. Clearly the balls selected are disjoint; we must check that (1.111) holds. If $B \in C$, then either $B = B_i$ for some i, or B intersects one of the B_i with $|B_i| \geq |B|$; if this were not the case, then B would have been chosen instead of the first ball B_k with $|B_k| < |B|$. Either way, $B \subset \tilde{B}_i$, so we have (1.111). It is easy to see that the result remains taking $\tilde{B}_i$ as the ball concentric with B_i and of $3 + \varepsilon$ times the radius, for any $\varepsilon > 0$; if C is finite we may take $\varepsilon = 0$. □

Mass distribution principle. Let μ be a mass distribution on F, and for some s there are numbers $c > 0$ and $\varepsilon > 0$ such that $\mu(U) \leq c|u|^s$ for all sets U with $|u| \leq \varepsilon$. Then $\mathcal{H}^s(F) \geq \mu(F)/c$ and

$$s \leq \dim_H F \leq \dim_B F \leq \overline{\dim}_B F.$$

Proof. If $\{U_i\}$ is any cover of F then $0 < \mu(F) \leq \mu(\bigcup_i U_i) \leq \sum_i \mu(U_i) \leq c \sum_i |U_i|^s$ by properties of measure and $\mu(U) \leq c|u|^s$. Taking infima, $\mathcal{H}^s_\delta(F) \geq \mu(F)/c$ is $\delta > 0$ if small enough, so $\mathcal{H}^s_\delta(F) \geq \mu(F)/c$. Since $\mu(F) > 0$ we get $\dim_H F \geq s$. □

The conclusion $\mathcal{H}^s_\delta(F) \geq \mu(F)/c$ is true if μ is the mass distribution on $\mathbb{R}^d$ and F is any subset.

Proposition 4 *Let μ be a mass distribution on $\mathbb{R}^d$, $F \subset \mathbb{R}^d$ a Borel set and let $c > 0$ a constant. Then,*

(a) *If $\overline{\lim\limits_{r \to 0}} \, \mu(B(x,r))/r^s < c$ for all $x \in F$ then $\mathcal{H}^s(F) \geq \mu(F)/c$,*

(b) *If $\overline{\lim\limits_{r \to 0}} \, \mu(B(x,r))/r^s > c$ for all $x \in F$ then $\mathcal{H}^s(F) \leq 2^s \mu(\mathbb{R}^d)/c$.*

Proof. (a) For each $\delta > 0$ let $F_\delta = \{x \in F : \mu(B(x,r)) < cr^s, \, \forall r \in (0,\delta]\}$. Let $\{U_i\}$ be a δ-cover of F and thus of F_δ. For each U_i containing a point $x \in F_\delta$, the ball $B(x, r = |U_i|)$ certainly contains U_i. By definition of F_δ, $\mu(U_i) \leq \mu(B) < c|U_i|^s$ so that $\mu(F_\delta) \leq \sum_i \{\mu(U_i) : U_i \cap F_\delta \neq \emptyset\} \leq c \sum_i |U_i|^s$. Since $\{U_i\}$ is any δ-cover of F, it follows that $\mu(F_\delta) \leq c\mathcal{H}^s_\delta(F) \leq c\mathcal{H}^s(F)$. But F_δ increases to F as δ decreases to 0, so $\mu(F) \leq c\mathcal{H}^s(F)$ by $\lim\limits_{\delta \to 0} \mu(F_\delta) = \mu(\bigcup_{\delta > 0} F_\delta)$.

(b) We prove a weaker version of (b) with 2^s replaced by 8^s, but the basic idea is similar. First, suppose that F is bounded. Fix $\delta > 0$ and let C be a family of balls $C = \{B(x,r) : x \in F, \, r \in (0,\delta], \, \mu(B) > cr^s\}$. Then by the hypothesis of (b) $F \subset \bigcup_{B \in C} B$. Apply the Covering Lemma 4 to the collection C such that $\bigcup_{B \in C} B \subset \bigcup_i \tilde{B}_i$ where $\tilde{B}_i$ is the ball concentric with B_i but of four times the radius. Thus $\{\tilde{B}_i\}$ is an 8δ-cover of F, and so

$$\mathcal{H}^s_{8\delta}(F) \leq \sum_i |\tilde{B}_i|^s \leq 4^s \sum_i |\tilde{B}_i|^s \leq 8^s c^{-1} \sum_i \mu(B_i) \leq 8^s c^{-1} \mu(\mathbb{R}^d).$$

Letting $\delta \to 0$, we get $\mathcal{H}^s(F) \leq 8^s c^{-1} \mu(\mathbb{R}^d) < \infty$. Finally, if F is unbounded and $\mathcal{H}^s(F) > 8^s c^{-1} \mu(\mathbb{R}^d) < \infty$, the $\mathcal{H}^s$-measure of some bounded subset of F will also exceed this value, contrary to the above. $\qquad\qquad\square$

It is immediate from Lemma 4 that if $\lim\limits_{r \to 0} \log \mu \frac{B(x,r)}{\log r} = s$ for all $x \in F$ then $\dim_H F = s$.

Proposition 5 *Let F be an s-set in $\mathbb{R}^d$. Then (a) $D^s(F,x) = \bar{D}^s(F,x) = 0$ for $\mathcal{H}^s$-almost all $x \notin F$, and (b) $2^{-s} \leq \bar{D}^s(F,x) \leq 1$ for almost all $x \in F$ (w.r.t. $\mathcal{H}^s$).*

Proof. (a) If F is closed and $x \notin F$ then $B(x,r) \cap F = \emptyset$ if $r > 0$ is small enough. Hence $\lim\limits_{r \to 0} \frac{\mathcal{H}^s(F \cap B(x,r))}{(2r)^s} = 0$. If F is not closed the proof is a little more involved and we omit it here.

(b) This follows from Proposition 4(a) by taking μ as a restriction of $\mathcal{H}^s$ to F, i.e., $\mu(A) = \mathcal{H}^s(F \cap A)$; then $\mathcal{H}^s(F_1) \geq \mathcal{H}^s(F)/c \geq \mathcal{H}^s(F_1)/c$, where $F_1 = \{x \in F : \bar{D}^s(F,x) < 2^{-s}c\}$. If $0 < c < 1$ this is only possible if $\mathcal{H}^s(F_1) = 0$. Thus $\bar{D}^s(F,x) \geq 2^{-s}$ holds for almost all $x \in F$. The upper bound follows in essentially the same way using Proposition 4(b). $\qquad\square$

Immediate consequence of Proposition 5(b) is that an irregular set has a lower density which is strictly less than 1 almost everywhere. We shall sometimes need to relate the densities of a set to choose the certain subsets. Let F be an s-set and let E be a Borel subset of F. Then

$$\frac{\mathcal{H}^s(F \cap B(x,r))}{(2r)^s} = \frac{\mathcal{H}^s(E \cap B(x,r))}{(2r)^s} + \frac{\mathcal{H}^s((F \backslash E) \cap B(x,r))}{(2r)^s}.$$

For almost all $x \in E$, we have $\frac{\mathcal{H}^s((F \backslash E) \cap B(x,r))}{(2r)^s} \xrightarrow[r \to 0]{} 0$ by Proposition 5(a), so letting $r \to 0$ gives

$$D^s(F,x) = D^s(E,x); \qquad \bar{D}^s(F,x) = \bar{D}^s(E,x)$$

for almost all $x \in E$ (w.r.t. $\mathcal{H}^s$). Thus, from the definitions of regularity, if E is a subset of an s-set F with $\mathcal{H}^s(E) > 0$, then E is regular if F is regular and E is irregular if F is irregular. In particular, the intersection of a regular and irregular set, being a subset of both, has measure zero.

In general it is quite involved to show that s-sets of fractal dimension are irregular, but in the case $0 < s < 1$ the following "annulus" proof is appealing.

Theorem 29 *Let F be an s-set in $\mathbb{R}^2$. Then F is irregular unless s is an integer.*

Proof. We show that F is irregular if $0 < s < 1$ by showing that the density $D^s(F, x)$ fails to exist almost everywhere in F. Assume the contrary: then there is a set $F_1 \subset F$ of positive measure where the density exists and therefore where $\frac{1}{2} < 2^{-s} \leq D^s(F, x)$ by Proposition 5(b). By Egorov's theorem, see [25], we find $r_0 > 0$ and Borel set $E \subset F_1 \subset F$ with $\mathcal{H}^s(E) > 0$ such that

$$\mathcal{H}^s(F \cap B(x, r)) > \frac{1}{2}(2r)^s \qquad (1.112)$$

for all $x \in E$ and $r < r_0$. Let $y \in E$ be a cluster point of E, i.e., a point y with other points of E arbitrarily close. Let η be a number with $0 < \eta < 1$ and let $A_{r,\eta}$ be the annulus $B(y, r(1+\eta)) \backslash B(y, r(1-\eta))$. Then

$$
\begin{aligned}
(2r)^{-s}\mathcal{H}^s(F \cap A_{r,\eta}) &= (2r)^{-s}\mathcal{H}^s(F \cap B(y, r(1+\eta))) \\
&\quad -(2r)^{-s}\mathcal{H}^s(F \cap B(y, r(1-\eta))) \to D^s(F, y)[(1+\eta)^s - (1-\eta)^s]
\end{aligned} \qquad (1.113)
$$

as $r \to 0$. For a sequence of values of r tending to 0, we may find $x \in E$ with $|x - y| = r$. Then $B(x, \frac{1}{2}r\eta) \subset A_{r,\eta}$ so by (1.112) $\frac{1}{2}r^s\eta^s < \mathcal{H}^s\left(F, B\left(x, \frac{1}{2}r\eta\right)\right) \leq \mathcal{H}^s(F, A_{r,\eta})$. Combining with (1.113) this implies that

$$2^{-(s+1)} \leq D^s(F, y)[(1+\eta)^s - (1-\eta)^s] = D^s(F, y)(2s\eta + O(\eta^2)).$$

Letting $\eta \to 0$ we see that this is impossible when $0 < s < 1$ and the result follows by contradiction. $\qquad \square$

1.5.2 Operations with fractals

Projections of fractals. Let L_θ be the line through the origin of $\mathbb{R}^2$ that makes an angle θ with the horizontal axis. Denote orthogonal projection onto L_θ by proj_θ, so that if F is a subset of $\mathbb{R}^2$, then $\mathrm{proj}_\theta F$ is the projection of F onto L_θ. Clearly, $|\mathrm{proj}_\theta x - \mathrm{proj}_\theta y| \leq |x - y|$ for all $x, y \in \mathbb{R}^2$, i.e., proj_θ is a Lipschitz mapping. Thus $\dim(\mathrm{proj}_\theta F) \leq \min\{\dim_H F, 1\}$ for any F and θ. As $\mathrm{proj}_\theta F$ is a subset of the line L_θ, its dimension cannot be more than 1.

Definition 34 For $s \geq 0$ the s-*potential* at a point $x \in \mathbb{R}^d$ due to the mass distribution μ on $\mathbb{R}^d$ is defined as $\Phi_s(x) = \int \frac{d\mu(y)}{|x-y|^s}$. If we are working in $\mathbb{R}^3$ and $s = 1$ then this is essentially the familiar Newtonian gravitational potential. The s-*energy* of μ is $I_s(\mu) = \int \Phi_s(x)d\mu(x) = \int \int \frac{d\mu(x)d\mu(y)}{|x-y|^s}$.

Theorem 30 *Let F be a subset of $\mathbb{R}^d$. (a) If there is a mass distribution μ on F with $I_s(\mu) < \infty$ then $\mathcal{H}^s(F) = \infty$ and $\dim_H F \geq s$. (b) If F is a Borel set with $\mathcal{H}^s(F) > 0$ then there exists a mass distribution μ on F with $I_t(\mu) < \infty$ for all $0 < t < s$.*

Proof. This can be found in [15, Section 4.3]. $\qquad \square$

Theorem 31 (Projection) *Let $F \subset \mathbb{R}^2$ be a Borel set.*
  (a) If $\dim_H F \leq 1$ then $\dim_H(\mathrm{proj}_\theta F) = \dim_H F$ for almost all $\theta \in [0, \pi)$.
  (b) If $\dim_H F > 1$ then $\mathrm{proj}_\theta F$ has positive length (as a subset of L_θ) and has dimension 1 for almost all $\theta \in [0, \pi)$.

Proof. This uses the potential characterization of Hausdorff dimension. If $s < \dim_H F \leq 1$ then (by Theorem 30) there exists a mass distribution μ on a compact subset of F with $0 < \mu(F) < \infty$ such that

$$\int_F \int_F \frac{d\mu(x)d\mu(y)}{|x-y|^s} < \infty. \tag{1.114}$$

For each θ we "project" the mass distribution μ onto the line L_θ to get a mass distribution μ_θ on $\text{proj}_\theta F$. Thus μ_θ is defined by the requirement that $\mu_\theta([a,b]) = \mu\{x : a \leq x \cdot \theta \leq b\}$ for each interval $[a,f]$, or equivalently, $\int_{-\infty}^{\infty} f(t)d\mu_\theta(t) = \int_F f(x \cdot \theta)d\mu(x)$ for each function $f \geq 0$. Hence θ is the unit vector in the direction θ, with x identified with its position vector and $x \cdot \theta$ the usual scalar product. Then

$$\int_0^x \left[\int_{-\infty}^{\infty} \int_{-\infty}^{\infty} \frac{d\mu_\theta(U)d\mu_\theta(v)}{|u-v|^s} \right] d\theta = \int_0^x \int_{F \times F} \frac{d\mu(x)d\mu(y)}{|x \cdot \theta - y \cdot \theta|^s} d\theta = \int_0^x \int_{F \times F} \frac{d\mu(x)d\mu(y)}{|(x-y) \cdot \theta|^s} d\theta = \int_0^x \frac{d\theta}{|\tau \cdot \theta|^s} \int_{F \times F} \frac{d\mu(x)d\mu(y)}{|x-y|^s} \tag{1.115}$$

for any fixed unit vector τ. The integral of $|(x-y) \cdot \theta|^{-s}$ with respect to θ depends only on $|x-y|$. If $s < 1$ then (1.115) is finite by virtue of (1.114) and that $\int_0^x \frac{d\theta}{|\tau \cdot \theta|^s} = \int_0^x \frac{d\theta}{|\cos(\tau-\theta)|^s} < \infty$. Hence $\int_F \int_F \frac{d\mu_\theta(u)d\mu_\theta(v)}{|u-v|^s} < \infty$ for almost all $\theta \in [0,\pi)$. By Theorem 30 (a) the existence of such a mass distribution μ_θ of $\text{proj}_\theta F$ implies that $\dim_H(\text{proj}_\theta F) \geq s$. This is true for all $s < \dim_H F$, so part (a) of the result follows. The proof of (b) follows similar lines, though Fourier transforms need to be introduced to show that projections have positive length. $\square$

These projection theorems are naturally generalized to high dimensions. The Grassmanian $G_{d,k}$ is the set of k-dimensional subspaces or k-*planes through the origin* in $\mathbb{R}^d$. These subspaces are parameterized by $k(d-k)$ coordinates (*generalized direction cosines*) so that we may refer to *almost all* subspaces in terms $k(d-k)$-dimensional Lebesgue measure. Write proj_Π for orthogonal projection onto k-plane Π.

Theorem 32 (High-dimensional projection) *Let $F \subset \mathbb{R}^2$ be a Borel set.*
(a) If $\dim_H \leq k$ then $\dim_H(\text{proj}_\Pi F) = \dim_\Pi F$ for almost all $\Pi \in G_{n,k}$.
(b) If $\dim_H > k$ then $\text{proj}_\Pi F$ has positive k-dimensional measure and has dimension k for almost all $\Pi \in G_{n,k}$.

Thus if $F \subset \mathbb{R}^3$, the plane projection of F is, in general, of dimension $\min\{2, \dim_H F\}$.

Theorem 33 *Let F be a regular 1-set in $\mathbb{R}^2$. Then $\text{proj}_\theta F$ has positive length except for at most one $\theta \in [0,\pi)$.*

In general, $\text{proj}_\theta F$ has positive length for all θ; there is an exceptional value θ only if F is contained in a set of parallel line segments.

Theorem 34 *Let F be an irregular 1-set in $\mathbb{R}^2$. Then $\text{proj}_\theta F$ has length zero for almost all $\theta \in [0,\pi)$.*

Corollary 7 *Let F be a 1-set in $\mathbb{R}^2$. If the regular part of F has $\mathcal{H}^1$-measure zero, then $\text{proj}_\theta F$ has length zero for almost all θ; otherwise it has positive length for all but at most one value of θ.*

Corollary 8 *A 1-set in $\mathbb{R}^2$ is irregular if and only if it has projections of zero length in at least two directions.*

As for example, the Cantor dust (Fig. 1.41) is an irregular 1-set. Moreover, there exist sets for which the projections in almost all directions are, to within length zero, anything that we care to prescribe.

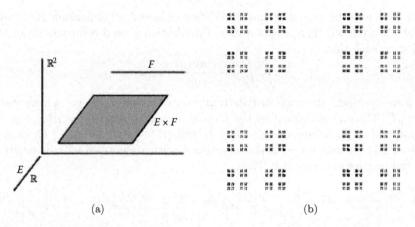

(a)   (b)

FIGURE 1.47: (a) The Cartesian product of a unit interval in $\mathbb{R}$ and a unit interval in $\mathbb{R}^2$. (b) The product $F \times F$, where F is the middle third Cantor set: $\dim_H F \times F = 2 \dim_H F = 2 \log 2 / \log 3$.

Theorem 35 *Let G_θ be a subset of L_θ for each $\theta \in [0, \pi)$ such that the set $\bigcup_{0 \le \theta < \pi} G_\theta$ is plane Lebesgue measurable. Then there exists a Borel set $F \subset \mathbb{R}^2$ such that*
(a) $proj_\theta F \supset G_\theta$ for all θ, and (b) the length $proj_\theta F \backslash G_\theta = 0$ for almost all θ.

Proof. Indicate the basic building block for such sets, which has been termed the *iterated Venetian blind* construction. Let E be a line segment of length λ; let ε be a small angle and k a large number. We replace E by k line segments of the lengths roughly λ/k, each at an angle ε to E and with end points equally spaced along E to form a new set E_1.

Repeat this process with each segment of E_1 to form a set E_2 consisting of k^2 line segments all of lengths about λ/k^2 and at angle 2ε to E. Continue in this way, to get E_r, a set of k^r segments all of lengths about λ/k^r and at angle $r\varepsilon$ to E. We stop when r is such that $r\varepsilon$ is, say, about $\frac{1}{4}\pi$. Comparing the projections of E_r with that of the original line segment E, we see that if $0 \le \theta < \frac{1}{2}\pi$ then $proj_\theta E_r$ are nearly the same since lines perpendicular to L_θ that cut E also cut E_r. If $-\frac{1}{4}\pi < \theta < 0$ then $proj_\theta E_r$ will have very small length, since most lines perpendicular to L_θ will pass straight between approximately angled *slats* of the construction. Thus the projections of E_r are similar to those of E, in certain directions, but are almost negligible in other directions.   □

This idea may be adapted to obtain sets with projections very close to G_θ in a narrow band of directions but with almost null projections in other directions. Unions of such sets for various small bands of directions give a set with approximately the required property. A limit of a sequence of sets which give increasingly accurate approximations leads to a set with the properties stated. This construction may be extended to high dimensions: there exists a set F in $\mathbb{R}^d$ such that almost all projections of F onto k-dimensional subspaces differ from prescribed sets by zero k-dimensional measure.

Products of fractals. Recall that if E is a subset of $\mathbb{R}^d$ and F is a subset of $\mathbb{R}^m$, the *Cartesian product*, or just *product* $E \times F = \{(x, u) \in \mathbb{R}^{d+m} : x \in E, y \in F\}$ – the set of points with first coordinate in E and second coordinate in F. Thus if E is a unit interval in $\mathbb{R}$, and F is a unit interval in $\mathbb{R}^2$, then $E \times F$ is a unit square in $\mathbb{R}^3$; see Fig. 1.47(a). Again, if F is the middle third Cantor set, then $F \times F$ is the *Cantor product* consisting of those points in the plane with both coordinates in F; see Fig. 1.47(b).

In the first example it is obvious that $\dim(E \times F) = \dim E + \dim F$ using the classical definition of dimension. This holds more generally, in the *smooth* situation, where E and F are smooth curves, surfaces of high-dimensional manifolds. Unfortunately, this equation is not always valid for *fractal* dimensions. For Hausdorff dimensions the best general result possible is an inequality $\dim_H(E \times F) \geq \dim_H E + \dim_H F$. Nevertheless, in many situations equality does not hold.

The proof of the product rule uses the Hausdorff measures on E and F to define a mass distribution $\mu(E \times F)$. Density bounds on E and F lead to estimates for μ suitable for a mass distribution method.

Proposition 6 *If $E \subset \mathbb{R}^d$, $F \subset \mathbb{R}^m$ are Borel sets with $\mathcal{H}^s(E), \mathcal{H}^t(F) < \infty$, then*

$$\mathcal{H}^{s+t}(E \times F) \geq c\mathcal{H}^s(E)\mathcal{H}^t(F) \tag{1.116}$$

where $c > 0$ depends only on s and t.

Proof. Assume that $E, F \subset \mathbb{R}$, so that $E \times F \subset \mathbb{R}^2$; the general proof is almost identical. If either $\mathcal{H}^s(E)$ or $\mathcal{H}^t(F)$ is zero, then (1.116) is trivial, so suppose that E is an s-set and F is a t-set, i.e., $0 < \mathcal{H}^s(E), \mathcal{H}^t(F) < \infty$. Define a mass distribution μ on $E \times F$ by utilizing the *product measure* of $\mathcal{H}^s(E)$ and $\mathcal{H}^t(F)$. Thus if $I, J \subset \mathbb{R}$, we define μ on the *rectangle* $I \times J$ by

$$\mu(I \times J) = \mathcal{H}^s(E \cap I)\mathcal{H}^t(F \cap J). \tag{1.117}$$

This defines a mass distribution μ on $E \times F$ with $\mu(\mathbb{R}^2) = \mathcal{H}^s(E)\mathcal{H}^t(F)$. By Proposition 5(b) we have

$$\overline{\lim_{r \to 0}} \mathcal{H}^s(E \cap B(x,r))(2r)^{-s} \leq 1 \quad \text{for } \mathcal{H}^s\text{-almost all } x \in E, \tag{1.118}$$

$$\overline{\lim_{r \to 0}} \mathcal{H}^t(F \cap B(y,r))(2r)^{-t} \leq 1 \quad \text{for } \mathcal{H}^t\text{-almost all } y \in F. \tag{1.119}$$

Of course, since we are concerned with subsets of $\mathbb{R}$, the "ball" $B(x,r)$ is just the interval of length $2r$ with midpoint x. By the definition of μ, both (1.118) and (1.119) hold for almost all $(x,y) \in E \times F$ (w.r.t. μ). Since $B((x,r),r) \subset B(x,r) \times B(y,r)$ we have

$$\mu(B((x,y),r)) \leq \mu(B(x,r) \times B(y,r)) = \mathcal{H}^s(E \cap B(x,r))\mathcal{H}^t(F \cap B(y,r)).$$

Thus

$$\frac{\mu(B((x,y),r))}{(2r)^{s+t}} \leq \frac{\mathcal{H}^s(E \cap B(x,r))}{(2r)^s} \cdot \frac{\mathcal{H}^t(F \cap B(y,r))}{(2r)^t}.$$

It follows, using (1.118) and (1.119), that $\overline{\lim}_{r \to 0}\mu(B((x,y),r))(2r)^{-(s+t)} \leq 1$ for almost all $(x,y) \in E \times F$ (w.r.t. μ). By Proposition 4(a), $\mathcal{H}^s(E \times F) \geq 2^{-(r+t)}\mu(E \times F) = 2^{-(r+t)}\mathcal{H}^s(E)\mathcal{H}^t(F)$. $\square$

Product formulas. These claim the following:
1. If $E \subset \mathbb{R}^d$, $F \subset \mathbb{R}^m$ are Borel sets then

$$\dim_H(E \times F) \geq \dim_H(E) + \dim_H(F). \tag{1.120}$$

2. For any sets $E \subset \mathbb{R}^d$ and $F \subset \mathbb{R}^m$ we have $\dim_H(E \times F) \leq \dim_H(E) + \overline{\dim}_B(F)$.
3. For any sets $E \subset \mathbb{R}^d$ and $F \subset \mathbb{R}^m$ $\overline{\dim}_B(E \times F) \leq \overline{\dim}_B E + \overline{\dim}_B F$.

Proof. 1. If s,t are any numbers with $s < \dim_H E$ and $t < \dim_H F$, then $\mathcal{H}^s(E) = \mathcal{H}^t(F) = \infty$. So, there are Borel sets $E_0 \subset E$ and $F_0 \subset F$ with $0 < \mathcal{H}^s(E_0), \mathcal{H}^t(F_0) < \infty$. By Proposition 6 $0 < \mathcal{H}^{s+t}(E \times F) \geq \mathcal{H}^{s+t}(E_0 \times F_0) \geq c\mathcal{H}^s(E_0)\mathcal{H}^t(F_0) > 0$. Hence, $\dim_H(E \times F) \geq s+t$. By choosing s and t arbitrarily close to $\dim_H(E)$ and $\dim_H(F)$, (1.120) follows.

2. For simplicity take $E \subset \mathbb{R}$ and $F \subset \mathbb{R}$. Choose numbers $s > \dim_H E$ and $t > \overline{\dim}_B F$. Then there is a number $\delta_0 > 0$ such that F may be covered by $N_\delta(F) \leq \delta^{-t}$ intervals of

length δ for all $\delta \leq \delta_0$. Let $\{U_i\}$ be any δ-cover of E by intervals with $\sum_i |U_i|^s < 1$. For each i, let $U_{i,j}$ be a cover of F by $N_{|U_i|}(F)$ intervals of length $|U_i|$. Then $U_i \times F$ is covered by $N_{|U_i|}(F)$ squares $\{U_i \times U_{i,j}\}$ of the side $|U_i|$. Thus $E \times F \subset \bigcup_i \bigcup_j (U_i \times U_{i,j})$, so that

$$\mathcal{H}^{s+t}_{\delta\sqrt{2}}(E \times F) \leq \sum_i \sum_j |U_i \times U_{i,j}|^{s+t} \leq \sum_i N_{|U_i|}(F) 2^{(s+t)/2} |U_i|^{s+t}$$

$$\leq 2^{(s+t)/2} \sum_i |U_i|^{-t} |U_i|^{s+t} < 2^{(s+t)/2}.$$

Letting $\delta \to 0$ gives $\mathcal{H}^{s+t}(E \times F) < \infty$ whenever $s > \dim_H E$ and $t > \overline{\dim}_B F$, so $\dim_H (E + F) \leq s + t$.

3. Just as in (1.117) – if E and F can be covered by $N_\delta(E)$ and $N_\delta(F)$ intervals of side δ, then $E \times F$ is covered by the $N_\delta(E)N_\delta(F)$ squares formed by products of these intervals. $\square$

Proposition 6 and Product formula 1 are valid for arbitrary (non-Borel) sets. It follows from (1.120) that the *Cantor product* $F \times F$, where F is the middle third Cantor set, has Hausdorff dimension at least $2\log 2/\log 3$ (Fig. 1.47(b)).

Corollary 9 *If* $\dim_H F = \overline{\dim}_B F$ *then* $\dim_H(E \times F) = \dim_H E + \dim_H F$.

Proof. Product formulas 1 and 2 give $\dim_H E + \dim_H F \leq \dim_H(E \times F) \leq \dim_H E + \overline{\dim}_B F$. $\square$

It is worth noting that the basic product inequality for upper box dimensions is opposite to that for Hausdorff dimensions.

Example 50 (Product with uniform Cantor sets) Let E, F be subsets of $\mathbb{R}$ with F a uniform Cantor set. Then $\dim_H(E \times F) = \dim_H E + \dim_H F$. Thus the *Cantor product* of the middle third Cantor set with itself has Hausdorff and box dimensions exactly $2\log 2/\log 3$. Similarly, if E is a subset of $\mathbb{R}$ and F is a straight line segment, then $\dim_H(E \times F) = \dim_H E + 1$.

Example 51 The *Cantor target* is the plane set given by $F' = \{(r, \theta) : r \in F, \ \theta \in [0, 2\pi]\}$ (in polar coordinates), where F is the middle third Cantor set. Show that $\dim F' = 1 + \log 2/\log 3$.
Solution. Set $f(x, y) = (x \cos y, x \sin y)$. Then f is a Lipschitz mapping and $F' = f(F \times [0, 2\pi])$. Thus

$$\dim_H F' = \dim_H f(F \times [0, 2\pi]) \leq \dim_H(F \times [0, 2\pi])$$

$$= \dim_H F + \dim_H[0, 2\pi] = \log 2/\log 3 + 1,$$

(if $f : F \to \mathbb{R}^d$ is a Lipschitz mapping then $\dim_H f(F) \leq \dim_H F$). But the restriction of f to $[\frac{2}{3}, 1] \times [0, \pi]$ is a bi-Lipschitz function, hence $\dim_H f(F) = \dim_H F$. Since $F' \supset f((F \cap [2/3, 1])) \times [0, \pi]$ then

$$\dim_H F' \geq \dim_H f((F \cap [\tfrac{2}{3}, 1])) \times [0, \pi] = \dim_H((F \cap [\tfrac{2}{3}, 1])) \times [0, \pi]$$
$$= \dim_H((F \cap [\tfrac{2}{3}, 1])) + \dim_H[0, \pi] = (\log 2/\log 3) + 1.$$

This argument requires only minor modification to show that F' is an s-set for $s = 1 + \log 2/\log 3$. $\square$

Example 52 Show that there exist sets $E, F \subset \mathbb{R}$ with $\dim_H E = \dim_H F = 0$ and $\dim_H(E \times F) \geq 1$.

Solution. Let $0 = m_0 < m_1 < \ldots$ be a rapidly increasing sequence of integers satisfying a condition to be specified below. Let E consist of those numbers in $[0, 1]$ with zero in the r-th decimal place whenever $m_k + 1 \leq r \leq m_{k+1}$ and k is even, and let F consist of those numbers with zero in the r-th decimal place if $m_k + 1 \leq r \leq m_{k+1}$ and k is odd. Looking at the first m_{k+1} decimal places for even k, there is an obvious cover of E by 10^{j_k} intervals of length $10^{-m_{k+1}}$, where $j_k = (m_2 - m_1) + \ldots + (m_k - m_{k-1})$. Then $\frac{\log 10^{j_k}}{-\log 10^{-m_{k+1}}} = j_k/m_{k+1}$ which tends to 0 as $k \to \infty$ provided that the m_k are chosen to increase sufficiently rapidly. So, $\dim_H E \leq \underline{\dim}_B E = 0$. If $0 < w < 1$ then we can write $w = x + y$ where $x \in E$ and $y \in F$; just take r-th decimal digit of $w \in E$ if $m_k + 1 \leq r \leq m_{k+1}$ and k is odd and $w \in F$ if k is even. The mapping $f : \mathbb{R}^2 \to \mathbb{R}$ given by $f(x, y) = x + y$ is Lipschitz, and so

$$\dim_H(E \times F) \geq \dim_H f(E \times F) \geq \dim_H(0, 1) = 1,$$

since for Lipschitz mapping f the inequality $\dim_H f(F) \leq \dim_H F)$ holds. $\qquad \square$

Intersections of fractals. The intersection of two fractals is often a fractal; it is natural to try to relate the dimension of this intersection to that of the original sets. It is immediately apparent that we can say almost nothing in the general case. For if F is bounded, there is a congruent copy F_1 of F such that $\dim_H(F \cap F_1) = \emptyset$ (take F and F_1 disjoint). We can say rather more, if we consider the intersection of F and a congruent copy in a *typical* relative position. To illustrate this, let F and F_1 be unit line segments in the plane. Then $F \cap F_1$ can be a line segment, but only in the exceptional situation when F and F_1 are collinear. If F and F_1 cross at an angle, then $F \cap F_1$ is a single point, but now $F \cap F_1$ remains a single point if F_1 is replaced by a nearby congruent copy. Thus, while (in general) $F \cap F_1$ contains at most one point, this situation occurs "frequently".

We can make this rather more precise. Recall that a right motion or direct congruence transformation σ of the plane transforms any set E to a congruent copy $\sigma(E)$ without reflection. The rigid motions may be parameterized by three coordinates (x, y, θ) where the origin is transformed to (x, y) and θ is the angle of rotation. Such parameterization provides a natural measure on the space of rigid motions, with the measure of a set A of rigid motions given by the three-dimensional Lebesgue measure of the (x, y, θ) parameterizing the motion in A. For example, the set of all rigid motions which map the origin to a point of the rectangle $[1, 2] \times [0, 3]$ has the measure $1 \times 3 \times 2\pi$. In the example with F a unit line segment, the set of transformations σ for which $F \cap \sigma(F)$ is a line segment has measure 0. But $F \cap \sigma(F)$ is a single point for a set of transformations of positive measure, in fact a set of measure 4.

Similar ideas hold in higher dimensions. In $\mathbb{R}^3$, *typically*, two surfaces intersect in a curve, a surface and a curve intersect in a point and two curves are disjoint. In $\mathbb{R}^d$, if smooth manifolds E and F intersect at all, then (in general) they intersect in a submanifold of dimension $\max\{0, \dim E + \dim F - d\}$. More precisely, if $\dim E + \dim F - d > 0$ then $\dim(E \cap \sigma(F)) = \dim E + \dim F - d$ for a set of rigid motions σ of positive measure, and is 0 for almost all other σ. Of course, σ is measured using the $\frac{1}{2}d(d + 1)$ parameters required to specify a rigid transformation of $\mathbb{R}^d$.

If we use Hausdorff dimension for fractals E and F then it is true, in general,

$$\dim_H(E \cap \sigma(F)) \leq \max\{0, \dim_H E + \dim_H F\} - d$$

and often

$$\dim_H(E \cap \sigma(F)) \geq \dim_H E + \dim_H F - d \qquad (1.121)$$

as σ ranges over a group G of transformations, such as the group of transformations, congruences or similarities. Of course, in general, this means for almost all σ and often means for a *set of σ of positive measure* with respect to a natural measure on the transformations in G. Generally, G can be parameterized by m coordinates in a direct way for some integer m and we can use Lebesgue measure on the parameter space $\mathbb{R}^m$. One may obtain upper bounds for $\dim_H(E \cap \sigma(F))$ when G is the group of translations; these bounds hold automatically for the larger groups of congruences and similarities. Recall that $F + x = \{x + y : y \in F\}$ denotes the translation of F by the vector x.

Proposition 7 *Let F be a Borel subset in $\mathbb{R}^2$. If $1 \leq s \leq 2$ then*

$$\int_{-\infty}^{\infty} \mathcal{H}^{s-1}(F \cap L_x) \, dx \leq \mathcal{H}^s(F). \tag{1.122}$$

Proof. Given $\varepsilon > 0$, let $\{U_i\}$ be a δ-cover of F such that $\sum_i |U_i|^s \leq \mathcal{H}_\delta^s(F) + \varepsilon$. Each U_i is contained in a square S_i of side $|U_i|$ with sides parallel to the coordinate axes.

Let χ_i be the indicator function of S_i, i.e., $\chi_i(x, y) = 1$ if $(x, y) \in S_i$ and $\chi_i(x, y) = 0$ if $(x, y) \notin S_i$. For each x, the sets $\{S_i \cap L_s\}$ form a δ-cover of $F \cap L_x$. So,

$$\mathcal{H}_\delta^{s-1}(F \cap L_x) \leq \sum_i |S_i \cap L_x|^{s-1} = \sum_i |U_i|^{s-2} |S_i \cap L_x| = \sum_i |U_i|^{s-2} \int \chi_i(x, y) dy.$$

Hence,

$$\int \mathcal{H}_\delta^{s-1}(F \cap L_x) \, dx \leq \sum_i |U_i|^{s-2} \iint \chi_i(x, y) dx \, dy = \sum_i |U_i|^s \leq \mathcal{H}_\delta^s(F) + \varepsilon.$$

Since $\varepsilon > 0$ is arbitrary, $\int \mathcal{H}_\delta^{s-1}(F \cap L_x) \, dx \leq \mathcal{H}_\delta^s(F) + \varepsilon$. Letting $\delta \to 0$ gives (1.122). $\square$

Corollary 10 *Let F be a Borel subset of $\mathbb{R}^2$. Then, for almost all x (w.r.t. one-dimensional Lebesgue measure) $\dim_H(F \cap L_x) \leq \max\{0, \dim_H F - 1\}$.*

Proof. Take $s > \dim_H F$, so that $\mathcal{H}^s(F) = 0$. If $s > 1$, formula (1.122) gives $\mathcal{H}^{s-1}(F \cap L_x) = 0$ and so $\dim_H(F \cap L_x) \leq s - 1$, for almost all x. $\square$

Theorem 36 *If $E, F \subset \mathbb{R}^d$ then $\dim_H(F \cap (F + x)) \leq \max\{0, \dim_H(E \times F) - d\}$ for all $x \in \mathbb{R}^d$.*

Proof. We prove this for $d = 1$, and the proof for $n > 1$ is similar, using dimensional analogy of Corollary 10. Let L_c be a line in the (x, y)-plane with equation $x = y + c$. Assuming that $\dim_H(E \times F) > 1$, it follows from Corollary 10 (rotating lines through 45^0 and changing notation slightly) that

$$\dim_H((E \times F) \cap L_c) \leq \dim_H(E \times F) - 1 \tag{1.123}$$

for almost all $c \in \mathbb{R}$. But a point $(x, x - c) \in (E \times F) \cap L_c$ and only if $x \in E \cap (F + c)$. In particular, $\dim_H(E \cap (F + c)) = \dim_H((E \times F) \cap L_c)$, so the result follows from (1.123). $\square$

Lower bounds for $\dim_H(E \cap \sigma(F))$ of the form (1.121) are rather harder to obtain. The main known results are contained in the following theorem.

Theorem 37 *Let $E, F \subset \mathbb{R}^d$ be Borel sets, and let G be a group of transformations on $\mathbb{R}^d$. Then*

$$\dim_H(E \cap \sigma(F)) \geq \dim_H E + \dim_H F - d$$

for a set of motions $\sigma \in G$ of positive measure in the following cases:

(a) G is the group of similarities, E and F are arbitrary sets.

(b) G is the group of rigid motions, E is arbitrary and F is a rectifiable curve, surface or manifold.

(c) G is the group of rigid motions, E and F are arbitrary, with either $\dim_H E > \frac{1}{2}(d+1)$ *or* $\dim_H F > \frac{1}{2}(d+1)$.

Proof. This uses the potential theoretic methods. The argument resembles that of Theorem 31, but various technical difficulties make it more complicated. Briefly, if $s < \dim_H E$ and $t < \dim_H F$, there are mass distributions $\mu(E)$ and $\nu(F)$ with energies $I_s(\mu)$ and $I_t(\nu)$ both finite. If ν happened to be absolutely continuous with respect to d-dimensional Lebesgue measure, i.e., if there were a function f such that $\nu(A) = \int_A f(x)\,\mathrm{d}x$ for each set A, then it would be natural to define a mass distribution η_σ on $E \cap \sigma(F)$ by

$$\eta_\sigma(A) = \int_A f(\sigma^{-1}(x))d\mu(x).$$

If we could show that $I_{s+t-d}(\eta_\sigma) < \infty$ for almost all σ then Theorem 30(a) would imply $\dim_H(E \cap \sigma(F)) \geq s + t - d$ if $\eta_\sigma(\mathbb{R}^d) > 0$. Unfortunately, when F is a fractal, ν is supported by a set of zero d-dimensional volume, so is anything but absolutely continuous mass distributions ν_δ supported by the δ-neighborhood of F. Then if

$$\nu_\delta(A) = \int_A f_\delta(x)\,\mathrm{d}x, \quad \eta_{\sigma,\delta}(A) = \int_A f_\delta(\sigma^{-1}(x))\,d\mu(x),$$

we can estimate $I_{s+t-d}(\eta_{\sigma,\sigma})$ and take the limit as $\delta \to 0$. Simplifying the integral $\int I_{s+t-d}(\eta_{\sigma,\delta})d\sigma$ isolates a term

$$\varphi_\delta(w) = \int\limits_{G_0} \int\limits_{\mathbb{R}^d} \nu_\delta(y)\nu_\delta(y + \sigma(w))dy\,d\sigma,$$

where integration with respect to σ is over the subgroup $G_0 \subset F$ which fixes the origin. Provided that

$$\varphi_\delta(w) \leq \text{const} \cdot |w|^{t-d} \tag{1.124}$$

for all w and δ, it may be shown that this constant c is independent of δ. Letting $\delta \to 0$, the measures $\eta_{\sigma,\delta}$ converge to measures η_σ on $E \cap \sigma(F)$, where $\int I_{s+t-d}(\eta_\sigma)d\sigma < \infty$ for almost all σ, so, by Theorem 30 (a), $\dim_H(E \cap \sigma(F)) \geq s + t - d$ whenever $\eta_\sigma(E \cap \sigma(F)) > 0$, which happens on a set of positive measure.

It may be shown that (1.124) holds if $I_t(\nu) < \infty$ in the cases (a), (b) and (c) listed above. This is easy to show for (a) and (b). Case (c) is more awkward, requiring the Fourier transform theory. $\qquad\square$

The condition that $\dim_H E$ or $\dim_H F$ is greater than $(d+1)/2$ in the case (c) is a curious consequence of the use of Fourier transforms. It is not known whether the theorem remains valid for the group of consequences if $d \geq 2$ and $d/2 < \dim_H E$, $\dim_H F \leq (d+1)/2$.

Example 53 Let $F \subset \mathbb{R}$ be the middle third Cantor set. For $\lambda, x \in \mathbb{R}$ write $\lambda F + x = \{\lambda y + x : y \in F\}$. Then $\dim_H(F \cap (F + x)) \leq \log 4/\log 3 - 1$ for almost all x, and $\dim_H(F \cap (\lambda F + x)) = \log 4/\log 3 - 1$ for a set of $(x, \lambda) \in \mathbb{R}^2$, of positive plane Lebesgue measure.

Solution. Since $\dim_H(F \times F) = \log 4/\log 3$, the stated dimensions follow from Theorems 36 and 37(a).

1.5.3   Fractal attractors in dynamical systems

We will illustrate various ways in which fractals can occur in dynamical systems. Let $D \subset \mathbb{R}^d$ be a subset (often $\mathbb{R}^d$ itself), and let $f : D \to D$ be a continuous mapping. Then f^k is the k-th iterate of f, so that

$$f^0(x) = x, \quad f^1(x) = f(x), \quad f^2(x) = f(f(x)), \quad \text{etc.}$$

Clearly $f^k(x) \subset D$ if $x \in D$.

An iterative scheme $\{f^k\}$ is called *discrete dynamical system*. We are interested in the behavior of the sequence of *iterates* or *orbits*, $\{f^k(x)\}_{k=1}^{\infty}$ for various initial points $x \in D$, particularly for large k. For example, if $f(x) = \cos x$, the sequence $f^k(x)$ converges to $0.739\ldots$ as $k \to \infty$ for any initial x: try repeatedly pressing the cosine button on a calculator and see! Sometimes the distribution of iterates appears almost random. Alternatively, $f^k(x)$ may converge to a *fixed point* $w \in D$, i.e., a point with $f(w) = w$. More generally, $f^k(x)$ may converge to an orbit of *period-p points* $\{w, f(w), \ldots, f^{p-1}(w)\}$, where p is the least positive integer with $f^p(w) = w$, in the sense that $|f^k(x) - f^k(w)| \to 0$ as $k \to \infty$. Sometimes, $f^k(x)$ may appear to move about at random, but always remaining close to a certain set, which may be fractal. Examine several ways in which such *fractal attractors* or *strange attractors* occur.

Roughly speaking, an attractor is a set to which all nearby orbits converge. Call a subset $F \subset D$ an attractor for f if F is a closed set that is *invariant* under f (i.e., with $f(F) = F$), such that the distance from $f^k(x)$ to F converges to zero as k tends to infinity for all x in an open set $V \supset F$. It is usual to require that F is minimal in the sense that it has no proper subset satisfying these conditions. Similarly, a closed invariant set F from which all nearby points not in F are iterated away is called a *repeller*. This is roughly equivalent to F being an *attractor* for the (perhaps multivalued) inverse f^{-1}. An attractor or repeller may just be a single point or a period $-p$ orbit. Even relatively simple maps f can have fractal attractors.

Note that $f(D) \subset D$ so that $f^k(D) \subset \ldots \subset f(D) \subset D$, so $\bigcap_{i=1}^{k} f^i(D) = f^k(D)$. Thus the set $F = \bigcap_{k=1}^{\infty} f^i(D)$ is invariant under f. Since $f^k(D) \subset \bigcap_{i=1}^{k} f^i(D)$ for all $x \in D$, the iterates $f^k(x)$ approach F as $k \to \infty$, and F might be an attractor. Often, if f has a fractal attractor or repeller F then f exhibits *chaotic* behavior on F. There are various definitions of chaos; f would certainly be regarded as chaotic on F if the following are true:

(a) The orbit $\{f^k(x)\}$ is dense in F for some $x \in F$;

(b) The periodic points of $f(x) \in F$, i.e., points for which $f^p(x) = x$ for some $p > 0$, are dense in F;

(c) f has *sensitive dependence on initial conditions*; that is, there is a number $\delta > 0$ such that for every $x \in F$ there are points $y \in F$ arbitrarily close to x such that $|f^k(x) - f^k(y)| \geq \delta$ for some k. Thus points that are initially closed to each other do not remain close under iterates of f.

Condition (a) implies that F cannot be decomposed into smaller closed invariant sets; (b) suggests a skeleton of regularity in the structure of F, and (c) reflects the unpredictability of iterates of points on F. In particular, (b) implies that accurate long-term numerical approximation to orbits of f is impossible, since a tiny numerical error will be magnified under iteration. Often conditions that give rise to fractal attractors also lead to chaotic behavior. Dynamical systems are naturally suited to computer investigation. Roughly speaking, attractors are the sets that are seen when orbits are plotted on a computer. For some initial points x one plots, say, 10000 iterates $f^k(x)$, starting at $k = 101$, say, to ensure that they are indistinguishable from any attractor. Under certain circumstances, a repeller in a dynamical system coincides with the attractor of related iterated function systems.

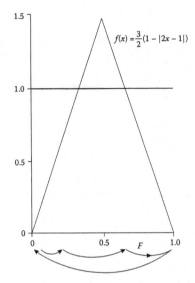

FIGURE 1.48: The tent map f; the iterates of a point are indicated by the arrows.

Example 54 The mapping $f : \mathbb{R} \to \mathbb{R}$ given by $f(x) = \frac{3}{2}(1 - |2x - 1|)$ is called the *tent map* because of the form of its graph (Fig. 1.48). Clearly, f maps $\mathbb{R}$ in a two-to-one manner onto $(-\infty, \frac{3}{2})$. Defining an iterated function system $S_1, S_2 : [0, 1] \to [0, 1]$ by contractions $S_1(x) = \frac{1}{3}x$, $S_2(x) = 1 - \frac{1}{3}x$, we see that $f(S_1(x)) = f(S_2(x)) = x$ for $0 \leq x \leq 1$. Thus S_1 and S_2 are the two branches of f^{-1}. So, there is a unique nonempty compact attractor $F \subset [0, 1]$ satisfying

$$F = S_1(F) \bigcup S_2(F) \tag{1.125}$$

which is given by $F = \bigcap_{k=0}^{\infty} S^k([0,1])$, writing $S(E) = S_1(E) \cup S_2(E)$ for any set E. Clearly the attractor F is the middle third Cantor set, with Hausdorff dimension $\log 2 / \log 3$. It follows from (1.125) that $f(F) = F$. To see that F is a repeller, observe that if $x < 0$ then $f(x) = 3x$, so $f^k(x) = 3^k x \to -\infty$ as $k \to \infty$. If $x > 1$ then $f(x) < 0$ and again $f^k(x) \to -\infty$. If $x \in [0, 1] \backslash F$ then for some k, we have

$$x \notin S^k[0, 1] = \cup \{S_{i_1} \circ \cdots \circ S_{i_k}[0, 1] : i_j = 1, 2\},$$

and so $f^k(x) \notin [0, 1]$, and again $f^k(x) \to -\infty$ as $k \to \infty$. All points outside F are iterated to $-\infty$ so F is a repeller. The chaotic nature of $f(F)$ is readily apparent. Denoting the points of F by $x_{i_1 i_2, \ldots}$ with $i_j = 1, 2$, $|x_{i_1 i_2, \ldots} - x_{i'_1 i'_2, \ldots}| \leq 3^{-k}$ if $i_1 = i'_1, \ldots, i_k = i'_k$. Since $x_{i_1 i_2, \ldots} = S_{i_1}(x_{i_2 i_3, \ldots})$, it follows that $f(x_{i_1 i_2 \ldots}) = x_{i_2 i_3 \ldots}$. Suppose that $(i_1, i_2, \ldots)$ is an infinite sequence with every finite sequence of 1 and 2 appearing as a consecutive block of terms; for example, $(1, 2, 1, 1, 1, 2, 2, 1, 2, 2, 1, 1, 1, 1, 1, 2, \ldots)$, where the spacing is just to indicate the form of the sequence. For each point $x_{i'_1 i'_2, \ldots} \in F$ and each integer q, we may find k such that $(i'_1, \ldots, i'_q) = (i_{k+1}, \ldots, i_{k+q})$. Then $|x_{i_{k+1} i_{k+2}, \ldots} - x_{i'_1 i'_2, \ldots}| < 3^{-q}$, so the iterates $f(x_{i_1 i_2, \ldots}) = x_{i_{k+1} i_{k+2}, \ldots}$ come arbitrarily close to each point of F for suitable large k, so that f has a dense orbit in F. Similarly, since $x_{i_1 \ldots i_k, i_1 \ldots i_k, i_1 \ldots}$ is a periodic point of period k, the periodic points of f are dense in F. The iterates have sensitive dependence on initial conditions, since $f^k(x_{i_1 \ldots i_k, 1 \ldots}) \in [0, \frac{1}{3}]$, but $f^k(x_{i_1 \ldots i_k, 1, \ldots}) \in [\frac{2}{3}, 1]$. Thus, F is a chaotic repeller for f. The study of f by its effect on points of F represented by sequences $(i_1, i_2, \ldots)$ is known as *symbolic dynamics*.

Scratching and folding transformation. One of the simplest planar dynamical systems with fractal attractor is Baker's transformation, so named because it resembles the

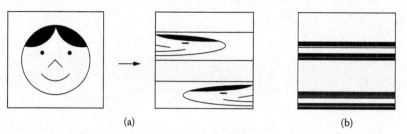

(a)   (b)

FIGURE 1.49: The Baker's transformation: (a) its effect on the unit square; (b) its attractor.

process of repeatedly stretching a piece of dough and folding in two. Let $E = [0, 1] \times [0, 1]$ be the unit space. For fixed $0 < \lambda < 1/2$ we define the *Baker's transformation* $f : E \to E$ by

$$f(x, y) = \begin{cases} (2x, 2y), & 0 \le x \le \frac{1}{2} \\ (2x - 1, \lambda y + \frac{1}{2}), & \frac{1}{2} < x \le 1. \end{cases} \tag{1.126}$$

This transformation is stretching E into $2 \times \lambda$ rectangle, cutting it into $1 \times \lambda$ rectangles and placing these above each other with a gap of $\frac{1}{2} - \lambda$ in-between (Fig. 1.49). Then $E_k = f^k(E)$ is a decreasing sequence of sets, with E_k comprising 2^k horizontal strips of height λ^k separated by gaps of at least $(\frac{1}{2} - \lambda)\lambda^{k-1}$. Since $F(E_k) = f^k(E)$, the compact limit set $F = \bigcap_{k=0}^{\infty} E_k$ satisfies $f(F) = F$. Strictly speaking, $f(F)$ does not include part of F in the left edge of the square E, a consequence of f being discontinuous. This has little effect on this study. If $(x, y) \in E$ then $f^k(x, y) \in E_k$, so $f^k(x, y)$ lies within distance $\lambda^k \in F$. Thus all points of E are attracted to F under iteration by f.

If the initial point (x, y) has $x = 0. a_1 a_2 \ldots$ in base 2, and $x \notin \{\frac{1}{2}, 1\}$, then $f^k(x, y) = (0. a_{k+1} a_{k+2} \ldots, y_k)$ where y_k is some point in the strip of E_k numbered $a_k a_{k-1} \ldots a_1$ (base 2) counting from the bottom with the bottom strip numbered 0. Thus when k is large the position of $f^k(x, y)$ depends largely on the base-2 digits a_i of x with i close to k arranging for $f^k(x, y)$ to be dense in F for certain initial (x, y), just as in the case of the tent map.

The Baker's transformation is rather artificial, being piecewise linear and discontinuous. It does serve to illustrate how the *stretching and cutting* procedure results in a fractal attractor.

The related process of *stretching and folding* can occur for continuous functions on plane regions. Suppose that $E = [0, 1] \times [0, 1]$ and f maps E in a one-to-one manner onto a horseshoe-shaped region $f(E) \in E$. Then f may be thought of as stretching E into a long thin rectangle, which is then bent in the middle. This figure is repeatedly stretched and bent by f so that $f^k(E)$ consists of an increasing number of side-by-side strips; see Fig. 1.50. We have $E \supset f(E) \supset \ldots$, and the compact set $F = \bigcap_{k=1}^{\infty} f^k(E)$ attracts all points of E. Locally, F is the product of a Cantor set and an interval.

Example of a *stretching and folding* transformation is a Hénon map $f : \mathbb{R}^2 \to \mathbb{R}^2$

$$f(x, y) = (y + 1 - ax^2, bx), \tag{1.127}$$

where a and b are constants, e.g., $a = 1.4$ and $b = 0.3$. For these values there is a quadrilateral D for which $f(D) \subset D$. This mapping has Jacobian $-b$ for all (x, y), so it contracts area at a constant rate throughout $\mathbb{R}^2$; to within a linear change of coordinates (1.127) is the most general quadratic mapping with this property. The transformation (1.127) may be decomposed into an area-preserving bend, a contraction, and a reflection, the net effect being *horseshoe-like* (Fig. 1.51). This leads us to expect f to have a fractal attractor, and this is borne out by computer pictures (Fig. 1.52).

Detailed pictures show banding indicative of a set that is locally the product of a line segment and a Cantor-like set. Numerical estimates suggest that the attractor has box

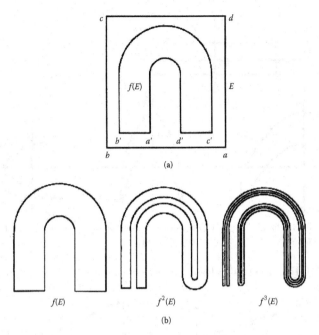

FIGURE 1.50: A horseshoe map. (a) The square E is transformed by stretching and bending to the horseshoe $f(E)$ with a, b, c, d mapped to a', b', c', d', respectively. (b) The iterates of E under f form a set that is locally a product of a line segment and a Cantor set.

dimension of about 1.26 when $a = 1.4$ and $b = 0.3$. Detailed analysis of the dynamics of the Hénon map is complicated. In particular the qualitative changes in behavior (bifurcations) that occur as a and b vary are highly intricate.

Many other types of *stretching and folding* are possible. Transformations can fold several times or even be many-to-one. For example, the ends of a horseshoe might cross. Such transformations often have fractal attractors, but their analysis tends to be difficult.

Multifractals. Now we expect that different parts of an attractor may be characterized by different values of the fractal dimensions. In such a situation, a single value of some fractal dimension is not sufficient to characterize the attractor adequately. Two quite different attractors might have the same correlation dimension, for example, but still differ widely in their "appearance". An object with a multiplicity of fractal dimensions is called a *multifractal*. We can visualize this object as a collection of overlapping fractal objects, each with its own fractal dimension. In fact, the multifractal description seems to be the appropriate description for many objects in nature (not just for attractors in state space). To some extent it has been the extension of the notion of fractals to multifractals that has led to the wide range of applications of fractals in almost all areas of science.

The scheme provides a distribution function, called $f(\alpha)$, which gives the distribution of *scaling exponents*, labeled by α. These notions can be understood by considering the partition of the attractor region of state space into a group of cells of size δ labeled by an index i, with $i = 1, 2, \ldots, N(\delta)$. Let a trajectory run for a long time and ask for the probability that the trajectory points fall in the i-th cell. That probability is defined as $p_i(\delta) = N_i/N$, where N_i is the number of trajectory points in the i-th cell and N is the total number of trajectory points. Alternatively, if we are considering an actual geometric object, we "cover" the object with cells of size δ and define the fraction (p_i) of the object's mass which falls within the i-th cell. Assume that $p_i(\delta_i)$ satisfies a scaling relation

$$p_i(\delta_i) = k\delta_i^{\alpha_i}, \tag{1.128}$$

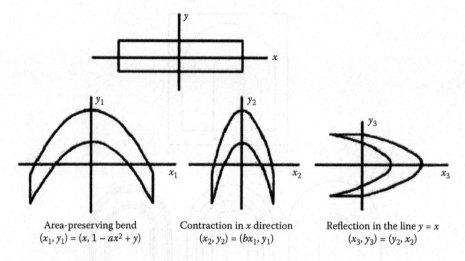

Area-preserving bend
$(x_1, y_1) = (x, 1 - ax^2 + y)$

Contraction in x direction
$(x_2, y_2) = (bx_1, y_1)$

Reflection in the line $y = x$
$(x_3, y_3) = (y_2, x_2)$

FIGURE 1.51: The Hénon map. Transformations of a rectangle.

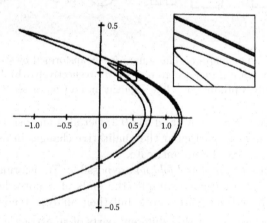

FIGURE 1.52: Iterates of a point under Hénon map (1.127) showing the form of attractor.

where k is some proportionality constant and δ_i is the size of the i-th cell. The value α_i is called the scaling index for i-th cell. As we make δ smaller, we increase the number of cells $N(\delta)$ and define the number of cells that have a scaling index in the range between α and $\alpha + d\alpha$. Call that number $n(\alpha)d\alpha$. The crucial assumption is that we expect the number of cells with α in the range $[\alpha, \alpha + d\alpha]$ to scale with the size of the cells with a characteristic exponent which we define as $f(\alpha)$. This characteristic exponent is formally defined by $n(\alpha) = K\delta^{-f(\alpha)}$. Interpret $f(\alpha)$ as a fractal dimension of the set of points with scaling index α. If we plot $f(\alpha)$ as a function of α, we get, for a multifractal, a graph like that shown in Fig. 1.53. The maximum value of $f(\alpha)$ corresponds to the average box-counting dimension of the object. For a truly one-dimensional attractor, such as the attractor for the logistic map at the period-doubling accumulation point, we would expect $f(\alpha)$ to be less than 1 since we know that the attractor is a fractal. For data from a two-dimensional mapping such as the Hénon map, we might expect $f(\alpha) < 2$.

Both the generalized dimensions introduced in the previous section and $f(\alpha)$ describe properties of the multifractal. We might reasonably expect that those quantities are related. In fact, in some cases, we compute $f(\alpha)$ by first finding the generalized dimensions and then following the procedure to be presented shortly. The connection between $f(\alpha)$ and D_q is

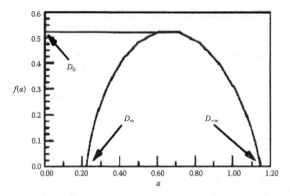

FIGURE 1.53: A plot of $f(\alpha)$ as a function of α for a two-scale Cantor (multifractal) set (see Fig. 1.54); the maximum value of $f(\alpha)$ is equal to the average box-counting dimension for the set.

established by looking at the probability sum

$$S_q(\delta) = \sum_{i=1}^{N(\delta)} p_i^q, \tag{1.129}$$

where D_q is the generalized (box-counting) dimension to be defined as

$$D_q = \lim_{R \to 0} \frac{\ln\left(\sum_{i=1}^{N(\delta)} p_i^q\right)}{(q-1)\ln\delta}. \tag{1.130}$$

The sum $S_q(\delta)$ is sometimes called the q-th order *partition function*, in analogy with partition function used in statistical mechanics. In order to make the connection to $f(\alpha)$, we write the probabilities in terms of α values to get the partition function (with some proportionally constant C)

$$S_q(\delta) = C \int \delta^{-f(\alpha)} \delta^{q\alpha} d\alpha. \tag{1.131}$$

The first factor in the integrand in (1.131) shows the number of cells which have index α, while the second factor is the q-th power of the probability associated with the index α. Since δ is supposed to be small, the numerical value of the integrand is largest when the exponent of δ, $q\alpha - f(\alpha)$, is smallest. That is, for each value of q, there is a corresponding value of α, which we call $\alpha_*(q)$ for which

$$[(q\alpha - f(\alpha))']_{\alpha=\alpha_*} = 0, \qquad [(q\alpha - f(\alpha))'']_{\alpha=\alpha_*} < 0. \tag{1.132}$$

From the first of equations (1.132), it follows $q = f'|_{\alpha=\alpha_*}$, which tells us that the slope of the $f(\alpha)$ curve is equal to 1 for the α value corresponding to $q = 1$. Remark that $f''|_{\alpha=\alpha_*} < 0$, which tells us that the $f(\alpha)$ curve must be concave downward. Returning to the evaluation of the integral, note that the integral value must be approximately the value of the integrand where it is a maximum multiplied by some proportionality constant. Thus, we can write

$$S_q(\delta) = C' \delta^{q\alpha_* - f(\alpha_*)}. \tag{1.133}$$

Recalling the definition of D_q in (1.130) and comparing it to (1.129) and (1.133), we see that

$$(q-1)D_q = q\alpha - f(\alpha), \tag{1.134}$$

where we have dropped the subscript "$*$" on the variable α.

In practice, for data generated by the trajectory of some model dynamical system or for data from some experiment, we often find D_q. Then we carry out various manipulations to

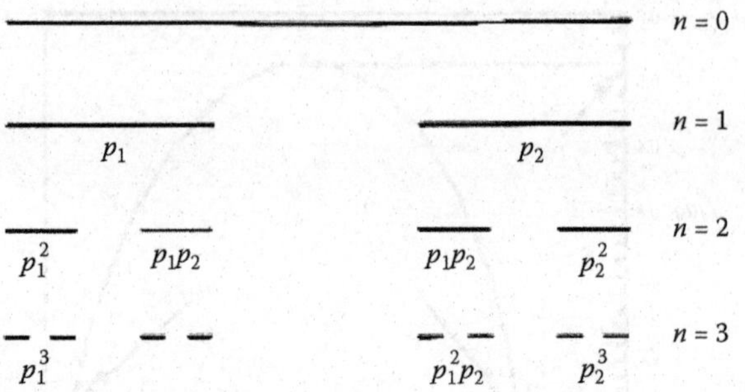

FIGURE 1.54: A sketch of the first few generations of the construction of a weighted Cantor set. For the $n = 3$ generation, only a few of the probabilities (weightings) have been indicated explicitly.

find for each q the corresponding values of α and $f(\alpha)$. To see how this works out, we first differentiate (1.134) by q,

$$\frac{d}{dq}[(q-1)D_q] = \alpha + q\frac{d\alpha}{dq} - \frac{df}{d\alpha}\frac{d\alpha}{dq} \qquad (1.135)$$

which, together with (1.134), yields

$$\alpha(q) = \frac{d}{dq}[(q-1)D_q], \qquad (1.136)$$

thus giving a value of α. Then we solve (1.134) for $f(\alpha)$:

$$f(\alpha) = q\frac{d}{dq}[(q-1)D_q] - (q-1)D_q. \qquad (1.137)$$

To summarize, once we have found D_q as a function of q, we can compute $f(\alpha)$ and α from (1.136) and (1.137). This change from the variables q and D_q to α and $f(\alpha)$ is an example of a *Legendre transformation*, used in the formalism of thermodynamics. There is a simple interpretation of various aspects of $f(\alpha)$ curve, which displays in a straightforward fashion some expected universal features.

To get some feeling for $f(\alpha)$, first compute this quantity for a simple probability distribution. It can be done by using (1.128) with a specified distribution function. Specifically, assume that the probability distribution function for $x \in [0,1]$ is given by $\rho(x) = kx^{-1/2}$, where k is a proportionality constant. The probability of finding the particle in segment of the length δ ($\delta \ll 1$) is given by the integral $p(\delta) = \int_{x_0}^{x_0+\delta} \rho(x)\,dx$ which gives

$$p(\delta) = \begin{cases} k\delta^{1/2}, & x_0 = 0 \\ k\delta, & x_0 > 0. \end{cases}$$

Since only one point ($x_0 = 0$) has the scaling index $\alpha = 1/2$, the corresponding fractal dimension $f(1/2) = 0$. On the other hand, a continuous one-dimensional segment of x values has the scaling index $\alpha = 1$ with the corresponding fractal dimension $f(1) = 1$.

For interesting examples involving various fractals generated by recursive procedures, it is useful to define a generalized partition function that allows for the possibility of cells of variable size. By evaluating this function for several simple examples, we will gain considerable insight into the workings of the $f(\alpha)$ curve. This generalized partition function is defined to be

$$\Gamma(q,\tau) = \sum_i p_i^q / \delta_i^\tau, \qquad (1.138)$$

where τ is chosen so that for a fixed value of q, $\Gamma(q, \tau) = 1$. The meaning of the parameter τ becomes apparent if we evaluate Γ for the special case of equal cell sizes $\delta_i = \delta$. Then taking the logarithm of both sides of (1.138) yields $\tau = \frac{\ln \sum_i p_i^q}{\ln \delta}$. Comparing this last equation with (1.138) gives us $\tau = (q - 1)D_q$. Thus, the generalized dimension D_q (or, more specifically, $(q - 1)D_q$) is that number which, for a given value of q, makes the generalized partition function equal 1. The actual numerical value of the partition function is not so important since for τ not equal to $(q - 1)D_q$, the partition function diverges to $\pm\infty$.

To see how this generalized partition function allows us to find $f(\alpha)$ for a recursively generated fractal, look at the case of a *weighted Cantor set*. The *canonical* middle-third Cantor set is generated by starting with a line segment $[0, 1]$ of the length 1 and eliminating the middle one-third, leaving two segments each of the length $\frac{1}{3}$. We can generalize this Cantor set by allowing the two newly generated line segments to have different weights, say p_1 and p_2 with $p_1 + p_2 = 1$. This weighting means that when we assign points to the two segments, a fraction p_1 of the points go to the segment on the left and p_2 go to the segment on the right at the first level of generation. At the second level, we have four segments: one with weight p_1^2, one with p_2^2, and two with $p_1 p_2$. Figure 1.54 shows the first few levels of recursion for this set. A moment's consideration should convince that the generalized partition function for the n-th generation can be written as

$$\Gamma_n(q, \tau) = \frac{1}{\delta^{n\tau}}(p_1^q + p_2^q)^n, \tag{1.139}$$

where $\delta = \frac{1}{3}$ for the *standard* Cantor set. Again, for a specified value q, the parameter τ is chosen so that the generalized partition function is equal to 1.

To evaluate the generalized partition function, we consider the binomial expression in (1.139) which can be expressed in the standard way as a sum of products of p_1 and p_2 raised to various powers:

$$\Gamma_n(q, \tau) = \frac{1}{\delta^{n\tau}} \sum_{k=0}^{n} \frac{n!}{(n-k)!k!} p_1^{qk} p_2^{q(n-k)}. \tag{1.140}$$

The factorial combination in (1.140) defines the number of line segments having the weight corresponding to a particular value k. Now comes the critical part of the argument. For large n, there is one term in the sum (1.140) that is largest and in fact dominates the sum. We can find the corresponding value of k (call k_*) by differentiating the natural logarithm of the summand with respect to k and setting the resulting derivative equal to 0. We use the natural logarithm because we can then use the Stirling's approximation: $\ln n! \simeq n \ln n - n$. If the logarithm has a maximum, so does the original function. When we carry out that differentiation, we find that k_* satisfies

$$n/k_* = 1 + (p_2/p_1)^q. \tag{1.141}$$

As an aside, we can find the value of τ by requiring that the associated term in the sum satisfy

$$\frac{1}{\delta^n} \frac{n!}{(n-k_*)!k_*!} p_1^{qk_*} p_2^{q(n-k_*)} = 1.$$

This expression can readily be solved for $\tau = (q - 1)D_q$. Figure 1.55 shows a plot of D_q as a function of q for the weighted Cantor set with $p_1 = 0.7$ and $p_2 = 0.3$. We make connection with $f(\alpha)$ by requiring that α be the exponent for the δ dependence of our dominant probability (weighting): $p_{k_*} = p_1^{k_*} p_2^{(n-k_*)} = \delta_n^\alpha$, where $\delta^n = 3^{-n}$ is the length of the segments constructed at the n-th generation. Hence,

$$\alpha = \frac{k_* \ln p_1 + (n - k_*) \ln p_2}{n \ln \delta}. \tag{1.142}$$

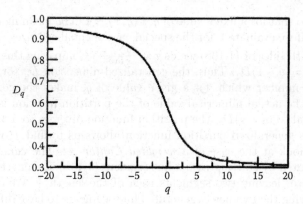

FIGURE 1.55: A plot of the generalized dimension D_q as a function of q for the weighted Cantor set with $p_1 = 0.7$ and $p_2 = 0.3$.

Note that α depends only on the ratio n/k_*. Similarly, we find $f(\alpha)$ by asking for the length dependence of the number of segments corresponding to the value k_*:

$$\frac{n!}{(n - k_*)! k_*!} = \delta_n^{-f(\alpha)}. \tag{1.143}$$

Using the Stirling approximation for the factorials and solving for $f(\alpha)$ yields

$$f(\alpha) = \frac{\ln(1 - k_*/n) + (k_*/n)\ln(n/k_* - 1)}{\ln \delta}, \tag{1.144}$$

which, like α, depends only on n/k_*. To determine the entire $f(\alpha)$ curve, for prescribed values of p_1 and p_2, we pick values of q (ranging from -40 to $+40$) and then find n/k_* from (1.141). Then we compute α and $f(\alpha)$ from (1.142) and (1.144); see Fig. 1.56(a) with the results for $p_1 = 0.7$ and $p_2 = 0.3$.

If $q \to \infty$ then by (1.140), $k_* \to n$. From (1.143), $f(\alpha) \to 0$ and we obtain the smallest value for α, namely, $\alpha_{\min} = \ln p_1 / \ln \delta = 0.324 \ldots$; see Fig. 1.56(a). On the other hand, for $q \to -\infty$, $f(\alpha) \to 0$, but we have the largest value of $f(\alpha)$ occuring for $q = 0$, for which we have $n/k_* = 2$. From (1.143), we find $f_{\max} = \ln 2 / \ln 3$, which is just the box-counting dimension D_0 for the Cantor set with $\delta = 1/3$. By (1.133), the largest value of α is equal to D_∞. Figure 1.55 shows how D_q approaches these values. Since $q \to \infty$ emphasizes the largest probability in the generalized partition function, $\alpha_{\min}$ corresponds to the most densely populated part of the fractal (the part with the largest probability weighting). At the other extreme, $\alpha_{\max}$ corresponds to $q \to -\infty$ and emphasizes that part of the fractal with the smallest probability, the least densely populated part. The interpretation carries over to all $f(\alpha)$ curves.

We come back to the example illustrated in Fig. 1.53. These results were computed for the so-called *two-scale Cantor set*, which is a combination of the asymmetric and weighted Cantor sets. At each stage of construction, we have two different segments δ_1 and δ_2 with two different weights p_1 and p_2. Using a method similar to that used earlier for the weighted Cantor set, it can be shown that it is relatively straightforward to find α and $f(\alpha)$ for this set. It can be also shown that $\alpha_{\max}$ is given by the larger of $\ln p_1 / \ln \delta_1$ or $\ln p_2 / \ln \delta_2$ and that $\alpha_{\min}$ is given by the smaller of those two ratios.

Consider another example, the $f(\alpha)$ distribution for the logistic map at the period-doubling accumulation point. The results are presented in Fig. 1.56(b). The logistic map data illustrate again the basic features of the $f(\alpha)$ distribution. The α values for which $f(\alpha)$ is not 0 lie between 0 and 1. We expect this for the logistic map because we can treat the attractor at the period-doubling accumulation point approximately as a Cantor set with two length scales ($1/\alpha_F$ and $1/\alpha_F^2$), where $\alpha_F \approx 2.502$ called as a Feigenbaum constant. The numerical value of $f(\alpha)$ at its peak is just the box-counting dimension about equal to 0.5388.... Consider, finally, an example of multifractal distributions. Look at the $f(\alpha)$

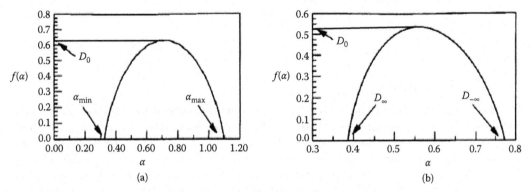

FIGURE 1.56: (a) A plot of $f(\alpha)$ for weighted Cantor set with $p_1 = 0.7$, $p_2 = 0.3$ and $\delta = \frac{1}{3}$. (b) A plot of $f(\alpha)$ distribution for the logistic map attractor at the period-doubling accumulation point.

distribution for data calculated from the sine-circle map

$$\theta_{n+1} = \theta_n + \Omega - \frac{K}{2\pi} \sin(2\pi\theta_n) \tag{1.145}$$

with parameter $K = 1$ and with *winding number* Ω (or *frequency-ratio parameter*) equal to the golden mean, $G = (\sqrt{5} - 1)/2 \approx 0.6180$. This set of conditions corresponds to the onset of chaotic behavior. The distribution is plotted in Fig. 1.57. The general behavior of the distribution is similar to that for the logistic map. The largest value of α can be expressed analytically as $\alpha_{\max} = \frac{\ln G}{\ln \beta^{-1}} \approx 1.8980$, where $\beta \approx 1.2885$ is the scaling factor in the neighborhood of $\theta = 0$ for the sine-circle map (1.145). At the other extreme, the smallest value for α is given by $\alpha_{\min} = \frac{\ln G}{\ln \beta^{-1}} \approx 0.6326$. The peak of the distribution occurs at $f(\alpha) = 1$, which is to be expected because the iterates of the sine-circle map completely cover the interval $\theta = [0, 2\pi]$, thus giving a box-counting exponent of 1.

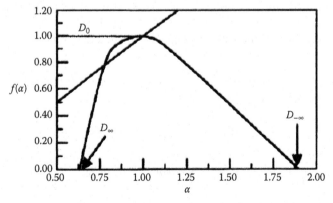

FIGURE 1.57: The $f(\alpha)$ distribution for data from the sine-circle map with $K = 1$, $G = 0.618\ldots$. The 45^0 line $f(\alpha)$ is shown for reference.

A variety of exponents have been analyzed using the $f(\alpha)$ formalism. This approach allows us to recognize universally classes among a diversity of dynamical systems. In practice, the correlation sum is used to compute the generalized dimensions D_q and $f(\alpha)$ is then found by using the Legendre transformations (1.135) and (1.136). Application of multifractals in turbulence is presented in Chapter 3.

1.6   Perturbations

Many problems faced today by physicists and applied mathematicians involve such difficulties as nonlinear governing equations, variable coefficients and nonlinear boundary conditions. Solution of these problems requires approximation and numerical methods, or a combination of both. Foremost among the approximation techniques is the systematic method of perturbations (asymptotic expansions) in terms of a small or a large parameter. This chapter is concerned with different perturbation techniques and their applications to ODE. According to perturbation techniques in use, the solution of the problem is represented by the first few terms of a perturbation expansion, usually the first two terms. The different techniques are described as formal procedures without any attempt at justifying them rigorously.

1.6.1   Regular perturbation theory

Basic estimations. Consider ODE $\dot{x} = f(x) + \varepsilon g(x, t, \varepsilon)$ $(x \in \mathcal{D} \subset \mathbb{R}^n)$, where $f, g \in C^r$ for some $r \geq 1$. Our aim is to describe the difference between the dynamics for $\varepsilon = 0$ and $\varepsilon > 0$.

Recall the fundamental theorem on ODEs, and the dependence of their solutions on parameters.

Theorem 38 *Let $F(x, t, \varepsilon) \in C^r(\mathcal{D}_0, \mathbb{R}^n)$, $r \geq 1$, for an open set $\mathcal{D}_0 \subset \mathbb{R}^n \times \mathbb{R} \times \mathbb{R}^p$. Then the solution of $\dot{x} = F(x, t, \varepsilon)$ with $x(t_0) = x_0$ is a C^r function of x_0, t_0, t and ε on its domain of existence.*

It follows that $x(t)$ admits an expansion of the form $x(t) = x_0(t) + \varepsilon x^1(t) + \varepsilon^2 x^2(t) + \ldots + \varepsilon^r x^r(t, \varepsilon)$ in some intervals of t and ε. In particular, $x_0(t)$ is a solution of the unperturbed equation $\dot{x} = f(x)$, and thus the solution of (1.148) stays close to $x_0(t)$ on compact time intervals.

A useful tool to estimate the time-dependence of the error $x(t) - x_0(t)$ is the following lemma.

Lemma 5 (Gronwall's inequality) *Let φ, α and β be continuous real-valued functions on $[a, b]$, with β nonnegative, and assume that*

$$\varphi(t) \leq \alpha(t) + \int_a^t \beta(s)\varphi(s)ds, \quad t \in [a, b]. \tag{1.146}$$

Then $\varphi(t) \leq \alpha(t) + \int\limits_a^t \beta(s)\alpha(s) \exp\left[\int\limits_s^t \beta(u)du\right]ds, \quad t \in [a, b].$

Proof. Let $R(t) = \int_a^t \beta(s)\varphi(s)ds$. Then $\varphi(t) \leq \alpha(t) + R(t)$ for all $t \in [a, b]$ and, since $\beta(t) \geq 0$,

$$R'(t) = \beta(t)\varphi(t) \leq \beta(t)\alpha(t) + \beta(t)R(t).$$

Let $B(s) = \int_a^s \beta(u)du$. Then $\frac{d}{ds}e^{-B(s)}R(s) = [R'(s) - \beta(s)R(s)]e^{-B(s)} \leq \beta(s)\alpha(s)e^{B(s)}$, and thus, integrating from a to t, we obtain $e^{-B(t)}R(t) \leq \int_a^t \beta(s)\alpha(s)e^{-B(s)}ds$. We complete the proof by multiplying this expression by $e^{B(t)}$ and inserting the result into (1.146). $\square$

Corollary 11 *Assume that f admits a uniform Lipschitz constant K in $\mathcal{D}$, and $\|g\|$ is uniformly bounded by M in $\mathcal{D} \times \mathbb{R} \times [0, \varepsilon_0]$. If $x(0) = x_0(0)$, then*

$$\|x(t) - x_0(t)\| \leq \frac{\varepsilon M}{K}(e^{Kt} - 1), \quad t \geq 0. \tag{1.147}$$

Proof. Let $y(t) = x(t) - x_0(t)$. Then $\dot{y} = f(x_0(y) + y(t)) - f(x_0(t)) + \varepsilon g(x(t), t, \varepsilon)$. Integrating between 0 and t, we obtain

$$y(t) = \int_0^t [f(x_0(s) + y(s)) - f(x_0(s))]ds + \varepsilon \int_0^t g(x(s), s, \varepsilon)\, ds.$$

Taking the norm, we arrive at the estimate

$$\|y(t)\| = \int_0^t \|f(x_0(s) + y(s)) - f(x_0(s))\|ds + \varepsilon \int_0^t \|g(x(s), s, \varepsilon)\|ds \leq \int_0^t K\|y(s)\|ds + \varepsilon Mt.$$

Applying Gronwall's inequality, we obtain (1.147), as desired. $\square$

The estimate (1.147) shows that for a given t, we can make the deviation $\|x(t) - x_0(t)\|$ small by taking ε small. Since e^{Kt} grows without bound as $t \to \infty$, we cannot make any statements on stability. Of course, (1.147) is only an upper bound. The following example shows that it cannot be improved without further assumptions on $f(x)$.

Example 55 Consider as a special case of (1.148) the equation $\dot{x} = Bx + \varepsilon g(t)$, where g is T-periodic in t. Assume that B is in diagonal form, with (possibly complex) eigenvalues $b_1, \ldots, b_n$. Then the solution can be written, for each component, as $x_j(t) = e^{b_j t}x_j(0) + \varepsilon \int_0^t e^{b_j(t-s)}g_j(s)ds$. If the g_j are written in Fourier series as $g_j(s) = \sum_{k \in \mathbb{Z}} c_{jk}e^{i\omega ks}$, $\omega = 2\pi/T$, we have to compute integrals of the form

$$\int_0^t e^{b_j(t-s)}e^{i\omega ks}\, ds = \begin{cases} \frac{e^{i\omega kt} - e^{b_j t}}{i\omega k - b_j} & \text{if} \quad b_j \neq i\omega k, \\ te^{b_j t} & \text{if} \quad b_j = i\omega k. \end{cases}$$

If $\operatorname{Re} b_j \neq 0$, solutions grow or decrease exponentially fast, just like for $\varepsilon = 0$. If $\operatorname{Re} b_j = 0$, the origin is a stable equilibrium point of the unperturbed system, while the perturbed system displays the phenomenon of *resonance*: If b_j is a multiple of $i\omega$, the deviation grows linearly with t; if not, it remains bounded, but the amplitude may become large if the denominator $i\omega k - b_j$ is small for some k.

Consider a perturbation of the integrable Hamiltonian of the form $H(\varphi, I, \varepsilon) = H_0(I) + \varepsilon H_1(I, \varphi, \varepsilon)$. The associated canonical equations are

$$\dot{\varphi}_j = \partial H_0/\partial I_j + \varepsilon \partial H_1/\partial I_j, \quad \dot{I}_j = -\varepsilon \partial H_1/\partial \varphi_j.$$

We can make the link with the system

$$\dot{x} = f(x) + \varepsilon g(x, t, \varepsilon), \quad x \in \mathcal{D} \subset \mathbb{R}^n. \tag{1.148}$$

A first possibility is to take $x = (\varphi, I)$, $f = (\partial H_0/\partial I, 0)$ and $g = (\partial H_1/\partial I, -\partial H_1/\partial \varphi)$. Then the function g is time-independent. There are other possibilities.

Example 56 Assume for instance that $\partial H_0/\partial I_1 \neq 0$ for all I. If $\partial H_1/\partial I_1$ is bounded, then $\varphi_1(t, \varepsilon)$ will be a monotonous on t for sufficiently small ε. Replace t by φ_1 and consider the $2n - 1$ equations

$$\begin{cases} \frac{d\varphi_k}{d\varphi_1} = \frac{\partial H_0/\partial I_k + \varepsilon\, \partial H_1/\partial I_k}{\partial H_0/\partial I_1 + \varepsilon\, \partial H_1/\partial I_1}, & k = 2, \ldots, n, \\ \frac{d\varphi_k}{d\varphi_1} = -\varepsilon \frac{\partial H_1/\partial \varphi_j}{\partial H_0/\partial I_1 + \varepsilon\, \partial H_1/\partial I_1}, & j = 1, \ldots, n. \end{cases}$$

We obtain a system of the form $\dot{x} = f(x) + \varepsilon\, g(x, t, \varepsilon)$ for $f = ((\partial H_0/\partial I)/(\partial H_0/\partial I_1), 0)$ and $x = (\varphi_2, \ldots, \varphi_n, I_1, \ldots, I_n)$. The remainder g depends periodically on φ_1 because φ_1 is an angle.

Example 57 Assume that $\gamma(t) = (q(t), p(t))$ is a T-periodic solution of a Hamiltonian system (which need not be integrable). The variable $y = x - \gamma(t)$ satisfies an equation of the form

$$\dot{y} = A(t) + g(y, t), \tag{1.149}$$

where $A(t) = \begin{pmatrix} \partial^2 H/\partial q\partial p & \partial^2 H/\partial p^2 \\ -\partial^2 H/\partial q^2 & -\partial^2 H/\partial q\partial p \end{pmatrix}\Big|_{\gamma(t)}$ and there are constants $M, \delta > 0$ such that $|g(y, t)| \leq M|y|^2$ for $|y| \leq \delta$. The equation (1.149) is not yet in the form we are looking for, but we are going to transform it. Let $U(t)$ be the solution of the problem

$$\dot{U} = A(t)U(t), \quad U(0) = 1. \tag{1.150}$$

Since the matrix A is T-periodic, by Floquet's theorem we have $U(t) = P(t)e^{tB}$ for some T-periodic matrix $P(t)$. By substitution into (1.150), we see that $\dot{P} = A(t)P(t) - P(t)B(t)$, and thus the change of variables $y = P(t)z$ yields the equation

$$\dot{z} = Bz + P^{-1}g(Pz, t). \tag{1.151}$$

Finally we curry out a rescaling $z = \varepsilon w$, which means that we zoom on a small neighborhood of the periodic orbit. Here ε is a small positive parameter, and the change of variables is defined for all $\varepsilon > 0$; thus we obtain the equation $\dot{w} = Bw + \varepsilon G(w, t, \varepsilon)$, having the form of (1.148) with

$$G(w, t, \varepsilon) = \frac{1}{\varepsilon^2} P^{-1} g(\varepsilon Pw, t), \quad \|G(w, t, \varepsilon)\| \leq \frac{1}{\varepsilon^2} \|P^{-1}\| \cdot \|M\| \cdot |\varepsilon Pw|^2 \leq M\|P^{-1}\| \cdot \|P\| \cdot |w|^2.$$

In fact, (1.149) and (1.151) can be considered as small perturbations of a linear system, deriving from a (possibly time-dependent) quadratic Hamiltonian. Since a quadratic form can be diagonalized, these linear equations are integrable, and thus the motion in a small neighborhood of a periodic orbit can be viewed as a perturbed integrable system.

Averaging and iterative methods. Corollary 11 gives a rather rough estimate on the difference between solutions of the perturbed and unperturbed equations. There exist a number of iterative methods that allow to improve these estimates, by replacing the unperturbed equation by a better approximation of the perturbed one, which still remains easier to solve.

Averaging. This method allows to deal with ODEs depending periodically on time, by comparing their solutions with a simpler autonomous equation. The method applies to systems of the form

$$\dot{x} = \varepsilon g(x, t, \varepsilon), \tag{1.152}$$

where g is T-periodic in t. We are interested in more general equations of the form $\dot{x} = f(x) + \varepsilon g(x, t, \varepsilon)$, but it is often possible to transform this equation into (1.152). Indeed, let $h(x, t, \varepsilon) \in C^r(\mathcal{D} \times \mathbb{R}/T\mathbb{Z} \times [0, \varepsilon_0], \mathbb{R}^n)$ be T-periodic in t, and invertible in x near $x = 0$.

Then the variable $y = h(x, t, \varepsilon)$ satisfies $\dot{y} = \partial_t h + \partial_x h f(x) + \varepsilon \partial_x h g(x, t, \varepsilon)$. Thus, if we manage to find a function h such that $\partial_t h + \partial_x h f(x) = O(\varepsilon)$, then y satisfies an equation of the form (1.152). This is, in fact, possible if the unperturbed equation has periodic solutions with a period close to T, i.e., near resonance.

Example 58 Consider the system of two equations $\dot{x} = B(\omega_0) x + \varepsilon g(x, t, \varepsilon)$ and $B(\omega_0) = \begin{pmatrix} 0 & 1 \\ -\omega_0^2 & 0 \end{pmatrix}$. The solution of the unperturbed equations is of the form

$$x_0(t) = e^{B(\omega_0) t} x_0(0), \qquad e^{B(\omega_0) t} = \begin{pmatrix} \cos \omega_0 t & \frac{1}{\omega_0} \sin \omega_0 t \\ -\omega_0 \sin \omega_0 t & \cos \omega_0 t \end{pmatrix},$$

and is thus periodic with period $2\pi/\omega_0$. Now let $\omega = 2\pi/T$ and take $h(x, t) = e^{-B(\omega) t} x$. We obtain that $\partial_t h + \partial_x h f(x) = e^{-B(\omega) t} [B(\omega_0) - B(\omega)] x$. Thus if ω is close to ω_0, that is, if $\omega^2 = \omega_0^2 + O(\varepsilon)$, then the expression above is indeed of order ε. More generally, taking $h(x, t) = e^{-B(\omega/k) t} x$ for some $k \in \mathbb{Z}$ allows to treat cases where $k^2 \omega^2 = k^2 \omega_0^2 + O(\varepsilon)$. This trick is known as the *Van der Pol transformation*.

We come back to (1.152). Assume that $g \in C^r(\mathcal{D} \times \mathbb{R}/T\mathbb{Z} \times [0, \varepsilon_0], \mathbb{R}^n)$, where $\mathcal{D}$ is a compact subset of $\mathbb{R}^n$ and $r \geq 2$. If M is an upper bound for $\|g\|$ in $\mathcal{D}$, then $\|x(t) - x(0)\| \leq \varepsilon M t$. Thus the trivial approximation $\dot{x} = 0$ of (1.152) is good to order ε for times of order 1 (this is, in fact, a partial case of Corollary 11 in the limit $K \to 0$). A better approximation is given by the averaged equation.

Definition 35 The *averaged system* associated with (1.152) is the autonomous equation

$$\dot{y}^0 = \varepsilon \langle g \rangle (y^0), \quad \langle g \rangle (x) = \frac{1}{T} \int_0^T g(x, t, 0) dt. \tag{1.153}$$

The averaging theorem shows that (1.153) is a good approximation of (1.152) up to time t of order $1/\varepsilon$.

Theorem 39 *There exists C^r change of coordinates $x = y + \varepsilon w(y, t)$, with $w(y, t)$ a T-periodic function of t, transforming (1.152) into*

$$\dot{y} = \varepsilon \langle g \rangle (y) + \varepsilon^2 g_1(y, t, \varepsilon), \tag{1.154}$$

with $g_1(y, t, \varepsilon)$ also T-periodic in t. Moreover,

• *if $x(t)$ and $y^0(t)$ are solutions of the original and averaged systems (1.152) and (1.153), respectively, with initial conditions $x(0) = y^0(0) + O(\varepsilon)$, then $x(t) = y^0(t) + O(\varepsilon)$ for t of order $1/\varepsilon$;*

• *if y^* is an equilibrium point of (1.153) such that $\frac{\partial}{\partial y} \langle g \rangle (y^*)$ has no zero eigenvalue, then (1.152) admits a T-periodic solution $\gamma_\varepsilon = y^* + O(\varepsilon)$ for sufficiently small ε; if, in addition, $\frac{\partial}{\partial y} \langle g \rangle (y^*)$ has no imaginary eigenvalue, then $\gamma_\varepsilon(t)$ has the same type of stability as y^* for sufficiently small ε.*

Proof. We build the change of variables explicitly. Split g into its averaged and oscillating parts:

$$g(x, t, \varepsilon) = \langle g \rangle (x) + g^0(x, t, \varepsilon), \quad \langle g^0 \rangle (x) = 0.$$

Inserting $x = y + \varepsilon w(y, t)$ into (1.152), we get

$$\dot{x} = (1 + \varepsilon \partial_y w) \dot{y} + \varepsilon \partial_y w(y, t) = \varepsilon \langle g \rangle (y + \varepsilon w) + \varepsilon g^0(y + \varepsilon w, t, \varepsilon),$$

and expanding $\dot{y}$ into Taylor series, we obtain $\dot{y} = \varepsilon[\langle g \rangle(y) + g^0(y, t, 0) - \partial_t w(y, t)] + O(\varepsilon)$, where the remainder of order ε^2 is a T-periodic function of t, the Taylor expansion of which can be computed if necessary. Now choosing $w(y, t) = \int_0^t g^0(y, s, 0)ds$ yields the desired form for $\dot{y}$. Indeed, w is T-periodic in t since g^0 has average zero. The difference $z(t) = y(t) - y^0(t)$ satisfies the equation $\dot{z} = \varepsilon[\langle g \rangle(y) - \langle g \rangle(y^0)] + \varepsilon^2 g_1(y, t, \varepsilon)$, and thus we can proceed as in the proof of Corollary 11. Since

$$z(t) = z(0) + \int_0^t \left\{ \varepsilon[\langle g \rangle(y(s)) - \langle g \rangle(y^0(s))] + \varepsilon^2 g_1(y(s), s, \varepsilon) \right\} ds,$$

taking K as a Lipschitz constant for $\langle g \rangle$ and M as a bound for $\|g_1\|$ (which must exist since $\mathcal{D}$ is compact), we arrive at the estimate

$$\|z(t)\| \leq \|z(0)\| + \int_0^t \varepsilon K \|z(s)\| ds + \varepsilon^2 M t,$$

and Gronwall's inequality provides us with the bound $\|z(t)\| \leq \|z(0)\| e^{\varepsilon K t} + \frac{\varepsilon M}{K}(e^{\varepsilon K t} - 1)$. Due to assumed $\|z(0)\| = O(\varepsilon)$, the error of the averaging approximation remains of order ε for $t \leq 1/(K\varepsilon)$.

Let φ_t denote the flow of the transformed equation (1.154), and let φ_t^0 denote the flow of the averaged equation (1.153) (φ^0 is obtained from φ by letting $g_1 = 0$, not by letting $\varepsilon = 0$). By Theorem 38, the flows may be Taylor-expanded in ε up to order ε^2. Setting $t = T$, we get the Poincaré maps for both systems

$$\begin{cases} \varphi_T^0(y) = P^0(y, \varepsilon) = y + \varepsilon P_1^0(y) + \varepsilon^2 P_2^0(y) \\ \varphi_T(y) = P(y, \varepsilon) = y + \varepsilon P_1(y) + \varepsilon^2 P_2(y). \end{cases}$$

By our estimate on $\|z(t)\|$, we must have $P_1(y) = P_1^0(y)$. If y^* is an equilibrium of (1.154), that is, if $\langle g \rangle(y^*) = 0$, it is a fixed point of P^0, and in particular $P_1^0(y^* = 0)$. Denote $\partial_y \langle g \rangle(y^*)$ by A, and obtain

$$\partial_y P^0(y^*, \varepsilon) = e^{\varepsilon A t} = 1 + \varepsilon A t + O(\varepsilon^2) \quad \Rightarrow \quad \partial_y P_1^0(y^*) = AT.$$

Now if we define the function $\Phi(y, \varepsilon) = P_1(y) + \varepsilon P_2(y, \varepsilon)$, the above relations imply that $\Phi(y^*, 0) = 0$ and $\partial_y \Phi(y^*, 0) = AT$. Thus if A has no vanishing eigenvalue, the implicit function theorem implies the existence, for small ε, of a fixed point $y^*(\varepsilon)$ of P. This fixed point corresponds to a periodic orbit γ_ε, which remains in an ε-neighborhood of y^* because of our estimate on $\|z\|$. Since $x = y + \varepsilon w(y, t)$, the corresponding x-coordinates of the periodic orbit are also ε-close to y^*. Finally, if A has no imaginary eigenvalue, then the linearization of P^0 at y^* has no eigenvalue on the unit circle, and this remains true, for sufficiently small ε, for the linearization of P at $y^*(\varepsilon)$. Thus the fixed points of P and P^0 have the same type of stability, and this property carries over to the periodic orbit. $\square$

The averaged approximation of a periodic system is considerably easier to study, since it does not depend on time. In order to find periodic orbits of the initial system, it is sufficient to find equilibrium points of the averaged system. The approximation is reliable on a larger time scale than the unperturbed system. It does not give any information on stability if the averaged system has an elliptic equilibrium.

Example 59 Consider a small perturbation of a two-degree-of-freedom integrable system, whose Hamiltonian depends only on one action:

$$H(\varphi_1, \varphi_2, I_1, I_2, \varepsilon) = H_0(I_1) + \varepsilon H_1(\varphi_1, \varphi_2, I_1, I_2, \varepsilon). \tag{1.155}$$

This situation arises, for instance, in the Kepler problem of a planet orbiting the Sun. If $\varepsilon = 0$, φ_1 rotates with constant speed $\Omega(I_1) = H_0'(I_1)$ as does the area swept by the planet in its orbit, while all other variables are fixed. If $\Omega(I_1) \neq 0$, the equations of motion may be written for sufficiently small ε as

$$\frac{d\varphi_2}{d\varphi_1} = \varepsilon \frac{\partial H_1/\partial I_2}{\Omega(I_1) + \varepsilon \partial H_1/\partial I_1}, \qquad \frac{dI_j}{d\varphi_1} = -\varepsilon \frac{\partial H_1/\partial \varphi_j}{\Omega(I_1) + \varepsilon \partial H_1/\partial I_1}, \qquad j = 1, 2. \quad (1.156)$$

The averaged version of this system is

$$
\begin{aligned}
d\varphi_2/d\varphi_1 &= \frac{\varepsilon}{\Omega(I_1)} \frac{1}{2\pi} \int_0^{2\pi} \partial H_1/\partial I_2(\varphi_1, \varphi_2, I_1, I_2, 0) d\varphi_1, \\
dI_2/d\varphi_1 &= -\frac{\varepsilon}{\Omega(I_1)} \frac{1}{2\pi} \int_0^{2\pi} \partial H_1/\partial \varphi_2(\varphi_1, \varphi_2, I_1, I_2, 0) d\varphi_1, \quad (1.157) \\
dI_1/d\varphi_1 &= -\frac{\varepsilon}{\Omega(I_1)} \frac{1}{2\pi} \int_0^{2\pi} \partial H_1/\partial \varphi_1(\varphi_1, \varphi_2, I_1, I_2, 0) d\varphi_1 = 0.
\end{aligned}
$$

Now we can make an important observation. Consider the *averaged Hamiltonian*

$$\langle H \rangle(\varphi_2, I_2; I_1, \varepsilon) = \frac{1}{2\pi} \int_0^{2\pi} H(\varphi_1, \varphi_2, I_1, I_2, \varepsilon) d\varphi_1 = H_0(I_1) + \varepsilon \langle H_1 \rangle(\varphi_2, I_2; I_1, \varepsilon). \quad (1.158)$$

The ordering of the variables in (1.158) is meant to be suggestive, since we will obtain a closed system of equations for φ_2, I_2. Indeed, the associated canonical equations are

$$
\begin{cases}
\dot\varphi_1 = \Omega(I_1) + \varepsilon \frac{\partial \langle H_1 \rangle}{\partial I_1}(\varphi_2, I_2; I_1, \varepsilon), & \dot I_1 = 0, \\
\dot\varphi_2 = \varepsilon \frac{\partial \langle H_1 \rangle}{\partial I_2}(\varphi_2, I_2; I_1, \varepsilon), & \dot I_2 = -\varepsilon \frac{\partial \langle H_1 \rangle}{\partial \varphi_2}(\varphi_2, I_2; I_1, \varepsilon).
\end{cases}
\quad (1.159)
$$

At first-order in ε, these equations are equivalent to (1.157). Conclude that for Hamiltonian systems of the form (1.156), averaging can be performed directly on the Hamiltonian, with respect to the "fast" variable φ_1. Note that I_1 and $\langle H \rangle$ are constants of the motion for the averaged system. By the averaging theorem, they vary at most by an amount of order ε during time intervals of order $1/\varepsilon$. Such quantities are called *adiabatic invariants*. Since the averaged Hamiltonian no longer depends on φ_1, the only dynamical variables are φ_2 and I_2, while I_1 plays the role of a parameter. The averaged Hamiltonian thus has one degree of freedom and is integrable. One can generalize the method to more degrees of freedom. In that case, the averaged system has one degree of freedom less than the original one. There also exist generalizations to systems with several fast variables.

The method of averaging of Hamiltonian systems with respect to fast variables is of major importance in applications. For instance, it allowed Laskar to integrate the motion of the Solar System over several hundred million years, while numerical integrations of the original system are only reliable on time spans of a few hundred to a few thousand years.

Iterative methods. The averaging procedure allows us to obtain a better approximation of the perturbed system, valid for longer times. A natural question is whether this method can be pushed further, in order to yield still better approximations. This is indeed possible to some extent.

Example 60 Consider ODE $\dot y = \varepsilon \langle g \rangle(y) + \varepsilon^2 g_1(y, t, \varepsilon)$ of the transformed system in Theorem 39. Assuming sufficient differentiability, and proceeding in the same way as when we constructed the change of variables $x = y + \varepsilon w(y, t)$, we can construct a new change of variables $y = y_2 + \varepsilon^2 w_2(y_2, t)$ such that

$$\dot y_2 = \varepsilon \langle g \rangle(y_2) + \varepsilon^2 \langle g_1 \rangle(y_2) + \varepsilon^3 g_2(y_2, t, \varepsilon).$$

The associated averaged system is $\dot{y}_2^0 = \varepsilon\langle g\rangle(y_2^0) + \varepsilon^2\langle g_1\rangle(y_2^0)$, and if K_2 and M_2 denote, respectively, a Lipschitz constant for $\langle g\rangle + \varepsilon\langle g_1\rangle$ and an upper bound on $\|g_2\|$, one obtains from Gronwall's inequality

$$\|y_2(t) - y_2^0(t)\| \leq (\|y_2(0) - y_2^0(0)\| + \varepsilon^2 M_2/K_2)e^{\varepsilon K_2 t}.$$

Thus for an appropriate initial condition, the error remains of order ε^2 for times smaller than $1/(K_2\varepsilon)$. This procedure can be repeated as long as the system is differentiable enough, and after $r - 1$ steps we get exact and approximated systems of the form

$$\begin{cases} \dot{y}_r = \varepsilon\langle g\rangle(y_r) + \cdots + \varepsilon^r\langle g_{r-1}\rangle(y_r) + \varepsilon^{r+1}g_r(y_r, t, \varepsilon), \\ \dot{y}_r^0 = \varepsilon\langle g\rangle(y_r^0) + \cdots + \varepsilon^r\langle g_{r-1}\rangle(y_r^0) + \varepsilon^{r+1}g_r(y_r, t, \varepsilon), \end{cases}$$

satisfying

$$\|y_r(t) - y_r^0(t)\| \leq (\|y_r(t) - y_r^0(t)\| + \varepsilon^r M_r/K_r)e^{\varepsilon K_r t}. \tag{1.160}$$

For sufficiently small $\|y_r(0) - y_r^0(0)\|$, this error remains of order ε^r for times of order $1/(K_r\varepsilon)$. One could thus wonder, if the initial system is analytic, whether this procedure can be repeated infinitely often, in such a way that the time-dependence is eliminated completely. This is not possible in general because the bounds M_r and K_r can grow rapidly with r (typically, like $r!$), so that the amplitude of the error (1.160) becomes large, while the time interval on which the approximation is valid shrinks to 0.

We are forced to conclude that the method of averaging provides very useful approximations on a given bounded time scale, up to some power in ε, but fails in general to describe the long-term dynamics, especially in the neighborhood of elliptic periodic orbits. We will return to this point in the next section.

Similar properties hold for Hamiltonian systems, for which these iterative methods were first developed. The main difference is that certain components of the averaged equations may vanish, e.g., (1.159), which implies that the averaged approximation may be valid for longer time intervals.

The method of matched asymptotic expansion. To describe the *method of matched asymptotic expansions*, we discuss the simple boundary value problem

$$\varepsilon\ddot{y} + \dot{y} + y(x) = 0, \tag{1.161a}$$

$$y(0) = \alpha, \quad y(1) = \beta. \tag{1.161b}$$

As $\varepsilon \to 0$, the equation (1.161a) reduces to

$$\dot{y} + y = 0, \tag{1.162}$$

which is a first-order equation that cannot satisfy in general both of the boundary conditions (1.161b). Hence one of these boundary conditions, $y(0) = \alpha$, must be dropped. This follows from its exact solution

$$y(x) = \frac{(\alpha e^{s_2 x} - \beta)e^{s_1 x} + (\beta - \alpha e^{s_1 x})e^{s_2 x}}{e^{s_2} - e^{s_1}}, \quad s_{1,2} = \frac{-1 \pm \sqrt{1 - 4\varepsilon}}{2\varepsilon}.$$

As $\varepsilon \to 0$ and for fixed $x \neq 0$

$$y(x) \to y^{out} := \beta e^{1-x}, \tag{1.163}$$

which is the solution of the reduced equation (1.162) subject to $y(1) = \beta$. The solution of the reduced equation (denoted by y^{out}) is called the *outer solution*. For small ε the solution of the reduced equation is close to the exact solution of (1.161a,b), except in a small interval at the end point $x = 0$ where the exact solution changes quickly in order to retrieve the boundary condition $y(0) = \alpha$ which is about to be lost. This small interval across which y changes very rapidly is called the *boundary layer* in fluid mechanics, the *edge layer* in solid mechanics and the *skin layer* in electrodynamics. To determine an expansion valid in the boundary layer, we magnify this layer using the *stretching transformation* $\xi = x/\varepsilon$. Under this transformation (1.161a) becomes $\frac{d^2y}{d\xi^2} + \frac{dy}{d\xi} + \varepsilon y = 0$, which for a fixed ξ and $\varepsilon \to 0$ reduces to $\frac{d^2y}{d\xi^2} + \frac{dy}{d\xi} = 0$. The general solution is

$$y(\xi) = A + Be^{-\xi},$$

where A and B are constants. Since this solution is valid in the boundary layer, it is valid at the origin; hence it must satisfy the boundary condition $y|_{x=0} = \alpha$. Since $\xi = 0$ corresponds to $x = 0$, we have $y|_{\xi=0} = \alpha$; hence $B = \alpha - A$, and thus $y(\xi) = A + (\alpha - A)e^{-\xi}$, which has an arbitrary constant A. Denote this solution by y^{inn} and call it the *inner solution* or *inner expansion*. To determine A, notice that $\lim_{x \to 0} y^{out}(x) = \beta e$. Moreover, from stretching transformation any small fixed x_0 corresponds to $\xi \to \infty$ as $\varepsilon \to 0$, and $\lim_{\xi \to \infty} y^{inn}(\xi) = A$. Thus these limits must represent the same value of y at a small $x = x_0 \neq 0$, hence $A = \beta e$. Hence

$$y^{inn} = \beta e + (\alpha - \beta e)e^{-\xi}. \tag{1.164}$$

In determining the outer and the inner expansions, we used two different limit processes – an *outer limit* defined by $y^{out} = \lim_{\varepsilon \to 0,\ x-\text{fixed}} y(x; \varepsilon)$ and *inner limit* defined by $y^{inn} = \lim_{\varepsilon \to 0,\ \xi-\text{fixed}} y(\varepsilon\xi; \varepsilon)$. The process of determining A is called *matching*, and we have used the *Van Dyke's matching principle*

$$\lim_{x \to 0} y^{out}(x; \varepsilon) = \lim_{\xi \to \infty} y^{inn}(\varepsilon\xi; \varepsilon), \tag{1.165}$$

which is equivalent to equating the inner limit of the outer solution denoted by $(y^{out})^{inn}$ to the outer limit of the inner solution $(y^{inn})^{out}$. An approximate solution to our original problem is given by (1.163) for x not near zero and by (1.164) near $x = 0$. To compute $y(x)$, one must switch from one solution to the other as x increases at some small value of x such as the value where the graphs of solutions y^{out} and y^{inn} may intersect. This switching is not convenient, and we form a single uniformly valid solution denoted y^{comp} and called the *composite solution*,

$$y^{comp} = y^{out} + y^{inn} - (y^{out})^{inn} = y^{out} + y^{inn} - (y^{inn})^{out}. \tag{1.166}$$

Since $((y^{out})^{inn})^{out} = (y^{out})^{inn} = (y^{inn})^{out} = ((y^{out})^{inn})^{inn}$, then we have

$$(y^{comp})^{out} = y^{out} + (y^{inn})^{out} - (y^{out})^{inn} = y^{out},$$

$$(y^{comp})^{inn} = (y^{out})^{inn} + y^{inn} - (y^{out})^{inn} = y^{inn}.$$

Thus the composite solution, which in our case looks like $y^{comp} = \beta e^{1-x} + (\alpha - \beta e)e^{-x/\varepsilon} + O(\varepsilon)$, is a good approximation in the outer region as the outer solution, and it is a good approximation in the inner region as the inner solution. This suggests that the composite solution is a uniform approximation over the whole interval of x including the gap between the outer and inner regions. The success of the matching may be due to the presence of an overlapping region in which both the outer and inner solutions are valid, hence there is no gap between the two regions.

Higher approximations and refined matching procedures. Determine higher approximations to the problem given by (1.161a,b). At first, we should start by determining the boundary conditions that must be dropped and determine the stretching transformation as a by-product. Then, we determine second-order inner and outer expansions and match them by using Van Dyke's matching principle. Finally, we form a uniformly valid composition.

If ε vanishes, (1.161a) reduces to the first-order equation (1.162), whose solution cannot satisfy both boundary conditions and, consequently, one of them must be dropped. At the end, where the boundary condition is dropped, y changes very rapidly from the solution of the reduced equation to the boundary value dropped at the end in a very small region called the *boundary layer* or *region of nonuniformity*. To determine if the boundary condition $y(1) = \beta$ must be dropped, we introduce the stretching transformation $\xi = (1-x)\varepsilon^{-\lambda}$, $\lambda > 0$ thereby transforming (1.161a) into

$$\varepsilon^{1-2\lambda}\frac{d^2y}{d\xi^2} - \varepsilon^{-\lambda}\frac{dy}{d\xi} + y(\xi) = 0. \qquad (1.167)$$

As $\varepsilon \to 0$, two limiting forms of (1.167) arise depending on the value λ.

Case $\lambda > 1$. $\frac{d^2y}{d\xi^2} = 0$ or $y^{inn} = A + B\xi$. Since the reduced equation (1.162) is assumed to be valid at $x = 0$,

$$y^{out} = \alpha e^{-x} \quad \text{(outer solution)}.$$

The matching principle (1.165) demands that $\lim_{\xi \to \infty} (A + B\xi) = \lim_{x \to 1} \alpha e^{-x}$, or $A = \alpha e^{-1}$ and $B = 0$. Hence $y^{inn} = \alpha e^{-1}$ (inner solution). Since this solution is valid at $x = 1$, it should satisfy the boundary condition $y|_{x=1} = \beta$, hence $\beta = \alpha e^{-1}$ which is not true in general. Therefore we discard this case because it does not permit the satisfaction of both boundary conditions.

Case $\lambda < 1$. $\frac{d^2y}{d\xi^2} - \frac{dy}{d\xi} = 0$ or $y^{inn} = A + Be^{\xi}$. Since the matching principle demands that $\lim_{\xi \to \infty} (A + Be^{\xi}) = \lim_{x \to 1} \alpha e^{-x}$, then $A = \alpha e^{-1}$ and $B = 0$. So this case must be discarded also because the application of the boundary condition $y|_{x=1} = \beta$ demands $\beta = \alpha e^{-1}$. Therefore the boundary layer cannot exist at $x = 1$, and hence the boundary condition $y|_{x=1} = \beta$ must not be dropped.

To investigate whether the boundary condition $y|_{x=0} = \alpha$ must be dropped, we introduce the stretching transformation $\xi = x\varepsilon^{-\lambda}$, $\lambda > 0$ thereby transforming (1.161a) into

$$\varepsilon^{1-2\lambda}\frac{d^2y}{d\xi^2} + \varepsilon^{-\lambda}\frac{dy}{d\xi} + y(\xi) = 0. \qquad (1.168)$$

In this case also three different limiting forms of (1.168) arise as $\varepsilon \to 0$ depending on λ.

Case $\lambda > 1$: $\frac{d^2y}{d\xi^2} = 0$ or $y^{inn} = A + B\xi$.

Case $\lambda < 1$: $\frac{dy}{d\xi} = 0$ or $y^{inn} = A$.

Case $\lambda = 1$: $\frac{d^2y}{d\xi^2} + \frac{dy}{d\xi} = 0$ or $y^{inn} = A + Be^{-\xi}$.

The first two cases must be discarded using arguments similar to those employed above. This leaves the third case which, by using $y(0) = \alpha$, leads to the inner solution $y^{inn} = A + (\alpha - A)e^{-\xi}$.

Since the matching principle demands $\lim_{\xi \to \infty} [A + (\alpha - A)e^{-\xi}] = \lim_{x \to 0} (\beta e^{1-x})$ or $A = \beta e$, then $y^{inn} = \beta e + (\alpha - \beta e)e^{-\xi}$. Therefore the boundary layer exists at $x = 0$, and the boundary condition $y(0) = \alpha$ cannot be imposed on the reduced equation (1.155). We have found as a by-product that the stretching transformation is $\xi = xe^{-1}$ as was chosen before, hence the region of nonuniformity is $x \sim O(\varepsilon)$.

Outer expansion. We seek an outer expansion for y in the form

$$y^{out}(x;\varepsilon) = \sum_{k=0}^{N-1} \varepsilon^k y_k(x) + O(\varepsilon^N) \tag{1.169}$$

using the *outer limit process* $\varepsilon \to 0$ keeping x to be fixed. Thus

$$y_0(x) = \lim_{\varepsilon \to 0,\ x\text{-fixed}} y(x;\varepsilon), \qquad y_m(x) = \lim_{\varepsilon \to 0,\ x\text{-fixed}} \frac{1}{\varepsilon^m}(y - \sum_{k=0}^{m-1} \varepsilon^k y_k(x)). \tag{1.170}$$

To determine this expansion we substitute (1.169) into (1.161a) and equate coefficients of powers of ε,

$$y_0' + y_0 = 0, \qquad y_k' + y_k = -y_{k-1}'', \quad k \geq 1. \tag{1.171}$$

As was mentioned in the previous section, this outer solution is valid everywhere except in the region $x \in O(\varepsilon)$, hence it must satisfy the boundary condition $y^{out}(1) = \beta$, which together with (1.169) yields $y_0(1) = \beta$, $y_k(1) = 0$, $k \geq 1$. The solution $(1.171)_1$ subject to $y_0(1) = \beta$ is $y_0 = \beta e^{1-x}$, while the solution of $(1.171)_2$ for $n = 1$ subject to $y_k(1) = 0$ is $y_1 = \beta(1-x)e^{1-x}$. Therefore

$$y^{out} = \beta[1 + \varepsilon(1-x)]e^{1-x} + O(\varepsilon^2). \tag{1.172}$$

Inner expansion. To determine an expansion valid near the origin, the stretching transformation $\xi = x\varepsilon^{-1}$ should be used to transform (1.161a) into

$$\frac{d^2 y^{inn}}{d\xi^2} + \frac{dy^{inn}}{d\xi} + \varepsilon y^{inn}(\xi) = 0 \tag{1.173}$$

and then be seek an inner expansion of y in the form

$$y^{inn}(\varepsilon\xi;\varepsilon) = \sum_{k=1}^{N-1} \varepsilon^k Y_k(\xi) + O(\varepsilon^N) \tag{1.174}$$

using the *inner limit process* $\varepsilon \to 0$ keeping $\xi = x/\varepsilon$ to be fixed. Thus

$$Y_0(\xi) = \lim_{\varepsilon \to 0,\ \xi\text{-fixed}} y(\varepsilon\xi;\varepsilon), \qquad Y_m(\xi) = \lim_{\varepsilon \to 0,\ \xi\text{-fixed}} \frac{1}{\varepsilon^m}[y(\varepsilon\xi;\varepsilon) - \sum_{k=0}^{m-1} \varepsilon^k Y_k(\xi)]. \tag{1.175}$$

To determine this expansion we substitute (1.174) into (1.173), equate coefficients of like power of ε, keeping in mind that ξ is the independent variable, and obtain

$$Y_0'' + Y_0' = 0, \qquad Y_k'' + Y_k' = -Y_{k-1}, \quad k \geq 1. \tag{1.176}$$

While this inner expansion satisfies the boundary condition at $x = 0$, it is not expected to satisfy in general the boundary condition at $x = 1$. Since $x = 0$ corresponds to $\xi = 0$, the boundary condition $y|_{x=0} = \alpha$ together with (1.174) gives $Y_0(0) = \alpha$, $Y_k(0) = 0$ for $k \geq 1$. The solution of $(1.176)_1$ subject to $Y_0(0) = \alpha$ is $Y_0 = \alpha - A_0(1 - e^{-\xi})$, while the solution of $(1.176)_2$ and subject to $Y_k(0) = 0$ for $k = 1$ is $Y_1 = A_1(1 - e^{-\xi}) - [\alpha - A_0(1 + e^{-\xi})]\xi$. Therefore

$$y^{inn} = \alpha - A_0(1 - e^{-\xi}) + \varepsilon\{A_1(1 - e^{-\xi}) - [\alpha - A_0(1 + e^{-\xi})]\xi\} + O(\varepsilon^2). \tag{1.177}$$

This inner expansion contains the arbitrary constants A_0 and A_1 which must be determined from matching with the outer expansion (1.172).

Refined matching procedures. The simplest possible form of matching the inner and outer expansion is $\lim_{x\to 0} y^{out} = \lim_{\xi\to\infty} y^{inn}$. This condition leads to the matching of the first terms in both the inner and the outer expansions, thereby giving $A_0 = \alpha - \beta e$. This matching principle cannot be used to match other than these first terms. In fact $\lim_{x\to 0} y^{out} = \beta e(1 + \varepsilon) + O(\varepsilon^2)$, while $\lim_{\xi\to\infty} y^{inn} = \alpha - A_0 + \varepsilon[A_1 - (\alpha - A_0)\xi] + O(\varepsilon^2)$. Since these last two expansions have to be equal according to the matching principle for all values of ξ:

$$A_0 = \alpha, \quad A_1 = \beta e(1/\varepsilon + 1),$$

which violates the assumption that $A_1 = O(1)$ which has been used in deriving (1.177).

More general forms of the matching condition are the following:

– The inner limit of the *outer limit* equals the outer limit of the *inner limit*.

– The inner expansion of the *outer expansion* equals the outer expansion of the *inner expansion* where these expansions are formed using the outer and the inner limit processes defined by (1.169), (1.170) and (1.174) through (1.175), respectively.

The higher order matching principle is

– The *m*-term inner expansion of the *k-term outer expansion* equals the *k*-term outer expansion of the *m-term inner expansion* where *m* and *n* are any two integers which may be equal or unequal.

Composite expansion. As was discussed, the outer expansion is not valid near the origin, while the inner expansion is not valid in general away from the region $x \in O(\varepsilon)$. To determine an expansion valid over the whole interval, a composite expansion y^{comp} should be formed according to (1.166). These two forms are equivalent since the matching principle requires $(y^{out})^{inn} = (y^{inn})^{out}$. Since $(y^{inn})^{inn} = y^{inn}$, $(y^{out})^{out} = y^{out}$, (1.166) implies that $(y^{comp})^{out} = y^{out}$, $(y^{comp})^{inn} = y^{inn}$. Therefore y^{comp} is a good approximation to $y(x)$ in the outer region as y^{out}, and it is also as good an approximation to $y(x)$ in the inner region as y^{inn}. Since $(y^{out})^{inn}$ is given by either $(y^{out})^{inn} = \beta e(1 + \varepsilon - \varepsilon\xi)$ or $(y^{inn})^{out} = (\alpha - A_0)(1 - x) + \varepsilon A_1$, then a composite expansion can be formed adding the outer expansion (1.172) and the inner expansion

$$y^{inn} = \beta e + (\alpha - \beta e)e^{-\xi} + \varepsilon\{\beta e(1 - e^{-\xi}) - [\beta e - (\alpha - \beta e)e^{-\xi}]\xi\} + O(\varepsilon^2)$$

and subtracting the inner expansion of the outer expansion $(y^{out})^{inn} = \beta e(1 + \varepsilon - \varepsilon\xi)$. This results in

$$y^{comp} = \beta[1 + \varepsilon(1 - x)]e^{1-x} + [(\alpha - \beta e)(1 + x) - \varepsilon\beta e]e^{-x/\varepsilon} + O(\varepsilon^2).$$

Second-order ODE with variable coefficients. Consider a first-order uniformly valid expansion for the solution of

$$\varepsilon\ddot{y} + a(x)\dot{y} + b(x)y = 0, \tag{1.178a}$$
$$y(0) = \alpha, \quad y(1) = \beta, \tag{1.178b}$$

where $\varepsilon \ll 1$ and $a(x)$, $b(x)$ are analytic functions in the interval $[0, 1]$. Equation (1.161a) can be obtained from (1.178a) by setting $a(x) = b(x) \equiv 1$. Indeed, if $\varepsilon = 0$, (1.178a) reduces to the first-order ODE

$$a(x)\dot{y} + b(x)y = 0,$$

whose solution cannot satisfy both boundary conditions; one of them must be dropped as a consequence.

As is shown below, the boundary condition that must be dropped depends on the sign of $a(x)$ in the interval $[0, 1]$. If $a(x) > 0$, the condition $y(0) = \alpha$ must be dropped, and an inner expansion near $x = 0$ must be developed and matched with the outer solution. If $a(x) < 0$, the condition $y(1) = \beta$ must be dropped, and an inner expansion near $x = 1$ must be obtained and matched with the outer expansion. If $a(x)$ changes sign in $[0, 1]$, y may change from oscillatory to exponentially growing or decaying across the zeros of $a(x)$. Such zeros are called *turning or transition points*.

To investigate whether $y(0) = \alpha$ must be dropped, i.e., the boundary layer is at the origin, we introduce again the stretching transformation $\xi = x\varepsilon^{-\lambda}$, $\lambda > 0$ to transform (1.178a) into

$$\varepsilon^{1-2\lambda}\frac{d^2y}{d\xi^2} + \varepsilon^{-\lambda}a(\varepsilon^\lambda\xi)\frac{dy}{d\xi} + b(\varepsilon^\lambda\xi)y(\xi) = 0. \tag{1.179}$$

As $\varepsilon \to 0$, (1.179) tends to

$$\frac{d^2y}{d\xi^2} = 0 \text{ if } \lambda > 1; \quad \frac{dy}{d\xi} = 0 \text{ if } \lambda < 1; \quad \frac{d^2y}{d\xi^2} + a(0)\frac{dy}{d\xi} = 0 \text{ if } \lambda = 1. \tag{1.180}$$

In order to be able to match the solutions of (1.180) with the outer solution as given by the reduced equation, we need the bounded solutions of (1.180) as $\xi \to \infty$. The bounded solutions of the first two cases are constants, hence they must be discarded because they lead to certain inconsistencies similar to those encountered in the previous section. Similarly, if $a(0) < 0$, the bounded solution of the last case (i.e., $\lambda = 1$) is a constant and must also be discarded. As a consequence, the boundary layer cannot exist at $x = 0$. If $a(0) > 0$, the general solution for $\lambda = 1$ is

$$y^{inn} = A + Be^{-a(0)\xi}, \tag{1.181}$$

which is bounded as $\xi \to \infty$ and contains two arbitrary constants. It is acceptable as an inner expansion because together with the solution of the reduced equation it can satisfy both boundary conditions.

To investigate whether $y(1) = \beta$ must be dropped, we introduce the stretching transformation $\eta = (1 - x)e^{-\lambda}$ ($\lambda > 0$) to transform (1.178a) into

$$\varepsilon^{1-2\lambda}\frac{d^2y}{d\eta^2} + \varepsilon^{-\lambda}a(1 - \varepsilon^\lambda\eta)\frac{dy}{d\eta} + b(1 - \varepsilon^\lambda\eta)y(\eta) = 0. \tag{1.182}$$

As $\varepsilon \to 0$, this equation tends to

$$\frac{d^2y}{d\eta^2} = 0 \text{ if } \lambda > 1; \quad \frac{dy}{d\eta} = 0 \text{ if } \lambda < 1; \quad \frac{d^2y}{d\eta^2} - a(1)\frac{dy}{d\eta} = 0 \text{ if } \lambda = 1. \tag{1.183}$$

The bounded solutions of (1.182) as $\eta \to \infty$ are constants if $\lambda \neq 1$ or $a(1) > 0$ when $\lambda = 1$, hence they must be discarded. If $a(1) < 0$ when $\lambda = 1$, the general solution of (1.183) is

$$y^{inn} = A + Be^{a(1)\eta},$$

which is bounded as $\eta \to \infty$. It is acceptable as an inner expansion because it contains two arbitrary constants which allow, when combined with the outer solution, to satisfy both boundary conditions.

In summary, the boundary layer exists at $x = 0$ if $a(x) > 0$, and $x = 1$ if $a(x) < 0$. Next we should determine a first-order uniformly valid expansion for the former case.

The case of $a(x) > 0$. In this case the first term of the outer expansion (1.169) is governed by

$$a(x)\dot{y} + b(x)y = 0, \quad y(1) = \beta,$$

which yields

$$y^{out} = \beta \exp\left[-\int_1^x b(t)/a(t)dt\right]. \tag{1.184}$$

The general solution for the first term of the inner expansion (1.174) is given by (1.181). Under transformation $\xi = x\varepsilon^{-\lambda}$ the boundary condition $y|_{x=0} = \alpha$ is transformed into $y|_{\xi=0} = \alpha$, hence

$$y^{inn} = \alpha - B + Be^{-\alpha(0)\xi}. \tag{1.185}$$

To match y^{out} and y^{inn}, we take $m = k = 1$ in the matching principle and proceed as follows. One-term inner expansion of *one-term outer expansion* is equal to $\beta \exp[-\int_1^0 \frac{b(t)}{a(t)}dt]$. One-term outer expansion of *one-term inner expansion* equals $\alpha - B$. By equating these two terms according to the matching principle, we have $B = \alpha - \beta \exp\left[-\int_1^0 b(t)/a(t)dt\right]$. Forming a composite expansion by adding (1.184) and (1.185) and subtracting $\alpha - B$ instead of one-term outer expansion of *one-term inner expansion,* we obtain

$$y^{comp} = \beta \exp[-\int_1^x b(t)/a(t)dt] + \left\{\alpha - \beta \exp[-\int_1^0 b(t)/a(t)dt]\right\} \exp[-\frac{a(0)x}{\varepsilon}] + O(\varepsilon).$$

Letting $a(x) = b(x) = 1$, we have $y^{comp} = \beta e^{1-x} + (\alpha - \beta)e^{-x/\varepsilon} + O(\varepsilon)$. Taking $a(x) = 2x+1$ and $b(x) = 2$, we obtain $y^{comp} = \frac{3\beta}{1+2x} + (\alpha - 3\beta)e^{-x/\varepsilon} + O(\varepsilon)$.

The straightforward expansion. Consider systems governed by equations having the form

$$\ddot{y} + f(y) = 0, \tag{1.186}$$

where, in general, $f(y)$ is a nonlinear function. It is convenient to shift the origin to location of the center, $y = y_0$, where $f(y_0) = 0$ and $f'(y_0) > 0$. Thus we let $x = y - y_0$, and (1.186) becomes $\ddot{x} + f(x + y_0) = 0$. Assuming f of class C^∞, we rewrite the above equation as

$$\ddot{x} + \sum_{k=1}^N \alpha_k x^k = 0, \quad \alpha_k = \frac{1}{k!}f^{(k)}(y_0). \tag{1.187}$$

The solution describes the response of the system to an initial disturbance. It is convenient to write the initial conditions (the initial position s_0 and the initial velocity v_0) in polar form using an amplitude and a phase according to

$$x(0) = s_0 \equiv a_0\cos\beta_0, \quad \dot{x}(0) = v_0 \equiv -a_0\omega_0\sin\beta_0, \tag{1.188}$$

where $\omega_0 = \sqrt{\alpha_1} = [f'(y_0)]^{1/2}$, $a_0 = [s_0^2 + (v_0/\omega_0)^2]^{1/2}$ and $\beta_0 = \cos^{-1}(s_0/\omega_0) = \sin^{-1}(-v_0/(a_0\omega_0))$. Assume that the solution of (1.187) can be represented by an expansion of the form

$$x(t;\varepsilon) = \varepsilon x_1(t) + \varepsilon^2 x_2(t) + \varepsilon^3 x_3(t) + \dots \tag{1.189}$$

Substituting (1.189) into (1.187) and setting the coefficients of each power of ε equal to zero, we obtain

$$\ddot{x}_1 + \omega_0^2 x_1 = 0 \quad (\text{order } \varepsilon) \tag{1.190a}$$
$$\ddot{x}_2 + \omega_0^2 x_2 = -\alpha_2 x_1^2 \quad (\text{order } \varepsilon^2) \tag{1.190b}$$
$$\ddot{x}_3 + \omega_0^2 x_3 = -2\alpha_2 x_1 x_2 - \alpha_3 x_1^3 \quad (\text{order } \varepsilon^3). \tag{1.190c}$$

In satisfying the initial conditions, there are the following alternatives:

1. One can substitute the expansion (1.189) into the initial condition (1.188) and collect coefficients of equal powers of ε. The result is

$$x_1(0) = a_0\cos\beta_0, \quad \dot{x}_1(0) = -\omega_0 a_0\sin\beta_0, \tag{1.191a}$$
$$x_k(0) = 0, \quad \dot{x}_k(0) = 0 \quad \text{for} \quad k \geq 2. \tag{1.191b}$$

Then one determines the constants of integration in x_1 such that (1.191a) is satisfied; and one includes the homogeneous solution in the expressions for x_k with $k \geq 2$, choosing the constants of integration such that (1.191b) is satisfied at each step.

2. One can ignore the initial conditions and the homogeneous solutions in all x_k ($k \geq 2$), until the last step. Then, considering the constants of integration in x_1 to be functions of ε, one expands the solution for x_1 in powers of ε and chooses such coefficients in the expansion that (1.188) is satisfied.

Initially it may appear that the second alternative is inconsistent because we stipulated that the x_k are independent of ε. As we demonstrate by an example, the two approaches are equivalent, yielding precisely the same result. We prefer the second approach because there is much less algebra involved and, in many cases, we are only concerned with steady-state responses, which frequently are independent of the initial conditions. The general solution of (1.190a) can be written in the form

$$x_1 = a\cos(\omega_0 t + \beta), \tag{1.192}$$

where a and β are constants. Following the first alternative, we let $a = a_0$ and $\beta = \beta_0$ in order to satisfy (1.191a). Following the second approach, we consider $a = a(\varepsilon)$ and $\beta = \beta(\varepsilon)$ to be functions of ε and at this point pay no regard to the initial conditions. Substituting x_1 of (1.192) into (1.191b) yields

$$\ddot{x}_2 + \omega_0^2 x_2 = -\alpha_2 a^2 \cos^2(\omega_0 t + \beta) = -\frac{1}{2}\alpha_2 a^2[1 + \cos(2\omega_0 t + 2\beta)],$$

where trigonometric identities were used to eliminate all products and powers of the cosines. This is a necessary step in all the subsequent perturbation methods. Due to discussion above, we have two choices for expressing x_2: either

$$x_2 = \frac{\alpha_2 a_0^2}{6\omega_0^2}[\cos(2\omega_0 t + 2\beta_0) - 3] + a_2 \cos(\omega_0 t + \beta_0) \tag{1.193a}$$

or

$$x_2 = \frac{\alpha_2 a_0^2}{6\omega_0^2}[\cos(2\omega_0 t + 2\beta_0) - 3], \tag{1.193b}$$

accordingly. Here a_2 and β_2 are additional constants of integration, independent of ε, chosen such that (1.191b) is satisfied. Thus, the first alternative gives

$$x = \varepsilon a_0 \cos(\omega_0 t + \beta_0) + \varepsilon^2\left\{\frac{a_0^2 \alpha_2}{6\omega_0^2}[\cos(2\omega_0 t + 2\beta_0) - 3] + a_2 \cos(\omega_0 t + \beta_0)\right\} + O(\varepsilon^3). \tag{1.194a}$$

Following the second alternative, we have

$$x = \varepsilon a \cos(\omega_0 t + \beta) + \varepsilon^2 \frac{a^2 \alpha_2}{6\omega_0^2}[\cos(2\omega_0 t + 2\beta) - 3] + O(\varepsilon^3). \tag{1.194b}$$

Now into (1.193a) we put $a = a(\varepsilon) = A_1 + \varepsilon A_2 + \dots$ and $b = b(\varepsilon) = B_0 + \varepsilon B + \dots$ Then

$$
\begin{aligned}
&\varepsilon a \cos(\omega_0 t + \beta)\\
&= (\varepsilon A_1 + \varepsilon^2 A_2 + \dots)\,[\cos(\omega_0 t + B_0)\cos(\varepsilon B_1 + \dots) - \sin(\omega_0 t + B_0)\cos(\varepsilon B_1 + \dots)]\\
&= \varepsilon A_1 \cos(\omega_0 t + \beta) + \varepsilon^2[A_2 \cos(\omega_0 t + \beta) - A_1 B_1 \sin(\omega_0 t + \beta)] + O(\varepsilon^3)\\
&= \varepsilon A_1 \cos(\omega_0 t + \beta) + \varepsilon^2(A_2^2 + A_1^2 B_1^2)^{1/2}\cos(\omega_0 t + \theta_2) + O(\varepsilon^3),
\end{aligned}
$$

where $\theta_2 = B_0 + \tan^{-1}(\frac{A_1 B_1}{A_2})$. Choose $A_1 = a_0$, $B_0 = \beta_0$, and A_2 and B_1 such that

$$(A_2^2 + A_1^2 B_1^2)^{1/2} = a_2, \quad \beta_0 + \tan^{-1}(A_1 B_1/A_2) = \beta_2.$$

Then (1.194a) and (1.194b) are equivalent. Thus either alternative can be used in the subsequent schemes, and either alternative can be used for higher-order approximations. Substituting (1.192) and (1.193b) into (1.190c) yields

$$\ddot{x}_3 + \omega_0^2 x_3 = \frac{\alpha_2^2 a^3}{3\omega_0^2} [3\cos(\omega_0 t + \beta) - \cos(\omega_0 t + \beta)\cos(2\omega_0 t + 2\beta)] - \alpha_3 a^3 \cos^3(\omega_0 t + \beta)$$

$$= \left(\frac{5\alpha_2^2}{6\omega_0^2} - \frac{3\alpha_3}{4}\right)a^3 \cos(\omega_0 t + \beta) - \left(\frac{\alpha_3}{4} + \frac{\alpha_2^2}{6\omega_0^2}\right)a^3 \cos(3\omega_0 t + 3\beta). \quad (1.195)$$

Any particular solution of (1.195) contains the term

$$\frac{10\alpha_2^2 - 9\alpha_3\omega_0^2}{24\omega_0^3} a^3 t \sin(\omega_0 t + \beta).$$

If the straightforward procedure is continued, the terms with $t^m \cos(\omega_0 t + \beta)$ and $t^m \sin(\omega_0 t + \beta)$ appear. Terms of this kind are called *secular terms*. Because of secular terms, (1.189) is not periodic. Moreover, x_3/x_1 and x_3/x_2 grow without bound as t increases. Thus x_3 does not always provide a small correction to x_1 and x_2. One says that expansion (1.189) is not uniformly valid as t increases.

Example 61 (Klein-Gordon equation) Consider nonlinear waves governed by

$$\partial_{tt}u - c^2\partial_{xx}u + \lambda^2 u = \varepsilon f(u, u_t, u_x). \quad (1.196)$$

If $\varepsilon = 0$, (1.196) admits solutions of the form

$$u = a\cos(k_0 x - \omega_0 t + \phi), \quad (1.197)$$

where a and ϕ are constants, and k_0 and ω_0 satisfy the dispersion relationship

$$\omega_0^2 = c^2 k_0^2 + \lambda^2. \quad (1.198)$$

For small but finite ε, we seek an expansion of the form

$$u = a\cos\psi + \varepsilon u_1(a, \psi) + \ldots, \quad (1.199)$$

where a is slowly varying function of both time and position according to

$$\partial_t a = \varepsilon A_1(a) + \varepsilon^2 A_2(a) + \ldots, \quad \partial_x a = \varepsilon B_1(a) + \varepsilon^2 B_2(a) + \ldots, \quad (1.200)$$

and ψ is a new phase variable which coincides with the phase of (1.197) for $\varepsilon = 0$,

$$\partial_t \psi = -\omega_0 + \varepsilon C_1(a) + \varepsilon^2 C_2(a) + \ldots, \quad \partial_x \psi = k_0 + \varepsilon D_1(a) + \varepsilon^2 D_2(a) + \ldots \quad (1.201)$$

In this case also, no u_k contains the fundamental $\cos\psi$.

Substituting (1.199) through (1.201)$_1$ into (1.196), using (1.198) and equating the coefficients of ε on both sides, we obtain

$$\lambda^2(\partial_{\psi\psi}^2 u_1 + u_1) = -2(\omega_0 A_1 + c^2 k_0 B_1)\sin\psi - 2a(\omega_0 C_1 + c^2 k_0 D_1)\cos\psi$$
$$+ f[a\cos\psi, a\omega_0\sin\psi, -ak_0\sin\psi]. \quad (1.202)$$

We will Fourier analyze f in terms of ψ as

$$f[a\cos\psi, a\omega_0\sin\psi, -ak_0\sin\psi] = g_0(a) + \sum_{m=1}^{\infty} [f_m(a)\sin m\psi + g_m(a)\cos m\psi].$$

Eliminating secular terms, we have

$$2\omega_0 A_1 + 2c^2 k_0 B_1 = f_1(a), \quad 2a(\omega_0 C_1 + c^2 k_0 D_1) = g_1(a). \tag{1.203}$$

Then the solution of (1.202) is

$$u_1 = \frac{g_0(a)}{\lambda^2} + \sum_{m=2}^{\infty} \frac{f_m(a)\sin m\psi + g_m(a)\cos m\psi}{\lambda^2(1-m^2)}.$$

Substituting for A_1, B_1, C_1 and D_1 from (1.200) through (1.201) into (1.203), we obtain

$$\partial_t a + \omega_0' \partial_x a = \varepsilon \frac{f_1(a)}{2\omega_0}, \quad \partial_t \beta + \omega_0' \partial_x \beta = \varepsilon \frac{g_1(a)}{2a\omega_0}, \tag{1.204}$$

where $\omega_0' = d\omega_0/dk_0$ the group velocity, and $\beta = \psi - k_0 x + \omega_0 t$. If $f_1 = 0$, then $a = h_1(x - \omega_0' t)$ and $\beta = \varepsilon(x + \omega_0' t)\frac{g_1(a)}{4a\omega_0\omega_0'} + h_2(x - \omega_0' t)$, where h_1 and h_2 are determined from the initial or boundary conditions. Equations (1.204) can be solved if a and β are functions of either time or position only.

The Lindstedt-Poincaré method. The idea is to introduce a new independent variable, say $\tau = \omega t$, where initially ω is an unspecified function of ε. The new governing equation will contain ω in the coefficient of the second derivative; this permits the frequency and amplitude to interact. One can choose the function ω in such a way as to eliminate the secular terms, i.e., to render the expansion periodic. Assume an expansion for ω:

$$\omega(\varepsilon) = \omega_0 + \varepsilon\omega_1 + \varepsilon^2\omega_2 + \dots, \tag{1.205}$$

where $\omega_1, \omega_2, \dots$ are unknown constants at this point. Let x be expanded in the form

$$x(t; \varepsilon) = \varepsilon x_1(\tau) + \varepsilon^2 x_2(\tau) + \varepsilon^3 x_3(\tau) + \dots, \tag{1.206}$$

where all x_j are independent of ε. Thus, the equation (1.187) becomes

$$(\omega_0 + \varepsilon\omega_1 + \dots)^2 \frac{d^2}{d\tau^2} (\varepsilon x_1 + \varepsilon^2 x_2 + \dots) + \sum_{k=1}^{N} \alpha_k (\varepsilon x_1 + \varepsilon^2 x_2 + \dots)^k = 0.$$

Equating the coefficients of like powers of ε to zero and recalling that $\alpha_1 = \omega_0^2$, we obtain

$$\frac{d^2 x_1}{d\tau^2} + x_1 = 0, \tag{1.207a}$$

$$\omega_0^2 \Big(\frac{d^2 x_2}{d\tau^2} + x_2\Big) = -2\omega_0\omega_1 \frac{d^2 x_1}{d\tau^2} - \alpha_2 x_1^2, \tag{1.207b}$$

$$\omega_0^2 \Big(\frac{d^2 x_3}{d\tau^2} + x_3\Big) = -2\omega_0\omega_1 \frac{d^2 x_2}{d\tau^2} - 2\alpha_2 x_1 x_2 - (\omega_1^2 + 2\omega_0\omega_2)\frac{d^2 x_1}{d\tau^2} - \alpha_3 x_1^3. \tag{1.207c}$$

We'll find the general solution of (1.207a) of the form $x_1 = a\cos\phi$, where $\phi = \tau + \beta$, with a and β being constants. Substituting $x_1 = a\cos\phi$ into (1.207b) leads to

$$\omega_0^2 \Big(\frac{d^2 x_2}{d\tau^2} + x_2\Big) = 2\omega_0\omega_1 a\cos\phi - \frac{1}{2}\alpha_2 a^2 (1 + \cos 2\phi). \tag{1.208}$$

Thus we must set $\omega_1 = 0$, or x_2 will contain the secular term $\omega_1\omega_0^{-1} a\tau\sin\phi$. Then disregarding the solution of the homogeneous equation, we write the solution of (1.208) as

$$x_2 = -\frac{\alpha_2 a^2}{2\omega_0^2}\Big(1 - \frac{1}{3}\cos 2\phi\Big). \tag{1.209}$$

Substituting for x_1 and x_2 into (1.207c) and recalling that $\omega_1 = 0$, we obtain

$$\omega_0^2 \left(\frac{d^2 x_3}{d\tau^2} + x_3 \right) = 2 \left(\omega_0 \omega_2 a - \frac{3}{8} \alpha_3 a^3 + \frac{5}{12} \frac{\alpha_2^2 a^3}{\omega_0^2} \right) \cos \phi - \frac{1}{4} \left(\frac{2\alpha_2^2}{3\omega_0^2} + \alpha_3 \right) a^3 \cos 3\phi.$$

To eliminate the secular term from x_3, we should put $\omega_2 = \frac{(9\alpha_3 \omega_0^2 - 10\alpha_2^2) a^2}{24\omega_0^3}$. Hence from $x = y - y_0$, (1.205), (1.206), $x_1 = a \cos \phi$, and (1.209), it follows

$$y = y_0 + \varepsilon a \cos(\omega t + \beta) - \varepsilon^2 \frac{a^2 \alpha_2}{2\alpha_1} \left[1 - \frac{1}{3} \cos(2\omega t + 2\beta) \right] + O(\varepsilon^3), \tag{1.210}$$

where

$$\omega = \sqrt{\alpha_1} \left(1 + \frac{9\alpha_3 \alpha_1 - 10\alpha_2^2}{24\alpha_1^2} \varepsilon^2 a^2 \right) + O(\varepsilon^3). \tag{1.211}$$

Carrying out the expansion to high order is cumbersome. One seldom has the courage to go beyond third-order unless the algebraic manipulations are performed by computer. Consequently, a recurrence algorithm was developed (e.g., in [33, 39]) by which the Fourier coefficients of solutions and the frequency corrections may be calculated rather than solving for the individual y_n.

Imposing the initial conditions (1.188), we have

$$\begin{cases} a_0 \cos \beta_0 = \varepsilon a \cos \beta - \varepsilon^2 \frac{a^2 \alpha_2}{2\alpha_1} \left(1 - \frac{1}{3} \cos 2\beta \right), \\ -\omega_0 a_0 \sin \beta_0 = -\varepsilon a \omega \sin \beta - \varepsilon^2 \frac{a^2 \alpha_2 \omega}{3\alpha_1} \sin 2\beta. \end{cases} \tag{1.212}$$

To solve (1.212), we should expand a and β in powers of ε and equate coefficients of like powers of ε. The result is $\varepsilon a = a_0 + \frac{\alpha_2 a_0^2}{12\alpha_1} (3 \cos \beta_0 + \cos 3\beta_0)$ and $\beta = \beta_0 - \frac{\alpha_2 a_0}{12\alpha_1} (9 \sin \beta_0 + \sin 3\beta_0)$. The solution is in agreement with the solution that can be derived by including the homogeneous solution in x_2 and satisfying the initial conditions at each level of approximation.

The Lindstedt-Poincaré procedure produced

(i) a periodic expression describing the motion of the system,

(ii) a frequency-amplitude relationship which is a direct consequence of requiring the expression to be periodic,

(iii) higher harmonic in the higher-order terms of the expression, and

(iv) a *drift* or *steady-streaming* term $-\frac{1}{2} \varepsilon^2 a^2 \alpha_2 \alpha_1^{-1}$.

Example 62 (The Duffing equation) The *Duffing equation* is an equation with small parameter ε at cubic nonlinearity for u. It has a form $\ddot{u} + u + \varepsilon u^3 = 0$. We apply Lindstedt-Poincaré technique to the Duffing equation. Under the transformation $t = s(1 + \varepsilon \omega_1 + \varepsilon^2 \omega_2 + \ldots)$ it becomes

$$\frac{d^2 u}{ds^2} + (1 + \varepsilon \omega_1 + \varepsilon^2 \omega_2 + \ldots)^2 (u + \varepsilon u^3) = 0. \tag{1.213}$$

Letting

$$u = \sum_{k=0}^{\infty} \varepsilon^k u_k(t), \tag{1.214}$$

and substituting (1.214) into (1.213), and then after equating coefficients of like powers of ε, we obtain

$$\begin{cases} \frac{d^2 u_0}{ds^2} + u_0 = 0, \\ \frac{d^2 u_1}{ds^2} + u_1 = -u_0^3 - 2\omega_1 u_0, \\ \frac{d^2 u_2}{ds^2} + u_2 = -3u_0^2 u_1 - 2\omega_1 (u_1 + u_0^3) - (\omega_1^2 + 2\omega_2) u_0. \end{cases} \tag{1.215}$$

The general solution of the first equation of the system (1.215) is

$$u_0 = a \cos(s + \varphi), \tag{1.216}$$

where a and φ are constants of integration. With (1.216), the second equation of (1.215) becomes

$$\frac{d^2 u_1}{ds^2} + u_1 = -\frac{1}{4}a^3 \cos 3(s + \varphi) - \left(\frac{3}{4}a^2 + 2\omega_1\right)a \cos(s + \varphi). \tag{1.217}$$

If a straightforward perturbation expansion is used, $\omega_k = 0$, i.e., $t = s$, then (1.217) reduces to

$$\frac{d^2 u_1}{ds^2} + u_1 = -\frac{1}{4}a^3(\cos 3s + 3 \cos s),$$

whose particular solution (for the initial value problem with trivial initial conditions and $\varphi = 0$)

$$u_1 = -\frac{3}{8}a^3 s \sin s + \frac{a^3}{32}(\cos 3s - \cos s)$$

contains a secular term. In order to avoid this secular term, ω_1 is chosen to eliminate the coefficient of $\cos(s + \varphi)$ on the right-hand side of (1.217). This condition determines ω_1 to be $\omega_1 = -\frac{3}{8}a^2$. Then the solution of (1.217) becomes $u_1 = \frac{1}{32}a^3 \cos 3(s + \varphi)$. Substituting for u_0, u_1 and ω_1 into the third equation of (1.215), we obtain

$$\frac{d^2 u_2}{ds^2} + u_2 = \left(\frac{51}{128}a^4 - 2\omega_2\right)a \cos(s + \varphi) + \text{NST},$$

where NST stands for terms that do not produce secular terms. Secular terms are eliminated if $\omega_2 = \frac{51}{256}a^4$. Therefore, $u = a \cos(\omega t + \varphi) + \varepsilon \frac{a^3}{32} \cos 3(\omega t + \varphi) + O(\varepsilon^2)$, where a and φ are constants of integration, and

$$\omega = \left(1 - \frac{3}{8}a^2 \varepsilon + \frac{51}{256}a^4 \varepsilon^2 + \ldots\right)^{-1} = 1 + \frac{3}{8}a^2 \varepsilon - \frac{51}{256}a^4 \varepsilon^2 + O(\varepsilon^3).$$

Example 63 (The Mathieu's equation) The Mathieu's equation may be an example for the studied perturbation methods to be applied. This has a form

$$\ddot{u} + (\delta + \varepsilon \cos 2t)u = 0. \tag{1.218}$$

The Mathieu's equation is a special case of Hill's equation which is a linear ODE with periodic coefficients. Equations similar to this appear in many problems in applied mathematics such as stability of a transverse column subjected to a periodic longitudinal load, stability of periodic solutions of nonlinear conservative systems, electromagnetic wave propagation in a medium with periodic structure, the lunar motion, and the excitation of certain electrical systems.

The qualitative nature of solutions of (1.218) can be described by using Floquet theory (see, for example, [17]). This equation has normal solutions of the form $u = e^{\gamma t}\phi(t)$, where ϕ is a periodic function with a period π or 2π, and γ may be real or complex depending on the values of the parameters δ and ε. Floquet theory shows that the transition curves in the $\delta - \varepsilon$ plane, separating stable from unstable solutions, correspond to periodic solutions of (1.218). Some of these curves are determined below by expanding both δ and $u(t, \varepsilon)$ as functions of ε. Thus we let

$$\delta = n^2 + \varepsilon\delta_1 + \varepsilon^2\delta_2 + \ldots, \qquad u(t) = u_0 + \varepsilon u_1 + \varepsilon^2 u_2 + \ldots, \tag{1.219}$$

where n is a nonnegative integer, and u_m/u_0 is bounded for all m in order that (1.219)$_1$ be a uniformly valid asymptotic expansion. Substituting (1.219) into (1.218), expanding, and equating coefficients of equal powers of ε, we obtain

$$\ddot{u}_0 + n^2 u_0 = 0, \tag{1.220a}$$

$$\ddot{u}_1 + n^2 u_1 = -(\delta_1 + \cos 2t)u_0, \tag{1.220b}$$

$$\ddot{u}_2 + n^2 u_2 = -(\delta_1 + \cos 2t)u_1 - \delta_2 u_0. \tag{1.220c}$$

The solution of (1.220a) is $u_0 = \begin{cases} \cos nt \\ \sin nt \end{cases} \quad n = 0, 1, 2, \ldots$

Next we determine the higher approximations for the cases $n = 0, 1$ and 2.

Case $n = 0$. In this case $u_0 = 1$, and (1.220a) becomes $\ddot{u}_1 = -\delta_1 - \cos 2t$. In order for (1.220b) to be uniformly valid asymptotic expansion, δ_1 must vanish, and $u_1 = \frac{1}{4} \cos 2t + c$ where c is a constant. With u_0 and u_1 known (1.220c) becomes $u_2 = -\delta_2 - \frac{1}{8} - c \cos 2t - \frac{1}{8} \cos 4t$. In order for u_2/u_0 to be bounded, δ_2 must be equal to $-\frac{1}{8}$, hence $\delta = -\frac{1}{8}\varepsilon^2 + O(\varepsilon^3)$.

Case $n = 1$. In this case $u_0 = \cos t$ or $\sin t$. Taking $u_0 = \cos t$, we find that (1.220b) becomes $\ddot{u}_1 + u_1 = -\left(\delta_1 + \frac{1}{2}\right) \cos t - \frac{1}{2} \cos 3t$. In order u_1/u_0 be bounded, δ_1 must be equal $-\frac{1}{2}$, and then $u_1 = \frac{1}{16} \cos 3t$. Equation (1.220c) then becomes $\ddot{u}_2 + u_2 = -\left(\frac{1}{32} + \delta_2\right) \cos t + \frac{1}{32} \cos 3t - \frac{1}{32} \cos 5t$. The condition that u_2/u_0 be bounded demands that $\delta_2 = -\frac{1}{32}$, hence $\delta = 1 - \frac{1}{2}\varepsilon - \frac{1}{32}\varepsilon^2 + O(\varepsilon^3)$. Notice that using $u_0 = \sin t$, we obtain the transition curve $\delta = 1 + \frac{1}{2}\varepsilon - \frac{1}{32}\varepsilon^2 + O(\varepsilon^3)$.

The method of multiple scales. The uniformly valid expansion given by (1.210) and (1.211) may be viewed as a function of two independent variables rather than a function of one. Namely we may regard x to be a function of t and $\varepsilon^2 t$. The underlying idea of the method of multiple scales is to consider the expansion representing the response to be a function of multiple independent variables, or scales, instead of a single variable. The method of multiple scales, though a little more involved, has advantages over the Lindstedt-Poincaré method. For example, it can treat damped systems conveniently.

Introduce new independent variables according to $T_k = \varepsilon^k t$ ($k = 0, 1, \ldots, m$). Assume that the solution of (1.187) can be represented by an expansion of the form

$$x(t; \varepsilon) = \varepsilon x_1(T_0, \ldots, T_m) + \varepsilon^2 x_2(T_0, \ldots, T_m) + \ldots + O(\varepsilon T_m). \tag{1.221}$$

Notice that the number of independent time scales should depend on the order to which the expansion is carried out. If the expansion is carried out to $O(\varepsilon^2)$, then T_0 and T_1 are needed. In this section we carry out the expansion to $O(\varepsilon^3)$, and hence we need T_0, T_1 and T_2. The derivatives with respect to t become expansions in terms of the partial derivatives with respect to T_k due to

$$\begin{cases} \frac{d}{dt} = \frac{dT_0}{dt}\partial_{T_0} + \frac{dT_1}{dt}\partial_{T_1} + \cdots = D_0 + \varepsilon D_1 + \cdots + \varepsilon^m D_m, \\ \frac{d^2}{dt^2} = D_0^2 + 2\varepsilon D_0 D_1 + \varepsilon^2 (D_1^2 + 2D_0 D_2) + \ldots \end{cases} \tag{1.222}$$

Substituting (1.221) and (1.222) into (1.187) and equating the coefficients of $\varepsilon, \varepsilon^2$, and ε^3 to zero, we get

$$D_0^2 x_1 + \omega_0^2 x_1 = 0, \tag{1.223a}$$

$$D_0^2 x_2 + \omega_0^2 x_2 = -2D_0 D_1 x_1 - \alpha_2 x_1^2, \tag{1.223b}$$

$$D_0^2 x_3 + \omega_0^2 x_3 = -2D_0 D_1 x_2 - (D_1^2 + 2D_0 D_2)x_1 - 2\alpha_2 x_1 x_2 - \alpha_3 x_1^3, \tag{1.223c}$$

where $\omega_0^2 = f'(y_0)$. It will be convenient to write the solution of (1.223a) in the form

$$x_1 = A \exp(i\omega_0 T_0) + \bar{A} \exp(-i\omega_0 T_0), \tag{1.224}$$

where $A = A(T_1, T_2)$ is an unknown complex function and $\bar{A}$ is the complex conjugate of A. The governing equations for A are obtained by requiring x_2 and x_3 to be periodic in T_0. Substituting (1.224) into (1.223b) leads to

$$D_0^2 x_2 + \omega_0^2 x_2 = -2i\omega_0 D_1 A \exp(i\omega_0 T_0) - \alpha_2 [A^2 \exp(2i\omega_0 T_0) + A\bar{A}] + c.c., \qquad (1.225)$$

where *c.c.* denotes the complex conjugate of the preceding terms. Any particular solution of (1.225) has a secular term containing the factor $T_0 \exp(i\omega_0 T_0)$ unless $D_1 A = 0$. Therefore A must be independent of T_1. With $D_1 A = 0$, the solution of (1.225) is

$$x_2 = \frac{\alpha_2 A^2}{3\omega_0^2} \exp(2i\omega_0 T_0) - \frac{\alpha_2}{\omega_0^2} A\bar{A} + c.c., \qquad (1.226)$$

where the solution of the homogeneous equation is not needed. Substituting for (1.224) and (1.226) for x_1 and x_2 into (1.223c) and recalling that $D_1 A = 0$, we obtain

$$D_0^2 x_3 + \omega_0^2 x_3 = -\left(2i\omega_0 D_2 A - \frac{10\alpha_2^2 - 9\alpha_3\omega_0^2}{3\omega_0^2} A^2 \bar{A}\right) \exp(i\omega_0 T_0)$$
$$-\frac{3\alpha_3\omega_0^2 + 2\alpha_2^2}{3\omega_0^2} A^3 \exp(3i\omega_0 T_0) + c.c.$$

To eliminate the secular term from x_3, we must put

$$2i\omega_0 D_2 A + \frac{9\alpha_3\omega_0^2 - 10\alpha_2^2}{3\omega_0^2} A^2 \bar{A} = 0. \qquad (1.227)$$

Solving equations having the form of (1.227), we shall write A in the polar form $A = \frac{1}{2}ae^{i\beta}$, where a and β are real functions of T_2. Substituting this A into (1.227) and separating the result into real and imaginary parts, we obtain

$$\omega a' = 0, \quad \omega_0 a\beta' + \frac{10\alpha_2^2 - 9\alpha_3\omega_0^2}{24\omega_0^2} a^3 = 0,$$

where $'$ denotes the derivative with respect to T_2. It follows that a is a constant and hence that $\beta = \frac{9\alpha_3\omega_0^2 - 10\alpha_2^2}{24\omega_0^3} a^2 T_2 + \beta_0$, where $\beta_0 = $ const. Returning to $A = \frac{1}{2}ae^{i\beta}$ and assuming $T_2 = \varepsilon^2 t$, we find

$$A = \frac{1}{2} a \exp\left(i \frac{9\alpha_3\omega_0^2 - 10\alpha_2^2}{24\omega_0^3} \varepsilon^2 a^2 t + i\beta_0\right). \qquad (1.228)$$

Substituting (1.224) and (1.226) accordingly for x_1 and x_2, and using (1.228), we get

$$x = \varepsilon a \cos(\omega t + \beta_0) - \varepsilon^2 \frac{a^2\alpha_2}{2\alpha_1}\left[1 - \frac{1}{3}\cos(2\omega t + 2\beta_0)\right] + O(\varepsilon^3),$$

where $\omega = \sqrt{\alpha_1}\left[1 + \frac{9\alpha_3\alpha_1 - 10\alpha_2^2}{24\alpha_1^2} \varepsilon^2 a^2\right] + O(\varepsilon^3)$ is in agreement with the solution (1.210) and (1.211), obtained by the Lindstedt-Poincaré procedure.

Example 64 (The Duffing equation, continue Example 62) Apply the derivative-expansion method (method of multiple scaling) to the Duffing equation

$$\ddot{u} + \omega_0^2 u + \varepsilon u^3 = 0. \qquad (1.229)$$

Assume that

$$u = \sum_{k=0}^{2} \varepsilon^k u_k(T_0, T_1, T_2) + O(\varepsilon^3), \qquad (1.230)$$

and denote

$$\frac{d}{dt} = D_0 + \varepsilon D_1 + \varepsilon^2 D_2 + \ldots, \qquad D_k = \partial_{T_k}. \tag{1.231}$$

Substituting (1.230) and (1.231) into (1.229) and equating coefficients of each power of ε to zero, we get

$$\begin{cases} D_0^2 u_0 + \omega_0^2 u_0 = 0, \\ D_0^2 u_1 + \omega_0^2 u_1 = -2D_0 D_1 u_0 - u_0^3, \\ D_0^2 u_2 + \omega_0^2 u_2 = -2D_0 D_1 u_1 - (2D_0 D_2 + D_1^2) u_0 - 3u_0^2 u_1. \end{cases} \tag{1.232}$$

The solution of the first equation of the system (1.232) is $u_0 = A(T_1, T_2)e^{i\omega_0 T_0} + \bar{A}(T_1, T_2)e^{-i\omega_0 T_0}$. The second equation of (1.232) then becomes

$$D_0^2 u_1 + \omega_0^2 u_1 = -(2i\omega_0 D_1 A + 3A^2 \bar{A})e^{i\omega_0 T_0} - A^3 e^{3i\omega_0 T_0} + c.c.$$

In order that u_1/u_0 be bounded for all T_0, terms that produce secular terms must be eliminated. Hence

$$2\,i\,\omega_0 D_1 A + 3A^2 \bar{A} = 0, \tag{1.233}$$

and the solution becomes $u_1 = B(T_1, T_2)e^{i\omega_0 T_0} + \frac{A^3}{8\omega_0^2}e^{3i\omega_0 T_0} + c.c.$ To solve (1.233), let $A = \frac{1}{2}ae^{i\varphi}$ with real a and φ, separate real and imaginary parts, and obtain $\partial_{T_1} a = 0$ and $-\omega_0 \partial_{T_1} \varphi + \frac{3a^2}{8} = 0$. Hence

$$a = a(T_2), \qquad \varphi = \frac{3}{8\omega_0}a^2 T_1 + \varphi_0(T_2). \tag{1.234}$$

Substituting for u_0 and u_1 into the third equation of (1.232) yields

$$D_0^2 u_2 + \omega_0^2 u_2 = \left[\frac{21}{8\omega_0^2}A^4 \bar{A} - 3BA^2\right]e^{3i\omega_0 T_0} - \frac{3}{8\omega_0^2}A^5 e^{5i\omega_0 T_0} - Q(T_1, T_2)e^{i\omega_0 T_0} + c.c.,$$

where $Q = 2i\omega_0 D_1 B + 3A^2 \bar{B} + 6A\bar{A}B + 2i\omega_0 D_2 A - \frac{15A^3 \bar{A}^2}{8\omega_0^2}$. Secular terms are eliminated if

$$B = 0, \qquad 2i\omega_0 D_2 A = \frac{15A^3 \bar{A}^2}{8\omega_0^2}. \tag{1.235}$$

With $Q = 0$, the solution is $u_2 = \frac{A^5}{64\omega_0^4}e^{5i\omega_0 T_0} - \frac{21A^4 \bar{A}}{64\omega_0^4}e^{3i\omega_0 T_0} + c.c.$ disregarding the homogeneous solution. Letting $A = \frac{1}{2}ae^{i\varphi}$ in (1.235)$_2$ and separating real and imaginary parts, we obtain

$$\partial_{T_2} a = 0, \qquad -\omega_0 \partial_{T_2} \varphi = \frac{15}{256\omega_0^2}a^4. \tag{1.236}$$

Equations (1.234) and (1.236) lead to a being a constant, hence $\varphi_0 = -\frac{15}{256\omega_0^3}a^4 T_2 + \chi$, with χ being a constant. Therefore, $\varphi = \frac{3}{8\omega_0}a^2 T_1 - \frac{15}{256\omega_0^3}a^4 T_2 + \chi$. Substituting u_0, u_1 and u_2 into (1.230), keeping in mind that $A = \frac{1}{2}a\exp(i\varphi)$ and expressing the result in terms of t, we obtain

$$u = a\cos(\omega t + \chi) + \varepsilon\frac{a^3}{32\omega_0^2}\left(1 - \varepsilon\frac{21a^2}{32\omega_0^2}\right)\cos 3(\omega t + \chi) + \varepsilon^2\frac{a^5}{1024\omega_0^4}\cos 5(\omega t + \chi) + O(\varepsilon^3) \tag{1.237}$$

with $\omega = \omega_0 + \varepsilon\frac{3a^2}{8\omega_0} - \varepsilon^2\frac{15a^4}{256\omega_0^3} + O(\varepsilon^3)$. In last terms of (1.237), ω_0 is replaced by ω with an error $O(\varepsilon^3)$.

Example 65 (The Mathieu's equation, continue Example 63) Due to the Floquet theory (see, for example, [17]) of linear ODEs with periodic coefficients, the $\delta - \varepsilon$ plane is divided into regions of stability and instability which are separated by transition curves along which u is periodic with a period of either π or 2π. In this example we find not only the transition curves but also the solutions, hence the degree of stability or instability. Let $\delta = \omega_0^2$ with positive ω_0, and assume that

$$u = u_0(T_0, T_1, T_2) + \varepsilon u_1(T_0, T_1, T_2) + \varepsilon^2 u_2(T_0, T_1, T_2) + \dots \qquad (1.238)$$

Different cases have to be distinguished depending on whether ω_0 is near or far away from an integer n.

Solution for ω_0 "far away" from an integer n. Express $\cos(2t)$ in terms of the time scale T_0 as $\cos(2T_0)$. Substituting (1.238) into (1.218) and equating the coefficients of ε^0, ε^1 and ε^2 to zero, we get

$$D_0^2 u_0 + \omega_0^2 u_0 = 0, \qquad (1.239a)$$
$$D_0^2 u_1 + \omega_0^2 u_1 = -2D_0 D_1 u_0 - u_0 \cos 2T_0, \qquad (1.239b)$$
$$D_0^2 u_2 + \omega_0^2 u_2 = -2D_0 D_1 u_1 - (D_1^2 + 2D_0 D_2)u_0 - u_1 \cos 2T_0. \qquad (1.239c)$$

Solution of (1.239a) is $u_0 = A(T_1, T_2)e^{i\omega_0 T_0} + \bar{A}(T_1, T_2)e^{-i\omega_0 T_0}$. Substituting for u_0 into (1.239b) yields

$$D_0^2 u_1 + \omega_0^2 u_1 = -2i\omega_0 D_1 A e^{i\omega_0 T_0} - \frac{1}{2}A e^{i(\omega_0+2)T_0} - \frac{1}{2}A e^{i(\omega_0-2)T_0} + c.c.$$

Since ω_0 is far away from 1, secular term is eliminated if $D_1 A = 0$ or $A = A(T_2)$. Then solution u_1 is

$$u_1 = \frac{1}{8(\omega_0+1)}A e^{i(\omega_0+2)T_0} - \frac{1}{8(\omega_0-1)}A e^{i(\omega_0-2)T_0} + c.c.$$

Substituting for u_0 and u_1 into (1.239c) yields

$$D_0^2 u_2 + \omega_0^2 u_2 = -2\left[i\omega_0 D_2 A - \frac{A}{16(\omega_0^2-1)}\right]e^{i\omega_0 T_0} - \frac{A}{16(\omega_0+1)}A e^{i(\omega_0+4)T_0}$$
$$+ \frac{A}{16(\omega_0-1)}A e^{i(\omega_0-4)T_0} + c.c. \qquad (1.240)$$

Since ω_0 is different from 1 or 2, secular terms will be eliminated if

$$D_2 A = -\frac{i}{16\omega_0(\omega_0^2 - 1)}A. \qquad (1.241)$$

If we let $A = \frac{1}{2}a \exp i\phi$ and separate real and imaginary parts, then obtain

$$\frac{da}{dT_2} = 0, \qquad \frac{d\phi}{dT_2} = -\frac{1}{16\omega_0(\omega_0^2 - 1)}.$$

Thus, $a = \text{const}$, and $\phi = -[16\omega_0(\omega_0^2 - 1)]^{-1}T_2 + \phi_0$ with $\phi_0 = \text{const}$. With (1.241) the solution of (1.240) is

$$u_2 = \frac{A e^{i(\omega_0+4)T_0}}{128(\omega_0+1)(\omega_0+2)} - \frac{A e^{i(\omega_0-4)T_0}}{128(\omega_0-1)(\omega_0-2)} + c.c.$$

Summarizing, to $O(\varepsilon^2)$, the solution for u is

$$u = a\cos(\omega t + \phi_0) + \frac{\varepsilon a}{8}\left(\frac{\cos\left[(\omega+2)t + \phi_0\right]}{\omega_0 + 1} - \frac{\cos\left[(\omega - 2)t + \phi_0\right]}{\omega_0 - 1}\right)$$

$$+ \frac{\varepsilon^2 a}{128}\left(\frac{\cos\left[(\omega + 4)t + \phi_0\right]}{(\omega_0 + 1)(\omega_0 + 2)} + \frac{\cos[(\omega - 4)t + \phi_0]}{(\omega_0 - 1)(\omega_0 - 2)}\right) + O(\varepsilon^3),$$

where $\omega = \omega_0 - \frac{\varepsilon^2}{16\omega_0(\omega_0^2 - 1)} + O(\varepsilon^3)$. This expansion is valid only when ω_0 is away from 1 and 2. As $\omega_0 \to 1$ or 2, $u \to \infty$. An expansion that is valid near $\omega_0 = 1$ is obtained in what follows.

Solution for ω_0 near 1. In this case we let

$$\delta = 1 + \varepsilon\delta_1 + \varepsilon^2\delta_2 + \dots, \tag{1.242}$$

with $\delta_1 = O(1)$ and $\delta_2 = O(1)$. Equation (1.242) modifies (1.239a-c) into

$$D_0^2 u_0 + \omega_0^2 u_0 = 0, \tag{1.243a}$$
$$D_0^2 u_1 + \omega_0^2 u_1 = -2D_0 D_1 u_0 - \delta_1 u_0 - u_0\cos 2T_0, \tag{1.243b}$$
$$D_0^2 u_2 + \omega_0^2 u_2 = -2D_0 D_1 u_1 - (D_1^2 + 2D_0 D_2)u_0 - \delta_2 u_0 - u_1\cos 2T_0. \tag{1.243c}$$

The solution of (1.243a) is $u_0 = A(T_1, T_2)e^{iT_0} + \bar{A}(T_1, T_2)e^{-iT_0}$. Substituting for u_0 into (1.243b) gives

$$D_0^2 u_1 + \omega_0^2 u_1 = -\left(2iD_1 A - \delta_1 A - \frac{1}{2}\bar{A}\right)e^{iT_0} - \frac{1}{2}Ae^{3iT_0} + c.c. \tag{1.244}$$

Secular terms with respect to the time scale T_0 will be eliminated if

$$D_1 A = \frac{i}{2}\left(\delta_1 A + \frac{1}{2}\bar{A}\right). \tag{1.245}$$

Then the solution of (1.244) is $u_1 = \frac{1}{16}(Ae^{3iT_0} + \bar{A}e^{-3iT_0})$. To solve (1.245), we assume that $A = A_{\text{Re}} + iA_{\text{Im}}$, where A_{Re} and A_{Im} are real, and separate real and imaginary parts to obtain $\frac{\partial A_{\text{Re}}}{\partial T_1} = \frac{1}{2}\left(\frac{1}{2} - \delta_1\right)A_{\text{Im}}$ and $\frac{\partial A_{\text{Im}}}{\partial T_1} = \frac{1}{2}\left(\frac{1}{2} + \delta_1\right)A_{\text{Re}}$. The solution of these equations is

$$A_{\text{Re}} = a_1(T_2)e^{\gamma_1 T_1} + a_2(T_2)e^{-\gamma_1 T_1}, \tag{1.246a}$$

$$A_{\text{Im}} = \frac{2\gamma_1}{1/2 - \delta_1}[a_1(T_2)e^{\gamma_1 T_1} + a_2(T_2)e^{-\gamma_1 T_1}], \tag{1.246b}$$

where $\gamma_1^2 = \frac{1}{4}\left(\frac{1}{4} - \delta_1^2\right)$. Here a_1 and a_2 are real valued functions of the time scale T_2, but to first approximation, a_1 and a_2 are constant. Equations (1.246a-b) show that A grows exponentially with T_1 (i.e., with t) if γ_1 is real or $|\delta_1| < \frac{1}{2}$, and A oscillates with T_1 if γ_1 is imaginary or $|\delta_1| \geq \frac{1}{2}$ (in this case the solution is written in terms of $\cos\gamma_1 T_1$ and $\sin\gamma_1 T_1$ to keep A_{Re} and A_{Im} to be real). Hence the boundaries (transition curves) that separate the stable from the unstable domains, emanating from $\delta = 1$, $\varepsilon = 0$, are given to a first approximation by $\delta_1 = \pm\frac{1}{2}$ or $\delta = 1 \pm \frac{1}{2}\varepsilon + O(\varepsilon^2)$. To determine a second approximation to u and the transition curves, we substitute u_0 and u_1 into (1.243c) and obtain

$$D_0^2 u_2 + u_2 = -[2iD_2 A + D_1^2 A + (\delta_2 + 1/32)A]e^{iT_0} + c.c. + \text{NST}.$$

The condition that must be satisfied for excluding secular terms is

$$2iD_2 A + D_1^2 A + (\delta_2 + 1/32)A = 0. \tag{1.247}$$

Since $A = A_{\mathrm{Re}} + iA_{\mathrm{Im}}$, (1.247) gives the following equations for A_{Re} and A_{Im} upon separation of real and imaginary parts $2\frac{\partial A_{\mathrm{Re}}}{\partial T_2} + \alpha A_{\mathrm{Im}} = 0$ and $-2\frac{\partial A_{\mathrm{Im}}}{\partial T_1} + \alpha A_{\mathrm{Re}} = 0$, where $\alpha = \gamma_1^2 + \delta_2 + \frac{1}{32}$. Replacing A_{Re} and A_{Im} by their expressions from (1.246a,b) and equating the coefficients of $\exp(\pm\gamma_1 T_1)$ to zero because they are functions of T_2, we obtain

$$2\frac{\partial a_1}{\partial T_2} + \frac{2\gamma_1}{1/2 - \delta_1}\alpha a_1 = 0, \quad -\frac{4\gamma_1}{1/2 - \delta_1}\frac{\partial a_1}{\partial T_2} + \alpha a_1 = 0,$$

$$2\frac{\partial a_2}{\partial T_2} + \frac{2\gamma_1}{1/2 - \delta_1}\alpha a_2 = 0, \quad \frac{4\gamma_1}{1/2 - \delta_1}\frac{\partial a_2}{\partial T_2} + \alpha a_2 = 0.$$

These equations lead to $\frac{da_1}{dT_2} = \frac{da_2}{dT_2} = 0$, or $a_1 = \mathrm{const}$ and $a_2 = \mathrm{const}$, and $\alpha = 0$ or $\delta_2 = -\gamma_1^2 - \frac{1}{32}$.

1.6.2 Singular perturbation theory

Slow manifolds. In this section, we will consider so-called *slow-fast systems* of the form

$$\varepsilon\dot{x} = f(x, y), \qquad \dot{y} = g(x, y), \tag{1.248}$$

where $x \in \mathbb{R}^n$, $y \in \mathbb{R}^m$, f and g are of class C^r, $r \geq 2$, in some domain $\mathcal{D} \subset \mathbb{R}^n \times \mathbb{R}^m$, and ε is a small parameter. The variable x is called *fast variable* and y is called *slow variable*, because $\dot{x}/\dot{y}$ can be of order ε^{-1}. Equation (1.248) is an example of a *singularly perturbed system*, because in the limit $\varepsilon \to 0$, it does not reduce to a differential equation of the same type, but to an algebraic-differential system

$$0 = f(x, y), \qquad \dot{y} = g(x, y). \tag{1.249}$$

Since this limiting system is not an ODE, it is not clear how solutions of (1.248) might be expanded into powers of ε, for instance. The set of points $(x, y) \in \mathcal{D}$ for which $f(x, y) = 0$ is called the *slow manifold*.

Assume that this manifold can be represented by an equation $x = x^*(y)$, then the dynamics of (1.249) on the slow manifold is governed by the equation

$$\dot{y} = g(x^*(y), y) \equiv G(y), \tag{1.250}$$

which is easier to solve than the original equation (1.248).

One of the questions in singular perturbation theory is: *What is the relation between solutions of the singularly perturbed system (1.248) and those of the reduced system (1.250)?* There is another way to study the singular limit $\varepsilon \to 0$. For the *fast time* $s = t/\varepsilon$, (1.248) can be written as

$$\dot{x} = f(x, y), \quad \dot{y} = \varepsilon g(x, y). \tag{1.251}$$

Taking the limit $\varepsilon \to 0$, we obtain the so-called *associated system*

$$dx/ds = f(x, y), \quad y = \mathrm{const}, \tag{1.252}$$

in which y plays the role of a parameter. The perturbed system in the form (1.251) can thus be considered as a modification of the associated system (1.252) in which the "parameter" y changes slowly in time. Note that the slow manifold $x = x^*(y)$ defines equilibrium points of the associated system.

Results from regular perturbation theory show that orbits of (1.251) will remain close to those of (1.252) for s of order 1, i.e., for t of order ε, but not necessarily for larger t. Our aim is to describe the dynamics for times t of order 1. We start by giving sufficient conditions

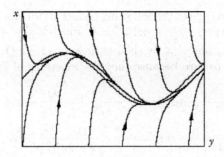

FIGURE 1.58: The heavy curve represents the slow manifold (1.254), while the light curves represent solutions of the slow/fast equation (1.253) for various initial conditions, and $\varepsilon = 0.07$. The solutions are quickly attracted by a small neighborhood of order ε of the slow manifold.

under which (1.250) and (1.252) indeed provide good approximations to different phases of the motion. Then we will discuss some cases in which these conditions are violated, and new phenomena occur.

In order to give an idea of what to expect in the general case, first discuss a solvable example.

Example 66 Consider the equation

$$\varepsilon \dot{x} = -x + \sin y, \quad \dot{y} = 1. \tag{1.253}$$

Then the slow manifold (Fig. 1.58) is given by the curve

$$x = x^*(y) = \sin y, \tag{1.254}$$

and the dynamics reduced to the slow manifold to be described by

$$x(t) = \sin y(t), \quad y(t) = y_0 + t. \tag{1.255}$$

The associated system for $s = t/\varepsilon$ is $\dot{x} = -x + \sin y$, $y = \text{const}$, and admits the solution

$$x(s) = (x_0 + \sin y)e^{-s} + \sin y. \tag{1.256}$$

Thus all solutions converge to the slow manifold $x = \sin y$ as $s \to \infty$.

We compare the approximate solutions to the solution of the original system (1.253) which can, in this case, be computed by the method of variation of the constant. It can be written in the form

$$x(t) = [x_0 - \bar{x}(0, \varepsilon)]e^{-t/\varepsilon}, \quad \bar{x}(t, \varepsilon) = \frac{\sin(y_0 + t) - \varepsilon \cos(y_0 + t)}{1 + \varepsilon^2},$$
$$y(t) = y_0 + t. \tag{1.257}$$

For small t, we have $\bar{x}(t, \varepsilon) = \sin(y_0) + O_1(\varepsilon) + O_2(t)$ so that $x(t) = (x_0 - \sin y_0)e^{-t/\varepsilon} + \sin y_0 + O_1(\varepsilon) + O_2(t)$, which is well approximated by the solution (1.256) of the associated system. As time t grows, the drift of y makes itself felt and the associated system no longer provides a good approximation. On the other hand, the factor $e^{-t/\varepsilon}$ decreases very fast. For $t = k\varepsilon|\ln \varepsilon|$, for instance, it is equal to ε^k, and for larger times, it goes to zero faster than any power of ε. The solution (1.257) thus becomes very close to $x(t) = \frac{\sin y - \varepsilon \cos y}{1 + \varepsilon^2}$, $y(t) = y_0 + t$, which lies at a distance of order ε from the slow manifold $x = \sin y$. Thus (1.255) reduced on the slow manifold provides a good approximation of (1.253) for $t \geq \varepsilon|\log \varepsilon|$. The A.Tikhonov's theorem gives sufficient conditions for this behavior for general systems: solutions are attracted, in a time of order $\varepsilon|\log \varepsilon|$, to a small neighborhood of order ε of a slow manifold.

The A.N. Tikhonov's theorem. Consider the slow-fast system (1.248) with $f \in C^r(\mathcal{D}, \mathbb{R}^n)$, $g \in C^r(\mathcal{D}, \mathbb{R}^m)$, $r \geq 2$. First, give conditions, which the slow manifold, defined implicitly by $f(x, y) = 0$, can be represented in the explicit form $x = x^*(y)$. Introduce the $n \times n$ matrix $A_x(y) = \partial_x f(x, y)$. Let $f(x_0, y_0) = 0$. The implicit function theorem states that if $\det A(y_0) \neq 0$, then there exists a neighborhood $\mathcal{N} = O(x_0, y_0)$ such that all solutions of $f(x, y) = 0$ in $\mathcal{N}$ can be written as $x = x^*(y)$. Here x^* is a function of class C^r, defined in a neighborhood $\mathcal{N}_0 = O(y_0) \subset \mathbb{R}^m$, and it satisfies

$$\partial_y x^*(y) = -[A_{x^*(y)}(y)]^{-1} \partial_y f(x^*(y), y). \tag{1.258}$$

We will abbreviate $A_{x^*(y)}(y)$ by $A(y)$. From now on we will always assume that (x, y) belongs to the open set $\mathcal{N}$, and that

1. $\det A(y)$ is bounded away from 0 whenever $(x^*(y), y) \in \mathcal{N}$;
2. the norms of f, g and all their mixed partial derivatives up to order 2 are bounded uniformly in $\mathcal{N}$.

We state a result describing how well the slow-fast system (1.248) is approximated by its limit when $\varepsilon \to 0$. The first versions of this result were proved independently by A.N. Tikhonov and I.S. Gradstein.

Theorem 40 *Let the eigenvalues $a_j(y)$ of $A(y)$ obey $\operatorname{Re} a_j(y) \leq -a_0$ ($y \in \mathcal{N}_0$, $j = 1, \ldots, n$) for some constant $a_0 > 0$. Let f and g (as well as their derivatives up to order 2) be uniformly bounded in $\mathcal{N}$. Then there exist constants $\varepsilon_0, c_0, c_1, c_2, c_3, c_4$, and $K_0 > 0$ such that the following hold for $0 < \varepsilon < \varepsilon_0$: any solution of (1.248) with the initial condition $(x_0, y_0) \in \mathcal{N}$ such that $\|x_0 - x^*(y_0)\| < c_0$ satisfies*

$$\|x(t) - x^*(y(t))\| \leq c_1 \varepsilon + c_2 \|x_0 - x^*(y_0)\| e^{-K_0 t / \varepsilon}. \tag{1.259}$$

Let $y^0(t)$ be the solution of the reduced system

$$\dot{y}^0 = G(y^0) \equiv g(x^*(y_0), y_0) \tag{1.260}$$

with initial condition $y^0(0) = y_0$. Let K_1 be a Lipschitz constant for G and $\mathcal{N}_0$. Then

$$\|y(t) - y^0(t)\| \leq c_3 \varepsilon e^{K_1 t} + c_4 \|x_0 - x^*(y_0)\| e^{-K_0 t / \varepsilon} \tag{1.261}$$

for all $t \geq 0$ such that $(x(s), y(s)) \in \mathcal{N}$ for $0 \leq s \leq t$. Roughly speaking, (1.259) implies

$$x(t) = x^*(y(t)) + O(\varepsilon), \quad t \geq \frac{1}{K_0} \varepsilon |\log \varepsilon|,$$

and thus $x(t)$ reaches a neighborhood of order ε of the slow manifold in a time of order $\varepsilon |\log \varepsilon|$, and stays there as long as this manifold is attracting. To determine $x(t)$ more precisely, we need to approximate $y(t)$ as well. This is done by (1.261), which implies

$$\begin{cases} y(t) = y^0 + O(\varepsilon), \\ x(t) = x^*(y^0(t)) + O(\varepsilon), \end{cases} \quad \text{for } \frac{1}{K_0} \varepsilon |\log \varepsilon| \leq t \leq \frac{1}{K_1}. \text{ Thus, if we can solve the reduced}$$

equation (1.260), then we can compute x and y up to errors of order ε and for times of order 1.

Proof. We will give a proof in the case $n = 1$, and comment on larger n at the end of the proof. For $n = 1$, $A(y) = a(y)$ is a scalar and $a(y) \leq a_0$ for all y.

Change of variables. The deviation of general solutions of (1.248) from the slow manifold $x = x^*(y)$ is characterized by the variable $z = x - x^*(y)$, which satisfies the equation

$$\varepsilon \dot{x} = \varepsilon \frac{d}{dt}(x - x^*(y)) = f(x^*(y) + z, y) - \varepsilon \partial_y x^*(y) g(x^*(y) + z, y).$$

The variables (y, z) belong to $\mathcal{N}' = \{(y, z) : (x^*(y) + z, y) \in \mathcal{N}\}$. We'll expand f in Taylor series to second order in z, and write the system in the form $\varepsilon \dot{z} = a(y)z + b(y, z) + \varepsilon w(y, z)$, $\dot{y} = g(x^*(y) + z, y)$. The nonlinear form $b(y, z) = f(x^*(y) + z, y) - a(y)z$ satisfies, by Taylor's formula and our assumption on the derivatives: $\|b(y, z)\| \leq M\|z\|^2$, $(y, z) \in \mathcal{N}'$ for some constant $M > 0$. The *drift term* is given by

$$w(y, z) = -\partial_y x^*(y) g(x^*(y) + z, y) = -[a(y)]^{-1} \partial_y f(x^*(y), y) g(x^*(y) + z, y);$$

see (1.258). Our assumption on f, g and a imply the existence of a constant $W > 0$ such that $\|w(y, z)\| \leq W$, $(\forall (y, z) \in \mathcal{N}')$. If w were absent, the slow manifold $z = 0$ would be invariant. The drift term acts like an inertial force, and may push solutions away from the slow manifold.

Proof of (1.259). The above bounds on a, b and w imply that

$$\begin{cases} \varepsilon \dot{z} \leq -a_0 z + M z^2 + \varepsilon W, & z \geq 0, \\ \varepsilon \dot{z} \geq -a_0 z - M z^2 - \varepsilon W, & z \leq 0, \end{cases}$$

whenever $(y, z) \in \mathcal{N}$. Thus if $\varphi(t) = |z(t)|$, we have $\varepsilon \dot{\varphi}(t) \leq -a_0 \varphi(t) + M \varphi^2(t) + \varepsilon W$. The function $\varphi(t)$ is not differentiable at those t for which $\varphi = 0$, but it is left and right differentiable at these points, so that this small inconvenience will not cause any problems. Now we take $c_0 = \frac{a_0}{2M}$, $c_1 = \frac{2W}{a_0}$, $\varepsilon_0 = \frac{a_0}{2Mc_1}$, so that $\varphi(0) = |z(0)| < \frac{a_0}{2M}$ and $c_1 \varepsilon < \frac{a_0}{2M}$ under hypothesis of the theorem. Define

$$\tau = \inf \left\{ t > 0 : \varphi(t) = \frac{a_0}{2M}, \text{ or } (y(t), z(t)) \notin \mathcal{N}' \right\}$$

with the convention that $\tau = \infty$ if the set on the right-hand side is empty. By continuity of solutions, this definition implies that $\tau < \infty \Rightarrow \varphi(\tau) = \frac{a_0}{2M}$ or $(y(\tau), z(\tau)) \in \partial \mathcal{N}'$. For $0 \leq t \leq \tau$, we have $\varepsilon \dot{\varphi}(t) \leq -\frac{a_0}{2} \varphi(t) + \varepsilon W$. We cannot apply Lemma 5 directly, because the coefficient of $\varphi(t)$ is negative. We can bound the difference between $\varphi(t)$ and the value we would obtain if this relation were to be an equality. Define a function $\psi(t)$ by $\varphi(t) = \frac{2\varepsilon}{a_0} W + \left(\varphi(0) - \frac{2\varepsilon}{a_0} W + \psi(t) \right) \exp(-a_0 t / 2\varepsilon)$. Differentiating with respect to time and using the inequality for $\dot{\varphi}$, we obtain that $\dot{\psi}(t) \leq 0$. Since $\psi(0) = 0$, we have

$$\varphi(t) \leq \frac{2\varepsilon}{a_0} W + \left(\varphi(0) - \frac{2\varepsilon}{a_0} W \right) \exp(-a_0 t / 2\varepsilon), \quad \forall t \in [0, \tau].$$

Our definition of c_0, c_1, ε_0 implies that $\varphi(t) < \frac{a_0}{2M}$ for $0 \leq t \leq \tau$. If $\tau < \infty$, we have $\varphi(\tau) < \frac{a_0}{2M}$ and thus necessarily $(y(\tau), z(\tau)) \in \partial \mathcal{N}'$. In other words, the above upper bound for $\varphi(t)$ holds for all t such that $(y(s), z(s)) \in \mathcal{N}'$ for $s \in [0, t]$. Since $\varphi(t) = \|x(t) - x^*(y(t))\|$, we have proved (1.259).

Proof of (1.261). We want to compare solutions of the equations

$$\dot{y} = g(x^*(y) + z, y) \equiv G(y) + z R(y, z), \qquad \dot{y}^0 = g(x^*(y^0), y^0) \equiv G(y^0).$$

Taylor's formula shows that $\|R(y, z)\| \leq M'$ for some M' uniformly in $\mathcal{N}'$. If $\eta(t) = y(t) - y^0(t)$, we find $\dot{\eta} = G(y^0 + \eta) - G(y^0) + z R(y^0 + \eta, z)$. Integrating from time 0 to t, using K_1 as a Lipschitz constant for G and (1.259), we obtain $|\eta(t)| \leq K_1 \int_0^t |\eta(s)| ds +$

$M'\big(c_1\varepsilon + c_2\|x_0 - x^*(y_0)\|e^{-K_0 t/\varepsilon}\big)$. Now the bound (1.261) follows from Gronwall's inequality and a short computation.

The case $n > 1$. In order to reduce the problem to a one-dimensional one, it is possible to use a so-called *Lyapunov function* $V : \mathbb{R}^n \to \mathbb{R}$, which decreases along orbits of the associated system. A possibility is to use a quadratic form $V(y, z) = z \cdot \mathbf{Q}(y)z$, for a suitable symmetric positive definite matrix $\mathbf{Q}(y)$, depending on $A(y)$. Then the function $\varphi(t) = V(y(t), z(t))^{1/2}$ obeys a similar equation as in the case $n = 1$. $\qquad\square$

Averaging. Another case of "potentially discontinuous" dependence on parameters is covered by the following averaging result.

Theorem 41 *Let $f : \mathbb{R}^n \times \mathbb{R}^2 \to \mathbb{R}^n$ be a continuous function which is τ-periodic with respect to its second argument t, and continuously differentiable with respect to its first argument. Let $\bar{x}_0 \in \mathbb{R}^n$ be such that $f(\bar{x}_0, t, \varepsilon) = 0$ for all t, ε. For $\bar{x} \in \mathbb{R}^n$ define $\bar{f}(\bar{x}, \varepsilon) = \int_0^\tau f(\bar{x}, t, \varepsilon)\,\mathrm{d}\tau$. If $\bar{f}'_x|_{x=0,\varepsilon=0}$ is a Hurwitz matrix, then, for small $\varepsilon > 0$, the equilibrium $x \equiv 0$ of the system*

$$\dot{x}(t) = \varepsilon f(x, t, \varepsilon) \tag{1.262}$$

is exponentially stable.

Proof. Consider the function $S : \mathbb{R}^n \times \mathbb{R}^2 \to \mathbb{R}^n$ which maps (x_0, ε) to $x(\tau)$, where $x(\cdot)$ is a solution of (1.262). It is sufficient to show that the derivative (Jacobian) $\dot{S}(\bar{x}, \varepsilon)$ of S with respect to its first argument, evaluated at $\bar{x}_0$ and $\varepsilon > 0$ sufficiently small, is a Schur matrix. Note first that, according to the rules on differentiating w.r.t. initial conditions, $\dot{S}(\bar{x}_0, \varepsilon) = \Delta(\tau, \varepsilon)$, where

$$\partial_t \Delta(t, \varepsilon) = \varepsilon f'_x(0, t, 0)\Delta(t, \varepsilon), \quad \Delta(0, \varepsilon) = \mathrm{id}.$$

Consider $\bar{D}(t, \varepsilon)$ defined by

$$\partial_t \bar{D}(t, \varepsilon) = \varepsilon f'_x(0, t, 0)\bar{\Delta}(t, \varepsilon), \quad \bar{\Delta}(0, \varepsilon) = \mathrm{id}.$$

Let $\delta(t)$ be the derivative of $\bar{\Delta}(t, \varepsilon)$ with respect to ε at $\varepsilon = 0$. Due to the rule for differentiating solutions of ODE with respect to parameters,

$$\delta(t) = \int_0^t f'_x(0, t_1, 0)\,\mathrm{d}t_1.$$

Hence $\delta(t) = \bar{f}'_x|_{x=0=z}$ holds by assumption of a Hurwitz matrix. On the other hand, $\Delta(\tau, \varepsilon) - \bar{\Delta}(\tau, \varepsilon) = o(\varepsilon)$. Combining this with $\bar{\Delta}(\tau, \varepsilon) = \mathrm{id} + \delta(\tau)\varepsilon + o(\varepsilon)$ yields

$$\Delta(\tau, \varepsilon) = \mathrm{id} + \varepsilon\delta(\tau) + o(\varepsilon).$$

Since $\delta(\tau)$ is a Hurwitz matrix, this implies that all eigenvalues of $\Delta(\tau, \varepsilon)$ have absolute value strictly less than one for all sufficiently small $\varepsilon > 0$. $\qquad\square$

Remark 7 Though the parameter dependence in Theorem 41 is continuous, the question asked is about the behavior at $t = \infty$, which makes system behavior for $\varepsilon = 0$ not a valid indicator of what will occur for sufficiently small $\varepsilon > 0$. Indeed, for $\varepsilon = 0$ the equilibrium $\bar{x}_0$ is not asymptotically stable.

Iterations and asymptotic series. In the proof of Theorem 40, we have used the fact that the original system (1.248) is transformed, by the translation $\bar{x} = x^*(y) + z$, into the system

$$\varepsilon\dot{z} = f(x^*(y) + z, y) - \varepsilon\partial_y x^*(y)g(x^*(y) + z, y), \quad \dot{y} = g(x^*(y) + z, y).$$

Write this system in the following way:

$$\varepsilon\dot{z} = A(y)z + b_0(y, z, \varepsilon) + \varepsilon w_0(y), \quad \dot{y} = g_0(y, z), \qquad (1.263)$$

where $A(y) = \partial_x f(x^*(y), y)$, and $w_0(y) = -\partial_y x^*(y)g(x^*(y), y)$ contains the terms which do not vanish at $z = 0$. The term b_0 contains all remainders of the Taylor expansion of f (at order 2) and g (at order 1), and is bounded in norm by a constant times $|z|^2 + \varepsilon|z|$.

We know that $z(t)$ is asymptotically of order ε. The reason why z does not go to zero is that the drift term $\varepsilon w_0(y)$ may push z away from zero. If we manage to increase the order of the drift term, then z may become smaller and we will have obtained a better approximation of solutions of (1.248).

We could use the fact that (1.263) is again a slow-fast system, where $\dot{z}$ vanishes on a slow manifold $z = z^*(y)$. Since $z^*(y)$ is difficult to compute, we will use an approximation of it. Consider the change of variables $z = z_1 + \varepsilon u_1(y)$ and $u_1(y) = -[A(y)]^{-1}w_0(y)$. Substituting into (1.263), we obtain

$$\varepsilon\dot{z}_1 = A(y)z_1 + b_0(y, z_1 + \varepsilon u_1(y)) - \varepsilon^2\partial_y u_1 g_0(y, z_1 + \varepsilon u_1(y)). \qquad (1.264)$$

We extract the terms which vanish for $z = 0$, which are of order ε^2. The system can be written as

$$\varepsilon\dot{z}_1 = A(y)z_1 + b_1(y, z_1, \varepsilon) + \varepsilon^2 w_1(y, \varepsilon), \quad \dot{y} = g_1(y, z_1, \varepsilon),$$

where w_1 denotes the right-hand side of (1.264) for $z_1 = 0$ and Taylor's formula shows again that $b_1(y, z, \varepsilon)$ is of order $\|z\|^2 + \varepsilon\|z\|$. As long as the system is sufficiently differentiable, we can repeat this procedure, increasing the order in ε of the drift term by successive changes of variables. Thus there is a transformation

$$x = x^*(y) + \varepsilon u_1(y) + \varepsilon^2 u_2(y) + \cdots + \varepsilon^r u_r(y) + z_r$$

which yields the form

$$\varepsilon\dot{z}_r = A(y)z_r + b_r(y, z_r, \varepsilon) + \varepsilon^{r+1}w_r(y, \varepsilon), \quad \dot{y} = g_r(y, z_r, \varepsilon). \qquad (1.265)$$

Proceeding as in the proof of Tikhonov's theorem, one can show that $z_r(t) = O(\varepsilon^{r+1})$ after a time of order $\varepsilon|\log\varepsilon|$. This implies that after this time, solutions of (1.248) satisfy

$$\begin{cases} x(t) = x^*(y(t)) + \varepsilon u_1(y(t)) + \cdots + \varepsilon^r u_r(y(t)) + O(\varepsilon^{r+1}), \\ \dot{y}(t) = g(x^*(y) + \varepsilon u_1(y) + \cdots + \varepsilon^r u_r(y), y) + O(\varepsilon^{r+1}) \end{cases}$$

for times of order 1. One might wonder whether these iterations may be pushed to $r = \infty$ if the system is analytic, in such a way that the remainder disappears. The answer is negative in general, because the amplitude of the drift term w_r tends to grow with r. We will illustrate this phenomenon.

Example 67 Consider the equations

$$\varepsilon\dot{x} = -x + h(y), \quad \dot{y} = 1, \qquad (1.266)$$

where we will assume that y is analytic in the strip $|\text{Im}\, y| \leq R$. This system can be solved exactly as

$$x(t) = x_0 e^{-t/\varepsilon} + \frac{1}{\varepsilon}\int_0^t e^{-(t-s)/\varepsilon}h(y_0 + s)ds, \quad y(t) = y_0 + t. \qquad (1.267)$$

Thus if $|h(y)|$ is uniformly bounded by M, we have

$$|x(t)| \leq |x_0|e^{-t/\varepsilon} + \frac{M}{\varepsilon}\int_0^t e^{-(t-s)/\varepsilon}ds = |x_0|e^{-t/\varepsilon} + M(1 - e^{-t/\varepsilon}).$$

Thus there is no question that $x(t)$ exists and is uniformly bounded for all $t \geq 0$, by a constant independent of ε. Now let us pretend that we do not know the exact solution of (1.267). The naive way to solve (1.266) is to look for a series of the form

$$x = x_0(y) + \varepsilon x_1(y) + \varepsilon^2 x_2(y) + \dots \tag{1.268}$$

It is unlikely that all solutions admit such a series representation, but if we obtain a particular solution of (1.266) which has this form, we can then determine the general solution. Substituting the series (1.268) into (1.266), and solving order by order, we obtain

$$
\begin{aligned}
0 &= -x_0(y) + h(y) &\Rightarrow\quad x_0(y) &= h(y), \\
(x_0)'(y) &= -x_1(y) &\Rightarrow\quad x_1(y) &= -h'(y), \\
(x_1)'(y) &= -x_2(y) &\Rightarrow\quad x_2(y) &= -h''(y),
\end{aligned}
$$

and so on. Thus x admits a formal series representation of the form

$$x(t) = h(y(t)) - \varepsilon h'(y(t)) + \dots + (-\varepsilon)^k h^{(k)}(y(t)) + \dots \tag{1.269}$$

Can this series converge? In certain cases yes, as shows the particular case $h(y) = \sin y$. In that case, the function $\bar{x}(t, \varepsilon)$ admits an expansion in ε which converges for $|\varepsilon| < 1$. In general, the derivatives $h^{(k)}(y)$ may grow with k. Cauchy's formula tells us that if h is analytic and bounded by M in a disc or radius R in a complex plane around y, then

$$h^{(k)}(y) = \frac{k!}{2\pi i} \int_{x-y|=R} \frac{h(z)}{(z-y)^{k+1}} dz \quad\Rightarrow\quad |h^{(k)}(y)| \leq MR^{-k}k!$$

Thus, $|h^{(k)}(y)|$ may grow like $k!$, which is too fast for the formal series (1.269) to converge.

A more subtle method consists in trying to simplify (1.266) by successive changes of variables. The translation $x = h(y) + z$ yields the equation $\varepsilon \dot{z} = -z + \varepsilon h'(y)$. The drift term $-\varepsilon h'(y)$ can be further decreased by the change of variables $z = -\varepsilon h'(y) + z_1$ and so on. In fact, the transformation

$$x = \sum_{j=0}^{k} (-\varepsilon)^j h^{(j)}(y) + z_k \tag{1.270}$$

results in the equation $\varepsilon \dot{z}_k = -z_k + (-\varepsilon)^{k+1} h^{(k+1)}(y)$. The solution is given by

$$z_k(t) = z_k(0)e^{-t/\varepsilon} - (-\varepsilon)^k \int_0^t e^{-(t-s)/\varepsilon} h^{(k+1)}(y_0 + s) ds.$$

Of course, the relations (1.270) and (1.271a) can be obtained from the exact solution (1.267) by k successive integrations by parts. The iterative method also works in cases in which we do not know an exact solution. In that case, the remainder $z_k(t)$ can be bounded as in the proof of Tikhonov's theorem.

We are looking for a particular solution admitting an expansion of the form (1.269). If we choose $x(0)$ in such a way that $z_k(0) = 0$, we obtain for each $k \in \mathbb{Z}$ a particular solution satisfying

$$x(t, \varepsilon) = \sum_{j=0}^{k} (-\varepsilon)^j h^{(j)}(y(t)) + z_k(t, \varepsilon), \tag{1.271a}$$

$$|z_k t(t, \varepsilon)| \leq \varepsilon^{k+1} M R^{-(k+1)}(k+1)! \tag{1.271b}$$

where M, R are independent of k. The first k terms of (1.271a) agree with formal solution (1.269). An expression of the form $x(t, \varepsilon) = \sum_{j=0}^{k} \varepsilon^j u_j(t) + \varepsilon^{k+1} r_k(t, \varepsilon)$, $(k \in \mathbb{Z})$, where r_k is bounded for all k, but the series does not necessarily converge, is called an *asymptotic series*. If the remainder grows like in (1.271b), this asymptotic series is called of type *Gevrey-1*; see [32]. This situation is common in singularly perturbed systems: bounded solutions exist, but they do not admit convergent series in ε.

Remark 8 We can determine at which order k to stop the asymptotic expansion (1.271a) in such a way that the remainder $\varepsilon^{k+1}|z_k|$ is as small as possible. This order will be approximately the one for which $k!\varepsilon^k$ is minimal. Recall Stirling's formula $k! \simeq k^k e^{-k}$. It is convenient to optimize the logarithm of $k!\varepsilon^k$. We have

$$\frac{d}{dk}\log(k^k e^{-k}\varepsilon^k) = \frac{d}{dk}(k\log k - k + k\log \varepsilon) = \log k + \log \varepsilon,$$

which vanishes for $k = 1/\varepsilon$ (we need to take an integer k but this hardly makes a difference for small ε). For this k, we have $k!\varepsilon^k \simeq (1/\varepsilon)^{1/\varepsilon}e^{-1/\varepsilon}\varepsilon^{1/\varepsilon} = e^{-1/\varepsilon}$. Conclude that the optimal truncation of the series is for $k = O(1/\varepsilon)$, which yields a remainder of order $e^{-1/\varepsilon}$. This function goes to zero faster than any power of ε when $\varepsilon \to 0$, and thus even though we cannot compute the solution exactly in general, we can compute it to a relatively high degree of precision. For general slow/fast systems one can show that if f and g are analytic in some open complex domain, then the iterative scheme leading to (1.265) can also be truncated in such a way that the remainder is exponentially small in ε. We will not prove this result here, but only mention that it heavily relies on Cauchy's formula.

Remark 9 There is another way which may help to understand why the asymptotic series (1.269) does not converge for general h. Assume that h is periodic in y, and write its Fourier series in the form

$$h(y) = \sum_{k\in\mathbb{Z}} h_k e^{iky}, \qquad h_k = \frac{1}{2\pi}\int_0^{2\pi} h(y)e^{-iky}dy. \qquad (1.272)$$

The formal series (1.269) will also be periodic in y, so that we may look for a solution of (1.266) of the form $x(y) = \sum_{k\in\mathbb{Z}} x_k e^{iky}$. Substituting in the equation for x, we obtain $\varepsilon ikx_k = -x_k + h_k$, $\forall k \in \mathbb{Z}$, and thus the periodic solution can be written as

$$x(y) = \sum_{k\in\mathbb{Z}} \frac{h_k}{1 + \varepsilon ik}e^{iky}. \qquad (1.273)$$

This Fourier series *converges* in a neighborhood of the real axis because the h_k decrease exponentially fast in $|k|$ (to see this, shift the integration path of h_k in (1.272) by an imaginary distance). Thus the equation admits a bounded periodic solution. In the plane of complex ε, the function (1.273) has poles at every $\varepsilon = -i/k$ for which $h_k \neq 0$. In the case $h(y) = \sin y$, there are only two poles at $\varepsilon = \pm i$, so that the series converges for $|\varepsilon| < 1$. If $h(y)$ has nonvanishing Fourier components with arbitrarily large k, the poles accumulate at $\varepsilon = 0$ and the radius of convergence of the series in ε is zero.

Dynamic bifurcation. We examine what happens to solutions of the slow-fast system (1.248), $\varepsilon \dot{x} = f(x,y)$, $\dot{y} = g(x,y)$, in the case where the assumptions of Tikhonov's theorem are violated. If the slow manifold is represented by $x = x^*(y)$, the main assumption of Tikhonov's theorem is that each eigenvalue of the matrix $A(y) = \partial_x f(x^*(y), y)$ has a strictly negative real part. This is equivalent to saying that $x^*(y)$ is an asymptotically stable equilibrium point of the associated system

$$dx/dt = f(x,y)\big|_{y=\text{const}}. \qquad (1.274)$$

A *dynamic bifurcation* occurs if the slow motion of y causes some eigenvalues of $A(y)$ to cross the imaginary axis. Then there may be a bifurcation in the associated system (1.274), and the main assumption of Tikhonov's theorem is no longer satisfied.

The two most generic cases (codimension 1 bifurcations of equilibria) are

• Hopf bifurcation: a pair of complex conjugate eigenvalues of A cross the imaginary axis; the slow manifold $x^*(y)$ continues to exist, but changes its stability.

• Saddle node bifurcation: an eigenvalue of $A(y)$ vanishes, and the slow manifold $x^*(y)$ exists in some cases (in fact, it has a fold).

We will discuss a few relatively simple examples of these two dynamic bifurcations.

Example 68 (Hopf bifurcation of slow-fast y-independent system) For the slow-fast system

$$\begin{cases} \varepsilon \dot{x}_1 = yx_1 - x_2 - x_1(x_1^2 + x_2^2), \\ \varepsilon \dot{x}_2 = x_1 + yx_2 - x_2(x_1^2 + x_2^2), \\ \dot{y}_1 = 1, \end{cases} \tag{1.275}$$

the associated system is

$$\begin{cases} dx_1/ds = yx_1 - x_2 - x_1(x_1^2 + x_2^2), \\ dx_2/ds = x_1 + yx_2 - x_2(x_1^2 + x_2^2), \\ y = \text{const.} \end{cases}$$

The slow manifold is given by $x^*(y) \equiv (0,0)$, and the matrix $A(y) = \begin{pmatrix} y & -1 \\ 1 & y \end{pmatrix}$ has eigenvalues $a_{\pm}(y) = y \pm 1$. Thus there is a Hopf bifurcation at $y = 0$. In fact, the complex variable $z = x_1 + ix_2$ satisfies $dz/ds = (y + i)z - |z|^2 z$, which becomes, in polar coordinates $z = re^{i\varphi}$,

$$dr/ds = yr - r^3, \quad d\varphi/ds = 1.$$

If $y \leq 0$, all solutions spiral to the origin $r = 0$, while for $y > 0$, the origin is unstable, and all solutions not starting in $r = 0$ are attracted by the periodic orbit $r = \sqrt{y}$.

Let us come back to the slow-fast system (1.275). By Tikhonov's theorem, solutions starting close to $x = 0$ at some negative y will reach a neighborhood of order ε of the origin after a time of order $\varepsilon |\log \varepsilon|$. What happens as y approaches zero? Since the origin becomes unstable, intuitively the solution will leave the neighborhood of the origin as soon as y becomes positive and approaches the periodic orbit. This assumption can be checked by computing the solutions. In the complex variable z, the slow-fast system (1.275) becomes

$$\varepsilon \dot{z} = (y + i)z - |z|^2 z, \quad \dot{y} = 1,$$

and in polar coordinates

$$\varepsilon \dot{r} = yr - r^3, \quad \varepsilon \dot{\varphi} = 1, \dot{y} = 1.$$

The solution with initial condition $(r, \varphi, y)(0) = (r_0, \varphi_0, y_0)$ can be written as

$$r(t) = r_0 c(t) e^{\alpha(t)/\varepsilon}, \quad \varphi(t) = \varphi_0 + t/\varepsilon, \quad y(t) = y_0 + t, \tag{1.276}$$

where

$$\alpha(t) = \int_0^t y(s)ds = y_0 t + \frac{1}{2}t^2, \quad c(t) = \Big(1 + \frac{2r_0^2}{\varepsilon} \int_0^t e^{\alpha(s)/\varepsilon} ds\Big)^{-1/2}.$$

Assume that we start with $y_0 < 0$. Then the function $\alpha(t)$ is negative for $0 < t < -2y_0$. For these t, the term $e^{\alpha(t)/\varepsilon}$ is small, while $c(t) \leq 1$. This implies that $r(t)$ is very small up to times t close to $-2y_0$. The bifurcation already happens at time $t = -y_0$. For $t > -2y_0$, the term $e^{\alpha(t)/\varepsilon}$ becomes large, but is counteracted by $c(t)$. In fact, it is not hard to show that $r(t)$ approaches $\sqrt{y(t)}$ as $t \to \infty$, which means that the solution (1.276) approaches the periodic orbit of the associated system.

In summary, the solution starting at a distance r_0 (independent of ε) from the origin

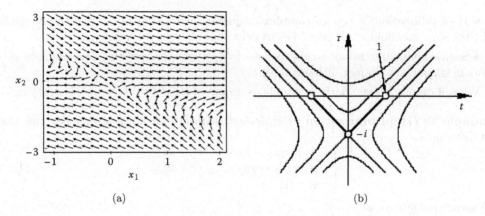

(a)   (b)

FIGURE 1.59: (a) A solution of equation (1.275) starting with a negative y_0. The Hopf bifurcation occurs at $y = 0$, but the solution approaches the limit cycle at $r = y^{1/2}$ only after $y = -y_0$. (b) Level lines of the function $\operatorname{Re}\alpha(t + i\tau)$. The function is small for large $|\tau|$ and large for large $|t|$. The integral $\Psi(t, t_0)$ is small only if t_0 and t can be connected by a path on which $\operatorname{Re}\alpha$ is non-increasing.

with $y_0 < 0$ is quickly attracted by the origin (in accordance with Tikhonov's theorem), but stays there until $y = -y_0$, although the actual bifurcation takes place at $y = 0$. Only after $y = -y_0$, the solution will approach the stable periodic orbit of the associated system, Fig. 1.59(a). This phenomenon is called a *bifurcation delay*.

Example 69 (Slow manifold dependent on y) A particularity of Example 68 is that the slow manifold does not depend on y, so that there is no drift term pushing solutions away from it. One might expect that the bifurcation delay is destroyed as soon as $x^*(y)$ depends on y. Modify (1.275) as follows:

$$\begin{cases} \varepsilon \dot{x}_1 = y(x_1 + y) - x_2 - (x_1 + y)\big[(x_1 + y)^2 + x_2^2\big], \\ \varepsilon \dot{x}_2 = (x_1 + y) + y x_2 - x_2\big[(x_1 + y)^2 + x_2^2\big], \\ \dot{y} = 1. \end{cases}$$

The slow manifold is given by $x^*(y) = (-y, 0)$. The associated system is the same as in Example 68, up to a translation of $x^*(y)$. There is a Hopf bifurcation at $y = 0$. The variable $z = (x_1 + y) + i x_2$ obeys

$$\varepsilon \dot{z} = (y + i)z - |z|^2 z + \varepsilon, \quad \dot{y} = 1. \tag{1.277}$$

The additional drift term stems from the y-dependence of the slow manifold.

In naive perturbation theory, one would look a particular solution of the form

$$z = z_0(y) + \varepsilon z_1(y) + \varepsilon^2 z_2(y) + \cdots \tag{1.278}$$

Inserting into (1.277) and solving order by order, we get

$$0 = (y + i)z_0(y) - |z_0(y)|^2 z_0(y) \;\Rightarrow\; z_0(y) = 0,$$
$$\dot{z}_0 = (y + i)z_1(y) - \big[(z_0(y))^2 \bar{z}_1(y) + 2|z_0(y)|^2 z_1(y)\big] + 1 \;\Rightarrow\; z_1(y) = -\tfrac{1}{y + i}$$
$$\cdots$$

Thus equation (1.277) seems to admit a particular solution

$$z_p(t) = -\frac{\varepsilon}{y(t) + i} + O(\varepsilon^2).$$

To obtain the general solution, we set $z = z_p + \zeta$, substitute in (1.277), and use the fact that z_p is a particular solution. The result is an equation for ζ of the form

$$\varepsilon\dot{\zeta} = (y + i)\zeta + O(\varepsilon^2|\zeta|) + O(|\zeta|^3). \tag{1.279}$$

The drift term has disappeared in the transformation, so that the right-hand side of (1.279) vanishes for $\zeta(t) \equiv 0$. Although this equation cannot be solved exactly, it should behave as in Example 68. To show this, assume that the system is started at time $t = t_0 = y_0$ in such a way that $y(0) = 0$. We use

$$\zeta(t) = \zeta_0 e^{\alpha(t,t_0)/\varepsilon} c(t), \quad \alpha(t, t_0) = i(t - t_0) + \frac{1}{2}(t^2 - t_0^2). \tag{1.280}$$

If the last three terms in (1.279) were absent, the solution would be given by (1.280) with $c(t) \equiv 1$. Inserting (1.280) into (1.279), we obtain a differential equation for c, which can be used to show that c remains of order 1 as long as $\operatorname{Re}\alpha(t, t_0) < 0$. Since $\operatorname{Re}\alpha(t, t_0) < 0$ for $-t_0 < t < t_0$, although the bifurcation happens at $t = 0$, we conclude that the bifurcation is delayed as in the previous example.

There is a major problem with this procedure. We already know that the asymptotic series (1.278) is unlikely to converge, but this does not necessarily imply that (1.277) has no solution of order ε. In this case, (1.277) *does not admit a solution of order ε for all times*. To see this, write the general solution implicitly as

$$z(t) = z_0 e^{\alpha(t,t_0)/\varepsilon} + \frac{1}{\varepsilon}\int_{t_0}^t e^{\alpha(t,s)/\varepsilon}\big[\varepsilon - |z(s)|^2 z(s)\big]ds.$$

We can try to solve this equation by iterations. Assume that $z_0 = O(\varepsilon)$. As long as z remains of order ε, we have $z(t) = z_0 e^{\alpha(t,t_0)/\varepsilon} + \int_{t_0}^t e^{\alpha(t,s)/\varepsilon}[1 + O(\varepsilon^2)]ds$. Its behavior is thus contained in the integral

$$\Psi(t, t_0) = \int_{t_0}^t e^{\alpha(t,s)/\varepsilon}ds. \tag{1.281}$$

As long as this integral remains of order ε, $z(t)$ is of order ε. If it becomes large, then z becomes large as well (though this has to be checked independently). The function Ψ can be evaluated by deformation of the integration path into the complex plane. The function $\alpha(t) := \alpha(t, 0)$ can be continued to the complex plane, and satisfies

$$\operatorname{Re}\alpha(t + i\tau) = -\tau + \frac{1}{2}(t^2 - \tau^2) = \frac{1}{2}\big[t^2 - (\tau + 1)^2 + 1\big].$$

The level lines of $\operatorname{Re}\alpha$ are hyperbolas centered at $t + i\tau = -i$. The integral (1.281) is small if we manage to find a path of integration connecting t_0 to t on which $\operatorname{Re}\alpha(t, s) = \operatorname{Re}\alpha(t) - \operatorname{Re}\alpha(s) \le 0$ for all s. This is only possible if $\operatorname{Re}\alpha(t) \le \operatorname{Re}\alpha(t_0)$, but this condition is not sufficient. In fact; see Fig. 1.59(b):

- if $t_0 \le 1$, such a path exists for all $t \le 1$;
- if $-1 \le t_0 \le 0$, such a path exists for all $t \le -t_0$;
- if $t_0 \ge 0$, such a path exists for all $t \ge t_0$.

Conclude that if $t_0 \le 0$, then $\Psi(t, t_0)$ remains small for all $t \le \min\{-t_0, 1\}$. As a consequence, $z(t)$ also remains small up to this time. The term $-t_0$ is natural, the same as in Example 68, but the term 1, called *maximal delay or buffer point*, is new. It is an effect of the drift term, and cannot be found by naive perturbation theory, because it is "hidden" in exponentially small terms. This example shows that one should be extremely cautious with asymptotic expansions near dynamic bifurcations.

1.7   Elements of tensor analysis

1.7.1   Transformations of coordinate systems

The choice of coordinate system is very important for general mathematical description of gas flow. The choice of this system depends on a concrete problem and therefore we have to know how to transform the governing equations from one coordinate system to another one; see [30, 49]. Although the laws of motion *may involve* coordinates, they *must be invariant* with respect to the choice of the coordinate system, which imposes known *limitations* on the form of mathematical descriptions of these laws.

Suppose that we have two coordinate systems ζ^i and η^i in $3d$ space and one-to-one C^1 relations between these coordinate systems are known,

$$\zeta^i = \zeta^i(\eta^1, \eta^2, \eta^3), \quad i = 1, 2, 3. \tag{1.282}$$

These may be rewritten in the inverse form as

$$\eta^i = \eta^i(\zeta^1, \zeta^2, \zeta^3), \quad i = 1, 2, 3, \tag{1.283}$$

where η^i are functions of class C^1.

Definition 36 Relations (1.282) and (1.283) are called the *transformations of coordinates*.

From (1.282) one obtains linear relations between local variations of coordinates

$$d\zeta^i = \frac{\partial \zeta^i}{\partial \eta^1} d\eta^1 + \frac{\partial \zeta^i}{\partial \eta^2} d\eta^2 + \frac{\partial \zeta^i}{\partial \eta^3} d\eta^3 = \frac{\partial \zeta^i}{\partial \eta^j} d\eta^j.$$

Here and hereafter we will imply the summation for repeating indices (Einstein summation convention). Define $a^i_j = \frac{\partial \zeta^i}{\partial \eta^j} d\eta^j$. Similarly we can write

$$d\eta^i = \frac{\partial \eta^i}{\partial \zeta^j} d\zeta^j \equiv b^i_j d\zeta^j. \tag{1.284}$$

Clearly, there exist simple relations between matrices $A = \|a^i_j\|$ and $B = \|b^i_j\|$,

$$A \cdot B = \mathrm{Id}, \quad a^i_j b^j_k = \frac{\partial \zeta^i}{\partial \eta^j} \frac{\partial \eta^j}{\partial \zeta^k} = \frac{\partial \zeta^i}{\partial \zeta^k} = \delta^i_k.$$

Here Id is the unit matrix and δ^i_k the Kronecker symbol.

Consider two points $M(\zeta^1, \zeta^2, \zeta^3)$ and $M'(\zeta^1 + \Delta\zeta^1, \zeta^2 + \Delta\zeta^2, \zeta^3 + \Delta\zeta^3)$ in coordinate system ζ^i. Introduce into consideration a new object, vector $dr = \overrightarrow{MM'}$, i.e., a pair of infinitesimally close points M and M' are taken in definite order (ordered pair) and with basis vector $dr/d\zeta^i = \mathbf{e}^i$, we can write $dr = d\zeta^1 \mathbf{e}_1 + d\zeta^2 \mathbf{e}_2 + d\zeta^3 \mathbf{e}_3$, where $d\zeta^i$ are the components of vector dr. The basis $\{\mathbf{e}_i\}$ in coordinate system $\{\zeta^i\}$ has the components $(1, 0, 0), (0, 1, 0), (0, 0, 1)$. In coordinate system $\{\eta^i\}$ vector dr has representation $dr = d\eta^1 \mathbf{e}'_1 + d\eta^2 \mathbf{e}'_2 + d\eta^3 \mathbf{e}'_3$. The components of vector dr as well as basis vectors depend on the coordinate system. Find the relations between $\{\mathbf{e}_i\}$ and $\{\mathbf{e}'\}$,

$$\mathbf{e}'_j = \frac{\partial \mathbf{r}}{\partial \eta^j} = \frac{\partial \mathbf{r}}{\partial \zeta^i} \frac{\partial \zeta^i}{\partial \eta^j} = \mathbf{e}_i a^i_j.$$

But for components $d\eta^j$ we have from (1.284) $d\eta^j = b_i^j d\zeta^i$. Hence, the basis vectors are transformed with matrix $A = \|a_j^i\|$ and the vector components are transformed with matrix $B = \|b_j^i\|$. The vector $\mathbf{r}$ is invariant with respect to transformation, $d\mathbf{r} = d\eta^j \mathbf{e}'_j = b_i^j d\zeta^i a_j^k \mathbf{e}_k = d\zeta^i \mathbf{e}_i$ due to $b_i^j a_j^k = d_i^k$.

Definition 37 The values which are transformed during coordinate transformation as basis vectors $\mathbf{e}_i$ (with matrix A) are called *covariant* values; the values which are transformed during coordinate transformation as vector components (with matrix B) are called *contravariant* values.

Further, all low indices will be associated with covariant values, and upper indices – with contravariant values. Because the vector is invariant with respect to the coordinate transformation, it is nature to postulate the following question: "Is it possible to introduce more complex objects which will be invariant with respect to the coordinate transformation?" The answer is positive.

Definition 38 One of the possible ways to do it is to use the new operator, called *polyadic products* of the basis vectors, which is designated by the sign $\otimes$.

Polyadic multiplication of vectors represents certain operations on vectors which lead to new objects (neither vectors nor scalar). For definition of this operator we have only to state its properties:

$$
\begin{aligned}
&1. &\mathbf{e}_i \otimes \mathbf{e}_j &\neq \mathbf{e}_j \otimes \mathbf{e}_i, \\
&2. &\mathbf{e}_i \otimes \mathbf{e}_j + \mathbf{e}_k \otimes \mathbf{e}_l &= \mathbf{e}_k \otimes \mathbf{e}_l + \mathbf{e}_i \otimes \mathbf{e}_j, \\
&3. &\mathbf{e}_i \otimes (\alpha \mathbf{e}_j + \beta \mathbf{e}_k) &= \alpha \mathbf{e}_i \otimes \mathbf{e}_j + \beta \mathbf{e}_i \otimes \mathbf{e}_k,
\end{aligned}
$$

where α, β are real numbers. Using these properties one can see how the new objects $\mathbf{e}_i \otimes \mathbf{e}_j$ are transformed during coordinate transformation: $\mathbf{e}'_i \otimes \mathbf{e}'_j = a_i^k \mathbf{e}_k \otimes a_j^l \mathbf{e}_l = a_i^k a_j^l (\mathbf{e}_k \otimes \mathbf{e}_l)$.

Definition 39 Using linear combination of polyadic product of the basis vectors we can introduce a new object $\mathbf{T} = T^{ij}(\mathbf{e}_i \otimes \mathbf{e}_j)$, which is called *tensor* of second order.

The main property of this new object is that it has to be invariant with respect to coordinate transformation. It means that $(T^{ij})'(\mathbf{e}'_i \otimes \mathbf{e}'_j) = T^{ij}(\mathbf{e}_i \otimes \mathbf{e}_j)$. To satisfy this condition the components T^{ij} during coordinate transformation have to transform by the way $(T^{ij})' = b_k^i b_l^j T^{kl}$. Determine the vector's length. To do it we introduce the space metric via scalar multiplication of the basis vectors $g_{ij} = \mathbf{e}_i \cdot \mathbf{e}_j$. In this case the square of the length of $d\mathbf{r}$ by definition is given by

$$(d\mathbf{r})^2 = ds^2 = (dr \cdot dr) = d\zeta^i \mathbf{e}_i d\zeta^j \mathbf{e}_j = g_{ij} d\zeta^i d\zeta^j. \tag{1.285}$$

Definition 40 The relation (1.285) is of quadratic form and it is called *fundamental quadratic form*.

Due to $d\mathbf{r}$ being invariant with respect to coordinate transformation we write $(d\mathbf{r})^2 = g'_{kl} d\eta^k d\eta^l$, and then $g'_{kl} = a_k^i a_l^j g_{ij}$. Thus, we can consider g_{ij} as the covariant components of the second-order tensor g.

Definition 41 Tensor g is called *fundamental metric tensor*. By definition of the scalar product, g is a symmetric tensor: $g_{ij} = g_{ji}$.

1.7.2   Covariant and contravariant derivatives

Consider a vector $\mathbf{w} = w^i \mathbf{e}_i$ in covariant basis. Our goal is to calculate the value $\partial \mathbf{w}/\partial \zeta^j$,

$$\frac{\partial \mathbf{w}}{\partial \zeta^j} = \frac{\partial w^k}{\partial \zeta^j} \mathbf{e}_k + w^k \frac{\partial \mathbf{e}_k}{\partial \zeta^j}. \tag{1.286}$$

Thus $\partial \mathbf{e}_k/\partial \zeta^j$ is also a vector which has its own components in the basis $\{\mathbf{e}_i\}$.

Definition 42 Suppose that

$$\frac{\partial \mathbf{e}_k}{\partial \zeta^j} = \Gamma_{kj}^i \mathbf{e}_i, \tag{1.287}$$

where coefficients Γ_{kj}^i are called Christoffel symbols. Using (1.287) the relation (1.286) may be rewritten as follows $\frac{\partial \mathbf{w}}{\partial \zeta^j} = \frac{\partial w^k}{\partial \zeta^j} \mathbf{e}_k + w^i \Gamma_{ij}^k \mathbf{e}_k = \nabla_j w^k \mathbf{e}_k$, where $\nabla_j w^k \equiv \frac{\partial w^k}{\partial \zeta^j} + w^i \Gamma_{ij}^k$.

By the definition, it is the covariant derivative of vector's contravariant component. Some properties of operator $\nabla_j w^k$ can be indicated.

1. For Cartesian coordinates, $\mathbf{e}_i = \text{const}$, $\partial \mathbf{e}_k/\partial \zeta^j = 0$, by (1.287): $\Gamma_{kj}^i = 0$, thus $\nabla_j w^k = \frac{\partial w^k}{\partial \zeta^j}$.

2. This operator is linear, $\nabla_j (\alpha w^k + \beta v^k) = \alpha \nabla_j w^k + \beta \nabla_j v^k$, where α, β are real or complex.

3. Operator ∇_j is the covariant derivative of the contravariant component of second-order tensor.

Let $H = H^{jk} \mathbf{e}_j \mathbf{e}_k$ be the second-order tensor. Then we can write

$$\frac{\partial H}{\partial \zeta^i} = \frac{\partial H^{jk}}{\partial \zeta^i} \mathbf{e}_j \mathbf{e}_k + H^{jk} \frac{\partial \mathbf{e}_j}{\partial \zeta^i} \mathbf{e}_k + H^{jk} \mathbf{e}_j \frac{\partial \mathbf{e}_k}{\partial \zeta^i} = \frac{\partial H^{jk}}{\partial \zeta^i} \mathbf{e}_j \mathbf{e}_k + H^{jk} \Gamma_{ji}^l \mathbf{e}_l \mathbf{e}_k + H^{jk} \Gamma_{ki}^l \mathbf{e}_j \mathbf{e}_l.$$

If in the second sum we change index l on index j and j on l we obtain

$$\frac{\partial H}{\partial \zeta^i} = \left(\frac{\partial H^{jk}}{\partial \zeta^i} + H^{lk} \Gamma_{li}^j + H^{jl} \Gamma_{li}^k \right) \mathbf{e}_j \mathbf{e}_k = \nabla_i H^{jk} \mathbf{e}_j \mathbf{e}_k.$$

Here

$$\nabla_i H^{jk} = \frac{\partial H^{jk}}{\partial \zeta^i} + H^{lk} \Gamma_{li}^j + H^{jl} \Gamma_{li}^k$$

is the covariant derivative of the contravariant component of second-order tensor.

4. This operator is a covariant differentiation of a product.

Because $v^j w^k$ can be considered as a contravariant component of second-order tensor, we can write

$$\nabla_i (v^i w^k) = \frac{\partial}{\partial \zeta^i} (v^j w^k) + v^l w^k \Gamma_{li}^j + v^j w^l \Gamma_{li}^k = w^k \nabla_i v^j + v^j \nabla_i w^k.$$

5. It is the covariant derivative of the covariant component of the vector.

If the vector is presented in covariant basis vectors $\mathbf{w} = w_j \mathbf{e}^j$, we have

$$\frac{\partial \mathbf{w}}{\partial \zeta^i} = \frac{\partial w_j}{\partial \zeta^i} \mathbf{e}^j + w_j \frac{\partial \mathbf{e}^j}{\partial \zeta^i}. \tag{1.288}$$

Note that $\mathbf{e}^j \cdot \mathbf{e}_k = g^{ij} \mathbf{e}_i \cdot \mathbf{e}_k = g^{ij} g_{ik} = \delta_k^j$. Differentiation of this relation gives us

$$\frac{\partial \mathbf{e}^j}{\partial \zeta^i} \mathbf{e}_k + \mathbf{e}^j (\Gamma_{ki}^l \mathbf{e}_l) = 0 \quad \Rightarrow \quad \frac{\partial \mathbf{e}^j}{\partial \zeta^i} \mathbf{e}_k = -\Gamma_{ki}^j.$$

Multiplying the last expression by $\mathbf{e}^k$ yields $\frac{\partial \mathbf{e}^j}{\partial \zeta^i} = -\Gamma_{ki}^j \mathbf{e}^k$. Therefore, (1.288) may be rewritten as

$$\frac{\partial \mathbf{w}}{\partial \zeta^i} = \left(\frac{\partial w_j}{\partial \zeta^i} - w_k \Gamma_{ji}^k\right)\mathbf{e}^j = \nabla_i w_j \mathbf{e}^j,$$

where $\nabla_i w_j = \frac{\partial w_j}{\partial \zeta^i} - w_k \Gamma_{ji}^k$ is a covariant derivative of the covariant component of the vector.

Note that $\nabla_i w_j$ and $\nabla_i w^j$ can be considered as covariant and mixed components of the second-order tensor $T = (\partial \mathbf{w}/\partial \zeta^i)\mathbf{e}^i = \nabla_i w_j \mathbf{e}^j \mathbf{e}^i = \nabla_i w^j \mathbf{e}_j \mathbf{e}^i$. But it means that there exists the following relation between components of the same tensor $\nabla_i w^j = g^{jk} \nabla_i w_k$. From the other side $w^j = g^{jk} w_k$ and $\nabla_i(g^{jk} w_k) = g^{jk} \nabla_i w_k$. It means $\nabla_i g^{jk} = 0$. Simultaneously, $\nabla_i g_{jk} = 0$. We advise the reader interested in more detailed study of the problems presented in this subsection to refer to the textbooks [2, 10].

1.7.3 Christoffel symbols and curvature tensor

Christoffel symbols have symmetry on low indices. Indeed,

$$\frac{\partial \mathbf{e}_j}{\partial \zeta^k} = \frac{\partial}{\partial \zeta^k}\left(\frac{\partial \mathbf{r}}{\partial \zeta^j}\right) = \frac{\partial}{\partial \zeta^j}\left(\frac{\partial \mathbf{r}}{\partial \zeta^k}\right) = \frac{\partial \mathbf{e}_k}{\partial \zeta^j},$$

and by definition of Christoffel symbols, $\Gamma_{jk}^i = \Gamma_{kj}^i$ [8]. To calculate these symbols, differentiate g_{im}:

$$\frac{\partial g_{im}}{\partial \zeta^k} = \frac{\partial}{\partial \zeta^k}(\mathbf{e}_i \cdot \mathbf{e}_m) = \frac{\partial \mathbf{e}_i}{\partial \zeta^k}\mathbf{e}_m + \frac{\partial \mathbf{e}_m}{\partial \zeta^k}\mathbf{e}_i.$$

Hence,

$$\frac{\partial g_{im}}{\partial \zeta^k} - \frac{\partial \mathbf{e}_m}{\partial \zeta^k}\mathbf{e}_i = \Gamma_{ik}^j \mathbf{e}_j \cdot \mathbf{e}_m = \Gamma_{ik}^j g_{jm}. \tag{1.289}$$

Analogously one can obtain

$$\frac{\partial g_{km}}{\partial \zeta^i} - \frac{\partial \mathbf{e}_m}{\partial \zeta^i}\mathbf{e}_k = \Gamma_{ki}^j g_{jm}. \tag{1.290}$$

Summing (1.289) and (1.290), considering the symmetry of Christoffel symbols $\Gamma_{ki}^j = \Gamma_{ik}^j$, and taking into account the relations $\frac{\partial \mathbf{e}_m}{\partial \zeta^k} = \frac{\partial \mathbf{e}_k}{\partial \zeta^m}$, $\frac{\partial \mathbf{e}_m}{\partial \zeta^i} = \frac{\partial \mathbf{e}_i}{\partial \zeta^m}$, we obtain $\frac{\partial g_{im}}{\partial \zeta^k} + \frac{\partial g_{km}}{\partial \zeta^i} - \frac{\partial g_{ik}}{\partial \zeta^m} = 2\Gamma_{ki}^j g_{jm}$. Multiplication of this expression on g^{jm} gives a useful explicit calculation formula for Γ_{ki}^j,

$$2\Gamma_{ki}^j = g^{jm}\left(\frac{\partial g_{im}}{\partial \zeta^k} + \frac{\partial g_{km}}{\partial \zeta^i} - \frac{\partial g_{ik}}{\partial \zeta^m}\right). \tag{1.291}$$

For orthogonal coordinate system formulas (1.291) may be simplified due to $g_{ij} = 0$, if $i \neq j$.

$$\Gamma_{ki}^k = \Gamma_{ik}^k = \frac{1}{2}g^{kk}\frac{\partial g_{kk}}{\partial \zeta^i}, \quad \Gamma_{ii}^k = -\frac{1}{2}g^{kk}\frac{\partial g_{ii}}{\partial \zeta^k}, \quad \Gamma_{kk}^k = \frac{1}{2}g^{kk}\frac{\partial g_{kk}}{\partial \zeta^k}.$$

For *cylindrical coordinates* we have

$$\zeta^1 = r, \quad \zeta^2 = \varphi, \quad \zeta^3 = z, \quad ds^2 = dr^2 + r^2 d\varphi^2 + dz^2, \quad g_{11} = 1, \quad g_{22} = 1, \quad g_{33} = 1.$$

For *spherical coordinates* we have $\zeta^1 = r$, $\zeta^2 = \phi$, $\zeta^3 = \theta$ and

$$ds^2 = dr^2 + r^2 d\phi^2 + r^2 \cos^2 \phi \, d\theta^2, \quad g_{11} = 1, \quad g_{22} = r^2, \quad g_{33} = r^2 \cos^2 \phi.$$

The *Riemannian curvature tensor* $R = (R^l_{ijk})$, that is, $R(e_i, e_j)e_k = R^l_{ijk}e_l$, has rang 4 and is given on the basis $(e_i = \partial_i)$ by

$$R(e_i, e_j)e_k = (\nabla_i\nabla_j - \nabla_j\nabla_i), e_k,$$

and these vanish in Euclidean space. In local coordinates (using Einstein's summation convention),

$$R^l_{ijk} = \frac{\partial \Gamma^l_{ik}}{\partial \zeta^j} - \frac{\partial \Gamma^l_{jk}}{\partial \zeta^i} + \Gamma^r_{ik}\Gamma^l_{jr} - \Gamma^r_{jk}\Gamma^l_{ir}.$$

We omit here the symmetries and other details about this important tensor, since we do not use them in the book. The contraction over the second and fourth (or equivalently over the first and the third) indices gives rise to a nontrivial symmetric tensor called *Ricci's tensor*:

$$\mathrm{Ric}_{ik} = R^l_{ilk}.$$

In local coordinates, $\mathrm{Ric}_{ik} = \frac{\partial \Gamma^l_{ik}}{\partial_l} - \frac{\partial \Gamma^l_{lk}}{\partial_i} + \Gamma^r_{ik}\Gamma^l_{lr} - \Gamma^r_{lk}\Gamma^l_{ir}$. The contraction of Ric produces the so-called *scalar curvature*: $S = \mathrm{Ric}^k_k$. Another relevant tensor is the so-called *Einstein's tensor*

$$G_{ik} = \mathrm{Ric}_{ik} - \frac{1}{2}g_{ik}S,$$

which is "divergence free" and plays a crucial role in General Relativity.

1.7.4 Integral formulas

To justify $3d$ Navier–Stokes equations for multicomponent chemical nonequilibrium gas flow we need some useful formulae from mathematical and tensor analyses; see [29]. Consider at time t the individual volume V of continuum media with surface Σ. At the moment $t+\Delta t$ the volume V transfers to volume V', and the surface Σ transfers to the surface Σ'. For the difference $V' - V$ we obtain

$$V' - V = \int_\Sigma u_n \Delta t\, d\sigma \ \Rightarrow\ \lim_{\Delta t\to 0}\frac{V'-V}{\Delta t} = \int_\Sigma u_n\, d\sigma,$$

where u_n is the normal to Σ component of the velocity. Next, by definition of div $\mathbf{u}$ we have div $\mathbf{u} = \lim_{\Delta t\to 0}\frac{V-V_0}{V_0\Delta t}$. Thus an infinitesimal small volume V_* may be written as $\lim_{\Delta t\to 0}\frac{V_*-V'_*}{\Delta t} = V_*$ div $\mathbf{u}$ or (using the first relation) $\lim_{\Delta t\to 0}\frac{V_*-V'_*}{\Delta t} = \int_\Sigma u_n\, d\sigma$. Consequently for V_* we get

$$\int_{\Sigma_*} u_n\, d\sigma = V_* \text{ div } \mathbf{u}.$$

This relation can be rewritten for a finite volume V in the form $\int_\Sigma u_n\, d\sigma = \int_V \int_0^t \text{div } \mathbf{u}\, d\mathbf{r}$. Simultaneously, we can write the following formula for an arbitrary differentiable vector A in a volume V:

$$\int_\Sigma (A\cdot\mathbf{n})\, d\sigma = \int_V \text{div } A\, d\mathbf{r}. \tag{1.292}$$

Let us obtain another useful formula for the value $\frac{d}{dt}\int_{V(t)} f(x, y, z, t)d\mathbf{r}$, where $f(x, y, z, t)$ is a function defined on an arbitrary frame $(x, y, z) \subset \mathbb{R}^3$ (including a tensor). By definition of derivative, we have

$$\frac{d}{dt}\int_{V(t)} f(\mathbf{r}, t)d\mathbf{r} = \lim_{\Delta t\to 0}\frac{1}{\Delta t}\Big[\int_{V(t+\Delta t)} f(\mathbf{r}, t+\Delta t)d\mathbf{r} - \int_{V(t)} f(\mathbf{r}, t)d\mathbf{r}\Big]$$
$$= \lim_{\Delta t\to 0}\Big\{\frac{1}{\Delta t}\int_{V(t)} [f(\mathbf{r}, t+\Delta t) - f(\mathbf{r}, t)]d\mathbf{r} + \frac{1}{\Delta t}\int_{V(t+\Delta t)-V(t)} f(\mathbf{r}, t+\Delta t)d\mathbf{r}\Big\}$$
$$= \int_{V(t)}\frac{\partial f}{\partial t}d\mathbf{r} + \int_\Sigma f\cdot u_n\, d\sigma,$$

because the volume $V' - V$ consists of a number of cylinders of volume $d\mathbf{r} = u_n \, d\sigma \, \Delta t$, and $f(\mathbf{r}, t + \Delta t) \to f(\mathbf{r}, t)$, $\Sigma' \to \Sigma$ for $\Delta t \to 0$. Using (1.292), we write

$$\frac{d}{dt} \int_{V(t)} f(\mathbf{r}, t) d\mathbf{r} = \int_{V(t)} \frac{\partial f}{\partial t} d\mathbf{r} + \int_{V(t)} \operatorname{div} (f \cdot \mathbf{u}) d\mathbf{r}. \tag{1.293}$$

Recollecting that

$$\frac{\partial f}{\partial t} + \operatorname{div} (f \cdot \mathbf{u}) = \frac{\partial f}{\partial t} + \nabla_i (f u^i) = \frac{\partial f}{\partial t} + u^i \nabla_i f + f \nabla_i u^i = \frac{df}{dt} + f \nabla_i u^i, \tag{1.294}$$

where $\frac{df}{dt} = \frac{\partial f}{\partial t} + u^i \nabla_i f$ is the full derivative of function $f(\mathbf{r}, t)$ with respect to time in arbitrary coordinate system, $\nabla_i u^i$ is the covariant derivative of the contravariant component of vector $\mathbf{u}$.

Using (1.294) the relation (1.293) may be rewritten as

$$\frac{d}{dt} \int_{V(t)} f(\mathbf{r}, t) d\mathbf{r} = \int_{V(t)} \left(\frac{df}{dt} + f \nabla_i u^i \right) d\mathbf{r}. \tag{1.295}$$

Let us apply (1.295) for the case $f = 1/V$. In this case, f depends only on t and for all t we have the following relations: $\int_{V(t)} \frac{1}{V} \, d\mathbf{r} = 1 \Rightarrow \frac{d}{dt} \int_{V(t)} \frac{1}{V} \, d\mathbf{r} = 0$, and consequently

$$\int_{V(t)} \left[\frac{d}{dt} V^{-1} + V^{-1} \operatorname{div} \mathbf{u} \right] d\mathbf{r} = 0. \tag{1.296}$$

It is important that (1.296) can be applied to all volume V of the moving media as well as for each part of this volume. Using (1.296) for small volume ΔV we obtain the following important relation:

$$\frac{d}{dt} \left(\frac{1}{\Delta V} \right) + \frac{1}{\Delta V} \operatorname{div} \mathbf{u} = 0.$$

1.8 Navier–Stokes equations for nonequilibrium gas mixture

In this section we will deduce three-dimensional Navier–Stokes equations for multicomponent chemical nonequilibrium gas mixtures.

1.8.1 Continuity, momentum and energy equations

Continuity equation. Now we will justify the governing equations for $3d$ multicomponent nonequilibrium gas flow in the framework of a curvilinear coordinate system. As a basis idea for justification we will use the general conservation laws of mass, momentum and energy.

First, assume that there are no chemical reactions in gas mixture. Thus the mass conservation law is satisfied at each moment of time t for each component (with mass m_i of density ρ_i and velocity $\mathbf{u}$) from N ones and the following relation can be written for arbitrary individual volume V:

$$\frac{d}{dt} m_i = 0 \Rightarrow \frac{d}{dt} \int_V \rho_i \, d\mathbf{r} = 0 \quad (i = 1, \ldots, N).$$

Using the integral formulas from Section 1.7.4, rewrite these N relations in the form

$$\int_V \left[\partial_t \rho_i + \text{div}(\rho_i \mathbf{u}_i) \right] d\mathbf{r} = 0 \quad (i = 1, \dots, N). \tag{1.297}$$

We call (1.297) as *integral form* of *mass conservation equations* or *continuity equations*.

Since (1.297) are satisfied for arbitrary volume V, we can write the continuous equations as [29]:

$$\partial_t \rho_i + \text{div}(\rho_i \mathbf{u}_i) = 0 \quad (i = 1, \dots, N).$$

Consider the general case when chemical reactions play a significant role. Then the individual mass m_i in individual volume V will be changed and we should take into account this effect also. Let $\dot{w}_i$ be the *total net mass production rate* of i-th species per unit time and per unit volume to chemical reactions proceeding in the gas mixture; we can neglect $\dot{w}_i$ when the chemical reaction is absent. Determination of these values is one of the main fundamental goals of chemistry and further these values are assumed to be known functions of density, temperature, pressure and concentration of each species; $\dot{w}_i$ has the dimensionality g/cm$^3 \cdot$ s. Hence the conservation law in this case has the following form:

$$\partial_t m_i = \int_V \dot{w}_i \, d\mathbf{r} \quad \Rightarrow \quad \frac{d}{dt} \int_V \rho_i \, d\mathbf{r} = \int_V \dot{w}_i \, d\mathbf{r} \quad (i = 1, \dots, N). \tag{1.298}$$

Thus, the continuity equations in *integral form* have a view:

$$\int_V \left[\partial_t \rho_i + \text{div}(\rho_i \mathbf{u}_i) - \dot{w}_i \right] d\mathbf{r} = 0 \quad (i = 1, \dots, N).$$

Since V is arbitrary, the *differential form* of the above is [29]

$$\partial_t \rho_i + \text{div}(\rho_i \mathbf{u}_i) - \dot{w}_i = 0 \quad (i = 1, \dots, N). \tag{1.299}$$

For further analysis (1.299) may be transferred to more convenient form. Because $\rho_i \mathbf{u}_i$ can be expressed via averaged velocity $\mathbf{u}$ of gas mixture and vector of diffusion flux $\mathbf{J}_i$,

$$\rho_i \mathbf{u}_i = \mathbf{J}_i + \rho_i \mathbf{u},$$

equations (1.299) can be rewritten as

$$\partial_t \rho_i + \text{div}\, \mathbf{J}_i + \text{div}(\rho_i \mathbf{u}) = \dot{w}_i \quad (i = 1, \dots, N). \tag{1.300}$$

Assuming that a full mass is conserved during chemical reactions $\sum_{i=1}^N \dot{w}_i = 0$, using $\sum_{i=1}^N \text{div}\, \mathbf{J}_i = 0$ and summing (1.300), one obtains a continuity equation for gas mixture as a whole

$$\partial_t \rho + \text{div}(\rho\, \mathbf{u}) = 0. \tag{1.301}$$

Multiplying (1.301) by mass concentration $c_i = \rho_i/\rho = m_i/m$, we obtain $\rho_i \, \text{div}\, \mathbf{u} = -\frac{\rho_i}{\rho} \partial_t \rho = -c_i \partial_t \rho$, and then, the convenient form for the continuity equation, for species i,

$$\rho\, \partial_t c_i + \text{div}\, \mathbf{J}_i = \dot{w}_i \quad (i = 1, \dots, N). \tag{1.302}$$

The source term in RHS of (1.302) is a complicated nonlinear exponential function of gas temperature.

Let the number of basis (or independent) components be equal to N_e. All other components are called reactions product A_i ($i = 1, \dots, N_r$, $N_r = N - N_e$). As example, consider fifth-components dissociated air ($N = 5$) with components N_2, O_2, NO, N, O. Basis components ($N_e = 2$) are N, O and the reaction products are N_2, O_2, NO. All reaction products A_i ($i = 1, \ldots, N_r$) can be represented via linear combinations of basis components A_i ($i = N_r + 1, \ldots, N$) under $A_i = \sum_{j=N_r+1}^{N} \nu_{ij} A_j$, where A_i are the chemical symbols, ν_{ij} – stoichiometric coefficients. For the case of air being chosen as the gas mixture, i.e., $N_2 = N + N$, $O_2 = O + O$, $NO = N + O$, the stoichiometric coefficients are $\nu_{14} = 2$, $\nu_{15} = 0$, $\nu_{24} = 0$, $\nu_{25} = 2$, $\nu_{34} = 1$, $\nu_{35} = 1$.

Definition 43 The following linear combination of mass concentration c_i

$$c_i^* = c_i + \sum_{k=1}^{N_r} \nu_{ki}(m_i/m_k)c_k, \quad (i = N_r + 1, \ldots, N) \tag{1.303}$$

is called *mass concentration of chemical element i*, and linear combination of diffusion fluxes $\mathbf{J}_i$

$$\mathbf{J}_i^* = \mathbf{J}_i + \sum_{k=1}^{N_r} \nu_{ki}(m_i/m_k)\mathbf{J}_k, \quad (i = N_r + 1, \ldots, N)$$

as *diffusion flux of chemical element*.

Following mass conservation law, full mass of chemical elements is not changed during chemical reactions and the following relations hold:

$$\dot{w}_i + \sum_{k=1}^{N_r} \nu_{ki}(m_i/m_k)\dot{w}_k = 0 \quad (i = N_r + 1, \ldots, N).$$

Multiplying first N_r equations (1.302) on $\nu_{ij}m_j/m_i$ ($i = N_r + 1, \ldots, N$), summing on i and then addding it to (1.302) for $j = N_r + 1, \ldots, N$, we obtain the following continuity equations for chemical elements:

$$\rho\, \partial_t c_j^* + \operatorname{div} \mathbf{J}_i^* = \dot{w}_i \quad (i = N_r + 1, \ldots, N).$$

For gas mixture flow with chemical reactions we have two characteristic time scales τ_g and τ_{ch}. The first of these scales is defined by gas dynamic characteristic velocity and linear scale accordingly, while the second one is only defined by character of homogeneous chemical reactions. Introducing these scales in our consideration we then obtain dimensionless form of the species continuity equation

$$\rho\, \partial_t c_i + \operatorname{div} \mathbf{J}_i = Dm \cdot \dot{w}_i \quad (i = 1, \ldots, N_r),$$

where $Dm = \tau_g/\tau_{\mathrm{ch}}$ is the *Damkohler number*. Depending on this number a chemical non-equilibrium gas flow can be classified as *frozen regime* (for $Dm \ll 1$) and *equilibrium regime* (for $Dm \gg 1$). From asymptotic point of view for frozen regime the chemical time scale τ_{ch} is huge in comparison with gas dynamic time scale τ_g. Thus as the first approximation the influence of the chemical reactions on the flow characteristic may be neglected, and the governing continuity equations have the form

$$\rho\, \partial_t c_i + \operatorname{div} \mathbf{J}_i = 0 \quad (i = 1, \ldots, N_r).$$

Opposite of this, for equilibrium regime the chemical time scale τ_{ch} is negligible by comparison with gas dynamic time scale τ_g. It means that for this regime as the first approximation the influence of gas dynamic flow on species concentrations c_i is completely determined by chemical equilibrium conditions, which can be written as follows:

$$\prod_{j=N_r+1}^{N} (x_j^{\nu_{ij}}/x_i) = K_{p_i}(T)/p^{\nu_i}, \quad \nu_i = \sum_{j=N_r+1}^{N} \nu_{ij} - 1, \quad i = 1, \ldots, N_r, \tag{1.304}$$

where the functions $K_{p_i}(T)$ are known (from chemistry) constants of equilibrium. Equations

(1.304) can be considered as algebraic integrals which may be used instead of N_r diffusion equation (1.302). Adding the system (1.303) of $N - N_r$ linear equations to (1.304) provides calculation of the values c_i (or x_i) as functions of P, T and c_i^*. Notice that different diffusion properties of species dependence of c_i on c_i^* can be very essential because the integrals $c_i^* = $ const, $j = N_r + 1, \ldots, N$ do not exist for a general case.

Momentum equations. To examine the motion of material continuum, the forces (as vector quantities) should be introduced for our consideration. In theoretical mechanics of material point we deal with *concentrated* forces while in continuum mechanics we face *distributed* forces. It means that the forces act on every part of a volume V or on every part of a surface Σ of continuous media, where the principal force vectors tend to zero when infinitesimal element of volume or surface tends to zero.

Definition 44 Forces distributed over a volume V are called *volume or mass forces.*

Denote by Φ the principal vector of mass forces acting on an element of mass Δm. Then the density $\mathbf{F}$ of the mass force at a given point is $\mathbf{F} = \lim_{\Delta m \to 0} \frac{\Phi}{\Delta m}$. In continuum mechanics, surface forces, i.e., forces which are distributed over the surface of continuous medium, play a more sufficient role than mass forces. Select the element $d\sigma$ of the surface Σ and introduce the surface force $d\mathbf{P} = \mathbf{p}\, d\sigma$, where $\mathbf{p} = \lim_{\Delta m \to 0} \frac{\Delta \mathbf{P}}{\Delta m}$ is the density of surface forces acting on $d\sigma$.

Definition 45 Imagine an arbitrary volume V and cut it into two parts V_1 and V_2 by cross-section Σ. Clearly the surface forces acting on Σ are different for different cross-sections Σ. Select some point M inside a volume V and consider at this point different elements of area $d\sigma$. Define the orientation of these elements by their normal vector $\mathbf{n}$. Denote by $d\mathbf{P}$ the resultant force acting from the side of the part of the medium in volume V_2 on the part of the medium in volume V_1 over area $d\sigma$ with normal $\mathbf{n}$. Further, assume that $d\mathbf{P} = \mathbf{p}_n\, d\sigma$ (with $\mathbf{p}_n$ being a finite vector which may be considered as a surface density of the forces of intersection between the parts divided by area $d\sigma$). Such surface forces may be introduced at any point of continuous medium. They are called the *forces of internal stress.*

At each point M of continuous medium there are infinitely many vectors $\mathbf{p}_n$ corresponding to infinitely many choices of area $d\sigma$ passing through M. The second Newton's law (momentum law) states that the sum of all external forces acting on inertial system of material points is equal to the exchange rate of linear momentum of the system. The same idea can be applied to individual volume V of gas mixture

$$\frac{d}{dt}\int_V \rho\, \mathbf{u}\, d\mathbf{r} = \int_V \rho\, \mathbf{F}\, d\mathbf{r} + \int_\Sigma \rho\, \mathbf{p}_n\, d\sigma. \tag{1.305}$$

Definition 46 We call (1.305) a *momentum equation*; it is a basic postulated dynamic relationship of continuous mechanics.

The stress $\mathbf{p}_n$ acting on any element $d\sigma$ can be expressed by the following linear combination: $\mathbf{p}_n = \mathbf{p}^i n_i = p^{ij}\mathbf{e}_j(\mathbf{e}_i \cdot \mathbf{n})$, where p^{ij} are contravariant components of the tensor $\mathbf{P}$ called the *internal stress tensor* [29]. By integral formulas (see Section 1.7.4) and continuity equation (1.301) for gas mixture as a whole, one can rewrite (1.305) in *integral form*

$$\int_V \rho(\partial_t u^i + u^j \nabla_j u^i)\, d\mathbf{r} = \int_V \rho F^i\, d\mathbf{r} + \int_V \nabla_j p^{ij}\, d\mathbf{r},$$

and in *differential form*

$$\partial_t u^i + u^j \nabla_j u^i = F^i + \frac{1}{\rho}\nabla_j p^{ij}. \tag{1.306}$$

To close the momentum equation, determine the dependence of internal stress tensor p^{ij} on the problem parameters. Assuming that chemical reactions have no influence on tensor $\mathbf{P}$, we write Navier–Stokes law,

$$p^{ij} = -pg^{ij} + g^{ij}\left(\zeta - \frac{2}{3}\mu\right) + 2\mu e^{ij},$$

where p is the pressure, g^{ij} are covariant components of the metric tensor, μ and ζ are dynamic and the second coefficient of viscosity, e^{ij} are covariant components of the strain rate tensor: $e^{ij} = \frac{1}{2}(\nabla^i u^j + \nabla^j u^i)$. Write (1.306) in vector form

$$\rho\frac{d\mathbf{u}}{dt} = \rho\,\mathbf{F} + \nabla_i \mathbf{p}^i, \tag{1.307}$$

where $\frac{d}{dt} \equiv \frac{\partial}{\partial t} + u^i\nabla_i$, $\mathbf{p}^i = p^{ji}\mathbf{e}_j$. Multiplying (1.307) on $\mathbf{u}$ one obtains the kinetic energy theorem,

$$\rho\frac{d}{dt}\left(\frac{1}{2}u^2\right) = \rho\mathbf{F}\cdot\mathbf{u} + u_j\nabla_i p^{ji} = \rho\mathbf{F}\cdot\mathbf{u} + \nabla_i(u_j p^{ji}) - p^{ji}\nabla_i u_j. \tag{1.308}$$

Thus, the differential of the kinetic energy density is equal to the sum of the element work densities done by external mass and external and internal surface forces.

Energy equation. Justification of energy equation is based on two fundamental laws of thermodynamics; they are called the first and the second thermodynamics laws. These laws deal with the concept of *internal state* and of *variables of state*. The number of the variables of state and the choice of these parameters differ for different models of continuum. Consider the example: from a macroscopic point of view, the state of uniform gas at rest is determined by altogether only two parameters, namely, the pressure p and the density ρ. Assume that for gas mixture which consists of N components, the internal energy U_m of small particle with mass m can be presented as $U_m = U_m(s, \rho, m_1, \ldots, m_N)$, where ρ is the particles density, s is entropy referred per unit mass, m_i is mass of i-th component. In fact it means that we assume the parameters ρ, s, m_i to be variable of state for multicomponent gas mixtures. Suppose that U_m is a differentiable function that can be written as

$$dU_m = \frac{\partial U_m}{\partial s}ds + \frac{\partial U_m}{\partial(1/\rho)}d(1/\rho) + \sum_{k=1}^{N}\frac{\partial U_m}{\partial m_k}dm_k. \tag{1.309}$$

Using (1.309), introduce the following functions:

$$m\theta(s, \rho, m_i) = \frac{\partial U_m}{\partial s}, \quad mp'(s, \rho, m_i) = \frac{\partial U_m}{\partial(1/\rho)}, \quad \mu_k(s, \rho, m_i) = \frac{\partial U_m}{\partial m_k}.$$

To clarify these definitions, remember that for perfect uniform ideal gas we have the relation for internal energy U_m: $dU_m = m[Tds - pd(1/\rho)]$, and that $\theta = T$, $p' = p$ for reversible process.

Definition 47 The quantity $\mu_k = \mu_k' - \theta s + p'/\rho$ is called the *chemical potential*.

The experimental data demonstrate that internal energy has additive property with respect to the mass. It means that for $s = $ const, $\rho = $ const the following relation is satisfied:

$$U_m(s, \rho, nm_1, \ldots, nm_N) = nU_m(s, \rho, m_1, \ldots, m_N).$$

For $n^{-1} = \sum m_i = m$ we obtain $U_m(s, \rho, m_i) = nU_m(s, \rho, c_i)$, where $c_i = m_i/m = \rho_i/\rho$.

In the case of the mixture of perfect gases with the same temperature T we have

$$U_m = \sum_{i=1}^{N} U_{m_i}(T) = m \sum_{i=1}^{N} c_{mi}\left[U_i^0 + \int_{T^0}^{T} c_{V_i}(T)dT\right].$$

Universal relation expressing the conservation energy law (the *first law of thermodynamics*) may be presented in the form

$$dU_m = -dA_{\text{surf}}^{(int)} + dQ^{(e)} + dQ^*,\qquad(1.310)$$

where $dQ^{(e)}$ is elementary heat flux into the volume V from outside, dQ^* is elementary flux from outside of other nonthermal forms of energy which differ from the work of microscopic mechanical forces (for example, Joule's heat), and $dA_{\text{surf}}^{(int)} = -\int_V \int_0^t p^{ij}\nabla_i u_j \, dt \, d\mathbf{r}$ is an elementary work done by internal surface stress on the volume V on infinitesimal displacement $d\mathbf{r}$. Because

$$\nabla_i u_j = \frac{1}{2}(\nabla_i u_j + \nabla_j u_i) + \frac{1}{2}(\nabla_i u_j - \nabla_j u_i) = e_{ij} + w_{ij},$$

and that the tensor p^{ij} is symmetric and the tensor w^{ij} is antisymmetric, we can write

$$dA_{\text{surf}}^{(int)} = -\left[\int_V p^{ij} e_{ij}\, d\mathbf{r} + \int_V p^{ij} w_{ij}\, d\mathbf{r}\right]dt = -\left(\int_V p^{ij} w_{ij}\, d\mathbf{r}\right)dt.\qquad(1.311)$$

The second law of thermodynamics says

$$T\, ds_m = dQ^{(e)} + dQ',\qquad(1.312)$$

where dQ' is *uncompensated heat* which is always higher or equal to zero. In reversible processes this term is equal to zero, i.e., $dQ' = 0$. For an infinitesimal small gas mixture particle with individual volume ΔV, using (1.312) we get $dQ^{(e)} + dQ^* = T\, ds_m + dQ^* - dQ'$ and replace this relation into (1.310),

$$dU_m = p^{ij} e_{ij}\Delta V\, dt + T\, ds_m + dQ^* - dQ'.\qquad(1.313)$$

Considering, additionally, that $p^{ij} = -pg^{ij} + \tau^{ij}$, the first term in the RHS of (1.313) reduces to

$$-p^{ij} e_{ij}\Delta V\, dt = -p\nabla_i u^i \frac{m}{\rho}\, dt = -p\frac{m}{\rho}(\text{div }\mathbf{u})\, dt = p\frac{m}{\rho}\frac{d\rho}{\rho} = -p\, d(m/\rho).$$

Here we used the fact that the particle mass m is not changed during the displacements and continuity equation is applicable, $\text{div } v\, dt = -(1/\rho)\, d\rho$. Using these formulas, (1.313) may be rewritten as follows:

$$dU_m = -pd(m/\rho)\tau^{ij} e_{ij}\Delta V\, dt + T\, ds_m + dQ^* - dQ'.$$

On the other hand, it can be written as

$$dU_m = \frac{\partial U_m}{\partial(m/\rho)}d(m/\rho) + \frac{\partial U_m}{\partial s_m}ds_m + \sum_{k=1}^{N}\mu_k dm_k.$$

Thus, we get the following definitions for pressure P and temperature T for irreversible processes:

$$P = -\frac{\partial U_m}{\partial(m/\rho)} = -\frac{\partial U}{\partial(1/\rho)},\quad T = \frac{\partial U_m}{\partial s_m}.$$

We can also write useful formula for $dQ' - dQ^*$:

$$dQ' - dQ^* = \tau^{ij} e_{ij}\Delta V\, dt - \sum_{k=1}^{N}\mu_k dm_k.\qquad(1.314)$$

From continuity equation for the i-th component the following relation can be written:

$$\rho\frac{d(m_i/m)}{dt} = \frac{\rho}{m}\frac{dm_i}{dt} = \frac{1}{\Delta V}\frac{dm_i}{dt} = \dot{w}_i - \text{div }\mathbf{J}_l,$$

and in this way, (1.314) may be transformed to

$$dQ' - dQ^* = \{\tau^{ij}e_{ij} + \sum\nolimits_{k=1}^{N}\mu_k(\text{div }\mathbf{J}_k - \dot{w}_k)\}\Delta V\, dt. \tag{1.315}$$

The value $\dot{w}_k$ can be presented via linear combination of all chemical reactions in the gas mixture,

$$\dot{w}_k = M_k\sum\nolimits_{\alpha=1}^{r}\nu_{k\alpha}\kappa_\alpha,$$

where r is the total number of homogeneous chemical reactions, $\nu_{k\alpha}$ known coefficients, M_k molecular mass of species k, and κ_α the rate of chemical reaction α. In this case, (1.315) has a form

$$dQ' - dQ^* = \{\tau^{ij}e_{ij} + \sum\nolimits_{k=1}^{N}\mu_k(\text{div }\mathbf{J}_k - M_k\sum\nolimits_{\alpha=1}^{r}\nu_{k\alpha}\kappa_\alpha)\}\Delta V\, dt,$$

and the second law of thermodynamics can be presented as

$$T\, ds_m = dQ^{(e)} + dQ^* + \{\tau^{ij}e_{ij} + \sum\nolimits_{k=1}^{N}\mu_k(\text{div }\mathbf{J}_k - M_k\sum\nolimits_{\alpha=1}^{r}\nu_{k\alpha}\kappa_\alpha)\}\Delta V\, dt.$$

Assume that the total external energy flux to the gas mixture particle of volume ΔV which moved with average velocity $\mathbf{u}$ can be represented in the form

$$dA^{(e)} + dQ^{(e)} + dQ^* = \Big[\sum_{k=1}^{N}\rho_k\mathbf{F}_k\mathbf{u}_k\Delta V + \sum_{k=1}^{N}\int_{\Sigma} p_k^{ij}n_j u_{ki}\, d\sigma - \int_{\Sigma} q_0^i n_j\, d\sigma\Big]\, dt$$

$$= \Big[\sum_{k=1}^{N}\rho_k\mathbf{F}_k\mathbf{u}_k\Delta V + \sum_{k=1}^{N}\int_{\Sigma} p_k^{ij}n_j u_{ki}\, d\sigma + \sum_{k=1}^{N}\mathbf{F}_k\mathbf{J}_k\Delta V + \nabla_j\Big(\sum_{k=1}^{N} p_k^{ij}J_{ki}/\rho_k\Big)\Delta V - \nabla_j q_0^j\Delta V\Big]\, dt.$$

$$\tag{1.316}$$

If we consider (1.316) and take into account the definition of $dA^{(e)}$,

$$dA^{(e)} = \sum\nolimits_{k=1}^{N}\rho_k F_k\mathbf{u}_k\Delta V\, dt + \int_{\Sigma} p_k^{ij}n_j u_i\, d\sigma\, dt,$$

then obtain the following formula for $dQ^{(e)} + dQ^*$:

$$dQ^{(e)} + dQ^* = \sum\nolimits_{k=1}^{N}F_k\mathbf{J}_k\Delta V\, dt + \nabla_j\Big(\sum\nolimits_{k=1}^{N} p_k^{ij}J_{ki}/\rho_k\Big)\Delta V\, dt - \nabla_j q_0^j\Delta V\, dt. \tag{1.317}$$

Substituting (1.311) for $dA_{\text{surf}}^{(int)}$ and (1.317) for $dQ^{(e)} + dQ^*$ into (1.310) gives the energy equation for internal energy equation U per unit mass of gas mixture as a whole,

$$\rho\frac{dU}{dt} = p^{ij}e_{ij} + \sum\nolimits_{k=1}^{N}\mathbf{F}_k\mathbf{J}_k - \text{div }\mathbf{J}_q, \tag{1.318}$$

where $\mathbf{J}_q = \mathbf{q}_0 - \sum_{k=1}^{N}\frac{p_k^{ij}}{\rho_k}J_{ki}$. Adding (1.308) to (1.318) leads to the energy equation for the total energy $H = \frac{1}{2}u^2 + U$ per unit mass:

$$\rho\frac{dH}{dt} = p^{ij}e_{ij} + u_j\nabla_i p^{ij} + \sum\nolimits_{k=1}^{N}\mathbf{F}_k\mathbf{J}_k + \rho\mathbf{F}\cdot\mathbf{u} - \text{div }\mathbf{J}_q. \tag{1.319}$$

Due to $p^{ij}e_{ij} + u_j\nabla_i p^{ij} = p^{ij}\nabla_i u_j + u_j\nabla_i p^{ij} = \nabla_i(p^{ij}u_j)$, (1.319) may be rewritten as follows:

$$\rho\frac{dH}{dt} = \nabla_i(p^{ij}u_j) + \sum_{k=1}^{N} \mathbf{F}_k\mathbf{J}_k + \rho\mathbf{F}\cdot\mathbf{u} - \operatorname{div}\mathbf{J}_q. \qquad (1.320)$$

For an important particular case when mass forces are equal to zero the last expression has a simple form $\rho\frac{dH}{dt} = \nabla_i(p^{ij}u_j) - \operatorname{div}\mathbf{J}_q$. How does one change the entropy of the mixture? From (1.317), which is the second law of thermodynamics, the total entropy changing can be obtained as follows:

$$\frac{ds_m}{dt} = \rho\Delta V\frac{ds}{dt} = m\frac{ds}{dt} = \Big(-\frac{1}{T}\operatorname{div}\mathbf{J}_q + \frac{1}{T}\sum_{k=1}^{N}\mu_k\operatorname{div}\mathbf{J}_k$$

$$+\frac{1}{T}\sum_{k=1}^{N}\mathbf{F}_k\mathbf{J}_k - \frac{1}{T}\sum_{k=1}^{N}\sum_{\alpha=1}^{r}\mu_k M_k\nu_{k\alpha}\dot{\kappa}_\alpha + \frac{1}{T}\tau^{ij}e_{ij}\Big)\Delta V. \qquad (1.321)$$

The value ds_m can be considered as the sum of external and internal terms $ds_m = ds_m^{(e)} + ds_m^{(i)}$. Using it, (1.321) may be rewritten as $\frac{ds_m}{dt} = \frac{ds_m^{(e)}}{dt} + \frac{ds_m^{(i)}}{dt}$. Denoting $ds_m = mds$, $ds_m^{(e)} = mds^{(e)}$, $ds_m^{(i)} = mds^{(i)}$, the equation above may be written as $\frac{ds_m}{dt} = \frac{ds^{(e)}}{dt} + \frac{ds^{(i)}}{dt}$. As the result we obtain

$$\rho\frac{ds^{(e)}}{dt} = -\operatorname{div}\frac{\mathbf{J}_q}{T} + \sum_{k=1}^{N}\operatorname{div}\frac{\mu_k\mathbf{J}_k}{T},$$

$$\rho\frac{ds^{(i)}}{dt} = \sigma = -\frac{1}{T^2}\mathbf{J}_q\nabla T - \sum_{k=1}^{N}\mathbf{J}_k\nabla\frac{\mu_k}{T} + \frac{1}{T}\sum_{k=1}^{N}\mathbf{F}_k\mathbf{J}_k - \frac{1}{T}\sum_{k=1}^{N}\sum_{\alpha=1}^{r}\mu_k M_k\nu_{k\alpha}\dot{\kappa}_\alpha + \frac{\tau^{ij}e_{ij}}{T}.$$
$$(1.322)$$

For more details for the problems presented in this section, see [11, 27, 31].

Main equations. Here we collect the main conservation equations described in this section.

Continuity equation for gas mixture as a whole [27]:

$$\partial_t\rho + \operatorname{div}(\rho\mathbf{u}) = 0. \qquad (1.323)$$

Continuity equation for each species of gas mixture:

$$\rho\,\partial_t c_i + \operatorname{div}\mathbf{J}_i = \dot{w}_i \quad (i = 1,\dots,N_r), \quad \rho\,\partial_t c_j^* + \operatorname{div}\mathbf{J}_j^* = 0 \quad (j = N_r+1,\dots,N). \qquad (1.324)$$

Momentum equation:

$$\rho\frac{d}{dt}\mathbf{u} = -\nabla p + \operatorname{div}\tau + \rho\mathbf{F}, \qquad (1.325)$$

with $\tau_{ij} = \delta_{ij}\big(\zeta - \frac{2}{3}\mu\big)\operatorname{div}\mathbf{u} + 2\mu e_{ij}$, $e_{ij} = \frac{1}{2}(\nabla_i u_j + \nabla_j u_i)$ and $\operatorname{div}\tau = \nabla^i\tau_{ij}$, where τ_{ij} is strain tensor, μ is viscosity coefficient and ζ is viscosity coefficient due to all-around compression (second viscosity).

Energy equation:

$$\rho\frac{dh}{dt} = \frac{dp}{dt} - \operatorname{div}\mathbf{J}_q + \Phi, \qquad (1.326)$$

with $h = \sum_{k=1}^{N} c_k h_k$, $h_k = h_k^0(T_0) + \int_{T_0}^{T} c_{p_k}dT$, $\mathbf{J}_q = \mathbf{q} + \sum_{k=1}^{N} h_k\mathbf{J}_k$ and $\Phi = \operatorname{div}(\tau\cdot\mathbf{u}) - \mathbf{u}\operatorname{div}\tau = \nabla^i(\tau_{ij}u^j) - u^j\nabla^i\tau_{ij}$. Sometimes it is more convenient to use energy equations with temperature T as unknown function. To do it we should use the following transformations:

$$dh = d\Big(\sum_{k=1}^{N} c_k h_k\Big) = \sum_{k=1}^{N}(c_k dh_k + h_k dc_k)$$

$$= \sum_{k=1}^{N} c_k c_{p_k} dT(c_k dh_k) + \sum_{k=1}^{N} h_k dc_k = c_p dT + \sum_{k=1}^{N} h_k dc_k. \qquad (1.327)$$

Considering (1.327) and (1.324), energy equation (1.326) may be rewritten in the form

$$\rho c_p \frac{dT}{dt} = \frac{dp}{dt} - \text{div } \mathbf{J}_q + \Phi - \sum_{k=1}^{N} h_k \dot{w}_k - \nabla T \cdot \sum_{k=1}^{N} c_{p_k} \mathbf{J}_k.$$

State equation:

$$p = \rho R_0 T \sum_{k=1}^{N} (c_k/m_k). \qquad (1.328)$$

1.8.2 Closing relations and transport coefficients

The above-listed differential equations have the simple useful first integrals:

$$\sum_{i=1}^{N} c_i = 1, \quad \sum_{i=1}^{N} J_i = 0.$$

The closing relations between fluxes and gradients of temperature, pressure and species concentrations can be represented in more convenient form [42]:

$$\rho \nabla x_i = \sum_{k=1}^{N} a_{ik} \frac{m^2}{m_i m_k} \left\{ (c_i \mathbf{J}_k - c_k \mathbf{J}_i) + \nabla(\ln T)(c_i D_k^T - c_k D_i^T) \right\} + \rho \nabla(\ln p)(c_i - x_i), \quad (1.329a)$$

$$\mathbf{q} = -\lambda \nabla T + R_A T \sum_{k,j=1}^{N} \frac{x_k a_{jk}}{m_j} D_j^T (\mathbf{J}_j/\rho_j - \mathbf{J}_k/\rho_k). \qquad (1.329b)$$

In practice, instead of a_{ij} the coefficients $D_{ij} = 1/a_{ij}$ are used, called *binary coefficients of diffusion*.

Let us calculate the total number of unknown functions and the number of equations in system (1.323)–(1.329b). The list of unknown functions is the following: three components of velocity $\mathbf{u}$, pressure p, density ρ, temperature T, N mass concentrations of species c_i, $3N$ components of vectors of diffusion fluxes $\mathbf{J}_i$. Thus, the number of unknown functions equals $4N + 6$. As for equations we have: one equation (1.323), N equations (1.324), three equations (1.325), one equation (1.326), one equation (1.328), $3N$ equations (1.329a). So, the number of equations is also $4N + 6$, thus our system is closed. The coefficients μ, λ, D_{ij}, etc. can be determined either from experimental data or from molecular gas theory. Here we present formulas from the molecular gas theory (for Lenard-Jones potential) for air:

$$\mu = \sum_{i=1}^{N} \frac{\mu_i}{c_i + G_i}, \quad \mu_i = 2.6693 \times 10^{-5} \frac{\sqrt{m_i T}}{\sigma_i^2 \Omega_i^{(2.2)*}},$$

$$\lambda = \sum_{i=1}^{N} \frac{\lambda_i c_i}{c_i + 1.065 G_i}, \quad \lambda_i = \frac{15}{4} \mu_i \cdot (R_A/m_i)(i = O, N),$$

$$\lambda_i = \frac{15}{4} \mu_i \cdot (R_A/m_i) \cdot (0.115 + 0.354 c_{pi}/R_A)(i = O_2, N_2, NO),$$

$$D_{ij} = 0.00268(T^{3/2}/(p\sigma_{ij}^2 \Omega_{ij}^{(1.1)*}))\sqrt{\frac{1}{2}(m_j + m_i)/m_i m_j}, \quad G_i = m_i \sum_{j=1, \, j\neq i}^{N} G_{ij} \frac{c_j}{m_j},$$

$$G_{ij} = \frac{1}{\sqrt{8}} \sqrt{m_j/(m_j + m_i)}[1 + (m_j/m_i)^{0.25}(\mu_i/\mu_j)^{0.5}]^2,$$

$$\Omega_{ij}^{(1.1)*} = 1.074 \cdot (kT/\varepsilon_{ij})^{-0.1604}, \quad \Omega_i^{(2.2)*} = 1.157 \cdot (kT/\varepsilon_i)^{-0.1472},$$

$$\sigma_{ij} = \frac{1}{2}(\sigma_i + \sigma_j), \quad \varepsilon_{ij} = \sqrt{\varepsilon_i \varepsilon_j},$$

where μ_i – viscosity of species i [g / cm · s], λ_i – thermal conductivity of species i [cal / cm · s · K], D_{ij} – binary diffusion coefficient [cm^2/s], m_i – molecular mass i [g / mol], p – pressure [atm], T – temperature [K], k – Boltzmann's constant, σ_i – the distance between particles of i-th species with interaction energy to be zero, ε_i – maximal interaction energy of i-th species; see the values for air in Table 1.5.

TABLE 1.5: The values ε_i/k and σ_i for air.

O_2	$\varepsilon/k = 106.7\text{K}$	$\sigma = 3.467$
N_2	$\varepsilon/k = 71.4\text{K}$	$\sigma = 3.798$
NO	$\varepsilon/k = 116.7\text{K}$	$\sigma = 3.492$
N	$\varepsilon/k = 71.4\text{K}$	$\sigma = 3.298$
O	$\varepsilon/k = 106.7\text{K}$	$\sigma = 3.050$

1.8.3　Boundary conditions

For a correct statement of the problem it is necessary to add to Navier–Stokes equations the boundary conditions. For external steady gas dynamic problems the boundary conditions should be given at the body surface and on the infinity. The boundary conditions at infinity are trivial:

$$\mathbf{u}\big|_{\mathbf{r}\to\infty} = \mathbf{V}_\infty, \quad p\big|_{\mathbf{r}\to\infty} = p_\infty, \quad T\big|_{\mathbf{r}\to\infty} = T_\infty, \quad c_i\big|_{\mathbf{r}\to\infty} = c_{i,\infty}. \tag{1.330}$$

The type of boundary conditions at the body surface depends on the problem parameters, e.g., on Reynolds number at infinity $Re_\infty = \rho_\infty V_\infty L_\infty/\mu_\infty$. First, consider a non-permeable surface. If Reynolds number at the body surface is rather large ($Re \geq 100$), the classical nonslip conditions are used:

$$u_\tau = 0, \quad u_n = 0, \quad T = T_w, \tag{1.331}$$

where u_τ and u_n are tangent and normal components of the velocity vector with respect to the body surface. The boundary conditions to diffusion equations are

$$(J_{in})_w = \dot{r}_i \quad (i = 1,\ldots,N_r), \qquad (J^*_{jn})_w = 0 \quad (j = N_r + 1,\ldots,N), \tag{1.332}$$

where $\dot{r}_i$ is the total net mass rate of production of species i-th per unit time per unit surface due to chemical reactions proceeding at the body surface. In fact these conditions present the mathematical formulation of the law which affirms that diffusion flux of the species is equal to the surface mass source of species due to heterogeneous surface chemical reactions.

If the body surface is permeable conditions (1.332) are modified as follows:

$$\begin{aligned}
(\rho u_n)_w(c_i - c_i^{(1)}) + (J_{in})_w &= \dot{r}_i \quad (i = 1,\ldots,N_r), \\
(\rho u_n)_w(c_j^* - c_j^{*(1)}) + (J^*_{jn})_w &= 0 \quad (j = N_r + 1,\ldots,N),
\end{aligned} \tag{1.333}$$

where $c_i^{(1)}$ and $c_j^{*(1)}$ are mass concentrations of i-th species and j-th chemical element in a gas mixture which is injected through the surface. In addition, instead of $(u_n)_w = 0$ the following condition is used:

$$(\rho u_n)_w = G,$$

where G is a known function. For low Reynolds numbers we should modify the boundary conditions for the flow velocity tangent component and temperature because for this regime the relations between velocity and temperature values at the body surface and their gradients for normal direction are true.

Definition 48 These links are called *slip conditions* and they can be presented in the form

$$u^\alpha = \frac{2-\vartheta}{\vartheta}\sqrt{\frac{\pi}{2}\frac{m}{R_A}}\frac{\mu}{\rho T^{1/2}}\frac{\partial u^\alpha}{\partial n}, \quad T_w = T_w^{(1)} + \frac{2-\nu}{\nu}\sqrt{\frac{\pi}{2}\frac{m}{R_A}}\frac{\lambda}{\rho(c_p + c_V)T^{1/2}}\frac{\partial T}{\partial n}, \tag{1.334}$$

where ϑ is the coefficient of diffusion reflectivity, ν is the coefficient of accommodation and $T_w^{(1)}$ is the body surface's temperature.

Let us compute the number of boundary conditions (1.330)–(1.334) and the order of the system of differential equations (1.323)–(1.329b). The boundary conditions contain $N + 5$ conditions (1.330) (at infinity), four conditions (1.331) and N conditions (1.332)–(1.333) (at the body surface). There are $2N + 9$ boundary conditions. As for equations we have first-order equation (1.323), N second-order equations (1.324), three second-order equations (1.325) and one second-order equation (1.326). The order of our system of differential equations is $2N + 9$, hence our boundary value problem is correct.

1.8.4   Deducing Navier–Stokes equation

In this section we deduce the *Navier–Stokes equations* (NSE) without influence of external forces. This assumption does not change the generality of the derivation of the equation but merely simplifies the form of the expressions which will allow the reader to focus his attention on the main.

In Section 1.8.1 we learned the existence of surface forces to describe fluid motion in addition to the volume force (such as gravity). The surface force is also termed as the stress and represented by a tensor. In fluid mechanics, the total stress tensor σ is the sum $\sigma = \sigma^{(p)} + \sigma^{(v)}$ of the pressure stress tensor $\sigma_{ij}^{(p)}$ and the viscous stress tensor $\sigma_{ij}^{(v)}$. The surface force is termed as the stress and represented by a tensor. A typical stress is the pressure $p\delta_{ij}$ (δ_{ij} represents *isotropic tensor*), which is written as

$$\sigma_{ij}^{(p)} = -p\delta_{ij}.$$

The pressure force acts perpendicularly to a surface element $\delta A(\mathbf{n})$ and is represented as $-pn_i dA$. The viscous stress (internal friction) $\sigma_{ij}^{(v)}$ has a tangential force to the surface $\delta A(\mathbf{n})$ as well. This tensor can be represented in the form

$$\sigma_{ij}^{(v)} = \mu\, E_{ij},$$

where the strain rate tensor E_{ij} is defined as follows:

$$E_{ij}(x,t) = \frac{1}{2}\Big(\frac{\partial u_i}{\partial x_j} + \frac{\partial u_j}{\partial x_i}\Big).$$

These components usually change with the coordinate $x = (x_1, x_2, x_3)$ and time t. The strain rate tensor E_{ij} expresses the rate at which the mean velocity changes in the medium as one moves away from the point x, except for the changes due to rotation of the medium about the point x as a rigid body, which do not change the relative distances of the particles and only contribute to the rotational part of the viscous stress via the rotation of the individual particles themselves.

For arbitrary viscous flow, tensor $\sigma^{(p)}$ is fully defined by denoting six scalar values. Consequently, it is necessary to involve a definite number of restrictions between components of stress tensor and velocity vector (with their partial derivations). The Newton law describes inter-restrictions between velocity field and the tensor of the field $d\mathbf{u}/d\mathbf{r}$. Assuming the motion direction along x_i, following Newton law, the friction force is equal to

$$p_{ij} = \mu\frac{\partial u_i}{\partial x_j}. \tag{1.335}$$

The fluids satisfying the law (1.335) are called *Newton fluids*.

Proposition 8 *Tensor $d\mathbf{u}/d\mathbf{r}$ can be presented as a sum of symmetric and antisymmetric tensors.*

Proof. Indeed, $\frac{d\mathbf{u}}{d\mathbf{r}} = S + T$, where $S_{ij} = \frac{1}{2}\left(\frac{\partial u_i}{\partial x_j} + \frac{\partial u_j}{\partial x_i}\right)$ is a symmetric (small deformation) tensor, and $T_{ij} = \frac{1}{2}\left(\frac{\partial u_i}{\partial x_j} - \frac{\partial u_j}{\partial x_i}\right)$ is an antisymmetric tensor. $\square$

Proposition 9 (Navier–Stokes equation) *Motion equation for viscous flow has the form*

$$\rho\frac{d\mathbf{u}}{dt} = \rho\mathbf{F} - \nabla p + \mu\Delta\mathbf{u}. \tag{1.336}$$

Proof. This consists of four parts based on four hypotheses.

Hypothesis 1. Antisymmetric tensor T describes quasi-solid motion of elementary fluid particles when all viscous forces equal zero. Thus, all components of tensor $\sigma^{(p)}$ depend only on components of the small deformation tensor S. Newton law formulates linear depending between these components only for the case of plane-parallel motion. So, one should postulate linear restriction between tensors $\sigma^{(p)}$ and S for arbitrary motion; also, $\sigma^{(p)} = AS + B$, where A depends only on the media properties, and B may depend on the first invariants of tensor S, i.e., on scalars, div $\mathbf{u}$ and $g^{ik}p_{ik}$.

Hypothesis 2 consists of the supposition of the isotropic media assumption, i.e., that physical properties of the media do not depend on the directions of the motion. Saying another, the main directions of tensors $\sigma^{(p)}$ and S must be collinear. Indeed, assume the contrary, i.e., let this condition not be satisfied. Then, we choose the main axis of tensor S as coordinate system and so nondiagonal elements S_{ij} $(i \neq j)$ will be zero. If the corresponding component is not equal to zero, we should ascribe it the sign plus or minus that leads to excretion of some preemptive directions that are contrary to the second hypothesis assumption. From this hypothesis, it follows that tensors A and B are diagonal in an arbitrary coordinate system. Moreover, their tensor ellipsoids have equal semi-axes, i.e., $A = aF$ and $B = bE$, where $a = a(\mu)$ depending on viscosity, and b is linear function of the first invariants of tensors $\sigma^{(p)}$ and S, $b = b_0 + b_1$ div $\mathbf{u} + b_2 g^{ij}p_{ij}$. We will find the restriction between tensors $\sigma^{(p)}$ and S in the form

$$\sigma^{(p)} = aS + \left(b_0 + b_1 \text{ div } \mathbf{u} + b_2 g^{ij}p_{ij}\right)E. \tag{1.337}$$

Hypothesis 3. In order to find coefficients a, b_0, b_1 and b_2, we involve an additional hypothesis: the law (1.337) must include Newton law as a partial case. Consider plane-parallel motion along axis x_i with $u_i = u_i(x_i, x_j)$, $u_j = 0$ $(j \neq i)$. Using (1.335) and (1.337), we obtain $p_{ij} = \frac{1}{2} a\left(\partial u_i/\partial x_j\right)$, that is, $a = 2\mu$. To define scalars b_0, b_1 and b_2, we should let the first invariants of tensors in both hand-sides of (1.337) be equal to each other,

$$g^{ij}p_{ij} = 2\mu \text{ div } \mathbf{u} + 3b_0 + 3b_1 \text{ div } \mathbf{u} + 3b_2 g^{ij}p_{ij}.$$

This equality must be satisfied for arbitrary form of the motion, i.e., must have a place for any values of $g^{ij}p_{ij}$ and div $\mathbf{u}$. Hence, equating corresponding addends, we obtain $b_0 = 0$, $b_1 = -\frac{2}{3}\mu$, $b_2 = \frac{1}{3}$. Formula (1.337) may be rewritten as

$$\sigma^{(p)} = 2\mu S + \left(\frac{1}{3} g^{ij}p_{ij} - \frac{2}{3}\mu \text{ div } \mathbf{u}\right)E. \tag{1.338}$$

It is known [27] that $g^{ij}p_{ij} = -3p$ for ideal inviscid flow.

Hypothesis 4. Assume that in a viscous case this is also true. Then, (1.338) has a view

$$\sigma^{(p)} = 2\mu S - \left(p + \frac{2}{3}\mu \text{ div } \mathbf{u}\right)E. \tag{1.339}$$

Formula (1.339) is called *general Newton law*. Equality (1.339) may be rewritten in scalar form

$$p_{ij} = \begin{cases} \mu(\partial u_i/\partial x_j + \partial u_j/\partial x_i) \\ 2\mu \, \partial u_i/\partial x_j - p - \frac{2}{3}\mu \text{ div } \mathbf{u}. \end{cases} \tag{1.340}$$

Let $\mathbf{w}$ be an acceleration of fluid particles, ρ density, $\mathbf{F}$ mass force's stress, $\mathbf{p}_n$ the surface force's stress. Using *d'Alembert principle* for exuding material system inside finite volume V bounded by surface Σ, we get

$$-\int_V \rho \mathbf{w}\, dV + \int_V \rho \mathbf{F}\, dV + \int_S \mathbf{p}_n\, d\Sigma = 0. \tag{1.341}$$

By the Gauss theorem, (1.341) is transformed as $\int_S \mathbf{p}_n d\Sigma = \int_V \operatorname{div} \sigma^{(p)} dV$, where $\operatorname{div} \sigma^{(p)} = \sum_i \partial p_i / \partial x_i$. Using (1.341) and arbitrary of the volume V, we obtain the general differential equation for fluid motion

$$\mathbf{w} = \mathbf{F} + \frac{1}{\rho} \operatorname{div} \sigma^{(p)}. \tag{1.342}$$

This equation is called a *motion equation in stress terms*. Equalities (1.340) give possibility to exclude components of tensor $\sigma^{(p)}$ from (1.342). Hence, the presentation for $\operatorname{div} \sigma^{(p)}$ has a view

$$\operatorname{div} \sigma^{(p)} = \sum_k \left(\frac{\partial p_{1k}}{\partial x_1} \mathbf{x}_k^0 + \frac{\partial p_{2k}}{\partial x_2} \mathbf{x}_k^0 + \frac{\partial p_{3k}}{\partial x_3} \mathbf{x}_k^0 \right) = \mathbf{x}_1^0 \sum \frac{\partial p_{k1}}{\partial x_k} + \mathbf{x}_2^0 \sum \frac{\partial p_{k2}}{\partial x_k} + \mathbf{x}_3^0 \sum \frac{\partial p_{k3}}{\partial x_k}. \tag{1.343}$$

Using (1.340) and (1.342) the stress equation (1.343) may be rewritten in scalar form:

$$\rho \partial_t u_1 = \rho \mathbf{F}_1 - \frac{\partial p}{\partial x_1} + 2\frac{\partial}{\partial x_1}\left(\mu \frac{\partial u_1}{\partial x_1}\right) + \frac{\partial}{\partial x_2}\left[\mu\left(\frac{\partial u_1}{\partial x_2} + \frac{\partial u_2}{\partial x_1}\right)\right] + \frac{\partial}{\partial x_3}\left[\mu\left(\frac{\partial u_1}{\partial x_3} + \frac{\partial u_3}{\partial x_1}\right)\right] - \frac{2}{3}\frac{\partial}{\partial x_1}\left(\mu \operatorname{div} \mathbf{u}\right),$$
$$\rho \partial_t u_2 = \rho \mathbf{F}_2 - \frac{\partial p}{\partial x_2} + 2\frac{\partial}{\partial x_1}\left[\mu\left(\frac{\partial u_1}{\partial x_2} + \frac{\partial u_2}{\partial x_1}\right)\right] + 2\frac{\partial}{\partial x_2}\left(\mu \frac{\partial u_2}{\partial x_2}\right) + \frac{\partial}{\partial x_3}\left[\mu\left(\frac{\partial u_2}{\partial x_3} + \frac{\partial u_3}{\partial x_2}\right)\right] - \frac{2}{3}\frac{\partial}{\partial x_2}\left(\mu \operatorname{div} \mathbf{u}\right),$$
$$\rho \partial_t u_3 = \rho \mathbf{F}_3 - \frac{\partial p}{\partial x_3} + 2\frac{\partial}{\partial x_1}\left[\mu\left(\frac{\partial u_1}{\partial x_3} + \frac{\partial u_3}{\partial x_1}\right)\right] + \frac{\partial}{\partial x_2}\left[\mu\left(\frac{\partial u_2}{\partial x_3} + \frac{\partial u_3}{\partial x_2}\right)\right] + 2\frac{\partial}{\partial x_3}\left(\mu \frac{\partial u_3}{\partial x_3}\right) - \frac{2}{3}\frac{\partial}{\partial x_3}\left(\mu \operatorname{div} \mathbf{u}\right). \tag{1.344}$$

For an inviscid case, and when additionally $\mu = \text{const}$, the system (1.344) is sufficiently simplified, and may be rewritten in vector form (1.336). $\qquad\square$

Equations (1.343) and (1.336) are called *Navier–Stokes equations*.

More detailed description of the stuff contained in the present section is given in [18].

1.8.5 Existence and uniqueness of solutions of the Navier–Stokes equation

In Section 1.8.4, we confined ourselves to the study of the NSE in the absence of external forces. In the present paragraph, we give results and present statement problems for the total case when external forces are not negligible. We would like to note that the problem of existence and uniqueness of solutions to the NSE have been occupied by mathematicians and theoretical physicists for about a hundred years. For this huge period for modern science, many articles and monographs have been published; considerable material was included in the manuscripts of textbooks, lecture courses, as well as in special courses and the preparation of graduate works and doctoral theses. We cannot even partially reveal this problem in one small paragraph. We only present our vision of problems with the enumeration of some already solved problems, and, most importantly, some of the main unsolved problems facing the researchers.

The *Navier-Stokes existence and smoothness problem* concerns the mathematical properties of solutions to the NSE, one of the pillars of fluid mechanics, such as with turbulence. These equations describe the motion of a fluid (that is, a liquid or a gas) in space. Solutions of the NSE are used in many practical applications. The evolution of NSE was considered in [45]. However, theoretical understanding of solutions to these equations is incomplete. In particular, solutions of the NSE often include turbulence, which remains one of the greatest unsolved problems in physics, despite its immense importance in science and engineering.

Even much more basic properties of solutions to NSE have never been proven. For the three-dimensional system of equations, and given some initial conditions, mathematicians have not yet proved that smooth solutions always exist, or that if they do exist, they have bounded energy per unit mass, see [52]. Since understanding the NSE is considered to be the first step to understanding the elusive phenomenon of turbulence, the Clay Mathematics Institute in May of 2000 made this problem one of its seven Millennium Prize problems in mathematics, see [16].

The NSE are a system of nonlinear PDEs for abstract vector fields of any size. In physics and engineering, they are used to build models the motion of liquids or non-rarefied gases (in which the mean free path is short enough so that it can be thought of as a continuum mean instead of a collection of particles) using continuum mechanics. The NSE are a statement of Newton's second law, with the forces modeled according to those in a viscous Newtonian fluid as the sum of contributions by pressure, viscous stress and an external body force. Since the setting of the problem proposed by the Clay Mathematics Institute is in three dimensions, for an incompressible and homogeneous fluid, only that case is considered below.

Let $\mathbf{u}(\mathbf{r}, t)$ be a 3d vector field, the velocity of the fluid, and let $p(\mathbf{r}, t)$ be the pressure of the fluid (more precisely, $p(\mathbf{r}, t)$ is the pressure divided by the fluid density, and the density is constant for this incompressible and homogeneous fluid). Then NSE are

$$\partial_t \mathbf{u} + (\mathbf{u} \cdot \nabla)\mathbf{u} = \nu \nabla^2 \mathbf{u} - \nabla p + \mathbf{Q}(\mathbf{r}, t), \qquad (1.345)$$

where ν is the kinematic viscosity, $\mathbf{Q}(\mathbf{r}, t)$ the external volumetric force. Since in 3d, there are three equations and four unknowns (three scalar velocities and the pressure), with the initial condition

$$\mathbf{u}(\mathbf{r}, 0) = \mathbf{u}_0(\mathbf{r}),$$

then a supplementary equation is needed. This extra equation is the continuity equation for incompressible fluids that describes the conservation of mass of the fluid:

$$\mathrm{div}\ \mathbf{u} = 0. \qquad (1.346)$$

Due to this last property, the solutions for the NSE are searched in the set of solenoidal ("divergence-free") functions. For this flow of a homogeneous medium, density and viscosity are constants. The pressure p can be eliminated by taking an operator rot (alternative notation: curl) of both sides of the NSE. In this case, the NSE reduce to the vorticity-transport equations.

Two settings: unbounded and periodic space. There are two settings for the Navier–Stokes existence and smoothness problem. The original problem is in the whole $\mathbb{R}^3$, which needs extra conditions on the growth behavior of the initial condition and solutions. In order to rule out the problems at infinity, the NSE be set in a periodic framework, which implies that they are no longer working on the whole space $\mathbb{R}^3$ but in the 3-dimensional torus $T^3 = \mathbb{R}^3 / \mathbb{Z}^3$. Each case will be treated separately.

Statement of the problem in the whole space. Indeed, for the 2d NSE there is no vorticity stretching term, and for this case one has global regularity. (This is still the case for the 2d Euler equation, $\nu = 0$). In particular, one can think of the 2d NSE as a spatially symmetric solution of the 3d equations. Global strong solutions for the 3d NSE with large data also exist with axial, rotational and helical symmetry. Ladyzhenskaya [26] proved the global existence of rotationally symmetric large strong solutions in domains Ω which are obtained by rotation about the x_3-axis of a planar domain Ω lying in the half-plane $\{x_2 = 0,\ x_1 > 0\}$ at a positive distance from the x_3-axis, assuming the angular components of the force $\mathbf{Q}$ and the initial data $\mathbf{u}_0$ do not depend on the angle of rotation ϕ. Ukhovskii-Iudovich [46] proved the global existence of large strong axially symmetric solutions in the whole space. By axial symmetry is meant that the solution is rotationally symmetric and its component in the x_3 direction is zero. Recently, Mahalov-Titi-Leibovich [28] established the existence of unique global strong solutions for large helically symmetric data. There, Ω is an infinite periodic pipe in the x_3-direction with Dirichlet boundary conditions on the sides. Helical symmetry means that the solution $\mathbf{u}(r, \phi, x_3)$ in cylindrical coordinates only depends on r and $n\phi + ax_3$, where $n \in \mathbb{Z} \setminus \{0\}$ and $a > 0$ are given parameters. As in 2d, the axially symmetric flow has no vorticity stretching. Global existence and regularity follow in these cases by estimating directly from the vorticity equation. On the other hand, for the 3d NSE with rotational or helical symmetry, the vorticity stretching term is nontrivial. Here, one can take advantage of the symmetry to reduce the spatial dimension of the problem and derive new Sobolev estimates.

Hypotheses and growth conditions. The initial condition $\mathbf{u}_0(\mathbf{r})$ is assumed to be smooth and divergence-free (see smooth function) such that, for every multi-index α and any $K > 0$, there exists a constant $C_{\alpha K} = C(\alpha, K) > 0$ such that

$$|\partial^\alpha \mathbf{u}_0(\mathbf{r})| \leq C_{\alpha K}/(1 + |\mathbf{r}|)^K, \quad \mathbf{r} \in \mathbb{R}^3. \tag{1.347}$$

The external force $\mathbf{Q}(\mathbf{r}, t)$ is assumed to be a smooth function as well, satisfying an analogous inequality (now the multi-index includes time derivatives as well):

$$|\partial_\mathbf{r}^\alpha \partial_t^m \mathbf{Q}(\mathbf{r}, t)| \leq C_{\alpha m K}/(1 + |\mathbf{r}| + t)^K, \quad (\mathbf{r}, t) \in \mathbb{R}^3 \times [0, \infty). \tag{1.348}$$

1. For physically reasonable conditions, the type of solutions expected are smooth functions that do not grow large as $|r| \to \infty$. More precisely, the following assumptions are made:

$$\mathbf{u}(\mathbf{r}, t) \in C^\infty(\mathbb{R}^3 \times [0, \infty)), \quad p(\mathbf{r}, t) \in C^\infty(\mathbb{R}^3 \times [0, \infty)). \tag{1.349}$$

2. There exists a constant $E > 0$ such that

$$\int_{\mathbb{R}^3} |\mathbf{u}(\mathbf{r}, t)|^2 d\mathbf{r} < E \text{ for all } t \geq 0 \quad \text{(bounded energy).} \tag{1.350}$$

Condition 1 implies that the functions are smooth and globally defined and condition 2 means that the kinetic energy of solutions is globally bounded.

Statement of the periodic problem; hypotheses. The functions sought now are periodic in the space variables of period 1. More precisely, let e_i be the unitary vector in the i-direction.

1. Then $\mathbf{u}(\mathbf{r}, t)$ is periodic in the space variables if for any $i = 1, 2, 3$, hence:

$$\mathbf{u}(\mathbf{r} + \mathbf{e}_i, t) = \mathbf{u}(\mathbf{r}, t), \quad (\mathbf{r}, t) \in C^\infty(\mathbb{R}^3 \times [0, \infty)) \tag{1.351}$$

$$\mathbf{u}_0(\mathbf{r} + \mathbf{e}_i) = \mathbf{u}_0(\mathbf{r}), \quad \mathbf{Q}_0(\mathbf{r} + \mathbf{e}_i, t) = \mathbf{Q}_0(\mathbf{r}, t), \quad (\mathbf{r}, t) \in C^\infty(\mathbb{R}^3 \times [0, \infty)). \tag{1.352}$$

In place (1.347) and (1.348), we assume that $\mathbf{u}_0(\mathbf{r})$ is smooth and that for any α, m, K:

$$|\partial_\mathbf{r}^\alpha \partial_t^m \mathbf{Q}(\mathbf{r}, t)| \leq C_{\alpha m K}/(1 + t)^K, \quad (\mathbf{r}, t) \in \mathbb{R}^3 \times [0, \infty). \tag{1.353}$$

2. Notice that this is considering the coordinates mod 1. This allows us to work not on the whole space $\mathbb{R}^3$ but on the quotient space $\mathbb{R}^3/\mathbb{Z}^3$, which turns out to be the 3-dimensional torus T^3.

3. Now the hypotheses can be stated properly. The initial condition $\mathbf{u}_0(\mathbf{r})$ is assumed to be a smooth and divergence-free function and the external force $\mathbf{Q}(\mathbf{r}, t)$ is assumed to be a smooth function as well. The type of solutions that are physically relevant are those who satisfy these conditions:

$$\mathbf{u}(\mathbf{r}, t) \in [C^\infty(T^3 \times [0, \infty))]^3, \quad p(\mathbf{r}, t) \in C^\infty(T^3 \times [0, \infty)).$$

4. There exists a constant $E > 0$ such that $\int_{T^3} |\mathbf{u}(\mathbf{r}, t)|^2 \, d\mathbf{r} < E$ for all $t \geq 0$ (bounded energy). Just as in the previous case, condition 3 implies that the functions are smooth and globally defined and condition 4 means that the kinetic energy of solutions is globally bounded. Moreover, it was shown in [22] that the periodicity problem

$$\partial_t \mathbf{u} + \mathbf{u} \cdot \nabla \mathbf{u} = -\nu \nabla^2 \mathbf{u} - \nabla p + \mathbf{Q}(\mathbf{r}, t), \quad \mathbf{r} \in \Omega, \ t \in \mathbb{R},$$

$$\text{div } \mathbf{u} = 0, \quad \mathbf{r} \in \Omega, \ t \in \mathbb{R},$$

$$\mathbf{u}|_{\partial\Omega} = 0, \quad t \in \mathbb{R}, \ \Omega \text{ bounded domain in } \mathbb{R}^m, \ \partial\Omega \text{ -smooth,}$$

may be formulated as an ordinary differential equation in the Hilbert space H_σ, which is defined as the closure of $C_{0,\sigma}^\infty$ in $L_2(\Omega)$

$$\frac{d}{dt} \mathbf{u} + A\mathbf{u} + P\mathbf{u} \cdot \nabla \mathbf{u} = P\mathbf{Q}(\mathbf{r}, t), \quad \mathbf{r} \in \Omega, \ t \in \mathbb{R},$$

$$\mathbf{u}(\mathbf{r}, t + \omega) = \mathbf{u}(\mathbf{r}, t), \quad t \in \mathbb{R} \tag{1.354}$$

by orthogonal decomposition $L_2(\Omega) = H_\sigma \oplus \{\nabla p : \ p \in H^1(\Omega)\}$.

Here, $C_{0,\sigma}^\infty = \{\phi \in C_0^\infty(\Omega) : \ \text{div } \phi = 0\}$, $H^m(\Omega)$ is the Sobolev space of vector-valued functions which are in $L_2(\Omega)$ together with their derivatives up to order m, $H_0^m(\Omega)$ is the completion of the set C^∞ in $H^m(\Omega)$, P is the orthogonal projection from $L_2(\Omega)$ onto H_σ. By the Stokes operator A there is denoted the Friedrichs extension of the symmetric operator $-P\Delta$ in H_σ with domain $D(A) = H^2(\Omega) \cap H_0^1(\Omega) \cap H_\sigma$. It was shown [22] that $D(A^{1/2}) = H_{0,\sigma}^1$, where $H_{0,\sigma}^1$ is the closure of $C_{0,\sigma}^\infty$ in $H^1(\Omega)$. As a result it was proven [22] that the solutions of the problem (1.354) exist and are unique.

Remarks in higher dimensions. If Ω is an exterior domain of $\mathbb{R}^m$ ($m > 3$) of class C^1, then for $\mathbf{a} \in L_{m-1}(\partial\Omega)$ the Stokes problem (1.357) in this case has a solution $\mathbf{u}_s \in L_m(\Omega)$ satisfying inequality $\|\mathbf{u}_s\|_{L_{m-1}(\partial\Omega)} \leq \gamma \|\mathbf{a}_{L_{m-1}(\partial\Omega)}\|$, see [38]. For $\mathbf{u} \in L_m(\Omega)$ consider the functional equation

$$\mathbf{u}' = \mathbf{u}_s + (\mathcal{K} + \mathcal{V})[\mathbf{u}] \tag{1.355}$$

in $L_m(\Omega)$, where the operator $\mathcal{K} + \mathcal{V}$ is defined in [9]. Taking into account that

$$\|(\mathcal{K} + \mathcal{V})[\mathbf{u}]\|_{L_m(\Omega)} \leq c \|\mathbf{u}\|_{L_m(\Omega)}^2,$$

and if $\|\mathbf{a}\|_{L_{m-1}(\Omega)}^2$ is sufficiently small, then $\mathbf{u}'$ is a contraction in a ball of $L_m(\Omega)$ and the fixed point of (1.355) is a smooth solution to the problem

$$\nu \nabla^2 \mathbf{u} - \mathbf{u} \cdot \nabla \mathbf{u} - \nabla p = 0, \quad \mathbf{r} \in \Omega,$$

$$\text{div } \mathbf{u} = 0, \quad \mathbf{r} \in \Omega,$$

$$\mathbf{u}|_{\partial\Omega} = \mathbf{a}, \quad \lim_{|\mathbf{r}| \to \infty} \mathbf{u}(\mathbf{r})|_{\partial\Omega} = 0 \tag{1.356}$$

in the m-dimensional exterior domain $\Omega = \mathbb{R}^m \setminus \cup_{i=1}^l \bar{\Omega}_i$, where Ω_i are bounded Lipschitz domains connected boundaries such that $\Omega_i \cup \Omega_j = \emptyset$ ($i \neq j$). In (1.356), $\mathbf{u}$ and p are the kinetic and pressure fields, and $\mathbf{a}$ is an assigned field on $\partial\Omega$. The linearized version of (1.356) yields the Stokes equations

$$\nu \nabla^2 \mathbf{u} - \nabla p = 0, \quad \mathbf{r} \in \Omega,$$

$$\text{div } \mathbf{u} = 0, \quad \mathbf{r} \in \Omega,$$

$$\mathbf{u}|_{\partial\Omega} = \mathbf{a}, \quad \lim_{|\mathbf{r}| \to \infty} \mathbf{u}(\mathbf{r})|_{\partial\Omega} = 0 \tag{1.357}$$

Note that if $\mathbf{a} \in L_3(\partial\Omega)$, then (1.357) has a solution $(\mathbf{u}_s, p_s)$ (see [38]) analytical in Ω such that $\mathbf{u}_s = O(r^{-2})$, $p_s = O(r^{-3})$ and which satisfies (1.357) in the sense of the non tangential convergence, i.e., there is a finite cone Γ such that $\lim \mathbf{u}(\mathbf{r}) = \mathbf{a}(\xi)$, for almost all $\xi \in \partial\Omega$, where $\Gamma_\xi \subset \Omega$ is a cone with vertex ξ, congruent to Γ. Moreover, if $\mathbf{u} + \mathbf{v} \in L_m(\Omega)$ is another solution to (1.356), then by Schwarz's and Sobolev's inequalities, we obtain

$$\nu \int_\Omega |\nabla \mathbf{w}|^2 \, d\mathbf{r} = \int_\Omega \mathbf{w} \cdot \nabla \mathbf{w} \cdot \mathbf{u} \, d\mathbf{r} \le \frac{m-1}{(m-2)\sqrt{m}} \|\mathbf{u}\|_{L_m(\Omega)} \cdot \|\nabla \mathbf{w}\|^2_{L_m(\Omega)}.$$

Hence, the above solution is unique in the ball $\|\mathbf{u}\|_{L_m(\Omega)} < \nu\sqrt{m}(m-2)/(m-1)$.

For domains of class C^1 one can repeat the argument (see the proof of [9, Theorem 1]) to see that for $\mathbf{a} \in L_{m-1}(\Omega)$ and fluxes obeying a condition of the type (1.345), then (1.345) has a solution $\mathbf{u} = \mathbf{u}_\varepsilon + \mathbf{v}_s + \mathbf{w}$, with $\mathbf{u}_\varepsilon$, $\mathbf{v}_s$ regular in Ω and $w \in D_0^{1,2}(\Omega)$. Here, $D_0(\Omega) \subset C(\mathbb{R}^m)$ is the subspace of continuous functions f such that $\sup_{|\mathbf{r}|>R} |f(\mathbf{r})| \to 0$ as $R \to \infty$. It is normed by $\|f\|_\infty = \sup_{\mathbf{r}\in\mathbb{R}^m} |f(\mathbf{r})|$. The domain $D^{1,2}(\Omega) = \{\phi \in L_1^{loc}(\Omega) : \|\nabla\phi\|_{L_2(\Omega)} < \infty\}$, and $D_0^{1,2}(\Omega)$ is the completion of $C_0^\infty(\Omega)$ with respect to $\|\nabla\phi\|_{L_2(\Omega)}$, where $L_1^{loc}(\Omega)$ are strongly measurable functions such that $\phi f \in L_p(\Omega)$ for all $\phi \in C_0^\infty(\Omega)$, where $C_0^\infty(\Omega)$ are functions that are smooth and compactly supported. Up to date, we do not have general results assuring that $\mathbf{w}$ is regular.

For domains of class $C^{1,1}$ (see, e.g., [5]) and boundary data in the Sobolev space $W^{1/4,4}(\partial\Omega)$ (see, e.g., [5]) problem (1.357) in bounded domains has been considered by several authors (see [19, p. 297] and the references therein). Here, $C^{k,\lambda}(\mathbb{R}^m)$ are functions $f \in C^k(\mathbb{R}^m)$ such that $\nabla^\alpha f \in C^\lambda(\mathbb{R}^m)$, $|\alpha| \le k$, and $W^{s,p}(\Omega)$ (where s is a positive integer) is L_p Sobolev space, normed by

$$\|\psi\|_{W^{s,p}} = \sum_{l=0}^s \sum_{|\alpha|=k} \|\nabla^\alpha \psi\|_p.$$

The steady-state boundary value problem for the NSE in 4D space was considered in [9].

Theorem 42 (existence and uniqueness, [9]) *If* $\mathbf{a} \in L_3(\partial\Omega)$ *and*

$$\mathcal{J} = \frac{1}{4\omega} \sum_{i=1}^m \left| \int_{\partial\Omega_i} (\mathbf{a} \cdot \mathbf{n}) \, d\mathbf{r}_i \right| \cdot \max_{\partial\Omega} |\mathbf{r} - \mathbf{r}_i|^{-2} < \nu,$$

where ω *is the measure of the unit sphere of* $\mathbb{R}^4$, $\mathbf{r}_i$ *is a fixed point on* Ω_i *and* $\mathbf{n}$ *is the outward (with respect to* Ω*) unit normal to* $\partial\Omega$*, then (1.356) has a solution* $(\mathbf{u}, p) \in [L_4(\Omega) \cap C^\infty(\Omega)] \times C^\infty(\Omega)$*. Moreover,* $\mathbf{u}$ *is unique in* $L_4(\Omega)$*, provided* $\|\mathbf{u}\|_{L_4(\Omega)} < \frac{4}{3}\nu$ *and*

$$\|\mathbf{u}_s\|_{L_4(\Omega)} + \frac{\|\mathbf{u}_s\|^2_{L_4(\Omega)}}{\frac{4}{3}\nu - \|\mathbf{u}_s\|_{L_4(\Omega)}} < \frac{4}{3}\nu, \tag{1.358}$$

where $\mathbf{u}_s$ *is the solution to (1.357) in* Ω *with the boundary data* $\mathbf{a}$*.*

Remark 10 1. Note that by inequality $\|\mathbf{u}_s\|_{L_m(\partial\Omega)} \le \gamma \|\mathbf{a}\|_{L_{m-1}(\partial\Omega)}$ and (1.358) is provided by $\gamma \|\mathbf{a}_s\|_{L_3(\partial\Omega)} < \frac{4}{3}\nu$ and

$$\gamma \|\mathbf{a}\|_{L_3(\partial\Omega)} + \frac{\gamma^2 \|\mathbf{a}\|^2_{L_3(\partial\Omega)}}{\frac{4}{3}\nu - \|\mathbf{a}\|_{L_3(\partial\Omega)}} < \frac{4}{3}\nu.$$

It was shown also in [40] that the proof of existence of a solution to (1.358) in $L_4(\Omega)$ requires only that the corresponding Stokes problem has a solution $\mathbf{u}_2 \in L_4(\Omega)$. Hence, it follows that if $\partial\Omega$ is of class $C^{1,1}$, then Theorem 42 can be extended to boundary data $\mathbf{a} \in W^{-1/4,4}(\partial\Omega)$, see [8*].

2. The Sobolev space $W^{s,p}$ of fractional order $s \in (0,1)$ consists of functions $f \in L_p(\Omega)$ such that

$$\|f\|_{W^{s,p}} = \left(\|f\|_{L_p(\Omega)}^p + \int_\Omega \frac{|f(x) - f(y)|^p}{|x - y|^{m+ps}}\, dx\, dy \right)^{1/p}.$$

The Sobolev space of negative order is defined by $H^{-k} = (H_0^k(\Omega))^*$, where $(*)$ means the conjugate space to $H_0^k(\Omega)$, and $H_0^k(\Omega)$ is the closure of the Sobolev space $H^k(\Omega)$, that is $W^{-s,p}(\Omega) = (W_0^{s,p}(\Omega))^*$.

Some partial results.

1. If the initial velocity $\mathbf{u}(\mathbf{r}, t)$ is sufficiently small, then the statement is true: *there are smooth and globally defined solutions to the NSE*, see [16].

2. Given an initial velocity $\mathbf{u}(\mathbf{r})$ there exists a finite time $T > 0$, depending on $\mathbf{u}_0(\mathbf{r})$ such that the NSE on $\mathbb{R}^3 \times (0, T)$ have smooth solutions $\mathbf{u}(\mathbf{r}, t)$ for different smooth $p(\mathbf{r}, t)$. It is not known if the solutions exist beyond that "blowup time" T, see [16].

3. In 2016 T. Tao published a finite time blowup result for an averaged version of the 3d NSE. He writes that the result formalizes a *supercriticality barrier* for the global regularity problem for the true NSE, and claims that the method of proof in fact hints at a possible route to establishing blowup for the true equations, see [43]. In this paper, a modification $\partial_t \mathbf{u} = \nu\nabla^2\mathbf{u} + \tilde{B}(\mathbf{u}, \mathbf{u})$ of this equation is considered, where $\tilde{B}$ is an averaged version of the bilinear operator B (where the average involves rotations, dilations and Fourier multipliers of order zero), and which also obeys the cancelation condition $\langle \tilde{B}(\mathbf{u}, \mathbf{u}), \mathbf{u} \rangle = 0$ (so that it obeys the usual energy identity). By analyzing a system of ODEs related to (but more complicated than) a dyadic Navier–Stokes model of Katz–Pavlovic [23], they construct an example of a smooth solution to such an averaged NSE which blows up in finite time.

4. The existence of global classical solutions to the 3d isentropic compressible NSE in a bounded domain was studied [53]. The considered problem in a cuboid domain is

$$\partial_t \rho + \operatorname{div}(\rho\mathbf{u}) = 0,$$
$$\partial_t(\rho\mathbf{u}) + \operatorname{div}(\rho\mathbf{u} \otimes \mathbf{u}) - \mu\nabla^2\mathbf{u} - (\mu + \lambda)\nabla(\operatorname{div}\mathbf{u}) + \nabla p(\rho) = 0, \qquad (1.359)$$

where $\rho \geq 0$, $\mathbf{u} = (u_1, u_2, u_3)$ and $p(\rho) = a\rho^\gamma$ (with $a > 0$ and polytropic coefficient $\gamma > 1$) represent the fluid density, velocity and pressure (dependent on fluid density ρ), respectively. The constant μ is the bulk viscosity coefficient satisfying the physical restrictions:

$$\mu > 0, \quad 2\mu + 3\gamma \geq 0.$$

The main purpose in this paper is to look for global classical solutions to (1.359) with initial data:

$$(\rho, \mathbf{u}(\mathbf{r}, t))|_{t=0} = (\rho_0, \mathbf{u}_0(\mathbf{r})), \quad \mathbf{r} \in [0, 1]^3,$$

and the boundary conditions

$$u_1 = \partial_1 u_2 = \partial_1 u_3 = 0 \text{ for } x_1 = 0 \text{ and } x_1 = 1,$$
$$u_2 = \partial_2 u_1 = \partial_2 u_3 = 0 \text{ for } x_2 = 0 \text{ and } x_2 = 1,$$
$$u_3 = \partial_3 u_1 = \partial_3 u_2 = 0 \text{ for } x_3 = 0 \text{ and } x_3 = 1. \qquad (1.360)$$

The main result is the following.

Theorem 43 (Existence and uniqueness of solution) *Given $M > 0$ and $\bar{\rho} > \rho \geq 0$, suppose that the initial data $(\rho_0, \mathbf{u}_0)$ satisfies (1.360), and*

$$0 \leq \inf \rho_0 \leq \sup \rho_0 \leq \bar{\rho}, \quad \|\nabla \mathbf{u}_0\|_{L_2}^2 = M,$$

$$\mathbf{u}_0 \in H^3, \quad (\rho_0 - \rho, p(\rho_0) - p(\rho)) \in H^3,$$

$$-\mu \nabla^2 \mathbf{u}_0 - (\mu + \lambda) \nabla(\text{div } \mathbf{u}_0) + \nabla p(\rho_0) = \rho_0 \, \mathbf{g}$$

for some $\mathbf{g}$ with $\sqrt{\rho_0}\, \mathbf{g}, \nabla g \in L_2$, $g^i|_{x_i=0,1} = 0$ and $i = 1, 2, 3$. Then there exists a constant $\varepsilon > 0$ depending on $\mu, \lambda, \bar{\rho}, \rho, M$ and Ω such that if $E_0 \leq \varepsilon$, then the problem (1.359)–(1.360) has a unique global classical solution $(\rho, \mathbf{u})$ satisfying for any $0 < \tau < T < \infty$ the conditions

$$0 \leq \rho(\mathbf{r}, t) \leq 2\bar{\rho} + 1, \quad \mathbf{r} \in \Omega, \ t \geq 0,$$

$$(\rho - \zeta, p(\rho) - p(\zeta)) \in C([0, T]; H^3),$$

$$\mathbf{u} \in C([0, T], H^3) \cap L_2([0, T]; H^4) \cap L_\infty([\tau, T]; H^4),$$

$$\partial_t \mathbf{u} \in L_\infty([0, T]; H^1) \cap L_2([0, T]; H^2) \cap L_\infty([\tau, T]; H^2) \cap H^1([\tau, T]; H^1),$$

$$\sqrt{\rho}\, \partial_t \mathbf{u} \in L_\infty([0, T]; L_2),$$

where the initial energy E_0 is defined as $E_0 = \int_\Omega (\frac{1}{2}\rho|\mathbf{u}|^2 + G(\rho_0))\, d\mathbf{r}$ with the potential energy density

$$G(\rho) = \rho \int_\zeta^\rho \frac{p(s) - p(\zeta)}{s^2}\, ds.$$

Moreover,

$$G(\rho) = \frac{1}{\gamma - 1} p(\rho), \quad \text{if } \zeta = 0,$$

$$c_1(\rho - \zeta)^2 \leq G(\rho) \leq c_2(\rho - \zeta)^2 \quad \text{if } \zeta > 0, \ 0 \leq \rho \leq \bar{\rho}$$

for some positive constants c_1 and c_2 depending on $\bar{\rho}$ and ζ.

Theorem 43 allows us precisely, through piecewise estimation, a priori to estimate established time uniform upper bounds for density under the assumption that the initial energy is small. The initial vacuum is also allowed. In addition, there were present higher-order regularity estimates.

NSE on Riemannian manifolds. When one moves from the cartesian to curvilinear coordinates, or more general, from Euclidean setting to Riemannian manifolds, the first question is how to write the equations. One should use not only metric tensor and Christoffel symbols (of the Levi–Civita connection ∇ associated with the metric) but also the curvature tensor, which obviously vanishes in Euclidean geometry. Also second-order differential operators become more complicated.

First, notice that the term $(\mathbf{u} \cdot \nabla)\mathbf{u}$ is the directional derivative of $\mathbf{u}$ in the direction of $\mathbf{u}$, that is generalized to $\nabla_\mathbf{u} \mathbf{u}$. Next question is how to interpret the viscosity operator (the Laplacian). The Laplace–Beltrami operator $\Delta := \text{div grad} = |g|^{-1/2} \partial_i(|g|^{1/2} g^{ij} \partial_j)$ acts on scalar functions, not on vector and tensor fields. There are known two natural generalizations of Δ: *Bochner Laplacian* $\Delta_H := \text{div } \nabla = -\nabla^* \nabla$ (where ∇^* is the adjoint to ∇) and *Hodge Laplacian* $\Delta_B := -(dd^* + d^*d)$ acting on differential forms. They are related by the *Bochner–Weitzenböck formula*

$$\nabla^* \nabla = d\, d^* + d^*d - \text{Ric}, \tag{1.361}$$

where Ric is the *Ricci curvature*. In the particular case of $\mathbb{R}^m$ (or when Ric $\equiv 0$), we have $\Delta_H = \Delta_B$.

Ebin and Marsden [12] indicated that when writing the NSE on an *Einstein manifold* (that is the Ricci curvature tensor is proportional to the metric tensor), the ordinary Laplacian should be replaced by $-\mathcal{L} := 2\mathrm{Def}^*\mathrm{Def}$ (a positive-semidefinite operator), where

$$(\mathrm{Def}\,\mathbf{u})_{ij} = \frac{1}{2}(\nabla_i u_j + \nabla_j u_i)$$

is the *deformation tensor*, and Def^* is its adjoint. The deformation tensor measures the degree to which the flow of a vector field $\mathbf{u}$ distorts the metric. Since $2\mathrm{Def}^*\mathrm{Def} = 2\,\mathrm{div}\,\mathrm{Def}$, a direct computation using (1.361) and a Ricci identity $\mathrm{Ric}^k_{j,\,k} = \frac{1}{2}S_{,j}$ (with scalar curvature S) gives us

$$\mathcal{L} = \Delta_H + 2\mathrm{Ric}.$$

They consider two types of compact space, a compact Riemannian manifold (M, g) with or without boundary, see [44]. The two types of fluid motion are modeled by NSE for incompressible fluids

$$\partial_t \mathbf{u} + \nabla_\mathbf{u}\mathbf{u} = \nu\,\mathcal{L}(\mathbf{u}) - \nabla p,$$

and by Euler equations for a perfect fluid

$$\partial_t \mathbf{u} + \nabla_\mathbf{u}\mathbf{u} = -\nabla p.$$

Unsolved problems. A fundamental problem in analysis is to decide whether such smooth, physically reasonable solutions exist for the NSE. To give reasonable leeway to solvers while retaining the heart of the problem, the Millennium Committee under Clay Mathematics Institute asks for a proof of one of the following four statements, see [16].

Conjectures in the whole space:

(A) *Existence and smoothness of the NSE solutions in* $\mathbb{R}^3$. Take $\nu > 0$ and $n = 3$. Let $\mathbf{u}_0(\mathbf{r})$ be a smooth, divergence-free vector field satisfying (1.347). Take $\mathbf{Q}(\mathbf{r}, t) \equiv 0$. Then there exist smooth functions $p(\mathbf{r}, t), \mathbf{u}(\mathbf{r}, t)$ on $\mathbb{R}^3 \times [0, \infty)$ that satisfy (1.345)–(1.346), (1.349) and (1.350).

(B) *Breakdown of the NSE solutions in* T^3. Take $\nu > 0$ and $n = 3$. Let $\mathbf{u}_0(\mathbf{r})$ be any smooth, divergence-free vector field satisfying periodic problem hypotheses (1.352) and growth conditions (1.353); we take $\mathbf{Q}(\mathbf{r}, t) \equiv 0$. Then there exist smooth functions $p(\mathbf{r}, t)$, $\mathbf{u}(\mathbf{r}, t)$ on $\mathbb{R}^3 \times [0, \infty)$ that satisfy (1.345)–(1.346), (1.349) and (1.351).

(C) *Breakdown of the NSE solutions in* $\mathbb{R}^3$. Take $\nu > 0$ and $n = 3$. Then there exist a smooth divergence-free vector field $\mathbf{u}(\mathbf{r})$ on $\mathbb{R}^3$ and a smooth $\mathbf{Q}(\mathbf{r}, t) \in \mathbb{R}^3 \times [0, \infty)$ satisfying (1.347) and (1.348), for which there exist no solutions $(p, \mathbf{u})$ of (1.345)–(1.346), (1.349) and (1.350) on $\mathbb{R}^3 \times [0, \infty)$.

(D) *Breakdown to the NSE solutions in* T^3. Take $\nu > 0$ and $n = 3$. Then there exist a smooth divergence-free vector field $\mathbf{u}_0(\mathbf{r})$ on $\mathbb{R}^3$ and a smooth $\mathbf{Q}(\mathbf{r}, t) \in \mathbb{R}^3 \times [0, \infty)$, satisfying periodic problem hypotheses and growth conditions (1.352) and (1.353), for which there exist no solutions $(p, \mathbf{u})$ of (1.345)–(1.346), (1.349) and (1.351) on $\mathbb{R}^3 \times [0, \infty)$.

These open problems are very important for Euler equations ($\nu = 0$), although these equations are not on the Clay Mathematics Institute's list of prize problems [16]. Let us sketch the main partial results for Euler and NSE, and conclude with a few remarks on the importance of the question.

In the 2d case, the analogues of assertions (A) and (B) have been known for a long time (Ladyzhenskaya [26]), also for the more difficult case of Euler equations. This gives no hint about the three-dimensional case, since the main difficulties are absent in two dimensions.

In the 3d case, (A) and (B) hold provided the initial velocity $\mathbf{u}_0$ satisfies a smallness condition. For initial data $\mathbf{u}_0(\mathbf{r})$ not assumed to be small, it is known that (A) and (B) hold (also for $\nu = 0$) if the time interval $[0, \infty)$ is replaced by a small time interval $[0, T)$, with T depending on the initial data. For a given initial $\mathbf{u}_0(\mathbf{r})$, the maximum allowable T is called the *blowup time*. Either (A) and (B) hold, or else there is a smooth, divergence-free $\mathbf{u}_0(\mathbf{r})$, for which the problem (1.345)–(1.346) has a solution with a finite blowup time. For the NSE ($\nu > 0$), if there is a solution with a finite blowup time T, then the velocity $u_i(\mathbf{r}, t)$, $i = 1, 2, 3$, becomes unbounded near the blowup time. Other unpleasant things may happen at the blowup time T, if $T < \infty$. For the Euler equations ($\nu = 0$), if there is a solution (with $\mathbf{Q} \equiv 0$, say) with finite blowup time T, then the vorticity $\omega(\mathbf{r}, t) = \operatorname{curl}(\mathbf{r}, t)$ satisfies [16]

$$\int_0^T \sup_{\mathbf{r} \in \mathbb{R}^3} |\omega(\mathbf{r}, t)| \, dt = \infty,$$

so that the vorticity blows up rapidly. Many numerical computations appear to exhibit blowup for solutions of the Euler equations, but the extreme numerical instability of the equations makes it very hard to draw reliable conclusions.

1.8.6 Relativistic Navier–Stokes equation

(a) **Derivation.** Starting with a brief introduction into the basics of relativistic fluid dynamics, we discuss our current knowledge of a relativistic theory of fluid dynamics in the presence of (mostly shear) viscosity. Derivations based on the generalized second law of thermodynamics, kinetic theory, and a complete second-order gradient expansion are reviewed. The resulting fluid dynamic equations are shown to be consistent for all these derivations, when properly accounting for the respective region of applicability, and can be applied to both weakly and strongly coupled systems. In its modern formulation, relativistic viscous hydrodynamics can directly be solved numerically. This has been useful for the problem of ultra-relativistic heavy-ion collisions, and one will review the setup and results of a hydrodynamic description of experimental data for this case. For non-ideal fluids, where dissipation can occur, the Euler equation generalizes to the NSE [27]

$$\partial_t u^i + u^k \frac{\partial u^i}{\partial x^k} = -\frac{\eta}{\rho} \frac{\partial \rho}{\partial x^i} - \frac{1}{\rho} \frac{\partial \tau^{ki}}{\partial x^k}, \quad \tau^{ki} = -\eta \left(\frac{\partial u^i}{\partial x^k} + \frac{\partial u^k}{\partial x^i} - \frac{2}{3} \delta^{ki} \frac{\partial u^l}{\partial x^l} \right) - \zeta \delta^{ki} \frac{\partial u^l}{\partial x^l},$$

where Latin indices denote the three space directions, e.g., $i = 1, 2, 3$. The viscous stress tensor τ_{ki} contains the coefficients for shear viscosity, η, and bulk viscosity, ζ, which produces a viscous effect associated with volume change and is independent of velocity. The non-relativistic Navier–Stokes equation is well tested and found to be reliable in many applications, so any successful theory of relativistic viscous hydrodynamics should reduce to it in the appropriate limit. For a relativistic system, the mass density $\rho(t, \mathbf{r})$ is not a good degree of freedom because it does not account for kinetic energy that may become sizable for motions close to the speed of light. Instead, it is useful to replace it by the total energy density $\varepsilon(t, \mathbf{r})$, which reduces to ρ in the non-relativistic limit. Similarly, $\mathbf{u}(t, \mathbf{r})$ is not a good degree of freedom because it does not transform appropriately under Lorentz transforms. Thus, it should be replaced by the Lorentz 4-vector for the velocity, $u^\mu \equiv dx^\mu / dT$, where Greek indices are related to Minkowski 4-space with metric $g_{\mu\nu} = \operatorname{diag}(1, -1, -1, -1)$ (the same symbol for the metric will also be used for curved space-times). The proper time increment dT is given by the line element,

$$(dT)^2 = g_{\mu\nu} dx^\mu dx^\nu = (dt)^2 - (d\mathbf{r})^2 = (dt)^2 (1 - |\mathbf{u}|^2),$$

where here and in the following, natural unit, speed of light, $c = 1$ will be used. This implies that

$$u^\mu = \frac{dt}{dT} \frac{x^\mu}{dt} = \frac{1}{\sqrt{1 - |\mathbf{u}^2|}} \begin{pmatrix} 1 \\ \mathbf{u} \end{pmatrix} = \gamma(\mathbf{u}) \begin{pmatrix} 1 \\ \mathbf{u} \end{pmatrix},$$

which reduces $u^\mu = (1, \mathbf{u})$ in the non-relativistic limit. In particular, one has $u^\mu = (1, \mathbf{0})$ if the fluid is locally at rest ("local rest frame"). Note that the 4-vector u^μ contains only three independent components since it obeys the relation $u^2 = g_{\mu\nu} u^\mu u^\nu = 1$, so one does not need additional equations when trading $\mathbf{u}$ for the fluid 4-velocity u^μ. To obtain the relativistic fluid dynamic equations, it is sufficient to derive the energy-momentum tensor $T^{\mu\nu}$ for a relativistic fluid, as will be shown below. The energy-momentum tensor of an ideal relativistic fluid (denoted as $T_0^{\mu\nu}$) has to be built out of the hydrodynamic degrees of freedom, namely two Lorentz scalars (ε, p) and one vector u^μ, as well as the metric tensor $g_{\mu\nu}$. Since $T^{\mu\nu}$ should be symmetric and transformed as a tensor under Lorentz transformations, the most general form allowed by symmetry is therefore

$$T_0^{\mu\nu} = \varepsilon(c_0 g^{\mu\nu} + c_1 u^\mu u^\nu) + p(c_2 g^{\mu\nu} + c_4 u^\mu u^\nu). \qquad (1.362)$$

In the local rest-frame, one requires the T_0^{00} component to represent the energy density ε of the fluid. Similarly, in the local rest frame, the momentum density should be vanishing $T_0^{0i} = 0$, and the spacelike components should be proportional to the pressure, $T_0^{ij} = p\delta^{ij}$ (see [27, Section 133]). Imposing these conditions onto the general form (1.362) leads to the equations

$$(c_0 + c_1)\varepsilon + (c_3 + c_4)p = \varepsilon, \quad -(c_0\varepsilon - c_2 p) = p,$$

which imply $c_0 = 0$, $c_1 = 1$, $c_2 = -1, c_3 = 1$, or $T_0^{\mu\nu} = \varepsilon u^\mu u^\nu - p(g^{\mu\nu} - u^\mu u^\nu)$. For later convenience, it is useful to introduce the tensor $\Delta^{\mu\nu} = g^{\mu\nu} - u^\mu u^\nu$. It has the properties $\Delta^{\mu\nu} u_\mu = \Delta^{\mu\nu} u_\nu = 0$ and $\Delta^{\mu\nu} \Delta_\nu^\alpha = \Delta^{\mu\alpha}$ (that is $\Delta^2 = \Delta$) and serves as a *projection operator* on the space orthogonal to the fluid velocity u^μ. In this notation, the energy-momentum tensor of an ideal relativistic fluid becomes $T_0^{\mu\nu} = \varepsilon u^\mu u^\nu - p\Delta^{\mu\nu}$. If there are no external sources, the energy-momentum tensor is conserved, $\partial_\mu T_0^{\mu\nu} = 0$. It is useful project these equations in the direction parallel $u_\nu \partial_\mu T_0^{\mu\nu}$ and perpendicular $\Delta_\nu^\alpha \partial_\mu T_0^{\mu\nu}$ to the fluid velocity. For the first projection, one finds

$$u_\nu \partial_\mu T_0^{\mu\nu} = u^\mu \partial_\mu \varepsilon + \varepsilon(\partial_\mu u^\mu) + \varepsilon u_\nu u^\mu \partial_\mu u^\nu - p u_\nu \partial_\mu \Delta^{\mu\nu} = (\varepsilon + p)\partial_\mu u^\mu + u^\mu \partial_\mu \varepsilon = 0, \quad (1.363)$$

where the identity $u_\nu \partial_\mu u^\nu = \frac{1}{2}\partial_\mu(u_\nu u^\nu) = 0$ was used. For the other projection one finds

$$\Delta_\nu^a \partial_\mu T_0^{\mu\nu} = \varepsilon u^\mu \Delta_\nu^a \partial_\mu u^\nu - \Delta^{\mu a} \partial_\mu p + p u^\mu \Delta_\nu^a \partial_\mu u^\nu = (\varepsilon + p)u^\mu \partial_\mu u^a - \Delta^{\mu a} \partial_\mu p. \quad (1.364)$$

Introducing the shorthand notations $D = u^\mu \partial_\mu$ and $\nabla^\alpha = \Delta^{\mu\alpha} \partial_\mu$ for the projection of derivatives parallel and perpendicular to u^μ, (1.363)–(1.364) can be written as

$$D\varepsilon + (\varepsilon + p)\partial_\mu u^\mu = 0, \qquad (1.365)$$

$$(\varepsilon + p)Du^\alpha - \nabla^\alpha p = 0. \qquad (1.366)$$

These are the fundamental equations for a relativistic ideal fluid. Their meaning becomes transparent in the non-relativistic limit: for small velocities $|\mathbf{u}| \ll 1$ one finds

$$D = u^\mu \partial_\mu \approx \partial_t + \mathbf{u} \cdot \boldsymbol{\partial} + O(|\mathbf{u}|^2), \quad \nabla^i = \Delta^{i\mu} \partial_\mu \approx \partial^i + O(|\mathbf{u}|).$$

So, D and ∇^i reduce to time and space derivatives, respectively. Imposing further a non-relativistic equation of state, where $\rho \ll \varepsilon$ and that energy density is dominated by mass density $\varepsilon \simeq \rho$, (1.365) becomes the continuity equation $\partial_t \rho + \mathrm{div}(\rho \mathbf{u}) = 0$, and (1.366) the non-relativistic Euler equation

$$\partial_t \mathbf{u} + (\mathbf{u} \cdot \boldsymbol{\partial})\mathbf{u} = -(\eta/\rho)\boldsymbol{\partial} p.$$

One thus recognizes the fluid dynamic equations (both relativistic and non-relativistic) to be identical to the conservation equations for the fluids energy-momentum tensor.

As a result, the relativistic equivalent to NSE or a more general Dissipative Relativistic Hydrodynamics Equation there would be an energy momentum tensor with the habitual view [27]:

$$T_{\mu\nu} = (\varepsilon + p)u_\mu u_\nu - p g_{\mu\nu} + \tau_{\mu\nu},$$

with the viscous stress tensor given by

$$\tau_{\mu\nu} = -\eta[\partial_\mu u_\nu + \partial_\nu u_\mu - u_\mu u^\alpha \partial_\alpha u_\nu - u_\nu u^\alpha \partial_\alpha u_\mu] - (\zeta - \frac{2}{3}\eta)(g_{\mu\nu} - u_\mu u_\nu)\partial_\alpha u^\alpha,$$

and the Conservation of Energy and Momentum which would lead to $\partial^\mu T_{\mu\nu} = 0$. Here, u^μ refers to the 4-velocity of the fluid, ε and p refer to the proper energy density and the pressure, respectively. The idea is to try to transpose the expressions you use in usual non-relativistic NSE and fix the terms via positivity of entropy production. The relativistic Euler equation can be obtained setting $\tau_{\mu\nu} = 0$, although some effort must be made to recover the Euler equation.

(b) **Existence and uniqueness of solutions.** Naturally, existence of solutions is not the only criteria to judge the reliability of a numerical simulations. Issues of discretization, numerical stability, etc., are also important. But these become relevant only when the underlying equations have (or are assumed to have) solutions. The modern theory admits existence and uniqueness of solutions and linear stability has been verified in the fluid's rest frame. As far as one knows, investigations addressing the stability of the equations of motion in a Lorentz boosted frame and coupling to Einstein's equations have not appeared in the literature for the energy-momentum tensor. Only recently it has been shown to yield a causal theory that admits existence and uniqueness of solutions, including when dynamically coupled to gravity, at least in the cases of non-rotational fluids or with restrictions on the initial data.

Finally, motivated by the rapid expansion and the highly anisotropic initial state of the quark-gluon plasma formed in heavy ion collisions, a new set of fluid dynamic equations have been studied defining the so-called anisotropic hydrodynamics formalism. This subject is still under development [41] and statements regarding stability, causality, and existence of solutions are not yet available. In [4] it was rigorously proven the existence, uniqueness, and causality of solutions to this viscous theory (in the full nonlinear regime) both in a Minkowski background and also when the fluid is dynamically coupled to Einstein's equations. In a later section, we establish the stability of solutions to the equations of motion in the linearized regime. This is the first time that such nontrivial statements can be rigorously made about viscous fluid dynamics in the relativistic regime since Eckart's first proposal [13].

We will focus on the case of conformal hydrodynamics, which provides the simplest set of assumptions regarding the properties of the underlying microscopic theory that can be used to study relativistic hydrodynamic phenomena. Consider Einstein's equations

$$R_{\mu\nu} - \frac{1}{2}R\,g_{\mu\nu} + \Lambda g_{\mu\nu} = 8\pi G T_{\mu\nu},$$

with energy-momentum tensor given by

$$T^{\mu\nu} = (\varepsilon + \frac{3}{4}\xi\varepsilon^{-1}\mathcal{D}\varepsilon)(u^\mu u^\nu + \frac{1}{3}\Delta^{\mu\nu}) - \eta\sigma^{\mu\nu} + \frac{3}{4}\lambda\varepsilon^{-1}(u^\mu g^{\alpha\nu}\mathcal{D}^\perp_\alpha\varepsilon + u^\nu g^{\alpha\mu}\mathcal{D}^\perp_\alpha\varepsilon), \quad (1.367)$$

where

$$\mathcal{D}\varepsilon = u^\mu\nabla_\mu\varepsilon + \frac{4}{3}\varepsilon\nabla_\mu u^\mu, \quad \mathcal{D}^\perp_\mu\varepsilon = \frac{4}{3}\varepsilon u^\lambda\nabla_\lambda u_\mu + \frac{1}{3}\nabla^\perp_\mu\varepsilon, \quad \sigma_{\mu\nu} = \nabla^\perp_\mu u_\nu + \nabla^\perp_\nu u_\mu - \frac{2}{3}\Delta_{\mu\nu}\nabla_\alpha u^\alpha,$$

and the shear tensor $\nabla_\mu^\perp = \Delta_\mu^\nu \nabla_\nu$ is the transverse covariant derivative, and η is a shear viscosity transport coefficient (for a conformal fluid $\eta \sim s \sim T^3$, with $s = \frac{4}{3}\varepsilon/T$ being entropy density). It is assumed that the equation $u^\mu u_\mu = -1$ is also part of the system. An initial data set $\mathcal{I}$ for this system consists of the usual initial conditions (Σ, g_0, κ) for Einstein's equations (Σ is a three-dimensional manifold endowed with a Riemannian metric g_0 and a symmetric two tensor κ), two scalar functions ε_0 and ε_1 on Σ (energy density and its time-derivative at the initial time), and two vector fields $\mathbf{u}_0$ and $\mathbf{u}_1$ on Σ (the initial values for the velocity and its time-derivative), such that the constraint equations are satisfied [47, Chapter 10]. For a conformal theory all transport coefficients are $\sim T^3$ so we can assume $\chi = a_1\eta$, $\lambda = a_2\eta$, with $a_{1,2}$ constants, and $\chi > 0$, $\lambda > 0$, which gives equations of motion for the hydrodynamic fields of second-order in derivatives. The meaning of "sufficiently regular" stated in the theorem is explained below.

Theorem 44 *Let $\mathcal{I} = (\Sigma, g_0, \kappa, \varepsilon_0, \varepsilon_1, u_0, u_1)$ be a sufficiently regular initial data set for Einstein's equations coupled to (1.367). Assume the following: Σ is compact with no boundary, $\varepsilon_0 > 0$, $\eta : (0,\infty) \to (0,\infty)$ is analytic, $a_1 = 4$ and $a_2 \geq 4$. Then*

(1) There exists a globally hyperbolic development M of $\mathcal{I}$.

(2) Let (g, ε, u) be a solution of Einstein's equations provided by the globally hyperbolic development M. For any $x \in M$ in the future of Σ, $(g(x), u(x), \varepsilon(x))$ depends only on $\mathcal{I}|_{i(\Sigma) \cap J^-(x)}$, where $J^-(x)$ is the causal past of x and $i : \Sigma \to M$ is the embedding associated with the globally hyperbolic development M.

Statement (1) means the Einstein's equations admit existence and uniqueness of solutions (uniqueness up to a diffeomorphism, as usual in general relativity). Statement (2) says that the system is causal. Σ is assumed compact and with no boundary for simplicity, as otherwise asymptotic and/or boundary conditions would have to be prescribed. The assumption $\varepsilon_0 > 0$ guarantees that (1.367) is well defined. Above, sufficiently regular means that the initial data belongs to appropriate *Gevrey spaces*. It is crucial to point out, however, that the causality of the equations does not depend on the use of Gevrey spaces, and it will automatically hold in any space of functions where existence and uniqueness can be established. The Gevrey space G^σ consists of the smooth functions f on $\mathbb{R}^n$ such that on every compact subset K there are constants C and R with

$$|D^\alpha f(x)| \geq CR^k k^{\sigma k}$$

for multi-indices α such that $|\alpha| = k$, and corresponding differential operators D^α; and x restricted to lie in the compact set K. When $\sigma = 1$, this class is the same as the class of analytic functions; for $\sigma > 1$ there are compactly supported functions in the class that are not identically zero.

Theorem 45 *Under assumptions $a_1 = 4$ and $a_2 \geq 4$, a statement similar to Theorem 44, i.e., existence, uniqueness, and causality holds for solutions of $\nabla_\mu T^{\mu\nu} = 0$, with $T^{\mu\nu}$ given by (1.367), in Minkowski background.*

Conditions $a_1 = 4$ and $a_2 \geq 4$ in Theorems 44 and 45 are technical, but we believe that they provide a wide range of values for applications in different situations of interest. Note that these are sufficient conditions, i.e., we are not saying (and we do not know) whether causality is lost if one of these two conditions is not satisfied.

1.9 Exercises

Determine a first-order uniformly valid expansion for a solution to the boundary value problems.

1. Find asymptotic solution to the boundary value problem

$$\varepsilon y'' + a(x)y' + b(x)y = 0, \quad y(0) = \alpha, \quad y(1) = \beta, \tag{1.368}$$

where $\varepsilon \ll 1$, and $a(x)$, $b(x)$ are analytic functions in the interval $x \in [0, 1]$.

Solution. If $\varepsilon = 0$ then (1.368) reduces to the first-order ODE

$$a(x)y' + b(x)y = 0, \tag{1.369}$$

whose solution cannot satisfy both boundary conditions, and one of them must be dropped as a consequence. Thus, the boundary condition that must be dropped depends on the sign of $a(x)$ in the interval $[0, 1]$. If $a(x) > 0$, $a(0) = \alpha$ must be dropped, and an inner expansion near $x = 0$ must be developed and matched with the outer solution. If $a(x) < 0$, $a(1) = \beta$ must be dropped and an inner expansion near $x = 1$ must be obtained and matched with the outer expansion. If $a(x)$ changes sign in $[0, 1]$, y may change from oscillatory to exponentially growing or decaying across the zeros of $a(x)$.

To investigate whether $y(0) = \alpha$ must be dropped (i.e., the boundary layer is at the origin), we introduce the stretching transformation

$$\zeta = x\varepsilon^{-\lambda}, \quad \lambda > 0 \tag{1.370}$$

to transform (1.368) into

$$\varepsilon^{1-2\lambda}\frac{d^2y}{d\zeta^2} + \varepsilon^{-\lambda}a(\varepsilon^\lambda\zeta)\frac{dy}{d\zeta} + b(\varepsilon^\lambda\zeta)y = 0. \tag{1.371}$$

As $\varepsilon \to 0$, (1.371) tends to

$$\begin{aligned}
d^2y/d\zeta^2 &= 0 \quad &&(\text{if} \quad \lambda > 1), \\
dy/d\zeta &= 0 \quad &&(\text{if} \quad \lambda < 1), \\
d^2y/d\zeta^2 + a(0)\, dy/d\zeta &= 0 \quad &&(\text{if} \quad \lambda = 1).
\end{aligned} \tag{1.372}$$

In order to be able to match the solutions to (1.371) with the outer solution as given by the reduced (1.369), we need the bounded solutions to (1.372) as $\zeta \to \infty$. The bounded solutions for the first two cases are constants. Similarly, if $a(0) < 0$, the bounded solution for the last case (i.e., when $\lambda = 1$) is a constant and must also be discarded; as a consequence, the boundary layer cannot exist at $x = 0$. If $a(0) > 0$, the general solution for $\lambda = 1$ is

$$y^{(i)} = A + Be^{-a(0)\zeta}, \tag{1.373}$$

which is bounded as $\zeta \to \infty$ and contains two arbitrary constants; hence it is acceptable as an inner expansion because together with the solution to the reduced (1.369) it can satisfy both boundary conditions.

To investigate whether $y(1) = \beta$ must be dropped, we introduce the stretching transformation $\eta = (1 - x)\varepsilon^{-\lambda}$, $\lambda > 0$. To transform (1.368) into $\varepsilon^{1-2\lambda}\frac{d^2y}{d\eta^2} + \varepsilon^{-\lambda}a(1 - \varepsilon^\lambda\eta)\frac{dy}{d\eta} + b(1 - \varepsilon^\lambda\eta)y = 0$. As $\varepsilon \to 0$, this equation tends to

$$\begin{aligned}
d^2y/d\eta^2 &= 0 \quad &&(\text{if } \lambda > 1), \\
dy/d\eta &= 0 \quad &&(\text{if } \lambda < 1), \\
d^2y/d\eta^2 + a(1)\, dy/d\eta &= 0 \quad &&(\text{if } \lambda = 1).
\end{aligned} \tag{1.374}$$

The bounded solutions of (1.374) as $\eta \to \infty$ are constants if $\lambda \neq 1$ or $a(1) > 0$ when $\lambda = 1$; hence they must be discarded. If $a(1) < 0$ when $\lambda = 1$, the general solution of (1.374) is $y^{(i)} = A + Be^{a(1)\eta}$, which is bounded as $\eta \to \infty$; hence it is acceptable as an inner expansion because it contains two arbitrary constants which allow, when combined with the outer solution, the satisfaction to both boundary conditions.

In summary, the boundary layer exists at $x = \begin{cases} 0 & \text{if} \quad a(x) > 0 \\ 1 & \text{if} \quad a(x) < 0. \end{cases}$ Next we determine a first-order uniformly valid expansion for the former case.

The case of $a(x) > 0$. In this case the first term of the outer expansion

$$y^{(0)}(x; \varepsilon) = \sum_{n=0}^{N-1} \varepsilon^n y_n(x) + O(\varepsilon^N) \tag{1.375}$$

is governed by $a(x)y' + b(x)y = 0$, $y(1) = \beta$, which yields

$$y^{(o)} = \beta \exp\left[-\int_1^x (b(t)/a(t))dt\right]. \tag{1.376}$$

The general solution for the first term of the inner expansion $y^{(i)}(\zeta; \varepsilon) = \sum_{m=0}^{N-1} \varepsilon^m Y_m(\zeta) + O(\varepsilon^N)$ is given by (1.373). Under the transformation (1.370) the boundary condition $y(x)|_{x=0} = \alpha$ is transformed into $y(\zeta)|_{\zeta=0} = \alpha$; hence

$$y^{(i)} = \alpha - B + Be^{-a(0)\zeta}. \tag{1.377}$$

To match $y^{(o)}$ and $y^{(i)}$, we take the m-term of inner expansion (1.377) and the n-term of outer expansion (1.375) to be equal unity and proceed as follows, one-term inner expansion of (one-term outer expansion) is equal

$$\beta \exp\left[-\int_1^0 (b(t)/a(t))dt\right] \tag{1.378}$$

one-term outer expansion of (one-term inner expansion) is equal

$$\alpha - B. \tag{1.379}$$

By (1.378) and (1.379) according matching principle,

$$\begin{array}{l} m \text{ - term inner expansion of the } n \text{ - term outer expansion} \\ n \text{ - term outer expansion of the } m \text{ - term inner expansion,} \end{array} \tag{1.380}$$

we have $B = \alpha - \beta \exp\left[-\int_1^0 (b(t)/a(t))dt\right]$. Forming a composite expansion by adding (1.376) and (1.377) and subtracting (1.379), we obtain

$$y^{(c)} = \beta \exp\left[-\int_1^x (b(t)/a(t))dt\right] + \left\{\alpha - \beta \exp\left[-\int_1^0 (b(t)/a(t))dt\right]\right\} \times \exp[-a(0)x/\varepsilon] + O(\varepsilon).$$

For the simple case when $a(x) = b(x) = 1$, we have $y^{(c)} = \beta e^{1-x} + (\alpha - \beta)e^{-x/\varepsilon} + O(\varepsilon)$. Taking $a(x) = 2x + 1$ and $b(x) = 2$, we obtain $y^{(c)} = \frac{3\beta}{1+2x} + (\alpha - 3\beta)e^{-x/\varepsilon} + O(\varepsilon)$.

2. The pressure distribution in an isothermal compressible film in an infinitely long slider bearing (Fig. 1.60(a)) is given by Reynolds equation, written in dimensionless quantities

$$(h^3 pp')' = \Lambda(ph)'. \tag{1.381}$$

Here the distance x, the film thickness h, and the pressure p have been made dimensionless with respect to the length L of the bearing in the flow direction, the film thickness T at the trailing edge, and the ambient pressure p_a, respectively. The bearing number Λ is given by $6\mu LU/p_a T^2$, where μ is the fluid viscosity and U is the velocity of the lower surface. The boundary conditions are $p|_{x=0} = p|_{x=1} = 1$. Find an asymptotic solution to this problem for large Λ using method of matched asymptotic expansions.

Solution. We seek an outer expansion of the form

$$p^{(0)} = p_0(x) + \Lambda^{-1}p_1(x) + \dots \tag{1.382}$$

Substituting this expansion into (1.381) and equating coefficients of like powers of Λ^{-1}, we obtain

$$(hp_0)' = 0, \quad (hp_1)' = (h^3 p_0 p_0')'.$$

Since these equations are of first order, one of the boundary conditions must be dropped, and the outer expansion is not valid near the boundary. Since h and Λ are positive, arguments show that the boundary condition at $x = 1$ must be dropped. Hence $p^{(0)} = 1$, which together with (1.382) gives $p_0(0) = 1$, $p_1(0) = 0$. Consequently, the solution for p_0 is $p_0 = h_0/h(x)$. Hence the solution for p_1 is $p_1 = h_0^2[h'(0) - h'(1)]/h(x)$. This outer expansion needs to be supplemented by an inner expansion (boundary layer solution) near $x = 1$. Introducing the stretching transformation $\zeta = (1 - x)\Lambda^\sigma$, $\sigma > 0$, we transform (1.381) into

$$\Lambda^{\sigma-1}\frac{d}{d\zeta}\left[h^3(1 - \zeta\Lambda^{-\sigma})p\frac{dp}{d\zeta}\right] = -\frac{d}{d\zeta}[h(1 - \zeta\Lambda^{-\sigma})p].$$

As $\Lambda \to \infty$ this equation reduces to

$$\begin{aligned}
\frac{d}{d\zeta}\left(p\frac{dp}{d\zeta}\right) &= 0 \quad (\text{if } \sigma > 1), \\
\frac{dp}{d\zeta} &= 0 \quad (\text{if } \sigma < 1), \\
\frac{d}{d\zeta}\left(p\frac{dp}{d\zeta}\right) &= -\frac{dp}{d\zeta} \quad (\text{if } \sigma = 1)
\end{aligned}$$

The first two cases must be discarded because the solution of the second one is a constant while the solution of the first case is a constant as $\zeta \to \infty$; hence in neither case can we satisfy the boundary conditions. In the third case a first integral is

$$p\frac{dp}{d\zeta} + p = A, \tag{1.383}$$

where A is a constant. The solution of (1.383) is

$$-\zeta = p + A\ln(p - A) + B, \tag{1.384}$$

where B is another constant. This solution thus represents the first term P_0 in the inner expansion

$$p^{(i)} = P_0(\zeta) + \Lambda^{-1}P_1(\zeta) + \dots \tag{1.385}$$

Hence it must satisfy the boundary condition $p|_{x=1} = 1$ or $p^{(i)}|_{\zeta=0} = 1$. This together with (1.385) leads to $P_0(0) = 1$ which, when substituted into (1.384), yields $B = -1 - A\ln(1 - A)$. Hence P_0 is given implicitly by

$$-\zeta = P_0 - 1 + A\ln[(A - P_0)/(A - 1)]. \tag{1.386}$$

To determine A we take $m = n = 1$ in matching principle (1.380) and proceed as follows

$$\text{one-term inner expansion of (one-term outer expansion)} = h_0 \tag{1.387}$$

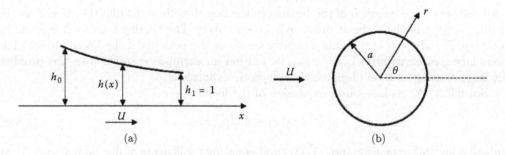

FIGURE 1.60: (a) Slider bearing. (b) Flow past a sphere.

one-term outer expansion of (one-term inner expansion) = A.   (1.388)

Equation (1.388) was obtained by expanding (1.386) for large ζ as a result of the implicit functional dependence of P_0 on ζ. Equating (1.387) and (1.388) according to matching principle, we have $A = h_0$; hence

$$-\zeta = P_0 - 1 + h_0 \ln \frac{h_0 - P_0}{h_0 - 1}.$$ (1.389)

A first-order uniformly valid composite expansion is then formed according to $y^{(c)} = y^{(0)} + y^{(i)} - \left(y^{(0)}\right)^{(i)} = \frac{h_0}{h(x)} + P_0(\zeta) - h_0 + O(\Lambda^{-1})$, where P_0 is given by (1.389).

3. The third exercise given here to show the difficulty arising from an infinite domain is the small Reynolds number, incompressible, uniform flow past a sphere. In spherical coordinate system shown in Fig. 1.60(b), full Navier–Stokes equations give the following dimensionless equation for the stream function

$$\psi(r,\theta): \quad (u_r = \psi_\theta/r^2 \sin\theta, \quad u_\theta = -\psi_r/r \sin\theta)$$

for axisymmetric flow

$$D^4\psi = \frac{Re}{r^2 \sin\theta}\left(\psi_\theta \frac{\partial}{\partial r} - \psi_r \frac{\partial}{\partial\theta} - 2\psi_r \cot\theta - 2\frac{\psi_\theta}{r}\right)D^2\psi,$$ (1.390)

where Reynolds number $Re = Ua/\nu$ (ν is the kinematic viscosity) and $D = \frac{\partial^2}{\partial r^2} + \frac{\sin\theta}{r^2}\frac{\partial}{\partial\theta}\left(\frac{1}{\sin\theta}\frac{\partial}{\partial\theta}\right)$. The boundary conditions at the sphere's surface call for the vanishing of the velocity: that is, in dimensionless form, $\psi(1,\theta) = \psi_r(1,\theta) = 0$. The condition that the flow is uniform gives

$$\psi(r,\theta) \to \frac{1}{2}r^2 \sin^2\theta, \quad r \to \infty.$$ (1.391)

Equations (1.390) through (1.391) form a well-posed problem[2] for the stream function ψ.

Solution. A. Let us seek a formal expansion of Poincaré type valid for small Re

$$\psi(r,\theta; Re) = \sum_{m=0}^{\infty} Re^m \psi_m(r,\theta).$$ (1.392)

Substituting (1.392) into (1.390) through (1.391), expanding for small Re, and equating coefficients of equal powers of Re lead to **Order Re^0**

$$D^4\psi_0 = 0,$$ (1.393a)

[2]Stokes G.G. On the effect of the internal friction of fluids on the motion of pendulums. – *Trans. Cambridge Phil. Soc.* V. 9, 1851, 8–106.

$$\psi_0(1,\theta) = \psi_{0r}(1,\theta) = 0, \tag{1.393b}$$

$$\psi_0(r,\theta) \to \frac{1}{2}r^2 \sin^2\theta, \quad r \to \infty. \tag{1.393c}$$

Order Re^1

$$D^4\psi_1 = \frac{Re}{r^2 \sin\theta}\left(\psi_{0\theta}\frac{\partial}{\partial r} - \psi_{0r}\frac{\partial}{\partial\theta} - 2\cot\theta\psi_{0r} - 2\frac{\psi_{0\theta}}{r}\right)D^2\psi_0, \tag{1.394a}$$

$$\psi_1(1,\theta) = \psi_{1r}(1,\theta) = 0, \tag{1.394b}$$

$$\psi_1(r,\theta) \to O(r^2), \quad r \to \infty. \tag{1.394c}$$

Equation (1.393c) suggests that ψ_0 has the form

$$\psi(r,\theta) \to f(r)\sin^2\theta.$$

Substituting this assumed form into (1.393a) leads to $f^{iv} - 4\frac{f''}{r^2} + 8\frac{f'}{r^3} - 8\frac{f}{r^4} = 0$, whose solution is

$$f = c_4 r^4 + c_2 r^2 + c_1 r + c_{-1} r^{-1}. \tag{1.395}$$

Boundary condition (1.393c) demands that $c_4 = 0$ and $c_2 = \frac{1}{2}$, while boundary conditions (1.393b) demand that $c_1 = -\frac{3}{24}$ and $c_{-1} = \frac{1}{4}$. Therefore

$$\psi_0 = \frac{1}{4}(2r^2 - 3r + 1/r)\sin^2\theta. \tag{1.396}$$

This solution was obtained by G.G. Stokes (1851). Substituting for ψ_0 from (1.396) into (1.394a) gives

$$D^4\psi_1 = -\frac{9}{4}\left(\frac{2}{r^2} - \frac{3}{r^3} + \frac{1}{r^5}\right)\sin^2\theta\cos\theta. \tag{1.397}$$

Equations (1.397) and boundary conditions (1.394b,c) suggest that a particular solution for ψ_1 has the form

$$\psi_1 = g(r)\cdot\sin^2\theta\cos\theta \tag{1.398}$$

with the form (1.398) for ψ_1, g should satisfy the following equation and boundary conditions

$$g^{iv} - 12\frac{g''}{r^2} + 24\frac{g'}{r^3} = -\frac{9}{4}\left(\frac{2}{r^2} - \frac{3}{r^3} + \frac{1}{r^5}\right), \tag{1.399a}$$

$$g(1) = g'(1) = 0, \quad g(r) = O(r^2), \quad r \to \infty. \tag{1.399b}$$

The general solution to the problem (1.399a,b) is $g = b_{-2}r^{-2} + b_0 + b_3 r^3 + b_5 r^5 - \frac{3}{16}r^2 + \frac{9}{32}r + \frac{3}{32}r^{-1}$.

The second of boundary conditions (1.399a) demands that $b_3 = b_5 = 0$. Even with this choice for b_3 and b_5, g does not behave properly as $r \to \infty$ because of the presence of the term $-\frac{3}{16}r^2$ in the particular solution for g. Clearly, no values for b_0 and b_{-2} can be found that correct this shortcoming. Moreover, no other complementary solution to (1.397) can be found which makes ψ_1 behave properly as $r \to \infty$. The first of boundary conditions (1.399b) demand that $b_0 = b_{-2} = -\frac{3}{32}$; hence

$$\psi_1 = -\frac{3}{32}(2r^2 - 3r + 1 - r^{-1} + r^{-2})\sin^2\theta\cos\theta. \tag{1.400}$$

Here again difficulty with the straightforward expansion arose because of the infinite domain. Two-term expansion for $Re \to 0$,

$$\psi = \frac{1}{4}(2r^2 - 3r + r^{-1})\sin^2\theta - \frac{3}{32}Re(2r^2 - 3r + 1 - r^{-1} + r^{-2})\sin^2\theta\cos\theta + O(Re^2),$$

satisfies the surface boundary conditions, but it does not satisfy the boundary condition at infinity. Thus, this expansion breaks down for large r and this breakdown is called *Whitehead's paradox* because[3] was the first to obtain this solution, although by iteration, and the first to point out its non-uniformity.

B. Let us seek an asymptotic solution to this problem small Re number using method of matched asymptotic expansions.

Oseen's Expansion: To investigate the source of this non-uniformity, we examine the relative magnitude of the convective terms neglected in the first subsection (A) of this exercise those retained (viscous). The right-hand side of (1.394a) shows that

$$\text{Neglected terms} = O(Re/r^2),$$ while the cross-product term retained in (1.393a) is

$$\frac{\partial^2}{\partial r^2}\left[\frac{\sin\theta}{r^2}\frac{\partial}{\partial\theta}\left(\frac{1}{\sin\theta}\frac{\partial}{\partial\theta}\right)\right]\psi_0 = O(r^{-3}).$$

Hence $\frac{\text{Neglected}}{\text{Retained}} = O(rRe)$, $r \to \infty$, and Stokes' expansion breaks down when r increases to $O(Re^{-1})$.

One can arrive at this conclusion by noting that the source of non-uniformity is the term $-\frac{3}{16}Rr^2\sin^2\theta\cos\theta$ in the particular integral that does not behave properly as $r \to \infty$. Consequently, Stokes' expansion is valid as long as this term is small compared to the term $-\frac{3}{4}r\sin^2\theta$ in ψ_0, and Stokes' expansion breaks down when these two terms are of the same order; that is, when $rRe = O(1)$.

The above arguments lead[4] to derive an approximation to the flow field valid everywhere. It is the first two terms of what called Oseen's expansion, derived using Oseen's limit process $Re \to 0$ with $\rho = rRe$ fixed. Note that ρ is a contracting rather than a stretching transformation. With this new variable (1.390) becomes

$$D^4\psi = \frac{Re^2}{\rho^2\sin\theta}\left(\psi_\theta\frac{\partial}{\partial\rho} - \psi_\rho\frac{\partial}{\partial\theta} - 2\psi_\rho\cot\theta - 2\frac{\psi_\theta}{\rho}\right)D^2\psi, \tag{1.401}$$

where $D = \frac{\partial^2}{\partial\rho^2} + \frac{\sin\theta}{\rho^2}\frac{\partial}{\partial\theta}\left(\frac{1}{\sin\theta}\frac{\partial}{\partial\theta}\right)$. Since the general solutions to PDEs are not known in general, it is more convenient and expedient to determine the inner and outer expansions term by term, employing the matching principle as a guide to the forms of these expansions. Since Stokes' solution (1.396) is uniformly valid, we can determine the first term in Oseen's expansion by taking Oseen's limit (1.396). To determine the form of the second term as well, we use the matching principle

one-term Stokes t (two-term Oseen)=two-term Oseen (one-term Stokes)
$$= \tfrac{1}{2}Re^{-2}\rho^2\sin^2\theta - \tfrac{3}{4}Re^{-1}\rho\sin^2\theta. \tag{1.402}$$

Thus, Oseen's expansion (denoted by $\psi^{(0)}$) must have the form $\psi^{(0)} = \tfrac{1}{2}Re^{-2}\rho^2\sin^2\theta + Re^{-1}\Psi_1(\rho,\theta) + \Psi_2(\rho,\theta) + \ldots$ and the first term is a consequence of the validity of Stokes' solution for all r. Substituting this expansion into (1.401) and equating the coefficients of Re on both sides, we obtain

$$\left(D^2 - \cos\theta\frac{\partial}{\partial\rho} + \frac{\sin\theta}{\rho}\frac{\partial}{\partial\theta}\right)D^2\Psi_1 = 0. \tag{1.403}$$

This equation is called Oseen's equation which he derived using physical arguments.

---

[3]Whitehead A.N., Second approximation to viscous fluid motion. *Quart. J. Math.*, V. 23, 1889, 143–152.

[4]Oseen C.W. Über die Stokessche Formel und über eine verwandte Aufgabe in der Hydrodynamik. – *Ark.* (1910).

To solve Oseen's equation we let

$$D^2\Psi_1 = \varphi \exp\left(\frac{1}{2}\rho\cos\theta\right) \tag{1.404}$$

and obtain

$$(D^2 - 1/4)\varphi = 0. \tag{1.405}$$

Rather than obtaining the general solution to (1.403) by solving (1.404) and (1.405) and then using the matching principle (1.402) to select the terms that match Stokes' solution. Since

$$D^2\psi^{(s)} = \frac{3}{2}r^{-1}\sin^2\theta + O(Re), \quad D^2\psi^{(0)} = Re^2 D^2\psi^{(o)} = ReD^2\Psi_1(\rho,\theta) + O(Re^2)$$

and the matching conditions demands that

one-term Stokes (one-term $D^2\psi^{(o)}$)=one-term Oseen (one-term $D^2\psi^{(s)}$),

we have

$$\text{one-term Stokes } (D^2\Psi_1) = \frac{3}{2}\rho^{-1}\sin^2\theta. \tag{1.406}$$

To satisfy this condition we seek a solution of (1.405) of the form $\varphi = f(\rho)\sin^2\theta$ so that $f'' - \left(2\rho^{-2} + \frac{1}{4}\right)f = 0$. The solution of f that does not grow exponentially with ρ is $f = A(1 + 2\rho^{-1})\exp(-\rho/2)$ so that $D^2\Psi_1 = A(1 + 2\rho^{-1})\sin^2\theta\exp\left[-\frac{1}{2}\rho(1 - \cos\theta)\right]$. Hence

$$\text{one - term Stokes } (D^2\Psi_1) = 2A(\rho Re)^{-1}\sin^2\theta.$$

Then (1.406) gives $A = \frac{3}{4}$, and $D^2\Psi_1$ becomes

$$D^2\Psi_1 = \frac{3}{4}(1 + 2\rho^{-1})\sin^2\theta\exp\left[-\frac{1}{2}\rho(1 - \cos\theta)\right]. \tag{1.407}$$

The particular integral of (1.407) is $\Psi_{1p} = \frac{3}{2}(1 + \cos\theta)\exp\left[-\frac{1}{2}\rho(1 - \cos\theta)\right]$ so that

$$\Psi_1 = \Psi_{1c} + \Psi_{1p} = \Psi_{1c} + \frac{3}{2}(1 + \cos\theta)\exp\left[-\frac{1}{2}\rho(1 - \cos\theta)\right].$$

Then, from (1.402) we obtain

$$\text{one-term Stokes } (D^2\Psi_{1c}) = -\frac{3}{2}(1 + \cos\theta).$$

Hence $\Psi_{1c} = -\frac{3}{2}(1 + \cos\theta)$, and Oseen's expansion becomes

$$\psi^{(o)} = \frac{1}{2}Re^{-2}\rho^2\sin^2\theta - \frac{3}{2}Re^{-1}(1 + \cos\theta)\left\{1 - \exp\left[-\frac{1}{2}\rho(1 - \cos\theta)\right]\right\} + O(1).$$

Second Term in Stokes' Expansion. Equations (1.396), (1.397) and (1.400) show that the second-order Stokes' expansion is

$$\psi^{(o)} = \frac{1}{4}(2r^2 - 3r + r^{-1})\sin^2\theta$$
$$+Re\left[-\frac{3}{32}(2r^2 - 3r + 1 - r^{-1} + r^{-2})\sin^2\theta\cos\theta + \psi_{1c}\right] + O(Re^2),$$

where ψ_{1c} is the complementary solution of (1.397). To determine ψ_{1c} we match two terms of $\psi^{(s)}$ with two terms of $\psi^{(0)}$ according to

two-term Stokes expansion: $\psi \sim \psi_0 + Re(\psi_{1p} + \psi_{1c})$

Rewritten in Oseen's variables: $\frac{1}{4}\left(2\frac{\rho^2}{Re^2} - 3\frac{\rho}{Re} + \frac{Re}{\rho}\right)\sin^2\theta$

$+Re\left[-\frac{3}{32}\left(2\frac{\rho^2}{Re^2} - 3\frac{\rho}{Re} + 1 - \frac{Re}{\rho} + \frac{Re^2}{\rho^2}\right)\right] \times \sin^2\theta\cos\theta + \psi_{1c}(\rho/Re, \theta).$

Two-term Oseen's expansion: $\frac{1}{2}Re^{-2}\rho^2\sin^2\theta + Re^{-1}\left[-\frac{3}{4}\left(\rho\sin^2\theta + \frac{1}{4}\rho^2\sin^2\theta\cos\theta\right)\right.$
$\left. + \text{ term } O(1) \text{ in } Re^2\psi_{1c}(\rho/Re, \theta)\right]$

$$\text{(1.408)}$$

Two-term Oseen's expansion:

$$\psi \sim \frac{1}{2}Re^{-2}\rho^2\sin^2\theta - \frac{3}{2}Re^{-1}(1 + \cos\theta) \times \left\{1 - \exp\left[-\frac{1}{2}\rho(1 - \cos\theta)\right]\right\},$$

rewritten in Stokes' variables:

$$\frac{1}{2}r^2\sin^2\theta - \frac{3}{2}Re^{-1}(1 + \cos\theta) \times \left\{1 - \exp\left[-\frac{1}{2}rRe(1 - \cos\theta)\right]\right\}.$$

Two-term Stokes' expansion: $\left(\frac{1}{2}r^2 - \frac{3}{4}r\right)\sin^2\theta + \frac{3}{16}r^2 Re\sin^2\theta(1 - \cos\theta).$ (1.409)

Equating (1.408) and (1.409) according to the matching principle, we have

$$\text{One-term Oceen } (\psi_{1c}) = \frac{3}{16}r^2\sin^2\theta.\qquad\text{(1.410)}$$

This suggests that $\psi_{1c} = f(r)\sin^2\theta$. Hence from (1.395): $f = c_4 r^4 + c_2 r^2 + c_1 r + c_{-1}r^{-1}$. Matching condition (1.410) demands that $c_4 = 0$ and $c_2 = \frac{3}{16}$, while the boundary conditions $\psi_1(1, \theta) = \psi_{1r}(1, \theta) = 0$ demand that $c_1 = -\frac{9}{32}$ and $c_{-1} = -\frac{3}{32}$. Therefore

$$\psi^{(s)} = \frac{1}{4}\left(2r^2 - 3r + \frac{1}{r}\right)\sin^2\theta$$
$$+\frac{3}{32}Re\left[(2r^2 - 3r + 1)\sin^2\theta - \left(2r^2 - 3r + 1 - \frac{1}{r} + \frac{1}{r^2}\right)\sin^2\theta\cos\theta\right] + O(Re^2)$$

4. Find the solution for forced Duffing oscillator and build its bifurcation diagrams.

Solution. We deal with forced Duffing oscillator. To compare the forced- and free-oscillation cases, consider an equation

$$\ddot{u} + u + \varepsilon(2\mu\dot{u} + \alpha u^3) = 2\varepsilon k\cos\Omega t\qquad\text{(1.411)}$$

for the case of primary resonance. In (1.411), ε is a small positive parameter, μ is a damping coefficient, α is the coefficient of the cubic nonlinearity, $2\varepsilon k$ is the amplitude of forcing, and Ω is the excitation frequency. The parameters μ, α, and k are independent of ε. Further, the excitation frequency is such that $\Omega = 1 + \varepsilon\sigma$, where the parameter σ is called the *external detuning*. By using the method of multiple scales [33], one obtains a first-order approximation for the solution of (1.411) as $u = a\cos(\Omega t - \gamma) + O(\varepsilon)$, where the amplitude a and phase γ are governed by

$$a' = -\mu a + k\sin\gamma, \quad a\gamma' = \sigma a - \frac{3}{8}\alpha a^3 + k\cos\gamma.\qquad\text{(1.412)}$$

The primes in (1.412) indicate derivatives with respect to the time scale $\tau = \varepsilon t$. The system (1.412) is a planar autonomous dynamical system. The fixed points (a_0, γ_0) of this system are given by

$$-\mu a_0 + k\sin\gamma_0 = 0, \quad \sigma a_0 - \frac{3}{8}\alpha a_0^3 + k\cos\gamma_0 = 0.\qquad\text{(1.413)}$$

From (1.413) we obtain so-called *frequency-response equation*

$$\left[\mu^2 + \left(\sigma - \frac{3}{8}\alpha a_0^2\right)^2\right]a_0^2 = k^2.\qquad\text{(1.414)}$$

Equation (1.414) is an implicit equation for the amplitude of the response a_0 as a function of the external detuning σ, i.e., the excitation frequency, and amplitude of the excitation k.

Let $\mathbf{X} = (p, q)$ and $\mathbf{F} = (F_x, F_y)$. The stability of (a_0, γ_0) is determined by the eigenvalues of the Jacobian matrix $D_{\mathbf{X}}\mathbf{F} = \begin{bmatrix} -\mu & -a_0\left(\sigma - \frac{3}{8}\alpha a_0^2\right) \\ a_0^{-1}\left(\sigma - \frac{9}{8}\alpha a_0^2\right) & -\mu \end{bmatrix}$. The corresponding eigenvalues λ_i are the roots of

$$\lambda^2 + 2\mu\lambda + \mu^2 + \left(\sigma - \frac{3}{8}\alpha a_0^2\right)\left(\sigma - \frac{9}{8}\alpha a_0^2\right) = 0. \tag{1.415}$$

From (1.415), we find that the sum of the eigenvalues is -2μ. This sum is negative because $\mu > 0$. Consequently, at least one of the two eigenvalues will always have a negative real part. This fact eliminates the possibility of pair of purely imaginary eigenvalues and, hence, a *Hopf bifurcation* (see Section 1.3.4). But static bifurcations can occur. To this end, we find that one of the eigenvalues vanishes when

$$\mu^2 + \left(\sigma - \frac{3}{8}\alpha a_0^2\right)\left(\sigma - \frac{9}{8}\alpha a_0^2\right) = 0. \tag{1.416}$$

For fixed $\mu > 0$, $\sigma > 0$, and $\alpha > 0$, there are shown the variation of a_0 with k in Fig. 1.61(a). In this bifurcation diagram, the solid and broken lines correspond to the stable and unstable fixed points of the both equations of (1.412), respectively. As the control parameter k is gradually increased from a zero value, a_0 follows the curve $A'H'B'$ until the critical value $k = k_2$ is reached. At $k = k_2$, a saddle-node bifurcation occurs, and, locally, there are no other solutions for $k > k_2$.

If we start from the point D' and decrease k gradually, a_0 follows the curve $D'C'E'$. At the critical point $k = k_1$, a saddle-node bifurcation occurs, and, locally, there are no other solutions for $k < k_1$. Points B' and E' are points of vertical tangencies. To this end, we find from (1.414) that

$$\frac{dk^2}{da_0^2} = \mu^2 + \left(\sigma - \frac{3}{8}\alpha a_0^2\right)\left(\sigma - \frac{9}{8}\alpha a_0^2\right),$$

which is 0 by virtue of (1.416). Recall that because saddle-node bifurcation points are locations of vertical tangencies, they are called *tangent bifurcations*. Further, because the geometry at such points, they are called *turning points* and *folds*. Yet another name for this bifurcation points is *limit points*.

Bifurcation diagrams, such as Fig. 1.61(a), are known as *force-response curves* because they show the variation of the response amplitude as a function of the forcing amplitude. Suppose that an experiment is conducted to construct Fig. 1.61(a). Then, as k is a gradually increased from a zero value, a_0 follows the curve $A'H'B'$ until the critical value $k = k_2$ is reached. Here, a jump occurs from the stable branch $A'H'B'$ to the stable branch $E'C'D'$. As k is increased beyond k_2, a_0 follows the curve $C'D'$. Consequently, as k is slowly increased, the state of the system (e.g., a_0) evolves continuously except at $k = k_2$, where it experiences a discontinuous (jump) or catastrophic change. Therefore, saddle-node bifurcations are examples of *discontinuous* or *catastrophic bifurcations*. If we start from the point D' and decrease k gradually, a_0 follows the curve $D'C'E'$. At $k = k_1$, a jump occurs from point E' to point H'. As k is decreased below k_1, a_0 follows the curve $H'A'$. Again, there is a discontinuous or catastrophic bifurcation at the saddle-node value $k = k_1$, at which the state-control function is discontinuous. In the range $k_1 < k < k_2$, the realized response depends on the direction of sweep of the control parameter. This phenomenon is called the *hysteresis phenomenon*.

In Fig. 1.61(a), for all values of $k < k_1$, there is only one stable fixed point of (1.412) in the (α, γ)-space. Hence, all evolutions in this space are attracted to the stable fixed point.

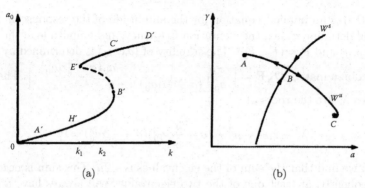

(a)   (b)

FIGURE 1.61: (a) Basins of attraction of the stable fixed point A and C in the (α, γ)-space. The stable manifold of the saddle point B divides this space into two regions. (b) Response amplitude a_0 versus the amplitude of the forcing k.

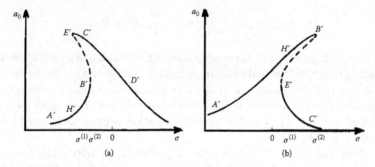

(a)   (b)

FIGURE 1.62: Bifurcation diagram by using σ as a control parameter: (a) $\alpha < 0$ and (b) $\alpha > 0$ (saddle-node bifurcations occur at B' and E').

Again, for all values of $k > k_2$, there is only one stable fixed point in the (α, γ)-space. In the interval $k_1 < k < k_2$, the two stable branches of solutions coexist. Therefore, this interval is referred to as an *interval of bistability*. In the bistability interval, there are two stable and one unstable fixed points for each value for each value of k. Let A, B, and C be the fixed points on the branches $A'H'B'$, $B'E'$ and $E'C'D'$, respectively. Then, evolutions in the (α, γ)-space approach either A or C as time $t \to \infty$, depending on the initial condition. In Fig. 1.61(b), it is shown a qualitative sketch of the global stable and unstable manifolds of B. One end of the unstable manifold of B are attracted to A, while the other end is attracted to C. The stable manifold partitions the (α, γ)-plane into two regions. The evolutions initiated in the region to the left of the stable manifold of B are attracted to A, while the evolutions initiated in the region to the right of the stable manifold are attracted to C. Remark that the state-space portraits for $k < k_1$ and $k_1 < k < k_2$ are *structurally different* because there is only one stable fixed point in one case and two stable and one unstable fixed points in the other case. Similarly, the state-space portraits for $k > k_2$ and $k_1 < k < k_2$ are *structurally different*.

Next, consider bifurcation diagrams when σ is used as a control parameter for $k > 0$ and $\mu > 0$. Figures 1.62 correspond to $\alpha < 0$ and $\alpha > 0$, respectively. Again, the solid and broken lines correspond to the stable and unstable fixed points of (1.412). For both $\alpha < 0$ and $\alpha > 0$, saddle-node bifurcations occur at $\sigma^{(1)}$ and $\sigma^{(2)}$. To verify if the bifurcation points are points of vertical tangency, we find from (1.414) that

$$\frac{d\sigma}{da_0^2} = \frac{3}{8}\alpha - \frac{\mu^2 + \left(\sigma - \frac{3}{8}\alpha a_0^2\right)^2}{2a_0^2\left(\sigma - \frac{3}{8}\alpha a_0^2\right)}. \tag{1.417}$$

Because (1.416) is satisfied at the bifurcation points, we substitute for μ^2 from (1.416) into (1.417). After simplifying, we find that $\frac{d\sigma}{da_0^2} = 0$.

Bifurcation diagrams such as Fig. 1.62 are called *frequency-response curves*. Further, the response displayed in Fig. 1.62(a) for $\alpha < 0$ is called a *softening-type response*, while the response displayed in Fig. 1.62 (b) for $\alpha > 0$ is called a *hardening-type response*. If experiments were conducted to construct the frequency-response curves, jumps would be observed at the saddle-node bifurcation points, which are catastrophic bifurcations.

5. Find the solution for parametrically excited Duffing oscillator, build its bifurcation diagram.

Solution. Its equation has a form

$$\ddot{u} + u + \varepsilon(2\mu\dot{u} + \alpha u^3 + 2ku\cos\Omega t) = 0. \tag{1.418}$$

The parameters μ, α, and k are all independent of ε, while the parameter Ω is such that $\Omega = 2 + \varepsilon\sigma$. This type of resonance, where the frequency of the parametric excitation is close to twice the natural frequency of the system, is called a *principal parametric resonance* (see, for example, [32]).

By using the method of multiple scales, one obtains the following first-order approximation for the solution of (1.418)

$$u = a\cos\left(\frac{1}{2}\Omega t - \frac{1}{2}\gamma\right) + O(\varepsilon), \tag{1.419}$$

where the amplitude a and phase γ are governed by

$$a' = -\mu a - \frac{1}{2}ka\sin\gamma, \qquad a\gamma' = \sigma a - \frac{3}{4}\alpha a^3 - \frac{1}{2}ka\cos\gamma, \tag{1.420}$$

and the prime indicates the derivative with respect to the time scale $\tau = \varepsilon t$. The first-order approximation can alternatively be expressed in the form

$$u = p\cos\left(\frac{1}{2}\Omega t\right) + q\sin\left(\frac{1}{2}\Omega t\right) + O(\varepsilon), \tag{1.421}$$

where

$$p' = -[\mu p + \frac{1}{2}(\sigma+k)q - \frac{3}{8}\alpha q(p^2 + q^2)], \quad q' = -[\mu q - \frac{1}{2}(\sigma-k)p + \frac{3}{8}\alpha p(p^2 + q^2)]. \tag{1.422}$$

It follows from (1.419) and (1.421) that

$$p = a\cos(\gamma/2), \quad q = a\sin(\gamma/2). \tag{1.423}$$

Equations (1.420) represent the polar form of the modulation equations, while (1.422) represent the Cartesian form of modulation equations. Observe that (1.420) are invariant under transformation $(a, \gamma) \Leftrightarrow (-a, \gamma)$, while (1.422) are invariant under transformation $(p, q) \Leftrightarrow (-p, -q)$.

Next, we examine the fixed points (a_0, γ_0) of (1.420). They are solutions of the algebraic system

$$\mu a_0 + \frac{1}{2}ka_0\sin\gamma_0 = 0, \quad -\sigma a_0 + \frac{3}{4}\alpha a_0^3 + \frac{1}{2}ka_0\cos\gamma_0 = 0. \tag{1.424}$$

There are two types of fixed points: (a) trivial fixed points corresponding to $a_0 = 0$ and (b) nontrivial fixed points corresponding to $a_0 \neq 0$. The corresponding fixed points (p_0, q_0) of

(1.422) can be found by using (1.423). We find from (1.424) that the nontrivial fixed points (a_0, γ_0) are given by

$$a_0^2 = \frac{4}{3\alpha}\left(\sigma \pm \sqrt{k^2 - 4\mu^2}\right), \quad \gamma_0 = \arcsin(-2\mu/k). \tag{1.425}$$

It is also obvious that the trivial fixed points share the symmetry property of the equations, while the nontrivial fixed points do not share this symmetry property. Hence, the trivial and nontrivial fixed points may be called as *symmetric* and *asymmetric solutions*.

Consider the stability of the trivial and nontrivial fixed points. For a trivial fixed point, (1.424) become identities. In such cases, where one or more of the state variables of the system assume a zero value and cause one or more of the fixed-point equations to become identities, it is not convenient to determine the stability of the fixed point from the polar form of the modulation equations. In these and other cases, the stability of a fixed point can be conveniently determined from the Cartesian form of the modulation equations. From (1.422), it can be found that the Jacobian matrix associated with (p_0, q_0) is given by

$$D_{\mathbf{X}}\mathbf{F} = \begin{bmatrix} -\mu + \frac{3}{4}\alpha p_0 q_0 & -\frac{1}{2}(\sigma + k) + \frac{3}{8}\alpha(p_0^2 + 3q_0^2) \\ \frac{1}{2}(\sigma - k) - \frac{3}{8}\alpha(p_0^2 + 3q_0^2) & -\mu - \frac{3}{4}\alpha p_0 q_0 \end{bmatrix}. \tag{1.426}$$

Note from (1.426) that the sum of the two eigenvalues is -2μ, which is negative. Hence, at least one of the two eigenvalues will always have a negative real part. This fact eliminates the possibility of a pair of purely imaginary eigenvalues and, hence, a *Hopf bifurcation*. Again, static bifurcation can occur. When k is used as a control parameter, the rank of the augmented matrix $[D_{\mathbf{X}}\mathbf{F}|\mathbf{F}_k]$ can be used to decide if a static bifurcation is a saddle-node bifurcation. To this end, we find from (1.422) that

$$\mathbf{F}_k = -\frac{1}{2}\left\{ \begin{array}{c} q \\ p \end{array} \right\}. \tag{1.427}$$

For the trivial fixed point $(p_0, q_0) = (0,0)$, (1.426) reduces to

$$D_{\mathbf{X}}\mathbf{F} = \begin{bmatrix} -\mu & -\frac{1}{2}(\sigma + k) \\ \frac{1}{2}(\sigma - k) & -\mu \end{bmatrix}$$

and its corresponding eigenvalues are the roots of $\lambda^2 + 2\mu\lambda + \mu^2 + \frac{1}{4}(\sigma^2 - k^2) = 0$. One of these eigenvalues is zero when $k = k_2 = \sqrt{4\mu^2 + \sigma^2}$. At $k = k_2$, (1.427) becomes $\mathbf{F}_k = \left\{ \begin{array}{c} 0 \\ 0 \end{array} \right\}$, and the rank of the augment matrix $[D_{\mathbf{X}}\mathbf{F}|\mathbf{F}_k]$ is one. Hence, the associated static bifurcation is not a saddle-node bifurcation.

For $\alpha > 0$ and $\sigma < 0$, it follows from (1.425) that nontrivial fixed points are possible only when $k > k_2$. They are given by

$$(a_{10}, \gamma_{10}) = (|a_0|, \gamma_0) \quad \text{and} \quad (a_{20}, \gamma_{20}) = (-|a_0|, \gamma_0), \tag{1.428}$$

where a_0^2 and γ_0 are defined by (1.425). The corresponding fixed points of (1.422) are

$$(p_{10}, \; q_{10}) = [a_{10}\cos(\gamma_{10}/2), \; a_{10}\sin(\gamma_{10}/2)],$$
$$(p_{20}, \; q_{20}) = [a_{20}\cos(\gamma_{20}/2), \; a_{20}\sin(\gamma_{20}/2)]. \tag{1.429}$$

In Fig. 1.63(a) it is shown the bifurcation diagram when k is used as a control parameter for $\alpha > 0$ and $\sigma < 0$. In this and following figures, the solid and broken lines correspond to the stable and unstable fixed points of (1.422), respectively. As k is gradually increased

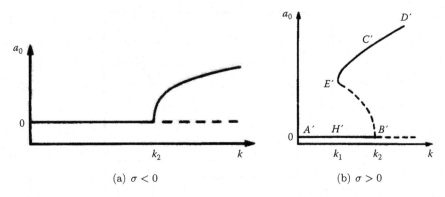

(a) $\sigma < 0$ (b) $\sigma > 0$

FIGURE 1.63: Bifurcation diagram constructed by using k as a control parameter for $\alpha > 0$.

from zero, the trivial fixed point remains stable until the critical value $k = k_2$ is reached. Here, a static bifurcation of the trivial fixed point occurs. For $k > k_2$, the trivial fixed point is a saddle point, and the stable nontrivial points (a_{10}, γ_{10}) and (a_{20}, γ_{20}) are created. The fixed point (a_{20}, γ_{20}) is not shown in Fig. 1.63(a). Both of the newly created fixed points are asymmetric solutions. Because the static bifurcation at $k = k_2$ leads to the creation of asymmetric solutions, so it is a *symmetry-breaking bifurcation*. Specifically, this bifurcation is a *supercritical pitchfork bifurcation*.

If an experiment were to be construct Fig. 1.63(a), a gradual transition from a trivial response amplitude to a nontrivial response amplitude would be observed at $k = k_2$. Consequently *supercritical pitchfork bifurcations* are continuous or safe bifurcations. For $\alpha > 0$ and $\sigma > 0$, it follows from (1.425) that nontrivial fixed points are possible for $k > k_1 = 2\mu$. At $k = k_1$, (1.428) and (1.429) become

$$(a_{10}, \gamma_{10}) = ((\frac{4\sigma}{3\alpha})^{1/2}, -\frac{1}{2}\pi), \quad (a_{20}, \gamma_{20}) = (-(\frac{3\sigma}{4\alpha})^{1/2}, -\frac{1}{2}\pi),$$

$$(p_{10}, q_{10}) = ((\frac{2\sigma}{3\alpha})^{1/2}, -(\frac{2\sigma}{3\alpha})^{1/2}), \quad (p_{20}, q_{20}) = (-(\frac{2\sigma}{3\alpha})^{1/2}, (\frac{2\sigma}{3\alpha})^{1/2}).$$

For (p_{10}, q_{10}), (1.426) reduces to $D_\mathbf{X}\mathbf{F} = \begin{bmatrix} -(\mu + \sigma/2) & -\mu + \sigma/2 \\ -(\mu + \sigma/2) & -\mu + \sigma/2 \end{bmatrix}$, which obviously

has a zero eigenvalue. Further, (1.427) becomes $\mathbf{F}_k = -\frac{1}{2}\begin{Bmatrix} q_{10} \\ p_{10} \end{Bmatrix}$. Because $\mathbf{F}_k$ does not

belong to the range of $D_\mathbf{X}\mathbf{F}$, the rank of the matrix $[D_\mathbf{X}\mathbf{F}|\mathbf{F}_k]$ is two. Hence, at $k = k_1$, the fixed point (p_{10}, q_{10}) experiences a saddle-node bifurcation. Due to symmetry, the fixed point (p_{20}, q_{20}) also experiences a saddle-node bifurcation. In Fig. 1.63(b) it is shown the bifurcation diagram when k is used as a control parameter for $\alpha > 0$ and $\sigma > 0$. Again, as k is gradually increased from zero, the trivial fixed point loses stability at $k = k_2$ due to a symmetry-breaking bifurcation. Here, locally, there are no other stable solutions for $k > k_2$, forcing the system to jump in a fast dynamic transient to C'. Therefore, we have a *subcritical pitchfork bifurcation*. To verify that this point is a point of vertical tangency, we find from (1.425) that $\frac{dk^2}{da_0^2} = \frac{3\alpha}{2}(\frac{3\alpha}{4}a_0^2 - \sigma)$, which is zero at $k = k_1$ because $a_0^2 = \frac{4}{3}(\sigma/\alpha)$.

Point out that in Figs. 1.63(a),(b), branches with different tangents meet at the pitchfork bifurcation points. The similar situation also occurs at transcritical bifurcation points. Hence, it may be referred to transcritical and pitchfork bifurcation points as *branch points*.

If an experiment were to be concluded for constructing Fig. 1.63(b), jumps would be observed at $k = k_2$ during a forward sweep at a control parameter and at $k = k_1$ during a reverse sweep of the control parameter. Therefore, the bifurcation at $k = k_1$ and k_2

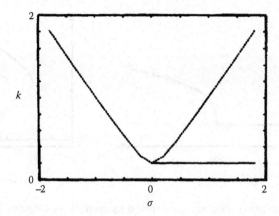

FIGURE 1.64: Bifurcation set in the k-σ plane.

are catastrophic bifurcations. Thus, subcritical pitchfork bifurcations are also catastrophic bifurcations. Figure 1.64 shows the loci of the possible bifurcation points of (1.422) in the space of the control parameters k and σ for $k > 0$ and $\mu = 0.1$. The broken and solid lines correspond to pitchfork and saddle-node bifurcation points, respectively. Here, we do not have any transcritical or Hopf bifurcation points.

The pitchfork bifurcation points fall on the curve $k^2 = 4\mu^2 + \sigma^2$. The saddle-node bifurcation points occur only for $\sigma > 0$, and they fall on the line $k = 2\mu$. When $\sigma = 0$, the pitchfork and saddle-node bifurcation points coalesce. Sets, such as Fig. 1.64, that consist of the different bifurcation points in the control-parameter space are called *bifurcation sets*. In other systems, one may need to use the numerical schemes to determine the loci of static and dynamic bifurcation points.

6. Solve the initial boundary value problem for the heat equation (it has been introduced in [24]).

Solution. Assume that the temperature field $u(x, t; \varepsilon)$ depends on one space variable x which ranges from 0 to $b(\varepsilon t)$, where b is a known function and ε is a small parameter. Thus b is a slowly varying function of t. The mathematical description of the problem is

$$u_t = u_{xx}, \quad 0 \le x \le b(\varepsilon t),$$
$$u(0, t) = \varphi(\varepsilon t), \quad u[b(\varepsilon t), t] = 0, \tag{1.430}$$
$$u(x, 0) = \psi(x), \quad 0 \le x \le b(0).$$

Solution. Changing from variable t to $\tau = \varepsilon t$, the first two equations of (1.430) become

$$\varepsilon u_\tau = u_{xx}, \quad 0 \le x \le b(\tau), \quad u(0, \tau) = \varphi(\tau), \quad u[b(\tau), \tau] = 0. \tag{1.431}$$

Since ε multiplies u_τ, the straightforward perturbation solution for small ε, keeping τ fixed, cannot satisfy in general the second initial condition in (1.430) and non-uniformity exists at and near $\tau = 0$. To describe the behavior of u near $t = 0$, a stretching transformation $t = \tau/\varepsilon$ is needed. As verified below, the special function $\exp[-g(\tau)/\varepsilon]$ with $g(\tau) \to \tau$ as $\tau \to 0$ describes the behavior of u in this region. Therefore, we assume that u has a uniformly valid asymptotic expansion of the form

$$u = \sum_{n=0}^{\infty} \varepsilon^n f_n(x, \tau) + e^{-g(\tau)/\varepsilon} \sum_{n=0}^{\infty} \varepsilon^n h_n(x, \tau).$$

Substituting this into the third boundary condition in (1.430) through the boundary conditions in (1.431) and equating to zero the coefficients of ε^n and $\varepsilon^n e^{-g(\tau)/\varepsilon}$, we obtain

$$f_{0,xx} = 0, \quad f_0(0, \tau) = \varphi(\tau), \quad f_0[b(\tau), \tau] = 0,$$
$$h_{0,xx} + g' h_0 = 0, \quad h_0(0, \tau) = 0, \quad h_0[b(\tau), \tau] = 0, \tag{1.432}$$
$$f_0(x, 0) + h_0(x, 0) = \psi(x), \quad 0 \le x \le b,$$

and for $n \geq 1$

$$
\begin{aligned}
f_{n,xx} &= f_{n-1,\tau}, \quad f_n(0,\tau) = f_n[b(\tau),\tau] = 0, \\
h_{n,xx} + g'h_n &= h_{n-1,\tau}, \quad h_n(0,\tau) = h_n[b(\tau),\tau] = 0, \\
f_n(x,0) &+ h_n(x,0) = 0,
\end{aligned} \tag{1.433}
$$

where g' stands for $dg/d\tau$. The solution to the upper problem in (1.432) is $f_0 = \varphi(\tau)[1 - x/b(\tau)]$. Since the boundary condition on h_0 are homogeneous, the equation for h_0 has a non-trivial solution if and only if g' is one of the eigenvalues $g'_k = [k\pi/b(\tau)]^2$. Then the corresponding normalized eigenfunctions are

$$
\chi_k = [2/b(\tau)]^{1/2} \sin \frac{k\pi x}{b(\tau)}.
$$

Hence $h_0 = a_0(\tau)\chi_k(x,\tau)$, where unknown function a_0 is determined from examination of the equation for h_1, with h_0 known the second problem in (1.433) for $n = 1$ becomes

$$
h_{1,xx} + g'_k h_1 = a'_0 \chi_k + a_0 \chi_{k,\tau}, \quad h_1(0,\tau) = h_1[b(\tau),\tau] = 0. \tag{1.434}
$$

Now we assume that h_1 can be expanded in terms of the eigenfunctions χ_k; that is

$$
h_1 = \sum_{s=1}^{\infty} c_s(\tau)\chi_s(x,\tau). \tag{1.435}
$$

Substituting (1.435) into (1.434) and using the fact that $\chi_{s,xx} = -g'_s \chi_s$, we obtain

$$
\sum_{s=1}^{\infty} (g'_k - g'_s)c_s \chi_s = a'_0 \chi_k + a_0 \chi_{k,\tau}.
$$

If we multiply this equation by χ_k and integrate from $x = 0$ to $b(\tau)$, the left-hand side vanishes because χ_k is orthogonal to χ_s for $k \neq s$, and $g'_s = g'_k$ when $k = s$. Therefore

$$
\int_0^{b(\tau)} (a'_0 \chi_k^2 + a_0 \chi_{k,\tau}\chi_k)\,dx = 0, \tag{1.436}
$$

which is the condition for solvability of problem (1.434). Since

$$
\int_0^{b(\tau)} \chi_k^2 \, dx = 1, \tag{1.437}
$$

then we have

$$
\frac{d}{d\tau} \int_0^{b(\tau)} \chi_k^2 \, dx = b'(\tau)\chi_k^2[b(\tau),\tau] + 2\int_0^{b(\tau)} \chi_k \chi_{k,\tau}\,dx = 0.
$$

Note that $\chi_k[b(\tau),\tau] = 0$, thus $\int_0^{b(\tau)} \chi_k \chi_{k,\tau}\,dx = 0$. Hence, (1.436) and (1.437) lead to $a_0 = \text{const}$. Therefore zero-order solution is

$$
u(x,\tau;\varepsilon) = \varphi(\tau)\left[1 - \frac{x}{b(\tau)}\right] + \sum_{k=1}^{\infty} a_k \left[\frac{2}{b(\tau)}\right]^{1/2} \sin \frac{k\pi x}{b(\tau)} \exp\left[-\frac{k^2\pi^2}{\varepsilon} \int_0^{\tau} b^{-2}(\xi)d\xi\right] + O(\varepsilon),
$$

where $a_k = [2/b(0)]^{1/2} \int_0^{b(0)} \{\psi(x) - \varphi(0)[1 - x/b(0)]\} \sin \frac{k\pi x}{b(\tau)} \, dx$ is constant.

Bibliography

[1] Abramowitz M. and Stegun I. A. *Handbook of Mathematical Functions*. Dover, New York, 1965, 830 pp.

[2] Arfken G., Weber H. *Mathematical Methods for Physicists*. Elsevier Inc., 2013, 1205 pp.

[3] Beigman I.L. and Gaissinski I. M. An analytical description of populations of highly excited levels. *J. Quant. Spectrosc. Radiat. Transfer*, 28(5), 1982, 441–454.

[4] Bemfica F.S., Disconzi M.M., and Noronha J. Causality and existence of solutions of relativistic viscous fluid dynamics with gravity. arXiv:1708.06255, 2017.

[5] Ben-Artzi M., Croisille J.-P. and Fishelov D. *Navier–Stokes Equations in Planar Domains*. London, 2013, 302 pp.

[6] Berglund N. *Geometrical Theory of Dynamical Systems*. ETH Zurich, 2001, 85 pp.

[7] Berglund N. *Perturbation Theory of Dynamical Systems*. ETH Zurich, 2001, 111 pp.

[8] Choquet-Bruhat Y. et al. *Analysis, Manifolds and Physics*. Elsevier, 1977, 544 pp.

[9] Coscia V. An existence and uniqueness theorem for the Navier–Stokes equations in dimension four. *Int. J. Applicable Analysis*, 2016, 1–6

[10] Dodson C. T. J. and Poston T. *Tensor Geometry*. Springer, 1991

[11] Dugdale J. S. *Entropy and its Physical Meaning*. Taylor & Francis, 1996.

[12] Ebin D.G. and Marsden J. Groups of diffeomorphisms and the motion of an incompressible fluid, *Ann. of Math.* 92 (1970) 102–163.

[13] Eckart C. The Thermodynamics of Irreversible Processes. III. Relativistic Theory of the Simple Fluid. *Phys. Rev.* 58, Issue 10, 1940, 919–923.

[14] Elbert A. Some recent results on the zeros of Bessel functions and orthogonal polynomials. *J. of Computational and Applied Math.* 133, Issues 1-2, 2001, 65–83.

[15] Falconer K. *Fractal Geometry* (3rd edition). Wiley, 2014, 398 pp.

[16] Fefferman C.L. Existence and Smoothness of the Navier–Stokes equation. Official statement of the problem, Clay Mathematics Institute, 6 pp.

[17] Gaissinski I. and Rovenski V. *Non-linear Models in Mechanics*. Lambert, 2010.

[18] Gaissinski I. and Kartvelishvili L. *Viscous Hypersonic Flow*. Verlag VDM, 2010, 220 pp.

[19] Galdi G.P. *An Introduction to the Mathematical Theory of the Navier–Stokes Equations. Steady-State Problems*. Springer, 2nd ed., 2011, 1018 pp.

[20] Heading J. *An Introduction to Phase Integral Method*. Wiley, New York, 1962, 160 pp.

[21] Hilborn R.C. *Chaos and Nonlinear Dynamics* (2nd edition). Oxford, 2000, 328 pp.

[22] Kato H. Existence of periodic solutions of the Navier–Stokes equations. *J. of Math Analysis and Applications*, 208, 1997, 141–157.

[23] Katz N.H. and Pavlovic N. Finite time blow-up for a dyadic model of the Euler equations. *Trans. AMS*, 357(2), 2004, 695–708.

[24] Keller H. *Numerical Methods for Two-Point Boundary Value Problems*. Blaisdell Publishing Co., Waltham, MA, 1968.

[25] Kolmogorov A., Fomin S. *Elements of Theory of Functions and Functional Analysis*, 1999.

[26] Ladyzhenskaya O.A. *The Mathematical Theory of Viscous Incompressible Flow*. Gordon and Breach Sci. Publishers, New York, 1969.

[27] Landau L.D. and Lifshitz E.M. *Course of Theor. Physics Vol. 6, Fluid Mechanics*, Elsevier, 1987.

[28] Mahalov A., Titi E.S. and Leibovich S. Invariant helical subspaces for the Navier–Stokes equations. *Arch. Rat. Mech. Anal.* 112, 1990, 193–222.

[29] Milne-Thomson L.M. *Theoretical Hydrodynamics*. Macmillan Company, 1996.

[30] Morita S. *Geometry of Differential Forms*. AMS, 2001.

[31] Moran M., Shapiro H. *Fundamentals of Engineering Thermodynamics*. Wiley & Sons, 2008.

[32] Nayfeh A.H. *Linear and Non-linear Structural Mechanics*. Wiley, 1997, 560 pp.

[33] Nayfeh A.H. *Introduction to Perturbation Techniques*. Wiley, 1981, 519 pp.

[34] Newman M.H. *Elements of the Topology of Plane Sets of Points*. Dover Publ., 1992.

[35] Oseledec V.I. A multiplicative ergodic theorem. Lyapunov characteristic numbers for dynamical systems, *Trans. Moscow Math Soc.*, 19, 1968, 197–231.

[36] Rand R.H. and Armbruster D. *Perturbation Methods, Bifurcation Theory and Computer Algebra*. Springer-Verlag, 1987, 243 pp.

[37] Ronveaux A. *Heun's Differential Equations*. Oxford University Press, 1995.

[38] Russo R. On Stokes' problem. In: Rannacher R., Sequeira A., eds. *Advances in Mathematical Fluid Mechanics*. Springer-Verlag, 2010, 473–511.

[39] Shivamoggi B.K. *Perturbation Methods for Differential Equations*. Birkhäuser, 2003, 354 pp.

[40] Sparrow C. *The Lorentz Equations: Bifurcations, Chaos, and Strange Attractors*. Springer, 1982, 278 pp.

[41] Strickland M. Anisotropic Hydrodynamics. *Acta Phys. Polon.* B, 45, 2014, 2355–2394.

[42] Suslov O.N. and Fateeva E.I. Analysis of multicomponent gas mixture flows with partial chemical equilibrium. *J. Fluid Dynamics*, 31(1), 1996, 97–106.

[43] Tao T. Finite time blowup for an averaged three-dimensional Navier–Stokes equation. *J. of the AMS*, 29(3), 2016, 601–674.

[44] Taylor M.E. *Partial Differential Equations III: Nonlinear Equations*, Springer, New York, 2011.

[45] Temam R. *Navier–Stokes Equations: Theory and Numerical Analysis*. AMS, Vol. 343; 1984; 408 pp.

[46] Ukhovskii, M.R. and Iudovich, V.I. Axially symmetric flows of ideal and viscous fluids filling the whole space. *J. Appl. Math. Mech.* 32, 1968, 52–62.

[47] Wald R.M. *General relativity*. University of Chicago Press, 2010.

[48] Watson G.N. A Treatise on the Theory of Bessel Functions. Cambridge Univ. Press, 1966, 76 pp.

[49] Weisstein E. W. *Coordinate System*. Wolfram Math. World, 2009.

[50] Whittaker E. and Watson G. *A Course of Modern Analysis*. Cambridge Univ. Press, 1999, 608 pp.

[51] Wiggins S. *Introduction to Applied Nonlinear Dynamical Systems and Chaos*. Springer, 2003.

[52] Younsi A. Regularity criteria for strong solutions to the 3d Navier–Stokes equations. *System Science and Control Engineering*, 3, 2015, 262–265.

[53] Yu H. and Zhao J. Global classical solutions to the 3d isentropic compressible Navier–Stokes equations in a bounded domain. *IOP Publ. Ltd. and LMS Nonlinearity*, 30, No. 1, 2016, 1–20.

Chapter 2

Models for Hydrodynamic Instabilities

There are about 20 instability types named after people, for example, Rayleigh–Taylor (occuring in density-stratified flows, astrophysics), Kelvin–Helmholtz (shear flow), Richtmyer–Meshkov (plasma physics [41], astrophysics [25]), Taylor–Couette (flow in rotating cylinder), Bénard instability (natural convection, Rayleigh-Bénard convection), Buneman (plasma physics), Darrieus–Landau (flame propagation), Kruskal–Shafranov (line-tied partial-toroidal plasmas), and Velikhov (nonequilibrium MHD).

In Chapter 2, we consider two types of them, Rayleigh–Taylor (Section 2.2) and Kelvin–Helmholtz (Section 2.3) instabilities with results by authors. We confined ourselves to the consideration of these two types of hydrodynamic instabilities for two reasons. First, Rayleigh–Taylor and Kelvin–Helmholtz instabilities are mostly often encountered in fluid mechanics under the absence of electric and magnetic fields. We also will not consider the theory of convection flow and flame propagation playing sufficient roles in the physics of combustion. The second reason is the limited volume of the present book, while each of even the basic instability types can be discussed indefinitely.

An exact solution to the problem of the motion of a thin round cylindrical shell due to different internal and external pressures is obtained. Analysis of nonlinear instabilities of Rayleigh–Taylor type in meridional cross-section is carried out. The linear perturbation model for Kelvin–Helmholtz instabilities is developed and applied to jets of low viscosity; the dispersion relation based on continuity and momentum conservation equations for viscous compressible jets is deduced.

2.1 Stability concepts

Assume that at every point $\mathbf{x} \in \mathbb{R}^3$ of the fluid, and at all times t, we can define the functions of density $\rho(\mathbf{x}, t)$, velocity $\mathbf{u}(\mathbf{x}, t)$, and pressure $p(\mathbf{x}, t)$, and that they vary smoothly (differentiable) over the fluid. Note that we do not deal with the dynamics of individual molecules. A small volume δV thus has a mass $\rho \, \delta V$ and a momentum $\rho \, \mathbf{u} \, \delta V$.

Definition 49 A *fluid particle* (i.e., material element) is one that moves with the fluid, so that its velocity is $\mathbf{u}(\mathbf{x}, t)$ and its position $\mathbf{x}(t)$ satisfy the differential equation $\dot{\mathbf{x}} = \mathbf{u}(\mathbf{x}, t)$. The rate of change of a quantity as seen by a fluid particle (the time derivative along particle trajectories) is called the *material derivative* and written D/Dt. It is given by the chain rule as $\frac{D}{Dt} \equiv \partial_t + \mathbf{u} \cdot \nabla$.

The *mass conservation law* (assuming commutativity of the operator, $\mathbf{u} \cdot \nabla = \nabla \cdot \mathbf{u}$, on scalar functions) is expressed as

$$\frac{D}{Dt} \rho \equiv \partial_t \rho + \mathrm{div}(\rho \, \mathbf{u}) = 0.$$

For an *incompressible* fluid, the density function is a positive constant, and the law reads

$$\text{div } \mathbf{u} = 0.$$

For a 2d flow, the above condition is automatically satisfied by $\mathbf{u} = \nabla \times (0, 0, \psi(x, y)) = (\partial_y \psi, -\partial_x \psi, 0)$, where $\psi(x, y)$ is called the *stream function*.

If axisymmetric flow is given in *cylindrical* coordinates (r, θ, z) by the Stokes *stream function*, $\psi(r, z)$, the incompressible condition $\text{div } \mathbf{u} = 0$ may be expressed as $\mathbf{u} = \nabla \times (0, \frac{\psi}{r}, 0) = (-\frac{1}{r}\partial_z \psi, 0, \frac{1}{r}\partial_r \psi)$.

Note that the *axisymmetric* property allows us to analyze a 3-dimensional object which is rotationally symmetric about an axis. The NSE for an incompressible flow have a view

$$\rho \frac{D}{Dt} \mathbf{u} \equiv \rho(\partial_t \mathbf{u} + (\mathbf{u} \cdot \nabla)\mathbf{u}) = -\nabla p + \mathbf{F} + \mu \nabla^2 \mathbf{u}, \tag{2.1}$$

where μ is the *viscosity* and $\mathbf{F}$ a *body force* (for example, a gravity, $\mathbf{F} = \rho \mathbf{g}$). In cylindrical coordinates, (2.1) for the velocity $\mathbf{u} = (u_r, u_\theta, u_z)$ becomes

$$\begin{cases} \rho\left(\frac{D}{Dt}u_r - \frac{u_\theta^2}{r}\right) = -\partial_r p + \mu\left(\nabla^2 u_r - \frac{u_r}{r^2} - \frac{2}{r^2}\partial_\theta u_\theta\right), \\ \rho\left(\frac{D}{Dt}u_\theta - \frac{u_r u_\theta}{r}\right) = -\frac{1}{r}\partial_\theta p + \mu\left(\nabla^2 u_\theta - \frac{u_\theta}{r^2} - \frac{2}{r^2}\partial_\theta u_r\right), \\ \rho\frac{D}{Dt}u_z = -\partial_z p + \mu\nabla^2 u_z, \end{cases}$$

where $\mathbf{u} \cdot \nabla = u_r \partial_r + u_\theta \frac{1}{r}\partial_\theta + u_z \partial_z, \quad \nabla^2 = \frac{1}{r}\partial_r(r\,\partial_r) + \frac{1}{r^2}\partial_\theta^2 + \partial_z^2.$

2.1.1 Boundary conditions

In order to determine the velocity $\mathbf{u}(\mathbf{x}, t)$ and the pressure $p(\mathbf{x}, t)$ in some region V, we need to determine the boundary conditions on the surface $S = \partial V$. The appropriate conditions to apply are that the velocity and the stress should be continuous across the whole interface.

Fluid/solid boundaries. For a stationary boundary of the fluid, it is sufficient to require that the velocity vanishes on the boundary,

$$\mathbf{u} = 0. \tag{2.2}$$

From (2.2) follows that the tangential velocity components are zero as well as the normal component. In *inviscid flow* only the normal velocity need be continuous at an interface, and a *slip velocity* must be permitted. The second derivative $\mu \nabla^2 \mathbf{u}$ in the NSE requires an extra boundary condition. A solid boundary can provide whatever stress is needed to support the fluid motion. The continuity in stress enables to calculate the force on the solid due to the fluid.

Fluid/fluid boundaries. These are more complicated, because the interface can move. It is a physical fact that an extra normal stress, due to the *surface tension*, acts on the interface. This extra stress takes the form $\gamma K(\mathbf{x})$ where $\gamma > 0$ is the surface tension constant, and K is the curvature of the fluid surface, which can be defined by $K = \mathbf{n} \cdot \nabla$, where $\mathbf{n}$ is the unit normal to the interface.

If one of the fluids is dynamically negligible, as often happens with a liquid/gas interface, then we can treat one fluid as having a constant pressure p_0 and neglect its motion. If such a surface is stationary, then the appropriate boundary conditions to apply on the other fluid are zero normal velocity and zero tangential stress. If the normal stress does not balance then the surface will not be stationary. More generally, if we describe the surface position at time t by the function $\zeta(\mathbf{x}, t) = 0$, then the boundary conditions can be written as

$$\frac{D}{Dt}\zeta = 0, \quad \mathbf{T} = -(\gamma K + p_0)\hat{\mathbf{n}}.$$

For inviscid flow, $\mu = 0$ holds, and the tangential stress condition is trivial.

2.1.2   Inviscid and high-Reynolds-number flow

When written in terms of non-dimensional variables, a parameter Re, known as the *Reynolds number* appears in the equations. Re essentially measures the relative importance of the inertial to the viscous forces, and is defined by $Re = \rho L u / \mu$, where L is a typical length-scale of the problem, and u a typical velocity magnitude.

For small values of Re, it can be proved that only one steady solution of the NSE exists, and that this flow is stable in the sense defined below. For high values of Re, there are many examples where more than one stable, steady solution exists. Flow instability is strongly linked with the existence of more than one solution. When $Re \gg 1$, it is tempting to neglect the viscous terms, setting $\mu = 0$. In this case, one of the boundary conditions must be omitted, usually allowing tangential slip. Some caution is necessary, as viscous boundary layers form near solid surfaces in which the velocity develops strong gradients so that the viscous term cannot be neglected. Boundary layers typically have thickness $L/Re^{1/2}$ and must remain thin for the *core/layer* structure to be valid. Inside a steady boundary layer, where x and y are measured parallel and normal to the boundary, the governing equations for $\mathbf{u} = (u_x, u_y, 0)$

$$\rho(u \, \partial_x u_x + u_y \, \partial_y u_y) = -\partial_x p + \mu \, \partial_y^2 u_x, \quad \partial_y p = 0, \quad \partial_x u_x + \partial_y u_y = 0.$$

These equations are *parabolic* which means they must be solved in the downstream direction. The pressure does not vary across the layer and is determined by the conditions at $y \to \infty$ which means the external potential flow. The boundary layer equations tend to be valid so long as the pressure gradient is *favourable*, which means $\partial_x p < 0$. If the pressure gradient is unfavourable, there is a strong likelihood that *separation* of the boundary layer will occur. This is manifested by the solution to the boundary layer equations developing a singularity. Separation completely alters the external flow, and leads for example to "stall" of aircraft.

The *vorticity equation* is obtained by taking *rot* of (2.1). Presenting the vorticity by $\boldsymbol{\omega} = \nabla \times \mathbf{u}$ we get

$$\rho(\partial_t \boldsymbol{\omega} + (\mathbf{u} \cdot \nabla) \, \boldsymbol{\omega} - (\boldsymbol{\omega} \cdot \nabla) \, \mathbf{u}) = \mu \nabla^2 \boldsymbol{\omega}. \tag{2.3}$$

For two-dimensional flow, if we write $\mathbf{u} = \nabla(0, 0, \psi(x, y, t))$ and $\boldsymbol{\omega} = (0, 0, \omega)$ then

$$\rho \frac{D}{Dt} \boldsymbol{\omega} = \mu \nabla^2 \boldsymbol{\omega}, \quad \omega = -\nabla^2 \psi.$$

Definition 50 A flow with zero vorticity, $\boldsymbol{\omega} \equiv 0$, is called *irrotational*.

Then we can introduce a velocity potential function ϕ, such that $\mathbf{u} = \nabla \phi$.

Inviscid flow. As there is no source terms in (2.3), vorticity can only be generated at boundaries. If $\mu = 0$ then a flow which is irrotational initially remains irrotational for all time. The *time-dependent Bernoulli theorem* states that for irrotational flows,

$$\partial_t \phi + \frac{1}{2} |\nabla \phi|^2 + gy + \frac{p}{\rho} = \text{const},$$

where $g \approx 9.81$ is the gravity acceleration. For high values of Re it is found experimentally [4, 11] that the fluid flows tend to become unsteady and highly chaotic, even though a simple steady flow could exist in theory. Turbulent flows have important practical implications but they are difficult to analyze. The manner in which the *transition to a turbulence* of a *laminar flow* occurs is an important topic. The first stage in this process is that the underlying steady flow becomes unstable. In this book we examine the *hydrodynamic stability*.

2.1.3   Basic definitions

For a given problem, we assume that the obtained solution $\mathbf{u} = \mathbf{U}(\mathbf{x})$ of governing equations is steady with a corresponding pressure distribution $p = P(\mathbf{x})$. Consider a small perturbation to the flow,

$$\mathbf{u} = \mathbf{U}(\mathbf{x}) + \varepsilon\mathbf{u}'(\mathbf{x}, t), \quad p = P(\mathbf{x}) + \varepsilon p'(\mathbf{x}, t),$$

where $\varepsilon > 0$ is a small constant. If $\varepsilon\,\mathbf{u}'$ remains small for all time, we say that the underlying flow is *stable*, whereas if it eventually becomes large no matter how small ε is, we say that the flow is *unstable*. Exact equations for the components $\mathbf{u}'$ and p' are the following:

$$\rho(\partial_t\mathbf{u}' + (\mathbf{U}\cdot\nabla)\,\mathbf{u}' + \varepsilon(\mathbf{u}'\cdot\nabla)\,\mathbf{u}') = -\nabla p' + \mu\nabla^2\mathbf{u}', \quad \mathrm{div}\,\mathbf{u} = 0.$$

The linear stability theory neglects the last term on the LHS, as $\varepsilon \to 0$. The resulting linear equation has solutions of the form $\mathbf{u}' = \hat{\mathbf{u}}(\mathbf{x})e^{st}$ for some vector function $\hat{\mathbf{u}}$ and constant s, and similarly $p' = \hat{p}(\mathbf{x})e^{st}$. This is because none of the coefficients depends on t as $\mathbf{U}$ is steady. The general solution to this problem will be a linear combination of all these particular solutions. The possible values of s can be regarded as *eigenvalues* of the system. Generally, these are complex:

$$s = s_{\mathrm{Re}} + i\,s_{\mathrm{Im}}, \quad e^{st} = e^{s_{\mathrm{Re}}t}[\cos(s_{\mathrm{Im}}t) + i\sin(s_{\mathrm{Im}}t)].$$

If $s_{\mathrm{Re}} < 0$ for all values of s we say that the flow is *stable*, and if $s_{\mathrm{Re}} > 0$ for at least one eigenvalue s then the flow is called *unstable*. If $s_{\mathrm{Re}} = 0$ for some eigenvalue, we say that the flow is *neutrally stable*, then the nonlinear terms may be important.

Surface stability. If the fluid has a free surface, it will deform in accordance with the normal stress associated with the perturbation velocity. Free surfaces can be unstable even at very small Re.

The above approach looks at perturbation modes with a fixed spatial structure and examines how they evolve in time, a process known as a *temporal stability*. An alternative approach, which is often appropriate, is to consider the spatial evolution of a localized disturbance in the flow. This disturbance may grow as it is an advected downstream, so that the place where the instability occurs is far away from the disturbance. This is known as *convective instability*. It is possible that the region of flow interest is too small for instability of a given initial magnitude to develop. If a disturbance at a given position leads to growth at that position this is known as *absolute instability*.

Consider **plane-wave representation**:

$$\mathbf{u} = \mathbf{u}_0 e^{i(\omega t - \mathbf{k}\cdot\mathbf{r})}, \tag{2.4}$$

where $\mathbf{u}_0$ is called an *initial amplitude*, $\mathbf{k}$ – a *wavevector*, its modulus $k = |\mathbf{k}|$ is a *wavenumber*, $\mathbf{r}$ the radius vector, ω the *frequency*. So the wavelength λ may be denoted as $\lambda = 2\pi/k$. The equation, which denotes dependency of the frequency from wavenumber, is called a *dispersion relation*. In cylindrical coordinates, $\mathbf{r} = (r, \theta, z)$, representation (2.4) has a form

$$\mathbf{u} = \mathbf{u}_0 e^{i(\omega t - k_\theta\theta - k_r r - k_z z)}. \tag{2.5}$$

In the case of axisymmetric flow, i.e., when $\mathbf{u}$ uniformly depends on θ, and (2.5) may be transformed to

$$\mathbf{u} = \mathbf{u}_0 e^{i(\omega t - n\theta - k_r r - k_z z)}$$

with integer n which is called a *perturbation mode*.

2.2 Rayleigh–Taylor instability

More than 100 years ago, Lord Rayleigh (1883) published his classical treatment[1] entitled *"Investigation of the Character of the Equilibrium of Incompressible Fluid Variable Density."* It was motivated as an illustration of the theory of cirrus clouds but it is not specific for this case. The work describes the linear stability eigenvalue problem for incompressible fluid under gravity. The general stability criterion for incompressible inviscid fluids is discussed and particular perturbation solutions for exponential density vibrations are derived. It is remarkable that even a transcendental dispersion relation, describing the stability of exponential transition profiles, was already fairly completely analyzed in this original work.

Rayleigh was probably the first who posed the stability problem in a general manner and recognized its principal significance for atmospheric stratification. At his time, stable surface waves had already been well-understood theoretically. The mathematical description of surface waves between superposed liquid of different densities may be largely due to Stokes (1847), see [29]. Some earlier observation of the density dependence of the oscillation period to be apparently made by B. Franklin (1762), see [29], who compared the behavior of oil-water and air-water interfaces. In later works, some of idealizations of the inviscid incompressible fluid model have been overcome. Schwarzschild[2] established the basic criterion for stability against convection in compressible atmospheres. Harrison [23] included the effect of viscosity in the treatment of superposed fluid layer and Rayleigh[3] explained the onset of Boussinesq convection by including viscosity and thermal conductivity.

Apart from these and other early studies of gravitational instabilities, the modern development has started with the work of Taylor (1950). In a companion works [32, 43], the basic physical importance of *Rayleigh–Taylor Instabilities* (RTI) for accelerated fluid layers could first be demonstrated. The theoretically predicted instability growth rates to be provided in excellent agreement with laboratory experiments. The subsequent stage of rising upstanding gas columns could be observed. The rate of rise of these gas columns could be explained by similar laws as for underwater bubbles and for bubbles rising in cylindrical tubes. These finding have been highly influential for the future developments.

2.2.1 Potential flow

An elementary description of RTI can be given in terms of potential flow theory. It provides a well-known framework for the study of surface modes and has been the basis of many classical stability results. Important aspects of this approach is that the most severe interfacial perturbations can be isolated by this model, giving usually an upper bound on instability growth. It pertains to nonlinear evolution in a relatively simple manner.

In this section, the basic equations governing potential flows with moving boundaries are outlined. First, the boundary-value problem for an arbitrary contact discontinuity between two fluids is introduced. Then, the mathematical model is specialized to linear perturbations about plane and spherically symmetric boundaries. The common aspects of the different geometries are to be emphasized by first deriving the general linearized boundary conditions and then performing the specific mode expansion.

---

[1] Rayleigh Lord, *Proc. London Math. Soc.*, 14, 1883, 170–177.

[2] Schwarzschild K. Die Sekularbeschleunigung des Mondes. *D.Ph. Georg-August-Univ. Gettingen*, 1906 (1905), p. 41.

[3] Rayleigh Lord, On convective currents in a horizontal layer of fluid when the higher temperature is on the under side. *Phil. Mag.*, 32, 1916, 529–546.

Let $\mathbf{u}(\mathbf{x}, t)$ be the velocity field of a uniform inviscid fluid of constant density ρ. In the RTI, fluid motions are considered to be two-dimensional and they arise from small initial perturbations of an equilibrium state where the fluid is at rest. Under these conditions, the flow can be assumed irrotational at least in the linear approximation. Inspection of the *linearized vorticity equation* shows [28] that the vorticity field is necessarily time-independent, $\partial_t(\nabla \times \mathbf{u}) = 0$. Even if the vorticity occurs in the initial conditions, it can be neglected after a short initial period in comparison with the unstable irrotational flow. Note that there is a basic difference to shear flow instabilities where the vorticity can be driven by the velocity shear of the basic flow. Incompressible irrotational motions are subject to the constraints

$$\operatorname{div} \mathbf{u} = 0, \quad \nabla \times \mathbf{u} = 0. \tag{2.6}$$

Alternatively, a scalar potential function ϕ can be introduced through the relations (see Definition 50)

$$\mathbf{u} = \nabla\phi, \quad \nabla^2\phi = 0. \tag{2.7}$$

The description of the motion of the fluid boundaries is partly a kinematical problem. Let $S(\mathbf{x}, t) = 0$ denotes the equation of a boundary surface moving with the fluid in a given flow field. Its evolution is then governed by the quasi-linear partial differential equation (PDE)

$$\partial_t S + (\mathbf{u} \cdot \nabla)S = 0. \tag{2.8}$$

Using the method of characteristics, solutions can be found by calculating the trajectories of surface particles. Here and in the following, it is understood without explicit notation that flow variables in surface equations are evaluated on the surface. The dynamics of the flow can be described under the conventional momentum conservation law for fluid particles. If each fluid particle is subjected to a conservative force field $-\nabla U$ and to a scalar pressure p, its motion is governed by the Euler equation

$$\rho \frac{D}{Dt}\mathbf{u} = -\nabla\mathbf{p} - \rho\nabla\mathbf{U}.$$

In this section we will mostly interested in constant accelerations of fluid layers. If a denotes the magnitude of the acceleration along the y-axis, one can set $U = ay$. Physically, this equation can describe either a fluid layer accelerated towards the positive y-axis or a fluid layer subject to gravitational acceleration along the negative y direction. Notice that the equivalence between gravity and inertial forces applies only under constant acceleration. For instance, the gravitational instability of a fluid sphere is different from the instability of spherically converging shell. Inserting $\mathbf{u} = \nabla\phi$ it follows immediately the integrated form

$$\rho(\partial_t\phi + \frac{1}{2}u^2 + U) + p = C(t). \tag{2.9}$$

This is a Bernoulli type equation, which defines the pressure variations inside the fluid. The function $C(t)$ can be chosen corresponding to a particular gauge of the velocity potential function ϕ.

The simplifying assumptions (2.6) apply only to the interior domain of a uniform inviscid fluid. Within the boundary layer between two potential flow regions one has to consider the complete conservation laws of fluid dynamics. It is well known that the theory of ideal fluid flow admits the existence of contact discontinuous [10] to be moving passively within the fluid. There is no mass flow across a contact discontinuity and, in the absence of a surface tension; it exerts no forces on its surroundings. More precisely, *contact discontinuities* are defined by the continuity conditions $[u_n] = 0$, $[p] = 0$, for the normal component of the fluid velocity $u_n = \partial\phi/\partial t$ and the pressure, respectively. The square brackets represent the jump of the argument across the boundary between two fluids; see Fig. 2.1.

Using Equations (2.8) and (2.9), these continuity conditions may be expressed as

$$\partial_t S + (\mathbf{u} \cdot \nabla)S = 0, \quad [\rho(\partial_t \phi + u^2/2 + U)] = [C(t)]. \tag{2.10}$$

The equations above often require the kinematical and dynamical boundary conditions, respectively. Note that $(2.10)_1$ has to be satisfied at the both sides of the boundary: thus representing actually two constraints. These imply the continuity of u_n since ∇S is directed along the surface normal and $S(\mathbf{x}, t)$ denotes the same function in the both fluids. The time function $[C(t)]$ in $(2.10)_2$ is usually chosen to be zero with possible exception for steady flow problems. The two potentials ϕ of the both fluids and the surface S have to be determined in accordance with these three boundary constraints. They determine the evolution of an arbitrary contact discontinuity in potential flow theory.

The perturbation problem is considerable importance in gaining an understanding of more complicated flow. A general approach is given under perturbing a known solution of high symmetry and studying the growth of surface perturbations in the linear approach.

Example 70 One may illustrate the foregoing remarks in the simplest case of a plane contact discontinuity. The unperturbed flow consists of two superposed fluid layers with an interface at $y = 0$ (see Fig. 2.1). Assume acceleration a along y-axis and a uniform flow velocity u in the x-axis. Small departures from the steady-state configuration are described by a potential perturbation $\delta\phi$ and a surface perturbation $S = \zeta(x, t) - y = 0$. The boundary conditions (2.10) are expanded about $y = 0$ and linearized in the perturbations. The procedure yields

$$\partial_t \zeta + u\partial_x \zeta - \partial_y \delta\phi = 0, \quad [\rho(\partial_t \delta\phi + u\partial_x \delta\phi + a\zeta)] = 0, \tag{2.11}$$

evaluated at the plane $y = 0$.

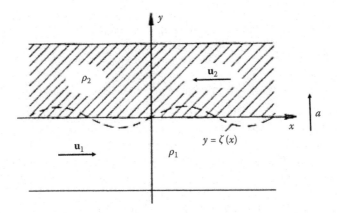

FIGURE 2.1: Instability of a plane contact discontinuity. Surface perturbations $y = \zeta(x)$ can grow by the *Kelvin–Helmholtz instability* (KHI) (see Section 2.3) if $\mathbf{u}_1 \neq \mathbf{u}_2$ or by RTI if $\rho_2 > \rho_1$ and acceleration a is applied in the direction toward the dense fluid.

Although the extension to the spherical symmetry is straightforward, a few basic differences should be noticed in advance. First, the acceleration can no longer be simply expressed by an equivalent gravity. Therefore it can be omitted the gravitational potential U and considered the acceleration problem explicitly. The treatment of the gravitational case is due to Kelvin and can be found in [7]. Second, the tangential flow discontinuity, as shown in Fig. 2.1, would be inconsistent with the basic assumption of irrotational flow. As a consequence, we restrict our attention to purely radial motions in the basic flow. This excludes, for instance, the instabilities of Taylor–Couette flow between rotating cylinders, a subject, which has also been extensively discussed elsewhere [7]. Finally, the incompressibility assumption requires a void inside the sphere. Therefore, we consider only the free-surface problem.

Now derive the free-surface conditions for the potential perturbations on the spherically symmetric boundaries. The derivations can be given in complete analogy for the plane ($\delta = 1$), cylindrical ($\delta = 2$), and spherical ($\delta = 3$) symmetry. More specifically, the unperturbed surface belongs to the $(\delta-1)$-dimensional sphere $r = R(t)$, where $R(t)$ describes an arbitrary radial motion. For $\delta = 1$ the radius r is identified with the y coordinate. The perturbed flow is written in the form

$$\phi = \phi_0(r, t) + \delta\phi, \quad S = R(t) + \zeta - r, \tag{2.12}$$

where $\phi_0(r, t)$ represents the spherically symmetric basic flow, $\delta\phi$ the perturbation, and ζ the corresponding surface displacement. Take into account a small non-uniformity in the applied pressure by writing $p = p_0(t) + \delta p$ for the surface pressure. The perturbation δp may be produced in the surrounding gas and enters in this model as a prescribed function.

First, we consider the dynamical boundary condition. At perturbed surface, Bernoulli's equation (2.9) assumes the form $\rho(\partial_t \phi + \frac{1}{2}u^2)|_{r=R+\zeta} + p_0 + \delta p = C(t)$. Expanding this equation in the perturbations and subtracting the radially symmetric part, one obtains the expression

$$\rho(\partial_t \delta\phi + u_0 \partial_r \delta\phi) + \rho\partial_r(\partial_t \phi_0 + u_0/2)\zeta + \delta p = 0, \tag{2.13}$$

evaluated on the unperturbed sphere $r = R$ with $u_0 = \partial_r \phi_0 = \dot{R}$. This equation can be further simplified by expressing the velocity potential in terms of the normalized coordinate $\eta = r/R$ and noting, that $\partial_t|_{\eta=1} = \partial_t|_{r=R} + \dot{R}\eta \, \partial_r|_{t=0}$. After this change of variables one obtains

$$\begin{cases} (\partial_t \delta\phi + u_0 \partial_r \delta\phi)|_{r=R} = \partial_t \delta\phi|_{\eta=1}, \\ \partial_r(\partial_t \phi_0 + \frac{1}{2}u_0)|_{r=R} = \partial_t u_0|_{\eta=1} = \ddot{R}. \end{cases} \tag{2.14}$$

Substituting (2.14) into (2.13) yields

$$\partial_t \delta\phi|_{\eta=1} + \ddot{R}\zeta + \delta p/p = 0. \tag{2.15a}$$

This equation has basically the same form as $(2.11)_2$ for plane geometry. In the co-moving frame, defined by the coordinate $\eta = 1$, the shell acceleration appears as an effective gravity. Differences between the geometries will arise, because of different forms of the potential perturbations inside a half-plane, a disc, and a sphere. Notice that both the acceleration and the pressure non-uniformity can have an arbitrary time-dependence. The second boundary condition is obtained by inserting the surface expression of (2.12) into the kinematical boundary condition $(2.10)_1$. The linearized equation becomes

$$\partial_t \zeta - (\partial_r u_0)\zeta - \partial_r \delta\phi = 0, \tag{2.15b}$$

on the sphere $r = R$. Comparing this equation with the corresponding equation $(2.11)_1$ in plane geometry, one can recognize an additional term for converging geometries. It arises from the radial variation of the unperturbed flow velocity. Such variation is required to conserve the mass flux across the sphere with area $\sim R^{\delta-1}$. Equations (2.15a,b) represent the basic perturbation equations for an accelerated interface, see Section 1.6.

Let us look for solutions of basic equations (2.7) that satisfy boundary conditions on a plane, a disc, and a sphere. The method of separation of variables provides well-known function systems that are known to be complete on these boundaries. We'll only give a brief summary of the basic potentials to be used in this work. The spherically symmetric flow depends only on the radial coordinate. The corresponding potentials are obtained from (2.7),

$$\phi_0 = \begin{cases} R\dot{R}\eta, & \text{for} \quad \delta = 1, \\ R\dot{R}\ln\eta, & \text{for} \quad \delta = 2, \\ -R\dot{R}/\eta, & \text{for} \quad \delta = 3. \end{cases}$$

They describe a uniform stream in the one-dimensional case and source flows in the two- and three-dimensional cases. The flow velocity

$$u_0 = \partial_r \phi_0 = \dot{R}\eta^{-(\delta-1)} \tag{2.16}$$

is directed radially and conserves the mass flux $\sim u_0 r^{(\delta-1)}$ through spherical shells.

The perturbations will be treated as being two-dimensional, depending on the Cartesian coordinates (x, y) in the plane and on the polar components (r, θ) in the cylindrical and spherical coordinates. In the linear approach, the assumption of two-dimensional flow represents only a restriction for the cylindrical geometry. Here, we neglect perturbations along the cylinder axis by restricting attention to simple flute modes. The symmetry of the plane and spherical geometries assures that normal mode growth will not depend on a second angle. For a plane boundary, one may consider a Fourier expansion with respect to the x coordinate. In the expansion of the complex velocity potential $\delta\phi = \delta\phi_{\text{Re}} + i\delta\phi_{\text{Im}}$, each term has the general form

$$\delta\phi(k) = (Ae^{-ky} + Be^{ky})e^{ikx}, \tag{2.17}$$

with a constant wavenumber $k \in \mathbb{R}$ and time-dependent amplitudes A and B. These potential perturbations will often be called surface modes, because they are exponentially damped toward the interior of the fluid. The boundary conditions at infinity are $A = 0$ for $y \to -\infty$ and $B = 0$ for $y \to \infty$.

A physical interpretation of surface modes may be given in terms of sound waves. Writing the *sound wave dispersion relation* in the familiar form, [30],

$$\omega^2 = c^2(k_x^2 + k_y^2),$$

where $\mathbf{k} = (k_x, k_y)$ is the wavevector, and $c > 0$ is the sound velocity, and solving for the normal component of the wavenumber, gives $k_y = \pm\sqrt{\omega^2/c^2 - k_x^2}$. For large sound velocities, $\omega^2/c^2 \ll 1$, wave propagation is no longer possible but instead of them damping modes can exist. Asymptotically, they approach the potential flow solutions. In cylindrical and spherical coordinates analogous expansions exist, but with the basic difference that the wavenumbers are no longer constants. Actually, the *arc* length $r\theta$ has to be identified with x coordinate, indicating wavenumbers of the form $k(r) = j/r$, where $j \in \mathbb{N}$. Assuming this dependence, the potential (2.17) may be generalized to the form [28],

$$\delta\phi = A(t)\exp\left(-\int_R^r k(z)dz\right)g_l(\theta) = A(t)(r/R)^{-j}g_l(\theta).$$

The angular parts $g_l(\theta)$ and the corresponding powers j can be found from the solution of the Laplace equation. There follows the familiar result,

$$g_l = \begin{cases} \sin(l\theta), & \text{for} \quad \delta = 2, \\ P_l(\theta), & \text{for} \quad \delta = 3, \end{cases}$$

where $P_l(\theta)$ are the Legendre polynomials. The possible j values are $j = \begin{cases} l, & \text{for} \quad \delta = 2, \\ 1 - l, & \text{for} \quad \delta = 3, \end{cases}$ corresponding to radially decaying and growing solutions, respectively. The above formulas provide the background of the potential flow model for RTI.

2.2.2 Plane boundaries

This section deals with the applications of potential flow theory to the stability of accelerated plane layers. At first, the stability of single plane interface is omitted, leading to the classical growth formula for RTI. The RTI is further illustrated for a variety different acceleration laws. The analysis of a single unstable interface is then extended to layers of finite width where Taylor's free-surface model is given.

Two of the most important principles of interfacial fluid instabilities can be inferred from the model of a single plane interface (Fig. 2.1). Consider a simple normal mode analysis of the stability of a plane interface. Normal mode perturbations with frequency ω and wavenumber k are taken proportional to the factor $\exp(-i\omega t + ikx)$. In linear systems, the frequency is subject to a dispersion relation $\omega = \omega(k)$. The system is linearly unstable, if there exists a positive growth rate $n = \text{Im}\,\omega$ for at least one real k.

The potential perturbation (2.17) with $A = 0$ in the lower and $B = 0$ in the upper half-plane is assumed. With this choice, k is restricted to positive values. Eliminating the surface displacement ζ from the boundary conditions (2.11), there follow the jump relations

$$[\partial_y \delta\phi/\tilde\omega] = 0, \quad [\tilde\omega\rho\delta\phi] - \rho a\partial_y\delta\phi/\tilde\omega = 0,$$

where $\tilde\omega = \omega - ku$. They form a linear system of equations for the potential amplitudes A and B. Non-vanishing solutions can only exist if the frequency satisfies the dispersion relation [28],

$$\omega = ku_s \pm i\sqrt{\alpha a k + (\beta v k)^2}, \tag{2.18}$$

with $u_s = \frac{\rho_1 u_1 + \rho_2 u_2}{\rho_1 + \rho_2}$, $v = \frac{1}{2}(u_2 - u_1)$, $\alpha = \frac{\rho_2 - \rho_1}{\rho_1 + \rho_2}$, $\beta = 2\sqrt{\rho_1\rho_2}(\rho_1 + \rho_2)$.

The interface is unstable if $\alpha a k + (\beta v k)^2 > 0$. Accordingly, instability can arise from acceleration towards the denser fluid ($\alpha > 0$) or from a tangential flow discontinuity ($v \neq 0$).

The special case of an unstable density discontinuity ($\alpha > 0$, $v = 0$) is called RTI. As a remark, the instability of a tangential flow discontinuity ($v \neq 0$, $a = 0$) is called *Kelvin–Helmholtz instability* (KHI, see Section 2.3). The corresponding growth rates are $n = \sqrt{\alpha a k}$, $n = k|v|$, respectively.

The dispersion relation (2.18) becomes particularly simple in a coordinate frame moving with the mean velocity u_s of both layers. In this frame, a purely growing mode with $\text{Re}(\omega) = 0$ is obtained.

(a) **Accelerated interfaces**. The stability of time-dependent accelerations cannot be described by constant growth rates. Instead, the perturbation equations have to be solved for the underlying surface motions and initial data. The boundary conditions (2.15a,b) will be specialized for a plane boundary of an accelerated fluid layer in the half-space $r > R(t)$. The potential perturbation can then be assumed of the form (2.17) with $B = 0$ and $y = r$. Noting $\eta = r/R$, the following relations hold:

$$\delta\phi = A(t)\exp(-kr + ikx) = A(t)\exp(-kR\eta + ikx),$$
$$\partial_r\delta\phi = -k\delta\phi, \quad \partial_t\left[(\partial_r\delta\phi)_{\,|\,r=R}\right] = -k\partial_t\delta\phi_{\,|\,\eta=1}. \tag{2.19}$$

With the help of (2.19), one can eliminate the potential from the boundary conditions (2.15a,b). Noting that $\partial_y u_0 = 0$ according to (2.16), one obtains the surface equation,

$$\partial_t^2\zeta - ak\zeta = \delta a. \tag{2.20}$$

It describes the growth of surface perturbations for a prescribed acceleration law $a(t) = \ddot R$. Non-uniform acceleration is described by the inhomogeneity $\delta a = k\delta p/p$. This term will be neglected until the discussion of non-uniformities (see below).

(b) **Constant acceleration**. The growth of the free-surface RTI under a constant acceleration a can be expressed by the growth rate $n = \sqrt{ak}$. The unstable modes can arise from an initial surface displacement ζ_0 or from an initial surface velocity $\partial_t \zeta_0$. The corresponding solution of (2.20) is given by

$$\zeta = \zeta_0 \cosh(nt) + n^{-1} \partial_t \zeta_0 \sinh(nt). \tag{2.21}$$

The RTI instability imposes principal limitations on the acceleration of foils by gas or ablation pressure. If a foil of thickness d has been accelerated over a distance $s = \frac{1}{2}at^2$, the growth increment becomes $nt = \sqrt{2ks} = \sqrt{2kdQ}$, where $Q = s/d$ is the basic dimensionless parameter foil stability. Depending on the most critical wavenumbers and the magnitude of the initial perturbations, the typical Q values seem to be limited to the interval $\simeq 5 - 15$. Although the instability mechanism is only dependent on the acceleration distance, the kinetic energy of the foil depends explicitly on the acceleration. Using the hydrostatic pressure law $p_0 = \rho ad$, the kinetic energy density of the foil is expressed in the form $E = \frac{1}{2}\rho u^2 = \rho a s = p_0 Q$.

(c) **Impulsive acceleration**. Another interesting limiting case is given by an impulsive acceleration law: $a = \Delta u \delta(t)$, where $\delta(t)$ denotes the delta-function and Δu the velocity increment imparted to the undisturbed foil. The solution of (2.20) predicts a constant perturbation velocity and corresponding linear amplitude growth

$$\zeta = \zeta_0 + (\partial_t \zeta_0 + k\Delta u \zeta_0)t. \tag{2.22}$$

The velocity increment of the perturbation depends on the initial amplitude and the wavenumber of the surface corrugations. The impulsive approximation requires that the acceleration time is much shorter than the exponential folding time of the unstable mode. The approximation applies mainly to long-wavelength modes with sufficiently small growth rates. Such modes can also reach large amplitudes $\zeta \sim k^{-1}$ before saturating nonlinearly.

(d) **Exponential acceleration law**. The finite duration of acceleration pulses limits the growth of long wavelength perturbations, having exponential folding times to be longer than the acceleration time. A simple example of a transient pressure pulse may be given under an exponential acceleration law

$$a(t) = a \exp(bt). \tag{2.23}$$

The constant a denotes an initial acceleration at $t = 0$ and the constant b can be positive for growing pulses or negative for decreasing pulses. In the later case, a useful interpolation formula between the limiting cases of constant and impulsive acceleration can be gained. In particular, an analytic expression for the asymptotic perturbation velocity imparted to the perturbation during the passage of the pulse will be obtained. Solutions of (2.20) with an exponential acceleration law (2.23) can be found by a variable substitution. Define $\tau = \tau_0 \exp(bt/2)$ as a new independent variable with initial value $\tau_0 = 2\sqrt{ak}/|b|$ that varies in the interval $0 < \tau < \tau_0$ for $b < 0$ and in the interval $\tau_0 < \tau < \infty$ for $b > 0$. Equation (2.20) becomes

$$\tau^2 \partial_\tau^2 \zeta + \tau \partial_\tau \zeta - \tau^2 \zeta = 0. \tag{2.24}$$

Two independent solutions of (2.24) are given by modified Bessel functions $I_0(\tau)$ and $K_0(\tau)$. Imposing initial conditions at $\tau = \tau_0$, we get

$$\zeta = \tau_0 \{[K_1(\tau_0)I_0(\tau) + K_0(\tau)I_1(\tau_0)]\zeta_0 + |K_0(\tau_0)I_0(\tau) + K_0(\tau)I_0(\tau_0)| \, \partial_t \zeta_0 / \sqrt{ak}\}. \tag{2.25}$$

The second bracket is positive for $\tau > \tau_0$ and negative for $\tau < \tau_0$. Its magnitude has to be taken because of a corresponding sign change of the constant b. If deriving (2.25) we have also used the relations $I_1 = \partial_\tau I_0$, $K_1 = -\partial_\tau K_0$ and $K_0 I_1 + K_1 I_0 = 1/\tau$ for the first derivatives and the Wronskian, respectively. The asymptotic limits of large and small τ values can be analyzed in more detail. Large τ values describe large growth rates and a corresponding slow time variation of the applied acceleration. Using for τ, $\tau_0 \gg 1$ the asymptotic expressions $I_n \to \sqrt{1/2\pi\tau}\exp(\tau)$ and $K_n \to \sqrt{1/2\pi\tau}\exp(-\tau)$, one obtains from (2.25)

$$\zeta = \sqrt{\tau_0/\tau}\left(\zeta_0 \cosh(\tau_0 - \tau) + \partial_\tau \zeta_0/\sqrt{ak}\sinh(|\tau_0 - \tau|)\right).$$

The opposite limit, $\tau \to 0$, describes the time asymptotic response to a transient acceleration pulse with $b < 0$. Asymptotically, the perturbations grow with constant velocity whose magnitude depends on the pulse duration. Assuming small arguments, the modified Bessel functions can be approximated in the form $I_0 \to 1$, $I_1 \to \tau/2$, $K_0 \to -\ln(\tau/2)$ and $K_1 \to 1/\tau$. Neglected I_0 in comparison with K_0 and noting that $\partial_\tau \ln\tau = b/2$, it can be obtained from (2.25) the asymptotic perturbation velocity $\partial_t \zeta = \sqrt{ak}I_1(\tau_0)\zeta_0 + I_0(\tau_0)\partial_t\zeta_0$. It increases monotonically with τ_0 that characterized the pulse length in comparison with the exponential folding time. Limiting forms for small and large parameter values are

$$\partial_t \zeta = \begin{cases} (ak/|b|)\zeta_0 + \partial_t\zeta_0, & \tau_0 \ll 1, \\ \left(\sqrt{ak}\zeta_0 + \partial_t\zeta_0\right)\sqrt{1/2\pi\tau_0}\exp(\tau_0), & \tau_0 \gg 1. \end{cases}$$

For short pulses or long wavelengths ($\tau_0 \ll 1$), we recover the result (2.22) for impulsive acceleration. The velocity increment $\Delta u = a/|b|$ is the time integral of $a(t)$ from $t = 0$ to $t \to \infty$. For long pulses or short wavelengths ($\tau_0 \gg 1$), the asymptotic velocity is substantially larger because of the exponential growth during the acceleration phase. A specific example

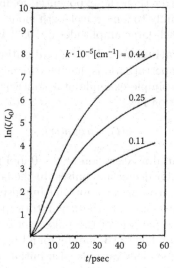

FIGURE 2.2: Evolution of a sinusoidal perturbation $\zeta = \zeta_0 \cos(kx)$ under a time-dependent acceleration law, $a(t) = 3.6 \times 10^{16}\exp(-0.1\,t/ps)\,[\text{cm/s}^2]$.

of the wavelength dependence of solution (2.25) can be recognized in Fig. 2.2. It shows the transition from exponential to linear amplitude growth for three different modes subject to the same acceleration pulse. The acceleration parameters are $a = 3.6 \times 10^{18}\,\text{cm/s}^2$ and $b = -10^{11}\,\text{s}^{-1}$, corresponding to a flight distance of $\sim 16\,\mu\text{m}$ in 55 ps.

(e) **Layers of finite width.** The model of a single free surface can readily be extended to fluid layers of finite width. Two important aspects of this treatment should be mentioned.

First, the RT growth rate is found independent of the layer width. Second, there can be strong interference effects for thin layers between the two surface modes developing at the front and at the rear side of the foil.

First, consider a plane fluid layer, $0 < r < d$, under a constant acceleration a toward a negative r axis. In addition to the solution (2.21) of the homogeneous equation, satisfying the initial conditions at $t = 0$, one has to consider a solution of the inhomogeneous equation with vanishing initial values. A potential perturbation will be assumed $\delta\phi = A_1 e^{-kr} + A_2 e^{k(r-d)}$. Define a vector $\mathbf{A} = (A_1, A_2)$, and a vector $\boldsymbol{\zeta}$ whose components are the displacements $\zeta_{1,2}$ at the surfaces $r = 0$ and $r = d$, respectively. Setting again $\delta p = 0$, the boundary conditions (2.15a,b), evaluated at both surfaces, become

$$\mathbf{M} \cdot \partial_t \mathbf{A} - a\boldsymbol{\zeta} = 0, \quad \partial_t \boldsymbol{\zeta} + k\mathbf{N} \cdot \mathbf{A} = 0, \tag{2.26}$$

where $\mathbf{M} = \begin{pmatrix} 1 & p \\ p & 1 \end{pmatrix}, \mathbf{N} = \begin{pmatrix} 1 & -p \\ p & -1 \end{pmatrix}, p = e^{-kd}$. Eliminating the displacement $\boldsymbol{\zeta}$ from (2.26) yields

$$\partial_t^2 \mathbf{A} = -ka\mathbf{M}^{-1} \cdot \mathbf{N} \cdot \mathbf{A} = -ka \begin{pmatrix} 1 & 0 \\ 0 & -1 \end{pmatrix} \cdot \mathbf{A},$$

where $\mathbf{M}^{-1} = \Delta^{-1} \begin{pmatrix} 1 & -p \\ -p & 1 \end{pmatrix}$ is the inverse and $\Delta = 1 - p^2$ the determinant of $\mathbf{M}$. The instability eigenvalue problem is already diagonal in the $\mathbf{A}$ representation. The normal mode solutions are

$$\mathbf{A} = A_1 \exp\left(\pm i\sqrt{ak}\,t\right) \begin{pmatrix} 1 \\ 0 \end{pmatrix}, \quad \mathbf{A} = A_2 \exp\left(\pm\sqrt{ak}\,t\right) \begin{pmatrix} 0 \\ 1 \end{pmatrix}.$$

These correspond to a stable surface wave arising from the rear side and an unstable RT mode arising from the front side of the foil, respectively. As already mentioned, the growth rates are independent of the layer width in this model.

The surface perturbations $\zeta_{1,2}$ are generally superpositions of both types of mode. Each surface is perturbed by the mode developing at the same surface and by an exponentially damped mode from the opposite surface. According to (2.26), this superposition can be expressed in the form $\boldsymbol{\zeta} = \mathbf{M} \cdot \mathbf{X}$, where the components of $\mathbf{X}$ are proportional to those of $\mathbf{A}$. Choosing static initial corrugations $\boldsymbol{\zeta}_0$, $\mathbf{X}_0$ at $t = 0$, the evolution equations can be written as,

$$\boldsymbol{\xi} = \mathbf{S} \cdot \boldsymbol{\zeta}_0, \quad \mathbf{X} = \mathbf{T} \cdot \mathbf{X}_0,$$

where $\mathbf{T} = \begin{pmatrix} c & 0 \\ 0 & C \end{pmatrix}$, $c = \cos\left(\sqrt{ak}\,t\right)$, $C = \cosh\left(\sqrt{ak}\,t\right)$, and

$$\mathbf{S} = \mathbf{M} \cdot \mathbf{T} \cdot \mathbf{M}^{-1} = \Delta^{-1} \begin{pmatrix} c - p^2 C & -p(c - C) \\ -p(c - C) & C - p^2 c \end{pmatrix}.$$

At late time, when the unstable mode dominates, the surface amplitudes satisfy the relations,

$$\zeta_1 = p\zeta_2, \quad \zeta_2 = \Delta^{-1}(\zeta_{20} - p\zeta_{10})C, \quad \zeta_2 - p\zeta_1 = (\zeta_{20} - p\zeta_{10})C.$$

At the unstable surface, the amplitude ξ_2 is increased by the factor $\Delta^{-1}(1 - p\zeta_{10}/\zeta_{20}) = \frac{1 - e^{-kd}\zeta_{10}/\zeta_{20}}{1 - e^{-2kd}}$ in comparison with the result for an infinitely thick layer. Especially for thin layers, this factor can describe an appreciable amplification of the initial amplitude ξ_{20}. Only the relative perturbation $\zeta_2 - p\zeta_1$ evolves exactly according to the single mode RTI. This behavior is an effect of mode interference in the initial state which will be discussed in more detail in the spherical shell problem.

2.2.3   Spherical boundaries

Spherical shapes lead to a number of modifications in the analysis of surface instabilities. The stability criterion becomes much more restrictive in the presence, where even an unaccelerated surface is unstable. Time-dependence of the basic flow is generic feature in these geometries. Conceptually, we no longer analyze the stability of an equilibrium of steady-state, but consider the evolution of symmetry perturbations for particular symmetric reference flows. In general, this stability problem can no longer be described by independent normal modes. Perturbation evolution is therefore much more dependent on particular initial data, shell motions and shell structures.

Symmetric shell motions. The motion of spherically cavities in an infinitely extended incompressible fluid plays an important role in the theory of cavitation bubbles. The following discussion is addressed to the dynamics of incompressible spherical shells of finite thickness. As in Section 2.2.1, a δ-dimensional surface is assumed, where plane ($\delta = 1$), cylindrical ($\delta = 2$), and spherical ($\delta = 3$) geometries are included. In this section, only unperturbed spherically symmetric basic flows are considered. The shell parameters are indicated in Fig. 2.3, consisting of a constant mass density, the inner radius $R_1(t)$, the outer radius $R_2(t)$, and the thickness $R_2(t) - R_1(t)$. For simplicity of notation, we'll often use the abbreviations $R(t) = R_1(t)$, $S(t) = R_1(t)/R_2(t)$ and $V(t) = \dot{R}_1(t)$. The shell is subject to an inside pressure $p_1(t)$ and the outside pressure $p_2(t)$. Its motion is therefore driven by the pressure difference $\Delta p = p_1 - p_2$. The dynamics of incompressible spherical shells follows from simple conservation laws for its mass and its energy. Alternatively, dynamical equations for the shell boundaries subject to the applied pressure can derived. Both approach will be outlined and then illustrated by specific examples. In the incompressible shell model, mass conservation governs the evolution of the shell thickness. One can therefore simply relate the motion of the outer surface to the motion of the inner surface using (2.16), the velocity of the outer surface is found to be

$$\dot{R}_2 = u_0(R_2) = VS^{\delta-1}. \tag{2.27}$$

The flux through the inner surface is equal to the flux through the outer surface and the mass integral

$$M = \rho \int_{R_1}^{R_2} r^{\delta-1}dr = \frac{\rho}{\delta}(R_2^\delta - R_1^\delta) \tag{2.28}$$

is conserved. The outer radius can therefore be expressed by use the inner radius and the corresponding initial values in the form

$$R_2 = (R_{20}^\delta - R_{10}^\delta + R_1^\delta)^{1/\delta}. \tag{2.29}$$

Noting (2.27) and (2.29), it is sufficient to examine the motion of one shell boundary, say R_1, only.

The study of shell dynamics is simplified by using the energy conservation law. Assume that the surface pressure is prescribed as a function of the shell radius without having an explicit time-dependence. It is then possible to derive the dynamics from a simple potential energy expression. Define the kinetic energy of the shell as $T = \frac{1}{2}\int_{R_1}^{R_2} \rho u_0^2 r^{\delta-1}dr$. Using (2.16), we get

$$T = \frac{1}{2}\rho V^2 \times \begin{cases} R_2 - R_1, & \delta = 1, \\ R^2 \ln S, & \delta = 2, \\ R^3(1 - S), & \delta = 3. \end{cases} \tag{2.30}$$

If the shell moves *ballistically*, i.e., if $\Delta p \equiv p_1 - p_2 = 0$, the kinetic energy T is conserved. More generally, one has to include a potential energy $W = W_0 - $

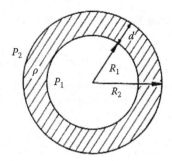

FIGURE 2.3: Incompressible shell of density ρ and thickness d.

$\int_0^1 (\dot{R}_1 R_1^{\delta-1} p_1 - \dot{R}_2 R_2^{\delta-1} p_2) dt$, arising from the work that is delivered to the shell by pressure forces acting on its surface. Making use of (2.27) and considering the pressure difference $\Delta p = p_1(R_1) - p_2(R_2) = \Delta p(R)$ as a known function of the inner radius, it follows

$$W = W_0 - \int_{R_0}^{R(t)} R^{\delta-1} \Delta p(R) dR. \tag{2.31}$$

The total energy $E = T + W$ is conserved during the shell motion. From this energy integral one obtains immediately the surface velocity V as a function of the shell radius R.

A somewhat more systematic treatment of shell motion may be based directly on the dynamical equation (2.9). Using (2.15a,b) and (2.16) into (2.9) and taking the difference between the boundary conditions at the inside and the outside shell surface, one can derive an expression for the pressure difference Δp. For a plane geometry, this equation is simply Newton's law

$$M\ddot{R} = \Delta p, \tag{2.32a}$$

applied to a fluid element with a real mass density $M = \rho d$. In cylindrical and spherical coordinates, the corresponding equations are

$$(R\ddot{R} + V^2)\ln(1/S) - \frac{1}{2}V^2(1 - S^2) = \Delta p/p, \tag{2.32b}$$

$$\left(R\ddot{R} + \frac{3}{2}V^2\right)(1 - S) - \frac{1}{2}V^2(1 - S^3) = \Delta p/p. \tag{2.32c}$$

These equations determine the motion for an arbitrarily prescribed pressure law. One can convince oneself that their energy integrals are identical with the energy conservation law given by (2.30) and (2.31). Simplifications can be gained for either thin or thick shells. In the limit of *thin shells* one can recover from (2.32a-c) the equation of motion for the mass element (2.28)

$$M\ddot{R} = R^{\delta-1} \Delta p \tag{2.33}$$

subject to the force $R^{\delta-1}\Delta p$. In the opposite limit of thick shell, (2.32b,c) reduce to the cavity equations

$$(R\ddot{R} + V^2)\ln(1/S) = \Delta p/p, \quad \text{when } \delta = 2,$$

$$\left(R\ddot{R} + \frac{3}{2}V^2\right) = \Delta p/p, \quad \text{when } \delta = 3.$$

Example 71 (Ballistic motion) The ballistic motion of an undriven shell ($\Delta p = 0$) is often realized after an initial acceleration phase when the applied pressure becomes negligibly small compared with the kinetic energy. It plays therefore an important role for the study of converging and diverging flows. Using (2.30), the conservation law for the total energy E becomes

$$E = T = \frac{1}{2}\rho V^2 R^3 (1 - S). \qquad (2.34)$$

To simply notation, we'll always choose units of lengths and time such that $V = 1$ at $R = 1$. The corresponding value S at $R = 1$ to be denoted as S_*. With this convention, ballistic motion to be governed by the following relations:

$$E = \frac{1}{2}\rho(1 - S_*), \qquad (2.35a)$$

$$V = \pm\left(\frac{1 - S_*}{1 - S}R^{-3}\right)^{1/2}, \qquad (2.35b)$$

$$\ddot{R} = -\frac{3}{2}(V^2/R)\left[1 - \frac{1}{3}(S + S^2 + S^3)\right], \qquad (2.35c)$$

where (2.34) and (2.32c) have been used. One can recognize a strong divergence of the surface velocity for converging flows. The surface acceleration is always directed toward the cavity, indicating RT stability at the inner shell surface. For cavities $(S, S_* \ll 1)$, the ballistic equation of motion can even be obtained explicitly as a function of time by integration of (2.35b). Choosing the initial condition $R(0) = 0$, the solution is found to be

$$R = \left(\frac{5}{2}|t|\right)^{2/5}, \quad V = \frac{2}{5}R/t, \quad \ddot{R} = -\frac{6}{25}T/t^2. \qquad (2.36)$$

Negative times describe imploding and positive times expanding cavities.

Example 72 (Acceleration) Another important case concerns the acceleration of a thin shell from rest toward the shell center. By (2.33), constant acceleration of thin shells requires a pressure law

$$\Delta p = -p_0(R_0/R)^2, \qquad (2.37)$$

where p_0 denotes the initial pressure applied to the outside surface at $R = R_0$. The pressure increases as R^{-2} to compensate for the decreasing surface area. Using (2.36) in (2.31), the potential energy becomes to $W = W_0 + p_0 R_0^2 (R - R_0)$. Assuming that the shell has been at rest, initially, the energy conservation law becomes $T + W - W_0 = 0$, yielding $\frac{1}{2}\rho V^2 R^3 (1 - S) - p_0 R_0^2 (R_0 - R) = 0$. Choose again the normalization convention $V = 1$, $S = S_*$ at $R = 1$ to obtain

$$p_0 = \frac{\rho(1 - S_*)}{2R_0^2(R_0 - 1)}, \quad V = -\left(\frac{1 - S_*}{1 - S}\frac{R_0 - R}{R_0 - 1}R^{-3}\right)^{1/2},$$

$$\ddot{R} = -\frac{1 - R_*}{(1 - S)R^3}\frac{1}{2(R_0 - 1)} - \frac{3}{2}\frac{V^2}{R}\left(1 - \frac{S + S^2 + S^3}{3}\right).$$

These equations determine the motion of an accelerated shell subject to the pressure law (2.37).

Cavities. We will discuss the stability of cylindrical and spherical cavities in an infinitely expanded fluid. The cavity model provides the some insight into the stability of converging flows in the absence of shell effects. A perturbation on a spherical cavity can be expanded into trigonometric functions $(\delta = 2)$ or Legendre polynomials $(\delta = 3)$ as in Section 2.2.1. Assume the single mode perturbations

$$\delta\phi = A(t)\exp\left(-\int_R^r k(z)dz\right)g_l(\theta) = A(t)\eta^{-j}g_l(\theta) \qquad (2.38)$$

with mode numbers $l \in \mathbb{N}$ and $j = l + \delta - 2$. Noting (2.16) and using that $\partial_t \left[(R \partial_t \delta \phi) \big|_{r=R} \right] = -j \partial_t \delta \varphi \big|_{\eta = 1}$, the boundary condition (2.15b) takes the form

$$\partial_t (R \partial_t \zeta) + (\delta - 1) \partial_t (\dot{R} \zeta) + j \partial_t \delta \phi \, {\big|}_{\eta = 1} = 0. \tag{2.39}$$

Eliminating $\delta \phi$ from (2.15a) and (2.39), yields

$$\partial_t^2 \zeta + \delta \frac{\dot{R}}{R} \partial_t \zeta - \frac{l - 1}{R} \ddot{R} \zeta = \frac{j}{R} \frac{\delta p}{p}. \tag{2.40}$$

Equation (2.40) governs the evolution of the surface displacement ζ for arbitrary radial motions and mode numbers. One can recognize similarities with (2.20), governing RTI in plane geometry, but also major differences because of the variation of the radius with time.

Firstly we attempt to give a more physical interpretation of the convergence effects described by (2.40). Defining $k = j/R$ and $m = \rho R^{\delta - 1}/k$, one can write (2.40) as an equation of motion

$$\partial_t (m \partial_t \zeta) - \frac{l - 1}{l + \delta - 2} m \ddot{R} k \zeta = R^{\delta - 1} \delta p, \tag{2.41}$$

with a time-dependent mass. Convergence effects are related to the variation of m and k with radius. These variations correspond to the time-dependent periodicity length $L = 2\pi R$ and surface area $\sim R^{\delta - 1}$ of the system. In addition, geometry effects are present for low l modes. They slightly reduce the numerical coefficient of the buoyancy term and lead to its elimination for $l = 1$. This mode describes translations of the sphere without fluid interchange across the sphere. Being interested in the evolution of asymmetries, we'll assume $l > 1$ in following. Also, the external pressure force in (2.41) will be omitted.

A new feature of converging flow consists in the possibility of amplification in the absence of acceleration. If a cavity implodes with a constant surface velocity V and zero acceleration, $\ddot{R} = 0$, the perturbed momentum $m \partial_t \zeta$ is conserved. This conservation law implies that the normalized amplitude $A = \zeta / R$ grows as a function of the convergence ratio $q = R_0/R$ accordingly to the formula

$$A = q \left(A_0 + \frac{1}{\delta - 1} (1 - q^{\delta - 1}) B_0 \right), \tag{2.42}$$

where $A_0 = \zeta_0 / R_0$ and $B_0 = \partial_t \zeta_0 / V$ denote initial values at $R = R_0$. For the static corrugations ($B_0 = 0$), the displacement ζ remains constant but the relative amplitude A still grows in proportion to the convergence ratio. If, on the other hand, $B_0 \neq 0$, the displacement itself diverges with the inverse of the surface area and A grows as q^δ. These convergence effects are independent of the mode number. If the surface is accelerated toward the cavity, surface oscillations can develop. Although the acceleration is stabilizing, it usually cannot prevent the amplification of the oscillation amplitudes. As an important example, we consider the ballistic equation (2.36) for three-dimensional cavities. Inserting this solution into (2.40) yields

$$\partial_t^2 \zeta + \frac{6}{5} t^{-1} \partial_t \zeta + \frac{6}{25} (l - 1) t^{-2} \zeta = 0. \tag{2.43}$$

Choosing $s = \ln q$ as the independent variable, the solutions to (2.43) can be expressed in the form $\zeta = e^{s/4} (C_1 e^{iks} + C_2 e^{-iks})$, with constants $C_{1,2}$ and $k = \frac{1}{4} \sqrt{25(l-1) - l} \simeq 1.25 \sqrt{l - 1}$. Except for $l = 1$, the exponent k is real. Taking the initial conditions into account and choosing the same notation as in (2.42), yields the result

$$A = q^{5/4} \left[A_0 \cos(ks) - \left(\frac{1}{4} A_0 + B_0 \right) k^{-1} \sin(ks) \right]. \tag{2.44}$$

It describes cavity oscillations whose frequencies increase with the square root of the mode number. This behavior is in accordance with the dispersion relation (2.18) for stable gravity modes. In addition, (2.44) leads to an amplification of the oscillation amplitudes with the convergence ratio. Notice that the amplification factor is independent of the mode number and considerably weaker than for unaccelerated motion. Acceleration can therefore partly compensate the convergence effects that are given in (2.42).

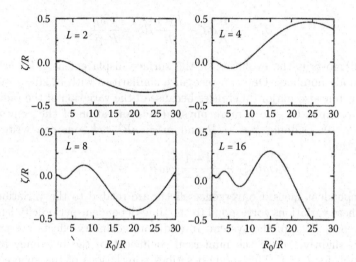

FIGURE 2.4: Surface oscillations of a ballistically imploding spherical cavity in an infinitely extended fluid. The relative surface displacement ζ/R has been calculated with initial amplitude $\zeta_0/R_0 = 0.01$ for different modes l.

The evolution of some cavity modes with $l = 2, 4, 8, 16$ is represented in Fig. 2.4. The figure shows the solution (2.44) for the initial amplitudes $A_0 = 0.01$, $B_0 = 0$ and for typical convergence ratios of $\simeq 25 - 30$. Oscillations can be recognized even for low l modes and the amplitudes can reach critical values of $\simeq 0.5 - 0.7$. Remark that the normalized amplitude $A = \zeta/R$ indicates stability failure when A approaches unity.

A general stability criterion for the homogeneous part of the differential equation (2.40) has been proved by Birkhoff [6]. It requires

$$\ddot{R} < 0, \quad R\ddot{R} + (2\delta - 1)\dot{R}\ddot{R} < 0, \tag{2.45}$$

for stability. According to the first condition the acceleration has to be directed toward the cavity which avoids RTI. If the acceleration is assumed constant and negative, the second condition becomes $\dot{R} > 0$. Accordingly, there exist no stable converging flows in this case. The stability criterion is an immediate consequence of the identity $\partial_t(\zeta^2 + p^2/f) = -(p/f)^2\partial_t f$, derived from (2.41) with $p = m\partial_t\zeta$ and $f = -\frac{l-1}{l+\delta-2}m^2 k\ddot{R}$. The displacement remains if both $f > 0$ and $\partial_t f > 0$. These conditions are identical to those stated in (2.45).

Shell perturbations. The spherical shell model can describe mutual dependences in the evolution of outside and inside surface instabilities. The outside RTI occurs during the acceleration phase. Its perturbations can penetrate the shell and cause cavity oscillations during the deceleration phase. For definiteness, we consider in the following a three-dimensional spherical shell whose boundaries $R_{1,2}$ are subject to displacements $\zeta_{1,2}$. The simpler notation of the dynamics equations $R(t) = R_1(t)$, $S(t) = R_1(t)/R_2(t)$ and $V(t) = \dot{R}_1(t)$ will also be used. Perturbations are chosen with angular part $P_l(\theta)$ and the corresponding potential

(2.38) is written in the form

$$\delta\phi = (R_1/r)^{l+1}A_1 + (r/R_2)^l A_2. \tag{2.46}$$

Imposing the boundary conditions (2.15a,b) on the potential solution (2.46) at both surfaces $r = R_{1,2}$, it follows the fourth-order system

$$R_1\partial_t\zeta_1 + 2\dot{R}_1\zeta_1 + (l+1)A_1 - lS^l A_2 = 0, \tag{2.47a}$$
$$R_2\partial_t\zeta_2 + 2\dot{R}_2\zeta_2 + (l+1)S^{l+1}A_1 - lA_2 = 0, \tag{2.47b}$$
$$\partial_t(A_1 + S^l A_2) + \ddot{R}_1\zeta_1 = 0, \tag{2.47c}$$
$$\partial_t(A_2 + S^{l+1}A_1) + \ddot{R}_2\zeta_2 = 0. \tag{2.47d}$$

These equations govern the evolution of shell perturbations for arbitrary radial motions.

In the numerical treatment, it is advantageous to choose the radius or the convergence ratio $q = R_0/R$ as the independent variable. The integration interval $1 < q < Q$ is then known in advance, if an implosion with initial radius R_0 and stagnation point radius $R_s = QR_0$ is considered. The radial dependence of the unperturbed shell velocities and accelerations follow simply from energy and mass conservation. It is therefore not necessary to solve for the explicit time-dependence of the motion. The transformation to the independent variable q can be achieved by noting that

$$\frac{d}{dq} = \frac{dt}{dq} \cdot \frac{d}{dt} = -\frac{R^2}{R_0 V} \cdot \frac{d}{dt} = -\frac{R_0}{q^2 V} \cdot \frac{d}{dt}.$$

As a minor drawback of this transformation, the system becomes singular at $V = 0$. One has therefore to initialize and determine the calculations with small nonzero shell velocity.

Thin shell model. Considerable simplification of the general shell problem can be gained in the limiting cases of thick ($S^l \ll 1$) and thin ($S \to 1$) shells. For thick shells, the outer surface can be considered at rest, while the inner surface is governed by the cavity equation (2.40). In the opposite limit of thin shells, the evolution of surface modes can be described by independent second-order equations for each shell boundary. The derivation of these equations by a systematic expansion procedure will be outlined there. If the shell thickness approaches zero, the boundary conditions at the inner and the outer surface become identical. To obtain a closed system of equations in this case, one has to consider an expansion up to first order in the parameter $\varepsilon = (R_2 - R_1)/R_1$. The motion of the outer surface can be related to the motion of the inner surface under the following first-order relations:

$$R_2 = (1+\varepsilon)R, \quad \zeta_{1,2} = C \pm \varepsilon D/2,$$
$$\dot{R}_2 = S^2\dot{R} = (1-2\varepsilon)\dot{R}, \quad \ddot{R}_2 = (1-2\varepsilon)\ddot{R} + 6\varepsilon\dot{R}^2/R. \tag{2.48}$$

Here, we used (2.27) for spherically symmetric flow and we introduced the mean and the relative displacements of the two surfaces

$$C = (\zeta_1 + \zeta_2)/2, \qquad D = (\zeta_1 - \zeta_2)/\varepsilon. \tag{2.49}$$

Both C and D are treated of order $O(1)$. Expanding now (2.47a-d) up to first order yields

$$R\dot{C} + 2\dot{R}C = -(l+1)A_1 + lA_2, \tag{2.50a}$$
$$\dot{A}_1 + \dot{A}_2 = -\ddot{R}C, \tag{2.50b}$$
$$R\dot{D} - \dot{R}D = -(l+1)(l+2)A_1 - l(l-1)A_2 - 6\dot{R}C, \tag{2.50c}$$
$$-(2l+1)\dot{A}_1 = \ddot{R}D + 3\dot{R}\dot{C} + (l+2)\ddot{R}C. \tag{2.50d}$$

Equations (2.50a-d) follow from the leading order of (2.48). Equations (2.50c,d) are obtained by taking the difference between the boundary conditions at the inner and outer surfaces. Thereby the following expansions have been used:

$$
\begin{aligned}
R_1 \partial_t \zeta_1 - R_2 \partial_t \zeta_2 &= \varepsilon(R\dot{D} - 3\dot{R}D - R\dot{C}), \\
\dot{R}_1 \zeta_1 - \dot{R}_2 \zeta_2 &= \varepsilon(\dot{R}D + 2\dot{R}C), \\
\ddot{R}_1 \zeta_1 - \ddot{R}_2 \zeta_2 &= \varepsilon[\ddot{R}D - (6\dot{R}^2/R - 2\ddot{R})C], \\
-(l+1)(1-S^{l+1})A_1 + l(S^l - 1)A_2 &= \varepsilon[-(l+1)^2 A_1 - l^2 A_2], \\
\partial_t[(1-S^{l+1})A_1 + (S^l - 1)A_2] &= \varepsilon[(2l+1)\dot{A}_1 + l\dot{R}C + 3(\dot{R}\dot{C} + 2C\dot{R}^2/R)].
\end{aligned}
$$

Eliminating A_1, A_2 from (2.50a-d), yields a set of two coupled second-order equations of the form $R\ddot{C} = \dot{R}D$ and $R\ddot{D} = \ddot{R}[3D + (l+2)(l-1)C]$, which can be decoupled by the variable transformation

$$
G = D + (l-1)C, \quad H = D - (l+2)C, \tag{2.51}
$$

yielding two independent second-order equations,

$$
R\ddot{G} - (l+2)\ddot{R}G = 0, \quad R\ddot{H} + (l-1)\ddot{R}H = 0, \tag{2.52}
$$

for G and H. The structure of these equations is analogous to the matrix equation in Section 2.2.2 (e):

$$
\partial_t^2 \mathbf{A} = -ka\mathbf{M}^{-1} \cdot \mathbf{N} \cdot \mathbf{A} = -ka\begin{pmatrix} 1 & 0 \\ 0 & -1 \end{pmatrix} \cdot \mathbf{A}.
$$

The present result is valid for arbitrary radial motions and mode numbers. To satisfy the initial conditions at the shell boundaries, we've to consider a superposition of the basic solutions. For definiteness, we define a system of fundamental solutions G_1, G_2, H_1, H_2 of (2.52) by *canonical initial conditions*

$$
G_{10} = H_{10} = 1, \quad \dot{G}_{10} = \dot{H}_{10} = 0, \quad G_{20} = H_{20} = 0, \quad \dot{G}_{20} = \dot{H}_{20} = 1. \tag{2.53}
$$

Setting $\mathbf{X} = \begin{pmatrix} G \\ H \end{pmatrix}$ and $\mathbf{T}_{12} = \begin{pmatrix} G_{12} & 0 \\ 0 & H_{12} \end{pmatrix}$, the solutions of (2.52) can be written as

$$
\mathbf{X} = \mathbf{T}_1 \cdot \mathbf{X}_0 + \mathbf{T}_2 \cdot \dot{\mathbf{X}}_0. \tag{2.54}
$$

This displacement vector ζ, as defined in (2.26), is related to the vector $\mathbf{X}$ by a linear but time-dependent transformation,

$$
\zeta = \mathbf{M} \cdot \mathbf{X}, \quad \mathbf{X} = \mathbf{M}^{-1} \cdot \zeta, \quad \dot{\mathbf{X}} = \mathbf{M}^{-1} \cdot \dot{\zeta} + \dot{\mathbf{M}}^{-1} \cdot \zeta. \tag{2.55}
$$

Using (2.49) and (2.51), one finds

$$
\mathbf{M} = \frac{1}{2l+1}\begin{pmatrix} 1+b & a-1 \\ 1-b & -a-1 \end{pmatrix}, \quad \mathbf{M}^{-1} = \frac{1}{\varepsilon}\begin{pmatrix} 1+a & a-1 \\ 1-b & -1-b \end{pmatrix}, \quad \dot{\mathbf{M}}^{-1} = \frac{3}{\varepsilon}\cdot\frac{\dot{R}}{R}\begin{pmatrix} 1 & -1 \\ 1 & -1 \end{pmatrix},
$$

with $a = \varepsilon(l-1)/2$ and $b = \varepsilon(l+2)/2$. The solution for ζ can be expressed in terms of the basic solutions (2.53). Using (2.54) and (2.55), one obtains

$$
\zeta = \mathbf{M} \cdot (\mathbf{T}_1 \cdot \mathbf{X}_0 + \mathbf{T}_2 \cdot \dot{\mathbf{X}}_0) = \mathbf{M} \cdot (\mathbf{T}_1 \cdot \mathbf{M}_0^{-1} + \mathbf{T}_2 \cdot \dot{\mathbf{M}}_0^{-1}) \cdot \zeta_0 + \mathbf{M} \cdot \mathbf{T}_2 \cdot \dot{\mathbf{M}}_0^{-1} \cdot \dot{\zeta}_0. \tag{2.56}
$$

As a specific example, consider a shell, initially at rest, with a corrugated outer surface.

Since $\dot{\mathbf{R}}_0 = \partial_t \zeta_0 = 0$, there is no contribution from the solution matrix $\mathbf{T}_2$ in this case. The contribution from $\mathbf{T}_1$ with $\zeta_{10} = 0$ yields

$$\zeta_1 = (\zeta_{20}/\Delta)[(1-a)(1+b_0)H_1 - (1+b)(1-a_0)G_1],$$
$$\zeta_2 = (\zeta_{20}/\Delta)[(1+a)(1+b_0)H_1 - (1-b)(1-a_0)G_1],$$

with $\Delta = (2l+1)\varepsilon_0$. The mode amplitudes are enhanced by the large factor Δ^{-1}. It determines the value from which the RTI actually grows.

Example 73 The thin shell approximation has led to the basic evolution equations (2.52) and (2.56). We examine their solutions for ballistic and accelerated shell motions.

The acceleration of a ballistically moving thin shell can be obtained from (2.35c) as $\ddot{R} = -3\varepsilon \dot{R}^2/R$. Being of order ε only, it can be neglected in the thin shell equations. With the help of (2.51) and (2.52), the ballistic motion can be described by the equations $\ddot{C} = \ddot{D} = 0$, $\ddot{R} = 0$. Accordingly, perturbations C and D grow with a constant velocity and vary linearly with a radius,

$$C = C_0 + \dot{C}_0(R - R_0)/\dot{R}, \qquad D = D_0 + \dot{D}_0(R - R_0)/\dot{R}. \tag{2.57}$$

Using an analogous notation as in (2.42), (2.44), the normalized displacement amplitudes follow from (2.48) and (2.57) to be

$$A_{12} = \frac{1}{2}q[(A_{10}+A_{20}) \pm (3q^2 - 2q^3)(A_{10}-A_{20}) - (1-q^{-1})(B_{10}+B_{20}) \pm (q^2-q^3)(B_{10}-B_{20})].$$

Here, we have used the scaling $\varepsilon = q^3 \varepsilon_0$ of the expansion parameter with the convergence ratio. One can recognize a sensitive dependence on the choice of initial conditions. Any perturbation of the shell thickness grows much faster than a mere distortion of the whole shell. Notice, that the result is limited due to the thin shell approximation ($\varepsilon \ll 1$) to moderate convergence ratios of the order $q^3 \leq 1/\varepsilon_0$. For larger convergence ratios the amplitude growth saturates and cavity oscillations, as described by (2.44), will develop.

Example 74 (The RTI of an accelerated thin shell) A constant acceleration a in the direction toward the shell center is assumed and the implosion time $\tau = \sqrt{2R_0/a}$ is used to form a dimensionless time variable $x = t/\tau$. The convergence ratio is related to the variable x by the expression $q = 1/(1-x^2)$. Typical acceleration distances are given by the values $q = 2$ or alternatively $x = 2^{-1/2} \simeq 0.71$. With x as the independent variable, the thin shell equations (2.52) become

$$(1-x^2)\ddot{G} + 2(l+2)G = 0, \qquad (1-x^2)\ddot{H} - 2(l-1)H = 0. \tag{2.58}$$

The different signs (in the equations for G and H) lead to qualitatively different solution behavior. While G describes stable surface oscillations, H has a purely growing solution branch. These two modes arise from the stable inner and from the RT unstable outer surface of the shell. We calculated some of the basic solutions G_1 and H_1, as defined by the initial conditions (2.53); see Fig. 2.5. The oscillation amplitudes of G_1 show no amplification and, actually, are even slightly decreasing with x. On the other hand, H_1 is monotonically growing and can lead to strong amplifications for large mode numbers.

For completeness, it is mentioned that the solutions of (2.58) can also be represented by Legendre functions $P_\nu(x)$, $Q_\nu(x)$ of degree ν. This representation is obtained by noting that $Y_1 = (1-x^2)\partial_x P_\nu(x)$ and $Y_2 = (1-x^2)\partial_x Q_\nu(x)$ are solutions of the differential equation

$$(1-x^2)\partial_x^2 Y + \nu(1+\nu)Y = 0. \tag{2.59}$$

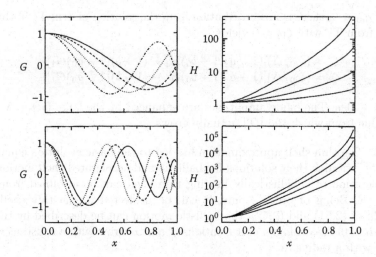

FIGURE 2.5: Evolution of the surface modes G and H of an accelerated thin shell. The time coordinates are dependent as $t = x\tau$. With increasing frequencies, the solutions for G correspond to the mode numbers $l = 2, 4, 8, 16, 40, 60, 80$. With increasing amplitudes, the solutions for H correspond to the mode numbers $l = 2, 4, 8, 16, 20, 30, 40, 50$.

TABLE 2.1: Comparison of the WKB approximation (2.60) with the numerical solution of (2.58). Values of the growing mode H_1 at $x = 0.71$ for different mode numbers l.

l	2	4	8	16
H_1 due to (2.60)	1.4	2.9	8.0	31
H_1 due to (2.58)	1.6	3.2	8.5	32

Solutions of (2.58) can be obtained by choosing the appropriate degree ν in (2.59). Unfortunately, the resulting degree is generally complex and the corresponding functions are not in common use.

It may also be interesting to compare the exact result with the simple Wentzel–Kramers–Brillouin (WKB) approximation [20],

$$G_1 \simeq (1 - x^2)^{1/4} \cos[\sqrt{2(1+2)}\, \text{arcsin}\, x],$$
$$H_1 \simeq (1 - x^2)^{1/4} \cosh[\sqrt{2(1-1)}\, \text{arcsin}\, x] = (1 - x^2)^{1/4}. \tag{2.60}$$

The validity of these expressions is generally limited to slowly varying coefficients in (2.58). Actually the WKB approximation proves fairly accurate for $x \leq 0.8$. The accuracy of the approximation may be appreciated from the comparison of exact and approximated results presented in Table 2.1.

2.2.4 Nonlinear perturbation theory

The nonlinear perturbation theories for the RTI have been intensively studied in the presence of surface tension. The prediction of the stability boundary between RTI and surface tension stabilization requires sophisticated perturbation methods. In a first treatment of this subject, a regular third-order expansion was given in [8]. Subsequently, singular perturbation methods, see Section 1.6.2, have been introduced in [37]. Thereby secular terms could be removed from expansion for the stable modes above the instability cutoff. The problem was reconsidered in [27, 35]. As a major result, it was found that finite-amplitude waves become nonlinearly unstable. While Rajappa's treatment apparently becomes invalid near the cutoff, Kiang and Nayfeh have obtained agreement in their nonlinear cutoff predictions. More recently, a different perturbation formula has been presented by Infeld and Rowlands [24], in making use of nonlinear traveling surface waves. Omitting the complications of surface tension, the present discussion is entirely based on regular perturbation theory. The purpose is to study the coupling to harmonics in an early stage of the instability evolution.

Bubble rise dynamics. RTI marks only onset of complex interpretation process, leading ultimately to the growth of mixing regions between neighboring fluids. On the other hand, the evolution of a free surface may be largely understood from first principles as an evolution of rising gas bubbles. The study of bubble dynamics, including various forms of bubble rise and bubble interactions, provides therefore a unique approach to a basic understanding of mixing-layer growth. The basic problem of the nonlinear free-surface instability consists in the determination of a potential $\phi(x, y, t)$ satisfying the boundary conditions

$$\left(\partial_t\phi + \frac{1}{2}u^2\right)\Big|_{y=\zeta(x,t)} + \zeta(x,t) = 0, \tag{2.61a}$$

$$\partial_t\zeta + u_1(x,\zeta,t)\partial_x\zeta - u_2(x,\zeta,t) = 0, \tag{2.61b}$$

at free surface $y = \zeta(x,t)$ at time t. Here, dimensionless variables by the length unit $1/k$ and the time unit $(ak)^{-1/2}$ have been used. In the perturbation approach, the boundary conditions are expanded about the undisturbed interface, leading to a sequence potential flow problem with boundary conditions on $y = 0$. In principle, each perturbation order can then be solved by a superposition of normal modes, being complete on a plane surface. The boundary condition (2.61a) can be rewritten in an alternative form that is particularly well suited for the perturbation approach. For this purpose, it can be used the shorthand notation

$$A(x,y,t)_{\,|\,y=\zeta} = 0, \quad A = \partial_t\phi + \frac{1}{2}u^2 + y, \tag{2.62}$$

and differentiate (2.61a) along the path of surface particles,

$$\frac{d}{dt}[A(x,y,t)_{\,|\,y=\zeta}] = (\partial_t A + u_1\partial_x A + u_2\partial_y A)_{\,|\,y=\zeta} = 0. \tag{2.63}$$

Substituting A in (2.63) from (2.62), we get a homogeneous boundary condition for the potential ϕ

$$\Phi(x,y,t)_{\,|\,y=\zeta(x,t)} = 0, \quad \Phi \equiv \partial_t^2\phi + \partial_y\phi + \partial_t u^2 + \mathbf{u}\cdot\nabla(\frac{1}{2}u^2). \tag{2.64}$$

We use Taylor series of $\Phi(x,y,t)$ with respect to y about $y = 0$, and represent both Φ and ζ by perturbation series,

$$\Phi = \Phi_{\,|\,y=0} + \partial_y\Phi_{\,|\,y=0}\zeta + \frac{1}{2}\partial_y^2\Phi_{\,|\,y=0}\zeta^2 + \cdots, \quad \Phi = \sum\nolimits_{n\in\mathbb{N}}\Phi_n\varepsilon^n, \quad \zeta = \sum\nolimits_{n\in\mathbb{N}}\zeta_n\varepsilon^n.$$

Ordering with respect to equal powers of ε yields up to third-order

$$\Phi_{\,|\,y=\zeta} = \varepsilon\Phi_1{}_{\,|\,y=0} + \varepsilon^2[\Phi_2 + (\partial_y\Phi_1)\zeta_1]_{\,|\,y=0} + \varepsilon^3\Big[\Phi_3 + (\partial_y\Phi_1)\zeta_2 + (\partial_y\Phi_2)\zeta_1 + \frac{1}{2}(\partial_y^2\Phi_1)\zeta_1^2\Big]\Big|_{y=0}. \tag{2.65}$$

The advantage of the homogeneous boundary condition (2.65) from the fact that the linear part Φ_1 vanishes identically, if it is required to vanish on the surface. This would not be true for the corresponding part A_1 of the original inhomogeneous boundary condition (2.62). As a result, the partial derivatives of Φ_1 can be dropped in (2.65), which greatly simplifies the evaluation of nonlinearities. Using (2.64) and (2.65), the boundary conditions at $y = 0$ for the first three orders of ϕ follow to be

$$\Phi_1 \;=\; \partial_t^2 \phi_1 + \partial_y \phi_1 = 0, \tag{2.66a}$$

$$\Phi_2 \;=\; \partial_t^2 \phi_2 + \partial_y \phi_2 + \partial_t u_1^2 = 0, \tag{2.66b}$$

$$\Phi_3 + (\partial_y \Phi_2)\zeta_1 \;=\; \partial_t^2 \phi_3 + \partial_y \phi_3 + 2\partial_t(\mathbf{u}_1 \cdot \mathbf{u}_2) + \mathbf{u}_1 \cdot \nabla(u_1^2/2)$$
$$+ \; \partial_y(\partial_t^2 \phi_2 + \partial_y \phi_2 + \partial_t u_1^2)\zeta_1 = 0. \tag{2.66c}$$

In addition to these boundary conditions, initial conditions have to be specified for ϕ_n and $\partial_t \phi_n$. We'll prescribe initial conditions on the first-order displacement ζ_1 and its velocity $\partial_t \zeta_1$. The higher orders are set equal to zero, initially: $\zeta_{n0} = \partial_t \zeta_{n0} = 0$ for $n > 1$. The surface displacement is defined by the original set of equations (2.61a,b). The expansion of these equations up to third-order is given by

$$\partial_t \phi_1 + \zeta_1 = 0, \quad \partial_t \zeta_1 - \partial_y \zeta_1 = 0, \tag{2.67a}$$

$$\partial_t \phi_2 + \zeta_2 + (\partial_t u_{y1})\zeta_1 + \frac{1}{2}u_1^2 = 0, \quad \partial_t \zeta_2 - \partial_y \phi_2 + u_{x1}\partial_x \zeta_1 - (\partial_y u_{y1})\zeta_1 = 0, \tag{2.67b}$$

$$\partial_t \phi_3 + \zeta_3 + (\partial_t u_{y1})\zeta_2 + (\partial_t u_{y2})\zeta_1 + \frac{1}{2}(\partial_t \partial_y u_{y1})\zeta_1^2 + \mathbf{u}_1 \cdot \mathbf{u}_2 + \frac{1}{2}(\partial_y u_1^2)\zeta_1 = 0, \tag{2.67c}$$

$$\partial_t \zeta_3 - \partial_y \phi_3 + u_{x2}\partial_x \zeta_1 + u_{x1}\partial_x \zeta_2 + (\partial_y u_{x1})(\partial_x \zeta_1)\zeta_1$$
$$- \left[(\partial_y u_{y1})\zeta_2 + (\partial_y u_{y2})\zeta_1 + \frac{1}{2}(\partial_y^2 u_{y1})\zeta_1^2\right] = 0. \tag{2.67d}$$

These equations have been commonly used in the perturbation approach. They determine both the potential and the displacement of each order simultaneously. In present treatment, the potential is first determined from the simpler set of boundary conditions (2.66a-c). The more complicated perturbation system (2.67a-d) is only used when explicit displacement expressions are required.

Harmonics. To discuss the dependence on initial conditions, we'll consider general sinusoidal flow perturbations arising either from the displacement or from the velocity of the initial surface. The first-order solution for the fundamental mode may be written in the form

$$\phi_1 \;=\; -[A \exp(t) + B \exp(-t)] \exp(-y) \cos x, \tag{2.68a}$$

$$\zeta_1 \;=\; [A \exp(t) - B \exp(-t)] \cos x, \tag{2.68b}$$

with arbitrary constants A and B. Using these expressions, the boundary condition (2.66b) for the second-order potential becomes $\partial_t^2 \phi_2 + \partial_y \phi_2 = -2[A^2 \exp(2t) - B^2 \exp(-2t)]$. With the help of (2.67b), the initial conditions $\zeta_{20} = 0$ and $\partial_t \zeta_{20} = 0$ can be expressed as $\partial_t \phi_{20} = -\left[A^2 + B^2 + \frac{1}{2}(A - B)^2 \cos 2x\right]$ and $\partial_y \phi_{20} = (A^2 - B^2) \cos 2x$, respectively. Solving these equations yields the second-order result

$$\phi_2 = -\frac{1}{2}(A^2 e^{2t} - B^2 e^{-2t}) - \frac{1}{2}\left[(A^2 - B^2)\cosh(\sqrt{2}t) + \frac{1}{\sqrt{2}}(A - B)^2 \sinh(\sqrt{2}t)\right]e^{-2y} \cos 2x,$$

$$\zeta_2 = -\frac{1}{2}\left[(Ae^t - Be^{-t})^2 - \sqrt{2}(A^2 - B^2)\sinh(\sqrt{2}t) - (A - B)^2 \cosh(\sqrt{2}t)\right]\cos 2x. \tag{2.69}$$

Notice the difference between different time-dependences in (2.69). In each perturbation order, there is a driven response, arising from the nonlinearity in the evolution equation, and a reactive response, resulting merely from specific initial conditions. In m-th order, the driven part will grow as e^{mt} but the reactive part only as $e^{\sqrt{m}t}$. In this sense, the nonlinear evolution appears rather universal and stable against normal mode perturbations $\sim e^{\sqrt{m}t}$. Neglecting reactive terms in the time-asymptotic limit, the dominant part of (2.69) follows to be

$$\phi_2 \to -\frac{1}{2}A^2 \exp(2t), \quad \zeta_2 \to -\frac{1}{2}A^2 \exp(2t)\cos 2x. \tag{2.70}$$

It is basically independent of the initial conditions, depending only on the unstable mode amplitude A. The asymptotic flow potential reduces to an unimportant time function, while the surface displacement becomes already modified by the first harmonic. Accordingly, simple-mode flow is consistent with nonlinear deformation of the surface. The negative sign of the first harmonic in (2.70)$_2$ describes the onset of bubble-spike asymmetry. The total displacement of the rising fluid ($x = 0$) is reduced in comparison with the total displacement of the falling fluid ($x = \pi$).

We will discuss two important cases. Choosing the initial conditions $A = B = V/2$, the evolution starts from a pure velocity perturbation. In this case, the first harmonic contribution to the velocity potential (2.69(a)) vanishes exactly. Up to second order, the corresponding solution is given by

$$\phi = -\frac{1}{4}V^2 \sinh(2t) - V\cosh t \exp(-y)\cos x, \tag{2.71a}$$

$$\zeta_2 = V\sinh t \cos x - \frac{1}{4}V^2[\cosh(2t) - 1]\cos(2x). \tag{2.71b}$$

We use this result to determine the third-order potential perturbation. Substituting (2.71a,b) into (2.66c) we obtain

$$\partial_2^t \phi_3 + \partial_y \phi_3 = -\mathbf{u}_1 \cdot \nabla(u^2/2) - (\partial_y \partial_t u_1^2)\zeta_1 = \frac{1}{4}[5\cosh(3t) - \cosh t]V^3 \cos x.$$

The third-order solution, corresponding to the initial conditions, $\zeta_{30} = -\partial_t \phi_{30} = 0$ and $\partial_t \zeta_{30} = \partial_y \phi_{30} = 0$, is found to be

$$\phi_3 = \left\{ \frac{5}{32}[\cosh(3t) - \cosh t] - \frac{1}{8}t\sinh t \right\}V^3 \exp(-y)\cos x.$$

Note that the nonlinear evolution can being consistently described by the fundamental mode of the flow potential up to third-order. Actually, the third-order provides merely a feedback to the time dependence of the flow amplitude.

To compare with previous work, we specialize to the more usual initial conditions $A = -B = Z/2$, describing static initial corrugations of amplitude Z. Combining (2.68a,b) and (2.69) yields

$$\phi = -Z\sinh t e^{-y}\cos x - \frac{1}{4}Z^2[\sinh(2t) + \sqrt{2}\sinh(\sqrt{2}t)e^{-2y}\cos(2x)],$$

$$\zeta = Z\cosh t \cos x - \frac{1}{4}Z^2[\cosh(2t) + 2\cosh(\sqrt{2}t) + 1]\cos(2x).$$

These solutions agree up to second-order with the results of [27]. Proceeding to third-order, the equations for the velocity potential become

$$\partial_t^2 \phi_3 + \partial_y \phi_3 = \left\{ -\sqrt{2}\partial_t[\sinh t \sinh(\sqrt{2}t)] + \frac{5}{4}\sinh(3t) + \frac{1}{4}\sinh(t) \right\}Z^3 \cos x,$$

$$\partial_y \phi_{30} = 0, \quad \partial_t \phi_{30} = -\frac{1}{8}Z^3[3\cos(3x) + \cos x].$$

Their solution includes a third-order contribution to the fundamental mode and an additional second harmonic

$$\phi_3 = \left[\tfrac{1}{8}t\cosh t + \tfrac{9}{32}\sinh t + \tfrac{5}{32}\sinh(3t) - \tfrac{\sqrt{2}}{2}\cosh(t)\sinh(\sqrt{2}t)\right]Z^3 e^{-y}\cos x$$
$$- \tfrac{\sqrt{3}}{8}Z^3\sinh(\sqrt{3}t)e^{-3y}\cos(3x).$$

The appearance of secular terms, as represented by the first member in the brackets, is typical for high-order perturbations. They can be removed by singular perturbation techniques, although this procedure has only basic justification for stable bounded solutions. For instance, the present term may be viewed as a nonlinear correction to the growth rate of the first-order solution $\phi_1 \to -Z\sinh\left((1 - \tfrac{1}{8}Z^2)t\right)e^{-y}\cos x$. This modification agrees with the rigorous derivations in [27, 35].

Nonlinear mode interactions. If the initial perturbation is a superposition of modes, beat waves can be excited under nonlinear mode coupling. The appearance of longer wavelengths in the evolution of multiple-mode problems is of considerable interest, since these modes can become dominant when the growth of the short-wavelength modes has reached nonlinear saturation. Unfortunately, the stage of saturation can no longer be described by the perturbation approach. To demonstrate the principle possibility of nonlinear mode coupling, we allow for a superposition of two modes in the first-order solution

$$\phi_1 = -m^{-1}F(t)e^{-my}\cos mx - n^{-1}G(t)e^{-ny}\cos nx,$$
$$\zeta_1 = m^{-1}\partial_t F(t)\cos mx + n^{-1}G(t)\cos nx.$$

In the second-order, there appears an additional mode with the difference mode number $l = m - n$, where $m > n > l$ is assumed. Its potential has the general form $\phi_2 = [b(t) + c_1\exp(\sqrt{l}t) + c_2\exp(-\sqrt{l}t)]\exp(-ly)\cos lx$, consisting of a driven solution $b(t)$ and the stable and unstable normal modes with amplitudes $c_{1,2}$. The boundary condition (2.66b) yields for b the differential equation $\partial_t^2 b - lb = -2\partial_t(FG)$. Noting that $\partial_t^2 F = mF$ and $\partial_t^2 G = nG$, the driven solution can be expressed in the form $b = -n^{-1}F\partial_t G$. From the initial conditions $\zeta_{20} = 0$ and $\partial_t^2\zeta_{20} = 0$ the constants $c_{1,2}$ are found

$$c_{1,2} = \frac{1}{4}\left[\pm(\sqrt{l}/nm)\partial_t F\partial_t G + n^{-1}F\partial_t G - m^{-1}G\partial_t F\right]\big|_{t=0},$$

where the upper sign refers to c_1, and the lower one to c_2. In this solution the freely evolving modes may be more important than the driven one because of the earlier saturation of short wavelength modes. This interpretation suggests, that the long-time behavior is dominated by the unstable mode with amplitude c_1. We remark, that such a mode can only exist if the initial surface displacement is nonzero, $\partial_t F_0 \neq 0$, or $\partial_t G_0 \neq 0$.

Least-squares approximation. With increasing amplitudes, high-order contribution can no longer be neglected in the perturbation series. The evolution of a large number of harmonics can be studied numerically by the method of least-squares approximation.

We first briefly outline the approximation method. According to the previous discussion, the n-th order perturbation solution has the general form

$$\phi(x, y, t) = \sum_{m=0}^{N} a_m(t)e^{-my}\cos mx, \tag{2.72}$$

where the amplitudes $a_m(t)$ contain contributions from various perturbation orders. Notice that this representation is not completely general. Even in the limit $N \to \infty$, it can fail to describe the full large-amplitude solution. The Fourier representation with respect to the x coordinate will only converge if the potential has no singularities on each line $y = $ const. It is therefore assumed, that the potential is analytic in the entire half-plane above the lowest surface point. This assumption becomes violated, if singularities appear in those regions of the half-plane that are not occupied by the fluid. Despite this principal limitation, it is of interest to see how far the nonlinear evolution can be described by the series (2.72) if high-order terms are included.

The method of least-squares approximation determines the mode amplitudes by minimizing the mean square error on the surface for an expansion of given order. The set of basis functions $f_m(x, y) = \exp(-my)\cos mx$ can be orthogonalized on the instantaneous surface $x = x(x_0, t)$, $y = (y_0, t)$ using the Gram–Schmidt orthogonalization procedure. This yields a new function $g_m(x, y, t)$ satisfying

$$(g_i, g_j) = \frac{2}{\pi}\int_0^\pi g_i(x(x_0,t), y(x_0,t), t)g_j(x(x_0,t), y(x_0,t), t)\,\mathrm{d}x_0 = \delta_{ij}.$$

If the series (2.72) is rewritten in terms of the function set $\{g_m\}$ as

$$\phi = \sum_{m=0}^N c_m(t)g_m(x, y, t), \tag{2.73}$$

the boundary condition (2.61a) assumes the general form

$$\sum_{m=0}^N \dot{c}_m g_m = R(x_0, \{c_m\}, t). \tag{2.74}$$

In this equation, the time-derivative $\dot{c}_m$ appears on the LHS only and R is a known function at time t,

$$R(x_0, \{c_m\}, t) = -\Big(\sum_{m=0}^N c_m\partial_t g_m + \frac{1}{2}u^2 + y\Big)\Big|_{x=x(x_0,t),\ y=y(x_0,t)}.$$

Taking projection of (2.74) on the basis functions and adding the equations of motions for the surface particles yields a closed system of ordinary differential equations

$$\dot{c}_m = (g_m, R), \quad c_m(0) = c_{m0}, \quad \dot{x}(x_0, t) = \partial_x\phi,$$
$$x(x_0, t) = x_0, \quad \dot{y}(x_0, t) = \partial_y\phi, \quad y(x_0, t) = \zeta_0. \tag{2.75}$$

Numerical solutions of the evolution equations (2.75) were obtained in [28]. An example is shown in Fig. 2.6, corresponding to the initial values $c_m(0) = 0$, $\zeta_0(x) = 0.1\cos x$. This calculation has been performed with $N = 11$ terms in the expansion (2.73) and with 100 particles distributed at equal distances over the initial surface. The normal mode spectrum at subsequent times is shown in Fig. 2.6(a). One can recognize that the rate of convergence of successive orders is quite fast at early times ($t = 0.6$ represented by square symbols) but it becomes increasingly worse in the course of time evolution. Note that the 10th harmonic amplitude grows from the order 10^{-11} up to the order 10^{-2} while the fundamental model grows only by somewhat more than a factor of 10. The decrease in the slope of the mode spectrum is particularly pronounced after the time $t = 3.0$ and leads computational failure after the time $t = 3.6$. Corresponding results for the evolution of the fluid interface are represented in Fig. 2.6(b). The heavy fluid above the interface is pushed under the pressure of a gas below the interface. It can be recognized that the sinusoidal initial perturbation develops into a rising bubble centered around $x = 0$ and into a falling spike centered around $x = \pi$. The accuracy of the calculation is believed to be good up to the time $t = 3.6$ where the bubble and spike amplitudes have reached the values 1.01 and -2.20, respectively.

Despite the satisfactory of bubble-spike asymmetry, it should be noticed that the perturbation method is faced with serious difficulties at larger amplitudes. In practice, calculations cannot be extended beyond a bubble-spike separation of at most half a wavelength. The

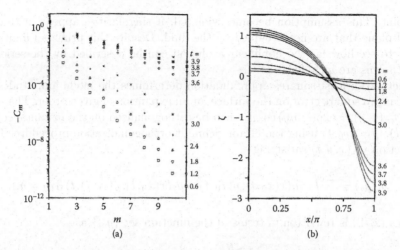

(a)

(b)

FIGURE 2.6: Nonlinear evolution of a sinusoidal surface displacement $\zeta = 0.1 \cos x$ in the free-surface RTI. The units are k^{-1} for the length and $1/\sqrt{ak}$ for the time. (a) Spectrum of the mode amplitudes c_m for the fundamental mode $m = 1$ and its first 10 harmonics $m = 2 - 11$. Subsequent times indicated by different marker symbols as defined at the right margin. (b) Surface displacement at corresponding times. The fluid occupies the region above the surface line and, for symmetry reason, only one-half of a wavelength is shown.

main reason is an enormous increase of the exponentials $\exp(-my)$ when the spike amplitude exceeds $y = -1$. In addition, the series may fail to converge when singularities in the bubble region are encountered.

Dynamics of a thin-wall cylindrical shell. Consider the radial scattering of the thin supper-conductive cylindrical shell in axial magnetic field of inductivity $\mathbf{B}_0$. The motion is governed by the dynamical equation of the form [14]:

$$\partial_t^2 \mathbf{r} = \frac{B^2(r, \psi, t)}{8\pi\rho} (\mathbf{e}_z \times \partial_\psi \mathbf{r}), \tag{2.76}$$

where $\mathbf{e}_z$ is the unit vector along z-axis, ρ is the density of the shell, $B(r, \psi, t)$ is the magnetic field in the point (r, ψ) at time t, ψ is the Lagrange coordinate, i.e., $dm = \rho \, d\psi$.

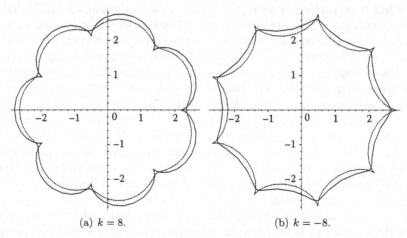

(a) $k = 8$.   (b) $k = -8$.

FIGURE 2.7: The form of the shell: (a) for $\mu = 1$ (solid line) and for $\mu = 0.5$ (dash line); (b) for $\mu = 0.5$ (solid line) and for $\mu = 1$ (dash line).

After substitution $Z = x + iy = r\exp(i\psi)$, (2.76) may be transformed to

$$\partial_t^2 Z = i\frac{B^2(Z,\psi,t)}{8\pi\rho}\,\partial_\psi Z$$

with initial conditions

$$Z(\psi,0) = \exp(i\psi) - \frac{\mu^*}{k}\exp(ik\psi), \quad \partial_t Z(\psi,0) = u_0[\exp(i\psi) - \frac{\mu(t)}{k}\exp(ik\psi)],$$

where u_0 is the initial velocity of the shell. We are looking for solution of the form

$$Z(\psi,t) = R(t)\Big[\exp(i\psi) - \frac{\mu(t)}{k}\exp(ik\psi)\Big]. \tag{2.77}$$

Thus, substituting (2.77) into (2.76) we obtain

$$\frac{d^2 R}{d\tau^2} = -aR(1-R^2)^{-2}, \quad \frac{d^2\mu}{d\tau^2} + \frac{2}{R}\frac{dR}{d\tau}\cdot\frac{d\mu}{d\tau} + \frac{1-k}{R}\cdot\frac{d^2 R}{d\tau^2} = 0$$

with boundary conditions $R(0) = R_0/R_B$, $\frac{dR}{d\tau}(0) = R_K/R_B$, $\mu(0) = \mu^*$ and $\frac{d\mu}{d\tau}(0) = 0$, where $a = B_0^2 t_0^2/\rho$, R_0 – the initial radius of the shell, R_K – maximal radial distance of the shells scattering, R_B – the radius of the solenoid creating the magnetic field ($R_B > R_K$), $t_0 = R_K/u_0$ – the characterizing time. Note, that if $|\mu(t^*)| \geq 1$, then the outer curve of the shell becomes self-crossed (Fig. 2.7) at the time $t \geq t^*$. This time t^* is called the *moment of the shells destroy*. Numerical calculations show that under conditions $B_0 = 0.2 - 0.7$ Tesla, $\mu^* = 0.95$, and for positive modes $k = 4, 8, 16, 32, 64$ (curves $1-5$ in Fig. 2.8) the increased perturbations do not become to the shells destroy. For the negative modes the shell is not to be destroyed only under very low initial perturbations amplitudes.

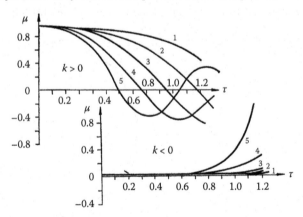

FIGURE 2.8: Temporal evolutions of value $\mu(\tau)$ for wave numbers $k > 0$ and $k < 0$.

Nonlinear dynamics of cylindrical liquid film. As a continuation of the previous section we consider Rayleigh–Taylor instability of a thin liquid film in the meridional plane (R, ϕ), assuming the flow to be cylindrically axisymmetric. The initial motion equation has nearly the similar view [18],

$$\rho\,\partial_{tt}^2\mathbf{r} = \Delta p(\mathbf{e}_z \times \partial_\psi\mathbf{r}), \tag{2.78}$$

where the density ρ and the pressure difference Δp between external and internal media are assumed to be constant, ψ is a Lagrange coordinate, i.e., $d\psi = \pi R_0^2\,d\phi$ per unit of the length along the z-axis, $\phi \in [0, 1]$ is dimensionless azimuth angle, R_0 is the initial generalized radius of the liquid film. Substitution $\mathbf{r} = [Z, 0]$, $Z = X + iY \in \mathbb{C}$ transforms (2.78) to the PDE for $Z(\phi, t)$,

$$\rho \, \partial_{tt}^2 Z = i \Delta p \, \partial_\psi Z.$$

Using the relation $d\psi = \pi R_0^2 \, d\phi$, the above equation may be rewritten as

$$\partial_{tt}^2 Z = i [\Delta p / (\pi \rho R_0^2)] \, \partial_\phi Z,$$

or equivalently, as the system of linear PDEs for real and imaginary parts of $Z(\phi, t)$,

$$\partial_{tt}^2 X = -[\Delta p / (\pi \rho R_0^2)] \, \partial_\phi Y, \quad \partial_{tt}^2 Y = [\Delta p / (\pi \rho R_0^2)] \, \partial_\phi X. \tag{2.79}$$

Generally, we assume $Z(\phi, 0) = X(\phi, 0) + iY(\phi, 0)$ to be given and

$$\partial_t Z(\phi, 0) = \Omega Z(\phi, 0).$$

Assuming that at the nozzle exit cross-section the cylindrical film rotates with angular velocity $\Omega = V_{\phi,0}/R_0$, where $V_{\phi,0}$ is the swirl velocity, and the radial velocity is zero, the initial conditions are given as follows:

$$Z_k(\phi, 0) = \exp(i\phi) - (\lambda_0/k) \exp(ik\phi), \quad \partial_t Z_k(\phi, 0) = \Omega \, Z_k(\phi, 0), \tag{2.80}$$

where $\lambda(0) = \lambda_0$, $|\lambda_0|$ being sufficiently smaller than 1, and $k \in \mathbb{Z} \setminus \{0\}$ a wave mode. Note that the second initial condition is different from the similar one in [14]. Let us find the solution to the problem (2.79)–(2.80) in the form; Fig. 2.9,

$$Z_k(\phi, t) = R(t)[\exp(i\phi) - (\lambda(t)/k) \exp(ik\phi)]. \tag{2.81}$$

Given $t \geq 0$ such that $|\lambda(t)| > 1$ (since $|\lambda(t_*)| = 1$) the graph $Z_k(\phi, t)$ becomes self-intersecting (a prolate hypocycloid or prolate epicycloid, depending on the sign of k, where $|k|$ is the number of branches). Figures 2.10 illustrate $|k-1|$-cusped epicycloidal and hypocycloidal (depending on the sign of k) cross-sections for $\lambda = 1$. It would appear natural to assume that the film lives up to an instant of time t_* such that $|\lambda(t_*)| = 1$.

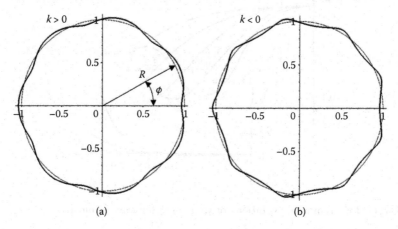

(a)   (b)

FIGURE 2.9: Epicycloidal and hypocycloidal cross-sections with $k = \pm 8$, $\varphi = 2\pi\phi$ and $|\lambda| < 1$.

The coefficient a may be represented in the form $a = \pi^{-1}(h_0/R_0)\mathrm{Lf}$, where $\mathrm{Lf} = \mathrm{Eu}/\mathrm{We}$ is the *Lefebvre number* named after Lefebvre by the authors [19]. Here $\mathrm{Eu} = \Delta p R_0/(2\sigma_*)$ is the Euler number, $\mathrm{We} = \rho h_0 V_{z0}/(2\sigma_*) > 0$ is the Weber number, $\sigma_* > 0$ is the surface tension coefficient, $h_0 > 0$ is the liquid film thickness. The Lefebvre number means the ratio of static and dynamic pressures. Substituting (2.81) into (2.79) gives the system

$$\partial_{\tau\tau}^2 \lambda + (2/\tilde{R})(\partial_\tau \tilde{R})(\partial_\tau \lambda) = a\lambda(1 - k), \quad \partial_{\tau\tau}^2 \tilde{R} = -a\tilde{R}, \quad a = \pi^{-1}\Delta p/(\rho V_{z0}^2) \tag{2.82}$$

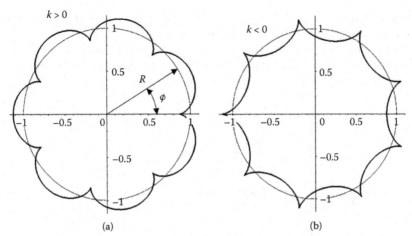

FIGURE 2.10: The perturbed form Z_k for epicycloidal and hypocycloidal cross-sections with $k = \pm 8$, $\varphi = 2\pi\phi$ and $|\lambda| = 1$.

with initial conditions (where $\tilde{R} = R/R_0$, $\tau = t/t_0$, $t_0 = R_0/V_{z0}$ being characteristic time of the motion, and $R(t) = R_0\tilde{R}(\tau)$)

$$\tilde{R}(0) = 1, \quad \partial_\tau \tilde{R}_{|\tau=0} = V_{\phi 0}/V_{z0} = V_\Omega, \tag{2.83a}$$

$$\lambda(0) = \lambda_0, \quad \partial_\tau \lambda_{|\tau=0} = 0. \tag{2.83b}$$

The system (2.82) splits onto two equations. The second one with its boundary conditions (2.83a) accepts analytical solution,

$$\tilde{R} = \begin{cases} \cos\sqrt{a}\tau + (V_\Omega/\sqrt{a})\sin\sqrt{a}\,\tau, & \Delta p > 0, \\ V_\Omega \tau + 1, & \Delta p = 0, \\ \cosh\sqrt{|a|}\tau + (V_\Omega/\sqrt{|a|})\sinh\sqrt{|a|}\,\tau, & \Delta p < 0. \end{cases} \tag{2.84}$$

Substitution (2.84) into the first equation of the system (2.82) gives

$$\partial^2_{\tau\tau}\lambda + F\partial_\tau\lambda = a(1-k)\lambda,$$

$$F = 2\sqrt{|a|} \times \begin{cases} \dfrac{b\cos\sqrt{a}\,\tau - \sin\sqrt{a}\,\tau}{\cos\sqrt{a}\,\tau + b\sin\sqrt{a}\,\tau}, & \Delta p > 0, \\ \dfrac{b\cosh\sqrt{|a|}\,\tau + \sinh\sqrt{|a|}\,\tau}{\cosh\sqrt{|a|}\,\tau + b\sinh\sqrt{|a|}\,\tau}, & \Delta p < 0, \end{cases} \tag{2.85}$$

where $b = V_\Omega/\sqrt{|a|}$. If the pressure drop is absent then (2.85) has a trivial solution, $\lambda = \lambda_0$. In general case, the problem (2.85) accepts analytical solution:

$$\lambda = \frac{\lambda_0}{\sqrt{|k|}} \begin{cases} \dfrac{1}{\cos\sqrt{a}\tau + b\sin\sqrt{a}\tau} \begin{cases} \sqrt{k}\cos\sqrt{ak}\,\tau + b\sin\sqrt{ak}\,\tau, & k > 0 \\ \sqrt{|k|}\cosh\sqrt{a|k|}\,\tau + b\sinh\sqrt{a|k|}\,\tau), & k < 0 \end{cases} & \Delta p > 0, \\ \sqrt{|k|}, \qquad\qquad k \neq 0, \quad \Delta p = 0, \\ \dfrac{1}{\cosh\sqrt{|a|}\,\tau + b\sinh\sqrt{|a|}\,\tau} \begin{cases} \sqrt{k}\cosh\sqrt{|a|k}\,\tau + b\sinh\sqrt{|a|k}\,\tau), & k > 0 \\ \sqrt{|k|}\cos\sqrt{ak}\,\tau + b\sin\sqrt{ak}\,\tau), & k < 0 \end{cases} & \Delta p < 0. \end{cases} \tag{2.86}$$

For a nonrotating jet, $V_\Omega = 0$, the solution (2.86) is transformed to the simple form,

$$\lambda = \lambda_0 \begin{cases} \cos\sqrt{ak}\,\tau/\cos\sqrt{a}\,\tau, & k > 0 \\ \cosh\sqrt{a|k|}\,\tau/\cos\sqrt{a}\,\tau, & k < 0 \end{cases} \quad \Delta p > 0, \\ 1, \qquad\qquad k \neq 0, \quad \Delta p = 0, \\ \begin{cases} \cosh\sqrt{|a|k}\,\tau/\cosh\sqrt{|a|}\,\tau, & k > 0 \\ \cos\sqrt{ak}\,\tau/\cosh\sqrt{|a|}\,\tau, & k < 0 \end{cases} \quad \Delta p < 0. \tag{2.87}$$

Let us study the modes $k = \pm 2, \pm 3, \pm 4, \pm 5, \pm 6, \ldots$ in a wide range temporary interval, $\tau > 100$, for low initial value of parameter $\lambda_0 = 0.05$. Figures 2.11–2.14 show dependence of λ on dimensionless time τ; the curves 3,4,6,8, and 16 correspond to the wave numbers $k = \pm 3, \pm 4, \pm 6, \pm 8, \pm 16, \ldots$, respectively (and the reader can find more illustrative figures in [19]).

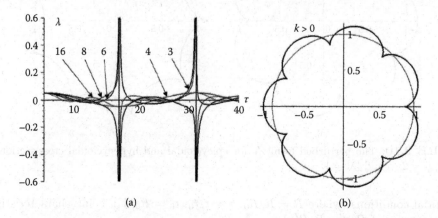

FIGURE 2.11: Function $\lambda(\tau)$ for $a = 0.04$, $b = 25$; modes $k = 3, 4, 6, 8, 16$.

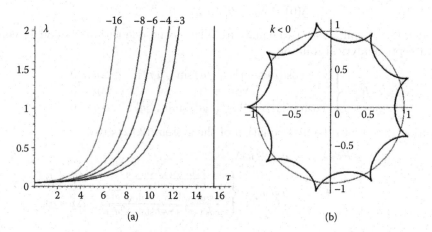

FIGURE 2.12: Function $\lambda(\tau)$ for $a = 0.04$, $b = 25$; modes $k = -3, -4, -6, -8, -16$.

For a positive value a, which corresponds to a positive pressure drop Δp and for wide range of parameter $0.8 - 25$, two positive modes $k = 4, 16$ are stable; Fig. 2.11, while modes $k = 4, 16$ are unstable. All modes with negative k are also unstable; Fig. 2.12. In the case of negative pressure drop corresponding to negative parameter the situation drastically changes – all negative modes decreased (Fig. 2.14), while all positive modes increased; Fig. 2.13. The lines $\lambda(\tau)$ are gathering with increasing $|a|$ as it plays a role of dimensionless frequency in (2.86) and (2.87). In the case of a nonrotating jet, when parameter $b = 0$, and at positive pressure drop all positive and negative modes become unstable. At negative pressure drop, all positive modes are unstable, and negative modes are stable. The rotating velocity affects jet stability so that only two of considered positive modes, $k = 4, 16$, become stable. The curves in Figs. 2.11–2.12 are periodic, their behavior is shown in the range of one and two periods.

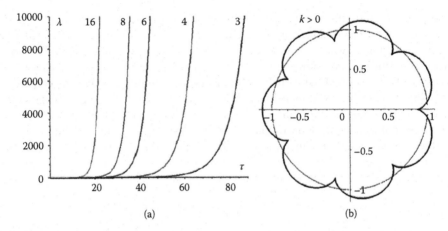

FIGURE 2.13: Function $\lambda(\tau)$ for $a = -0.04$, $b = 25$; modes $k = 3, 4, 6, 8, 16$.

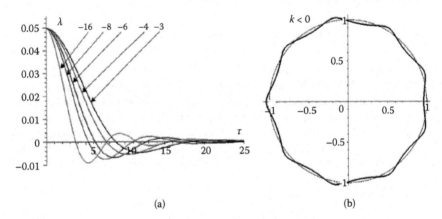

FIGURE 2.14: Function $\lambda(\tau)$ for $a = -0.04$, $b = 25$; modes $k = -3, -4, -6, -8, -16$.

2.2.5 Inhomogeneous fluids

The potential flow model describes the motion of fluid layers with piecewise constant densities. In the following, we'll consider incompressible fluids with continuous density variations. A basic difference arises from the possibility of vorticity generation in the presence of noncolinear pressure and density gradients. The evolution of these rotational modes and their relationship to the surface modes of potential flow theory is the main subject of this section.

Incompressible flow model. Incompressible motion of an inhomogeneous fluid in an external gravitational potential $U(x)$ are governed by the set of basic equations

$$\frac{D}{Dt}\rho = 0, \quad \text{div } \mathbf{u} = 0, \tag{2.88a}$$

$$\frac{D}{Dt}\mathbf{u} = -\frac{1}{\rho}\nabla p - \nabla U. \tag{2.88b}$$

In this model, the density be assumed constant along particle trajectories, but it can be different for different fluid particles. We first discuss the equilibrium states in this model. Any steady incompressible flow is subject to the condition div $\mathbf{u} = 0$. It excludes the possibility of mass flow along the density gradient. We will therefore restrict attention to the hydrostatic equilibrium

$$\mathbf{u} = 0, \quad \partial_t \rho = 0, \quad \nabla p = -\rho\nabla U.$$

Both the pressure and the density have to be constant on equipotential surfaces. In fact, their gradients are directed along the surface normal direction ∇U according to the relations $\nabla U \times \nabla p = 0$ and $\nabla \times \nabla p = -\nabla \times (\rho \nabla U) = -\nabla \rho \times \nabla U = \nabla U \times \nabla \rho = 0$. The condition of hydrostatic equilibrium assumes therefore the form $p(U) = -\int \rho(U)dU$, where the function $\rho(U)$ can be prescribed arbitrarily. The incompressible fluid equations can be further simplified in the case of two-dimensional flow. To satisfy the incompressibility condition, the velocity field is first to be expressed by the rot of a vector potential ψ, $\mathbf{u} = \nabla \times \psi$. If the flow is uniform along a particular direction specified by the unit vector $\mathbf{b}$, it can be described by the single component $\psi = \mathbf{b} \cdot \psi$, according to the relations $\mathbf{b} \cdot \nabla \equiv 0$, $\mathbf{b} \times \mathbf{u} = \mathbf{b} \times (\nabla \times \psi) = \nabla \psi$ and $\mathbf{u} = \nabla \psi \times \mathbf{b}$. The function ψ is known as the stream function because it is constant along steady stream lines. This property follows from the identity

$$\mathbf{u} \cdot \nabla \psi = (\nabla \psi \times \mathbf{b}) \cdot \nabla \psi = \mathbf{b} \cdot (\nabla \psi \times \nabla \psi) = 0.$$

The rot of the velocity field is generally nonzero, $\nabla \times \mathbf{u} = -\nabla^2 \psi \, \mathbf{b}$. According to (2.88b), its evolution is governed by the vorticity equation

$$\frac{D}{Dt}(\nabla^2 \psi) = \partial_t \nabla^2 \psi + \{\nabla^2 \psi, \psi\} = \rho^{-2}\{p, \rho\}. \tag{2.89}$$

The brackets $\{f, g\}$ are signed $\{f, g\} \equiv \mathbf{b} \cdot (\nabla f \times \nabla g) = \partial_x f \, \partial_y g - \partial_y g \, \partial_x f$, with $\mathbf{b}$ being normal to the (x, y)-plane. Noncolinear pressure and density gradients are responsible for vorticity generation in inhomogeneous media. Irrotational potential flow is basically limited to homogeneous conditions. In this spectral context, the stream function ψ is harmonic $(\nabla^2 \psi = 0)$ and it is related to the velocity potential ψ by the Cauchy–Riemann differential equations $\mathbf{u} = \nabla \phi = \nabla \psi \times \mathbf{b}$. To describe the fluid motion in terms of the variables ρ and ψ only, it is necessary to eliminate the pressure (2.88a,b) and (2.89) yields

$$\rho(\partial_t \nabla^2 \psi + \{\nabla^2 \psi, \psi\}) + \nabla \rho \cdot (\partial_t \nabla \psi + \{\nabla^2 \psi, \psi\}) = \{\rho, U\}, \tag{2.90a}$$

$$\partial_t \rho + \{\rho, \psi\} = 0. \tag{2.90b}$$

These equations still describe the full nonlinear evolution of two-dimensional incompressible flow.

Consider an arbitrary hydrostatic equilibrium. Small perturbations can be described by linearizing (2.90a,b) with respect to ψ. This yields

$$\mathrm{div}(\rho \partial_t^2 \psi) + \{\{\rho, \psi\}, U\} = 0, \tag{2.91}$$

where ρ denotes the unperturbed equilibrium density. The double brackets are proportional to the applied gravitational acceleration $\mathbf{g} = \nabla U$. More explicitly, it can be written as $\{\{\rho, \psi\}, U\} = \mathbf{g} \cdot [\nabla \times (\nabla \rho \times \nabla \psi)]$.

Stability eigenvalue problem. The gravitational instability of fluid layers with variable density was first analyzed by Rayleigh (1883). The classical eigenvalue problem and various extensions have been reviewed in great detail in [7]. The present discussion is mainly concerned with more recent developments. While (2.91) is valid for arbitrary potentials $U(x, y)$, we specialize in the physically important case of $U = ay$, describing acceleration along the y direction. Assuming normal mode perturbations, $\psi \sim \exp[i(kx - \omega t)]$, (2.91) becomes

$$\partial_\eta^2 \psi + \beta \partial_\eta \psi - (1 + \lambda\beta)\psi = 0, \tag{2.92}$$

where $\eta = ky$, $\beta = \partial_\eta \rho / \rho$, and $\lambda = ak/\omega^2$. This is Rayleigh's famous eigenvalue equation of the normal mode growth rate and frequencies of accelerated fluid layers. For a given density profile, the possible eigenvalues λ are determined by boundary conditions imposed on the solution ψ. The most common boundary conditions are

$$\lim_{|\eta|\to\infty} \psi(\eta) = 0, \tag{2.93a}$$

$$\psi_{\,|\,\eta=\eta_i} = 0, \tag{2.93b}$$

$$(\partial_\eta \psi - \lambda\psi)_{\,|\,\eta=\eta_i} = 0. \tag{2.93c}$$

They apply to (a) infinite systems with stable boundary regions and to bounded systems with (b) rigid walls or (c) free surfaces, respectively. In addition, internal interfaces with discontinuous density variations may be present. Multiplying (2.92) by ρ and integrating across the interface layer, one finds the continuity conditions

$$[\psi] = 0, \quad [\rho(\partial_\eta \psi - \lambda\psi)] = 0, \tag{2.94}$$

with the same notation as in Section 2.2.1.

To illustrate the relationship between the different sets of boundary conditions, we consider an arbitrary fluid layer $0 < \eta < h$ and a boundary medium with constant densities ρ_1 for $\eta < 0$ and ρ_2 for $\eta > h$. According to the infinite medium boundary conditions (a), the solution inside the boundary regions becomes

$$\psi = \left\{ \begin{array}{ll} \psi(0)\exp(\eta), & \eta < 0, \\ \psi(h)\exp(h-\eta), & \eta > h. \end{array} \right. \tag{2.95}$$

The rigid- and free-surface problems can be recovered by the following procedures:

High-density limit. We apply the continuity conditions (2.94) to the interfaces at $\eta = 0$ and $\eta = h$ and perform the limit $\rho_{1,2} \to \infty$. Using (2.95) for the solutions inside the boundary medium, one is left with constraints $(1-\lambda)\psi(0) = 0$ and $(1+\lambda)\psi(h) = 0$ on the boundary values of the solution inside the fluid layer. These constraints can be satisfied in three different ways under setting

$$\psi(0) = \psi(h) = 0, \tag{2.96a}$$

$$\psi(0) = 0 \quad \text{and} \quad \lambda = -1, \tag{2.96b}$$

$$\psi(h) = 0 \quad \text{and} \quad \lambda = 1, \tag{2.96c}$$

respectively. The first case recovers the rigid wall condition (2.93a). The other cases yield two further modes with eigenvalues $\lambda = \pm 1$. The vanishing of the solution at one boundary only can always be satisfied, by taking an appropriate linear combination of the two independent solutions to (2.92). These additional modes correspond to the free-surfaces of the boundary medium at $\eta = 0$ and $\eta = h$ with respect to the fluid layer of low density.

Low-density limit. Taking the opposite limit $\rho_{1,2} \to 0$, the second continuity condition of (2.93a) reduces simply to the free-surface condition (2.93c). In contrast to the previous case, the low-density limit is equivalent to the free-surface problem. Here one can also find free-surface modes with eigenvalues $\lambda = \pm 1$. The corresponding solutions are simply $\psi \sim \exp(\pm 1)$, satisfying both the differential equation (2.92) and the boundary conditions (2.93c). Remark that these free-surface modes exist for any density profile of the fluid layer. This confirms and extends the potential flow result of Section 2.2.2 (e) that free-surface growth rates are independent of the layer width. Apparently, growth reductions by inhomogeneity effects require partial stabilization of these modes.

Exponential density variation. As an explicit example, we discuss the stability of fluid layers with exponential density variations. This profile model permits explicit solutions for the evolution of internal modes and demonstrates their connection with the surface modes of potential flow theory. Particular attention is devoted to the discussion of density transition layers in an infinitely extended fluid and to its basic limits of short and long wavelength perturbations.

Example 75 (Model equations) Inside the layer $0 < \eta < h$, the density profile is assumed of the form $\rho = \rho_1 \exp(\beta\eta)$ with a constant gradient $\beta > 0$. The general solution of (2.92) becomes

$$\psi = A \exp(q_1\eta) + B \exp(q_2\eta), \tag{2.97}$$

with $q_{1,2} = -\beta/2 \pm z$, $z = v + iw = \sqrt{1 + \lambda\beta + \beta^2/4}$. Since the eigenvalues λ are real, the branch of the square root is chosen on the positive real axis ($v > 0$, $w = 0$) or on the positive imaginary axis ($v = 0$, $w > 0$), depending on the sign of the discriminant. The point $z = 0$ is excluded because the solution (2.97) would be incomplete in this case. The boundary conditions restrict the possible z values. The corresponding growth rates $n = \text{Im}(\omega)$ can then be obtained as

$$\frac{n^2}{ak} = -\frac{1}{\lambda} = \frac{\beta}{1 + \beta^2/4 - z^2}. \tag{2.98}$$

In what follows, we will discuss in succession the normal mode solutions corresponding to the boundary conditions (2.93a-c).

Example 76 (Rigid boundaries) Using (2.97) and (2.98), one finds the linear system

$$A + B = 0, \quad \exp(q_1h)A + \exp(q_2h)B = 0.$$

The allowable solutions are given by $B = -A$ and $z = iw_m$, where $w_m = m\pi/h$, $m = 1, 2, 3, \ldots$ denotes a positive integer. The corresponding growth rates and eigenfunctions follow to be

$$\frac{n^2}{ak} = \frac{\beta}{1 + \beta^2/4 + w_m^2}, \quad \psi = \exp(-\beta\eta/2)\sin(w_m\eta). \tag{2.99}$$

To discuss the physical dependences of this result, we introduce the gradient scale length L, the layer thickness d, a density increment C, and a parameter p by relations

$$\beta = 1/kL, \quad h = kd, \quad C = \frac{1}{2}\beta h = \frac{1}{2}\ln[\rho(h)/\rho(0)], \quad p = \frac{1}{2}\sqrt{1 + (\pi m/C)^2}.$$

The growth rate, given by (2.99), can then be expressed in the form

$$\frac{n^2}{ak} = \frac{kL}{(kL)^2 + p^2}, \quad \frac{n}{\sqrt{a/L}} = kL/((kL)^2 + p^2)^{1/2}.$$

Remark that n increases linearly for $kL \to 0$ and approaches the constant value $\sqrt{a/L}$ for $kL \to \infty$. In both limits, it stays below the free-surface value. This reduced growth is due to the effect of finite gradients for short-wavelength modes and due to the stabilizing influence of rigid boundaries for long-wavelength modes. In the intermediate range of k values the ratio $f = n^2/ak$ reaches a maximum given by $k_m L = p$, $k_m d = 2pC$ and $f_m = 1/2p$. Note that f_m cannot be larger than 1 and frequently will be smaller. For instance, choosing the lowest mode number $m = 1$ and fairly large density ratio $\exp(\beta h) = 50$, the above parameters are given by $k_m L = 0.95$, $k_m d = 3.7$, $f_m = 0.53$. In the presence of surface modes, these internal modes will be negligible for most applications.

Example 77 (Free boundaries) Using (2.98) together with (2.97), it follows

$$(q_1 - \lambda)A + (q_2 - \lambda)B = 0, \quad (q_1 - \lambda)\exp(q_1 h)A + (q_2 - \lambda)\exp(q_2 h)B = 0. \quad (2.100)$$

The possible solutions to (2.100) are restricted by $(q_1 - \lambda)(q_2 - \lambda)[\exp(q_1 h) - \exp(q_2 h)] = 0$. From vanishing of the first two factors, the free-surface modes $\lambda = \pm 1$, $\psi = \exp(\pm \eta)$ are obtained. Vanishing of the third factor leads again to the eigenvalues (2.99) of the rigid boundary case. Now, the eigenfunctions are shifted in phase $\psi_m = \exp(-\beta \eta/2)[\beta w_m \cos(w_m \eta) - (1 - w_m^2 - \beta^2/4)\sin(w_m \eta)]$. Remark that the exponential profile is inversion invariant satisfying $\rho(\eta) = \frac{1}{\rho(-\eta)}$.

Example 78 (Transition layers) Assume homogeneous boundary regions with densities ρ_1 for $\eta < 0$ and $\rho_2 = \rho_1 \exp(\beta h)$ for $\eta > h$. In contrast to the previous cases, this model can demonstrate the transition from the surface modes in long wavelength limit to the internal modes in the short wavelength limit. Using (2.95) and (2.97), the continuity conditions (2.94) at the interfaces $\eta = 0$ and $\eta = h$ yield $(1 - q_1)A + (1 - q_2)B = 0$ and $(1 + q_1)\exp(q_1 h)A + (1 + q_2)\exp(q_2 h)B = 0$. Setting the determinant of the coefficient matrix equal to zero, the variable z is found restricted by

$$\tanh(zh) = \frac{2z}{\beta^2/4 - (1 + z^2)}. \quad (2.101)$$

For each solution (2.101), the eigenfunctions can be expressed in the form

$$\psi = \exp(-\chi \eta/2)[z \cosh(z\eta) + (1 + \beta/2)\sinh(z\eta)].$$

For short wavelengths, one has the scaling $\beta \to 0$, $z = O(\beta)$, $h = O(\beta^{-1})$. In this limit (2.101) reduces to the form $\tan(wh) = 0$. We recover the growth rates of the internal modes (2.99). Since the wavelengths are assumed short, the result becomes independent of the boundary conditions used.

The long wavelength limit requires a more careful treatment. Define the small expansion parameter $\varepsilon = 2/\beta$ and set $x = \varepsilon z$, $C = \beta h/2 = h/\varepsilon$. Using this scaling, (2.98) and (2.101) become

$$\frac{n^2}{ak} = \frac{2\varepsilon}{1 + x^2 + \varepsilon^2}, \quad (2.102a)$$

$$2\varepsilon x = (1 - x^2 - \varepsilon^2)\tanh(Cx). \quad (2.102b)$$

Expanding the solution x to (2.102a,b) in powers of ε one obtains $x = 1 - a\varepsilon$, $1 - x^2 = 2a\varepsilon + b\varepsilon^2$ and $\tanh(Cx) = \alpha - \delta C a\varepsilon$, with

$$\alpha = \tanh(C) = \frac{\exp(\beta h) - 1}{\exp(\beta h) + 1} = \frac{\rho_2 - \rho_1}{\rho_2 + \rho_1}, \quad \delta = \partial_C \tanh(C) = \cosh^{-2}(C) = 1 - \alpha^2.$$

Comparing equal powers of ε in (2.102b) yields $\alpha a = 1$ and $\alpha b = \alpha - 2/\alpha + 2\delta C/\alpha^2$. Inserting these expansion coefficients in (2.102a), it follows the growth rate expression $\frac{n^2}{ak} = \alpha - \frac{2}{\beta}(\frac{1}{\alpha} - \alpha)(C - \alpha)$. It determines the maximum growth rate of the diffuse boundary model in the long wavelength limit.

Example 79 (Single-mode model) We illustrate mixing motions in the 2d case RTI. Consider the passive motion of fluid particles in the field of a single internal mode. Although the model gives no consistent dynamical description of nonlinear evolution, it is based on exact particle trajectories and conservation of both the mass and the total energy of the system.

The passive motion of particles in a single internal RT mode can be described by the system

$$\psi = \psi_0 \cos(px) \sin(qy), \quad u_1 = \partial_y \psi, \quad u_2 = -\partial_x \psi. \tag{2.103}$$

The stream function follows from (2.99) in the limit of weak density gradients. The time dependence of the amplitude ψ_0 is arbitrary. To calculate the particle motions in this flow field, it is convenient to transform to the variables $\xi = px$, $\eta = qy + \pi/2$ and $\tau = pq \int_0^1 \psi_0 dt$. Denoting the derivative with respect to τ by a "dot", the transformed equations become

$$\psi = -\psi_0 \cos(\xi) \cos(\eta), \quad \dot{\xi} = \cos(\xi) \sin(\eta), \quad \dot{\eta} = -\sin(\xi) \cos(\eta). \tag{2.104}$$

The stream line $\psi(\xi, \eta) = \psi(\xi_0, \eta_0)$ through some reference point (ξ_0, η_0) is given by the equation

$$\cos(\xi) \cos(\eta) = \cos(\xi_0) \cos(\eta_0) = C. \tag{2.105}$$

The stream lines are periodic and form closed orbits inside the vortex cell: $-\pi/2 < \xi < \pi/2$, $-\pi/2 < \eta < \pi/2$. The motion of fluid particles along these stream lines can be described by the equations

$$\dot{\xi} = \pm \text{sign}(\eta) \sqrt{\cos^2 \xi + C}, \quad \dot{\eta} = \mp \text{sign}(\xi) \sqrt{\cos^2 \eta + C},$$

where the upper sign corresponds to clockwise and the lower to counterclockwise rotation. According to (2.104), these signs alternate for subsequent cells both in the directions ξ and η. Since ξ is related to η by the stream line equation (2.105), it is sufficient to solve the equation for η. The solution can be expressed in the form $\sin \eta = \sqrt{\mu} \cdot \text{sn}(\tau + \tau_0, \mu)$ and $\sin \eta_0 = \sqrt{\mu} \cdot \text{sn}(\tau_0, \mu)$, where $\mu = 1 - C^2$, and sn denotes the *elliptic Jacobi functions*. In the present model, mixing is basically a result of the different periods of revolution for particles on different stream lines. The variable τ is given by

$$\tau = 4 \int_0^{\pi/2} \frac{dx}{\sqrt{1 - \mu \sin^2 x}}.$$

It approaches 2π for the innermost ($\mu = 0$) and ∞ for the outermost stream line ($\mu = 1$) of the cell.

The time dependence of the flow amplitude has not yet been specified. In principle, it can be chosen in such a way that global energy conservation results. The increased kinetic energy is then limited by the available potential energy of the fluid layer. We illustrate this procedure for the initial small amplitude regime. Setting $\psi_0 = \exp(\int n \, dt)$, the variations of the kinetic and potential energies are given by

$$\frac{dT}{dt} = \int dx \int \left(\frac{1}{2} u^2 \partial_t \rho + n\rho u^2 \right) dy,$$

$$\frac{dU}{dt} = \int dx \int ay \, \partial_t \rho \, dy = \int dx \int ay[-\partial_y(\rho u^2)] \, dy = \int dx \int a\rho u^2 \, dy,$$

where $n^2 = a \frac{\partial_y \rho_0}{\rho_0} \frac{p^2}{p^2 + q^2}$ is the linear growth rate expression. In linear theory the following relations hold: $\int \rho u_2 \, dx = -\frac{1}{n} \partial_y \rho_0 \int (u_2)^2 \, dx$ and $\int u^2 \partial_t \rho \, dx = O(|u|^3)$. The requirement of energy conservation

$$\frac{d}{dt}(T + U) = n\rho_0 \int dx \int [(u_1)^2 + (u_2)^2] \, dy - \frac{a}{n} \partial_y \rho_0 \int dx \int (u_2)^2 \, dy = 0,$$

yields, together with (2.103). It agrees with the weak gradient limit of the result (2.99) of the normal mode analysis.

2.2.6 Viscous fluids

Viscosity presents another source for vorticity generation in incompressible fluids. In contrast to the rotational modes in inviscid fluids, viscosity acts on already homogeneous fluids and it is responsible for the dissipation of kinetic energy on small spatial scales. This property is particularly important in boundary layer flows, in the evolution of homogeneous fluid turbulence, and in the present context, as a damping mechanism for fluid instabilities.

The grow rates $n = (\alpha a k)^{1/2}$, $n = k|u|$ (see Section 2.2.2) for the surface instabilities in inviscid fluids are increasing indefinitely with the wavenumber. This lack of a finite maximum growth rate in the inviscid fluid description can be overcome by the inclusion of viscosity. Concerning RTI, the first treatment of viscosity effects has been given in [23]. Viscous stresses can often be related linearly to the spatial derivatives of the velocity field. Within this framework, the stress tensor of an incompressible viscous fluid can be assumed of the form [30]

$$\sigma_{ij} = -p\delta_{ij} + \frac{1}{2}\mu(\partial_{x_j}u_i + \partial_{x_i}u_j), \tag{2.106a}$$

where δ_{ij} denotes the unit tensor and μ the coefficient of viscosity. The corresponding fluid equations for incompressible motion of a nonuniform viscous fluid are given by

$$\text{div } \mathbf{u} = 0, \quad \frac{D}{Dt}\rho = 0, \quad \frac{D}{Dt}\mathbf{u} = \frac{1}{\rho}\text{div } \mathbf{u} - \nabla U.$$

They replace the system (2.88a) in the presence of viscosity. In the special case of a homogeneous fluid layer with the kinematic viscosity $\nu = \mu/\rho$, the equation of motion simplifies to the well-known NSE

$$\frac{D}{Dt}\mathbf{u} = -\nabla\left(\frac{p}{\rho} + U\right) + \nu\nabla^2\mathbf{u}. \tag{2.106b}$$

Remark on its basic role for the theory of homogeneous fluid turbulence [2, pp. 47–53].

As an immediate consequence of viscous stresses, tangential flow discontinuities do not exist. Instead, one has to deal with viscous boundary layers where the flow velocity changes suddenly but in a continuous manner. Normally accelerated viscous shear layers can be represented in the form

$$\nabla U = (0, a, 0), \quad \mathbf{u} = (u_0(y), 0, 0), \quad \rho = \rho_0(y), \quad p = p_0(y) + cx. \tag{2.107a}$$

While the density profile $\rho_0(y)$ is arbitrary, the velocity and pressure profiles are given by

$$u_0(y) = -\int \frac{cy + d}{\mu}dy, \quad p_0(y) = -a\int \rho_0(y)dy. \tag{2.107b}$$

These are strictly parallel steady-state flows of the system (2.106a-c) with arbitrary integration constants c and d. In the absence of flow ($u_0 = 0$), one recovers the hydrostatic equilibrium state of the inviscid theory as presented in Section 2.2.5 (a). In the presence of flow ($u_0 \neq 0$), viscosity imposes a constraint on the velocity profile according to (2.107b). In the past, particular attention has been devoted to KH-type instabilities of unaccelerated shear flows (see Section 2.3).

(a) **Perturbations.** If each steady flow variable X_0 is subjected to a two-dimensional perturbation $\delta X = \delta X(y)\exp(-i\omega t + ikx)$ in the (x, y)-plane, (2.106a-c) yield the linear perturbation system

$$\delta(\partial_y u_y) + ik\delta(u_x) = 0, \tag{2.108a}$$

$$-i\tilde{\omega}\delta p + \rho_0'\delta(u_y) = 0, \tag{2.108b}$$

$$\rho_0(-i\tilde{\omega}\delta(u_x) + u_0'\delta(u_y)) = ik(-\delta p + 2ik\mu\delta(u_x)) + \partial_y\{\mu(\delta(\partial_y u_x) + ik\delta(u_y))\}, \tag{2.108c}$$

$$\rho_0(-i\tilde{\omega})\delta(u_y) = -a\delta p - \delta p' + ik\mu(ik\delta(u_x) + \delta(\partial_y u_x)) + 2\partial_y(\mu\delta(\partial_y u_y)), \tag{2.108d}$$

where $\tilde{\omega} = \omega - ku_0$ and the prime denotes differentiation with respect to the y coordinate.

In deriving (2.108a-d), it was assumed that viscosity perturbations are negligibly small. If Δ denotes the shear layer width and q a characteristic perturbation wavenumber, this approximation requires

$$\delta\mu/\mu \ll \delta u'/u_0 \sim q\Delta\delta u/u_0. \qquad (2.109)$$

It applies so-called *Boussinesq fluids*, having negligible variations in their thermodynamic properties, and also to pore hydrostatic equilibrium with $u_0 = 0$. In general, (2.109) appears rather restrictive. If it is not satisfied strong viscous heating $\sim \delta\mu(u_0')^2$ and corresponding mass diffusion will ensue. These effects could probably only be reconciled with each other in a more advanced compressible theory and they are therefore neglected for the present purposes. The perturbations of homogeneous steady-state flows with constant densities and velocities can be obtained most early from the vorticity equation,

$$[-i\tilde{\omega} + \nu(k^2 - \partial_y^2)]\nabla \times \delta\mathbf{u} = 0. \qquad (2.110)$$

It follows immediately from the rot of the linearized (2.106b) by noting that $\delta p = 0$ for $\tilde{\omega} \neq 0$ according to (2.108b). For unstable modes with complex frequencies, the condition $\tilde{\omega} \neq 0$ can be assumed without restriction. By (2.110), vorticity perturbations are generated by viscosity. In addition, irrotational potential flow perturbations can exist as in the absence of viscosity. The corresponding wavenumbers along the y-axis are given by

$$\pm\sqrt{k^2 - i\tilde{\omega}/\nu} \quad \text{for} \quad \nabla \times \delta\mathbf{u} \neq 0; \quad \pm k \quad \text{for} \quad \nabla \times \delta\mathbf{u} = 0. \qquad (2.111)$$

Consider two superposed homogeneous fluid layers 1 and 2 in the half-spaces $y < 0$ and $y > \Delta$, respectively, separated by an inhomogeneous shear layer of width Δ. In the upper and lower half-spaces, the perturbations that converge at infinity ($|y| \to \infty$) can be written in the form

$$\begin{aligned}
\delta u_x &= A\exp(-ky) + B\exp(-q_2 y), \quad \delta u_y = iA\exp(-ky) + i(k/q_2)B\exp(-q_2 y), \\
\delta p/p_2 &= (\tilde{\omega}_2/k)A\exp(-ky), \quad y > \Delta;
\end{aligned} \qquad (2.112a)$$

and, respectively,

$$\begin{aligned}
\delta u_x &= C\exp(ky) + D\exp(q_1 y), \quad \delta u_y = -iC\exp(ky) - i(k/q_1)D\exp(-q_1 y), \\
\delta p/p_1 &= (\tilde{\omega}_1/k)C\exp(ky), \quad y < 0.
\end{aligned} \qquad (2.112b)$$

In (2.112a,b) the wavenumber of the vorticity perturbation is defined to be $q = (k^2 - i\tilde{\omega}/\nu)^{1/2}$ and the branch of the square root is chosen such that $\text{Re}(q) > 0$ to ensure convergence at infinity.

(b) Jump conditions. In the inviscid instability theory, the unstable interface between different fluid layers can be simply idealized as a co-moving contact surface where the continuity conditions (2.112a,b) are know to be satisfied. The jump conditions

$$[\partial_y \delta\phi/\tilde{\omega}] = 0, \quad [\tilde{\omega}\rho\delta\phi] - \rho a\partial_y \delta\phi/\tilde{\omega} = 0 \qquad (2.113)$$

are then found (Section 2.2.2) to hold for inviscid surface modes in the linear approximation. Unfortunately, a viscous boundary layer cannot be treated in an analogous way as a constant surface between two viscous fluids. The corresponding continuity conditions would be $[\mathbf{u}] = 0$, $[\boldsymbol{\sigma}\cdot\mathbf{n}] = 0$, where the brackets denote jumps across the contact surface and $\mathbf{b}$ is the surface normal unit vector of the surface element. These conditions cannot be satisfied in general by the basic flow (2.107a,b) if the jumps are evaluated between the boundaries $y = 0$ and $y = \Delta$ of the shear layer. Similar arguments apply to perturbations. Vortex motion will lead to non-uniformities in the shear layer width and viscous stresses will contribute to the momentum balance between both fluids. It should therefore be clearly recognized that the viscous boundary layer plays an active role in the evolution of perturbations in neighboring fluids.

The matching relations between the outer solutions (2.112a,b) have to be derived from the viscous flow equations. This can be accomplished in the long wavelength approximation where the layer thickness, $\Delta \to 0$, the steady flow gradients ρ_0' and u_0' diverge. These terms have to be balanced by the highest derivatives δp, $\delta p'$, $\delta u_x''$ and $\delta u_y''$ of each flow variable in the perturbation system (2.112a,b). The variation of the lower derivatives δp, $\delta u_x'$, $\delta u_y'$ will remain bounded and that of the functions δu_1, δu_2 will even be continuous across the layer. Using these properties and integrating (2.108a-d) across an infinitely thin layer $0 < y < \Delta$, the following jump conditions are obtained

$$[\delta u_x] = 0, \tag{2.114a}$$

$$[\delta u_y] = 0, \tag{2.114b}$$

$$[\mu(\delta u_x' + ik\delta u_y)] = J_1\delta u_y, \tag{2.114c}$$

$$[2\mu\delta u_y' - \delta p] = J_2\delta u_y, \tag{2.114d}$$

where $J_1 = \lim\limits_{\Delta \to 0} \int_0^\Delta \rho_0 u_0' \, dy$, $J_2 = \lim\limits_{\Delta \to 0} a \int_0^\Delta \frac{\rho'}{i\tilde{\omega}} \, dy$. The first two conditions require the continuity of the perturbed flow at the unperturbed interface. The displacement $\zeta = \delta u_2/(-i\tilde{\omega})$ of the shear layer boundaries is therefore no longer continuous as in (2.113). Instead one has the relationship $u_{01}\zeta_1 = u_{02}\zeta_2$. For symmetric layers with $u_{01} = -u_{02}$ it leads to opposite displacements of equal magnitudes $\zeta_1 = \zeta_2$. The last two conditions include the source terms $J_{1,2}$ for surface instabilities. In particular, the viscous RTI is obtained from a density discontinuity with $u_0 = 0$, $J_1 = 0$, and $J_2 = a[\rho]/i\omega$, and the viscous KHI follows from a flow discontinuity with $\rho_0 = \text{const}$, and $J_2 = 0$.

(c) Viscous surface instabilities. The normal mode frequencies and growth rates of RTI in viscous fluid can be obtained from the linear set of equations (2.112a,b) and (2.114a-d). In explicit form these equations become

$$A + B = C + D, \quad iA + i(k/q_2)B = -iC - i(k/q_1)D,$$
$$\mu_2(-kA - q_2B) - \mu_1(kC + q_1D) - \{(\mu_2 - \mu_1)k + iJ_1\}\{A + (k/q_2)B\} = 0, \tag{2.115}$$
$$2k(\mu_2 - \mu_1)(A + B) + J_2(A + (k/q_2)B) - (i\tilde{\omega}_2\rho_2/k)A' + (i\tilde{\omega}_1\rho_1/k)C = 0.$$

Solving the first equation for D and the second for B, one can rewrite (2.115) in the form

$$D = A + B - C, \quad B = d_1A + d_2C, \quad (a_{12} + d_1b_{12})A + (c_{12} + d_2b_{12})C = 0, \tag{2.116}$$

where the coefficients are defined by the expressions

$$a_1 = -2k\mu_2 - (q_1 - k)\mu_1 - iJ_1, \quad a_2 = 2k(\mu_2 - \mu_1) - i\tilde{\omega}_2\rho_2/k + J_2,$$
$$b_1 = -\mu_2q_2 - \mu_1q_1 - (k^2/q_2)(\mu_2 - \mu_1) - i(k/q_2)J_1,$$
$$b_2 = 2(\mu_2 - \mu_1)k + (k/q_2)J_2, \quad c_1 = \mu_1(q_1 - k), \quad c_2 = i\tilde{\omega}_1\rho_1/k,$$
$$d_1 = -(q_2/k)(q_1 + k)/(q_1 + q_2), \quad d_2 = -(q_2/k)(q_1 - k)/(q_1 + q_2).$$

The fast member in (2.116) represents two separate equations, one for the first and one for the second subscript. The solubility condition for these equations assumes the form

$$(a_1 + b_1d_1)c_2 - (a_2 + b_2d_1)c_1 + (a_1b_2 - a_2b_1)d_2 = 0. \tag{2.117}$$

It determines the possible complex eigenfrequencies $\omega = \omega(k)$ of normal mode perturbations.

Since the general expression (2.117) is still complex, we'll restrict attention to velocity and density discontinuities.

(d) Viscous RTI. We will illustrate the RTI of viscous fluids for the simplest case of an accelerated free surface. Assuming $u_1 = \rho_1 = \mu_1 = 0$ for the lower fluid and omitting the subscript 2 for the upper fluid, the general dispersion relation (2.117) reduces to the simpler form

$$a_1 b_1 - a_2 b_2 = 0, \tag{2.118}$$

where the coefficients are given by

$$a_1 = -2k\mu, \quad a_2 = 2k\mu - i\rho\tilde{\omega}/k + J_2, \quad J_2 = -ia\rho/\tilde{\omega},$$
$$b_1 = -\mu(q + k^2/q), \quad b_2 = 2k\mu + J_2 k/q.$$

From (2.118) it follows $J_2\mu(q^2 - k^2) - 4qk^2\mu^2 + \mu(q^2 + k^2)(2k\mu - i\rho\tilde{\omega}/k) = 0$. Eliminating q with the help of (2.111) yields

$$4\sqrt{1 - i\tilde{\omega}/\nu k^2} = (2 - i\tilde{\omega}/\nu k^2)^2 - a/\nu^2 k^3. \tag{2.119}$$

To determine the instability growth rate (2.119) has to be solved for $\tilde{\omega}$ together with the constraint $\mathrm{Re}(q) > 0$ for bounded solutions. Without proof, it is noted that only a single unstable branch exists which has a real growth rate $n = -i\tilde{\omega}$. For a further discussion of this solution, it is convenient to define dimensionless variables by setting

$$\sigma = (\nu/a^2)^{1/3} n, \quad \tilde{k} = (\nu^2/a)^{1/3} k, \quad Q = \tilde{k}^{-3}, \quad s = \sigma/\tilde{k}^2, \quad y = \sqrt{1 + s}. \tag{2.120}$$

In terms of σ and $\tilde{k}$, Equation (2.119) becomes

$$\Lambda(\sigma, \tilde{k}) \equiv 4\sqrt{1 + \sigma/\tilde{k}^2} - (2 + \sigma/\tilde{k}^2)^2 + \tilde{k}^{-3} = 0. \tag{2.121}$$

In the limits of small and large wavenumbers, one can readily obtain the asymptotic expressions

$$\tilde{k} \le 0.1: \quad \sigma = \sqrt{\tilde{k}} - \tilde{k}^2, \quad n = \sqrt{ak} - 2\nu k^2;$$
$$\tilde{k} \ge 2.5: \quad \sigma = 1/2\tilde{k}, \quad n = a/2\nu k, \tag{2.122}$$

for the instability growth rate, respectively. They are in excellent agreement with the exact numerical solution within the given wavenumber limits [7]. A small wavenumbers', viscosity becomes negligible and the inviscid growth rate $n = \sqrt{\alpha a k}$ is approached. At large wavenumbers, the RTI is strongly damped by viscosity. In the range of intermediate wavenumbers, the growth rate curve assumes a maximum, where $\frac{d\sigma}{d\tilde{k}} = -\partial_{\tilde{k}}\Lambda/\partial_\sigma\Lambda = 0$. By (2.120)–(2.121), the condition for the maximum growth rate has the form

$$\Lambda = Q - (y - 1)(y^3 + y^2 + 3y - 1) = 0,$$
$$\tilde{k}^{-1}\partial_x\Lambda = -3Q + 4(y - 1)(y^3 + y^2 + y - 1/y) = 0. \tag{2.123}$$

These equations specify the maximum in terms of the variables Q and y. Eliminating Q from (2.123) yields a single equation for y,

$$y^4 + y^3 - 5y^2 + 3y = 4 \tag{2.124}$$

with one real solution for $y > 1$. Numerically, one finds $y_m = 1.706$, $Q_m = 0.491$, $\tilde{k}_m = 0.491$ and $\sigma_m = 0.460$, where (2.124), (2.123) and (2.120) have been used and m refers to the growth rate maximum. The maximum viscous growth rate is about $\frac{2}{3}$ of the corresponding inviscid growth rate $\sigma_m = 0.657\tilde{k}_m^{1/2}$.

The free-surface model of the viscous RTI can be extended similarly to two fluids with arbitrary densities ρ_1 and ρ_2, and viscosities ν_1 and ν_2. Using appropriately scaled variables, and defining dimensionless variables $\sigma = (\bar{\nu}/\alpha^2 a^2)^{1/3} n$ and $\tilde{k} = (\bar{\nu}/\alpha a)^{1/3} k$ in terms of the *Atwood number* (ratio) $\text{At} = |\rho_2 - \rho_1|/(\rho_2 + \rho_1)$ and the average viscosity $\bar{\nu} = (\rho_1 \nu_1 + \rho_2 \nu_2)/(\rho_2 + \rho_1)$, the variation of the growth rate maximum (2.16) with density and viscosity is limited as $0.478 \leq \tilde{k}_m \leq 0.526$, $0.448 \leq \sigma_m \leq 0.5$.

(e) Spatial amplification. If a fluid layer moves to its surface, TRI can become convectively unstable. This situation may arise for a steady-state flow around free-surface bubbles. Perturbations that become excited at the topside of the bubbles will be carried with the flow along the bubble surface, eventually reaching large amplitudes in the region of the failing spikes. A moving plane layer with a free surface is considered and both inviscid and viscous amplification factors are derived. The present criteria for convective instabilities arose from studies in plasma physics and have been reviewed in a comprehensive form in [5].

First, illustrate the role of convective amplification in the inviscid instability theory. If the fluid layer moves along the x-axis with a constant velocity u, the dispersion relation of the free-surface RTI becomes

$$\omega = ku + i\sqrt{ak}. \tag{2.125}$$

The frequency is simply Doppler shifted with respect to the familiar part, being valid in the test frame of the fluid. Because of the horizontal motion of the fluid layer, a local surface perturbation will only be present for a transient time at each fixed position in the laboratory frame. Mathematically, this behavior is expressed by the fact that the dispersion relation (2.125) has no branch-points with $\partial \omega/\partial k = 0$ in the physical complex half-plane $\text{Re}\,k > 0$. Such instabilities are commonly called *convective*. It is customary to describe convective instabilities wavenumbers and real frequencies. Solving therefore (2.17) for k yields

$$k = (a/u^2)[\omega u/a - 1/2 \mp i(\omega u/a - 1/4)^{1/2}]. \tag{2.126}$$

Complex wavenumbers are indicative of instability if the additional constraints

$$\text{Re}\,k > 0, \quad \text{Im}\,k > 0 \text{ for } \text{Im}\,\omega \to +\infty; \quad \text{Im}\,k < 0 \text{ for } \text{Im}\,\omega = 0 \tag{2.127}$$

are satisfied. The first condition defines the physical half-plane where the potential perturbations (2.17) are spatially damped inside the fluid. Comparing with (2.126), one obtains a lower cut-off frequency

$$\omega > a/2u \tag{2.128}$$

for convective amplification. The wavenumbers (2.126) of above the cut-off (2.128) are complex and the upper sign corresponds to a spatially growing mode in the x direction.

A convectively unstable system can be distinguished from a merely impenetrable medium by the second condition in (2.127). If $\text{Im}\,k$ changes its sign as $\text{Im}\,\omega$ varies from $+\infty$ to 0, spatially bounded initial perturbations can evolve into unbounded spatially growing modes. This criterion holds only for the upper sign in (2.126), showing that amplification occurs only in the flow direction.

The spatial growth increment of the inviscid theory, $\text{Im}(k\,x)$, increases indefinitely with the frequency. Actually, high-frequency oscillations will become strongly damped by viscosity. This can be shown in the limit of high viscosities where (2.122) yields the asymptotic result

$$\omega = ku + ia/2\nu k, \quad k = \omega/u - ia/2\nu\omega. \tag{2.129}$$

The complex wavenumber describes a convectively unstable mode, but its imaginary part decreases with increasing oscillation frequencies. For a rough estimate of the frequency of maximum amplification, one may look for the crossing point of the asymptotic growth increments. Equating the amplification factors of (2.126) and (2.129), one finds $\omega_m \sim ua^{1/3}/\nu^{2/3}$

and $k_m \sim (1/u)a^{2/3}/\nu^{1/3}$. A more complete description of the viscosity effects on the convectively unstable mode could be based on the general dispersion relation (2.119). It would require an analysis of the complex k roots of this equation when $\tilde{\omega} = \omega - ku$ is substituted and ω is taken real. At last we'll notice that the polynomial form is of degree seven and overview can only be gained by numerical means.

2.3 Kelvin–Helmholtz instability

The Kelvin–Helmholtz instability (KHI) often accompanies flows where shear is present, such as the case near the inversion of a cloud topped planetary boundary layer. Just as heat and moisture can have large jumps across the inversion, as discussed in *cloud top entrainment instability* (CTEI), which is a contour plot of liquid water, i.e., cloud water, from a two-dimensional simulation of CTEI. At every point in space, the air can only hold so much water in vapor form, up to 100% relative humidity, and the rest is in liquid form. In the contour plots the black region is above the cloud top and has no liquid water, i.e., it is undersaturated. In the plots, there is a light gray ramp with black being no cloud water and dark (black) being the maximum amount. Everywhere you see the same color, it has the same liquid water amount; we can have jumps in the mean horizontal velocity across the inversion layer as well, i.e., shear. When two parallel streams of different velocities are adjacent to each other, the flow can be unstable to perturbations, even infinitesimal ones. The flow forms convective patterns called *Kelvin cat's eyes*, shown below (Fig. 2.15(a); photo taken from *An Album of Fluid Motion*).

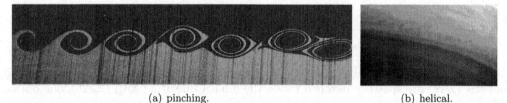

(a) pinching.   (b) helical.

FIGURE 2.15: (a) Kelvin cat's eyes. (b) Observation of KHI on the planet Saturn formed at the interaction of two bands on the planet's atmosphere.

In the atmosphere, the inversion layer makes a smooth transition of both velocity and density (stably stratified with lighter on top of heavier fluid) across the thickness of the inversion. The important measure of how much shear there is, and whether to expect instability, is the Richardson number, a ratio of buoyancy forces (stable stratification) versus convective forces (shear instability). If this number is less than a critical value, the KHI will ensue, spreading the thickness of the inversion and increasing the Richardson number in the process. The cat's eyes form individually, then pair up, as the inversion thickness continues to grow. The process will fizzles out if the Richardson number crosses the critical value. The spreading of the inversion layer can cause a cloud layer to evaporate, as the above inversion conditions being spread into the cloudy surface layer are much drier. The theory can be used to predict the onset of instability and transition to turbulent flow in fluids of different densities ρ_1, ρ_2 moving of various speeds u_1, u_2. Helmholtz studied the dynamics of two fluids of different densities when a small disturbance such as a wave is introduced at the boundary connecting fluids.

The Kelvin–Helmholtz billows occur fairly frequently in the atmosphere, with wavelengths up to a few kilometers (as, for example, see Fig. 2.15(b)). As they induce vertical air motion, they sometimes cause billow clouds. They can be seen on the top of straight-form clouds, with the cloud droplets acting as a tracer (like my animation above). They may also be observed when they occur in the clear air, by using high power radar or by making measurements from aircraft-borne instruments. Such observations have linked Kelvin–Helmholtz billows to *clear-air turbulence* making their prediction important for aviation safety. They also act to mix across considerably vertical depths (up to around 1 km) and dissipate energy. Estimates of the energy dissipated by billows at high levels in the troposphere indicate that they could have a significant impact on the larger-scale dynamics.

The KHI results from velocity shears between two media. These media need not even be of different densities. Any time there is a non-zero curvature, the flow of one fluid around another will lead to a slight centrifugal force which in turn leads to a change in pressure thereby amplifying the ripple. The most familiar example of this is wind blowing over calm water. Tiny dimples in the smooth surface will quickly be amplified to small waves and finally to frothing *white caps*. As with RTI, any sort of surface tension will hinder KHI. If there is some restoring force F_g, the instability will arise if

$$\Delta u^2 \geq \frac{2(\rho_1 + \rho_2)}{\rho_1 \rho_2}[F_g(\rho_2 - \rho_1)]^{1/2}.$$

For water waves, surface tension stabilizes the interface until the wind reaches $u > 650$ cm/s (12.5 knots). For astrophysical applications, magnetic field tension supplies a stabilizing force. The stability requirement is $\Delta u \leq 2u_A = 2B/(\mu_0\rho)^{1/2}$, where B is the magnetic field's induction and μ_0 is the magnetic conductivity.

In the evolved stages of KHI, cyclonic cat's eye structures are formed. When combined with RTI, KHI tends to form *mushroom caps* (Fig. 2.16) on the end of Rayleigh–Taylor fingers [26]. Since the growth rate of these fingers is proportional to their cross-section, Kelvin–Helmholtz tends to slow down finger growth.

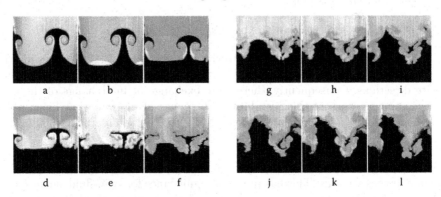

FIGURE 2.16: The experimental images of mushroom caps.

A disturbance, generating from the KH instabilities in free jets, propagates along the recompression shock to the top boundary layer and out into the top shear layer of the jet. Along this path, it pulses the flow at this oscillation frequency recent large eddy simulation (LES) computations on the 60 million-node grid have reinforced the proposed mechanism generating the 34 kHz signals [36, p. 319]. Because shear layers are unstable, they are susceptible to a range of inputs. The vortex train found in the bottom shear layer results from the merging of the two streams of different velocities and the splitter plate (i.e., vortex shedding). This is an exciting theory because it suggests that a single point (the splitter plate) modulates a vastly large spatial domain everywhere.

The KHI occurs in many natural and laboratory plasmas, in complex configurations with magnetic shear and density and temperature jumps across the velocity. The addition of compressibility effects leads to notable complexities since the dispersion relation yields a polynomial of tenth degree that depends on six independent parameters. Then, the results of the incompressible theory are usually employed while interpreting experimental data since in this approximation the stability condition has an analytical form, and it is assumed that compressibility tends to stabilize the Kelvin–Helmholtz modes, an assumption based on Newcomb's well-known theorems. Nevertheless these theorems cannot be applied in our case because perturbations with negative energy are possible in configurations with mass flow, and in consequence new instabilities may develop if new modes appear due to compressibility. In this chapter we show that the effect of compressibility destabilizes the plasma for low relative velocities in cases that are stable in the incompressible limit.

2.3.1   Instability of annular incompressible jet

First, consider the incompressible case, which will be sufficient to describe the appearance of typical motions. It is known from analysis of small perturbations of the gas dynamical equations that the linear modes split into entropy-vortex and acoustic modes. This indicates that there is only weak interaction between entropy-vortex motions and acoustic waves if the former are slow compared with the speed of sound. Note that we are not interested in turbulence in the bulk of uniform fluid. Therefore, $\nabla \times \mathbf{u} = 0$ everywhere outside density jumps (assuming the density is uniform outside such jumps). In this case, the motion outside density jump is potential ($\mathbf{u} = \nabla \phi$ and $\nabla^2 \phi = 0$). Since the equations in ϕ are elliptic, the perturbation potential falls off exponentially with distance from the density jump surface (if there are no boundaries nearby). Such discontinuous are usually discussed in connection with KHI. The appearance of this discontinuity, or jump, for KHI is shown in Fig. 2.18. In the periodic case, segments with positive and negative vorticity alternate. The simplest model of such distribution is [40]

$$\phi = -u_0 k^{-1} \text{sign}(y) \cos(kx) \exp(-k|y|), \quad u = u_0 \cos(kx) \exp(-k|y|),$$

where u is the velocity component normal to the boundary, Fig. 2.17(a). This distribution of the velocity v results in alternating streams that cross the initial position of the boundary in opposite directions. Consequently, there is an exchange or interchange of the materials in contact in the presence of this instability. The types of motions that bring about the exchange are fundamentally different in the case of KHI: the boundary twists in a spiral, so that cat's eyes are formed, Fig. 2.15(a).

Example 80 (Typical KHI) Consider the breakup process of an annular liquid jet (or sheet) in moving gas streams. This has practical significance for twin-fluid, such as air-assist or air-blast, atomization. By exposing a relatively slow-moving liquid sheet to high-velocity air (or gas) streams, the growth of disturbances at the gas-liquid interfaces is enhanced to cause the liquid jet to breakup more rapidly into ligaments and then droplets, forming a spray. The rate of disturbance amplification is mainly because of the complex influence of surface tension force and aerodynamic interaction between the liquid and gas phases. This type of liquid atomization process is used in many practical power generation and propulsion systems. The understanding of the mechanism of atomization is still far from complete. The primary objective of this example is to investigate the fundamental mechanism of the interfacial instability of an annular viscous liquid jet subject to both inner and outer gas streams of unequal velocities. The instability of a stationary viscous annular liquid sheet with unequal gas velocities for the inner and outer gas streams was formulated in [12]. Two

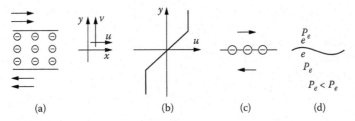

(a)   (b)   (c)   (d)

FIGURE 2.17: Mechanism for growth of perturbations amplitude in the case of KHI. (a) Layer of thickness δ in which the tangential velocity u changes sign. (b) The dependence $u(y)$. The vertical component of the velocity is $v = 0$. In the layer δ, there is a gradual variation of u. (c) We come to the tangential discontinuity in the limit $\delta \to 0$. (d) The development of KHI is due to the fact that at a curve boundary, the velocity decreases and the pressure p_e grows in the region of expansion of the flow e, in accordance with the Bernoulli integral, while the opposing pressure p_c decreases due to the increase of the velocity in the region c where the flow is compressed.

dispersion relations corresponding to each interface were derived. Only cases for inviscid liquids were examined in their study.

Small disturbances are assumed in the derivation of the dispersion relation and the equation for the amplitude ratio of initial disturbances at two gas-liquid interfaces by considering unequal gas velocities on both sides of an annular viscous liquid jet. As shown in Fig. 2.18, the annular jet is assumed semi-infinitely long and with inner radius a, outer radius b of annular sheet, constant density p_{lj}, and viscosity μ_{lj}. The jet moves at a uniform axial velocity U_{lj}^* with its two sides exposed to inviscid gas streams of velocity U_{ig} inside of the jet and U_{og} outside of the jet. The gas density is denoted by ρ_g. Both, gas and liquid velocities are presumed to be small compared with the velocity of sound. Therefore, both gas and liquid can be considered incompressible. Furthermore, gravity is neglected. The pressure field is constant within the liquid and gas phases, respectively, and it has a jump across the two liquid-gas interfaces, because of the effect of σ. When disturbances set in, resulting in the interface deformation and deviation away from the equilibrium configuration of the annular liquid jet, the flow field is disturbed with the perturbed flow velocity $\mathbf{u}$ and pressure p superimposed on the base flow velocity $\bar{\mathbf{U}}$ and pressure $\bar{P}$. Then, in a cylindrical coordinate system (z, r, θ) the perturbed flow fields become $U_\alpha = \bar{U} + u_\alpha$, $\mathbf{u} = (u_\alpha, v_\alpha, 0)$ and $P_\alpha = \bar{P}_\alpha + p_\alpha$, where $\bar{U}_* = \bar{U}/\bar{U}_{lj}$ is the dimensionless velocity of base flow (hereinafter the indices "$*$" will be omitted), $\mathbf{u}$ is the change of velocity vector induced under perturbation, $\mathbf{u}, \bar{\mathbf{U}}$ and $p, \bar{P}$ are the velocities and pressures, accordingly, of perturbed and base flow; the subscript $\alpha = lj, ig$ and $\alpha = oj$ corresponds to the liquid jet, the inner, and outer gas streams, respectively. The base flow quantities are given by

$$\bar{\mathbf{U}} = (\bar{U}_{lj}, 0, 0), \quad \bar{\mathbf{U}}_{ig} = (\bar{U}_{ig}, 0, 0), \quad \bar{\mathbf{U}}_{og} = (\bar{U}_{og}, 0, 0),$$

$$\bar{p}_{lj} = \bar{\mathbf{P}}_{ig} - \sigma/a = F_{og} + \sigma/b.$$

The equations governing the motion of the perturbed flow are the conservation of mass and momentum, which become, upon linearization,

$$\nabla u_\alpha = 0, \tag{2.130a}$$

$$p_\alpha(\partial_t + U_n\partial_z)u_\alpha = -\nabla p_\alpha + \mu_\alpha \nabla^2 u_\alpha. \tag{2.130b}$$

The boundary conditions that the solutions of the previous governing equations have to satisfy are the kinematic and dynamic conditions at the inner and outer interfaces, which are represented by $r = a + \eta_{ig}(z, t)$ and $r = b + \eta_{og}(z, t)$, respectively. In linear stability theory, these conditions need not be applied at the disturbed liquid-gas interfaces. Rather, they can be linearized in the same manner as done previously for the governing equations. Then the linearized boundary conditions can be applied at the unperturbed interfaces. Because the interfaces are material surfaces, the kinematic boundary conditions at $r = a$ and $r = b$ are, respectively,

$$v_{lj} = \partial_t \eta_{ig} + U_{lj}\partial_z \eta_{ig} \quad \text{and} \quad v_{lj} = \partial_t \eta_{og} + U_{lj}\partial_z \eta_{og}, \quad (2.131a)$$

$$v_{ig} = \partial_t \eta_{ig} + U_{ig}\partial_z \eta_{ig} \quad \text{and} \quad v_{og} = \partial_t \eta_{og} + U_{og}\partial_z \eta_{og}. \quad (2.131b)$$

The dynamic condition implies that
  – the shear stress must vanish at the interfaces because of the inviscid assumption for the gas phase,
  – normal stresses across the interfaces must be continuous with allowance for the effect of σ.
These conditions can be expressed as follows:

$$\mu_{lj}(\partial_r u_{lj} + \partial_z v_{lj})|_{r=a} = 0, \quad \mu_{lj}(\partial_r u_{lj} + \partial_z v_{lj})||_{r=b} = 0, \quad (2.132a)$$

$$p_{ig} - p_{lj} + 2\mu_{lj}\partial_r v_{lj} = -\sigma(a^{-2}\eta_{ig} + \partial_z^2 \eta_{ig}) \quad \text{at } r = a, \quad (2.132b)$$

$$p_{og} - p_{lj} + 2\mu_{lj}\partial_r v_{lj} = -\sigma(b^{-2}\eta_{og} + \partial_z^2 \eta_{og}) \quad \text{at } r = b, \quad (2.132c)$$

Further, in the ambient gas phase, the effects of disturbances should physically remain bounded, whether it is at the centerline or far away from the liquid jet. That is,

$$u_{ig}, p_{ig} - \quad \text{is bounded as} \quad r \to 0,$$

$$u_{og}, p_{og} - \quad \text{is bounded as} \quad r \to \infty.$$

The solutions to the governing equations are sought in terms of the normal mode in the

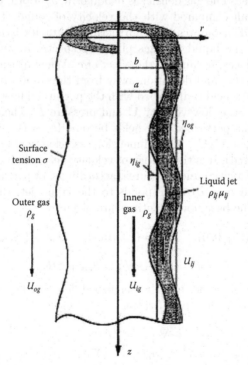

FIGURE 2.18: Schematic of annular liquid jet.

form [31]:

$$(u_\alpha, p_\alpha, \eta_{ig}, \eta_{og}) = [\tilde{u}_\alpha(r), \tilde{p}_\alpha(r), \varepsilon_{ig}\varepsilon_{og}] \exp(\Omega t + i\kappa z), \quad (2.133)$$

where $\alpha = lj$, ig and og, ε_{ig} and E, are the amplitudes of initial disturbances at the inner and outer interfaces, and are regarded to be much smaller than the inner and outer radius as well as the thickness of the annular liquid jet. The axial wave number of the disturbance κ is related to the disturbance wavelength Λ under the relation $\kappa = 2\pi/\lambda$. The complex frequency Ω has a real and imaginary part. The real part Ω_{Re} represents the rate of growth or decay of the disturbance with time, the imaginary part Ω_{Im} is equal to 2π times of the disturbance frequency, and $-\Omega_{Im}/\kappa$ represents the wave propagation velocity of the disturbance.

Substituting (2.133) into the governing equations (2.130a,b) yields the required general solution with a set of unknown integration constants, which can be determined by the boundary conditions (2.131a)–(2.132c) and the two limiting conditions (2.133) and (2.134a). The solutions are

$$
\begin{aligned}
u_{ig} &= -\varepsilon_{ig}\frac{I_0(\kappa r)}{I_1(\kappa a)}(\kappa U_{ig} - \Omega)e^{i(\kappa z - \Omega t)}, \\
v_{ig} &= i\varepsilon_{ig}\frac{I_1(\kappa r)}{I_1(\kappa a)}(\kappa U_{ig} - \Omega)e^{i(\kappa z - \Omega t)}, \\
p_{ig} &= \varepsilon_{ig}\frac{\rho_g I_0(\kappa r)}{k I_1(\kappa a)}(\kappa U_{ig} - \Omega)^2 e^{i(\kappa z - \Omega t)}
\end{aligned}
\tag{2.134a}
$$

for inner gas flow, and

$$
\begin{aligned}
u_{og} &= \varepsilon_{og}\frac{K_0(\kappa r)}{K_1(\kappa b)}(\kappa U_{og} - \Omega)e^{i(\kappa z - \Omega t)}, \\
v_{og} &= i\varepsilon_{og}\frac{K_1(\kappa r)}{K_1(\kappa b)}(\kappa U_{og} - \Omega)e^{i(\kappa z - \Omega t)}, \\
p_{og} &= \varepsilon_{og}\frac{\rho_g K_0(\kappa r)}{\kappa K_1(\kappa b)}(\kappa U_{og} - \Omega)^2 e^{i(\kappa z - \Omega t)}
\end{aligned}
\tag{2.134b}
$$

for the outer gas flow, and

$$
\begin{aligned}
u_{lj} &= \frac{1}{r}\partial_r\phi = [A_1 S I_0(Sr) - A_2 S K_0(Sr) + A_3\kappa I_0(\kappa r) - A_4\kappa K_0(\kappa r)]e^{i(\kappa z - \Omega t)}, \\
v_{lj} &= -\frac{1}{r}\partial_z\phi = -i\kappa[A_1 I_1(Sr) + A_2 K_1(Sr) + A_3 I_1(\kappa r) + A_4 K_1(\kappa r)]e^{i(\kappa z - \Omega t)}, \\
p_{lj} &= -\rho_{lj}(\kappa U_{lj} - \Omega)[A_3 I_0(\kappa r) - A_4 K_0(\kappa r)]e^{i(\kappa z - \Omega t)}
\end{aligned}
\tag{2.134c}
$$

for the liquid flow. Here

$$
A_1 = \frac{2i\kappa\nu_{lj}[\varepsilon_{og}K_1(Sa) - \varepsilon_{ig}K_1(Sb)]}{[I_1(Sa)K_1(Sb) - I_1(Sb)K_1(Sa)]}, \quad A_2 = \frac{2i\kappa\nu_{lj}[\varepsilon_{ig}I_1(Sb) - \varepsilon_{og}I_1(Sa)]}{[I_1(Sa)K_1(Sb) - I_1(Sb)K_1(Sa)]},
$$

$$
A_3 = \frac{i\nu_{lj}(S^2 + \kappa^2)[\varepsilon_{ig}K_1(\kappa b) - \varepsilon_{og}K_1(\kappa a)]}{\kappa[I_1(\kappa a)K_1(\kappa b) - I_1(\kappa b)K_1(\kappa a)]}, \quad A_4 = \frac{i\nu_{lj}(S^2 + \kappa^2)[\varepsilon_{og}I_1(\kappa a) - \varepsilon_{ig}I_1(\kappa b)]}{\kappa[I_1(\kappa a)K_1(\kappa b) - I_1(\kappa b)K_1(\kappa a)]}.
$$

Substituting the above solutions of each flow (2.134a-c) into the dynamic boundary conditions at the two interfaces, (2.132b) and (2.132c) lead to two equations with two unknowns ε_{ig} and ε_{og}. Since these two equations are homogeneous, the coefficients of the equations must satisfy a condition for nontrivial solutions to exist. That is, the determinant of the coefficient matrix must vanish, which gives the following dispersion relation between Ω and κ:

$$
\begin{aligned}
&\left\{(S^2 + \kappa^2)^2\Delta_1\Delta_4 + 4\kappa^3 S\Delta_3\Delta_6 + \frac{1}{\nu_l^2}\left[\frac{\kappa\sigma}{\rho_{lj}}\left(\frac{1}{a^2 - \kappa^2}\right) - \frac{\rho_g}{\rho_{lj}}(\kappa U_{ig} - \Omega)^2\frac{I_0(\kappa a)}{I_1(\kappa a)}\right]\right. \\
&\left. - \frac{2i\kappa}{a\nu_{lj}}(\kappa U_l - \Omega)\right\} \times \left\{(S^2 + \kappa^2)^2\Delta_1\Delta_4 + 4\kappa^3 S\Delta_3\Delta_6 + \frac{2i\kappa}{b\nu_{lj}}(\kappa U_{lj} - \Omega)\right. \\
&\left. + \frac{1}{\nu_{lj}^2}\left[\frac{\kappa\sigma}{\rho_{lj}}\left(\frac{1}{b^2} - \kappa^2\right) - \frac{\rho_g}{\rho_{lj}}(\kappa U_{ig} - \Omega)^2\frac{K_0(\kappa a)}{K_1(\kappa a)}\right]\right\} - \frac{1}{ab}\left[4\kappa^3\Delta_3 - \frac{(S^2 + \kappa^2)^2}{\kappa}\Delta_4\right] = 0,
\end{aligned}
\tag{2.135}
$$

where $\rho_g = \rho_{ig}$ is the gas density, I_0, I_1 are the modified Bessel functions of the first kind, and K_0, K_1 are the modified Bessel functions of the second kind of orders 0 and 1, respectively. The values S, $\Delta_1, \Delta_2, \Delta_3, \Delta_4, \Delta_5, \Delta_6$ are defined as $S = \sqrt{\kappa^2 + i(\kappa U_{lj} - \Omega)/\nu_{lj}}$ and

$$\Delta_1 = I_0(\kappa a)K_1(\kappa b) + K_0(\kappa a)I_1(\kappa b), \qquad \Delta_2 = I_0(\kappa b)K_1(\kappa a) + K_0(\kappa b)I_1(\kappa a),$$
$$\Delta_3 = [I_0(\kappa a)K_1(\kappa b) - I_1(\kappa b)K_1(\kappa a)]^{-1}, \quad \Delta_4 = [I_1(\kappa a)K_1(\kappa b) - I_1(\kappa b)K_1(\kappa a)]^{-1},$$
$$\Delta_5 = [I_0(Sb)K_1(Sa) + K_0(Sb)I_1(Sa)], \qquad \Delta_6 = -[I_0(Sa)K_1(Sb) + K_0(Sa)I_1(Sb)].$$

As a part of solution, the amplitude ratio of the initial disturbances at the inner and outer interfaces becomes

$$\frac{\varepsilon_{ig}}{\varepsilon_{og}} = \frac{G_3}{aG_4}, \quad G_3 = -4\kappa^3\Delta_3 + \frac{(S^2+\kappa^2)^2}{\kappa}\Delta_4,$$
$$G_4 = (S^2 + \kappa^2)^2\Delta_1\Delta_4 + \frac{\kappa\sigma}{\rho_{lj}\nu_{lj}^2}\left[\frac{1}{(a)^2} - \kappa^2\right] + \frac{\rho_g}{\rho_{lj}\nu_{lj}^2}\frac{I_0(\kappa a)}{I_1(\kappa a)}(\kappa U_{ig} - \Omega)^2 \qquad (2.136a)$$
$$-\frac{2i\kappa}{a\nu_{il}}(\kappa U_{lj} - \Omega) - 4\kappa^3 S\Delta_3\Delta_6,$$

or

$$\frac{\varepsilon_{og}}{\varepsilon_{ig}} = \frac{G_5}{bG_6}, \quad G_6 = -4\kappa^3\Delta_3 + \frac{(S^2+\kappa^2)^2}{\kappa}\Delta_4,$$
$$G_4 = (S^2 + \kappa^2)^2\Delta_2\Delta_4 + \frac{\kappa\sigma}{\rho_{lj}\nu_{lj}^2}\left[\frac{1}{(b)^2} - \kappa^2\right] + \frac{\rho_g}{\rho_{lj}\nu_{lj}^2}\frac{K_0(\kappa b)}{K_1(\kappa b)}(\kappa U_{og} - \Omega)^2 \qquad (2.136b)$$
$$-\frac{2i\kappa}{a\nu_{il}}(\kappa U_{lj} - \Omega) - 4\kappa^3 S\Delta_3\Delta_5.$$

The dimensional variables in (2.135) are transformed to non-dimensional form as follows: length scale is denoted by $r_h = (b - a)/2$; velocity scale U_h is denoted by U_{li}; time is normalized by r_h/U_h; *Reynolds number* $Re = U_h r_h/\nu_{lj}$; *Weber number* $\mathrm{We} = \rho_{lj}U_h^2 r_h/\sigma$; wave number $k = \kappa r_h$; wave frequency $\omega = \Omega r_h/U_h$; $l = S r_h = \sqrt{k^2 + iRe(kU_h - \omega)}$; the geometric sizes become normalized by r_h, i.e., $a = a/r_h$, $b = b/r_h$; density ratio $\rho = \rho_g/\rho_{lj}$. Thus, the relation (2.135) takes the dimensionless form:

$$\left\{ (l^2 + k^2)^2\Delta_1\Delta_4 + 4k^3 l\Delta_3\Delta_6 + \left[\frac{k}{\mathrm{We}}\left(\frac{1}{a^2} - k^2\right) - \rho(\omega + ikU_{ig})^2\frac{I_0(ka)}{I_1(ka)}\right]Re^2 \right.$$
$$-i\frac{2k}{a}Re(kU_h - \omega)\right\}\left\{(l^2+k^2)^2\Delta_2\Delta_4 - 4k^3 l\Delta_3\Delta_5 + \left[\frac{K}{\mathrm{We}}\left(\frac{1}{b^2} - k^2\right) - \rho(\omega + ikU_{og})^2\frac{K_0(kb)}{K_1(kb)}\right]$$
$$i\frac{2k}{b}Re(kU_h - \omega)\} - \frac{1}{ab}\left[\frac{1}{k}(l^2+k^2)^2\Delta_4 - 4k^3\Delta_3\right]^2 = 0, \qquad (2.137)$$

where Δ_1, Δ_2, Δ_3, Δ_4, Δ_5, Δ_6 are defined as

$$\Delta_1 = I_0(ka)K_1(kb) + K_0(ka)I_1(kb), \qquad \Delta_2 = I_0(kb)K_1(ka) + K_0(kb)I_1(ka),$$
$$\Delta_3 = [I_0(la)K_1(lb) - I_1(lb)K_1(la)]^{-1}, \quad \Delta_4 = [I_1(ka)K_1(kb) - I_1(kb)K_1(ka)]^{-1},$$
$$\Delta_5 = [I_0(lb)K_1(la) + K_0(lb)I_1(la)], \qquad \Delta_5 = -[I_0(la)K_1(lb) + K_0(la)I_1(lb)].$$

As a part of the solution, the amplitude ratios (2.136a,b) of initial disturbances at the inner and outer interfaces become

$$\frac{\varepsilon_{ig}}{\varepsilon_{og}} = -\frac{(l^2 + k^2)^2\Delta_4/k - 4k^3\Delta_3}{a[(l^2 + k^2)^2\Delta_1\Delta_4 + 4k^3l\Delta_3\Delta_6 - i(2k/a)Re(kU_h - \omega) + B_{ig}]}, \qquad (2.138a)$$

or

$$\frac{\varepsilon_{og}}{\varepsilon_{ig}} = -\frac{(l^2 + k^2)^2\Delta_4/k - 4k^3\Delta_3}{b[(l^2 + k^2)^2\Delta_2\Delta_4 + 4k^3l\Delta_3\Delta_5 - i(2k/b)Re(kU_h - \omega) + B_{og}]}, \qquad (2.138b)$$

where

$$B_{ig} = Re^2\left[\frac{k}{\mathrm{We}}\left(\frac{1}{a^2} - k^2\right) - \rho(kU_{ig} - \omega)^2\frac{I_0(ka)}{I_1(ka)}\right],$$
$$B_{og} = Re^2\left[\frac{k}{\mathrm{We}}\left(\frac{1}{b^2} - k^2\right) - \rho(kU_{og} - \omega)^2\frac{K_0(kb)}{K_1(kb)}\right].$$

The relation for inviscid annular jet can be obtained from (2.137) under $Re \to \infty$

$$\left\{(kU_h^2 - \omega)^2 \Delta_1 - \frac{1}{\Delta_4}\left[\rho(kU_{ig} - \omega)^2 \frac{I_0(ka)}{I_1(ka)} + \frac{k}{\text{We}}\left(\frac{1}{a^2} - k^2\right)\right]\right\}$$
$$\times \left\{(kU_h^2 - \omega)^2 \Delta_2 - \frac{1}{\Delta_4}\left[\rho(kU_{og} - \omega)^2 \frac{K_0(kb)}{K_1(kb)} + \frac{k}{\text{We}}\left(\frac{1}{b^2} - k^2\right)\right]\right\} - \frac{(kU_h - \omega^4)}{k^2 ab} = 0,$$

$$\frac{\varepsilon_{ig}}{\varepsilon_{og}} = \frac{(kU_h - \omega)^2}{ka\left\{(kU_h - \omega)^2 \Delta_1 - \left[\rho(kU_h - \omega)^2 \frac{I_0(ka)}{I_1(ka)} + k(a^{-2} - k^2)/\text{We}\right]\Delta_4^{-1}\right\}}, \qquad (2.139a)$$

or

$$\frac{\varepsilon_{og}}{\varepsilon_{ig}} = \frac{(kU_h - \omega)^2}{kb\left\{(kU_h - \omega)^2 \Delta_2 - \left[\rho(kU_h - \omega)^2 \frac{K_0(kb)}{K_1(kb)} + k(b^{-2} - k^2)/\text{We}\right]\Delta_4^{-1}\right\}}. \qquad (2.139b)$$

The temporal linear instability analysis [42] is carried out to examine the effects of various inner U_{ig} and/or outer gas velocities U_{og} as well as fluid properties on the jet instability. By using Muller's method [34] the dispersion relation (2.137) is solved for temporal modes of unstable wave growth rate ω_{Re} the positive real part of complex wave frequency representing the degree of jet instability. Similar to our earlier study of liquid jets in stationary gas medium two families of unstable solutions, parasinuous and paravaricose, to be obtained. The results shown here are all for a fixed nozzle geometry of $a = 40.12$ (and, hence, $b = 42.12$), which corresponds to an annular nozzle of $2a = 9.525 \times 10^{-3}$m and $2b = 10.0 \times 10^{-3}$m used in the experiment [42] of a water-air system.

By Fig. 2.19(a) that the growth rate for parasinuous disturbances decreases along with the dominant wave number as the inner gas velocity increases from 0 to 1,or the velocity difference $\Delta U_{ig} = |U_{ig} - 1|$ between the gas and liquid phases as the inner interface decreases from 1 to 0. Further increase in U_{ig} up to 2 causes or to increase, since the velocity difference U_{ig} increases from 0 to 1. The unstable wave number range does not change significantly for $0.5 \leq U_{ig} \leq 2$. When U_{ig} is larger than 2, ω_{Re} increases considerably while the dominant wave number increases relatively slowly. Clearly, the velocity difference across the interface enhances the jet instability and extends the unstable wave regime to higher wave numbers. The growth rate at a higher gas velocity (e.g., $U_{ig} = 1.7$) is larger than that at a lower one (e.g., $U_{ig} = 0.3$), even though the velocity difference across each interface is the same. Hence, it indicates that not only the velocity difference across each interface, but also the absolute gas and liquid velocity themselves affect the breakup processes of annular liquid jets. Calculations of disturbance energy implies that a shift in the disturbance frequency and phase angle of gas pressure fluctuations to be responsible for this behavior.

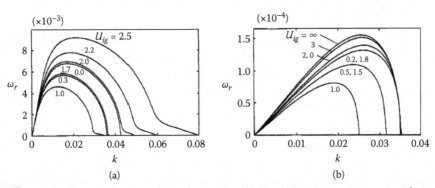

FIGURE 2.19: Wave growth rate for different velocities of inner gas stream with stationary outer gas medium; $a = 40.12$, $Re = 4112$, $We = 19.25$, $p = 0.00129$, $a = 40.12$, and $U_{0g} = 0$; (a) parasinuous, and (b) paravaricose mode.

As for the paravaricose mode; see Fig. 2.19(b), ω_{Re} also increases with ΔU_{ig} but approaches a certain limit. This indicates that when the gas velocity is sufficiently large, a further increase in the gas velocity has little effect on the growth rate of the paravaricose mode. This is because the paravaricose mode is always related primarily with the smaller of the two velocity differences across the two interfaces, whereas the parasinuous mode is always associated with the larger velocity difference across the interfaces. By Fig. 2.19 that the unstable wave growth rate for the parasinuous mode is approximately two orders of magnitude larger than that for the corresponding paravaricose mode, indicating that, in practice, the parasinuous unstable waves will outgrow paravaricose ones and predominate the jet breakup processes.

When the inner gas stream is stationary ($U_{ig} = 0$), but U_{og} is varied, ω_{Re} exhibits the same trends as discussed previously. The comparison of these two sets of parasinuous growth rates are shown in Fig. 2.20 at low and high gas velocities, respectively. At low gas velocities ($U_{ig} \leq 2$ or $U_{og} \leq 2$), by Fig. 2.20(a) that ω_{Re}, shown by the dashed curves for $U_{ig} = 0$, is larger than that given by the solid curves for $U_{og} = 0$, for a comparable gas velocity on the other side of the liquid jet. For either $U_{ig} \geq 2$ or $U_{og} \geq 2$ and the gas on the other side of the liquid jet stationary, Fig. 2.20(b) shows that ω_{Re} for $U_{og} = 0$ is larger than that for $U_{ig} = 0$. By Fig. 2.20, the larger wave growth rate always occurs when the velocity difference across the inner surface $\Delta U_{ig} = |U_{ig} - 1|$ is larger than that across the outer interface $\Delta U_{og} = |U_{0g} - 1|$. This implies that to promote jet instability, a gas stream applied to the outer interface is more effective than when applied to the inner surface when the gas velocity is less than twice the liquid velocity, whereas the gas stream with a velocity higher than twice the liquid velocity should be exposed to the inner interface.

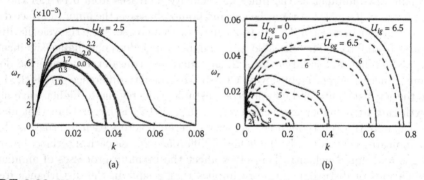

FIGURE 2.20: Wave growth rate of the parasinuous mode for different velocities of gas stream on either inner or outer side of the jet; $a = 40.12$, $Re = 4112$, $We = 19.25$, $p = 0.00129$, $a=40.12$: (a) low and (b) high gas velocity.

The initial amplitude ratio shown in Fig. 2.21, calculated from (2.138a) or (2.138b) under the same conditions as those for the wave growth rate in Fig. 2.20, shows that at larger wave numbers, the initial amplitude at the inner interface ε_{ig} is larger than that at the outer interface ε_{og} for $U_{ig} > U_{og}$ whereas ε_{og} is larger than ε_{ig} for $U_{og} > U_{ig}$. This implies that the larger initial amplitude occurs at the interface that is subject to a gas stream of higher velocity compared with the other side of the jet for short waves. Note that this result is obtained for the case where only one interface is exposed to a moving gas stream. At smaller wave numbers, the initial amplitude ratio ($\varepsilon_{ig}/\varepsilon_{og}$) or ($\varepsilon_{og}/\varepsilon_{ig}$) approaches a fixed value $\frac{\varepsilon_{ig}}{E} + \frac{b}{a}/[1 - p(1 - b^2/a^2)]$, which is the limit of (2.137) as the wave number $k \to 0$: When the gas-to-liquid density ratio is much smaller than 1 and b/a is of order of 1, approaches $\varepsilon_{ig}/\varepsilon_{og}$ approaches b/a, i.e., 1.05 for the present case.

A linear instability analysis has been carried out for an annular viscous liquid jet exposed to both inner and outer gas streams of unequal velocities. Muller's method is used to solve the dispersion relation for the temporal mode of unstable solutions. The results show that not only the velocity difference across each interface, but also the absolute velocity of each fluid phase is important for jet instability, although the effect of absolute velocity is secondary compared to that of relative velocity. Co-flowing gas at high velocities is found to significantly improve atomization performance, while a low-velocity co-flowing gas stream has little effect on jet instability when compared with the case where a gas stream of the same velocity is applied only on one side of the liquid jet. A high-velocity gas inside of the annular liquid jet promotes the jet breakup processes more than gas outside of the jet with equivalent velocity. For equal liquid and gas velocities, surface tension, liquid and gas density exhibit effects completely opposite to those with velocity discontinuity across interfaces. The viscous damping effect on jet instability always exists for the cases with and without the velocity differences at high Weber numbers.

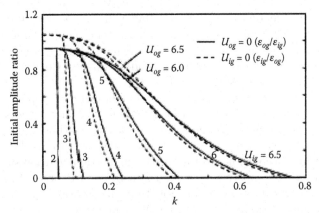

FIGURE 2.21: Amplitude ratio of initial disturbances at the inner and outer interfaces for the parasinuous mode, the conditions as in Fig. 2.20(b).

Atomization performance is enhanced by an increase in the liquid inertia, density ratio, and relatively large gas velocity, and hindered by the effect of surface tension and liquid viscosity.

2.3.2 Rotating jets

Stability of cylindrical jets may be presented as a contribution to the study of the atomization phenomena of liquid jets. These phenomena occur in many applications of hydrodynamics and often they are either desirable or, on contrary, undesirable. A fine atomization of the fuel, for instance, is asked for in some types of burner, whereas the atomization of the condensate flowing in a thin layer within a tube through which transported vapor can be very inconvenient. This chapter analysis contains only some of the conditions for which atomization may be expected.

We will resort to the application of the hydrodynamic stability theory in which infinitesimally small perturbations are superimposed upon a stationary flow field. Assume that the stable perturbations do not give rise to atomization of the jet, i.e., that atomization occurs only if unstable perturbations are present. This does not mean that every unstable perturbation will cause atomization. The reason for this lies in the assumption that the perturbations are infinitesimally small, since, after a finite interval of time, unstable perturbations will have grown to finite quantities and the linear theory of hydrodynamic stability does not apply to this state. According to this theory, therefore, a flow field disturbed by unstable perturbations may adopt another stationary state without the occurrence of atomization. In present article we will neglect this difficulty, which can be solved only by studying the effect of finite perturbations. In that case, nonlinear differential equations must be solved, whereas the investigation of infinitesimally small perturbations leads to linear differential equations only. Therefore we will deal only with the latter.

The hydrodynamic stability theory will be applied to cylindrical liquid jets under the action of surface tension and inertial effects of the jet. In a single case we will take into account the liquid viscosity or the inertial effects of the inviscid ambient medium, e.g. the air surrounding the jet. Thus, in the most cases we neglect the liquid viscosity and the inertial effects of the air, while it is supposed that the ambient air assumed be inviscid in all cases.

In general case, when the liquid viscosity and inertial effects of the surrounding air are neglected, the undisturbed jet will be a rotating one. This means a generalization of some of well-known results by Lord Rayleigh and those of Weber [44]. These authors described non-rotating cylindrical jets with free surface. Rayleigh[4] treated the dynamics of revolving fluids, but in this case the fluid moved between two fixed coaxial cylinders. The method we will use here has the advantage over those of Rayleigh and Weber of being straightforward, because it is an application of the standard method of the theory of hydrodynamic stability, which has been explained, for instance, by Lin [33].

As regards the nature of the perturbations that have to be applied to the given flow field, in general they should not have rotational symmetry. Many authors, including Weber and Lin, do apply rotationally symmetric perturbations. Although Lin wrote that that existing work deals only with the case of rotational symmetry, it must be said that Rayleigh had already taken into account non-rotationally symmetric perturbations in inviscid case. In present work it will be apparent that for rotating jets the non-rotationally symmetric perturbations are in some cases more unstable than the rotationally symmetric ones. This stresses the importance of taking non-rotationally symmetric perturbations into account.

Three types of jet will be investigated, in which the liquid fills the space defined by $r \leq R_0$, $R_1 \leq r \leq R_2$, $r \geq R_0$. These jets are called, accordingly, entire jet, annular jet and annular infinitely thick jet. In the case of entire jet, we will trace the influence of the liquid viscosity (in the case of non-rotating jet) or the inertial effects of the surrounding air (for inviscid jet).

Inviscid entire jet. For the case of inviscid entire jet and no external forces the continuity and the motion equations may be written as follows

$$\text{div}\,\mathbf{V} = 0, \quad \frac{\partial \mathbf{V}}{\partial t} + \frac{1}{2}\nabla(\mathbf{V}\cdot\mathbf{V}) - \mathbf{V}\times(\nabla\times\mathbf{V}) = -\frac{1}{\rho}\nabla P. \tag{2.140}$$

Let U, V and W be the components of the velocity with reference to axes r, φ and z of cylindrical coordinate system, with z-axis along the axis of the jet. Choose the unperturbed flow field be potential

$$U = 0, \quad V = \Omega/r, \quad W = W_0 \quad (0 \leq r \leq R_0), \tag{2.141}$$

where Ω and W_0 are constants. The flow given by (2.141) satisfies (2.140), and it follows

$$P = -\frac{1}{2}\rho(\Omega/r)^2 + \text{const.} \tag{2.142}$$

We have to fulfill the boundary condition at the free surface $r = R_0$. Because R_0^{-1} is the mean curvature of the free surface, the boundary condition is $P_{|\,r=R_0} = \sigma/R_0 + P_e$, where σ/R_0 is the difference between the internal and external pressure caused by the surface tension, P_e is the external pressure, and σ is the dynamic surface tension; subscript $(_e)$

---

[4]Rayleigh Lord, On the dynamics of revolving fluids. *Proc. Roy. Soc. London*, V. A93, 1917, 148–154.

indicates external (with respect to liquid) medium. Thus, assuming P_e be a constant, we can find the constant in (2.142) as follows

$$\text{const} = \frac{1}{2}\rho(\Omega/R_0)^2 + \sigma/R_0 + P_e. \tag{2.143}$$

Proposition 10 *In the case of inviscid entire jet with the boundary condition (2.143), the dispersion equation has a form*

$$\left(\omega - \frac{n\Omega}{R_0^2} - kW_0\right)^2 = \left[\frac{\sigma}{\rho}\left(\frac{n^2-1}{R_0^2} + k^2\right) - \frac{\Omega^2}{R_0^3}\right]k\frac{I_n'(kR_0)}{I_n(kR_0)}, \tag{2.144}$$

where ' means differentiation.

Proof. Let us apply theory of perturbations

$$U = \bar{U} + u, \quad V = \bar{V} + v, \quad W = \bar{W} + w, \quad P = \bar{P} + p. \tag{2.145}$$

Here $\bar{U}$, $\bar{V}$, $\bar{W}$ and $\bar{P}$ are the unperturbed solutions to (2.141) and (2.142). Substituting (2.145) in (2.140) and using (2.141) we obtain

$$\frac{1}{r}\partial_r(ru) + \frac{1}{r}\partial_\varphi v + \partial_z w = 0, \tag{2.146a}$$

$$\partial_t u + u\partial_r u + \frac{1}{r}(\frac{\Omega}{r} + v)\partial_\phi u + (W_0 + w)\partial_z u - \frac{1}{r}(\frac{\Omega}{r} + v)^2 = -\frac{1}{\rho}\partial_r(P + p),$$
$$\partial_t v + u\partial_r v + \frac{1}{r}(\frac{\Omega}{r} + v)\partial_\phi v + (W_0 + w)\partial_z v - \frac{uv}{r} = -\frac{1}{\rho}\partial_\phi(P + p), \tag{2.146b}$$
$$\partial_t w + u\partial_r w + \frac{1}{r}(\frac{\Omega}{r} + v)\partial_\phi w + (W_0 + w)\partial_z w = -\frac{1}{\rho}\partial_z(P + p).$$

Because of (2.142), the terms $-\Omega^2/r^3$ in the left-hand member of the first equation of (2.146b) and $-\rho^{-1}\partial_r P$ in the right-hand member cancel out; and so do $-\rho^{-1}\partial_\varphi P$ and $-\rho^{-1}\partial_z P$ in the second and third equations of (2.146b). Thus the remaining equations (2.146a) and (2.146b) contain only terms of the first or second order in u, v, w and p. Assuming the perturbations u, v, w, p to be infinitesimally small, we may neglect, in linear approach, the second order terms. We get

$$\partial_t u + \frac{\Omega}{r^2}\partial_\phi u + W_0\partial_z u - \frac{2\Omega}{r^2}v = -\frac{1}{\rho}\partial_r p,$$
$$\partial_t v + \frac{\Omega}{r^2}\partial_\phi v + W_0\partial_z v = -\frac{1}{\rho}\partial_\phi p, \tag{2.147}$$
$$\partial_t w + \frac{\Omega}{r^2}\partial_\phi w + W_0\partial_z w = -\frac{1}{\rho}\partial_z p.$$

To find the solutions of (2.146a) and (2.147) we note that t, φ and z do not occur explicitly in the equations, whereas r occurs explicitly in the left-hand member of (2.147) and in (2.146a). In view of this we will find the solutions of the plane wave representation

$$u = u^*(r)e^{-i\omega t + in\varphi + ikz}, \quad v = v^*(r)e^{-i\omega t + in\varphi + ikz},$$
$$w = v^*(r)e^{-i\omega t + in\varphi + ikz}, \quad p = p^*(r)e^{-i\omega t + in\varphi + ikz}, \tag{2.148}$$

where $u^*(r)$, $v^*(r)$, $w^*(r)$ and $p^*(r)$ are functions of r only, and n, called mode, is integer. We require the wave number k to be a real constant. This means that we have analyzed the perturbations into their Fourier components so far as φ and z concerned. In the general case the frequency ω will be a complex constant. Obviously the perturbations of view (2.148) are unstable if imaginary part of frequency $\omega_{Im} > 0$ and stable if $\omega_{Im} \leq 0$; so we have to look for those modes (n, k) which give rise to a complex value ω with a positive imaginary part. From the solutions (2.148) to (2.146a) and (2.147), the boundary condition for the pressure and the kinematic boundary condition, it follows that there is a function F with $F(\omega, n, k) = 0$

called as characteristic or dispersion equation. Substituting (2.148) into (2.146a), (2.147), and dividing resulting equations by $e^{-i\omega t + in\phi + ikz}$, we obtain

$$\frac{1}{r}\frac{d}{dr}(ru^*(r)) + i\frac{n}{r}\left[v^*(r) + \frac{kr}{n}w^*(r)\right] = 0, \tag{2.149a}$$

$$i(-\omega + \frac{n\Omega}{r^2} + kW_0)u^*(r) - \frac{2\Omega}{r^2}v^*(r) = -\frac{1}{\rho}\partial_r p^*,$$
$$i(-\omega + \frac{n\Omega}{r^2} + kW_0)v^*(r) = -\frac{1}{\rho}i\frac{n}{r}p^*, \tag{2.149b}$$
$$i(-\omega + \frac{n\Omega}{r^2} + kW_0)v^*(r) = -\frac{1}{\rho}ikp^*.$$

Eliminating $p^*(r)$ from the first two and the last two equations of (2.149b), we get

$$\frac{d}{dr}(rv^*(r)) = inu^*(r), \quad v^*(r) = \frac{n}{kr}w^*(r) \tag{2.150}$$

accordingly. Then, substituting the expressions above into (2.149a), we obtain

$$\frac{d^2w^*}{dr^2} + \frac{1}{r}\frac{dw^*}{dr} - \left(k^2 + \frac{n^2}{r^2}\right)w^*(r) = 0. \tag{2.151}$$

Equation (2.151) is the modified Bessel equation, and its solutions are modified Bessel functions of the first and third kinds of order n, so we find, using (2.150) and one of equations (2.149b):

$$u^* = -iB\frac{dI_n}{d(kr)} - iC\frac{dK_n}{d(kr)}, \quad v^* = \frac{n}{kr}BI_n(kr) + \frac{n}{kr}CK_n(kr),$$
$$w^* = BI_n(kr) + CK_n(kr), \quad p^* = \frac{\rho}{k}(\omega - \frac{n\Omega}{r^2} - kW_0)[BI_n(kr) + CK_n(kr)], \tag{2.152}$$

where B and C are constants which have to be chosen to satisfy the boundary conditions.

If $r = R_0 + \delta$ defines the disturbed boundary for entire jet, the boundary condition for the pressure reads

$$P = \sigma K + P_e, \quad r = R_0 + \delta, \tag{2.153}$$

where K is mean curvature of the disturbed boundary; it has to be positive for unperturbed surface. Following the theory of hydrodynamic stability consistently we need to know K only up to terms of the first order in δ because δ is assumed to be infinitesimally small. Using $K = \text{div}(\nabla G/|\nabla G|)$, where $G(r, \varphi, z) = 0$ denotes the surface for which K has to be determined, it follows,

$$K = \frac{1}{R_0} - \frac{\delta}{R_0^2} - \frac{1}{R_0^2}\partial_\varphi^2\delta - \partial_z^2\delta \tag{2.154}$$

up to terms of the first order in δ. Expanding (2.153) up to terms of the first order in the perturbations, it follows from (2.142), (2.143) and (2.154) that terms of zero order cancel. Introduce for δ an expression of the same type as for $u^*(r)$, $v^*(r)$, $w^*(r)$ and $p^*(r)$,

$$\delta = \delta^*(r)e^{-i\omega t + in\varphi + ikz}. \tag{2.155}$$

1. First, consider the simplest case when influence of surrounded air not taken into account. Consider the case where the perturbations in the external air pressure are zero, i.e., $p_e = 0$. Substituting (2.155) in the appropriate equations and recalling P_e to be constant, we find for entire jet, using (2.142), (2.145), (2.148), (2.153)–(2.155) and neglecting the air influence,

$$\rho\frac{\Omega^2}{R_0^3}\delta^* + p^*(R_0) = \sigma\left(\frac{n^2 - 1}{R_0^2} + k^2\right)\delta^*,$$

where the left-hand member follows from expanding (2.142) up to terms of the first order.

Displacement of the boundary by an amount $d\delta/dt$ is related to the velocity u as $u =$

$\partial_t \delta + (\bar{U}, \bar{V}, \bar{W}) \nabla \delta$ up to terms of the first order, as may be derived with the aid of differential geometry. In present case it follows from (2.141) and (2.142) that

$$u = \partial_t \delta + \frac{\Omega}{R_0^2} \, \partial_\varphi \delta + W_0 \, \partial_z \delta. \tag{2.156}$$

The required characteristic equation follows from representation (2.155)–(2.156), where the value C, in (2.152), has been put equal to zero due to singularity K_n at $r \to 0$. The result is the relation (2.144) that completes the proof. □

We may take n and $k \geq 0$. Then $I_n'/I_n > 0$, so that we have unstable perturbations if only

$$n^2 - 1 + k^2 R_0^2 - \rho \Omega^2 / \sigma R_0 < 0 \tag{2.157}$$

for only in that case has one of the values of ω, which are roots of (2.157) with imaginary part > 0.

Introducing the dimensionless parameters

$$q = kR_0, \quad s = \rho \Omega^2 / \sigma R_0, \quad b^2 = (\rho R_0^3 / \sigma) \omega_{Re}^2. \tag{2.158}$$

Equation (2.144) is transformed as

$$\left[b^2 - \frac{\rho R_0^3}{\sigma} \left(\omega_{Re} - n \frac{\Omega}{R_0^2} - kW_0 \right)^2 - 2i \sqrt{\frac{\rho R_0^3}{\sigma}} \left(\omega_{Re} - n \frac{\Omega}{R_0^2} - kW_0 \right) b \right] = (s + 1 - n^2 - q^2) q \frac{I_n'(q)}{I_n(q)}.$$

If the condition $s + 1 - n^2 - q^2 > 0$ is valid, it follows,

$$b^2 = (s + 1 - n^2 - q^2) \frac{q I_n'(q)}{I_n(q)}. \tag{2.159}$$

From (2.159) it can be concluded:

(a) The perturbations are only unstable if k is within a finite interval which has one end point in $k = 0$ and if n takes the value 0 and all positive integer values lower than a finite positive integral number, i.e., $n^2 < s + 1$.

(b) For non-rotating jet, when $\Omega = 0$, n must be equal zero and $0 < kR_0 = q < 1$ if instability is required. In other words: non-rotating jet is stable for all positive n perturbing modes [29, 44]. Rayleigh derived his formula from considerations of jets potential and kinetic energy.

(c) The higher values Ω and n, and the longer the intervals of k give rise to unstable modes.

(d) By (2.159), the surface tension plays an important role. The value ω_{Im} increases proportionally to the square root of the surface tension if $\Omega = 0$.

(e) The surface tension enters into the formulae together with the density of the jets liquid medium. This is obvious because these are the only physical quantities containing the dimension of mass in their dimension formulae.

2. If the influence of surrounding air may not be assumed negligible it is necessary to introduce assumptions on undisturbed airflow. As for the jet itself we choose a potential flow

$$\bar{U}_e = 0, \quad \bar{V}_e = \Omega_e / r, \quad \bar{W}_e = W_{e0} \quad (R_0 \leq r < \infty),$$

where Ω_e and W_{e0} are constant. For u_e, v_e, w_e, whose meanings are obvious, and p_e is chosen again of the type (2.147), analogous calculations give the following characteristic equation

$$\left(\omega - \frac{n\Omega}{R_0^2} - kW_0 \right)^2 - \frac{\rho_e}{\rho} \left(\omega - \frac{n\Omega_e}{R_0^2} - kW_{e0} \right)^2 \frac{I_n'(kR_0)K_n(q)}{I_n(kR_0)K_n'(q)}$$
$$= \left[\frac{\sigma}{\rho} \left(\frac{n^2 - 1}{R_0^2} + k^2 \right) - \frac{\Omega^2}{R_0^3} - \frac{\rho_e}{\rho} \frac{\Omega_e^2}{R_0^3} \right] k \frac{I_n'(q)}{I_n(q)}, \tag{2.160}$$

where in the solutions for the case of air, correspond to (2.152) B has now to be taken equal zero due to singularity I_n at infinity. If ρ_e, in (2.160), is put equal zero, we obtain (2.144) as we should.

Consequence. For a special case when $\Omega = \Omega_e$ and $W_0 = W_{e0}$, the mean velocity field is continuous at $r = R_0$, and (2.160) becomes

$$\frac{\rho R_0^3}{\sigma}\Big(\omega - \frac{n\Omega}{R_0^2} - kW_0\Big)^2 = \frac{[k^2R_0^2 + (n^2 - 1) - \frac{\rho\Omega^2}{\sigma R_0}(1 - \frac{\rho_e}{\rho})]kR_0\frac{I'_n(q)}{I_n(q)}}{1 - \frac{\rho_e}{\rho}\frac{I'_n(q)K_n(q)}{I_n(q)K'_n(q)}}. \tag{2.161}$$

Using (2.158) we can analogously transform (2.161) to dimensionless form

$$b^2 - \Big[\frac{\rho R_0^3}{\sigma}\Big(\omega_{Re} - \frac{n\Omega}{R_0^2} - kW_0\Big)^2 + 2i\sqrt{\frac{\rho R_0^3}{\sigma}}\Big(\omega_{Re} - \frac{n\Omega}{R_0^2} - kW_0\Big)b\Big]$$

$$= \frac{[s(1 - \rho_e/\rho) + 1 - n^2 - q^2]\frac{qI'_n(q)}{I_n(q)}}{1 - \frac{\rho_e}{\rho}\frac{I'_n(q)K_n(q)}{I_n(q)K'_n(q)}}.$$

Since $\rho_e \ll \rho$, we have unstable perturbations if only $s(1 - \rho_e/\rho) + 1 - n^2 - q^2 > 0$, that leads to

$$b^2 = \Big[s\Big(1 - \frac{\rho_e}{\rho}\Big) + 1 - n^2 - q^2\Big] \cdot \Big[1 - \frac{\rho_e}{\rho}\frac{I'_n(q)K_n(q)}{I_n(q)K'_n(q)}\Big]^{-1}\frac{qI'_n(q)}{I_n(q)}.$$

We arrive at the result in [9].

Viscous entire jet (non-rotating case). We single out the case where undisturbed jet is non-rotating and the influence of surrounded air may be neglected. Again as in previous section taking the case of incompressible liquid without external forces acting on it, we are based on the following continuous and conservation equations of motion (Navier–Stokes equations):

$$\text{div}\,\mathbf{U} = 0, \quad \partial_t\mathbf{U} + \frac{1}{2}\nabla(\mathbf{U}\cdot\mathbf{U}) - \mathbf{U}\times(\nabla\times\mathbf{U}) = -\frac{1}{\rho}\nabla P - \nu\nabla\times\nabla\times\mathbf{U}, \tag{2.162}$$

where $\nu = \mu/\rho$ is the coefficient of viscosity. The undisturbed motion is given by

$$\bar{U} = \bar{V} = 0, \quad \bar{W} = W_0 \quad (0 \le r \le R_0), \tag{2.163}$$

where W_0 is a constant. If we apply the method in Section 2.2.1 to the present case, as will be done below, we get ODEs of the fourth order, whereas we got ODE of the second order before. Note that the solution is uniquely determined because there are two boundary conditions,

$$\partial_z u + \partial_r w = 0, \quad r\partial_r(u/r) + \frac{1}{r}\partial_\phi u = 0. \tag{2.164}$$

Equations (2.164) are valid at the disturbed boundary $r = R_0 + \delta$ and express the fact that the shearing stresses are zero at the boundary. If we expand (2.164) up to terms of the first order, we may take $r = R_0$ in (2.164).

Proposition 11 *In the case of inviscid entire jet with the boundary condition (2.164) the dispersion equation has a view*

$$\begin{aligned}(\omega - kW_0 &+ 2i\nu k^2)^2 I_0(kR_0)I_1(lR_0) = 4(i\nu)^2 k^3 l I_1(kR_0)I_0(lR_0)\\ &+ \Big[\frac{\sigma}{\rho}\Big(-\frac{1}{R_0^2} + k^2\Big)k + 2i\nu\frac{k}{R_0}(\omega - kW_0)\Big]I_1(kR_0)I_1(lR_0)\end{aligned} \tag{2.165}$$

with l satisfied the relation $l^2 = k^2 - (i/\nu)(\omega - kW_0)$, $\mathrm{Re}(l) \ge 0$.

Proof. Applying the method from Section 2.2.2 to (2.162) and (2.163), we find for perturbation equations

$$-i(\omega - kW_0)\begin{pmatrix} u \\ v \\ w \end{pmatrix} = -\frac{1}{\rho}\nabla p - \nu \nabla \times \nabla \times \begin{pmatrix} u \\ v \\ w \end{pmatrix}, \quad \mathrm{div}\begin{pmatrix} u \\ v \\ w \end{pmatrix} = 0, \qquad (2.166)$$

where (u, v, w) is a vector field. Applying operator divergence to the upper equation (2.166), we get

$$\Delta p = 0. \qquad (2.167)$$

Substituting plane wave representation $p = p^*(r)\exp(-i\omega t + in\phi + ikz)$ we find from (2.167) an ODE for p

$$\left[\frac{d^2}{dr^2} + \frac{1}{r}\frac{d}{dr} - \left(\frac{n^2}{r^2} + k^2\right)\right]p^*(r) = 0.$$

So

$$p^*(r) = BI_n(kr) + CK_n(kr) \qquad (2.168)$$

with B and C to be constants. To find differential equation for u, v, w we put

$$\begin{pmatrix} u \\ v \\ w \end{pmatrix} = \nabla\Phi + \begin{pmatrix} u^0 \\ v^0 \\ w^0 \end{pmatrix}, \qquad (2.169)$$

$$-i(\omega - kW_0)\Phi = -p/\rho. \qquad (2.170)$$

Substituting (2.169) and (2.170) into (2.166) we obtain, using (2.167)

$$-i(\omega - kW_0)\begin{pmatrix} u^0 \\ v^0 \\ w^0 \end{pmatrix} = -\nu\nabla \times \nabla \times \begin{pmatrix} u^0 \\ v^0 \\ w^0 \end{pmatrix}, \quad \mathrm{div}\begin{pmatrix} u^0 \\ v^0 \\ w^0 \end{pmatrix} = 0. \qquad (2.171)$$

We again use plain wave representations

$$u^0 = u^{0*}(r)e^{-i\omega t + in\phi + ikz}, \quad v^0 = v^{0*}(r)e^{-i\omega t + in\phi + ikz}, \quad w^0 = w^{0*}(r)e^{-i\omega t + in\phi + ikz}, \quad (2.172)$$

and from (2.171) derive, on defining

$$l^2 = k^2 - \frac{i}{\nu}(\omega - kW_0), \quad Re(l) \geq 0 \qquad (2.173)$$

that

$$\frac{1}{r}\frac{d^2}{dr^2}[ru^{0*}(r)] + \frac{in}{r}v^{0*}(r) + ikw^{0*}(r) = 0, \qquad (2.174)$$

$$\begin{aligned} \left(\frac{d^2}{dr^2} + \frac{1}{r}\frac{d}{dr} - \frac{n^2+1}{r^2} - l^2\right)u^{0*}(r) - \frac{2in}{r^2}v^{0*}(r) &= 0, \\ \left(\frac{d^2}{dr^2} + \frac{1}{r}\frac{d}{dr} - \frac{n^2+1}{r^2} - l^2\right)v^{0*}(r) + \frac{2in}{r^2}u^{0*}(r) &= 0, \\ \left(\frac{d^2}{dr^2} + \frac{1}{r}\frac{d}{dr} - \frac{n^2}{r^2} - l^2\right)w^{0*}(r) &= 0. \end{aligned} \qquad (2.175)$$

The last equation of (2.175) gives the solution for $w^{0*}(r)$:

$$w^{0*}(r) = DI_n(lr) + EK_n(lr), \qquad (2.176)$$

with D and E to be constants. We eliminate $w^{0*}(r)$ from (2.175) by multiplying by r^2, then

$$\left\{r^2\left(\frac{d^2}{dr^2} + \frac{1}{r}\frac{d}{dr} - \frac{n^2+1}{r^2} - l^2\right) + 2n\right\}\left\{r^2\left(\frac{d^2}{dr^2} + \frac{1}{r}\frac{d}{dr} - \frac{n^2+1}{r^2} - l^2\right) - 2n\right\}u^{0*}(r) = 0. \qquad (2.177)$$

Because we may interchange the operators between figure scratches, solutions to (2.177) may be found from equations below

$$\left\{ r^2 \left(\frac{d^2}{dr^2} + \frac{1}{r}\frac{d}{dr} - \frac{n^2+1}{r^2} - l^2 \right) \pm 2n \right\} u^{0*}(r) = 0$$

or

$$r^2 \left[\frac{d^2}{dr^2} + \frac{1}{r}\frac{d}{dr} - \frac{(n\pm 1)^2}{r^2} - l^2 \right] u^{0*}(r) = 0. \tag{2.178}$$

Thus

$$u^{0*}(r) = P I_{n-1}(lr) + Q K_{n-1}(lr) + R I_{n+1}(lr) + S K_{n+1}(lr), \tag{2.179}$$

where P, Q, R, S are constants. From the first of (2.175) we find

$$v^{0*}(r) = i[P I_{n-1}(lr) + Q K_{n-1}(lr) - R I_{n+1}(lr) - S K_{n+1}(lr)]. \tag{2.180}$$

Now the solutions (2.176), (2.178), (2.180) have to fulfill (2.174). Substituting them in (2.174) and using

$$I_n(z) = \frac{n+1}{z} I_{n+1}(z) + I'_{n+1}(z) = -\frac{n-1}{z} I_{n-1}(z) + I'_{n+1}(z),$$
$$K_n(z) = -\frac{n+1}{z} K_{n+1}(z) - K'_{n+1}(z) = \frac{n-1}{z} K_{n-1}(z) - K'_{n+1}(z),$$

we find $P I_n(lr) - Q K_n(lr) + R I_n(lr) - S K_n(lr) + i\frac{k}{l}[D I_n(lr) + E K_n(lr)] = 0$. Thus, due to independently $I_n(lr)$ and $K_n(lr)$, we get

$$P + R = -i(k/l)D, \quad Q + S = i(k/l)E. \tag{2.181}$$

Now if we reduce (2.179) and (2.180) to expressions with Bessel functions of order n, we find defining $P - R = -iF$, $Q - S = iG$ and using (2.181),

$$u^{0*}(r) = -i\left(\frac{nF}{lr} I_n(lr) + \frac{k}{l} D I'_n(lr) + \frac{nG}{lr} K_n(lr) + \frac{k}{l} K'_n(lr) \right),$$
$$v^{0*}(r) = F I'_n(lr) + \frac{nD}{lr}\frac{k}{l} I_n(lr) + G K'_n(lr) + \frac{nE}{lr}\frac{k}{l} K_n(lr). \tag{2.182}$$

Finally, from (2.150)–(2.152), (2.176), (2.182), using plane wave representations,

$$u = u^*(r)e^{-i\omega t+in\varphi+ikz}, \quad v = v^*(r)e^{-i\omega t+in\varphi+ikz}, \quad w = w^*(r)e^{-i\omega t+in\varphi+ikz}, \tag{2.183}$$

we find

$$u^*(r) = -i\Big[\frac{kB}{\rho(\omega-kW_0)} I'_n(kr) + \frac{kC}{\rho(\omega-kW_0)} K'_n(kr) + \frac{k}{l} D I'_n(lr) \\ + \frac{k}{l} E K'_n(lr) + \frac{nF}{lr} I_n(lr) + \frac{nG}{lr} K_n(lr) \Big], \tag{2.184a}$$

$$v^*(r) = \frac{nB}{\rho(\omega-kW_0)r} I_n(kr) + \frac{nC}{\rho(\omega-kW_0)r} K_n(kr) \\ + \frac{nk}{l^2 r} D I_n(lr) + \frac{nk}{l^2 r} E K_n(lr) + F I'_n(lr) + G K'_n(lr), \tag{2.184b}$$

$$w^*(r) = \frac{kB}{\rho(\omega-kW_0)} I_n(kr) + \frac{kC}{\rho(\omega-kW_0)} K_n(kr) + D I_n(lr) + E K_n(lr). \tag{2.184c}$$

Equations (2.168) and (2.184a-c) are valid for all three types of jet. We will restrict ourselves to the entire jet, so that, in view of singularity $K_n(z)$ at $z \to 0$, we have $C = E = G = 0$. Therefore we still have the constants B, D, F to be chosen to satisfy the boundary conditions (see Section 2.1.3)

$$P|_{r=R_0+\delta} = \sigma K + P_e + 2p\nu\, \partial_r U, \quad (\partial_z U + \partial_r W)\big|_{r=R_0+\delta} = 0,$$
$$(r\partial_r(V/r) + \frac{1}{r}\partial_\rho U)\big|_{r=R_0+\delta} = 0. \tag{2.185}$$

Using (2.144), (2.155), (2.168), (2.183) and expression for curvature, $K = \text{div}(\nabla G/|\nabla G|)$ we find, if we expand (2.185) up to terms of the first order in perturbations,

$$
\begin{aligned}
&p^*(R_0) = \sigma\left(\tfrac{n^2-1}{R_0^2} + k^2\right)\delta^* + 2\rho\nu u^{*\prime}(R_0), \\
&iku^*(R_0) + w^{*\prime}(R_0) = 0, \\
&R_0\left[\tfrac{v^*(R_0)}{R^0}\right]' + i\tfrac{nu^*(R_0)}{R_0} = 0,
\end{aligned}
\tag{2.186}
$$

where again ($'$) means differentiation. The kinematic boundary conditions read $u^*(R_0) = -i(\omega - kW_0)\delta^*$, as follows from (2.180), (2.183) and (2.155), so that we find, from (2.186),

$$
p^*(R_0) = -\sigma\left(\frac{n^2-1}{R_0^2} + k^2\right)\frac{u^*(R_0)}{i(\omega - kW_0)} + 2\rho\nu\frac{du^*(R_0)}{dr}.
$$

Now substitute (2.168), (2.184a-c) in the second and third (2.186). We have three homogeneous equations in B, D, F; thus their determinant has to be equal zero. From this we find the characteristic equation $\det H = 0$, where H is a matrix with the columns

$$
\begin{bmatrix}
\frac{\sigma}{\rho}\left(\frac{n^2-1}{R_0^2} + k^2\right)\frac{kI_n'(kR_0)}{(\omega-kW_0)^2} - \frac{2i\nu k^2}{\omega-kW_0}I_n''(kR_0) - I_n(kR_0) \\
\frac{2k^2}{\rho(\omega-kW_0)}I_n'(kR_0) \\
\frac{2nk^2}{\rho(\omega-kW_0)}\left[\frac{I_n(kR_0)}{kR_0}\right]'
\end{bmatrix},
$$

$$
\begin{bmatrix}
\frac{\sigma}{\rho}\left(\frac{n^2-1}{R_0^2}+k^2\right)\frac{\rho kI_n'(lR_0)}{l(\omega-kW_0)} - 2i\rho\nu kI_n''(lR_0) \\
\frac{k^2+l^2}{l}I_n'(lR_0) \\
2nk\left[\frac{I_n(lR_0)}{lR_0}\right]'
\end{bmatrix},
\quad
\begin{bmatrix}
\frac{\sigma}{\rho}\left(\frac{n^2-1}{R_0^2}+k^2\right)\frac{\rho nI_n(lR_0)}{lR_0(\omega-kW_0)} - 2i\rho\nu ln\left[\frac{I_n(lR_0)}{lR_0}\right]' \\
\frac{nk}{lR_0}I_n(lR_0) \\
\frac{n^2}{lR_0^2}I_n(lR_0) + l^2R_0\left[\frac{I_n'(lR_0)}{lR_0}\right]'
\end{bmatrix}.
$$

The equation $\det H = 0$ is the characteristic for entire non-rotating viscous jet under inertial effects of surrounding air to be neglected. Consideration only pinching perturbed modes ($n = 0$), which may be done because high modes with $n > 0$ are probably stable for non-rotating disturbed jet, leads

$$
\begin{vmatrix}
\frac{\sigma}{\rho}\left(k^2 - \frac{1}{R_0^2}\right)\frac{kI_0'(kR_0)}{(\omega-kW_0)^2} & \frac{\sigma}{\rho}\left(k^2 - \frac{1}{R_0^2}\right)\frac{\rho kI_0'(lR_0)}{l(\omega-kW_0)} \\
-\frac{2i\nu k^2}{\omega-kW_0}I_0''(kR_0) - I_0(kR_0) & -2i\rho\nu kI_0''(lR_0) \\
\frac{2k^2}{\rho(\omega-kW_0)}I_0'(kR_0) & \frac{k^2+l^2}{l}I_0'(lR_0)
\end{vmatrix} = 0.
\tag{2.187}
$$

Using (2.173) and well-known differential equations for I_0, (2.187) are reduced to (2.165). $\square$

This result is the same as that derived by Rayleigh[5] and Weber [44]. Consider two limit cases, when the jets viscosity is very low or very high, provided $q = kR_0 \neq 0$ and $q \neq 1$.

1. For the case of high velocity we may assume $\omega - kW_0$ to be finite. This is also probable from a physical point of view because perturbations with high frequencies will be damped out. Therefore, introducing the abbreviation

$$
y = \frac{\omega - kW_0}{\nu}R_0^2,
\tag{2.188}
$$

we note that $y \to 0$ with the viscosity tends to infinity and l tends to k follow (2.172). Substituting $y = \nu^{-m}$, $(m > 0)$ in (2.187), using (2.188) gives that $m = 2$ for $q \neq 0$, 1.

[5]Rayleigh Lord, On the stability of cylindrical fluid surfaces. *London, Edinburgh and Dublin Phil. Mag. and J. of Sci.*, 5th Series, V. 34, 1892, 145–154, 177–180.

Put $y = c_1 \nu^{-2} + c_2 \nu^{-4} + \ldots$, where c_i are constants, and after simple manipulations using trivial formulas for Bessel functions, we get

$$\omega_{\mathrm{Re}} = kW_0 + O(\nu^{-5}), \quad (q \neq 0, 1);$$
$$\omega_{\mathrm{Im}} = \frac{\sigma(1-q^2)}{2\rho\nu R_0} \frac{I_1^2(q)}{q^2 I_0^2(q) - I_1^2(q)(1+q)^2} - \frac{\sigma^2}{\rho^2 \nu^3} \frac{(1-q^2)^2 I_1^3(q)}{8q[q^2 I_0^2(q) - I_1^2(q)(1+q^2)]^3}$$
$$\times [2q I_0^2(q) I_1(q) - q^2 I_0^3(q) - q I_1^3(q) + q^2 I_0(q) I_1^2(q) + I_0(q) I_1^2(q)] + O(\nu^{-5}). \tag{2.189}$$

For infinitely high viscosity ν, it follows (2.189) that the instability range for q is, as before, $0 < q < 1$ due to $q^2 I_0^2(q) - I_1^2(q)(1+q)^2 > 0$, as follows from Bessel functions properties. Hence, $\omega_{Im} \propto \nu^{-1}$. Neglecting $O(\nu^{-5})$, we see that the phase velocity of the perturbed modes is equal W_0.

2. For the case of low velocity and when $\omega_{Im} > 0$, i.e., if perturbations are unstable, Taylor expansion, after substitution $y = c_1 \nu^{-1} + c_2 \nu^{-1/2} + c_3 \nu^0 + \ldots$ in (2.187), using (2.188) and well-known formulas of asymptotic expansions for Bessel functions, yields

$$\omega_{\mathrm{Re}} = kW_0 + O(\nu^{3/2});$$
$$\omega_{\mathrm{Im}} = \left[\frac{\sigma}{\rho R_0^3}(1-q)^2 q \frac{I_1(q)}{I_0(q)}\right]^{1/2} - \frac{q}{R_0^2} \frac{[2q I_0(q) - I_1(q)]}{I_0(q)} \nu + O(\nu^{3/2}) \quad (q \neq 0, 1). \tag{2.190}$$

To derive (2.190) use has been made of the fact that assuming ω_{Im} and $\mathrm{Re}(l)$ to be positive, $\mathrm{Re}(lR_0) = \mathrm{Re}\left(\sqrt{q^2 - iy}\right) \to +\infty$ as $\nu \to 0$. Formula (2.190) shows that the instability range for q is again $0 < q < 1$, due to transformed to $2I_0(q) - I_1(q) > 0$. Note that for very low viscosity case, (2.190) transformed to (2.156) if $A = n = 0$.

Inviscid annular jet. Firstly, we will consider the limit case of *infinitely thick* inviscid jet. As in Section 2.2.2, we assume jet to be inviscid and incompressible and the external forces to be zero. Undisturbed flow is again given by $\bar{U} = 0$, $\bar{V} = \Omega/r$, $\bar{W} = W_0$ and $(R_0 \leq r < \infty)$ where Ω, W_0 are constants. For the pressure we get the same expression (2.142), and for the boundary condition $P|_{r=R_0} = -\sigma/R_0 + P_e$, where P_e is the pressure of the air within the annular jet. Assuming P_e to be a constant, we get with (2.142)

$$\frac{1}{2}\rho(\Omega/R_0)^2 - \sigma/R_0 + P_e = \mathrm{const}. \tag{2.191}$$

The derivation of (2.142) and (2.191) is analogous to the derivation of (2.142) and (2.143) and reception of the characteristic (dispersion) relation is likewise analogous to that in Section 2.2.2. Neglecting the inertial effects of surrounding air, we find (compare with (2.156))

$$\left(\omega - \frac{n\Omega}{R_0^2} - kW_0\right)^2 = \left[\frac{\sigma}{\rho}\left(\frac{n^2-1}{R_0^2} + k^2\right) - \frac{\Omega^2}{R_0^3}\right] k \frac{K_n'(kR_0)}{K_n(kR_0)}, \tag{2.192}$$

where B has to be taken equal zero in (2.151) due to singularity I_n at infinity. Equation (2.192) can be also obtained from (2.160) by interchanging ρ and ρ_e, Ω and Ω_e, W_0 and W_{0e} and than under let $\rho_e \to 0$ after multiplication obtained equation by ρ_e. For non-negative modes (n, k) follow $K_n'(kR_0)/K_n(kR_0) < 0$ we have instability only if (compare with (2.157)) $n^2 - 1 + k^2 R_0^2 + \rho\Omega^2/\sigma R_0 < 0$. The conclusion is the same as in Section 2.2.2 except for the point **(c)**. Instead of **(c)** we get:

(c)′ The higher value Ω, the lower the values n and the shorter the intervals of k give rise to unstable modes. If A is sufficiently high, the jet is even stable against all modes (n, k). The critical value of A for this case is determined by $s \equiv \frac{\rho\Omega^2}{\sigma R_0} = 1$.

Proposition 12 *In the case of liquid annular jet within $R_1 \leq r \leq R_2$ the dispersion relation may be written as*

$$A_n(b^* - nS^*\gamma^2)(b^* - nS^*)^2 + B_n q^*(b^* - nS^*)^2[(n^2-1)\gamma^2 + q^{*2} + S^{*2}\gamma^3]$$
$$-C_n q^*(b^* - nS^*\gamma^2)^2(n^2 - 1 + q^{*2} - S^{*2})$$
$$+D_n q^{*2}[(n^2-1)\gamma^2 + q^{*2} + S^{*2}\gamma^3](n^2 - 1 + q^{*2} - S^{*2}) = 0, \tag{2.193}$$

where $\gamma = R_2/R_1$, $b^{*2} = \frac{\rho R_2^3}{\sigma}(\omega - kW_0)^2$, $q^* = kR_2$ and $S^{*2} = \rho\Omega^2/\sigma R_2$.

Proof. In this case jet fills the space within $R_1 \leq r \leq R_2$. As in Sections 2.2.2 and 2.2.3, we assume the liquid viscosity and inertial effects of surrounded air to be negligible, and the liquid to be incompressible with the external forces to be zero. Undisturbed flow field is taken by

$$\bar{U} = 0, \quad \bar{V} = \Omega/r, \quad \bar{W} = W_0 \quad (R_1 \leq r \leq R_2).$$

Thus for the pressure we get as before, expression (2.142), but we arrive to two boundary conditions: $P|_{r=R_1} = -\sigma/R_1 + P_{e1}$ and $P|_{r=R_2} = \sigma/R_2 + P_{e1}$, where P_{e1} and P_{e2} are the external pressures at $r = R_1$ and $r = R_2$. Thus, from (2.142) it follows

$$-\frac{1}{2}\rho\Omega^2/R_1^2 + \text{const} = -\sigma/R_1 + P_{e1}, \quad -\frac{1}{2}\rho\Omega^2/R_2^2 + \text{const} = \sigma/R_2 + P_{e2}. \quad (2.194)$$

Equations (2.194) are only possible if $P_{e2} - P_{e1} = -\frac{1}{2}\rho\Omega^2\left(\frac{1}{R_2^2} - \frac{1}{R_1^2}\right) - \sigma\left(\frac{1}{R_2} - \frac{1}{R_1}\right)$. Analysis for this case is analogous to that used in Sections 2.2.2 and 2.2.3 with definitions $\gamma = R_2/R_1$, $b^{*2} = \frac{\rho R_2^3}{\sigma}(\omega - kW_0)^2$, $q^* = kR_2$ and $S^{*2} = \rho\Omega^2/\sigma R_2$,

$$\begin{aligned}
A_n &= I_n(q^*)K_n(q^*/\gamma) - K_n(q^*)I_n(q^*/\gamma), \\
B_n &= -I_n(q^*)K_n'(q^*/\gamma) + K_n(q^*)I_n'(q^*/\gamma), \\
C_n &= I_n'(q^*)K_n(q^*/\gamma) - K_n'(q^*)I_n(q^*/\gamma), \\
D_n &= -I_n'(q^*)K_n'(q^*/\gamma) + K_n'(q^*)I_n'(q^*/\gamma)
\end{aligned} \quad (2.195)$$

we find the characteristic equation (2.193). $\qquad\square$

The equation above is of the forth degree on b^*. First, consider the simplest case, when jet is non-rotating, i.e., when $\Omega = S^* = 0$. Then (2.195) becomes to an equation of the second degree on b^{*2},

$$\begin{aligned}
A_n b^{*4} + B_n q^* b^{*2}[(n^2 - 1)\gamma^2 + q^{*2}] - C_n q^* b^{*2}(n^2 - 1 + q^{*2}) \\
+ D_n q^{*2}[(n^2 - 1)\gamma^2 + q^{*2}](n^2 - 1 + q^{*2}) = 0.
\end{aligned} \quad (2.196)$$

Equation (2.196) has positive discriminant

$$\begin{aligned}
\Delta = \{B_n q^*[(n^2 - 1)\gamma^2 + q^{*2}] - C_n q^*(n^2 - 1 + q^{*2})\}^2 \\
-4(A_n D_n - B_n C_n)q^{*2}[(n^2 - 1)\gamma^2 + q^{*2}](n^2 - 1 + q^{*2})
\end{aligned}$$

and so these expressions are positive whatever the signs may be of $(n^2 - 1)\gamma^2 + q^{*2}$ and $n^2 - 1 + q^{*2}$ because $A_n D_n - B_n C_n = -\gamma/q^{*2} < 0$ and because A_n, B_n, C_n, D_n are positive for $\gamma > 1$.

For the partial case if $\gamma = 1$, then $\gamma = 1$, $A_n = D_n = 0$ and $B_n = C_n = 1/q^*$; and the derivatives of A_n, B_n, C_n, D_n are positive with respect to γ due to positivity of I_n, I_n', I_n'', K_n, $-K_n'$ and K_n''. Here n and q^* have been taken positive. From these considerations it follows that (2.196) has only real roots and that it has negative roots only if $(n^2 - 1)\gamma^2 + q^{*2} < 0$ and $n^2 - 1 + q^{*2} \leq 0$. Thus we have only negative roots if $n = 0$ and we have the same conclusion as before; the non-rotating jet is stable for perturbations with $n > 0$. For rotational case and pinching perturbed modes, $n = 0$, we also get a quadratic equation on b^{*2}, and a reasoning similar to that employed above shows that this equation has negative roots only if

$$-\gamma^2 + q^{*2} + S^{*2}\gamma^3 < 0, \quad -1 + q^{*2} - S^{*2} < 0, \quad (2.197)$$

where in (2.197) either one of the inequalities signs has to be replaced by less-equal ones.

If we take together the above results we receive: the characteristic equation (2.195) has complex roots b^*, with positive imaginary part only if

$$q^{*2} < \max\{1 + S^{*2}, \gamma^2 - S^{*2}\gamma^2\}, \quad n = 0, \quad A \geq 0. \tag{2.198}$$

Consider the case $S^{*2} = 0$, which occur when we are dealing with non-rotating jet. If the jet is very thin, i.e., if $\gamma \simeq 1$, we find, using differential equation for I_0, K_0 and for the Wronskian of these functions, $A_0 = \gamma - 1 + \ldots$, $B_0 = 1/q^* + \ldots$ and $D_0 = \gamma - 1 + \ldots$ and it then follows from (2.195), that for $n = 0$, the characteristic equation has an approximate view:

$$(\gamma - 1)b^{*4} + 2(1 - q^{*2})b^{*2} - (\gamma - 1)\big[(1 - q^{*2})^2 - S^{*4}\big]q^{*2} = 0. \tag{2.199}$$

Its roots are, for $S^* = 0$,

$$b^{*2} = -2\frac{1 - q^{*2}}{\gamma - 1} + \ldots \qquad b^{*2} = -\frac{1}{2}(\gamma - 1)(1 - q^{*2}) + \ldots,$$

and from (2.199) it follows that there is instability only in the range $0 < q^* < 1$. Then, $b^*_{\text{Im}} \to \infty$, and hence non-rotating jet is unstable in the range $0 < q^* < 1$ if it is very thin.

Consider another limit case $\gamma \to \infty$, that means $R_1 \to 0$, or, in another words, when annular jet tends to entire one. Then, it follows from Taylor's expansions of I_0 and K_0, we get $A_0 = I_0(q^*)\ln\gamma + \ldots$, $B_0 = I_0(q^*)\gamma(q^*)^{-1} + \ldots$, $C_0 = I_1(q^*)\ln\gamma + \ldots$, and $D_0 = I_1(q^*)\gamma(q^*)^{-1} + \ldots$ The characteristic equation for $n = 0$ then becomes

$$I_0(q^*)\ln\gamma \cdot b^{*4} + I_0(q^*)\gamma^2\ln\gamma \cdot b^{*2}(1 - S^{*2}\gamma) + I_1(q^*)\ln\gamma \cdot q^*b^{*2}(1 - q^{*2} + S^{*2})$$
$$+ I_1(q^*)\gamma^3 \cdot q^*(1 - S^{*2}\gamma)(1 - q^{*2} + S^{*2}) = 0. \tag{2.200}$$

Neglect the third term in (2.200) with respect to the second, and if $b^* \to \infty$, the fourth term with respect to the second, or if b^* tends to finite value, the first term with respect to the second. Thus we find for $S^* = 0$:

$$b^{*2} = -\frac{\gamma^3}{\ln\gamma} + \ldots, \qquad b^{*2} = (-1 + q^{*2})\frac{q^*I_1(q^*)}{I_0(q^*)} + \ldots \tag{2.201}$$

The second expression in (2.201) is the same as was obtained for entire jet (see (2.159)), and from the first expression it follows, using (2.198), that for $S^* = 0$

$$b^*_{\text{Im}} = \gamma^{3/2}\ln^{-1/2}(\gamma) + \ldots, \quad 0 < q^* < \gamma.$$

It follows that in interval $0 < q^* < \gamma$, b^*_{Im} tends to infinity as $\gamma \to \infty$ or $\gamma \to 1$. This means that b^*_{Im} has somewhere a minimum value for every fixed q^* in that interval when $\gamma \in (1, \infty)$.

If for instance q^* is near to zero it follows, for $n = 0$, that $A_0 = \ln\gamma + \ldots$, $B_0 = \gamma/q^* + \ldots$, $C_0 = 1/q^* + \ldots$ and $D_0 = \frac{1}{2}(\gamma - 1) + \ldots$ and the quadratic equation becomes

$$\ln\gamma \cdot b^{*4} - \gamma b^{*2}(\gamma^2 - S^{*2}\gamma^3) + b^{*2}(1 + S^{*2}) + \frac{1}{2}(\gamma - 1/\gamma)q^{*2}(\gamma^2 - S^{*2}\gamma^3)(1 + S^{*2}) = 0.$$

Proceeding as before, we find

$$b^{*2} = -\frac{\gamma(\gamma^2 - S^{*2}\gamma^3) + (1 + S^{*2})}{\ln\gamma} + \ldots, \quad b^{*2} = O(q^{*2}) \tag{2.202}$$

and for $S^* = 0$ it follows, from (2.202), that b^*_{Im} takes its minimum value to be found as a root of transcendental equation $\ln\gamma^3 = 1 + 1/\gamma^3$.

Considering of rotating jet with $S^{*2} > 0$ and $\gamma \approx 1$ (very thin jet) gives

$$b^{*2} = -2\frac{1 - q^{*2}}{\gamma - 1} + \ldots, \quad b^{*2} = -\frac{1}{2}(\gamma - 1)(1 - q^{*2})q^{*2}[1 - S^{*4}(q^{*2} - 1)^{-2}] + \ldots$$

The last gives two branches for b^*_{Im} if $S^* < 1$. The first branch is valid in the range $0 < q^* < (1 + S^{*2})^{1/2}$, and b^*_{Im} then tends to infinity in the interval $0 < q^* < 1$ and to zero in the interval $0 < q^* < (1 + S^{*2})^{1/2}$. The second branch is valid in the range $0 < q^* < (1 + S^{*2})^{1/2}$, and b^*_{Im} then tends to zero.

For $S^* < 1$, we are left with the first branch only. It can be postulated the same conclusion as for $S^{*2} = 0$. For very thick jet, $\gamma \to \infty$, we can find, from (2.202) $b^{*2} = \frac{S^{*2}\gamma^4}{\ln \gamma}$, and again the solution is the same as for entire jet. Hence, there is no disagreement as far as b^*_{Im} is concerned, but for b^*_{Re} we find the second solution. Finally, we come to the conclusion that non-rotationally symmetric disturbances may give rise to more unstable modes than rotationally symmetric ones.

2.3.3 Supersonic viscous jet

Supersonic gaseous jets have been studied experimentally and theoretically for a century. Their scales vary from centimeters in the laboratory to millions of light years in radio galaxies. The nozzle pressure can be greater than (underexpanded jet), equal to (matched jet), or less than (overexpanded jet) the atmospheric pressure. In any of these cases, nearly periodic arrays of the shock waves are found be embedded in the flow. If the Mach number is high enough, the interesting shocks form a transverse Mach stem (in slabs) or Mach disk (in axial jets). In aerodynamics, the existence of these periodic internal shocks has been known since the beginning of supersonic flight. One sees the long axisymmetric jet with periodic bright and dark spots [1]. The bright spots can be caused by either secondary recombination of fuel in the high temperature regions following the internal shocks or by chemiluminescence. The observations of supersonic jets point to a very similar underlying hydrodynamics flow. Knots of high pressure, density, and temperature, caused by internal shocks, seem to be ubiquitous in the supersonic jets. These internal shocks typically do not disrupt the jet, but the very strong shock at the head of the jet (the "hot spot") does terminate the flow. The head of the jet is often embedded in an amorphous lobe and indeed the jet itself is often surrounded by a "cocoon" of low surface brightness emission. Finally, the jets seem remarkably stable. The shear layers on the surface of the jets are subject to KHI, which can grow to nonlinear amplitude. The large wavelength instabilities (pinching and kinking modes) threaten global jet stability, while small wavelength instabilities should lead to turbulent mixing. In spite of this, supersonic jets are able to propagate to large distances.

Supersonic jets in pressure balance with an external medium are unstable to surface perturbation analogous to the KHI of vortex sheet ([12], [20] – [22], etc). Pinching and helical wave modes are growing perturbations with maximally unstable wavelength longer than or on the order of the instantaneous jet's radius, and previous work has provided estimates of wave phase velocity and maximum growth rate [21]. In addition to these wave modes, a cylindrical jet with sharp velocity discontinuity is unstable to an infinite number of other wave modes with progressively shorter, maximally unstable wavelength and a progressively more rapid growth rate. These growing waves convert directed flow energy into wave energy in the jet and external medium. Because initial growth of surface perturbation is rapid, several authors have investigated ways of slowing the rate of growth which is decreased as the amplitude of helical wave increases [3], and growth of perturbations to the jet surface can be linear rather than exponential if a jet expands rapidly enough [21]. Short wavelength perturbations to a jet surface can be stabilized by the presence of velocity shear between jet and external medium [38]. While it may be possible to completely stabilize a jet by a combination of jet expansion and velocity shear, true jet conditions are not yet well known, and it seems likely that some instability will be present. For this reason we should consider the behavior of long wavelength and short wavelength perturbations to a jet and consider the effect of the resulting growing wave modes on a jet.

Dispersion equation. The problem to be considered is the stability of a constantly expanding jet of compressible fluid moving through a medium of compressible fluid. It will be assumed that a sharp discontinuity in velocity exists at the boundary between two fluids and that the pressure is continuous across the boundary. The fact that a sharp boundary is assumed means that the results are accurate in the more general case if the boundary layer has scale size much less than the wavelength of perturbations to the boundary. The continuity and momentum equations have a form:

$$\partial_t \rho + \operatorname{div}(\rho\,\mathbf{u}) = 0, \qquad \rho[\partial_t \mathbf{u} + (\mathbf{u}\cdot\nabla)\mathbf{u}] = -\nabla P + \mu\nabla^2\mathbf{u}, \qquad (2.203)$$

where P is a static pressure. To analyze a diverging flow of a zero opening angle, we use cylindrical coordinates in which the axial axis is along the centerline of the diverging jet and radial axis is across the jet. In this coordinate system, the boundary between jet and external medium is given by $r = R$. The initial flow velocity (at $z = 0$) is $u_{0z} = u$ and $u_{0r} = 0$ for a jet with opening angle equal to zero. The jet is assumed to be axisymmetric and nonrotational, $u_{0\phi} = 0$.

Note that densities and pressures satisfy the time-independent equations of continuity and momentum, and for the case of constant velocity, it follows that the time-independent densities and pressures are proportional to z^{-1} in an isothermal jet and external medium.

(a) The *linearized* continuity and momentum equations (2.203) are (see Sections 1.6, 1.8),

$$\partial_t \rho' + \rho_0 \operatorname{div}\mathbf{u}' + \mathbf{u}_0\cdot\nabla\rho' + (\rho'\operatorname{div}\mathbf{u}_0 + \mathbf{u}'\cdot\nabla\rho_0) = 0, \qquad (2.204a)$$

$$\rho_0\big(\partial_t\mathbf{u}' + (\mathbf{u}_0\cdot\nabla)\mathbf{u}'\big) + \big((\rho_0\mathbf{u}' + \rho'\mathbf{u}_0)\cdot\nabla\big)\mathbf{u}_0 = -\nabla P' + \mu\nabla^2\mathbf{u}', \qquad (2.204b)$$

where $\rho = \rho_0 + \rho'$, $P = P_0 + P'$ and $\mathbf{u} = \mathbf{u}_0 + \mathbf{u}'$.

In cylindrical coordinates, denote $\mathbf{u}_0 = (u_{0r}, 0, u_{0z})$ (a nonrotational jet) and $\mathbf{u}' = (u'_r, u'_\phi, u'_z)$. The solutions of the time-independent equations inside high-velocity jets ($r < R$) are $\rho_0 = \rho_J(r,z)$, $P_0 = P_0(r,z)$ and $\mathbf{u}_0 = (0, 0, u_J)$. In the external medium (ambient atmosphere, i.e., for $r > R$), these quantities are constant: $\rho_0 = \rho_a = \mathrm{const}$, $P_0 = P_a = \mathrm{const}$ and $\mathbf{u}_0 = (0, 0, u_a)$. Equations (2.204a,b) may be written in the following form:

$$\partial_t\rho' + u_{0r}\Big(\partial_r\rho' + \frac{1}{r}\rho'\Big) + u_{0z}\partial_z\rho' + \rho'\Big(\frac{1}{r}u_{0r} + \partial_r u_{0r} + \partial_z u_{0z}\Big)$$

$$+\rho_0\Big(\frac{\partial_r(r u'_r)}{r} + \partial_z u'_z + \partial_\varphi u'_\varphi\Big) + u'_r\partial_r\rho_0 + u'_z\partial_z\rho_0 = 0, \quad (2.205a)$$

$$\rho_0(\partial_t u'_r + u_{0z}\partial_z u'_r) = -\partial_r P' + \mu\Big[\frac{1}{r}\partial_r(r\partial_r u'_r) + \frac{\partial^2_{\phi\phi} u'_r}{r^2} + \partial^2_{zz}u'_r - \frac{u'_r}{r^2} - \frac{2}{r^2}\partial_\phi u'_\phi\Big], \quad (2.205b)$$

$$\rho_0(\partial_t u'_\phi + u_{0z}\partial_z u'_\phi) = -\frac{1}{r}\partial_\phi P' + \mu\Big[\frac{1}{r}\partial_r(r\partial_r u'_\phi) + \frac{\partial^2_{\phi\phi} u'_\phi}{r^2} + \partial^2_{zz}u'_\phi - \frac{u'_\phi}{r^2} + \frac{2}{r^2}\partial_\phi u'_r\Big], \quad (2.205c)$$

$$\rho_0(\partial_t u'_z + u_{0z}\partial_z u'_z) = -\partial_z P' + \mu\Big[\frac{1}{r}\partial_r(r\partial_r u'_z) + \frac{1}{r^2}\partial^2_{\phi\phi} u'_z + \partial^2_{zz}u'_z\Big]. \quad (2.205d)$$

Follow universal gaseous law, $P/\rho = a^2$ (where a is the sound velocity), we obtain that for adiabatic perturbation, $dP'/d\rho' = a^2$. Hence

$$\partial_r P' = a^2\partial_r\rho', \quad \partial_\phi P' = a^2\partial_\phi\rho', \quad \partial_z P' = a^2\partial_z\rho'.$$

(b) Assume the velocity and density perturbations $(\mathbf{u}', \rho')$ of the form $f(r, \phi, z; t) = F(r)e^{i(kz+n\phi-\omega t)}$, namely,

$$u'_r = u_{r1}(r)e^{i(kz+n\phi-\omega t)}, \ u'_\phi = u_{\phi 1}\Psi(r)e^{i(kz+n\phi-\omega t)}, \ u'_z = u_{z1}\Psi(r)e^{i(kz+n\phi-\omega t)}, \quad (2.206a)$$

$$\rho' = \rho_1\Psi(r)e^{i(kz+n\phi-\omega t)}. \quad (2.206b)$$

Note that (2.205a-c) contain the differential $\partial/\partial t + u_{0r}\partial/\partial r + u_{0z}\partial/\partial z$, which for (2.206a,b) transforms to $-i\omega + iku_{0z} + 0 \cdot \partial/\partial r$. This allows us to pick out the velocity u_{0z} and then introduce it as a condition at the boundary. Thus, within the jet fluid and external medium, we may set $u_{0z} = 0$ and substitute (2.206a,b) in (2.205a) to obtain

$$\left(u_{z1}(r)\partial_z\rho_0 - i\omega\rho_1\right)\Psi(r) + \rho_0\left[i\left(ku_{z1} + n\frac{1}{r}u_{\phi 1}\right)\Psi(r) + \frac{1}{r}u_{r1}(r) + \partial_r u_{r1}(r)\right] + u_{r1}(r)\partial_r\rho_0 = 0. \quad (2.207)$$

Since for the majority of gaseous jets $\varepsilon = \frac{\nu k}{\omega/k} \ll 1$, one may solve (2.207) by perturbation methods with a small parameter ε.

(c) Denote $\mathbf{u}^{(0)} = (u_{r1}^{(0)}, u_{\phi 1}^{(0)}, u_{z1}^{(0)})$, where

$$u_{r1}^{(0)} = -\frac{i}{\omega}\left(\frac{\rho_1}{\rho_0}\right)a^2\,\Psi'(r), \quad u_{\phi 1}^{(0)} = \frac{n}{\omega}\left(\frac{\rho_1}{\rho_0}\right)\frac{1}{r}a^2, \quad u_{z1}^{(0)} = \frac{k}{\omega}\left(\frac{\rho_1}{\rho_0}\right)a^2 \quad (2.208)$$

are solutions of (2.205b-c) with zero viscosity. We shall find a function u_{r1} obeying (2.207), of the form

$$u_{r1}(kr) = u_{r1}^{(0)} + \varepsilon \cdot (\nabla^2\mathbf{u}^{(0)})_r, \quad (2.209)$$

where

$$(\nabla^2\mathbf{u}^{(0)})_r := \partial^2_{kr,kr}u_{r1}^{(0)} + \frac{1}{kr}\partial_{kr}u_{r1}^{(0)} + \frac{1}{(kr)^2}\partial^2_{\phi,\phi}u_{r1}^{(0)} + \partial^2_{kz,kz}u_{r1}^{(0)} - \frac{u_{r1}^{(0)}}{(kr)^2} - \frac{2}{(kr)^2}\partial_\phi u_{\phi 1}^{(0)}.$$

Substituting (2.208) and (2.209) for $\varepsilon = 0$ in (2.207), we find the differential equation for Ψ,

$$r^2\Psi''(r) + r\,\Psi'(r) + \left(\beta^2 k^2 r^2 - n^2\right)\Psi(r) = 0, \quad (2.210)$$

where $\beta = \sqrt{\omega^2/(k^2a^2) - 1}$. The general solution of the Bessel equation (2.210) has the form

$$\Psi(r) = c_1 J_n(\beta kr) + c_2 Y_n(\beta kr),$$

where J_n and Y_n are Bessel and Neumann functions, respectively. For approximate calculation of Bessel functions, $J_\nu(x)$, positive roots may be used for formulas (1.18) and (1.19) from Section 1.1.5. Substituting this result for Ψ in (2.208), one may find $u_{r1}^{(0)}(kr)$,

$$u_{r1}^{(0)} = -ia\frac{\rho_1}{\rho_0}\sqrt{1 - \left(\frac{ka}{\omega}\right)^2}[c_1 J_n'(\beta kr) + c_2 Y_n'(\beta kr)].$$

From the above we obtain the solution u_{r1}:

$$
\begin{aligned}
u_{r1} = {}& -ia\frac{\rho_1}{\rho_0}\sqrt{1 - \left(\frac{ka}{\omega}\right)^2}\Big\{\Psi'(r) + \frac{\nu k^2}{\omega}\Big[\beta^2\Psi'''(r) + \frac{\beta}{kr}\Psi''(r) \\
& - ((n^2 + 2in + 1)/(kr)^2 + 1)\Psi'(r)\Big]\Big\}.
\end{aligned}
$$

(d) This result combined with (2.206b) for ρ' allows us to transform (2.206a) for the velocity perturbation in the r direction as

$$
\begin{aligned}
u'_r = {}& ia\frac{\rho'}{\rho_0}\sqrt{1 - \left(\frac{ka}{\omega}\right)^2}\Big\{\Psi'(r) + \frac{\nu k^2}{\omega}\Big[\beta^2\Psi'''(r) + \frac{\beta}{kr}\Psi''(r) \\
& - ((n^2 + 2in + 1)/(kr)^2 + 1)\Psi'(r)\Big]\Big\}/\Psi(r). \quad (2.211)
\end{aligned}
$$

Note that (2.211) for $r < R$ becomes

$$
\begin{aligned}
u'_r &= ia_J \frac{\rho'_J}{\rho_J} \sqrt{1 - (ka_J/\omega_J)^2} \Big\{ J'_n(\beta_J kr) + \frac{\nu_J k^2}{\omega_J} [\beta_J^2 J'''_n(\beta_J kr) \\
&\quad + \frac{\beta_J}{kr} J''_n(\beta_J kr) - ((n^2 + 2in + 1)/(kr)^2 + 1) J'_n(\beta_J kr)] \Big\} / J_n(\beta_J kr), \quad (2.212a)
\end{aligned}
$$

and for $r > R$ becomes

$$
\begin{aligned}
u'_r &= ia_a \frac{\rho'_a}{\rho_a} \sqrt{1 - (ka_a/\omega_a)^2} \Big\{ H_n^{(1)'}(\beta_a kr) + \frac{\nu_a k^2}{\omega_a} [\beta_a^2 H_n^{(1)'''}(\beta_a kr) \quad (2.212b) \\
&\quad + \frac{\beta_a}{kr} H_n^{(1)''}(\beta_a kr) - ((n^2 + 2in + 1)/(kr)^2 + 1) H_n^{(1)'}(\beta_a kr)] \Big\} / H_n^{(1)}(\beta_a kr),
\end{aligned}
$$

where $H_n^{(1)} = J_n + iY_n$ is the Hankel function describing outward propagating disturbances. In (2.212a,b) we denote $\beta_a = \left(\frac{\omega_a^2}{k^2 a_a^2} - 1\right)^{1/2}$ and $\beta_J = \left(\frac{\omega_J^2}{k^2 a_J^2} - 1\right)^{1/2}$, where the subscript (a) refers to quantities in the ambient atmosphere (external medium) and the subscript (J) refers to quantities in the jet measured in the jet reference frame. Displacement of the boundary $(r = R)$ between two fluids by an amount $\delta' = \delta_1 e^{i(kz + n\phi - \omega t)}$ is related to the velocity by $\partial_t \delta' = -i\omega \delta' = u'_r$.

(e) Thus, we may write the displacement using (2.212a,b) as

$$
\begin{aligned}
-i\omega_J \delta' &= ia_J \frac{\rho'_J}{\rho_J} \sqrt{1 - \left(ka_J/\omega_J\right)^2} \Big\{ J'_n(\beta_J kr) + \frac{\nu_J k^2}{\omega_J} [\beta_J^2 J'''_n(\beta_J kr) + \frac{\beta_J}{kr} J''_n(\beta_J kr) \\
&\quad - ((n^2 + 2in + 1)/(kr)^2 + 1) J'_n(\beta_J kr)] \Big\} / J_n(\beta_J kr), \\
-i\omega_a \delta' &= ia_a \frac{\rho'_a}{\rho_a} \sqrt{1 - \left(ka_a/\omega_a\right)^2} \Big\{ H_n^{(1)'}(\beta_a kr) + \frac{\nu_a k^2}{\omega_a} [\beta_a^2 H_n^{(1)'''}(\beta_a kr) \\
&\quad + \frac{\beta_a}{kr} H_n^{(1)''}(\beta_a kr) - ((n^2 + 2in + 1)/(kr)^2 + 1) H_n^{(1)'}(\beta_a kr)] \Big\} / H_n^{(1)}(\beta_a kr).
\end{aligned}
$$

Using the transformations for frequencies to the frame of reference of the internal and the external medium (i.e., $\omega_J = \omega - ku_J$ and $\omega_a = \omega - ku_a$) and the fact that $\rho'_J a_J^2 = \rho'_a a_a^2$ (since the pressure is continuous across the boundary) allows us to present the dispersion equation as

$$
\begin{aligned}
&\frac{J'_n(\beta_{in} X) + \Upsilon_\nu^{in} \{ \beta_{in}^2 J'''_n(\beta_{in} X) + (\beta_{in}/X) J''_n(\beta_{in} X) - G(X) J'_n(\beta_{in} X) \}}{H_n^{(1)'}(\beta_{ex} X) + \Upsilon_\nu^{ex} [\beta_{ex}^2 H_n^{(1)'''}(\beta_{ex} X) + (\beta_{ex}/X) H_n^{(1)''}(\beta_{ex} X) - G(X) H_n^{(1)'}(\beta_{ex} X)]} \\
&\times \frac{\beta_{in} H_n^{(1)}(\beta_{ex} X)}{\beta_{ex} J_n(\beta_{in} X)} = \eta^{-1} \frac{\rho_J}{\rho_a} \left(\frac{\Phi - M_J}{\Phi/\eta^{1/2} - M_a} \right)^2, \quad (2.213)
\end{aligned}
$$

where

$$
X = kR, \quad G(X) = \frac{n^2 + 2in + 1}{X^2} + 1, \quad \eta = (a_a/a_J)^2, \quad \Phi = \frac{\omega}{ka_J},
$$

$$
\Upsilon_\nu^{ex} = \frac{\nu_a k}{a_a} (\Phi/\eta^{1/2} - M_a)^{-1}, \quad \Upsilon_\nu^{in} = \frac{\nu_J k}{a_J} (\Phi - M_J)^{-1},
$$

$$
\beta_{ex} = \sqrt{(\Phi/\eta^{1/2} - M_a)^2 - 1}, \quad \beta_{in} = \sqrt{(\Phi - M_J)^2 - 1}.
$$

The values M_J and M_a are the mach numbers of the jet at the nozzle exit and of the ambient atmosphere, respectively; a_J and a_a– the sound velocities; γ_J and γ_a – specific heat ratios;

$\nu_J = \mu_J/\rho_J$ and $\nu_a = \mu_a/\rho_a$ – viscosities; ρ_J and ρ_a – nonperturbed density of the jet and of the ambient atmosphere, respectively. For the case of zero viscosity, our equation (2.213) transformed to simple form, see [21].

$$\frac{J'_n(\beta_{in}X)H_n^{(1)}(\beta_{ex}X)}{J_n(\beta_{in}X)H_n^{(1)'}(\beta_{ex}X)} \cdot \frac{\beta_{in}}{\beta_{ex}} = \eta^{-1}\frac{\rho_J}{\rho_a}\left(\frac{\Phi - M_J}{\Phi/\eta^{1/2} - M_a}\right)^2. \tag{2.214}$$

Remark 11 For the case of a diverging flow of constant opening angle (for example, conical nozzle exit with angle θ) we should use spherical coordinates in which the polar axis is along the center of the diverging jet and radial lines are along streamlines of the flow. In this coordinate system, the boundary between jet and external medium is given by $\theta = \Theta$, and the cone angle of the diverging jet is 2Θ. After analytical manipulations similar to those that were done in [21], we'll obtain the same dispersion equation (2.213), but with choice of the values: $M_a^{(con)} = M_a \cos\theta$ instead of M_a; $X = kR[1 + (z/R)\tan\theta]$ instead of $X = kR$. Besides that instead β_{in} and β_{ex} we have to choose the values

$$\beta_{in}^* = \sqrt{(\Phi - M_J)^2 - (1 + n^2/k^2 - 2i/k)}, \quad \beta_{ex}^* = \sqrt{(\Phi/\eta^{1/2} - M_a)^2 - (1 + n^2/k^2 - 2i/k)}.$$

Our dispersion equation (2.213) is fair only under condition $\nu_{\{J,a\}}k/(\omega/k) \ll 1$.

Asymptotic analysis of instabilities. The dispersion equation (2.213) can be expanded in the limit of small and large arguments of the Bessel and Hankel functions. It is instructive to begin with the expansion for small arguments $|\xi| \ll 1$ and $|\zeta| \ll 1$, where $\xi = \beta_{in}kR$, and $\zeta = \beta_{ex}kR$, i.e., the long-wavelength limit.

1. For the case $n = 0$, which is a pinching wave mode, we use the following expansions of the Bessel and Hankel functions:

$$J_0(\xi) \approx 1 - \frac{\xi^2}{4}, \quad J'_0(\xi) \approx -\frac{\xi}{2}\left(1 - \frac{\xi^2}{8}\right), \quad J''_0(\xi) \approx -\frac{1}{2}\left(1 - \frac{3}{8}\xi^2\right), \quad J'''_0(\xi) \approx \frac{\xi}{8}\left(3 - \frac{5}{12}\xi^2\right);$$

$$\begin{aligned}
H_0^{(1)}(\zeta) &\approx 1 + i\frac{2}{\pi}(\ln(\tfrac{1}{2}\zeta) + C) + o(\zeta),\\
H_0^{(1)'}(\zeta) &\approx -\tfrac{1}{2}\zeta - i\frac{2}{\pi}[\tfrac{1}{4}(2C - 1)\zeta + \tfrac{1}{2}\zeta\ln(\tfrac{1}{2}\zeta) - \zeta^{-1}] + o(\zeta),\\
H_0^{(1)''}(\zeta) &\approx -\tfrac{1}{2} - i\frac{2}{\pi}[\tfrac{1}{2}\ln(\tfrac{1}{2}\zeta) + \tfrac{1}{4}(2C + 1) + \zeta^{-2}] + o(\zeta),\\
H_0^{(1)'''}(\zeta) &\approx \tfrac{3}{8}\zeta + i\frac{2}{\pi}[\tfrac{3}{8}\zeta\ln(\tfrac{1}{2}\zeta) + \tfrac{1}{2}\zeta^{-1}(4\zeta^{-2} - 1) + \tfrac{1}{32}(12C - 5)\zeta] + o(\zeta),
\end{aligned} \tag{2.215}$$

where $C \approx 0.577$ is Euler's constant, to write an expression for the pinching wave mode along wavelength. In the limit $|\xi| \ll 1$ and $|\zeta| \ll 1$, the dispersion relation (2.213) becomes

$$\frac{\left(\frac{\beta_{in}}{\beta_{ex}}\right)\left[(\beta_{in}^2 + 1)\Upsilon_\nu^{in} - 1\right]\left[1 + i\frac{2}{\pi}(\ln(\frac{\xi}{2}) + C)\right]\frac{\xi}{2}}{\frac{\xi}{2}[\Upsilon_\nu^{ex}(\beta_{ex}^2 + 1) - 1] + i\frac{2}{\pi}(\frac{1}{\xi})[1 - (1 + \frac{1}{2}\beta_{ex}^2)\Upsilon_\nu^{ex}]} = \eta^{-1}(\frac{\rho_J}{\rho_a})\frac{\Phi - M_J}{\Phi/\eta^{1/2} - M_a}.$$

After some manipulations we'll obtain

$$\eta\frac{\gamma_J}{\gamma_a}\left(\frac{\Phi - M_J}{\Phi/\eta^{1/2} - M_a}\right)^2 = \frac{1}{2}\xi^2[(\beta_{in}^2 + 1)\Upsilon_\nu^{in} - 1]\frac{\ln\zeta + C}{1 - (\frac{1}{2}\beta_{ex}^2 + 1)\Upsilon_\nu^{ex}}. \tag{2.216}$$

For the case of low jet viscosity we have $\left|\Upsilon_\nu^{in}\right| \ll 1$ and $\left|\Upsilon_\nu^{ex}\right| \ll 1$. In the case of a pinching wave mode $n = 0$ the dispersion equation (2.216) under condition $\nu_{\{J,a\}} \to 0$ transforms to the following equation, which is similar to the equation in [12]

$$\eta^{-1}\frac{\rho_J}{\rho_a}\left(\frac{\Phi - M_J}{\Phi/\eta^{1/2} - M_a}\right)^2 = -\frac{1}{2}\xi^2\left[\ln\left(\frac{\zeta}{2}\right) + C\right].$$

An analytical solution of this equation can be derived for $\eta(\gamma_J/\gamma_a) \ll 1$ (but $kR(\gamma_a/\gamma_J)/\eta \ll 1$) adopting the iterative method. Then at zero order we obtain the double root $\Phi^{(0)} = M_J - \eta^{1/2}M_a$, and the first-order growth rate comes out to be:

$$\Phi_{\mathrm{Im}} \approx \begin{cases} \dfrac{kR\Phi^{(0)}}{\sqrt{2(\rho_J/\rho_a)\eta}} \left| \ln\left(\dfrac{\Phi^{(0)}}{\sqrt{(\rho_J/\rho_a)\eta}}kR\right) \right|^{1/2}, & \Phi^{(0)} < \dfrac{\sqrt{(\rho_J/\rho_a)\eta^{-1}}}{kR} \\ 0, & \Phi^{(0)} > \dfrac{\sqrt{(\rho_J/\rho_a)\eta^{-1}}}{kR}, \end{cases} \tag{2.217}$$

where subscripts $(^{\mathrm{Re}})$, $(^{\mathrm{Im}})$ indicate real or imaginary part, respectively. A maximum growth rate $(\Phi_{\mathrm{Im}})_{\max} \approx [(1/2e)\ln(1/\sqrt{e})]^{1/2}$ is reached for the relative Mach number

$$\tilde{M} = M_J^{\max} - \eta^{1/2}M_a \approx (1/\sqrt{e})\sqrt{(\rho_J/\rho_a)\eta^{-1}}/(kR),$$

and its value appears to be almost independent of all parameters, but the corresponding $M_J^{\max}$ (the maximal value of M_J) decreases with $(\rho_J/\rho_a)\eta^{-1}$. Just above this critical Mach number growth rate probably falls zero: this range of M_J is again near the limit of validity of the asymptotic condition $|\beta_{ex}kR| \ll 1$ since

$$|\beta_{ex}kR|\sqrt{(\rho_J/\rho_a)\eta^{-1}} \sim |\Phi^{(0)}| \sim M_J.$$

2. In the case of $n = 1$ (helical wave mode), the appropriate long-wavelength expansions of the Bessel and Hankel functions are

$$J_1(\xi) \approx \left(1 - \frac{\xi^2}{8}\right)\frac{\xi}{2}, \quad J_1'(\xi) \approx \frac{1}{2}\left(1 - \frac{3}{8}\xi^2\right), \quad J_1''(\xi) \approx -\frac{\xi}{8}\left(3 - \frac{5}{12}\xi^2\right), \quad J_1'''(\xi) \approx -\frac{1}{8}\left(3 - \frac{5}{4}\xi^2\right);$$

$$\begin{aligned} H_1^{(1)}(\zeta) &\approx \tfrac{1}{2}\zeta + i\tfrac{2}{\pi}\left[\tfrac{1}{2}\zeta(\ln(\tfrac{1}{2}\zeta) + C) - \zeta^{-1}(1 + \tfrac{1}{2}\zeta^2)\right] + o(\zeta), \\ H_1^{(1)'}(\zeta) &\approx \tfrac{1}{2} + i\tfrac{2}{\pi}\left[\tfrac{1}{2}(\ln(\tfrac{1}{2}\zeta) + C) + \zeta^{-2} + \tfrac{1}{4}\right] + o(\zeta), \\ H_1^{(1)''}(\zeta) &\approx -\tfrac{3}{8}\zeta - i\tfrac{2}{\pi}\left[\tfrac{3}{8}\zeta(\ln(\tfrac{1}{2}\zeta) + C - \tfrac{5}{12}) + 2\zeta^{-3}(1 - \tfrac{1}{4}\zeta^2)\right] + o(\zeta), \\ H_1^{(1)'''}(\zeta) &\approx -\tfrac{3}{8} - i\tfrac{2}{\pi}\left[\tfrac{3}{8}((\ln\tfrac{1}{2}\zeta) + C + \tfrac{7}{12}) - \tfrac{1}{2}\zeta^{-4}(12 - \zeta^2)\right] + o(\zeta). \end{aligned} \tag{2.218}$$

In the limit $|\xi| \ll 1$ and $|\zeta| \ll 1$, the dispersion equation (2.213) becomes

$$\eta^{-1}\frac{\rho_J}{\rho_a}\left(\frac{\Phi - M_J}{\Phi/\eta^{1/2} - M_a}\right)^2 = -\frac{1 - 2(1 + i)\Upsilon_\nu^{\mathrm{in}}/(kr)^2}{1 + 2(1 - i)\Upsilon_\nu^{\mathrm{ex}}/(kr)^2}. \tag{2.219}$$

Note that this dispersion equation under condition $\nu_{\{J,a\}} \to 0$ transforms into the equation similar to one in [21]

$$\eta^{-1}\frac{\rho_J}{\rho_a}\left(\frac{\Phi - M_J}{\Phi/\eta^{1/2} - M_a}\right)^2 = -1.$$

For long-wavelength limit $kR < 1$ (but $kR \gg \mathrm{Re}_J^{-1/2}$) Equation (2.219) has an analytical solution

$$\Phi_{\mathrm{Re}} \approx \frac{M_J + \eta^{1/2}(\rho_a/\rho_J)M_a}{1 + \rho_a/\rho_J}, \quad \Phi_{\mathrm{Im}} \approx \left(\frac{\rho_a}{\rho_J}\right)^{1/2}\frac{M_J + \eta^{1/2}(\rho_a/\rho_J)M_a}{1 + \rho_a/\rho_J}.$$

For long waves, when $kR \ll \mathrm{Re}_J^{-1/2}$ the solution of this equation is

$$\Phi_{\mathrm{Re}} \approx \frac{[2 - \sqrt{3}(\rho_a/\rho_J)^{\frac{1}{3}}] \cdot [2M_J - \sqrt{3}\eta^{\frac{1}{2}}(\rho_a/\rho_J)^{\frac{1}{3}}M_a] + \eta^{\frac{1}{2}}(\rho_a/\rho_J)^{\frac{2}{3}}M_a}{\left[2 - \sqrt{3}\eta(\rho_a/\rho_J)^{\frac{1}{3}}\right]^2 + (\rho_a/\rho_J)^{\frac{2}{3}}}, \tag{2.220a}$$

$$\Phi_{\mathrm{Im}} \approx \eta^{\frac{1}{2}}(\rho_a/\rho_J)^{\frac{1}{3}}\frac{M_a\{[2 - \sqrt{3}(\rho_a/\rho_J)^{\frac{1}{3}} + 2\sqrt{3} - 4(\rho_a/\rho_J)^{-\frac{1}{3}}] - 2\eta^{-\frac{1}{2}}M_J\}}{\left[2 - \sqrt{3}(\rho_a/\rho_J)^{1/3}\right]^2 + (\rho_a/\rho_J)^{2/3}}. \tag{2.220b}$$

These perturbations grow under condition

$$M_J < \left\{\left[1 - \frac{1}{2}(\rho_a/\rho_J)^{1/3}\right] + \left[\sqrt{3} - 2(\rho_a/\rho_J)^{-1/3}\right]\right\}\eta^{1/2}M_a.$$

In the case of a helical mode ($n = 1$) both conditions $|\Upsilon_v^{\text{in}}| \ll 1$ and $|\Upsilon_v^{\text{in}}|/(kr)^2 \gg 1$ in the dispersion equation (2.219) may be executed together. Really, using transformation $\tilde{\Phi} = \Phi - \eta^{1/2}M_a$ and $\tilde{M} = M_J - M_a$, Equation (2.219) becomes

$$\frac{\rho_J}{\rho_a}\left(\frac{\tilde{\Phi} - \tilde{M}}{\tilde{\Phi}}\right)^2 = -\frac{1 - 2(1+i)\tilde{\Upsilon}_v^{\text{in}}/(kR)^2}{1 + 2(1-i)\tilde{\Upsilon}_v^{\text{ex}}/(kr)^2}, \tag{2.221}$$

where

$$\tilde{\Upsilon}_v^{\text{in}} = \tilde{M}kR/[Re_J \cdot (\tilde{\Phi} - \tilde{M})], \quad \tilde{\Upsilon}_v^{\text{ex}} = \eta^{1/2}\tilde{M}kR/(Re_a\tilde{\Phi}),$$

$Re_J = (u_J - u_a)R/\nu_J$ and $Re_a = (u_J - u_a)R/\nu_a$ – the *Reynolds numbers* in the jet and in the ambient atmosphere, respectively, for the relative flow. This transformation is equivalent to the transformations for frequencies to the frame of the external medium, i.e., $\omega = \omega_a$ and $\omega_J = \omega - k(u_J - u_a)$. The conditions $|\tilde{\Upsilon}_v^{\text{in}}| \ll 1$ and $|\tilde{\Upsilon}_v^{\text{in}}|/(kr)^2 \gg 1$ are equivalent to

$$\left|\frac{u_\omega - (u_J - u_a)}{u_J - u_a}\right|\tilde{M}Re_J \cdot kR \ll 1 \ll \left|\frac{u_\omega - (u_J - u_a)}{u_J - u_a}\right|\tilde{M}Re_J/kR. \tag{2.222}$$

They are executed in two cases: in the trivial case of very long wavelength perturbations, when $Re_J \cdot kR \ll 1$, and in the resonance case with $|u_\omega - (u_J - u_a)|/(u_J - u_a) < 1/(\tilde{M}Re_J)$. Then the condition (2.222) is equivalent to the trivial long-wavelength limit $kR \ll 1$. In the resonance case, (2.221) becomes

$$\frac{\rho_J}{\rho_a}\frac{(\tilde{\Phi} - \tilde{M})^3}{\tilde{\Phi}^2}Re_J \cdot kR = 2\sqrt{2}\exp(i\pi/4).$$

Using the appropriate expansion of the Bessel and Hankel functions for $|\xi| > 1$ and $|\zeta| > 1$,

$$J_n(\xi) \approx \sqrt{2/(n\xi)}\cos\left(\xi - \frac{1}{4}\pi(2n+1)\right), \quad H_n^{(1)}(\eta) \approx \sqrt{2/(n\xi)}\left(1 + i(4n^2 - 1)/\eta\right)e^{i(\eta - \frac{1}{4}\pi(2n+1))},$$

and known recurrent formulas

$$J_n'(\xi) = \frac{n}{\xi}J_n(\xi) - J_{n+1}(\xi), \quad H_n^{(1)\prime}(\zeta) = \frac{n}{\zeta}H_n^{(1)}(\zeta) - H_{n+1}^{(1)}(\zeta),$$

$$J_n''(\xi) = \frac{n(n-1)}{\xi^2}J_n(\xi) - \frac{2n+1}{\xi}J_{n+1}(\xi) + J_{n+2}(\xi),$$

$$H_n^{(1)\prime\prime}(\zeta) = \frac{n(n-1)}{\zeta^2}H_n^{(1)}(\zeta) - \frac{2n+1}{\zeta}H_{n+1}^{(1)}(\zeta) + H_{n+2}^{(1)}(\zeta),$$

$$J_n'''(\xi) = \frac{n(n-1)(n-2)}{\xi^3}J_n(\xi) - \frac{3n^2}{\xi^2}J_{n+1}(\xi) + \frac{3(n+1)}{\xi}J_{n+2}(\xi) - J_{n+3}(\xi),$$

$$H_n^{(1)\prime\prime\prime}(\zeta) = \frac{n(n-1)(n-2)}{\zeta^3}H_n^{(1)}(\zeta) - \frac{3n^2}{\zeta^2}H_{n+1}^{(1)}(\zeta) + \frac{3(n+1)}{\zeta}H_{n+2}^{(1)}(\zeta) - H_{n+3}^{(1)}(\zeta),$$

we get for short wavelength perturbations, $kR > 1$, the dispersion relation (from (2.221)) in the form

$$\frac{A_{\text{Re}}^+ + iA_{\text{Im}}}{A_{\text{Re}}^- - iA_{\text{Im}}} = \exp\left[i\left(\xi - \frac{2n+1}{4}\pi\right)\right], \tag{2.223}$$

where

$$A_{\text{Re}}^{\pm} = \beta_{in}[1+\Upsilon_{\nu}^{in}(1-\beta_{in}^2)](\Phi/\eta^{1/2}-M_a)^2 \pm \beta_{ex}\eta^{-1}(\rho_J/\rho_a)[1-\Upsilon_{\nu}^{ex}(1+\beta_{ex}^2)](\Phi-M_J)^2,$$

$$A_{\text{Im}} = (n/kR)t\{(\Phi/\eta^{1/2}-M_a)^2 + \eta^{-1}(\rho_J/\rho_a)(\Phi-M_J)^2$$
$$+ \Upsilon_{\nu}^{in}[\beta_{in}^2 n(3n+4)-1](\Phi/\eta^{1/2}-M_2)^2 - \Upsilon_{\nu}^{ex}\eta^{-1}(\rho_J/\rho_a)[\beta_{ex}^2 n(3n-4)+1](\Phi-M_J)^2\}.$$

In the case of pinching mode $n = 0$ for short wavelength perturbations, Equation (2.223) may be transformed to equation $\frac{A_{\text{Re}}^+}{A_{\text{Re}}^-} = \frac{\sqrt{2}}{2}(1 - i)\exp(i\xi)$. Dispersion relation (2.223) under condition $\nu_{\{J,a\}} \to 0$ may be transformed to the equation (similar to the equation in [21]), which is valid for all $n \geq 0$:

$$\frac{\beta_{in}L_1 + \beta_{ex}L_2 + i\frac{n}{kR}(L_1+L_2)}{\beta_{in}L_1 - \beta_{ex}L_2 - i\frac{n}{kR}(L_1+L_2)} = \exp\left[i(\xi - \frac{2n+1}{4}\pi)\right],$$

where $L_1 = (\Phi/\eta^{1/2} - M_a)^2$ and $L_2 = \eta^{-1}(\rho_J/\rho_a)(\Phi - M_J)^2$.

Numerical results. We continue investigation of the $n = 0$ pinching wave mode and $n = 1$ helical wave mode by numerical methods in order to find the wavenumber at which growth of these wave modes is a maximum, and how maximum growth is related to viscosity of the jet and external medium, i.e., to Mach number ($M_\infty \equiv M_a$) of the external flow. We concentrate on these two wave modes since they are the most likely to have directly observable consequences. This should be the case because longer wavelengths will be able to grow to larger amplitudes before nonlinear effects become important. As it was shown in [21], wave modes with high values of n travel around the jet more obliquely with respect to the flow and for a given wavelength value $\lambda_{r,z}$ at constant polar angle ϕ, high n corresponds to shorter true wavelength $\lambda_{r,z}$, and we expect the amplitudes of these wave modes to be too small to be directly observable.

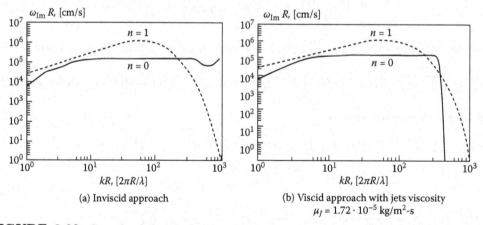

(a) Inviscid approach

(b) Viscid approach with jets viscosity
$\mu_J = 1.72 \cdot 10^{-5}$ kg/m^2-s

FIGURE 2.22: Growth of the pinching and helical waves as a function of the wave number $k = 2\pi R/\lambda$, normalized to the jet radius.

During numerical calculations we have assumed pressures of $P_J = P_{in} = 1.6$ atm, $P_\infty \equiv P_a = P_{ex} = 0.35$ atm, and the densities $\rho_J = 0.28$ kg/m^3, $\rho_a = 0.55$ kg/m^3, and specific heat ratios $\gamma_J = 1.29$, and $\gamma_a = 1.4$. This particular choice of pressures, densities and specific heat ratios is somewhat arbitrary, but represents conditions that might exist, respectively, in the plume's medium and in the external medium at altitude $H = 8$ km. The choice of pressure affects the results only in the value of the sound speeds ($a_J \simeq 1000$ m/s, and $a_a \simeq 375$ m/s) and calculated using pressure and some choice of density in the external medium and density ratio $\eta \simeq 0.141$ between jet and external medium. While the maximum growth rate is related to this choice of pressure because the maximum is a function of the jet velocity $u_{in} = M_J a_J$, the shapes of the curves and the wavelength at which maximum growth occurs, are functions of M_J and η independent of P. Equations (2.213) and (2.214) have the form

$$f_n(X, \Phi) = 0, \qquad (2.224)$$

where $n = 0, 1$ and the complex-valued function $f_n = f_n^{\mathrm{Re}} + i f_n^{\mathrm{Im}}$ is considered for two versions relative parameter v: zero and small, respectively. One may assume $\Phi = \Phi^{\mathrm{Re}} + i \Phi^{\mathrm{Im}}$ to be an implicit function of $X \in I$, where, by a physical sense, $I = [0.01, 10]$. Rewrite (2.224) as the nonlinear system of two real equations that will be solved numerically:

$$f_n^{\mathrm{Re}}(X, \Phi^{\mathrm{Re}}, \Phi^{\mathrm{Im}}) = 0, \quad f_n^{\mathrm{Im}}(X, \Phi^{\mathrm{Re}}, \Phi^{\mathrm{Im}}) = 0. \qquad (2.225)$$

By a physical sense we restrict our attention to the ranges $0 \le \Phi^{\mathrm{Re}} \le 5$ and $-0.4 \le \Phi^{\mathrm{Im}} \le 0.4$. The solutions (roots) of (2.225) for $X \in I$ are presented as intersections $R_x = \alpha_x^{\mathrm{Re}} \cap \alpha_x^{\mathrm{Im}}$ of the curves $\alpha_x^{\mathrm{Re}} = \{f_n^{\mathrm{Re}}(X, \Phi^{\mathrm{Re}}, \Phi^{\mathrm{Im}}) = 0\}$ and $\alpha_x^{\mathrm{Im}} = \{f_n^{\mathrm{Im}}(X, \Phi^{\mathrm{Re}}, \Phi^{\mathrm{Im}}) = 0\}$. For studying the roots' branches, one may graph two curves α_x^{Re}, α_x^{Im} in the domain $I_\Phi = [0, 5] \times [-0.4, 0.4]$ of the $(\Phi^{\mathrm{Re}}, \Phi^{\mathrm{Im}})$-plane. The topology of these curves is complicated. Since Φ^{Im}-range is small with respect to Φ^{Re}-range, the curve α_x^{Im} has oscillations in Φ^{Im}-direction that are grouped inside vertical strips, and the curve α_x^{Re} intersects each of these strips. Since the most sensitive perturbations are of low frequency (Φ^{Re}), we select solution $(\Phi^{\mathrm{Re}}(X), \Phi^{\mathrm{Im}}(X))$ in the strip with minimal abscissa. For a few number of isolated values of X we obtain bifurcations of the roots – the part of the curve α_x^{Re} degenerates to a small cycle and we jump to other components of α_x^{Im}. It also appears when the root's ordinate $\Phi^{\mathrm{Im}}(X)$ comes out of the prescribed range $[-0.4, 0.4]$.

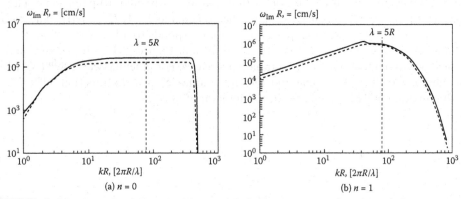

FIGURE 2.23: Growth of the pinching and helical waves as a function of the wave number $k = 2\pi R/\lambda$, normalized to the jet radius for two cases of ambient atmospheric flow. Solid line -$M_\infty = 2.4$, dashed line -$M_\infty = 0$. Jets viscosity -$\mu_J = 1.72 \times 10^{-5} \frac{\mathrm{kg}}{\mathrm{m^2 s}}$.

First, show that growth rate of both the pinching wave mode and the helical wave mode under plumes and external medium viscosities to be neglected – the inviscid case. Figure 2.22 shows us the behavior of the growth rate of these two wave modes for the internal Mach number $M_J = 3.0$ and for the external Mach number $M_\infty = 2.4$. For helical mode the value of the perturbations phase velocity $u_{ph}^\omega = \omega_{\mathrm{Im}} R$ has asymptotic $\sim kR$, see (2.220a,b), at $kR \ll 1$, while the pinching mode of perturbations phase velocity has asymptotic $\sim (kr)^2$, see (2.217). At the same time the helical mode perturbations phase velocity has maximum at $0.2 < kR < 1$ and strongly decreases until zero at $kR \gg 1$, while for the pinching mode the value $u_{ph}^\omega \to$ const at $kR \gg 1$.

Figure 2.23 shows us that behavior of the value u_{ph}^ω weakly depends on the jet's viscosity (under condition that we consider low viscosity jets), while the pinching oscillations

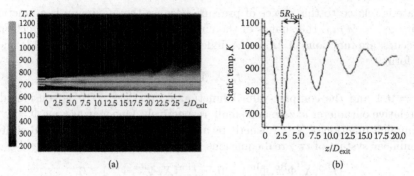

FIGURE 2.24: Contour (a) and graphic of jet's temperature along jet's centerline (b) numerical calculations under $D_{exit} = 2R$.

disappear at $kR > 3$, i.e., short wavelength oscillations die down for even too low viscosity jets. Figure 2.24 presents good consistent of our numerical results and the analytical results of instability research. Indeed (see Figs. 2.23 and 2.24), the perturbation waves with wavelength equal to $2.5D_{exit} - 3.0D_{exit}$, where $D_{exit} = 2R$ is the nozzle exit diameter. Note that the maximum of u_{ph}^{ω} (for helical mode) corresponds to $kR \simeq 0.8$, i.e., to the wavelength $\lambda \simeq 2.5D_{exit}$.

Figures 2.15 and 2.26 present the parameterized curves $c(X) = (\Phi^{Re}(X), \Phi^{Im}(X))$ for the versions of (2.225) under discussion: with zero and small viscosity parameter, with pinching and helical waves. In all these cases, by a physical sense, we divide the domain I onto three intervals $I_1 = [0.01, 0.1]$, $I_2 = [0.1, 1]$ and $I_3 = [1, 10]$. On each of these intervals we take a sufficiently dense net of points (for example, a regular 10-point net). Thus the parameterized curves $c(X)$ consist of three parts denoted by I, II and III. The value of $M_\infty = 2.4$ is used as practically upper bound; it corresponds to all curves in Fig. 2.15. The solid lines in Fig. 2.26 are the same as the solid lines in Figures 2.17(a,b), correspondingly. For helical waves the function $\Phi^{Im}(X)$ rapidly increases. For pinching waves $\Phi^{Im}(X)$ reaches its maximum ≈ 0.18 at $X \approx 0.05$ and for large values $X > 5$ the function $\Phi^{Im}(X) \to 0$, while the function $\Phi^{Re}(X)$ oscillates in the interval of $[3, 4]$.

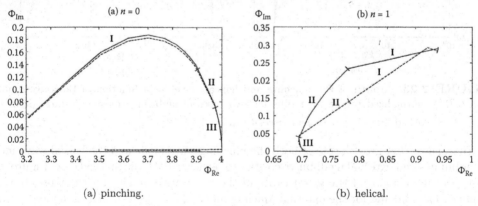

FIGURE 2.25: Phase trajectories of (2.213) for the pinching (a) and helical (b) waves for $M_\infty = 2.4$. Solid line – for viscous approach (with $\mu_J = 1.72 \times 10^{-5} \frac{kg}{m^2 s}$), and dashed line – for inviscid approach; zones: I – $X \in [0.01, 0.1]$, II – $X \in [0.1, 1.0]$, and III – $X \in [1.0, 10.0]$.

The dispersion equation (2.213) for linear perturbing model of viscous compressible jet has been developed. Its simplified cases coincide with various forms of dispersion equation obtained by other authors. In resonance cases the above equation (2.213) is reduced to simple algebraic equations. Moreover, (2.213) allows to study dependence of the complex function $\Phi = \omega/ka_J$ (which is nondimensional phase velocity) on the parameter $X = kR$ and to present numerical results in a graphical way.

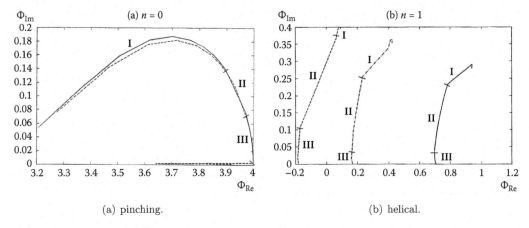

(a) pinching.   (b) helical.

FIGURE 2.26: Phase trajectories for (2.213) for the pinching (a) and helical (b) waves for $\mu_J = 1.72 \times 10^{-5} \frac{\text{kg}}{\text{m}^2\text{s}}$. Solid line – $M_\infty = 2.4$, dashed dots – $M_\infty = 1.0$, and dashed line – $M_\infty = 0.1$; zones: I – $X \in [0.01, 0.1]$, II – $X \in [0.1, 1.0]$, and III – $X \in [1.0, 10.0]$.

It can be watched that pinching and helical wave modes increase and the maximum of their oscillations corresponds to the wavelengths longer than or on the order of the instantaneous jet's radius. For helical mode the value of the perturbations phase velocity $u_{ph}^\omega = \omega_{Im} R$ has asymptotic $\sim kR$, while the pinching mode of these perturbations has asymptotic $\sim (kr)^2$ for $kR \ll 1$. The helical mode perturbations phase velocity has maximum in interval $0.2 < kR < 1$ and decreases rapidly to zero for $kR > 1$, while the pinching mode of the phase velocity goes to nonzero constant for $kR \gg 1$. The phase velocity weakly depends on the jets viscosity (for low viscosities), while the pinching oscillations disappear for $kR > 3$ in the viscous case by comparison with inviscid case, i.e., short wavelength oscillations die down even for the case of low viscosities. The analytical results obtained in present paper are of a good consistence with numerical results.

2.3.4 Supersonic viscous jet with Gaussian sound velocity distribution

Assumption of the sound velocity radial distribution $a(r)$ may be considered only as a first approximation. In reality, there is a fairly strong jump at the boundary between jet and the ambient media. Exact analytical solution of such problems is very difficult if it is possible. Therefore, as the next step, it makes sense to choose the real distribution obtained from numerical experiments. From many numerical experiments we obtained the radial distribution which shape has a strong Gaussian peak on the boundary layer of the jet while it is nearly constant in the rest of the layer [15, 16]. This dependence can be approximated with sufficient accuracy by the model function of the form:

$$a = a_0\left[1 + \frac{1}{2}e^{-(r-R_0)^2/\sigma^2}\right]. \tag{2.226}$$

Basic equations. To develop the model for KHI we are based again on continuity, momentum and energy equations of fluid mechanics [30, 39, 45]. Continuity and momentum equations have a form [30]:

$$\frac{\partial \rho}{\partial t} + \text{div}(\rho \mathbf{u}) = 0, \qquad \frac{\partial \mathbf{u}}{\partial t} + (\mathbf{u} \cdot \nabla)\mathbf{u} = \frac{1}{\rho}\nabla P + \nu \nabla^2 \mathbf{u}, \tag{2.227}$$

where $(x, y, z) \in \mathbb{R}^3$, $\nu = \mu/\rho$ – kinematic viscosity, $\rho(x, y, z, t)$ – density, $\mathbf{u}(x, y, z, t) = (u_x, u_y, u_z)$ – fluid velocity, $P(x, y, z, t)$ – pressure. In the more simple inviscid case the value of viscosity is assumed to be zero $\mu = 0$. For such low viscous media as gaseous, we can assume $\mu = \mu_0 = \mathrm{const}$.

Energy equation has the following form [30, 45]:

$$\frac{\partial}{\partial t}\left(\frac{1}{2}\rho u^2 + \rho\varepsilon\right) = -\nabla \cdot \left[\rho\mathbf{u}\left(\varepsilon + \frac{1}{2}u^2\right) + P\mathbf{u}\right] + \mu_0[c_P\nabla^2 T - \Pi : \nabla\mathbf{u}] \qquad (2.228)$$

where Π is the viscous stress tensor

$$\left(\Pi : \nabla\mathbf{u}\right)_{ik} = \frac{\partial u_i}{\partial x_k} + \frac{\partial u_k}{\partial x_i} - \frac{2}{3}\delta_{ik}\frac{\partial u_j}{\partial x_j}, \qquad (2.229)$$

and ε – *internal energy* per unit mass, c_P – *heat capacity* under constant pressure P. For low viscous media we can neglect the viscous stress tensor. Hence, we can rewrite the energy equation (2.228) as

$$\frac{\partial}{\partial t}\left(\frac{1}{2}\rho u^2 + \rho\varepsilon\right) = -\nabla \cdot \left[\rho\mathbf{u}\left(\varepsilon + \frac{1}{2}u^2\right) + P\mathbf{u}\right] + \mu_0 c_p\nabla^2 T. \qquad (2.230)$$

Proposition 13 *For adiabatic flow, the internal energy ε is given by $\varepsilon = \frac{A}{\gamma-1}T$, where A – universal gas constant per unit mass, $\gamma = c_P/c_V$ – isentropic ratio. So, the energy equation may be rewritten as*

$$\frac{\partial}{\partial t}\left(\frac{A}{\gamma-1}\rho T + \frac{1}{2}\rho u^2\right) = -\nabla \cdot \left[\rho\mathbf{u}\left(\frac{A}{\gamma-1}T + \frac{1}{2u^2}\right) + P\mathbf{u}\right] + \mu_0 c_p\nabla^2 T$$

Proof. Consider the ideal gas equation satisfied thermodynamic relation [45]

$$V = A\rho T, \qquad (2.231a)$$

$$T\,ds = d\varepsilon + P\,dV, \qquad (2.231b)$$

where s – specific entropy. For adiabatic flow, we will use adiabatic condition

$$\frac{ds}{dt} = 0, \quad s = c_V \ln(P/\rho^\gamma) + \mathrm{const} \qquad (2.232)$$

where c_V – heat capacity under constant volume V. Hence, (2.232) may be rewritten in the form

$$\frac{1}{P}\frac{dP}{dt} + \frac{\gamma}{V}\frac{dV}{dt} = 0. \qquad (2.233)$$

Using thermodynamic relation (2.231b) and adiabatic condition (2.232), we obtain

$$\frac{d\varepsilon}{dt} = -P\frac{dV}{dt}. \qquad (2.234)$$

Equations (2.231a) and (2.233) give

$$P\frac{dV}{dt} + V\frac{dP}{dt} = A\frac{dT}{dt}, \quad V\frac{dP}{dt} + \gamma P\frac{dV}{dt} = 0. \qquad (2.235)$$

Equations (2.234) and (2.235) lead to

$$P\frac{dV}{dt} = \frac{A}{\gamma-1}\frac{dT}{dt}. \qquad (2.236)$$

Substituting (2.236) in (2.234) gives $\frac{d\varepsilon}{dt} = \frac{A}{\gamma-1}\frac{dT}{dt}$. Integrating (2.229), we obtain

$$\varepsilon = \frac{A}{\gamma-1}T + \text{const.} \qquad (2.237)$$

Since ε is the inner energy of the gas, when the temperature T is zero (Kelvin), the gaseous micro-particles (molecules, atoms, or for plasmas, electrons and ions) have zero velocity and so zero kinetic energy. Hence, we assume this constant to be zero, const $= 0$. We find from (2.237) that $\varepsilon = \frac{A}{\gamma-1}T$. $\qquad\square$

Linearization. Now we use the perturbation method we write expansions:

$$\mathbf{u} = \mathbf{u}_0 + \mathbf{u}', \quad \rho = \rho_0 + \rho', \quad P = P_0 + P', \quad T = T_0 + T'. \qquad (2.238)$$

Remark 12 We use the perturbation theory under approach that perturbed values $\mathbf{u}'$, ρ', P', T' are smaller than unperturbed background (phone) variables $\mathbf{u}_0$, ρ_0, P_0, T_0 (stationary solution); this definition is true because an unperturbed media may be considered as a phone where the perturbations to be considered, so the products of perturbed variables are smaller (low values of the second order) than the products of phone's variables and perturbed ones.

First, consider linearized continuity, momentum and energy equations.

Proposition 14 *Linearized continuity and momentum equations (2.227) have the form*

$$\frac{\partial \rho'}{\partial t} + \rho_0 \operatorname{div} \mathbf{u}' + \mathbf{u}_0 \nabla \rho' + (\rho' \operatorname{div} \mathbf{u}_0 + \mathbf{u}' \nabla \rho_0) = 0, \qquad (2.239a)$$

$$\rho_0 \left(\frac{\partial \mathbf{u}'}{dt} + (\mathbf{u}_0 \cdot \nabla)\mathbf{u}' \right) + ((\rho_0 \mathbf{u}' + \rho' \mathbf{u}_0) \cdot \nabla)\mathbf{u}_0 = -\nabla P' + \mu_0 \nabla^2 \mathbf{u}'. \qquad (2.239b)$$

Proof. Substituting (2.238), without $T = T_0 + T'$, in (2.227), we obtain

$$\frac{\partial}{\partial t}(\rho_0 + \rho') + \operatorname{div}[(\rho_0 + \rho')(\mathbf{u}_0 + \mathbf{u}')] = 0,$$
$$(\rho_0 + \rho')\left[\frac{\partial}{\partial t}(\mathbf{u}_0 + \mathbf{u}') + ((\mathbf{u}_0 + \mathbf{u}') \cdot \nabla)(\mathbf{u}_0 + \mathbf{u}')\right] = -\nabla(P_0 + P') + \mu_0 \nabla^2(\mathbf{u}_0 + \mathbf{u}') = 0. \qquad (2.240)$$

Neglect the following products of the perturbations in (2.240) due to our limitation of first order perturbed terms, see Remark 12: $\rho'\mathbf{u}'$, $\operatorname{div}(\rho'\mathbf{u}')$, $(\mathbf{u}' \cdot \nabla)\mathbf{u}'$, $(\rho'\mathbf{u}' \cdot \nabla)\mathbf{u}_0$, $\rho'\mathbf{u}_0\nabla\mathbf{u}'$, $(\rho'\mathbf{u}'\cdot\nabla)\mathbf{u}'$. We also use the solutions ρ_0, $\mathbf{u}_0$, P_0 to unperturbed equations

$$\frac{\partial \rho_0}{\partial t} + \operatorname{div}(\rho_0 \mathbf{u}_0) = 0, \quad \frac{\partial \mathbf{u}_0}{\partial t} + (\mathbf{u}_0 \cdot \nabla)\mathbf{u}_0 = -\frac{1}{\rho_0}\nabla P_0 + \nu_0 \nabla^2 \mathbf{u}_0.$$

For steady-state case we obtain (2.239a,b) immediately from (2.240). $\qquad\square$

Proposition 15 *Linearized energy equation (2.230) has a view*

$$\frac{A}{\gamma-1}\rho_0\frac{\partial T'}{\partial t} + \rho_0 \mathbf{u}_0 \cdot \frac{\partial \mathbf{u}'}{\partial t} + \left(\frac{A}{\gamma-1}T + \frac{1}{2}u_0^2\right)\frac{\partial \rho'}{\partial t} =$$
$$-\nabla \cdot \mathbf{u}_0\left[\frac{A\rho_0}{\gamma-1}T' + P' + \left(\frac{AT_0}{\gamma-1} + \frac{1}{2}u_0^2\right)\rho'\right] - \nabla \cdot \mathbf{u}'\left[\left(\frac{3}{2}\rho_0 u_0^2 + P_0 + \frac{A\rho_0 T_0}{\gamma-1}\right)\right] + \mu_0 c_P \nabla^2 T'. \qquad (2.241)$$

Proof. Substituting (2.238) in (2.230) we obtain

$$\frac{\partial}{\partial t}\left[\frac{A}{\gamma-1}(\rho_0 + \rho')(T_0 + T') + \frac{1}{2}(\rho_0 + \rho')(\mathbf{u}_0 + \mathbf{u}')^2\right]$$
$$= -\nabla \cdot \left\{(\rho_0 + \rho')(\mathbf{u}_0 + \mathbf{u}')\left[\frac{A}{\gamma-1}(T_0 + T') + \frac{1}{2}(\mathbf{u}_0 + \mathbf{u}')^2\right]\right\}$$
$$- \nabla \cdot [(P_0 + P')(\mathbf{u}_0 + \mathbf{u}')] + \mu_0 c_p \nabla^2(T_0 + T').$$

Neglect products of the perturbed values $\rho'T'$, u'^2, $\rho'u'^2$, $\rho'\mathbf{u}'$, $P'\mathbf{u}'$, $\mathbf{u}'T'$ because they are the second-order values (Remark 12). We use unperturbed equation

$$\frac{\partial}{\partial t}\left(\frac{A}{\gamma-1}\rho_0 T_0 + \frac{1}{2}\rho_0 u_0^2\right) = -\nabla\cdot\left[\rho_0\mathbf{u}_0\left(\frac{A}{\gamma-1}T_0 + \frac{1}{2}u_0^2\right) + P_0\mathbf{u}_0\right] + \mu_0 c_p \nabla^2 T_0$$

to obtain

$$\frac{\partial}{\partial t}\left[\frac{A}{\gamma-1}(\rho_0 T' + \rho' T_0) + \rho_0\mathbf{u}_0\cdot\mathbf{u}' + \frac{1}{2}\rho' u_0^2\right]$$
$$= -\nabla\cdot\left[\frac{A\rho_0 T_0}{\gamma-1}T' + \mathbf{u}_0 P' + \left(\frac{A\rho T_0}{\gamma-1} + \frac{1}{2}u_0^2\right)\rho'\mathbf{u}_0 + \left(\frac{AT_0}{\gamma-1} + \frac{P_0}{\rho_0} + \frac{3}{2}u_0^2\right)\rho_0\mathbf{u}'\right] + \mu_0\nabla^2 T'.$$
$$(2.242)$$

Equation (2.242) leads to (2.241). $\qquad\square$

To analyze a diverging flow of an opening angle equal zero, we apply cylindrical coordinates (r, ϕ, z), in which the axial axis is along the centerline of the diverging jet and radial axis is across the jet. The cylindrical boundary between jet and external medium is given $r = R_0$. Denote $\mathbf{u}_0 = (u_{0r}, u_{0\phi}, u_{0z})$, $\mathbf{u}' = (u_r', u_\phi', u_z')$ and assume $\rho_0 = \rho_0(r, z)$, $P_0 = P_0(r, z)$, $T_0 = T_0(r, z)$.

Proposition 16 ([15, 17]) *Equations (2.239a,b) in cylindrical coordinates have the form*

$$\frac{\partial\rho'}{\partial t} + \left(u_r'\frac{\partial\rho_0}{\partial r} + u_z'\frac{\partial\rho_0}{\partial z}\right) + \rho_0\left(\frac{u_r'}{r} + \frac{\partial u_r'}{\partial r} + \frac{1}{r}\frac{\partial u_\phi'}{\partial\phi} + \frac{\partial u_z'}{\partial z}\right) = 0,$$
$$\frac{\partial u_r'}{\partial t} = -\frac{1}{\rho_0}\frac{\partial P'}{\partial r} + \nu_0\left(\frac{\partial^2 u_r'}{\partial r^2} + \frac{1}{r^2}\frac{\partial^2 u_r'}{\partial\phi^2} + \frac{\partial^2 u_r'}{\partial z^2} + \frac{1}{r}\frac{\partial u_r'}{\partial r} - \frac{2}{r^2}\frac{\partial u_\phi'}{\partial\phi} - \frac{u_r'}{r^2}\right),$$
$$\frac{\partial u_\phi'}{\partial t} = -\frac{1}{\rho_0 r}\frac{\partial P'}{\partial\phi} + \nu_0\left(\frac{\partial^2 u_\phi'}{\partial r^2} + \frac{1}{r^2}\frac{\partial^2 u_\phi'}{\partial\phi^2} + \frac{\partial^2 u_\phi'}{\partial z^2} + \frac{1}{r}\frac{\partial u_\phi'}{\partial r} + \frac{2}{r^2}\frac{\partial u_r'}{\partial\phi} - \frac{u_\phi'}{r^2}\right),$$
$$\frac{\partial u_z'}{\partial t} = -\frac{1}{\rho_0}\frac{\partial P'}{\partial z} + \nu_0\left(\frac{\partial^2 u_z'}{\partial r^2} + \frac{1}{r^2}\frac{\partial^2 u_z'}{\partial\phi^2} + \frac{\partial^2 u_z'}{\partial z^2} + \frac{1}{r}\frac{\partial u_z'}{\partial r}\right).$$
$$(2.243)$$

Proof. We use formulas from vector analysis for *divergence, gradient* and *Laplace operator* in cylindrical coordinates to translate linearized continue and momentum equations (2.239a,b). Note that

$$\nabla\cdot\mathbf{u} = u_r'/r + \partial u_r'/\partial r + (1/r)\partial u_\phi'/\partial\phi + \partial u_z'/\partial z,$$

$$\nabla\rho' = \left(\frac{\partial\rho'}{\partial r}, 0, \frac{\partial\rho'}{\partial z}\right), \quad \nabla\rho_0 = \left(\frac{\partial\rho_0}{\partial r}, 0, \frac{\partial\rho_0}{\partial z}\right) \quad \left(\text{with } \frac{\partial\rho'}{\partial\phi} = \frac{\partial\rho_0}{\partial\phi} = 0\right),$$

$$\nabla\cdot\mathbf{u}_0 = \frac{u_{0r}}{r} + \frac{\partial u_{0r}}{\partial r} + \frac{1}{r}\frac{\partial u_{0\phi}}{\partial\phi} + \frac{\partial u_{0z}}{\partial z}.$$

Substituting $\nabla\cdot\mathbf{u}$, $\nabla\rho'$, $\nabla\rho_0$ in (2.239a), we obtain

$$\frac{\partial\rho'}{\partial t} + \left(u_{0r}\frac{\partial\rho'}{\partial r} + u_{0z}\frac{\partial\rho'}{\partial z}\right) + \left(u_r'\frac{\partial\rho_0}{\partial r} + u_z'\frac{\partial\rho_0}{\partial z}\right)$$
$$+\rho_0\left(\frac{u_r'}{r} + \frac{\partial u_r'}{\partial r} + \frac{1}{r}\frac{\partial u_\phi'}{\partial\phi} + \frac{\partial u_z'}{\partial z}\right) + \rho'\left(\frac{u_{0r}}{r} + \frac{\partial u_{0r}}{\partial r} + \frac{1}{r}\frac{\partial u_{0\phi}}{\partial\phi} + \frac{\partial u_{0z}}{\partial z}\right).$$
$$(2.244)$$

Also we use the following:

$$((\rho_0\mathbf{u}' + \rho'\mathbf{u}_0)\cdot\nabla)\mathbf{u}_0 = \Big\{(\rho_0 u_r' + \rho' u_{0r})\frac{\partial u_{0r}}{\partial r} + (\rho_0 u_\phi' + \rho' u_{0\phi})\frac{1}{r}\frac{\partial u_{0r}}{\partial\phi} + (\rho_0 u_z' + \rho' u_{0z})\frac{\partial u_{0r}}{\partial z},$$
$$(\rho_0 u_r' + \rho' u_{0r})\frac{\partial u_{0\phi}}{\partial r} + (\rho_0 u_\phi' + \rho' u_{0\phi})\frac{1}{r}\frac{\partial u_{0\phi}}{\partial\phi} + (\rho_0 u_z' + \rho' u_{0z})\frac{\partial u_{0\phi}}{\partial z},$$
$$(\rho_0 u_r' + \rho' u_{0r})\frac{\partial u_{0z}}{\partial r} + (\rho_0 u_\phi' + \rho' u_{0\phi})\frac{1}{r}\frac{\partial u_{0z}}{\partial\phi} + (\rho_0 u_z' + \rho' u_{0z})\frac{\partial u_{0z}}{\partial z}\Big\},$$
$$\nabla P = \left(\frac{\partial P'}{\partial r}, \frac{1}{r}\frac{\partial P'}{\partial\phi}, \frac{\partial P'}{\partial z}\right).$$

Then, substituting $\mathbf{u}_0 \cdot \nabla \mathbf{u}'$, $\nabla P'$, $((\rho_0 \mathbf{u}' + \rho' \mathbf{u}_0) \cdot \nabla)\mathbf{u}_0$ in (2.239b) we obtain

$$\rho_0\left(\frac{\partial u'_r}{\partial t} + u_{0r}\frac{\partial u'_r}{\partial r} + u_{0\phi}\frac{1}{r}\frac{\partial u'_r}{\partial \phi} + u_{0z}\frac{\partial u'_r}{\partial z}\right) + (\rho_0 u'_r + \rho' u_{0r})\frac{\partial u_{0r}}{\partial r} + (\rho_0 u'_\phi + \rho' u_{0\phi})\frac{1}{r}\frac{\partial u_{0r}}{\partial \phi}$$
$$+(\rho_0 u'_z + \rho' u_{0z})\frac{\partial u_{0r}}{\partial z} = -\frac{\partial P'}{\partial r} + \mu_0\left(\frac{\partial^2 u'_r}{\partial r^2} + \frac{1}{r^2}\frac{\partial^2 u'_r}{\partial \phi^2} + \frac{\partial^2 u'_r}{\partial z^2} + \frac{1}{r}\frac{\partial u'_r}{\partial r} - \frac{2}{r^2}\frac{\partial u'_\phi}{\partial \phi} - \frac{u'_r}{r^2}\right),$$
$$\rho_0\left(\frac{\partial u'_\phi}{\partial t} + u_{0r}\frac{\partial u'_\phi}{\partial r} + u_{0\phi}\frac{1}{r}\frac{\partial u'_\phi}{\partial \phi} + u_{0z}\frac{\partial u'_\phi}{\partial z}\right) + (\rho_0 u'_r + \rho' u_{0r})\frac{\partial u_{0\phi}}{\partial r} + (\rho_0 u'_\phi + \rho' u_{0\phi})\frac{1}{r}\frac{\partial u_{0\phi}}{\partial \phi}$$
$$+(\rho_0 u'_z + \rho' u_{0z})\frac{\partial u_{0\phi}}{\partial z} = -\frac{1}{r}\frac{\partial P'}{\partial \phi} + \mu_0\left(\frac{\partial^2 u'_\phi}{\partial r^2} + \frac{1}{r^2}\frac{\partial^2 u'_\phi}{\partial \phi^2} + \frac{\partial^2 u'_\phi}{\partial z^2} + \frac{1}{r}\frac{\partial u'_\phi}{\partial r} + \frac{2}{r^2}\frac{\partial u'_r}{\partial \phi} - \frac{u'_\phi}{r^2}\right),$$
$$\rho_0\left(\frac{\partial u'_z}{\partial t} + u_{0r}\frac{\partial u'_z}{\partial r} + u_{0\phi}\frac{1}{r}\frac{\partial u'_z}{\partial \phi} + u_{0z}\frac{\partial u'_z}{\partial z}\right) + (\rho_0 u'_r + \rho' u_{0r})\frac{\partial u_{0z}}{\partial r} + (\rho_0 u'_\phi + \rho' u_{0\phi})\frac{1}{r}\frac{\partial u_{0z}}{\partial \phi}$$
$$+(\rho_0 u'_{0z} + \rho' u_{0z})\frac{\partial u_{0z}}{\partial z} = -\frac{\partial P'}{\partial z} + \mu_0\left(\frac{\partial^2 u'_z}{\partial r^2} + \frac{1}{r^2}\frac{\partial^2 u'_z}{\partial \phi^2} + \frac{\partial^2 u'_z}{\partial z^2} + \frac{1}{r}\frac{\partial u'_z}{\partial r}\right).$$

$$(2.245)$$

Since the jet is assumed to be rotational symmetric about the axial axis we have $u_{0\phi} = 0$. For simplification we will consider the flow with only one nonzero initial velocity component $u_{0z} = u$, while the radial and azimuthal component we assume to be zero, i.e., $u_{0r} = u_{0\phi} = 0$ at $z = 0$. Moreover, we may also let $u_{0z} = 0$ and then to account the difference u as additional term for the boundary conditions. Finally, we conclude that linearized continuity and momentum equations (2.239a,b) in cylindrical coordinates using (2.244) and (2.245) and $\mathbf{u}_0 = 0$, are (2.243). $\qquad \square$

Proposition 17 *Energy equation (2.241), in cylindrical coordinates, has the view:*

$$\frac{A}{\gamma-1}\left(\rho_0\frac{\partial T'}{\partial t} + T_0\frac{\partial \rho'}{\partial t}\right) = -\frac{P_0 u'_r}{r} - u'_r\frac{\partial P_0}{\partial r} - P_0\frac{\partial u'_r}{\partial r} - \frac{P_0}{r}\frac{\partial u'_\phi}{\partial \phi} - u'_z\frac{\partial P_0}{\partial z} - P_0\frac{\partial u'_z}{\partial z}$$
$$-\frac{A}{\gamma-1}\left(\frac{\rho_0 T_0 u'_r}{r} + T_0 u'_r\frac{\partial \rho_0}{\partial r} + \rho_0 u'_r\frac{\partial T_0}{\partial r} + \rho_0 T_0\frac{\partial u'_r}{\partial r}\frac{\rho_0 T_0}{r}\frac{\partial u'_\phi}{\partial \phi} + T_0 u'_z\frac{\partial \rho_0}{\partial z} + \rho_0 u'_z\frac{\partial T_0}{\partial z} + \rho_0 T_0\frac{\partial u'_z}{\partial z}\right)$$
$$+\mu_0 c_P\left(\frac{1}{r}\frac{\partial T'}{\partial r} + \frac{\partial^2 T'}{\partial r^2} + \frac{1}{r^2}\frac{\partial^2 T'}{\partial \phi^2} + \frac{\partial^2 T'}{\partial z^2}\right).$$

$$(2.246)$$

Proof. Using (2.242) and (2.243), we will transform linearized energy equation (2.242). First, we deal with LHS term of (2.242):

$$\frac{A\rho_0}{\gamma-1}\frac{\partial T'}{\partial t} + \rho_0\left(u_{0r}\frac{\partial u'_r}{\partial t} + r u_{0\phi}\frac{\partial u'_\phi}{\partial t} + u_{0z}\frac{\partial u'_z}{\partial t}\right) + \left[\frac{AT_0}{\gamma-1} + \frac{1}{2}(u_{0r}^2 + r u_{0\phi}^2 + u_{0z}^2)\right]\frac{\partial \rho'}{\partial t}.$$

$$(2.247)$$

The RHS of (2.242) we break into six parts as follows:

$$\nabla \cdot (\rho_0 \mathbf{u}_0 T') = \left(\frac{\rho_0 u_{0r} T'}{r} + u_{0r} T'\frac{\partial \rho_0}{\partial r} + \rho_0 T'\frac{\partial u_{0r}}{\partial r} + \rho_0 u_{0r}\frac{\partial T'}{\partial r}\right)$$
$$+\frac{1}{r}\left(u_{0\phi} T'\frac{\partial \rho_0}{\partial \phi} + \rho_0 T'\frac{\partial u_{0\phi}}{\partial \phi} + \rho_0 u_{0\phi}\frac{\partial T'}{\partial \phi}\right) + \left(u_{0z} T'\frac{\partial \rho_0}{\partial z} + \rho_0 T'\frac{\partial u_{0z}}{\partial z} + \rho_0 u_{0z}\frac{\partial T'}{\partial z}\right),$$
$$\nabla \cdot (\rho_0 \mathbf{u}_0 \rho') = \left(\frac{\rho_0 u_{0r} \rho'}{r} + T_0 \rho'\frac{\partial u_{0r}}{\partial r} + u_{0r} T_0 \rho'\frac{\partial T_0}{\partial r} + T_0 u_{0r}\frac{\partial \rho'}{\partial r}\right)$$
$$+\frac{1}{r}\left(T_0 \rho'\frac{\partial u_{0\phi}}{\partial \phi} + u_{0\phi} T_0 \rho'\frac{\partial T_0}{\partial \phi} + T_0 u_{0\phi}\frac{\partial \rho'}{\partial \phi}\right) + \left(T_0 \rho'\frac{\partial u_{0z}}{\partial z} + u_{0z} T_0 \rho'\frac{\partial T_0}{\partial z} + T_0 u_{0z}\frac{\partial \rho'}{\partial z}\right),$$
$$\nabla \cdot (\mathbf{u}_0 u_{0z}^2 \rho') = \frac{1}{r}\left[\frac{\partial}{\partial r}(r u_{0r} u_{0z}^2 \rho') + \frac{\partial}{\partial \phi}(u_{0\phi} u_{0z}^2 \rho') + \frac{\partial}{\partial z}(r u_{0z} u_{0z}^2 \rho')\right],$$
$$\nabla \cdot (\rho_0 u_0^2 \mathbf{u}') = \left(\frac{\rho_0 u_{0z}^2 u'_r}{r} + u_{0z}^2 u'_r\frac{\partial \rho_0}{\partial r} + 2\rho_0 u'_r u_{0z}\frac{\partial u_{0z}}{\partial r} + \rho_0 u_{0z}^2\frac{\partial u'_r}{\partial r}\right)$$
$$+\frac{1}{r}\left(u_{0z}^2 u'_r\frac{\partial \rho_0}{\partial \phi} + 2\rho_0 u'_r u_{0z}\frac{\partial u_{0z}}{\partial \phi} + \rho_0 u_{0z}^2\frac{\partial u'_r}{\partial \phi}\right) + \left(u_{0z}^2 u'_r\frac{\partial \rho_0}{\partial z} + 2\rho_0 u'_r u_{0z}\frac{\partial u_{0z}}{\partial z} + \rho_0 u_{0z}^2\frac{\partial u'_r}{\partial z}\right),$$
$$\nabla \cdot (P_0 \mathbf{u}') = \left(\frac{u'_r P_0}{r} + u'_r\frac{\partial P_0}{\partial r} + P_0\frac{\partial u'_r}{\partial r}\right) + \frac{1}{r}\left(u'_\phi\frac{\partial P_0}{\partial \phi} + P_0\frac{\partial u'_\phi}{\partial \phi}\right) + \left(u'_z\frac{\partial P_0}{\partial z} + P_0\frac{\partial u'_z}{\partial z}\right),$$
$$\nabla \cdot (\rho_0 T_0 \mathbf{u}') = \left(\frac{u'_r \rho_0 T_0}{r} + u'_r T_0\frac{\partial \rho_0}{\partial r} + u'_r \rho_0\frac{\partial T_0}{\partial r} + \rho_0 T_0\frac{\partial u'_r}{\partial r}\right)$$
$$+\frac{1}{r}\left(u'_\phi T_0\frac{\partial \rho_0}{\partial \phi} + u'_\phi \rho_0\frac{\partial T_0}{\partial \phi} + \rho_0 T_0\frac{\partial u'_\phi}{\partial \phi}\right) + \left(u'_z T_0\frac{\partial \rho_0}{\partial z} + u'_z \rho_0\frac{\partial T_0}{\partial z} + \rho_0 T_0\frac{\partial u'_z}{\partial z}\right).$$

Substitute the first term of (2.247) and all six parts of the second and third terms into (2.242) and use the basic assumption $\mathbf{u}_0 = 0$ to deduce (2.246). $\qquad \square$

Plane wave expansion. We will find solution in the *plane wave expansion* form for the quantities:

$$u'_r = u_{r1}(r)e^{kz+n\phi-\omega t}, \quad u'_\phi = u_{\phi1}\Psi(r)e^{kz+n\phi-\omega t}, \quad u'_r = u_{z1}\Psi(r)e^{kz+n\phi-\omega t},$$
$$\rho' = \rho_1(r)\Psi(r)e^{kz+n\phi-\omega t}, \quad T' = T_1\Psi(r)e^{kz+n\phi-\omega t}, \tag{2.248}$$

where $n \in \mathbb{N}$ is the wave number, ω – frequency, $u_{r1}(r)$, $\Psi(r)$ – smooth functions.

Proposition 18 *Continuity and momentum equations (2.243), in plane wave expansion, have view:*

$$\Psi(r)\left(u_{z1}\frac{\partial\rho_0}{\partial z}-i\omega\rho_1\right)+u_{r1}(r)\frac{\partial\rho_0}{\partial z}+\rho_0\left[\Psi(r)(iku_{z1}+in\frac{u_{\phi1}}{r})+\frac{u_{r1}(r)}{r}+\frac{du_{r1}}{dr}\right]=0, \tag{2.249a}$$

$$u_{r1}(r)+\frac{i}{\omega}\frac{\rho_1}{\rho_0}\frac{\partial P'}{\partial\rho'}\Psi'(r)-\frac{i}{\omega}\nu_0(\nabla^2\mathbf{u}')_r=0, \tag{2.249b}$$

$$u_{\phi1}-\frac{n}{\omega}\frac{\rho_1}{\rho_0}\frac{1}{r}\frac{\partial P'}{\partial\rho'}-\frac{i}{\omega}\nu_0\frac{1}{\Psi(r)}(\nabla^2\mathbf{u}')_\phi=0, \tag{2.249c}$$

$$u_{z1}-\frac{k}{\omega}\frac{\rho_1}{\rho_0}\frac{\partial P'}{\partial\rho'}-\frac{i}{\omega}\nu_0\frac{1}{\Psi(r)}(\nabla^2\mathbf{u}')_z=0. \tag{2.249d}$$

Proof. Note that (2.244) and (2.245) contain differential $(\partial\cdot/\partial t)+(u_{0r}\partial\cdot/\partial t)+(u_{0z}\partial\cdot/\partial t)$ which transformed to $i\omega+iku_{0z}+(0\cdot\partial\cdot/\partial z)$. This allows us to neglect the velocity u_{0z} and then to account it at the boundary conditions. Thus within the jet fluid and the external medium, we set $u_{0z}=0$ and substitute (2.248) in (2.244) and (2.245) to obtain (2.249a) and (2.249b-c). □

It is impossible to find an exact solution to the system (2.249a-c). Nevertheless, there is a possibility to solve it using the perturbation method due for gaseous jets, the dimensionless value

$$\varepsilon_1 = \nu_0 k^2/\omega \ll 1. \tag{2.250}$$

We will find the solution to (2.249a-c) in the form

$$u_{r1}(r) = u_{r1}^{(0)} + \varepsilon_1(\nabla^2\mathbf{u}^{(0)})_r$$
$$(\nabla^2\mathbf{u}^{(0)})_r = \frac{\partial^2 u_{r1}^{(0)}}{\partial(kr)^2} + \frac{1}{(kr)^2}\frac{\partial^2 u_{r1}^{(0)}}{\partial\phi^2} + \frac{\partial^2 u_{r1}^{(0)}}{\partial(kz)^2} + \frac{1}{kr}\frac{\partial u_{r1}^{(0)}}{\partial(kr)} - \frac{2}{(kr)^2}\frac{\partial u_{\phi1}^{(0)}}{\partial\phi} - \frac{1}{(kr)^2}u_{r1}^{(0)} \tag{2.251}$$

where $\mathbf{u}^0 = (u_{r1}^{(0)}, u_{\phi1}^{(0)}, u_{z1}^{(0)})$ is the solution to the system (2.249a,b), but with zero viscosity $\nu_0 = 0$.

Proposition 19 *Energy equation (2.246), in plane wave expansion, has a view:*

$$\mu_0 c_P T_1\left[\Psi'' + \frac{1}{r}\Psi' - \Psi(r)\left(\frac{n^2}{r^2}+k^2\right)\right] + i\omega\frac{A}{\gamma-1}\Psi(r)(\rho_0 T_1+\rho_1 T_0)$$
$$-c_0\rho_0^\gamma\frac{\gamma}{\gamma-1}\left[\frac{u_{r1}(r)}{r}+\frac{du_{r1}}{dr}+i\Psi(r)(\frac{nu_{\phi1}}{r}+u_{z1}k)\right]=0, \tag{2.252}$$

where c_0 is a constant that relates units of pressure and density.

Proof. Substituting (2.248) in (2.246) we obtain

$$i\omega\frac{A}{\gamma-1}\Psi(r)(\rho_0 T_1+\rho_1 T_0)-u_{r1}(r)\left[\frac{1}{r}\left(P_0+\frac{A}{\gamma-1}\rho_0 T_0\right)+\partial_r P_0+T_0\partial_r\rho_0+\rho_0\partial_r T_0\right]$$
$$-\frac{du_{r1}}{dr}\left(P_0+\frac{A}{\gamma-1}\rho_0 T_0\right)-\Psi(r)\left[i(n\frac{u_{\phi1}}{r}+ku_{z1})(P_0+\frac{A}{\gamma-1}\rho_0 T_0)+T_0\partial_r\rho_0+\rho_0\partial_r T_0\right]$$
$$+\mu_0 c_P T_1\left[\Psi'' + \frac{1}{r}\Psi' - \Psi(r)\left(\frac{n^2}{r^2}+k^2\right)\right]=0. \tag{2.253}$$

Since $P_0 = c_0\rho_0^\gamma$ and $P_0 = A\rho_0 T_0$ for the stationary solution, it follows that

$$
\begin{aligned}
\partial_r P_0 &= \frac{\partial P_0}{\partial \rho_0}\frac{\partial \rho_0}{\partial r} = c_0\gamma\rho_0^{\gamma-1}\frac{\partial \rho_0}{\partial r} = c_0\gamma\rho_0^\gamma\partial_r \ln \rho_0, \\
T_0\,\partial_r\rho_0 &= \frac{c_0}{A}\rho_0^{\gamma-1}\frac{\partial \rho_0}{\partial r} = \frac{c_0}{A}\rho_0^\gamma\partial_r \ln \rho_0, \\
\rho_0\,\partial_r T_0 &= \rho_0\frac{c_0}{A}\frac{\partial \rho_0}{\partial r}(\gamma-1)\rho_0^{\gamma-2} = \frac{c_0(\gamma-1)}{A}\rho_0^\gamma\partial_r \ln \rho_0, \\
T_0\frac{\partial \rho_0}{\partial z} &= \frac{c_0}{A}\rho_0^{\gamma-1}\frac{\partial \rho_0}{\partial z} = \frac{c_0}{A}\rho_0^\gamma\partial_z \ln \rho_0, \\
\rho_0\,\partial_z T_0 &= \rho_0\frac{c_0}{A}\frac{\partial \rho_0}{\partial z}(\gamma-1)\rho_0^{\gamma-2} = \frac{c_0(\gamma-1)}{A}\rho_0^\gamma\partial_z \ln \rho_0.
\end{aligned}
\tag{2.254}
$$

Now we will formulate a useful lemma which helps to simplify expressions in (2.254).

Lemma 6 *The function $\ln \rho_0$ is a weak function.*

Proof. For this purpose we estimate terms in square brackets in (2.253), i.e.,

$$
\frac{1}{r}\left(P_0 + \frac{A}{\gamma-1}\rho_0 T_0\right) + \frac{\partial P_0}{\partial r} + T_0\frac{\partial \rho_0}{\partial r} + \rho_0\frac{\partial T_0}{\partial r}.
\tag{2.255}
$$

The last three terms in (2.255) are $\sim \rho_0^\gamma\partial(\ln \rho_0)/\partial r$, and hence it will be enough to estimate one ratio

$$
\frac{\partial P_0/\partial r}{r^{-1}(P_0 + \frac{A}{\gamma-1}\rho_0 T_0)} = \frac{r}{\rho_0}\frac{A^{-1}\rho_0^\gamma}{(P_0/\rho_0)(1+1/A)}\partial_r \ln \rho_0 = r\frac{\gamma}{A+1}\frac{\rho_0^{\gamma-1}}{a_0^2}\partial_r \ln \rho_0.
$$

Then

$$
r\frac{\gamma}{A+1}\frac{\rho_0^{\gamma-1}}{a_0^2}\partial_r \ln \rho_0 \le \gamma r\frac{\rho_0^{\gamma-2}}{a_0^2}\partial_r\rho_0.
\tag{2.256}
$$

The aim of our estimation is to find the characteristic length l of the boundary layer, where the dumping of the density is maximal. The reason of the boundary layer appearance is the inner dissipative processes – viscosity and thermo-conductivity of the gas. We compare inertial forces with the forces caused by viscosity μ_0 and thermo-conductivity κ. For U being a velocity scaling, and R being a characteristic size of the flow (in the radial direction), then the inertial term is $\sim \rho U^2/R$. The viscous term $\mu_0\nabla^2\mathbf{u}$ in the momentum equations may be estimated as $\mu_0 U^2/R^2$; and thus, $\frac{1}{Re} = \frac{\nu_0}{UR} \sim \frac{l}{R}\frac{c}{U}$. Similarly, by comparing the conductive heat transfer with the mechanical one, we get

$$
\frac{1}{Pe} = \frac{\kappa}{\rho_0 c_P U R} = \frac{\chi}{UR} \sim \frac{l}{R}\frac{a}{U},
$$

where c_P – the heat capacity under constant pressure; Pe is the Peckle number, which in gases is nearly equal to Reynolds number Re [45], since for gases, molecular thermo-conductivity $\chi = \kappa/\rho_0 c_P$ is almost equal to kinematic viscosity $\nu_0 = \mu_0/\rho_0$. Thus,

$$
l \sim \frac{U}{a_0}\frac{R}{Re} = \frac{M\cdot R}{Re},
\tag{2.257}
$$

where $a_0 = \sqrt{\gamma P_0/T_0}$ – sound velocity, and M – Mach number. Substituting (2.257) in (2.256) gives

$$
\begin{aligned}
\frac{\partial P_0/\partial r}{r^{-1}(P_0 + \frac{A}{\gamma-1}\rho_0 T_0)} &\le \gamma r\frac{\rho_0^{\gamma-2}}{a_0^2}\frac{\partial\rho_0}{\partial r} \approx \gamma R\frac{\rho_0^{\gamma-2}}{a_0^2}\frac{\delta\rho_0}{l} \le \gamma\frac{\rho_0^{\gamma-2}\delta\rho_0}{a_0^2}\frac{R}{M}\frac{Re}{R} \\
&= \gamma\frac{\rho_0^{\gamma-2}\delta\rho_0}{a_0^2}\frac{RU}{\nu_0 M} = \gamma\frac{\rho_0^{\gamma-1}\delta\rho_0}{a_0}R\frac{(U/a_0)}{\mu_0} = \frac{\gamma}{a_0}\frac{\rho_0^{\gamma-1}\delta\rho_0}{\mu_0}R.
\end{aligned}
\tag{2.258}
$$

Now, estimate the value $\delta \rho_0$. Using Bernoulli equation $P_0 + \rho_0 u_0^2 = P_a + \rho_a u_a^2$ and adiabatic relation $P = c_0 \rho^\gamma$, we obtain $\rho_0^\gamma - \rho_a^\gamma = \gamma_a M_a^2 \rho_a^{\gamma a} - \gamma M^2 \rho_0^\gamma$. Here the subscript 'a' signs the ambient atmospheric properties. Hence the damp of the density at the boundary of the jet and the ambient atmosphere may be estimates, under assumption $\gamma \approx \gamma_a$, as $\delta \rho_0 = |\rho_a - \rho_0| \simeq \rho_0 |1 - \left(\frac{1+\gamma M^2}{1+\gamma M_a^2}\right)^{1/\gamma}|$.

For supersonic high temperature jets with parameters $a_0 \simeq 2000 \text{m / s}, \gamma \simeq 1.2 - 1.5$, $R < 0.5\text{m}$, $M \simeq 3$, $M_a \simeq 2.4 - 2.6$, $\rho_0 \simeq 0.1 \text{kg / m}^3$, $\mu_0 \simeq 1.7 \times 10^{-5} \text{kg} \cdot \text{m / s}$, we finally obtain

$$\frac{\partial P_0/\partial r}{r^{-1}\left(P_0 + \frac{A}{\gamma-1}\rho_0 T_0\right)} \leq \frac{\gamma}{a_0} R \frac{\rho_0^\gamma}{\mu_0} \left|1 - \left(\frac{1+\gamma M^2}{1+\gamma M_a^2}\right)^{1/\gamma}\right|. \tag{2.259}$$

The last term may be estimated between 0.15 and 0.2. Estimation (2.259) was obtained for very thin boundary layer, which corresponds more native with the width of the shock wave front. In fact, the real width of the boundary layer is not so thin and consequently the real gradient $\partial \rho_0/\partial r \ll \delta \rho_0/l_{\text{shock}}$, with $l_{\text{shock}} \sim MR/Re$ follow estimation (2.257). Note that we used rough estimate for U, in (2.258), which denotes Reynolds number. In fact we should take into account only its radial component Re_r, or U_r, which is much less than their total values. Really, in the nozzle of conical form with the angle $\theta \leq 12^0$, $U_r/U \leq \sin(\theta/2) \leq 0.1$. On the other hand the velocity radial component appears even in strongly cylindrical nozzle due to difference of the jet and ambient atmospheric pressures, i.e.,

$$U_r = (\delta P_0/\rho_0)^{1/2} < (P_0/\rho_0)^{1/2} = \gamma^{-1/2} a_0 = \gamma^{-1/2} U/M.$$

For $\gamma = 1.4$ and $M = 3.0$, we estimate the ratio $U_r/U \leq (U_r/U)_{\text{conic nozzle}} + (U_r/U)_{\delta P} \simeq 0.38$. Hence

$$\frac{\partial P_0/\partial r}{r^{-1}\left(P_0 + \frac{A}{\gamma-1}\rho_0 T_0\right)} \leq \frac{\gamma}{a_0} R \frac{\rho_0^\gamma}{\mu_0} \left|1 - \left(\frac{1+\gamma M^2}{1+\gamma M_a^2}\right)^{1/\gamma}\right| (U_r/U).$$

The last term is estimated between 0.05 and 0.08. Hence, $\ln \rho_0$ is a weak function. □

Thus, follow the lemma, $\frac{\partial \ln \rho_0}{\partial r} \ll 1$ and $\frac{\partial \ln \rho_0}{\partial z} \ll 1$, and we put $\partial \ln \rho_0/\partial r = \partial \ln \rho_0/\partial z = 0$ in (2.254). A simple calculation shows that using $P_0 = c_0 \rho_0^\gamma$ and $P_0 = A\rho_0 T_0$ for unperturbed solution we obtain

$$P_0 + \frac{A}{\gamma-1}\rho_0 T_0 = c_0 \rho_0^\gamma \frac{\gamma}{\gamma-1}. \tag{2.260}$$

Substituting (2.260) in (2.253) we find that energy equation (2.246) reads (2.252). □

Boundary value problem for radial function. We deduce the second order linear ODE for a function Ψ and then study two possibilities for the sound velocity: constant or of Gaussian (bell) form. We solve the boundary value problem for Ψ bounded and given for $r < R_0$ and $r > R_0$, then we glue the two solutions on the boundary layer $r = R_0$ between the internal and external media.

Second-order ODE. Zero-order solutions $\mathbf{u}^0 = \left(u_{r1}^{(0)}, u_{\phi1}^{(0)}, u_{z1}^{(0)}\right)$ may be found from expressions (2.249b-c) under $\nu_0 = 0$

$$u_{r1}^{(0)} = -\frac{i}{\omega}\frac{\rho_1}{\rho_0}\frac{\partial P'}{\partial \rho'}\Psi'(r), \quad u_{\phi1}^{(0)} = \frac{n}{\omega}\frac{\rho_1}{\rho_0}\frac{1}{r}\frac{\partial P'}{\partial \rho'}, \quad u_{z1}^{(0)} = \frac{k}{\omega}\frac{\rho_1}{\rho_0}\frac{\partial P'}{\partial \rho'}. \tag{2.261}$$

Substituting (2.261) in (2.249a) and (2.252) gives

$$\Psi'' + \left[\frac{1}{r} + \left(\frac{\partial P'}{\partial \rho'}\right)^{-1}\frac{\partial}{\partial r}\left(\frac{\partial P'}{\partial \rho'}\right)\right]\psi' + \left[\left(\frac{\partial P'}{\partial \rho'}\right)^{-1}\omega^2 - \left(k^2 + \frac{n^2}{r^2}\right)\right]\Psi(r) = 0, \tag{2.262a}$$

$$c_0 \rho_0^\gamma \gamma \frac{1}{\omega}\frac{\rho_1}{\rho_0}\frac{\partial P'}{\partial \rho'}\left(\Psi'' + \frac{1}{r}\psi'\right) + \left[A\omega(\rho_0 T_1 + \rho_0 T_0) - \left(k^2 + \frac{n^2}{r^2}\right)c_0 \rho_0^\gamma \gamma \frac{1}{\omega}\frac{\rho_1}{\rho_0}\frac{\partial P'}{\partial \rho'}\right]\Psi(r) = 0. \tag{2.262b}$$

Thus $\left(\frac{\partial P'}{\partial \rho'}\right)^{-1} \frac{\partial}{\partial r}\left(\frac{\partial P'}{\partial \rho'}\right) = \frac{\partial}{\partial r}\left(\ln \frac{\partial P'}{\partial \rho'}\right)$, that simplifies (2.262a)

$$\Psi'' + \left[\frac{1}{r} + \frac{\partial}{\partial r}\left(\ln \frac{\partial P'}{\partial \rho'}\right)\right]\psi' + \left[\left(\frac{\partial P'}{\partial \rho'}\right)^{-1}\omega^2 - \left(k^2 + \frac{n^2}{r^2}\right)\right]\Psi(r) = 0.$$

Proposition 20 *The following formula holds* $\partial P'/\partial \rho' = \gamma A T_0$.

Proof. The differentials dP' and $d\rho'$ may be written

$$dP' = \frac{\partial P'}{\partial r}dr + \frac{\partial P'}{\partial \phi}d\phi + \frac{\partial P'}{\partial z}dz, \quad d\rho' = \frac{\partial \rho'}{\partial r}dr + \frac{\partial \rho'}{\partial \phi}d\phi + \frac{\partial \rho'}{\partial z}dz.$$

Using well-known rules from mathematical analysis

$$\frac{\partial P'}{\partial r} = \frac{\partial P'}{\partial \rho'}\frac{\partial \rho'}{\partial r}, \quad \frac{\partial P'}{\partial \phi} = \frac{\partial P'}{\partial \rho'}\frac{\partial \rho'}{\partial \phi}, \quad \frac{\partial P'}{\partial z} = \frac{\partial P'}{\partial \rho'}\frac{\partial \rho'}{\partial z}, \quad \frac{\partial P'}{\partial t} = \frac{\partial P'}{\partial \rho'}\frac{\partial \rho'}{\partial t},$$

one obtains

$$\frac{dP'}{d\rho'} = \frac{\partial P'}{\partial \rho'}. \tag{2.263}$$

Using (2.263) and plane expansion representations for ρ' and P',

$$\rho' = \rho_1 \Psi(r)e^{i(kz+n\phi-\omega t)}, \quad P' = P_1 \Psi(r)e^{i(kz+n\phi-\omega t)}, \tag{2.264}$$

we get

$$\frac{\partial P'}{\partial \rho'} = \frac{dP'}{d\rho'} = \frac{dP'/dt}{d\rho'/dt} = \frac{P_1}{\rho_1}. \tag{2.265}$$

Using formula ideal gas (2.231a) under adiabatic condition

$$P = c_0\rho^\gamma \tag{2.266}$$

for unperturbed values P_0, ρ_0, we obtain

$$AT_0 = c_0\rho_0^{\gamma-1}. \tag{2.267}$$

Substituting plane wave represented values P', ρ' of view (2.264) in (2.265), using (2.231a) and neglecting the second order term $\rho'T'$, one can find $P_1 = A(\rho_1 T_0 + \rho_0 T_1)$. Substituting (2.264) in (2.266), using binomial theorem, remaining only linear terms and accounting adiabatic relation (2.266) for unperturbed quantities, we obtain $P_1 = \gamma c_0 \rho_0^{\gamma-1}\rho_1$. Hence,

$$P_1/\rho_1 = \gamma c_0\rho_0^{\gamma-1}. \tag{2.268}$$

Substitution (2.267) in (2.268) gives

$$P_1/\rho_1 = \gamma A T_0. \tag{2.269}$$

Substituting (2.269) in (2.265), we finally obtain $\frac{\partial P'}{\partial \rho'} = \gamma A T_0$. $\square$

Proposition 21 *The radial function* $\Psi(r)$ *satisfies to linear ODE of the form*

$$\Psi'' + \frac{1}{r}\Psi' + \left(\beta(r)^2 - \frac{n^2}{r^2}\right)\Psi = 0, \tag{2.270}$$

where

$$\beta^2(r) = \omega^2/a^2(r) - k^2. \tag{2.271}$$

Proof. Follow Proposition 20, (2.262a) may be rewritten as

$$\Psi'' + \left[\frac{1}{r} + \frac{\partial}{\partial r}(\ln \gamma A T_0)\right]\Psi' + \left[\frac{\omega^2}{\gamma A T_0} - \left(k^2 + \frac{n^2}{r^2}\right)\right]\Psi = 0.$$

Note, that

$$\frac{\partial}{\partial r}(\ln \gamma A T_0) = \gamma c_0 \frac{\partial}{\partial r}(\ln \rho_0^{\gamma-1}) = \gamma c_0 (\gamma - 1)\frac{\partial}{\partial r}(\ln \rho_0) = 0,$$

due to $\ln \rho_0$ is a weak function follow to the lemma. Hence,

$$\Psi'' + \frac{1}{r}\Psi' + \left[\frac{\omega^2}{a^2(r)} - \left(k^2 + \frac{n^2}{r^2}\right)\right]\Psi = 0. \tag{2.272}$$

Denoting $\beta^2(r) = \omega^2/a^2(r) - k^2$, we complete the proof. $\square$

Proposition 22 *Equations (2.262a) and (2.262b) are equivalent.*

Proof. Dividing (2.262b) by $c_0 \gamma \rho_0^\gamma (\rho_1/\rho_0)\partial P'/\partial \rho'$ gives

$$\Psi'' + \frac{1}{r}\Psi' + \left[\frac{A\omega^2(\rho_0 T_1 + \rho_1 T_0)}{c_0 \gamma \rho_0^\gamma (\rho_1/\rho_0)(\partial P'/\partial \rho')} - \left(k^2 + \frac{n^2}{r^2}\right)\right]\Psi = 0. \tag{2.273}$$

Follow Proposition 20, we get

$$P_1/\rho_1 = \gamma c_0 \rho_0^{\gamma-1}, \quad P_1 = A(\rho_0 T_1 + \rho_1 T_0) = \gamma c_0 \rho_0^\gamma (\rho_1/\rho_0).$$

Hence, (2.273) may be rewritten in the form

$$\Psi'' + \frac{1}{r}\Psi' + \left[\omega^2\left(\frac{\partial P'}{\partial \rho'}\right)^{-1} - \left(k^2 + \frac{n^2}{r^2}\right)\right]\Psi = 0.$$

Again, using definition $a_0 = \sqrt{\gamma A T_0}$, we obtain

$$\Psi'' + \frac{1}{r}\Psi' + \left[\frac{\omega^2}{a_0^2} - \left(k^2 + \frac{n^2}{r^2}\right)\right]\Psi = 0,$$

which in view of (2.271) is transformed to the view (2.270). $\square$

Remark 13 (a) By Proposition 22, we may work with the continuity equation instead of the energy equation. Thus we find the solution of (2.249b) in the form (2.251) with condition (2.250). (b) This does not contradicts to physics because energy and mass are related according to Einstein's equation $E = mc^2$, where m is mass and c is the speed of light.

Reduction of equation for radial functions to Heun equation. For the sound velocity of Gaussian form (2.226), note that this distribution is a good approximation of the profile of $a(r)$ [16, 17].

Proposition 23 *Equation (2.270), for $a(r_1)$ of the form (2.226), has the canonical form, see (1.22):*

$$\tilde{\Psi}'' + \left(\frac{1 + B_0}{r_1} - B_1 - 2r_1\right)\tilde{\Psi}' + \left[(B_2 - B_0 - 2) - \frac{B_1}{2r_1}(1 + B_0)\right]\tilde{\Psi} = 0,$$

$$\Psi(r_1) = r_1^{B_0/2}\exp\left[-\frac{1}{2}(B_1 r_1 + r_1^2)\right]\tilde{\Psi}, \tag{2.274}$$

where (denoting R_0 – jet's throat radius)

$$B_0 = 2n, \quad B_1 = 2b\omega^2/a_0^2 R_0 \alpha^3 \beta_1^3, \quad B_2 = \alpha^{-2} + \tfrac{1}{4}B_1^2, \quad B_3 = 0, \quad r_1 = \alpha\beta_1 r,$$
$$\beta_1^2 = c\omega^2/a_0^2 - k^2, \quad b = \tfrac{800}{27}\ln 50, \quad c = b + \tfrac{4}{9}, \quad \alpha = -A_4^{1/4}, \quad A_4 = b\omega^2/(a_0^2 R_0^2 \beta_1^4). \tag{2.275}$$

Proof. We will find the solutions to (2.272) by two steps:

1. Assume that the sound velocity has a Gaussian form. 2. Estimate constant $\sigma > 0$. Calculations show that Gaussian peak (maximum) at the boundary layer has a width $\sigma \simeq 0.1R_0$, i.e.,

$$r - R_0 = R_0/10. \tag{2.276}$$

Substituting (2.276) in $\frac{1}{2}e^{-(r-R_0)^2/\sigma^2}$ gives $\frac{1}{2}\exp\left[-(\frac{1}{10}R_0)^2/\sigma^2\right] \simeq \frac{1}{100}$. Hence,

$$\sigma^2 \simeq \frac{1}{100\ln 50}R_0^2. \tag{2.277}$$

Substituting (2.277) in (2.276) gives $a(r) = a_0\{1 + \frac{1}{2}\exp\left[-100\ln 50(r/R_0 - 1)^2\right]\}$. The last value may be rewritten in more convenient form

$$a^{-2}(r) = a_0^{-2}\left\{1 + \frac{1}{2}\exp\left[-100\ln 50(r/R_0 - 1)^2\right]\right\}^{-2}. \tag{2.278}$$

Denoting $x = -100\ln 50(r/R_0 - 1)^2$ and representing the term $\exp\left[-100\ln 50(r/R_0 - 1)^2\right]$ through Taylor series $e^x \approx 1 + x$ (under neglecting high-order terms due to thickness of the boundary layer), we obtain $a^{-2}(r) = a_0^{-2}\left(\frac{3}{2}\right)^{-2}\left(1 + \frac{1}{3}x\right)^{-2}$. Then, using the binomial theorem as follows $\left(1 + \frac{1}{3}x\right)^{-2} \approx 1 - \frac{2}{3}x$, we obtain, from (2.278),

$$a^{-2}(r) \approx \frac{4}{9}a_0^{-2}\left(1 - \frac{2}{3}x\right) = \frac{4}{9}a_0^{-2}\left[1 + \frac{200\ln 50}{3}(r/R_0 - 1)^2\right]. \tag{2.279}$$

Substituting (2.279) in (2.270) gives $\beta(r) = \sqrt{\frac{4}{9}\frac{\omega^2}{a_0^2}\left[1 + \frac{200\ln 50}{3R_0^2}(r - R_0^2)\right] - k^2}$. Equation (2.270) is reduced to linear ODE, of the form,

$$\Psi'' + \frac{1}{r}\Psi' + \left\{\frac{4\omega^2}{9a_0^2}\left[1 + \frac{200\ln 50}{3R_0^2}(r - R_0)^2\right] - k^2 + \frac{n^2}{r^2}\right\}\Psi = 0.$$

The last equation may be rewritten as

$$\Psi'' + \frac{1}{r}\Psi' + \left\{\frac{b\omega^2}{a_0^2 R_0^2}r^2 - \frac{2b\omega^2}{a_0^2 R_0}r + \frac{c\omega^2}{a_0^2} - k^2 + \frac{n^2}{r^2}\right\}\Psi = 0. \tag{2.280}$$

Simultaneously, for (2.280) we introduce $\beta_1 = \sqrt{c\omega^2/a_0^2 - k^2} = k\sqrt{c\omega^2/(a_0^2 k^2) - 1}$. Replacing variable r by $r_1 = \alpha\beta_1 r$ in (2.280) we obtain for $\Psi(r_1)$:

$$\Psi'' + \frac{1}{r}\Psi' + \left\{\frac{b\omega^2}{a_0^2 R_0^2\alpha^4\beta_1^4}r_1^2 - \frac{2b\omega^2}{a_0^2 R_0\alpha^3\beta_1^3}r_1 + \frac{1}{\alpha^2} + \frac{n^2}{r_1^2}\right\}\Psi = 0. \tag{2.281}$$

One can see that linear ODE (2.281) for $\Psi(r_1)$ is of the representative form of *Biconfluent Heun equation* (BHE) if we take

$$A_0' = -n^2, \quad A_1' = 0, \quad A_2' = 1/\alpha^2, \quad A_3' = -\frac{2b\omega^2}{a_0^2 R_0\alpha^3\beta_1^3}, \quad A_4' = \frac{b\omega^2}{a_0^2 R_0^2\alpha^4\beta_1^4}.$$

Assuming $A_4' = -1$ we obtain $\alpha = (-A_4)^{1/4}$ where $A_4 = b\omega^2/a_0^2 R_0^2\beta_1^4$. Equation (2.281) became to its canonical form (2.274) if we put $r_1 = r$, $B_3 = 0$, let be coefficient from (2.275) and let be additionally

$$\begin{aligned} A_0' &= -\tfrac{1}{4}B_0^2 = -n^2, \quad A_1' = -\tfrac{1}{2}B_3 = 0, \quad A_2' = B_2 - \tfrac{1}{4}B_1 = 1/\alpha^2, \\ A_3' &= -B_1 = -2b\omega^2/a_0^2 R_0\alpha^3\beta_1^3, \quad A_1' = 2b\omega^2/a_0^2 R_0^2\alpha^4\beta_1^4 = -1. \end{aligned} \tag{2.282}$$

We solve (2.282) in order to obtain the coefficients B_0, B_1, B_2, B_3 as in (2.275).

The general solution of (2.274) is the function $N(B_0, B_1, B_2, 0; r_1)$, see Section 1.1.5. $\square$

Remark 14 One may consider approximation of order $k > 1$, $e^x \approx 1 + x + \cdots + x^k/k!$. Then, we will get from (2.278) more general representation, where, in our case, $A_0 = -1$ and $A_1 = A_3 = A_5 = \ldots = 0$. General canonical form depending on m may be presented by

$$x = -100 \ln 50 (r/R_0 - 1)^2, \quad m = 2k.$$

Dispersion equation. The KHI model developed in this paragraph, generalizes [16]. In what follows, $H_n^{(1)}$ is the Hankel function and $\widetilde{B}_n$ is the Heun function. Namely, $\widetilde{B}_n(r_1) = r_1^{B_0/2} e^{-\frac{1}{2}(B_1 r_1 + r_1^2)} C_1 N_n(B_0, B_1, B_2, 0; r_1)$, where N_n is the general solution of (1.22) in Section 1.1.5,

$$\widetilde{\Psi}(r_1) = C_1 N_n(B_0, B_1, B_2, 0; r_1) + C_2 \Big[N_n(B_0, B_1, B_2, 0; r_1) \ln r_1 + \sum_{n \in \mathbb{N}} d_n r_1^{n - B_0} \Big].$$

Also, let ρ_{in} and ρ_{ex} be unperturbed densities of the jet and ambient atmosphere media.

Theorem 46 *The complex function $\Phi = \omega/(k a_{in})$ (i.e., the ratio of phase velocity ω/k and the jets sound velocity a_{in}) satisfies to the following dispersion equation:*

$$\frac{\widetilde{B}_n'(\beta_{in}X) + \Upsilon_\nu^{in}\Big\{ \beta_{in}^2 \widetilde{B}_n'''(\beta_{in}X) + \frac{\beta_{in}}{X} \widetilde{B}_n''(\beta_{in}X) - \frac{1}{X^2} \widetilde{B}_n'(\beta_{in}X) \Big\}}{H_n^{(1)'}(\beta_{ex}X) + \Upsilon_\nu^{ex}\Big\{ \beta_{ex}^2 H_n^{(1)'''}(\beta_{ex}X) + \frac{\beta_{in}}{X} H_n^{(1)''}(\beta_{ex}X) - \frac{1}{X^2} H_n^{(1)'}(\beta_{ex}X) \Big\}}$$
$$\times \frac{\beta_{in} H_n^{(1)}(\beta_{ex}X)}{\beta_{ex} \widetilde{B}_n(\beta_{in}X)} = \frac{\rho_{in}}{\eta \rho_{ex}} \Big(\frac{\Phi - M_{in}}{\Phi/\sqrt{\eta} - M_{ex}} \Big)^2, \tag{2.283}$$

where all denotation, except the function $\widetilde{B}_n$, are the same as in previous section.

Remark 15 For inviscid case, when $\nu_{in} = \nu_{ex} = 0$, (2.283) is reduced to a simple form:

$$\frac{H_n^{(1)}(\beta_{ex}X)\widetilde{B}'(\beta_{in}X)}{\widetilde{B}(\beta_{ex}X)H_n^{(1)'}(\beta_{ex}X)} \cdot \frac{\beta_{in}}{\beta_{ex}} = \frac{\rho_{in}}{\eta \rho_{ex}} \Big(\frac{\Phi - M_{in}}{\Phi/\sqrt{\eta} - M_{ex}} \Big)^2. \tag{2.284}$$

Proof. Write the canonical form (2.274) as

$$\widetilde{\Psi}'' + \Big(\frac{1 + B_0}{r_1} - B_1 - 2r_1 \Big) \widetilde{\Psi}' + \Big[(B_2 - B_0 - 2) - \frac{B_1}{2r_1}(1 + B_0) \Big] \widetilde{\Psi}(r_1) = 0,$$

where $r_1 = \alpha \beta_1 r$, $\alpha = -A_4^{1/4}$, $\beta_1 = k\sqrt{c\omega^2/(a_0^2 k^2) - 1}$. From the boundary condition at $r_1 = 0$ we get

$$N_n(B_0, B_1, B_2, 0; r_1) \ln r_1 + \sum_{n \in \mathbb{N}} d_n r_1^{b - B_0} \to \infty.$$

Hence, $C_2 = 0$, and $\widetilde{\Psi}_n(r_1) = C_1 N_n(B_0, B_1, B_2, 0; r_1)$. Substituting this value into $\widetilde{B}_n = r_1^{B_0/2} e^{-\frac{1}{2}(B_1 r_1 + r_1^2)} \widetilde{\Psi}_n(r_1)$ gives $\Psi_n(r_1) = C_1 \widetilde{B}_n(r_1)$. Substitution this into (2.261) gives

$$u_{r1}^{(0)} = -i \frac{\rho_1}{\rho_0} a_0 g(kr) C_1 t_{in} \widetilde{B}_n'(r_1)$$

for $r < R_0$, where $r_1 = \beta_1 kr$, $\beta_1 = \alpha\sqrt{c\omega^2/(a_0^2 k^2) - 1}$, and

$$t_1 = \sqrt{1 - a_0^2 k^2/(c\omega^2)}, \quad g(kr) = 1 + \frac{1}{2} e^{-100 \ln 50(kr/kR_0 - 1)^2}, \quad c = \frac{800}{27} \ln 50 + \frac{4}{9}.$$

Recall that $u_{r1}(r) = u_{r1}^{(0)} + \varepsilon_1 (\nabla^2 \mathbf{u}^{(0)})_r$, where

$$(\nabla^2 \mathbf{u}_0)_{kr} = \frac{\partial^2 u_{r1}^{(0)}}{\partial (kr)^2} + \frac{1}{(kr)^2} \frac{\partial^2 u_{r1}^{(0)}}{\partial (k\phi)^2} + \frac{\partial^2 u_{r1}^{(0)}}{\partial (kz)^2} - \frac{1}{kr} \frac{\partial u_{r1}^{(0)}}{\partial (kr)} - \frac{2}{(kr)^2} \frac{\partial u_{\phi 1}^{(0)}}{\partial (k\phi)} - \frac{u_{r1}^{(0)}}{(kr)^2}$$
$$= -\frac{i}{\omega} \frac{\rho_1}{\rho_0} \Big\{ \frac{\partial^2 [a_0^2(kr)\widetilde{B}_n'(r_1)]}{\partial (kr)^2} + \frac{1}{kr} \frac{\partial [a_0^2(kr)\widetilde{B}_n'(r_1)]}{\partial (kr)} - \frac{1}{(kr)^2} [a_0^2(kr)\widetilde{B}_n'(r_1)] \Big\}.$$

Follow Proposition 20 and (2.268)

$$\frac{\partial a_0^2}{\partial r} = \frac{\partial}{\partial r}(\gamma A T_0) = c_0 \gamma \frac{\partial}{\partial r} \rho_0^{\gamma-1} = c_0 \gamma(\gamma-1)\rho_0^{\gamma-1}\frac{\partial \ln \rho_0}{\partial r} \ll 1,$$
$$\frac{\partial^2 a_0^2}{\partial r^2} = c_0 \gamma(\gamma-1)\frac{\partial}{\partial r}\left(\rho_0^{\gamma-1}\partial_r \ln \rho_0\right) \ll 1.$$

Hence,

$$(\nabla^2 \mathbf{u}_0)_{kr} \simeq -\frac{i}{\omega}\frac{\rho_1}{\rho_0}a_0^2(kr)\left[\beta_1^2 \widetilde{B}_n'''(r_1) + \frac{\beta_1 \widetilde{B}_n''(r_1)}{kr} - \frac{\widetilde{B}_n'(r_1)}{(kr)^2}\right],$$

and so

$$u_{r1} = -i\frac{\rho_1}{\rho_0}a_0 g(kr)t_1\left\{\widetilde{B}_n'(r_1) + \varepsilon_1\left[\beta_1^2 \widetilde{B}_n'''(r_1) + \frac{\beta_1}{kr}\widetilde{B}_n''(r_1) - \frac{\widetilde{B}_n'(r_1)}{(kr)^2}\right]\right\}. \tag{2.285}$$

Substituting (2.285) in the first expansion (2.248) gives

$$u_r' = -i\frac{\rho_1}{\rho_0}a_0 g(kr)t_1\left\{\widetilde{B}_n'(r_1) + \varepsilon_1\left[\beta_1^2 \widetilde{B}_n'''(r_1) + \frac{\beta_1}{kr}\widetilde{B}_n''(r_1) - \frac{\widetilde{B}_n'(r_1)}{(kr)^2}\right]\right\}e^{-i(kz+n\phi-\omega t)}. \tag{2.286}$$

Using the plane wave expansion for perturbed density: $\rho_1 = \frac{\rho'}{\widetilde{B}_n(r_1)}e^{-i(kz+n\phi-\omega t)}$, (2.286) may be rewritten as follows

$$u_r' = -i\frac{\rho'}{\rho_0}a_0 g(kr)t_1\left\{\widetilde{B}_n'(r_1) + \varepsilon_1\left[\beta_1^2 \widetilde{B}_n'''(r_1) + \frac{\beta_1}{kr}\widetilde{B}_n''(r_1) - \frac{\widetilde{B}_n'(r_1)}{(kr)^2}\right]\right\}/\widetilde{B}_n(r_1).$$

Again, for the jets medium (internal), $r \leq R_0$ the last relation becomes

$$u_r' = -i\frac{\rho_{in}'}{\rho_{in}}a_{in}g(kr)t_{in} \times$$

$$\times\left\{\widetilde{B}_n'(\beta_{in}kr) + \varepsilon_{in}\left[\beta_{in}^2 \widetilde{B}_n'''(\beta_{in}kr) + \frac{\beta_{in}}{kr}\widetilde{B}_n''(\beta_{in}kr) - \frac{\widetilde{B}_n'(\beta_{in}kr)}{(kr)^2}\right]\right\}/\widetilde{B}_n(r_1) \tag{2.287}$$

where $t_{in} = \sqrt{1-(ka_{in}/c^{1/2}\omega_{in})^2}$. For the external (ambient atmosphere) medium, $r > R_0$ the solution is the same as in the case described in the Section 2.3.3, i.e., it has a form

$$\Psi(\beta_0 kr) = C_3 J_n(\beta_0 kr) + C_4 Y_n(\beta_0 kr).$$

Again, assuming $C_4 = iC_3$, $\Psi(\beta_0 kr) = C_3 H_n^{(1)}(\beta_0 kr)$, and for the internal medium, $r \leq R_0$ we obtain

$$u_r' = -ia_{ex}\frac{\rho_{ex}'}{\rho_{ex}}t_{ex}\left\{H_n^{(1)'}(\beta_{ex}kr) + \varepsilon_{ex}\left[\beta_{ex}^2 H_n^{(1)'''}(\beta_{ex}kr) + \frac{\beta_{ex}}{kr}H_n^{(1)''}(\beta_{ex}kr)\right.\right.$$
$$\left.\left. - \frac{1}{(kr)^2}H_n^{(1)'}(\beta_{ex}kr)\right]\right\}/H_n^{(1)}(\beta_{ex}kr),$$

where $t_{ex} = \sqrt{1-(ka_{ex}/\omega_{ex})^2}$. Displacement of the boundary $r = R_0$ between two flows by an amount $\delta' = \delta_1 e^{i(kz+n\phi-\omega t)}$ is related to the perturbed velocity by $\frac{\partial \delta'}{\partial t} = -i\omega\delta_1 e^{i(kz+n\phi-\omega t)} = -i\omega\delta'$. Thus the displacement may be written, follow (2.286), (2.287), and accounting $g(X) = 1$,

$$\delta' = \frac{1}{\omega_{in}}\frac{\rho_{in}'}{\rho_{in}}a_{in}t_{in}\frac{\widetilde{B}_n'(\beta_{in}X) + \varepsilon_{in}[\beta_{in}^2 \widetilde{B}_n'''(\beta_{in}X) + \frac{\beta_{in}}{X}\widetilde{B}_n''(\beta_{in}X) - \frac{\widetilde{B}_n'(\beta_{in}X)}{X^2}]}{\widetilde{B}_n(\beta_{in}X)},$$

$$\delta' = \frac{a_{ex}\rho_{ex}'}{\omega_{ex}\rho_{ex}}t_{ex}\frac{H_n^{(1)'}(\beta_{ex}X) + \varepsilon_{ex}[\beta_{ex}^2 H_n^{(1)'''}(\beta_{ex}X) + \frac{\beta_{ex}}{X}H_n^{(1)''}(\beta_{ex}X) - \frac{1}{X^2}H_n^{(1)'}(\beta_{ex}X)]}{H_n^{(1)}(\beta_{ex}X)}.$$

Comparing these two expressions we obtain

$$\frac{1}{\omega_{in}}\frac{\rho'_{in}}{\rho_{in}}a_{in}t_{in}\frac{\tilde{B}'_n(\beta_{in}X)+\varepsilon_{in}[\beta^2_{in}\tilde{B}'''_n(\beta_{in}X)+\frac{\beta_{in}}{X}\tilde{B}''_n(\beta_{in}X)-\frac{\tilde{B}'_n(\beta_{in}X)}{X^2}]}{\tilde{B}_n(\beta_{in}X)}$$
$$=\frac{a_{ex}\rho'_{ex}}{\omega_{ex}\rho_{ex}}t_{ex}\frac{H^{(1)'}_n(\beta_{ex}X)+\varepsilon_{ex}[\beta^2_{ex}H^{(1)'''}_n(\beta_{ex}X)+\frac{\beta_{ex}}{X}H^{(1)''}_n(\beta_{ex}X)-\frac{1}{X^2}H^{(1)'}_n(\beta_{ex}X)]}{H^{(1)}_n(\beta_{ex}X)}.$$

Follow the pressure to be continuous across the boundary, i.e., $\rho'_{in}a^2_{in}=\rho'_{ex}a^2_{ex}$, we obtain, from the equation above, the *dispersion relation* as (2.283)

$$\frac{\tilde{B}'_n(\beta_{in}X)+\Upsilon^{in}_\nu\left\{\beta^2_{in}\tilde{B}'''_n(\beta_{in}X)+\frac{\beta_{in}}{X}\tilde{B}''_n(\beta_{in}X)-\frac{1}{X^2}\tilde{B}'_n(\beta_{in}X)\right\}}{H^{(1)'}_n(\beta_{ex}X)+\Upsilon^{ex}_\nu\left\{\beta^2_{ex}H^{(1)'''}_n(\beta_{ex}X)+\frac{\beta_{ex}}{X}H^{(1)''}_n(\beta_{ex}X)-\frac{1}{X^2}H^{(1)'}_n(\beta_{ex}X)\right\}}$$
$$\times\frac{\beta_{in}H^{(1)}_n(\beta_{ex}X)}{\beta_{ex}\tilde{B}_n(\beta_{in}X)}=\frac{\rho_{in}}{\eta\rho_{ex}}\left(\frac{\Phi-M_{in}}{\Phi/\sqrt{\eta}-M_{ex}}\right)^2. \tag{2.288}$$

Similarly to Theorem 46, denoting

$$M_{in}=\frac{u_{in}}{a_{in}},\ M_{ex}=\frac{u_{ex}}{a_{ex}},\ \beta_{in}=\alpha\sqrt{c(\Phi-M_{in})^2-1},\ \beta_{ex}=\alpha\sqrt{(\Phi/\eta^{1/2}-M_{in})^2-1},$$
$$\Upsilon^{in}_\nu=\frac{\nu_{in}k}{a_{in}(\Phi-M_{in})},\ \Upsilon^{ex}_\nu=(\nu_{ex}k/a_{ex})/(\Phi/\eta^{1/2}-M_{ex}),$$
$$\alpha=(-A_4)^{1/4},\ A_4=b\omega^2/a^2_0R^2_0\beta^4_1,\ \beta_1=\sqrt{c\omega^2/a^2_0-k^2},$$

we find

$$\alpha=b^{1/4}\sqrt{i(\Phi-M_{in})/X\left[c(\Phi-M_{in})^2-1\right]}. \tag{2.289}$$

Substituting (2.289) in $\beta_{in}=\alpha\sqrt{c(\Phi-M_{in})^2-1}$, we obtain $\beta_{in}=b^{1/4}\sqrt{i(\Phi-M_{in})/X}$, so (2.288) becomes to (2.283).   □

Asymptotic analysis of instabilities. The *dispersion relation* (2.283) can be expanded in the limit of small and large arguments of the Heun and Hankel functions. We start with the expansion for $|\xi|\ll1,\ |\zeta|\ll1$, i.e., for the long-wavelength limit.
    (a) *Pinching mode* ($n=0$).

Proposition 24 *In limit* $|\xi|\ll1$, $|\zeta|\ll1$, *for low viscosity, the dispersion relation (2.283) becomes*

$$\left(\frac{\Phi-M_{in}}{\Phi-\sqrt{\eta}M_{ex}}\right)^2=\frac{1}{2}(\tilde{\beta}_{in}X)^2\frac{\rho_{ex}}{\rho_{in}}\ln\frac{2}{\beta_{ex}X}, \tag{2.290}$$

where

$$M_{in}=u_{in}/a_{in},\quad M_{ex}=u_{ex}/a_{ex},\quad\tilde{\beta}_{in}=\sqrt{c(\Phi-M_{in})^2-1},$$
$$\beta_{ex}=\sqrt{(\Phi/\eta^{1/2}-M_{ex})^2-1},\quad c=\frac{800}{27}\ln50+\frac{4}{9}.$$

Proof. Start again with the following expansions for Hankel functions, (2.215). Now we should apply first order expansions for Heun functions,

$$\tilde{B}_0(\xi)=\tilde{B}_0(0)+\tilde{B}'_0(0)\xi+o(\xi),\quad\tilde{B}'_0(\xi)=\tilde{B}'_0(0)+\tilde{B}''_0(0)\xi+o(\xi),$$
$$\tilde{B}''_0(\xi)=\tilde{B}''_0(0)+\tilde{B}'''_0(0)\xi+o(\xi),\quad\tilde{B}'''_0(\xi)=\tilde{B}'''_0(0)+\tilde{B}^{(4)}_0(0)\xi+o(\xi). \tag{2.291}$$

Our purpose is to find the coefficients $\tilde{B}_0(0)$, $\tilde{B}'_0(\xi)$, $\tilde{B}''_0(0)$, $\tilde{B}'''_0(0)$, $\tilde{B}^{(4)}_0(0)$. Recall that

$$\tilde{B}_n(\xi)=\xi^{B_0/2}e^{-\frac{1}{2}(B_1\xi+\xi^2)}C_1N_n(B_0,B_1,B_2,0;\xi),$$

where from (2.275) follows $B_0=2n$. So, for the case $n=0$ we obtain

$$\tilde{B}_0(\xi)=e^{-\frac{1}{2}(B_1\xi+\xi^2)}C_1N_n(0,B_1,B_2,0;\xi). \tag{2.292}$$

Using results in Section 1.1.5, we get $N_0(0, B_1, B_2, 0; \xi) = \sum_{j=0}^{\infty} \tilde{a}_j \xi^j / (j!)^2$, where

$$\tilde{a}_0 = 1, \quad \tilde{a}_1 = \tfrac{1}{2} B_1,$$
$$\tilde{a}_{j+2} = \left[(j+1) B_1 + \tfrac{1}{2} B_1\right] \tilde{a}_{j+1} - (j+1)^2 [B_2 - 2(j+1)] \tilde{a}_j, \quad j \geq 2.$$

In particular,

$$\tilde{a}_2 = \tfrac{3}{4} B_1^2 - B_2 + 2, \quad \tilde{a}_3 = \tfrac{15}{8} B_1^3 - \tfrac{9}{2} B_1 B_2 + 13 B_1,$$
$$\tilde{a}_4 = \tfrac{105}{16} B_1^4 - \tfrac{45}{2} B_1^2 B_2 + \tfrac{172}{2} B_1^2 + 9 B_2^2 - 72 B_2 + 108.$$

Here, B_1 and B_2 are given by (2.275). From (2.292), we find

$$\widetilde{B}_0(0) = C_1 \tilde{a}_0 + C_1,$$
$$\widetilde{B}_0'(0) = C_1(-\tfrac{1}{2} B_1 \tilde{a}_0 + \tilde{a}_1) = 0,$$
$$\widetilde{B}_0''(0) = C_1 \left[(\tfrac{1}{4} B_1^2 - 1) \tilde{a}_0 - B_1 \tilde{a}_1 + \tfrac{1}{2!} \tilde{a}_2\right] = C_1 b_1$$
$$\widetilde{B}_0'''(0) = -C_1 \left\{ \tfrac{3}{4} B_1 \tilde{a}_2 + [\tfrac{1}{2}(\tfrac{1}{4} B_1^2 - 1) - 1] B_1 \tilde{a}_0 - 3(\tfrac{1}{4} B_1^2 - 1) \tilde{a}_1 - \tfrac{1}{6} \tilde{a}_3 \right\} = b_2 C_1,$$
$$\widetilde{B}_0^{(4)}(0) = C_1 \left[\tfrac{1}{3} B_1 \tilde{a}_3 - (\tfrac{3}{4} B_1^2 - 3) \tilde{a}_2 + (\tfrac{1}{2} B_1^2 - 6) B_1 \tilde{a}_1 - (\tfrac{1}{16} B_1^3 - \tfrac{11}{4} B_1^2 - \tfrac{1}{4} B_1 + 3) \tilde{a}_0 \right.$$
$$\left. - \tfrac{1}{4!} \tilde{a}_4\right] = C_1 b_3,$$

where

$$N_0(0, B_1, B_2, 0; 0) = \tilde{a}_0, \quad N_0'(0, B_1, B_2, 0; 0) = \tilde{a}_1, \quad N_0''(0, B_1, B_2, 0; 0) = \tfrac{1}{2} \tilde{a}_2,$$
$$N_0'''(0, B_1, B_2, 0; 0) = \tfrac{1}{6} \tilde{a}_3, \quad N_0^{(4)}(0, B_1, B_2, 0; 0) = \tfrac{1}{24} \tilde{a}_4.$$

Thus, $b_1 = \tfrac{1}{2}(\tfrac{1}{4} B_1^2 - B_2)$, $b_2 = \tfrac{2}{3} B_1$ and $b_3 = \tfrac{3}{2}(\tfrac{1}{64} B_1^4 - \tfrac{1}{8} B_1^2 B_2 + \tfrac{1}{4} B_2^2 + 1)$, and so $b_1 = -\tfrac{1}{2} \alpha^{-2}$ with $\alpha = \tfrac{\sqrt{2}}{2} A_4^{1/4} (1 + i)$. Substituting $\widetilde{B}_1^{(k)}(0)$ $(k = 0, \ldots, 4)$ in (2.291) we obtain

$$\widetilde{B}_0(\xi) = C_1 + o(\xi), \quad \widetilde{B}_0'(\xi) = C_1 b_1 \xi + o(\xi),$$
$$\widetilde{B}_0''(\xi) = C_1(b_1 + b_2 \xi) + o(\xi), \quad \widetilde{B}_0'''(\xi) = C_1(b_2 + b_3 \xi) + o(\xi). \tag{2.293}$$

In the case $n = 0$, dispersion relation (2.283) becomes

$$\frac{\widetilde{B}_0'(\beta_{in} X) + \Upsilon_\nu^{in} \left\{ \beta_{in}^2 \widetilde{B}_0'''(\beta_{in} X) + \frac{\beta_{in}}{X} \widetilde{B}_0''(\beta_{in} X) - \frac{1}{X^2} \widetilde{B}_0'(\beta_{in} X) \right\}}{H_0^{(1)'}(\beta_{ex} X) + \Upsilon_\nu^{ex} \left\{ \beta_{ex}^2 H_0^{(1)'''}(\beta_{ex} X) + \frac{\beta_{in}}{X} H_0^{(1)''}(\beta_{ex} X) - \frac{1}{X^2} H_0^{(1)'}(\beta_{ex} X) \right\}}$$
$$\times \frac{\beta_{in} H_0^{(1)}(\beta_{ex} X)}{\beta_{ex} \widetilde{B}_0(\beta_{in} X)} = \frac{\rho_{in}}{\eta \rho_{ex}} \left(\frac{\Phi - M_{in}}{\Phi / \sqrt{\eta} - M_{ex}}\right)^2. \tag{2.294}$$

For long wavelength perturbations, $|\xi| \ll 1$, $|\zeta| \ll 1$ substitution (2.215) and (2.293) in (2.294) yields

$$\frac{i \frac{2}{\pi} \xi^2 \left[b_1 + 2 b_2 (\Upsilon_\nu^{in} / X^2) \xi\right] (\ln \tfrac{1}{2} \zeta + C)}{\left\{ -\tfrac{1}{2} + i \frac{2}{\pi} [\tfrac{1}{2} \ln \tfrac{1}{2} \zeta + \tfrac{1}{4}(2C - 1) + \zeta^{-2}] - i \frac{1}{\pi} (\Upsilon_\nu^{ex} / X^2) \right\} \zeta^2} = \frac{\rho_{in}}{\eta \rho_{ex}} \left(\frac{\Phi - M_{in}}{\Phi / \sqrt{\eta} - M_{ex}}\right)^2$$

In inviscid case, the equation above becomes

$$\frac{i \frac{2}{\pi} b_1 \xi^2 (\ln \tfrac{1}{2} \zeta + C)}{-\tfrac{1}{2} \zeta^2 + i \frac{2}{\pi} \left[\tfrac{1}{2} \zeta^2 \ln \tfrac{1}{2} \zeta + \tfrac{1}{4} \zeta^2 (2C - 1) + 1\right]} = \frac{\rho_{in}}{\eta \rho_{ex}} \left(\frac{\Phi - M_{in}}{\Phi / \sqrt{\eta} - M_{ex}}\right)^2$$

Accounting $|\xi| \ll 1$, $|\zeta| \ll 1$, it leads to

$$\frac{1}{\eta} \frac{\rho_{in}}{\rho_{ex}} \left(\frac{\Phi - M_{in}}{\Phi / \sqrt{\eta} - M_{ex}}\right)^2 = b_1 (\beta_{in} X)^2 \left(\ln \frac{1}{2} \beta_{ex} X + C\right). \tag{2.295}$$

Substitution $b_1 = -\tfrac{1}{2} \alpha^{-2}$, $\alpha = \tfrac{\sqrt{2}}{2} A_4^{1/4} (1 + i)$, $\beta_{in} = \alpha \sqrt{c(\Phi - M_{in})^2 - 1}$, and $\beta_{ex} = $

$\sqrt{(\Phi/\sqrt{\eta} - M_{ex})^2 - 1}$, and accounting $\ln \frac{1}{2}\beta_{ex} X + C \approx -\ln \frac{2}{\beta_{ex} X}$ leads to transformation (2.290) into (2.295). $\qquad\square$

Calculation for (2.290), for $X \ll 1$, shows two roots: the first, $\Phi^{Re} \simeq 3.0$, $\Phi^{Im} \simeq 0.05$ and the second, $\Phi^{Re} \simeq 0.9$, $\Phi^{Im} \simeq 0.05$ (for $X = 0.01$). Notice that for $\Phi^{Re} \simeq 0.9$, LHS tends to infinity. The first root may be estimated as follows from Remark 15.

Remark 16 Similarly to the case $n = 0$, RHS of (2.290) is sufficiently less of unity if $X \ll 1$. Thus we can use the iterative method that has been used in [15, 16] to find Φ^{Re} and Φ^{Im}. Again, the double root of zero order is $\Phi^{(0)} = M_{in}$, and in the first order, we obtain

$$\Phi^{Re} \simeq \frac{M_{in} - \sqrt{\eta}M_{ex}}{1 + \frac{1}{2}\frac{\rho_{ex}}{\rho_{in}}X^2 \ln \frac{2}{X\sqrt{(M_{in}/\sqrt{\eta}-M_{ex})^2-1}}} + \sqrt{\eta}M_{ex},$$

$$\Phi^{Im} \simeq \frac{(M_{in} - \sqrt{\eta}\,M_{ex})], X}{1 - \frac{1}{2}\frac{\rho_{ex}}{\rho_{in}}X^2 \ln\left[\frac{1}{2}X\sqrt{(M_{in}/\sqrt{\eta}-M_{ex})^2-1}\right]}\sqrt{\frac{1}{2}\frac{\rho_{ex}}{\rho_{in}}}\sqrt{\frac{2}{X[(M_{in}/\sqrt{\eta}-M_{ex})^2-1]^{1/2}}}.$$

For $X \ll 1$, again $\frac{\rho_{ex}}{\rho_{in}}X^2 \ln\left[\frac{1}{2}X\sqrt{(M_{in}/\sqrt{\eta} - M_{ex})^2 - 1}\right] \ll 1$, and we can conclude that

$$\Phi^{Re} \sim M_{in}, \quad \Phi^{Im} \sim \left(\frac{\rho_{ex}}{\rho_{in}}\right)^{1/2}(M_{in} - \sqrt{\eta}M_{ex})\left[(M_{in}/\sqrt{\eta} - M_{ex})^2 - 1\right]^{-1/4}X^{1/2}. \quad (2.296)$$

Simple calculations give $\Phi^{Im} \approx 0.046$ at the point $X = 0.01$ that helps to choose the starting point of the curve for numerical study.

(b) *Helical mode* ($n = 1$).

Proposition 25 *In limit* $|\xi| \ll 1$, $|\zeta| \ll 1$, *for low viscosity, the relation (2.283) becomes*

$$\frac{1}{\eta}\frac{\rho_{in}}{\rho_{ex}}\left(\frac{\Phi - M_{in}}{\Phi - \sqrt{\eta}M_{ex}}\right)^2 = -1. \quad (2.297)$$

Proof. For $n = 1$ the appropriate long wavelength expansions of Hankel functions are (2.218). Again we apply the first order expansions for Heun functions

$$\tilde{B}_1(\xi) = \tilde{B}_1(0) + \tilde{B}_1'(0)\xi + o(\xi), \quad \tilde{B}_1'(\xi) = \tilde{B}_1'(0) + \tilde{B}_1''(0)\xi + o(\xi),$$
$$\tilde{B}_1''(\xi) = \tilde{B}_1''(0) + \tilde{B}_1'''(0)\xi + o(\xi), \quad \tilde{B}_1'''(\xi) = \tilde{B}_1'''(0) + \tilde{B}_1^{(4)}(0)\xi + o(\xi) \qquad (2.298)$$

Recall that

$$\tilde{B}_n(\xi) = \xi^{B_0/2}e^{-\frac{1}{2}(B_1\xi + \xi^2)}C_1 N_n(B_0, B_1, B_2, 0; \xi)$$

where, from (2.275), follows $B_0 = 2n$. So, for the case $n = 1$, we obtain

$$\tilde{B}_1(\xi) = \xi e^{-\frac{1}{2}(B_1\xi + \xi^2)}C_1 N_n(2, B_1, B_2, 0; \xi). \quad (2.299)$$

Using results in Section 1.1.5, we get $N_1(2, B_1, B_2, 0; \xi) = \sum_{j=0}^{\infty} \tilde{a}_j \frac{2!}{(j+2)!}\frac{\xi^j}{j!}$, where

$$\tilde{a}_0 = 1, \quad \tilde{a}_1 = \frac{3}{2}B_1, \quad \tilde{a}_{j+2} = [(j+1)B_1 + \frac{3}{2}B_1]\tilde{a}_{j+1} - (j+1)(j+3)[B_2 - 2(j+2)]\tilde{a}_j,$$

for $j \geq 0$. In particular, $\tilde{a}_2 = \frac{15}{4}B_1^2 - 3B_2 + 12$ and $\tilde{a}_3 = \frac{105}{8}B_1^3 - \frac{45}{2}B_1 B_2 + 114 B_1$. Here, B_1 and B_2 are given in (2.275). Using (2.299) we find

$$\tilde{B}_1(0) = 0, \quad \tilde{B}_1'(0) = C_1\tilde{a}_0 = C_1, \quad \tilde{B}_1''(0) = C_1(\frac{2}{3}\tilde{a}_1 - B_1\tilde{a}_0) = 0,$$
$$\tilde{B}_1'''(0) = -C_1\left[\frac{1}{4}\tilde{a}_2 - B_1\tilde{a}_1 + \left(\frac{3}{4}B_1^2 - 3\right)\tilde{a}_0\right] = C_1\hat{b}_1,$$
$$\tilde{B}_0^{(4)}(0) = C_1\left[\frac{1}{15}\tilde{a}_3 - \frac{1}{2}B_1\tilde{a}_2 + (B_1^2 - 4)\tilde{a}_1 - \left(\frac{1}{2}B_1^2 - 6\right)B_1\tilde{a}_0\right] = C_1\hat{b}_2,$$

where

$$N_1(2, B_1, B_2, 0; 0) = \tilde{a}_0, \quad N_1'(2, B_1, B_2, 0; 0) = \tfrac{1}{3}\tilde{a}_1,$$
$$N_1''(2, B_1, B_2, 0; 0) = \tfrac{1}{12}\tilde{a}_2, \quad N_1'''(2, B_1, B_2, 0; 0) = \tfrac{1}{60}\tilde{a}_3,$$
$$N_1^{(4)}(2, B_1, B_2, 0; 0) = \tfrac{1}{360}\tilde{a}_4,$$

and $\hat{b}_1 = \tfrac{3}{16}(B_1^2 - 4B_2)$, $\hat{b}_2 = \tfrac{8}{5}B_1$. Substituting $\tilde{B}_1^{(k)}(0)$ $(k = 0, \ldots, 4)$ in (2.298) we get

$$\tilde{B}_1(\xi) = C_1\xi + o(\xi), \quad \tilde{B}_1'(\xi) = C_1 + o(\xi),$$
$$\tilde{B}_1''(\xi) = C_1\hat{b}_1\xi + o(\xi), \quad \tilde{B}_1'''(\xi) = C_1(\hat{b}_1 + \hat{b}_2\xi) + o(\xi). \tag{2.300}$$

In the case $n = 1$, the dispersion relation (2.283) becomes

$$\frac{\tilde{B}_1'(\beta_{in}X) + \Upsilon_\nu^{\text{in}}\left\{\beta_{in}^2 \tilde{B}_1'''(\beta_{in}X) + \frac{\beta_{in}}{X}\tilde{B}_1''(\beta_{in}X) - \frac{1}{X^2}\tilde{B}_1'(\beta_{in}X)\right\}}{H_1^{(1)'}(\beta_{ex}X) + \Upsilon_\nu^{\text{ex}}\left\{\beta_{ex}^2 H_1^{(1)'''}(\beta_{ex}X) + \frac{\beta_{in}}{X}H_1^{(1)''}(\beta_{ex}X) - \frac{1}{X^2}H_1^{(1)'}(\beta_{ex}X)\right\}}$$
$$\times \frac{\beta_{in}H_1^{(1)}(\beta_{ex}X)}{\beta_{ex}\tilde{B}_1(\beta_{in}X)} = \frac{\rho_{in}}{\eta\rho_{ex}}\left(\frac{\Phi - M_{in}}{\Phi/\sqrt{\eta} - M_{ex}}\right)^2. \tag{2.301}$$

Substituting (2.218) and (2.300) in (2.301), in view $|\xi| \ll 1$, $|\zeta| \ll 1$, yields

$$\frac{\rho_{in}}{\eta\rho_{ex}}\left(\frac{\Phi - M_{in}}{\Phi/\sqrt{\eta} - M_{ex}}\right)^2 = -\frac{1 - \Upsilon_\nu^{\text{in}}/X^2}{1 + 3\Upsilon_\nu^{\text{ex}}/X^2}. \tag{2.302}$$

For low viscosity, $\Upsilon_\nu^{\text{in}} \ll 1$, $\Upsilon_\nu^{\text{ex}} \ll 1$, and (2.302) tends to (2.297). □

Remark 17 Similarly to Section 2.3.4 $(n = 1)$, we find Φ^{Re} and Φ^{Im} as a solution to (2.297). Again for very long wavelength perturbations, when $X^2 \leq \Upsilon_\nu^{(\text{in,ex})}$, Equation (2.302) becomes

$$\frac{\rho_{in}}{\eta\rho_{ex}}\left(\frac{\Phi - M_{in}}{\Phi/\sqrt{\eta} - M_{ex}}\right)^2 = \frac{\Upsilon_\nu^{\text{in}}}{3\Upsilon_\nu^{\text{ex}}}$$

Substitution Υ_ν^{in}, Υ_ν^{ex} denoting in Theorem 46, in the equation above gives

$$\Phi = \frac{3^{1/3}M_{in} - M_{ex}\eta^{1/2}[(\rho_{ex}/\rho_{in})(\nu_{in}/\nu_{ex})]^{1/3}}{3^{1/3} - [(\rho_{ex}/\rho_{in})(\nu_{in}/\nu_{ex})]^{1/3}}.$$

Numerical results. Consider a jet with parameters corresponding to the plume media at altitude $H = 8$ km. These values are $a_{in} \simeq 1000$ m/s, $a_{ex} \simeq 375$ m/s, $\rho_{in} \simeq 0.28$ kg/m^3, $\rho_{ex} \simeq 0.55$ kg/m^3, $\mu_{in} \simeq 1.58 \cdot 10^{-5}$ kg/m $\cdot$ s, $\mu_{in} \simeq 1.72 \cdot 10^{-5}$ kg/m $\cdot$ s, $M_{in} = 3$, $M_{ex} = 2.4$, $\gamma_{in} \simeq 1.29$, $\gamma_{ex} \simeq 1.40$, $P_{in} = 1.6$ atm, and so the ratio between jets and ambient atmosphere densities is equal to $\eta \simeq 0.14$. We investigate pinching and helical modes by the numerical approach in order to find the wavenumber at which growth of these modes is maximal, and how the maximum growth is related to viscosity of the jet and of external medium. The study of high number modes $(n = 2, 3, \ldots)$ is more complicated; so we limit our research to low modes $(n = 0, 1)$. Generally, the radial cross-section of a jet has a form of epicycloid and hypocycloid with the number of petals equal to n. The dispersion equation (2.283) has the form

$$f_n(X, \Phi) = 0, \tag{2.303}$$

where $f_n = f_n^{Re} + if_n^{Im}$. Let $\Phi = \Phi_n^{Re} + i\Phi_n^{Im}$ be an implicit function of $X \in I$, whereby a physical sense $I = [0.01, 10]$. Then (2.303) reads as a system of two real equations

$$f_n^{Re}(X, \Phi_n^{Re}, \Phi_n^{Im}) = 0, \quad f_n^{Im}(X, \Phi_n^{Re}, \Phi_n^{Im}) = 0. \tag{2.304}$$

This system was solved numerically. By a physical sense we restrict our attention to the ranges $0 \leq \Phi^{Re} \leq 5$, and $-0.4 \leq \Phi^{Im} \leq 2.5$ since in this range the behavior of the solutions (roots) may be watched. The solutions of (2.304) for $X \in I$ are presented as intersections $R_X = \alpha_X^{Re} \cap \alpha_X^{Im}$ of the curves $\alpha_X^{Re} = f_n^{Re}(X, \Phi^{Re}, \Phi^{Im}) = 0$, $\alpha_X^{Im} = f_n^{Im}(X, \Phi^{Re}, \Phi^{Im}) = 0$. To study the root's branches, one may graph two curves α_X^{Re}, α_X^{Im} in the domain $I_\Phi = [0, 5] \times [-0.4, 2.5]$ at the plane (Φ^{Re}, Φ^{Im}); the curves topology is complicated. The curve α_X^{Re} intersects each of these strips, the curve α_X^{Im} has oscillations in Φ^{Re}-direction which groped inside vertical strips; the reason is that Φ^{Im}-range to be small with respect to Φ^{Re}-range. Since the most sensitive perturbations are of low frequency $|\Phi^{Re}|$, we select solution $(\Phi^{Re}(X), \Phi^{Im}(X))$ in the strip with minimal $|\Phi^{Im}|$. For a few number of isolated values X we obtain bifurcations of the roots – the part of the curve α_X^{Re} degenerates to a small cycle and we jump other component of α_X^{Im}. It also appears when the root's ordinate $\Phi^{Im}(X)$ comes out of the prescribed range [-0.4, 2.5].

The dispersion equation for linear approach of viscous term has been developed in (2.295). Its simplified cases coincide with various forms of dispersion equation obtained by other authors. In [15] there is study resonance cases but for constant sound velocity. Analysis of dispersion equations allows us to study dependence of the nondimensional phase velocity $\Phi = \omega/ka_{in}$ (complex function) on the parameter X. It was shown that pinching and helical modes increased and the maximum of their oscillations corresponds to the higher wavelengths than to the jet's radius. The analytical results show a good consistency with numerical results. Continue our investigation of the $n = 0$ and $n = 1$ cases when the sound velocity is of *Gaussian form*. First, we show that growth rate for the pinching and helical modes under plumes and external medium viscosities to be neglected – inviscid case. Figures 2.27(b) and 2.28(a) show us the behavior of the growth rate of these two wave modes for the internal Mach number $M_{in} = 3.0$ and for external Mach number $M_{ex} = 2.4$. For the pinching mode the value of the phase velocity u_{ph}^ω has asymptotic $\sim X$ (2.295) at $X \ll 1$ and the phase velocity is increased in almost straight line with increasing of the wave number k, while the phase velocity of the helical mode is increased as well with X. The analytical results show a good consistency with numerical results. Continue our investigation of the $n = 0$ and $n = 1$ cases when the sound velocity is of *Gaussian form*. First, we show that growth rate for pinching and helical modes under plumes and external medium viscosities to be neglected – inviscid case. Figure 2.28(b) shows that behavior of u_{ph}^ω weakly depends on the jet's viscosity (under condition of low viscosity jets). Since we get the same figure for $n = 0$ – inviscid case we can claim that the low viscosity does not affect the phase velocity of the jet while for $n = 1$, the figures are different. The phase velocity $u_{ph}^\omega = \omega/(ka_{in}) = \omega^{Im}R$ (see the end of Section 2.3.3) is increased more slowly, with X, than for zero viscosity. Notice that the result shown in Fig. 2.28(b) is limited by $X \approx 0.3$ due to appearance of singularities there.

Figure 2.29 shows the pinching and helical modes phase velocities u_{ph}^ω of the jet of sufficiently higher viscosity $(\mu_{in} = 9.64 \cdot 10^{-5}\,\text{kg/m} \cdot \text{s})$ since there can occur about 40% of alumina (Al_2O_3). The phase velocity weakly depends on the jet's viscosity (for low viscosities) and has a maximum at $X \simeq 0.4$ and it is decreased rapidly for $X \gg 1$ by comparison with inviscid case, Fig. 2.27(b), i.e., short wavelength oscillations die down in the case of low viscosities. Figure 2.29(a) shows a similar behavior of the wavelength oscillations, in the case of value of viscosity $\mu_{in} = 9.64 \cdot 10^{-5}\,\text{kg/m} \cdot \text{s}$, as in Fig. 2.27(b).

We introduce spectral graphs (Figs. 2.30–2.31) for dispersion equations (2.283) – viscous case and (2.284) – inviscid case for $n = 0$ and $n = 1$ cases. Spectral graphs show that for low viscosity, (2.283) has asymptotic solutions described in Remark 16, (2.296). Hence, the asymptotic equation's roots are $\Phi^{Re} \simeq 3.0$ and $\Phi^{Im} \simeq 0.04$ at $X = 0.01$, that is confirmed by the iterative method on the interval $0.01 \leq X \leq 0.1$, i.e., the topology of the curves is preserved. These graphs show that the number of roots increases rapidly with X, here Φ^{Im} is increased while Φ^{Re} is remained $\simeq 3.0$, i.e., there it occurs the dissipation of the perturbed

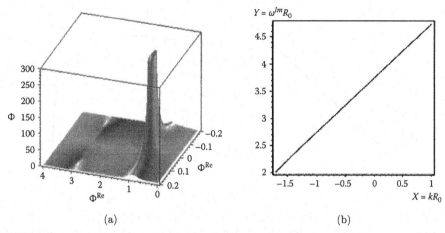

(a)   (b)

FIGURE 2.27: (a) Geometry of the function Φ. (b) Growth of pinching $(n = 0)$ mode as a function of $X = kR_0 = (\lambda/2\pi R_0)^{-1}$ for Gaussian form of sound velocity, for zero viscosity and for $\mu_{in} = 1.58 \cdot 10^{-5} kg/m^2 - s$; $\lambda/2\pi R_0$ is a ratio of the waves length to the jets radius at the inlet.

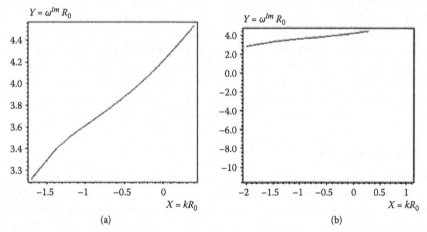

(a)   (b)

FIGURE 2.28: Growth of helical $(n = 1)$ mode as a function of $X = kR_0 = (\lambda/2\pi R_0)^{-1}$ for Gaussian form of sound velocity; $\lambda/2\pi R_0$ is a ratio of the wave's length to the jet's radius at the inlet. (a) Zero viscosity, $\mu = 0$. (b) Viscosity $\mu = 9.64 \cdot 10^{-5}$ kg/m^2-s.

waves to a high number of microwaves. This process starts from $X \geq 0.6$; the value $X \simeq 0.6$ may be considered as the point of wave dissipation. The developed microwave flow may be considered as turbulence of the jet's boundary layer (Fig. 2.30) started at the point $X \simeq 0.6$. Our model is not valid for developed turbulent flow which is a subject of special study. In this stage we can conclude that for $n = 0$ the low viscosity does not affect on the topology of the roots; it influences only when dissipation is begun. Spectral graphs for the helical mode are similar to the graphs for the pinching mode; hence the results described above are true also for $n = 1$ (Fig. 2.30), for both the inviscid and viscous cases. At $X \simeq 0.08$ we can see only one root $\Phi^{Re} \simeq 3.0$ and $\Phi^{Im} \simeq 1.4$. At $X = 1$ we have two roots, one is $\Phi^{Re} \simeq 3.0$, $\Phi^{Im} \simeq 1.2$ and second $-$ $\Phi^{Re} \simeq 3.0$, $\Phi^{Im} \simeq 1.2$. For $X = 0.3$ one may see the rapid growth of the number of roots. The *starting point* $X \simeq 0.6$ may be considered as the *point of wave dissipation*. Similar to the pinching mode we can conclude that low viscosity affects only when the dissipation is begun.

Remark 18 For inviscid jet and $n = 1$; see Fig. 2.31, we omit the plot for low viscous

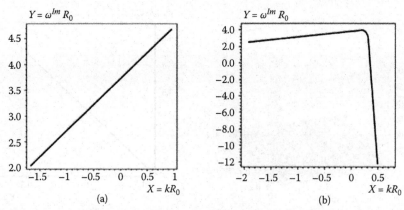

FIGURE 2.29: (a) Growth of pinching ($n = 0$) mode as a function of $X = kR_0 = (\lambda/2\pi R_0)^{-1}$ for Gaussian form of sound velocity; (b) growth of helical ($n = 1$) mode as a function of $X = kR_0 = (\lambda/2\pi R_0)^{-1}$ for Gaussian form of sound velocity.

case. There are x, y-axes represent dimensionless frequencies Φ^{Im} and Φ^{Re}, accordingly. The fact that an expanding jet is unstable to surface perturbations can be understood as resulting from a decrease in the angle θ_r as compared with the angle θ_z for a jet of constant cross-section. Jet expansion allows shorter wavelength perturbations to grow with this rapid growth rate. Equation (2.288) is appropriate for shorter wavelengths on an expanding jet because a shorter wavelength propagates with the same angle on the expanding jet, as does a somewhat longer wavelength on the cylindrical jet. The restriction on growth of shorter wavelength perturbations is in fact a restriction on the propagation angle $\theta_{r,z}$.

2.3.5   Relativistic jet

It is widely believed that extragalactic radio sources are produced by jets of material ejected from a power source at the nucleus of these objects. The morphology of the resulting radio sources to be probably controlled by several different processes. For example, the radio source 3C 31 has been modeled by assuming that jets of material are expelled from an active galaxy which is gravitationally bound to a nearby companion galaxy. A second process involves precession of the central power source, as in the case for the galactic source SS 433. In this situation the ejecta can probably be treated as consisting of discrete components with dynamics like that of ram-pressure-confined plasma clouds. Thus, it can be considered initially straight jets and investigate the properties of steady flow of material through an external medium for which there is little or no precessional motion of the central power source. In particular, we wish to study hydrodynamic stability of radio jet.

Perturbations to a well-directed flow may prove to be responsible for the overall structure of many radio sources. Long wavelength instability is capable of disrupting well-directed flow and of dissipating jets energy by sideways motion of the jet. The numerical calculations [21] showed the stability of a cylindrical jet of constant cross section whose density was less than that of the surrounding medium. Such a jet is subject to exponentially growing or at best algebraically growing perturbations at wavelengths longer than the jets radius and cannot propagate very far. Observation of jets shows that they expand as they leave the nucleus. We'll restrict out attention to the case of a jet of constant cone angle and velocity and we'll assume that both jet and external medium be isothermal. While it is not likely that jets are isothermal, observations of number of sources show that surface brightness does not fall as fast as might be expected if expansion were adiabatic. It seems likely that some of the

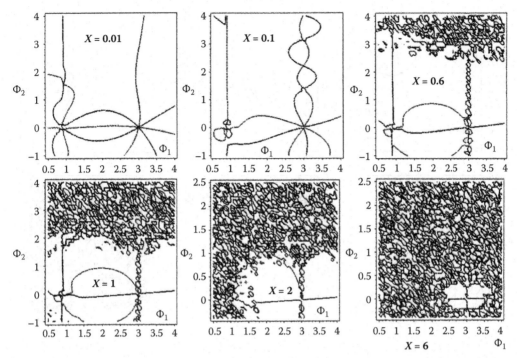

FIGURE 2.30: Spectral graphs for roots of (2.288) for the viscous case, $n = 0$ and for sound velocity of Gaussian form. The angle $\theta_{r,z}$ is decreased as n and $\lambda_{r,z}$ are increased. At constant angle ϕ (cylindrical geometry) one finds that growing perturbations appear restricted to longer and longer wavelengths $\lambda_{r,z}$ as $\theta_{r,z}$ is restricted to small angles. Analogous to the vortex sheet, we expect this to occur as the flow Mach number to be increased.

flow energy is thermalised, and the jets temperature probably varies slowly. If flow energy is much greater than the energy lost to thermalization, then the flow velocity will also vary slowly. Thus, these assumptions are reasonable compromise between likely jet behavior and a mathematically solvable problem. So, the obtained results are relevant to any jet in which the opening angle, jets velocity, and sound speeds do not change rapidly.

Dispersion equation. We start from the equations of continuity and momentum for a relativistic fluid, which perturbations to each fluid are such that the perturbed velocity in each fluid relative to the rest of the fluid is much less than c, the speed of light:

$$\partial_t \rho + \mathrm{div}((\rho + P/c^2)\mathbf{u}) = 0, \tag{2.305a}$$

$$(\rho + P/c^2)[\partial_t \mathbf{u} + (\mathbf{u} \cdot \nabla)\mathbf{u}] = -\nabla P. \tag{2.305b}$$

To analyze a diverging flow of constant opening angle, it can be used spherical coordinates in which the polar axis is again (Section 2.3.5) along the center of the diverging jet and radial lines are along streamlines of the flow. In this coordinate system, the boundary between jet and external medium is given by $\theta = \Theta$, and the cone angle of the diverging jet is 2Θ. The flow velocity is $u_{0r} = u$ and $u_{0\theta} = 0$ for a jet with constant opening angle. The jet is assumed to be axisymmetric and nonrotational, $u_{0\phi} = 0$. Densities and pressures satisfy the time-independent equations of continuity and momentum. For the case of constant cone angle and constant velocity, it follows that the time-independent densities and pressures are

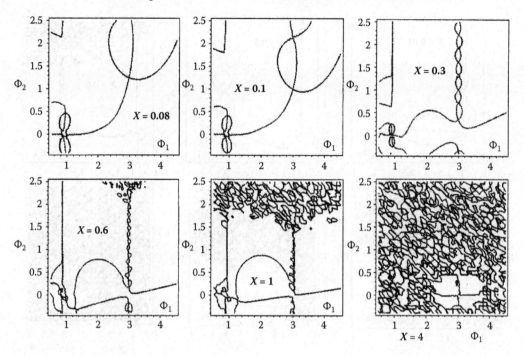

FIGURE 2.31: Spectral graphs for roots of (2.288) for inviscid case, $n = 1$ and for sound velocity of Gaussian form.

proportional to r^{-2} in an isothermal jet and external medium. The linearized (2.305a,b) are

$$\partial_t \rho' + \left(\rho_0 + \frac{P_0}{c^2}\right)\operatorname{div}\mathbf{u}' + \mathbf{u}_0\cdot\nabla\left(\rho' + \frac{P'}{c^2}\right) + \left(\rho' + \frac{P'}{c^2}\right)\operatorname{div}\mathbf{u}_0 + \mathbf{u}'\cdot\nabla\left(\rho_0 + \frac{P_0}{c^2}\right) = 0, \quad (2.306a)$$

$$\left(\rho_0 + \frac{P_0}{c^2}\right)(\partial_t\mathbf{u}' + \mathbf{u}_0\cdot\nabla\mathbf{u}') + \left[\left(\rho_0 + \frac{P_0}{c^2}\right)\mathbf{u}' + \left(\rho' + \frac{P'}{c^2}\right)\mathbf{u}_0\right]\cdot\nabla\mathbf{u}_0 = -\nabla P', \quad (2.306b)$$

where $\rho = \rho_0 + \rho'$, $P = P_0 + P'$, and $\mathbf{u} = \mathbf{u}_0 + \mathbf{u}'$, and it may be assumed that the sound speed is much less than the speed of light. In spherical coordinates, $\rho_0 = \rho_{in}(r)$, $P_0 = P_0(r)$ and $\mathbf{u}_0 = (u_{0r}, u_{0\theta}, u_{0\phi})$ with $u_{0r} \equiv u$, $u_{0\theta} = 0$ and $u_{0\phi} = 0$, are solutions of the dime-independent equations inside the jet given $\rho_0 = \rho_{ex}(r)$, $P_0 = P_0(r)$ and $\mathbf{u}_0 = 0$ in the external medium. Equations (2.306a,b) may be written as

$$\begin{aligned} &\partial_t\rho' + u_{0r}\left(\partial_r\rho' + \frac{1}{c^2}\partial_r P'\right) + u'_r\partial_r\left(\rho_0 + \frac{P_0}{c^2}\right) \\ &+\left(\rho_0 + \frac{P_0}{c^2}\right)\left[\frac{1}{r^2}\partial_r(r^2 u'_r) + \frac{1}{r\sin\theta}\partial_\theta(u'_\theta\sin\theta) + \frac{1}{r\sin\theta}\partial_\phi u'_\phi\right] = 0, \end{aligned} \quad (2.307)$$

$$\left(\rho_0 + \frac{P_0}{c^2}\right)(\partial_t u'_\phi + u_{0r}\partial_r u'_\phi) = -\frac{1}{r\sin\theta}\partial_\phi P', \quad (2.308a)$$

$$\left(\rho_0 + \frac{P_0}{c^2}\right)(\partial_t u'_\theta + u_{0r}\partial_r u'_\theta) = -\frac{1}{r}\partial_\theta P', \quad (2.308b)$$

$$\left(\rho_0 + \frac{P_0}{c^2}\right)(\partial_t u'_r + u_{0r}\partial_r u'_r) = -\partial_r P'. \quad (2.308c)$$

Assume the velocity and density perturbations (analogously to the previous section) of the plane wave form,

$$u'_r = u_{r1}\Psi(\theta)e^{i(kr + n\phi - \omega t)}, \quad u'_\phi = u_{\phi 1}\Psi(\theta)e^{i(kr + n\phi - \omega t)}, \quad u'_\theta = u_\theta(\theta)e^{i(kr + n\phi - \omega t)}, \quad (2.309a)$$

$$\rho' = \rho_1\Psi(\theta)e^{i(kr + n\phi - \omega t)}. \quad (2.309b)$$

Provided the term $\frac{1}{c^2}\partial_r P' = \frac{1}{c^2}\partial_{\rho'} P'\partial_r \rho' = \frac{a^2}{c^2}\partial_r \rho' \ll \partial_r \rho'$, i.e., a is much less than the speed of light c, we find that (2.307) and (2.308a,b) contain the differential $\partial_t + u_{0r}\partial_r$, which becomes $-i\omega + iku_{0r}$. This allows to transform away the velocity u_{0r}, and then to introduce it as a condition at the boundary. Thus, within the jet fluid and external medium, we may set $u_{0r} = 0$ and insert (2.309a,b) into (2.307) and (2.308a,b) to obtain

$$-i\omega\rho_1\Psi(\theta) + (\rho_0 + \tfrac{P_0}{c^2})\big[iku_{r1}\Psi(\theta) + \tfrac{2u_{r1}}{r}\Psi(\theta) + \partial_r u_{r1}\Psi(\theta) + \tfrac{1}{r}\partial_r u_\theta \Psi(\theta) \\ + \cos\theta\tfrac{u_\theta}{r} + in\tfrac{u_{\phi 1}}{r\sin\theta}\Psi(\theta)\big] + u_{r1}\Psi(\theta)\partial_r(\rho_0 + \tfrac{P_0}{c^2}) = 0, \tag{2.310a}$$

$$u_{\phi 1} = \frac{n}{r\sin\theta}\frac{\rho_1}{\omega(\rho_0 + P_0/c^2)}\frac{\partial P'}{\partial\rho'}, \tag{2.310b}$$

$$u_\theta = -\frac{i}{r}\frac{\rho_1}{\omega(\rho_0 + P_0/c^2)}\frac{\partial P'}{\partial\rho'}\Psi'(\theta), \tag{2.310c}$$

$$u_{r1} = k\frac{\rho_1}{\omega(\rho_0 + P_0/c^2)}\frac{\partial P'}{\partial\rho'}. \tag{2.310d}$$

The last equations (2.310b-c) may be used to obtain a differential equation for $\Psi(\theta)$:

$$\sin^2\theta\frac{\partial^2}{\partial\sin^2\theta}\Psi(\theta) + \sin\theta\frac{\partial}{\partial\sin\theta}\Psi(\theta) + \Big[\beta^2(r)\frac{\sin^2\theta}{\cos^2\theta} - \frac{n^2}{\cos^2\theta}\Big]\Psi(\theta) = 0, \tag{2.311}$$

where

$$\beta^2(r) \equiv \frac{\omega^2 r^2}{(\partial P'/\partial\rho')} - k^2 r^2 + i\Big[2kr + \frac{kr^2}{(\partial P'/\partial\rho')}\partial_r\Big(\frac{\partial P'}{\partial\rho'}\Big)\Big]. \tag{2.312}$$

Equation (2.311) is exact but does not have a simple analytic solution. If it is assumed that $\sin^2\theta \ll 1$, then we may use the approximation $\cos^{-2}\theta \simeq 1 + \sin^2\theta$ to write (2.311) as

$$\Psi''(x) + \frac{1}{x}\Psi'(x) + \Big\{\beta^2(r) - n^2 - \frac{n^2}{x^2}\Big\}\Psi(x) = 0, \tag{2.313}$$

where $x = \sin\theta$. Equation (2.313) is valid provided $x \ll 1$. Equation (2.313) becomes Bessel's equation if $\beta(r)$ is a constant independent of r. For adiabatic perturbation, the sound speed in the fluid is

$$a^2 = \Gamma \cdot (\partial P'/\partial\rho'),$$

where Γ is specific heat ratio, and (2.312), becomes

$$\beta^2(r) = \omega^2 r^2/a^2 - k^2 r^2 + i(2kr + (kr^2/a^2)\partial_r(a^2)),$$

For jet and external medium to be isothermal, the sound speed is independent of r, and

$$\beta^2(r) = \omega^2 r^2/a^2 - k^2 r^2 + 2ikr,$$

which can be independent of r provided ω, and k are inversely proportional to r. In this case (2.313) may be written as

$$\beta_n^2 x^2(\partial_{\beta_n x, \beta_n x}^2\Psi(x)) + \beta_n x(\partial_{\beta_n x}\Psi(x)) + (\beta_n^2 x^2 - n^2)\Psi(x) = 0, \tag{2.314}$$

where $\beta_n^2 \equiv \omega^2 r^2/a^2 - k^2 r^2 - n^2 + 2ikr$, and ωr, and kr to be constants. Equation (2.314) is Bessel's equation and it has the solution $\Psi(x) = C_1 J_n(\beta_n x) + C_2 Y_n(\beta_n x)$, where J_n and Y_n are Bessel and Neumann functions, respectively,

$$u_\theta = -\frac{1}{r}\Big[\frac{\rho_1 a^2}{\omega(\rho_0 + P_0/c^2)}\Big]\beta_n\cos\theta\Big[C_1 J_n'(\beta_n x) + C_2 Y_n'(\beta_n x)\Big],$$

where the primes denote derivatives with respect to the argument. The result combined with (2.309b) for ρ', allows to write (2.309a) for the θ-component of the velocity perturbation as

$$u'_\theta = -\frac{i}{r}\left[\frac{\rho' a^2}{\omega(\rho_0 + P_0/c^2)}\right]\beta_n\cos\theta\,\frac{C_1 J'_n(\beta_n x) + C_2 Y'_n(\beta_n x)}{C_1 J_n(\beta_n x) + C_2 Y_n(\beta_n x)}. \tag{2.315}$$

Assume that the boundary between jet and external medium is $\theta = \Theta$, $x = X \equiv \sin\Theta$, with the result that (2.315) for $x < X$ becomes

$$u'_\theta = -\frac{i}{r}\left[\frac{\rho'_{in} a^2_{in}}{\omega_{in}(\rho_{in} + P_0/c^2)}\right]\beta^{in}_n\cos\theta\,\frac{J'_n(\beta^{in}_n x)}{J_n(\beta^{in}_n x)}, \tag{2.316a}$$

and for $x > X$ becomes

$$u'_\theta = -\frac{i}{r}\left[\frac{\rho'_{ex} a^2_{ex}}{\omega_{ex}(\rho_{ex} + P_0/c^2)}\right]\beta^{ex}_n\cos\theta\,\frac{H^{(1)'}_n(\beta^{ex}_n x)}{H^{(1)}_n(\beta^{ex}_n x)}, \tag{2.316b}$$

where $H^{(1)}_n = J_n + iY_n$ is the Hankel function describing outward propagating disturbances. In (2.316a,b)

$$\beta^{in}_n = \left(\frac{\omega^2_{in} r^2}{a^2_{in}} - k^2_{in} r^2 - n^2 + 2ik_{in}r\right)^{1/2}, \quad \beta^{ex}_n = \left(\frac{\omega^2_{ex} r^2}{a^2_{ex}} - k^2_{ex} r^2 - n^2 + 2ik_{ex}r\right)^{1/2},$$

where the subscript (in) refers to quantities in the jet measured in the jet reference frame, and the subscript (ex) refers to the quantities in the external medium. Displacement of the boundary between the two fluids by an amount $\eta' = \eta_1 \exp[i(kr + n\phi - \omega t)]$ is related to the velocity by $\partial_t \eta' = -i\omega\eta' = u'_\theta$. Thus, we may write the displacement using (2.316a,b) as

$$-i\omega_{in}\eta' = -\frac{i}{r}\left[\frac{\rho'_{in} a^2_{in}}{\omega_{in}(\rho_{in} + P_0/c^2)}\right]\beta^{in}_n\cos\Theta\,\frac{J'_n(\beta^{in}_n x)}{J_n(\beta^{in}_n x)},$$

$$-i\omega_{ex}\eta' = -\frac{i}{r}\left[\frac{\rho'_{ex} a^2_{ex}}{\omega_{ex}(\rho_{ex} + P_0/c^2)}\right]\beta^{ex}_n\cos\Theta\,\frac{H^{(1)'}_n(\beta^{ex}_n x)}{H^{(1)}_n(\beta^{ex}_n x)},$$

which leads to the following eigenvalue equation for adiabatic perturbations to the surface of discontinuity:

$$\frac{\rho'_{in} a^2_{in}\beta^{in}_n}{\omega^2_{in}(\rho_{in} + P_0/c^2)}\frac{J'_n(\beta^{in}_n X)}{J_n(\beta^{in}_n X)} = \frac{\rho'_{ex} a^2_{ex}\beta^{ex}_n}{\omega^2_{ex}(\rho_{ex} + P_0/c^2)}\frac{H^{(1)'}_n(\beta^{ex}_n X)}{H^{(1)}_n(\beta^{ex}_n X)}. \tag{2.317}$$

Only if β be a constant, the eigenvalue equation (2.317) independent of r. This suggests that we define $(\omega, k) \equiv (\omega_{ex}r, k_{ex}r)$ as the appropriate variables. Using these variables and transforming (again, likely in Section 2.3.5) wavenumbers to the frame of reference of the external medium, i.e. $\omega_{in}r = \gamma(\omega - ku)$ and $k_{in}r = \gamma[k - u(\omega/c^2)]$, where $\gamma = (1 - u^2/c^2)^{-1/2}$, and using the fact that $\rho'_{in} a^2_{in} = \rho'_{ex} a^2_{ex}$ (due to the pressure being continuous across the boundary), it allows us to rewrite (2.317) as

$$\frac{J'_n(kX\xi_n)H^{(1)}_n(kX\zeta_n)}{J_n(kX\xi_n)H^{(1)'}_n(kX\zeta_n)}\frac{\xi_n}{\zeta_n} = \gamma^2\eta\frac{(\Phi - M_{in})^2}{\Phi^2},$$

where

$$\xi_n \equiv \gamma\sqrt{(\Phi - M_{in})^2 - (1 - \beta^2/M_{in}\Phi)^2 - n^2/(\gamma^2 k^2) + 2i/(\gamma k)(1 - \beta^2/M_{in}\Phi)},$$

$$\zeta_n \equiv \vartheta^{-1/2}\sqrt{\Phi^2 - (1 + n^2/k^2 - 2i/k)^2\vartheta},$$

$$\Phi = \frac{\omega}{ka_{in}}, \quad a_{in,ex} = \Gamma_{in,ex}\frac{P_0}{\rho_{in,ex}}, \quad M_{in} = u/a_{in}, \quad \beta = u/c,$$

$$\eta = \frac{\rho_{in} + P_0/c^2}{\rho_{ex} + P_0/c^2}, \quad \vartheta = \frac{a^2_{ex}}{a^2_{in}} = \frac{\Gamma_{ex}}{\Gamma_{in}}\frac{\rho_{in}}{\rho_{ex}}.$$

Asymptotic analysis of instabilities. For analytical study the behavior of perturbations, consider again the cases of small and large arguments $|kX\xi_n| < 1$, $|kX\zeta_n| < 1$, and $|kX\xi_n| > 1$, $|kX\zeta_n| > 1$. The pinching mode $n = 0$, when $|kX\xi_0| \ll 1$, $|kX\zeta_0| \ll 1$, gives the dispersion relation of the

$$\gamma^2 \eta \frac{(\Phi - M_{in})^2}{\Phi^2} \simeq -\frac{1}{2} k^2 X^2 \xi_0^2 \left[\ln \left(\frac{1}{2} kX\zeta_0 \right) + C \right], \tag{2.318}$$

where $C \approx 0.577$ is Euler constant. In the regime of wavelength much longer than the jets radius, the term $\ln(\frac{1}{2}kX\zeta_0 + C)$ is on the order $(i - 1)$ and, thus, relation (2.318) may be written $\gamma^2 \eta (\Phi - M_{in})^2 \simeq \alpha'_0 k^2 X^2 \gamma^2 (i/k)(i-1)\Phi^2$, where α'_0 is a numerical factor relatively insensitive to the value of $kX\zeta_0$, and we have used $\xi_0^2 \simeq 2i\gamma^2/k$. It can be shown that $\Phi = \Phi_{Re} + i\Phi_{Im}$ is complex and has $\Phi_{Im} > 0$ when $M_{in} > \Phi_{Re} > M_{in} - 1$, i.e., when $u > \omega_{Re}/k > u - a_{in}$. This is the expected result for a wave mode driven by the Bernoulli effect whose phase velocity $u_{ph} = \omega_{Re}/k$ must be such that material travels subsonically in the wave frame. For long wavelengths $\Phi \simeq M_{in}$ it can be found that [21]

$$\Phi \simeq M_{in} + i\alpha_0 \eta^{-1/2} k^{1/2} X M_{in}, \tag{2.319}$$

where α_0 is of order unity. Thus at long wavelengths the rate of growth $\omega_{Im} \simeq \alpha_0 \eta^{-1/2} k^{3/2} X u$, and the phase velocity is $u_{ph} \leq u$. In the limit of short wavelengths, it can be found, from (2.319), that the expression for the $n = 0$ pinching wave mode is

$$\xi_0 \simeq \frac{\pi}{4kX} - \frac{i}{2kX} \ln \left[\frac{\Phi^2 \xi_0 + \gamma^2 \eta (\Phi - M_{in})^2 \zeta_0}{\Phi^2 \xi_0 - \gamma^2 \eta (\Phi - M_{in})^2 \zeta_0} \right]. \tag{2.320}$$

If (2.320) is squared, then at least approximately

$$(\Phi - M_{in})^2 \simeq 1 + \frac{(\frac{1}{4}\pi^2 - A_{Re}^2)}{4k^2 X^2} - i\frac{\pi}{8k^2 X^2} A_{Re}, \tag{2.321}$$

where

$$A_{Re} \equiv \mathrm{Re} \left(\ln \frac{\Phi^2 \xi_0 + \gamma^2 \eta (\Phi - M_{in})^2 \zeta_0}{\Phi^2 \xi_0 - \gamma^2 \eta (\Phi - M_{in})^2 \zeta_0} \right). \tag{2.322}$$

For short wavelengths we obtain $\Phi_{Re} \simeq M_{in} \pm 1$, and (2.321) for perturbation becomes

$$\Phi \simeq (M_{in} \pm 1) \mp i\frac{\pi}{16k^2 X^2} A_{Re}. \tag{2.323}$$

Under condition $\mp A_{Re} > 0$, the perturbation is growing. It is not difficult to show that when $M_{in} + 1 \geq \Phi_{Re} \geq M_{in} - 1$, $\xi_0 \simeq \left[(\Phi - M_{in})^2 - 1 \right]^2$ is imaginary, and the magnitude of the argument in Equation (2.322) is equal to unity with the result that $A_{Re} \simeq 0$. On the other hand, when $\Phi_{Re} = M_{in} \pm \varepsilon$, where $\varepsilon \geq 1$, it can be shown that

$$A_{Re} \simeq 2\beta'_1 \eta (1 - \vartheta M_{in}^{-2})^{1/2} (\vartheta^{1/2} M_{in})^{-1}, \tag{2.324}$$

where it should be used the estimates $\Phi^2 \simeq M_{in}^2$, $(\Phi - M_{in})^2 \simeq \varepsilon^2$, $\xi_0 \simeq \left[(\Phi - M_{in})^2 - 1 \right]^{1/2}$ and $\zeta_0 \simeq (M_{in}^2 - \vartheta)^{1/2} \vartheta^{-1/2}$ in (2.322), and $\beta'_1 = \varepsilon^2 (\varepsilon^2 - 1)^{1/2}$. Provided $\vartheta/M_{in}^2 < 1$, Equation (2.324) is valid and $A_{Re} > 0$, implying that

$$\Phi \simeq M_{in} - 1 + i\beta'_1 (1 - \vartheta/M_{in}^2) \eta / (8\vartheta^{1/2} M_{in} k^2 X^2).$$

Thus, at short wavelengths the rate growth is

$$\omega_l \simeq \beta'_1 \left[\left(1 - \vartheta/M_{in}^2 \right)^{1/2} \eta / (8\vartheta^{1/2} M_{in} k^2 X^2) \right] ku,$$

and the phase velocity is $u_{ph} \leq u - a_{in}$. These short wavelength perturbations are no longer driven by the Bernoulli effect. Perturbations with phase velocity $u_{ph} \geq u + a_{in}$, ($\Phi_{Re} \geq M_{in} + 1$), are dumped.

The $n = 1$ wave mode will distort a cylindrical jet in a helical pattern. This wave mode along with all other wave modes with $n > 0$ is more affected by jet expansion. In the long wavelength limit, i.e., when $|kX\xi_1| \ll 1$ and $|kX\zeta_1| \ll 1$, we obtain relation [21]

$$\Phi \simeq \left(\frac{\gamma^2\eta}{1 + \gamma^2\eta} \pm i \frac{\gamma\eta^{1/2}}{1 + \gamma^2\eta} \right) M_{in}. \qquad (2.325)$$

In the other limit of analytic interest, i.e., when $|kX\xi_1| \gg 1$ and $|kX\zeta_1| \gg 1$, we obtain an expression for the $n = 1$ helical wave mode, analogous to relation (2.321) for the pinching wave mode,

$$(\Phi - M_{in})^2 \simeq 1 + k^{-2} + \left[\frac{9}{4}\pi^2 - (A_{Re}^{(1)})^2 \right] / (4k^2 X^2) - i \frac{3\pi}{8k^2 X^2} A_{Re}^{(1)}. \qquad (2.326)$$

In (2.326), $k^{-1} \equiv (k_{ex}r)^{-1} = (\lambda/2\pi R) < 1$, where R is the initial jets radius, and

$$A_{Re}^{(1)} = \text{Re} \left(\ln \frac{\Phi^2\xi_1 + \gamma^2\eta(\Phi - M_{in})^2\zeta_1 + (i/kX)[\gamma^2\eta(\Phi - M_{in})^2 + \Phi^2]}{\Phi^2\xi_1 - \gamma^2\eta(\Phi - M_{in})^2\zeta_1 - (i/kX)[\gamma^2\eta(\Phi - M_{in})^2 + \Phi^2]} \right).$$

For short wavelengths we find that $\Phi_{Re} \simeq M_{in} \pm 1$, and (2.326) describing a short wavelength helical perturbations becomes $\Phi \simeq (M_{in} \pm 1) \mp i \frac{3\pi}{16k^2 X^2} A_{Re}^{(1)}$, which is identical to (2.323) for the pinching mode. Under condition $\mp A_{Re}^{(1)} > 0$ the perturbations growing. It is easy to see that as $kX \gg 1$, $\xi_1 \to \xi_0$, and $\zeta_1 \to \zeta_0$ with the result that $A_{Re}^{(1)} \to A_{Re}$ in this limit. Thus, we expect growing perturbations when $\Phi_{Re} \leq M_{in} - 1$, with growth rate decreasing rapidly as (kX) increases. Only as (kX) decreases toward unity will there be significant differences in the solutions for the different value modes.

All perturbations to the jet boundary with $n > 1$ may be described as fluting wave modes. Unlike the helical wave mode, these modes do not displace the jet axis and, unlike the pinching wave mode, these modes will not significantly alter the cross sectional area of the jet. The fluting wave modes have a wavelength $\lambda_\phi = 2\pi R/n$ around the jet circumference. This wavelength serves to describe n nodes around the jet circumference, which twist along the jet. On a jet of constant cross section, a node rotates through 2π radians over a length $l_r = 2\pi n/k$, where k is the wavenumber of the perturbation. Thus, for large wavenumber perturbations a node rotates through 2π radians in shorter length l_r. At constant ϕ, fluting appears as a sinusoidal oscillation with wavelength $l_r = 2\pi/k$. One is particularly interested in any modification to the growth rate and wavelength range of growing perturbations that may be result of jet broadening. On a jet of constant cross section it was found that when $|kX\xi| \ll 1$ and $|kX\zeta| \ll 1$ the rate of growth and wave phase velocity are given by (2.325) for $n > 0$.

One can see what the mathematics mean by imaging an unrolled cylinder and an unrolled constantly expanding cone, Fig. 2.32(a). In general, each wave mode n can be represented as a wave propagating with angle $\sin(\theta_{r,z}) = (2\pi R/n)/[\lambda_{r,z}^2 + (2\pi R/n)^2]^{1/2}$ with respect to the axis of the cylinder or cone. One can immediately see from the figure, which drawn to represent the helical wave mode ($n = 1$), that the wavelength λ_r of a wave which propagates at a constant angle with respect to the axis increases proportional to the radius of the cone. As n to be increased, the angle $\theta_{r,z}$ decreases for a fixed value of λ_r or λ_z. The true wavelength of a wave is then given by $\Lambda_{r,z} = \lambda_{r,z} \sin(\theta_{r,z} + \Theta)$. For the cylinder, i.e., if $\Theta = 0$ then $\Lambda_z = (2\pi R/n)/[1 + (2\pi R/n\lambda_z)^2]^{1/2}$. Thus, for the case $n > 0$, the true wavelength $\Lambda_{r,z} = \lambda_{r,z}$. The effect of the cylindrical coordinates as opposed to plane parallel geometry is to restrict perturbations to wavelengths shorter than about the jet circumference.

The fact that an expanding jet appears to be unstable to surface perturbations can be understood as resulting from a decrease in the angle θ_r as compared with the angle θ_z for a jet of constant cross section. Jet expansion allows shorter wavelength perturbations to grow with this rapid growth rate. Equation (2.325) is appropriate for shorter wavelengths on an expanding jet because a shorter wavelength propagates with the same angle on the expanding jet, as does a somewhat longer wavelength on the cylindrical jet. The restriction on growth of shorter wavelength perturbations is in fact a restriction on the propagation angle $\theta_{r,z}$. The angle $\theta_{r,z}$ is decreased as n is increased and as $\lambda_{r,z}$ is increased. At constant angle ϕ (cylindrical coordinates) one finds that growing perturbations appear restricted to longer and longer wavelengths $\lambda_{r,z}$ as $\theta_{r,z}$ is restricted to smaller and smaller angles. Analogous to the vortex sheet, we expect this to occur as the flow Mach number to be increased.

Numerical analysis. In performing the numerical calculations, it was chosen a pressure of $P_{ex} = P_{1n} = 1.38 \times 10^{-12}$dynes/cm^2. This particular choice of pressure is somewhat arbitrary but represents conditions that might exist in the interstellar medium in the outer regions of elliptical galaxies or in the intergalactic medium external to these galaxies [20, 21]. For example, this is the pressure of an intergalactic or interstellar medium with total particle number density of 10^{-4} cm^{-3} and temperature of 10^8 K. The choice of pressure affects the results only in the value of the sound speeds calculated using this pressure and some choice of density in the external medium and density ratio η between jet and external medium. While the maximum growth rate is related to this choice of pressure because the maximum is a function of the jet velocity $u = M_{in}a_{in}$, the shapes of the curves and the wavelength at which maximum growth occurs are functions of M_{in} and η independent of P.

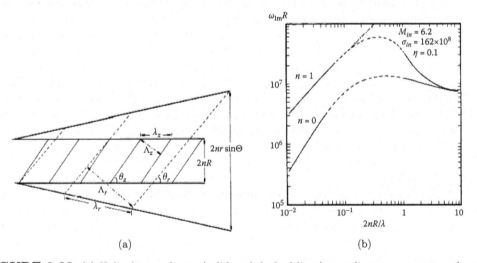

(a)　　　　　　　　　　　　　　　(b)

FIGURE 2.32: (a) Helical wave fronts (solid and dashed lines) traveling at constant angle with respect to the flow on an unrolled cylinder and cone, respectively. If $\lambda_r = \lambda_z$ at $2\pi r \sin\Theta = 2\pi R$ then $\theta_r < \theta_z$ and the wavelength λ_r increases linearly with the radius $r \sin\Theta$. The true wavelength between wavefronts is Λ_r and Λ_z. (b) Growth of the pinching ($n = 0$) and helical ($n = 1$) waves as a function of the wavenumber $k = 2\pi R/\lambda$, which is normalized to the jet radius. The growth rate is $\omega_{Im}R$ in units of s^{-1}. The jet density is $1/10$ the density in the surrounding medium.

Figure 2.32(b), see [21], shows behavior of the growth rate of two wave modes $n = 0$ and $n = 1$ for internal Mach number of 6.2. Since the jet is less dense than the surrounding medium, the sound speed in the external medium is less than that internal to the jet, and the Mach number of the flow in the external medium is higher by a factor $\eta^{-1/2} \simeq \sqrt{10}$. This figure shows the result for a jet with proton number density $N_{in} = 3.16 \times 10^{-5}$cm^{-3} and sound speed $a_{in} = 1.62 \times 10^8$sm s^{-1}. The jet velocity is 1×10^9cm s^{-1}. Since $\eta = 0.1$, we chose $N_{ex} = 3.16 \times 10^{-4} \text{sm}^{-3}$ and the external medium temperature of 1.58×10^7K. The dashed lines represent regions in which the approximations used for the Bessel and Hankel functions are not entirely valid and only represent the likely behavior of the growth rates in these wavelength regimes. The vertical scale is the growth rate ω_{Im} in s^{-1} times the jet radius R in cm, and the horizontal scale is the wavenumber $k \equiv k_{ex} R = 2\pi R/\lambda$ and is normalized to the jet radius R. The numerical results presented in Fig. 2.32(b) have been obtained under assumption the jet to be constantly expanding with opening angle $2\Theta = 0.16$ rad. Because this slow jet expansion will not significantly alter the growth rate or the wavelength at which the growth rate is a maximum, all results may be applied to a jet of constant radius R. To facilitate use of these growth rates for situation other than those considered in this figure, the imaginary part of (2.325) is included in the figure as a dashed and doted line. Inspection of Fig. 2.32(b) tells us that the maximum growth rate of the helical wave mode is reduced from what would be predicted, using (2.325) combined with the wavenumber at which this peak occurs, by about a factor of 2. The maximum rate of growth of the pinching mode is less than that of the helical wave. The growth rate of the pinching mode is not sharply peaked and forms a plateau over a fairly broad range of wavelengths. This result was found in [13].

An analysis of KHI for high Mach number jet flow shows that jet expansion allows wave modes with $n > 0$ to have higher growth rates at shorter wavelength than is true if the jet radius is constant. The pinching ($n = 0$) and helical ($n = 1$) wave modes are potentially capable of changing the flow in observable fashion: the pinching mode through quasi-periodic intensity enhancement of a jet, and the helical mode through helical twisting of the jet radius and increases as the radius increases. At short wavelengths the growth rate of these wave modes decreases rapidly. Numerical study shows that the growth rates decreases at least as k^{-2}. At long wavelength the growth rate of pinching and helical wave modes increases as $k^{3/2}$ and k, respectively. The wave modes can be relatively sharply peaked in growth rate at a particular wavelength. By [20, 21, 22], the helical wave mode grows faster at wavelength $\lambda_h \simeq 1.6 \cdot (M_{in}/\eta^{1/3})R$. The growth rate is sharply peaked for all Mach numbers and jet densities. The growth rate of the pinching mode is not so sharply peaked. As the Mach number of the flow is increased for constant jet density, a plateau forms in the growth rate and a wavelength at which maximum growth occurs is not well defined. For a given Mach number, an increase in jet density yields a more sharply peaked growth rate at wavelength $\lambda_p \simeq 0.6\,(M_{in}/\eta^{1/3})R$. Maximum growth of pinching is always less than the helical wave but becomes comparable as M_{in} is decreased or η is increased.

A comparison of the stability of relativistic and nonrelativistic jets. Jet flows in extragalactic radio sources and those associated with the galactic superluminals could have relativistic velocities for a significant fraction of their length. For each of the pinching, helical, triangular, rectangular, and higher order normal Kelvin–Helmholtz's modes, there is one surface and many body waves. Axisymmetric simulations, like the ones discussed here, only allow waves of the pinch mode, since higher order modes require non-axisymmetry. Note that the higher order modes have smaller growth lengths (or faster growth rates) than the pinch mode, so axisymmetric jets remain essentially stable for longer distances than in simulations where higher order modes are allowed.

In Fig. 2.33 we show schlieren plots, in which darker regions indicate a larger density gradient, for all of the simulations as the bow shock nears the right edge of the grid ($z \simeq 40R$). Typical jet features include a bow shock, a terminal shock or Mach disk of the jet flow, a cocoon of material lighter than the jet that has passed through this shock, and internal structure in the form of biconical shocks along the jet. In the smaller Lorentz factor relativistic simulations, there is some curvature of the terminal shock, which is the very thin line close to the leading edge of the jet and almost perpendicular to the jet axis in the schlieren images. There is some difference in this curvature between the simulations,

TABLE 2.2: Simulation parameters

									Power	Trust	Stability
Run[a]	Γ	η	$\mathcal{M}$	γ	η	M_{in}	M_{ex}	η	M_{in}	M_{ez}	η_x^c
A[b] ...	5/3	0.10	6	1.005	0.11	6.1	18.1	0.11	6.0	18.1	
E...	5/3	0.11	8	2.5	1.01	10.2	9.9	0.70	8.3	9.9	8.8
B...	5/3	0.16	8	5.0	6.70	12.6	4.9	3.96	9.4	9.9	41.8
C...	5/3	0.56	8	10.0	73.1	18.0	2.1	55.5	12.8	9.9	83.0
D...	4/3	0.57	15	10.0	74.4	25.4	2.9	57.1	18.0	9.9	93.1

[a] Relativistic simulations with $\mathcal{M} = \gamma\beta/\gamma_s\beta_s$, $M_{in} = 6.0$, $M_{ext} = 19.0$, where $\beta = u/c$, $\gamma \equiv (1 - \beta^2)^{-1/2}$ is the *Lorentz factor*, and subscript "s" refers to the sound speed, i.e., $\beta_s = a_{in}/c$, $\gamma_s \equiv (1 - \beta_s^2)^{-1/2}$; [b] a single nonrelativistic equivalent of this slow jet was run with $\eta \equiv \rho_{in}/\rho_{ex} = 0.1$, [c] estimated from setting the wave speed of the maximum growing modes to its nonrelativistic equivalent with appropriate values from the relativistic simulations.

particularly between runs E and B (Table 2.2), and in general the shape and curvature of the terminal shock varies with time. In some of the simulations, and in particular run C (Table 2.2), there is a secondary shock roughly $10 - 20R$ behind and parallel to the bow shock, which is caused by shocked ambient material expanding supersonically backward. This secondary shock is weaker or nonexistent in run D, a consequence of slower expansion accompanying the softer adiabatic index. Some of the schlieren plots show a nonphysical reflection of the bow shock at the outer radial boundary, generated by the edge of the numerical grid. The strength of this feature is somewhat exaggerated in a schlieren plot, and the reflected bow shock does not interact significantly with any of the more important jet structures. Despite the appearance that the reflection coincides with an enhancement in cocoon width in run B, we have determined that these coincident structures are unrelated. In this case, the expansion of the cocoon occurs before the reflection arrives.

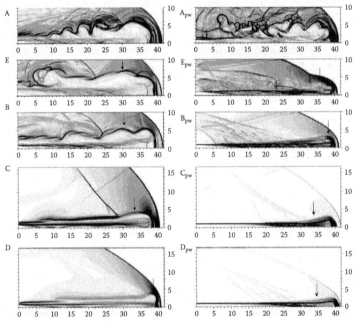

FIGURE 2.33: Schlieren plots of the relativistic and nonrelativistic (with subscriptions "pw") sets of simulations.

From the schlieren images in Fig. 2.33, we see that the morphology of the jets changes with the Lorentz factor in a similar manner for all three sets of simulations. The two codes do not give identical results for the relativistic simulation with a nonrelativistic jet flow speed and for its nonrelativistic equivalent. While the cocoon width is similar, the detailed structure of the cocoon with the relativistic code is more regular and appears somewhat underesolved. These differences within the cocoon lead to the less regularly spaced biconical shocks within the jet in the nonrelativistic equivalent. Given these differences between numerical schemes for the same simulation, we will focus primarily on large-scale differences such as in the cocoon, jet head, and bow shock between a relativistic simulation and its nonrelativistic equivalents.

An important result of these comparisons is that the velocity field of nonrelativistic jet simulations cannot be scaled up to give the spatial distribution of Lorentz factors seen in relativistic simulations. Such a scaling substantially under-predicts the size of the region of a significant Lorentz factor. Since the local Lorentz factor is the primary factor determining the brightness of the source, this suggests that a nonrelativistic simulation cannot yield the proper intensity distribution from a relativistic jet. Other general results are that relativistic simulations and their nonrelativistic equivalents have similar ages (in dynamical time units, R/a_{ex}) and efficiencies, and that jets with a larger Lorentz factor have a smaller cocoon size. The ages and efficiencies in the nonrelativistic simulations with equal thrust are a particularly good match to the relativistic simulations. Also, for both relativistic simulations and their nonrelativistic equivalents with the heat ratio $\Gamma = 5/3$, the widths of the cocoon radii predicted from simple models frequently match a density minimum within the cocoon, and therefore are adequate estimators of cocoon centers. Due to the limited range of estimated cocoon widths in all of our relativistic and nonrelativistic sets of simulations, these simple models do not discriminate well between relativistic and nonrelativistic flow speeds. Slices of pressure near the symmetry axis vary similarly in both relativistic and nonrelativistic simulation sets. Even in the nonrelativistic simulations, these data slices show fewer and smaller deviations as the Lorentz factor in the equivalent relativistic simulation is increased.

As an example of using these simulations to observations, we apply our results to estimate the age, jet to ambient density, and jet to ambient enthalpy of galaxies **Cygnus A**. Although none of these estimates can determine if the jet in **Cygnus A** is relativistic or nonrelativistic, they confirm independent estimates of these parameters. Our results suggest simulation that the jet is light with $\eta \simeq 10^{-4}$, and possibly cold if the jet has a relativistic flow speed $\eta_r \simeq 10^{-4}$. More detailed summary of the results given in this section and the latest research are presented in original works [15, 16, 17].

2.4   Exercises

1. What is the surface tension?
2. How does the surface tension influence on the instability increment?
3. Deduce dispersion relation for inviscid entire jet with undisturbed velocity field.
Hint: Follow methodology described in Section 2.3.2.
4. Deduce dispersion relation for viscous entire jet with undisturbed velocity field.
Hint: Follow methodology described in Section 2.3.2.
5. Deduce dispersion relation for inviscid annular jet with undisturbed velocity field.
Hint: Follow methodology described in Section 2.3.2.
6. Deduce dispersion relation for viscous annular jet with undisturbed velocity field.
Hint: Follow methodology described in Section 2.3.2.

Bibliography

[1] Avital G., et al. Experimental and computational study of infrared emission from under-expanded Rocket exhaust plumes. *J. of Thermophysics and Heat Transfer*, 15, 2001, 377–383.

[2] Batchelor G. K. *The Theory of Homogeneous Turbulence*. Cambridge Univ. Press, 1953, 197 pp.

[3] Benford G. Stability of galactic radio jets. *Astrophysical J.*, 257(3), pt. 1, 1981, 792–802.

[4] Benjamin R. F. Experimental observations of shock stability and shock-induced turbulence. In: *Advances in Compressible Turbulent Mixing*, ed. WP Dannevik, AC Buckingham, 1992, 341–348.

[5] Bers A. Space-time evolution of plasma instabilities – absolute and convective. In *Handbook of Plasma Physics*, vol 1, Eds. Rosenbluth M. N. and Sagdeev R. Z., North-Holland, Amsterdam, 1984, 451 pp.

[6] Birkhoff G. Note on Taylor instability. *Quart. Appl. Math.*, 12, 1954, 306–318.

[7] Chandrasekhar S. *Hydrodynamic and Hydromagnetic Stability*. Oxford Univ. Press, Oxford, 1961.

[8] Chang C. T. Dynamic instability of accelerated fluids. *Phys. Fluids*, 2, 1969, 656–663.

[9] Christiansen R. M. and Hixson A. N. Breakup of a liquid jet in a denser liquid. *Ind. Eng. Chem.*, V. 49, 1957, 1017–1024.

[10] Courant R. and Friedrichs K. *Supersonic Flow and Shock Waves*. Springer, Berlin, 1979, 464 pp.

[11] Dimonte G. and Schneider M. Turbulent Rayleigh–Taylor instability experiments with strong radiatively driven shocks. *Phys. Plasmas*, 4, 1997, 4347–4357.

[12] Ferrari A., Trussoni E. and Zaninetti L. Relativistic Kelvin–Helmholtz instabilities in extragalactic radio sources. *Astronumy and Astrophysics*, 64, 1978, 43–52.

[13] Fiedler R. and Jones T. Stability of magnetized astrophysical jets. *Bull AAS*, 12, 1980, 854–854.

[14] Gaissinski I., Kalinin A. and Stepanov A. Impact of magnetic field on braking the particles. *Sov. J. Appl. Mechan. and Techn. Phys.*, 6, 1979, 40–46.

[15] Gaissinski I. and Rovenski V. Kelvin–Helmholtz instability model for supersonic viscous jet. *Int. J. of Pure and Applied Math.*, 24, No. 4, 2005, 537–551.

[16] Gaissinski I., Kelis O. and Rovenski V. Kelvin–Helmholtz instability model for supersonic adiabatic jet. *Int. J. of Pure and Applied Math.*, 44, N. 2, 2008, 235–248.

[17] Gaissinski I., Kelis O. and Rovenski V. *Hydrodynamic Instability Analysis*. Verlag Dr. Müller, 2009.

[18] Gaissinski I., Levy Y., Rovenski V. Sherbaum V. Rotational liquid film interacted with ambient gaseous media, *Geometry and Applications*, Springer Proc. in Math. and Statistics, 72, 2014, 199–224.

[19] Gaissinski I., Levy Y., Kutikov D., Sherbaum V. and Rovenski V. Operation study of miniature air-blast atomizer under very low liquid pressures, *Int. J. Turbo Jet Eng.*, 2017, in press.

[20] Hardee P. E. On the configuration and propagation of jets in extragalactic radio sources. *Astrophysical J.*, 234, 1979, 47–55.

[21] Hardee P. E. Helical and pinching instability of supersonic expanding jets in extragalactic radio sources. *Astrophysical J.*, 257, 1982, 509–526.

[22] Hardee P. E. Effects of the Kelvin–Helmholtz surface instability on supersonic jets. *Astrophysical J.*, 269, 1983, 94–101.

[23] Harrison W. J. The influence of viscosity on the oscillations of superposed fluids. *Proc. London Math. Soc. 2nd Ser.*, 6, 1908, 396–405.

[24] Infeld E. and Rowlands G. Simple model for the nonlinear evolution of the Rayleigh–Taylor instability. *Phys. Rev. Lett.*, 60, 1988, 2273–2275.

[25] Inogamov N. A. A role of Rayleigh–Taylor and Richtmyer–Meshkov instabilities in astrophysics: An Introduction. *Astrophys. Space Phys.*, 1999, 10, 1–335.

[26] Kadau K. et al. Nanohydrodynamics simulations: An atomistic view of the Rayleigh–Taylor instability. *Proc. Natural Acad. Sci. USA*, 101(16), 2004, 5851–5855.

[27] Kiang R. L. Nonlinear theory of inviscid Taylor instability near the cutoff wavenumber. *Phys. Fluids*, 12, 1969, 1333–1339.

[28] Kull H. J. Theory of the Rayleigh–Taylor Instability. *Phys. Rep.*, North-Holland, 206, 1991, 197–325.

[29] Lamb H. *Hydrodynamics*. Reprint of the 1932 sixth edition. Cambridge University Press, Cambridge, 1993, 738 pp.

[30] Landau L. D. and Lifshitz E. M. *Fluid Mechanics, Course of Theoretical Physics*, vol. 6. Pergamon Press, Oxford, 1959.

[31] Lee J. and Chen, L. Linear stability analysis of gas-liquid interface, *AIAA J.*, 29, 1991, 1589–1595.

[32] Lewis D. J. The Instability of liquid surfaces when accelerated in a direction perpendicular to their planes. II. *Proc. R. Soc. A*, 202, 1950, 81–96.

[33] Lin C. C. *The Theory of Hydrodynamic Stability*. Corrected reprinting Cambridge University Press, New York 1966, 155 pp.

[34] Muller D. E. A Method for solving algebraic equations using an automatic computer. *Mathematical Tables and Aids to Computation*, 10, 1956, 208–215.

[35] Nayfeh A. H. On the nonlinear Lamb–Taylor instability. *J. Fluid Mech.*, 38, 1969, 619–631.

[36] Pollard A., Castillo L., Danaila L. and Glauser M. *Whither Turbulence and Big Data in the 21st Century?* Springer-Verlag, 2017, 574 pp.

[37] Rajappa N. R. and Amaranath T. Nonlinear theory of Taylor instability of supposed fluids of different densities in three dimensions. *Quart. J. Mech. Appl. Math.*, 30, 1977, 131–142.

[38] Ray T. P. The effects of a simple shear layer on the growth of Kelvin–Helmholtz instabilities. *Monthly Notices of the Royal Astronomical Society*, 198, 1982, 617–625.

[39] Reynolds S. B., et. al. Buoyant radiolobes in viscous intracluster medium. *Astron. Soc.*, 2004, 1–12.

[40] Ramshaw J. D. Simple model for linear and nonlinear mixing at unstable fluid interfaces with variable acceleration. *Phys. Rev. E*, 58, 1998, 5834–5840.

[41] Richtmyer R. D. Taylor instability in shock acceleration of compressible fluids. *Commun. Pure and Appl. Math.*, 13, 1960, 297–319.

[42] Shen J. and Li. X. Breakup of annular viscous liquid jets in two gas streams. *J. Propulsion and Power*, 12, 1996, 752–759.

[43] Taylor G. The Instability of liquid surfaces when accelerated in a direction perpendicular to their Planes. I *Proc. R. Soc. A*, 201, 1950, 192–196.

[44] Weber C. On breakup of liquid jets, *Z. Angew. Math. Mech.*, V. 11, 1931, 136–154.

[45] Zel'dovich Y. B. and Raizer Y. P. *Physics of shock waves and high-temperature hydrodynamics phenomena.* Academy of Science, USSR, Moscow, 1966.

Chapter 3

Models for Turbulence

In fluid dynamics, *turbulence* (or turbulent flow) means a flow regime characterized by chaotic, stochastic property changes. This includes low momentum diffusion, high momentum convection, and rapid variation of pressure and velocity in space and time. Flow that is not turbulent is called laminar flow. The Reynolds number characterizes whether flow conditions lead to laminar or turbulent flow; e.g., a pipe flow with a Reynolds number above about 2300 will be turbulent. The statistical description of turbulent flow was suggested by A. Kolmogorov; this description is known to be approximate at best. At very low speeds the flow is laminar and the flow is smooth (though it may involve vortices on a large scale). As the speed increases, at some point the transition is made to turbulent chaotic flow. In turbulent flow, unsteady vortices appear on many scales and interact with each other. Drag due to boundary layer skin friction increases. The structure and location of boundary layer separation often changes, sometimes resulting in a reduction of overall drag. Since laminar-turbulent transition is governed by Reynolds number, the same transition occurs if the size of the object gradually increases, or the viscosity of the fluid decreases, or if the density of the fluid increases. Turbulence causes the formation of eddies which are defined by the Kolmogorov length scale and a turbulent diffusion coefficient. In large bodies of water like oceans this coefficient can be found using Richardson's four-third power law and it is governed by the random walk principle. Consider main models of turbulent flow processes.

Chaos theory is a field of study in mathematics and natural sciences about the behavior of dynamical systems that are highly sensitive to initial conditions. This sensitivity is popularly referred to as the butterfly effect. Small changes in initial conditions (such as those due to rounding errors in numerical computation) yield widely diverging outcomes for chaotic systems, rendering long-term prediction impossible in general. Chaotic behavior can be observed in many natural systems, such as the weather. Related results on the topic of nonlinear differential equations, were carried out by G. Birkhoff [10], A. Kolmogorov [53, 54], M. Cartwright and J. Littlewood [13], and S. Smale [96]. Except for Smale these studies were all directly inspired by physics: the three-body problem in the case of Birkhoff, turbulence and astronomical problems in the case of Kolmogorov, and radio engineering in the case of Cartwright and Littlewood. Study of the critical point beyond which a system creates turbulence was important for Chaos theory, analyzed by physicist L. Landau who developed the Landau–Hopf theory of turbulence. D. Ruelle [92] and F. Takens [102] later predicted, against Landau, that fluid turbulence could develop through a strange attractor, a main concept of chaos theory. Feigenbaum notably discovered the universality in chaos, permitting an application of chaos theory to many different phenomena. A. Libchaber and J. Maurer [65] presented experimental observation of the bifurcation cascade that leads to chaos and turbulence in convective Rayleigh–Benard systems.

The idea of similarity in physical space, for example, expressed it explicitly, but the force of Kolmogorov's theory [53, 54] seems to have oriented attention in later years to wavenumber space. The balance seems to have been reverting to real space in the last two decades or so, but, as noted by B. Mandelbrot [77], turbulent structures are still largely described in terms of Euclidean objects such as blobs, rods, slabs, and ribbons, and occasionally in terms of more imaginative, but unhelpful, metaphors such as melons, beans, spaghetti, lettuce, and the like; on the other hand, reality [35] calls for "in-between" patterns. B. Mandelbrot pointed out that scale similarity cannot be built up in traditional descriptions without introducing dubious assertions, whereas scale similarity is the natural attribute of fractals.

It is a fact of experience almost too familiar to notice that dissipative physical systems subject to weak steady driving approach states of dynamic equilibrium that are independent of initial condition. As the strength of the driving is increased, these systems typically undergo a sequence of transitions – the details depending on the system – and arrive eventually at behavior that may be described as chaotic or turbulent. The turbulent motion is not entirely without regularity, but the regularity is statistical in character and appears only when long-term time averages are examined. Ideally, the mechanisms producing the transition from steady to chaotic behavior, and the detailed nature of the motion in the chaotic regime, should be deducible directly from the equations of motion for the system in question, i.e., the Navier–Stokes or Boussinesq equations in the case of classical kinds of fluid systems. Direct attacks on these equations meet with overwhelming difficulties. On the one hand, control over the analytic properties of the equations is not yet good enough either to prove or to disprove the existence of regular solutions for all times and arbitrary regular initial data. On the other hand, it seems quite hopeless to try to compute explicit analytic solutions with chaotic behavior, to say nothing of computing, from first principles, the statistical distribution describing the behavior of typical solutions. To circumvent the difficulties of a direct approach, a number of oblique lines of attack have been developed. One of these approaches is known as the strange attractor theory of turbulence.

Sir H. Lamb once said [43]: "...There are two matters on which I hope enlightenment. One is quantum electrodynamics and the other is turbulence of fluids. About the former, I'm really rather optimistic." Possibly Lamb's pessimism about turbulence had been short-sighted. There exist signs that the two issues that concerned Lamb are not disconnected. The connections were brewing for some while, and began to take clearer form recently. This promises renewed vigor in the intellectual pursuit for understanding this long-standing problem. This is not due to some outstanding developments of new tools in the theoretical study of turbulence, but rather due to developments in neighboring fields. The great successes of the theory of critical and chaotic phenomena and popularity of nonlinear physics of classical systems attracted efforts that combined the strength of fields like quantum field theory and condensed matter physics leading to renewed optimism about the solubility of the problem of turbulence. It seems that this area of research will have a renaissance of rapid growth that promises excitement into 20th century. Fluid turbulence, which is highly complex, chaotic and vertical flow that is characteristic of all fluids under large stresses, is a paramount example of these phenomena that are immensely challenging to the physicist and the mathematician alike.

3.1 Symmetries and conservation laws

3.1.1 Euler and Navier–Stokes equations

The history of fluid mechanics starts with L. Euler who was invited by Frederick the Great to Potsdam in 1741. As a true theorist, he began by trying to understand the laws of fluid motion, and in 1755 discovered the Newton's laws for a fluid (of constant *density*) which in modern notation read [62]

$$\partial_t \mathbf{u} + (\mathbf{u} \cdot \nabla)\,\mathbf{u} = -\nabla p/\rho, \tag{3.1}$$

where $\mathbf{u} = \mathbf{u}(\mathbf{r}, t)$ and $p = p(\mathbf{r}, t)$ are the fluid velocity and pressure at the spatial point $\mathbf{r}$ and at time t. The LHS of this *Euler equation* is just the material time derivative of the momentum, and the RHS is the force represented as the gradient of the pressure imposed on the fluid. This equation predicts (for a given pressure) velocities that are much higher than anything observed. One missing idea was the viscous dissipation ν that is the friction per of one portion of fluid against neighboring ones. The appropriate term was added to (3.1) by C. Navier in 1827 and by G. Stokes in 1845, see [62]. The result is known as *Navier–Stokes equations* (NSE),

$$\partial_t \mathbf{u} + (\mathbf{u} \cdot \nabla) \mathbf{u} = -\nabla p / \rho + \nu \nabla^2 \mathbf{u}, \tag{3.2}$$

where $\nu = \mu / \rho$ is the kinematic viscosity (about 10^{-2} and $0.15 \, \mathrm{cm}^2/\mathrm{sec}$ for water and air, respectively, at room temperature). Without this term the kinematic energy $u^2/2$ is conserved, but with it such energy is dissipated and turned into heat. The effect of the term $\nu \nabla^2 \mathbf{u}$ is to stabilize and control the nonlinear kinematic energy conserving by (3.1).

In mathematics, a *dynamical system* is a quadruplet $(\Omega, \mathcal{A}, P, G_t)$; the *probability space* Ω consists of all positive continuous $\mathbf{u}$; the *time-shift* is the map $G_t : \mathbf{u} \to \mathbf{u}(t)$ to be satisfied semi-group property $G_0 = I$, $G_t G_{t'} = G_{t+t'}$, and conserves the *probability*: $P(G_t^{-1} A) = P(A)$ for all $t \geq 0$ and $A \in \mathcal{A}$, where $\mathcal{A}$ is a family of measurable subsets of Ω. For the incompressible case, the fluid flow is governed by non-uniform NSE

$$\begin{cases} \partial_t \mathbf{u} + (\mathbf{u} \cdot \nabla) \mathbf{u} = -\nabla p / \rho + \nu \nabla^2 \mathbf{u} + \mathbf{f}, \\ \operatorname{div} \mathbf{u} = 0, \\ \mathbf{u}_{\,|\,t=0} = \mathbf{u}_0 \quad \text{(boundary conditions).} \end{cases} \tag{3.3}$$

The initial condition is chosen in a suitable space of functions satisfying the boundary conditions and the incompressibility constraint. The force $\mathbf{f}$ is assumed to be time independent. Hence G_t also maps $\mathbf{u}_s$ into $\mathbf{u}_{s+t}$. As for a probability measure P on Ω, it is invariant under the time shift.

The existence of the time-shift G_t and an *invariant measure* P are, in general, only conjectures. In three dimension case, there is no theorem guaranteeing the existence and uniqueness of solutions to NSE. For chaotic systems, rigorous proofs are available only for simple finite-dimensional models.

First, its invariant measure fills the whole available space $[0, 1]$. In contrast, it is typical for dissipative systems in finite dimensions to have their invariant measure concentrated on an attractor with zero Lebesgue measure and with fractal structure. A well-known instance is the *Henon map* $(x, y) \mapsto (y + 1 - ax^2, \; bx)$.

Second, it is typical for dissipative dynamical systems to have more than one attractor and therefore more than one invariant measure [36]. Each attractor has an associated *basin*. The statistical properties of solutions will then depend on to which basin the initial condition belongs. Thus, not only may the detailed behavior of orbits be unpredictable (because of the sensitivity to the initial conditions), but even their *statistical properties may be unpredictable*, insofar as it may be impossible to determine to which basin the initial condition belongs. Translated into meteorological vocabulary, this is equivalent to stating that not only the weather but also the climate may be unpredictable.

In order to study incompressible flow, we rewrite (3.2) and the continuity equation of (3.3) as

$$\partial_t u_i + \sum_j u_j \partial_j u_i = -\partial_i p / \rho + \nu \sum_j \partial_{jj} u_j, \qquad \sum_i \partial_i u_i = 0. \tag{3.4}$$

Here we used the notation: $\partial_j \equiv \frac{\partial}{\partial x_j}$, $\partial_{ij} \equiv \frac{\partial^2}{\partial x_i \partial x_j}$. In order to achieve maximum symmetry, it is advantageous not to have any boundaries. Thus, assume that the fluid fills the whole space $\mathbb{R}^3$. The unboundedness of the space $\mathbb{R}^3$ leads to some mathematical difficulties. We'll therefore often assume *periodic boundary conditions* in the space variables $\mathbf{r} = (x, y, z)$:

$\mathbf{u}(x + nL,\ y + mL,\ z + qL) = \mathbf{u}(x, y, z)$, for all $x, y, z \in \mathbb{R}$ and $n, m, q \in \mathbb{Z}$. The real $L > 0$ is called the *period*. It is then enough to consider the restriction of the flow to a *periodicity box* $B_L = \{0 \leq x,\ y,\ z \leq L\}$. The case of a fluid in the unbounded space $\mathbb{R}^3$ may be recovered by letting $L \to \infty$.

The space of L-periodic functions $\mathbf{u}(\mathbf{r})$ satisfying condition div $\mathbf{u} = 0$ will be considered, and it should be supplemented by prescribing a suitable norm. We'll refrain from this because our purpose is not to derive fully rigorous results: the present state of mathematics for three-dimensional NSE makes it unreasonable to set higher standards. With periodic boundary conditions, it is easy to eliminate the pressure from NSE, as it will be shown. This is rather an elementary exercise which gives us an opportunity to introduce useful notation. Taking the divergence of $(3.4)_1$ and using $(3.4)_2$ we get

$$\sum_{i,j} \partial_i (u_j \partial_j u_i) = \sum_{i,j} \partial_{ij}(u_i u_j) = -\sum_j \partial_{jj} p/\rho = -\nabla^2 p/\rho. \tag{3.5}$$

Equation (3.5) is an instance of the *Poisson equation*: $\nabla^2 p = \sigma$. This equation can be solved in the class of L-periodic functions provided that $\sigma(\mathbf{r})$ has zero average (angular brackets denote *space averages*):

$$\langle \sigma \rangle = \frac{1}{L^3} \int_{B_L} \sigma(\mathbf{r}) d\mathbf{r} = 0. \tag{3.6}$$

Obviously, the function $\sigma = -\sum_{i,j} \partial_{ij}(u_i u_j)$, being made of space-derivatives of periodic functions, possesses the solvability property (3.6). The solution of the Poisson equation is readily obtained by going from the *physical space*, $\mathbf{r}$-space, to the *Fourier space*, $\mathbf{k}$-space, using Fourier series:

$$\sigma(\mathbf{r}) = \sum_k e^{i\mathbf{k} \cdot \mathbf{r}} \hat{\sigma}_{\mathbf{k}}, \quad p(\mathbf{r}) = \sum_k e^{i\mathbf{k} \cdot \mathbf{r}} \hat{p}_{\mathbf{k}},$$

where $\mathbf{k} \in \left(\frac{2\pi}{L}\mathbb{Z}\right)^3$. The Fourier coefficients are given by $\hat{\sigma}_{\mathbf{k}} = \langle e^{-i\mathbf{k} \cdot \mathbf{r}} \sigma(\mathbf{r}) \rangle$, $\hat{p}_{\mathbf{k}} = \langle e^{-i\mathbf{k} \cdot \mathbf{r}} p(\mathbf{r}) \rangle$. Notice that by (3.6), $\hat{\sigma}_0$ vanishes. It follows from the Poisson equation that $\hat{p}_{\mathbf{k}} = -\frac{\hat{\sigma}_{\mathbf{k}}}{k^2}$, where $k \neq 0$ is the modulus of the *wavevector* $\mathbf{k}$. The coefficient $\hat{p}_0$ is arbitrary. Indeed, the solution of the Poisson equation is defined up to an additive constant. Adding a constant to the pressure does not change NSE. This (not quite unique) solution will be denoted in the physical space as $\nabla^{-2}\sigma$. In the physical space, it is a non-local operator (its explicit expression involves a convolution).

After the pressure has been eliminated by solving the Poisson equation, NSE may be rewritten as

$$\partial_t u_i + \sum_l (\delta_{il} - \partial_{il}\nabla^{-2}) \sum_j \partial_j (u_j u_l) = \nu\nabla^2 u_i. \tag{3.7}$$

It is sufficient to impose the divergence condition $\sum_j \partial_j u_j = 0$ at $t = 0$, since (3.7) will propagate this condition to all times. An alternative way to eliminate the pressure is to work with the *vorticity*

$$\omega = \nabla \times \mathbf{u}. \tag{3.8}$$

Taking "rot" of (3.2), and using identity $\nabla(u^2) = 2\,(\mathbf{u} \cdot \nabla)\mathbf{u} + 2\,\mathbf{u} \times (\nabla\mathbf{u})$, we get the *vorticity equation*:

$$\partial_t \omega = \nabla \times (\mathbf{u} \times \omega) + \nu\nabla^2\omega. \tag{3.9}$$

If we try to rewrite (3.9) in terms, solely, of the vorticity field, we first must solve (3.8) for the velocity. This can be done by taking the "rot" of (3.8) and solving the Poisson equation. Hence, the same "non-local" operator ∇^{-2} appears as in the velocity formalism (3.7).

3.1.2 Symmetries

The term *symmetry* is often used for a discrete or continuous invariance group of a dynamical system.

Definition 51 Let Gr denote a group of transformations acting on space-limited functions $\mathbf{u}(\mathbf{r}, t)$, which are spatially periodic and divergence free. A group Gr is a *symmetry group of NSE* if, for all $\mathbf{u}$ which are solutions of NSE, and all $g \in$ Gr, the function $g(\mathbf{u})$ is also a solution.

Hereafter we give a list of well-known symmetries of NSE.

Proposition 26 *The following take place for NSE:*

   a) Space-translations, $g_{\rho}^{\text{space}} : (t, \mathbf{r}, \mathbf{u}) \mapsto (t, \mathbf{r} + \boldsymbol{\rho}, \mathbf{u}), \quad \boldsymbol{\rho} \in \mathbb{R}^3;$

   b) Time-translations, $g_{\tau}^{\text{space}} : (t, \mathbf{r}, \mathbf{u}) \mapsto (t + \tau, \mathbf{r}, \mathbf{u}), \quad \tau \in \mathbb{R};$

   c) Galilean transformations, $g_{u}^{Gal} : (t, \mathbf{r}, \mathbf{u}) \mapsto (t + \tau, \mathbf{r} + \mathbf{U}t, \mathbf{u} + \mathbf{U}), \quad \mathbf{u}, \mathbf{U} \in \mathbb{R}^3;$

   d) Parity, $\mathbf{P} : (t, \mathbf{r}, \mathbf{u}) \mapsto (t, -\mathbf{r}, -\mathbf{u});$

   e) Rotations, only for $L \to \infty$, $g_{A}^{\text{rot}} : (t, \mathbf{r}, \mathbf{u}) \mapsto (t + \tau, \mathbf{A}\mathbf{r}, \mathbf{A}\mathbf{u}), \quad \mathbf{A} \in SO(\mathbb{R}^3);$

   f) Scaling, only for inviscid case $\nu = 0$, $g_{\mathbf{A}}^{\text{rot}} : (t, \mathbf{r}, \mathbf{u}) \mapsto (\lambda^{1-h}t, \lambda\mathbf{r}, \lambda^h \mathbf{u}), \quad \lambda \in \mathbb{R}^+, h \in \mathbb{R}.$

Concerning the notation in Proposition 26, note that it is simpler to write, for example, $(t, \mathbf{r}, \mathbf{u}) \mapsto (t, \mathbf{A}\mathbf{r}, \mathbf{A}\mathbf{u})$ than the equivalent statement $\mathbf{u}(\mathbf{r}, t) \mapsto \mathbf{A}\mathbf{u}(\mathbf{A}^{-1}\mathbf{r}, t)$. We omitted transformations for pressure p because it can be eliminated from NSE. By (3.5), the pressure is transformed as u^2.

Proof and comments. The space-and time- translation symmetries are obvious. As for Galilean transformations, observe that when we substitute $\mathbf{u}(\mathbf{r} - \mathbf{U}t, t) + \mathbf{U}$ for $\mathbf{u}$ in NSE, there is a cancelation of terms for $\partial_t \mathbf{u}$ and $\mathbf{u} \cdot \nabla \mathbf{u}$. Under parity, all the terms in NSE change sign (in particular, $\nabla \mapsto -\nabla$). Observe also that the symmetry $\mathbf{u} \mapsto -\mathbf{u}$ is not consistent with the equations, except when nonlinear term is negligible. Arbitrary (continuous) rotational invariance is not consistent with periodic boundary conditions since the latter single out certain directions, so that only a discrete subset of rotations is permitted. As for the scaling transformations, when t is changed into $\lambda^{1-h}t$, $\mathbf{r}$ into $\lambda\mathbf{r}$, and $\mathbf{u}$ into $\lambda^h \mathbf{u}$, all terms in NSE are multiplied by λ^{2h-1}, except the viscous term which is multiplied by λ^{h-2}. Thus, for finite viscosity, only $h = -1$ is permitted. The corresponding symmetry is then equivalent to the well-known *similarity principle* of fluid dynamics, because the scaling transformations are then seem to preserve the Reynolds number. If we ignore the viscous term or merely let it tend to zero, as may be justified at very high Reynolds numbers, then we find that there are *infinitely many* scaling groups, labeled by their *scaling exponent* h, which can be any real number.   □

3.1.3 Conservation laws

In mechanics, it is customary to discuss conservation laws together with their symmetries. For *conservative* systems describable by a Lagrangian function there is Noether's theorem which gives a rationale for this association. The theorem states that for each symmetry there is a corresponding conservation law. For instance, momentum conservation corresponds to the invariance of the Lagrangian under space-translations. Such results are not directly relevant for turbulence since NSE are dissipative (by comparison with the Euler equation, obtained under $\nu = 0$, which is conservative and possesses various Lagrangian formulations). Still, it is useful to discuss conservation laws involving an integration over the whole volume occupied by fluid. Other more local conservation laws, such as the conservation of circulation, may be even more important but have found surprisingly few applications to turbulence theory. Periodic boundary conditions are assumed as in the previous section. Angular brackets are used to denote averages over the fundamental periodicity box:

$$\langle f \rangle = \frac{1}{L^3} \int_{B_L} f(\mathbf{r}) \, d\mathbf{r},$$

where $f(\mathbf{r})$ is an arbitrary L-periodic function. (In what follows all functions are L-periodic). We list some useful identities which are readily proved by performing integrations by parts

$$\langle \partial_t f \rangle = 0, \tag{3.10a}$$

$$\langle (\partial_i f) g \rangle = -\langle f(\partial_i g) \rangle, \tag{3.10b}$$

$$\langle (\nabla^2 f) g \rangle = -\langle (\partial_i f)(\partial_i g) \rangle, \tag{3.10c}$$

$$\langle \mathbf{u} \cdot (\nabla \times \mathbf{v}) \rangle = \langle (\nabla \times \mathbf{u}) \mathbf{v} \rangle, \tag{3.10d}$$

$$\langle \mathbf{u} \cdot \nabla^2 \mathbf{v} \rangle = -\langle (\nabla \times \mathbf{u}) \cdot (\nabla \times \mathbf{v}) \rangle, \quad \operatorname{div} \mathbf{v} = 0. \tag{3.10e}$$

We list the main known conservation laws. We include relations, such as the energy balance equation, which become conservation laws only when the viscosity is set equal to zero.

Proposition 27 *Let $\mathbf{u}(\mathbf{r}, t)$ be a solution of NSE, and $\boldsymbol{\omega} = \nabla \times \mathbf{u}$. Then the conservation laws are satisfied:*

$$\frac{d}{dt} \langle \mathbf{u} \rangle = 0 \quad \text{(momentum)}, \tag{3.11a}$$

$$\frac{d}{dt} \langle \frac{1}{2} u^2 \rangle = -\frac{1}{2} \nu \langle \sum_{i,j} (\partial_i u_j + \partial_j u_i)^2 \rangle = -\nu \langle |\boldsymbol{\omega}|^2 \rangle \quad \text{(energy)}, \tag{3.11b}$$

$$\frac{d}{dt} \langle \frac{1}{2} \mathbf{u} \cdot \boldsymbol{\omega} \rangle = -\nu \langle \boldsymbol{\omega} \cdot \nabla \times \boldsymbol{\omega} \rangle \quad \text{(helicity)}. \tag{3.11c}$$

Proof. Momentum conservation is proved by observing that the advection term $\sum_j u_j \partial_j u_i$ in $(3.4)_1$ can be written as $\sum_j \partial_j (u_j u_i)$. Thus in NSE, all terms different from $\partial_t u_i$ are spatial derivatives of periodic functions. Hence (3.11a) follows from (3.10a). For the energy balance relation (3.11b), we multiply $(3.4)_1$ by u_i and use $(3.4)_2$ to obtain

$$\frac{1}{2} \sum_i \partial_t (u_i u_i) + \frac{1}{2} \sum_{i,j} \partial_j (u_j u_i u_i) = -\sum_i u_i \partial_i p / \rho + \nu \sum_i u_i \nabla^2 u_i,$$

from which (3.11b) follows by use of $(3.4)_2$ and (3.10a,b,e). For the helicity balance relation (3.11c), we start from the vorticity equation (3.9) and take the scalar product with $\mathbf{u}$, average and observe that by (3.10d): $\frac{d}{dt} \langle \mathbf{u} \cdot \boldsymbol{\omega} \rangle = 2 \langle \mathbf{u} \cdot (\partial_t \boldsymbol{\omega}) \rangle$. Thus, we obtain

$$\frac{1}{2} \frac{d}{dt} \langle \mathbf{u} \cdot \boldsymbol{\omega} \rangle = \langle \mathbf{u} \cdot \nabla \times (\mathbf{u} \times \boldsymbol{\omega}) \rangle + \nu (\mathbf{u} \cdot \nabla^2 \boldsymbol{\omega}),$$

from which the helicity relation (3.11c) follows by use of (3.10d,e).   □

Introduce some important notations.

Definition 52 *Call E the mean energy, H – the helicity and Ω – the enstrophy (the term enstrophy was coined by C.Leith by analogy with energy; the word "mean" is often omitted):*

$$E \equiv \langle \frac{1}{2} |\mathbf{u}|^2 \rangle, \quad \Omega \equiv \langle \frac{1}{2} |\boldsymbol{\omega}|^2 \rangle, \quad H \equiv \langle \frac{1}{2} \mathbf{u} \cdot \boldsymbol{\omega} \rangle, \quad H_\omega \equiv \langle \frac{1}{2} \boldsymbol{\omega} \cdot \nabla \times \boldsymbol{\omega} \rangle.$$

The energy and helicity balance equations may thus be written as

$$\frac{d}{dt} E = -2\nu \Omega, \quad \frac{d}{dt} H = -2\nu H_\omega.$$

The quantity H_ω might be called the *mean vortical helicity*. The *mean energy dissipation* (per unit mass) $\varepsilon \equiv -\frac{d}{dt} E$, is one of most frequently used quantities in turbulence studies.

Remark 19 1. In deriving the conservation laws, we assumed that the velocity and the pressure fields are sufficiently smooth to permit all necessary manipulations, such as integration by parts, derivations of products, etc. This sort of smoothness is generally conjectured to hold for any finite positive viscosity. For the solutions of the Euler equation ($\nu = 0$) it may not hold and the energy conservation may break down. Increasingly weak smoothness conditions ensuring energy conservation for the Euler equation have been obtained in [25, 100] and so on.

2. Note that $\nu|\omega|^2$ and the *local dissipation* $\varepsilon_{loc} = \frac{1}{2}\nu \sum_{i,j} (\partial_i u_j + \partial_j u_i)^2$ have the same space average. They actually differ by a term proportional to the Laplacian of pressure. Indeed, by (3.5) we get

$$\nabla^2 p/\rho = \frac{1}{4}\sum_{i,j}(\partial_i u_j - \partial_j u_i)^2 - \frac{1}{4}\sum_{i,j}(\partial_i u_j + \partial_j u_i)^2.$$

Only the quantity ε_{loc} which involves the rate-of-strain tensor (the symmetric part of the velocity gradient), deserves to be a local dissipation. Indeed, in a region of quasi-uniform vorticity there is an almost solid rotation of the fluid and hence no dissipation.

3. The energy balance equation plays a critical role in proving the existence "in the large" (for all times) for the three-dimensional NSE. Unfortunately, uniqueness in the large is proven only in two-dimensional case. In two dimensions, there is indeed an additional balance equation for the enstrophy:

$$\frac{d}{dt}\Omega = -2\nu P, \qquad P \equiv \langle \frac{1}{2}|\nabla \times \omega|^2\rangle, \tag{3.12}$$

where P is called the *mean palinstrophy*, and (3.12) also has applications for *2d turbulence*.

3.2 Anomalous scaling exponents

Studies of the universal small scale structure of turbulence can be classified broadly into two classes.

First there is a large body of phenomenological models that by attempting to achieve agreement with experiments reached important insights on the nature of the cascade or the statistics of the turbulent fields [36]. In particular there appeared influential ideas, following [76], about the fractal geometry of highly turbulent fields which allow scaling properties that are sufficiently complicated to include also non-Kolmogorov scaling. By introducing multifractals one can accomodate the nonlinear dependence of ζ_n and n, [37]. These models are not derived on the basis of the equations of fluid mechanics; one is always left with uncertainties about the validity or relevance of these models.

The second class of approaches is based on the equations of fluid mechanics. Typically one acknowledges the fact that fluid mechanics is a (classical) field theory and resorts to field theoretic methods in order to compute statistical quantities. Even though there had been a continuous effort during almost 50 years in this direction, the analytic derivation of the scaling laws for $K_{\varepsilon\varepsilon}(R)$ and $S_n(R)$ from (3.2) and the calculation of the numerical value of the scaling exponents μ and ζ_n have been among the most elusive goals of theoretical research. Why did it turn out to be so difficult?

3.2.1   Multifractal models

There is a very simple way to modify the phenomenological model, so as to incorporate a form of intermittency. The idea of so called the β-model is: at each stage of Richardson cascade, the number of *daughters* of a given *mother-eddy* is chosen such that the fraction of volume occupied is decreased by the factor β ($0 < \beta < 1$), which is an adjustable parameter of the model. Otherwise, nothing is changed in the presentation of the cascade. In the β-model, the fraction p_l of the space which is "active," i.e., within a *daughter-eddy* of size $l = r^n l_0$ it decreases as a power of l,

$$p_l = \beta^n = \beta^{\frac{\ln(l/l_0)}{\ln r}} = (l/l_0)^{3-D}, \quad 3 - D \equiv \ln \beta / \ln r.$$

The notation for D is justified by the observation that it can be interpreted as a *fractal dimension* [27, 76]. This statement is made intuitively by Fig. 3.1(a) which shows within a unit cube three objects, a point, a curve and a surface having, respectively, the dimension D (in the ordinary sense of manifold) of zero, one and two. We are interesting the question, what is the probability p_l that a ball of small radius l, whose center is chosen in the cube with a random uniform distribution, will insert such an object. The answer follows immediately from the geometric construction in Fig. 3.1(a). For a surface, the center of the intersecting ball has to be within a sandwich of thickness $2l$; for a curve, it has to be within a sausage of radius l and for the point it has to be within a ball of radius l. Thus, in all cases

$$p_l \sim l^{3-D}, \quad l \to 0. \tag{3.13}$$

The probability would scale the same way if little cubes of size l were used instead of balls. If the embedding space is d-dimensional instead of 3-dimensional, (3.13) becomes $p_l \sim l^{d-D}$, $l \to 0$. The quantity $d - D$ is called the *codimension*. We derive its scaling laws by adapting the standard K41 [53] phenomenology. Observe that it is justified to define the eddy turnover time t_l as l/v_l, since the formation of smaller eddies is supposed to be important only within active eddies. Active eddies of size $\sim l$ fill only a fraction p_l of the total volume; thus, the energy per unit mass associated with motion on scale $\sim l$ is $E_l \sim v_l^2 p_l = v_l^2 (l/l_0)^{3-D}$. The energy flux Π_l' from scales $\sim l$ to smaller scales is $\sim E_l/t_l$. Thus,

$$\Pi_l' \sim \frac{v_l^3}{l} (l/l_0)^{3-D}. \tag{3.14}$$

Assume as usual that for right Reynolds number there is an inertial range in which the energy flux is independent of l: $\Pi_l' \sim \varepsilon \sim v_0^3/l_0$. This relation is a consequence of (3.14) evaluated for $l \sim l_0$. From (3.14) and the relation $\Pi_l' \sim \varepsilon \sim v_0^3/l_0$ we obtain the following:

$$v_l \sim v_0 (l/l_0)^{1/3 - (3-D)/3}, \quad t_l \sim l/v_l \sim \frac{l_0}{v_l} (l/l_0)^{2/3 + (3-D)/3}. \tag{3.15}$$

Equation $(3.15)_1$ may be viewed as the statement that the velocity field has the scaling exponent

$$h = \frac{1}{3} - \frac{1}{3}(3 - D) \tag{3.16}$$

on the set J of fractal dimension D on which the cascade accumulates. This reformulation will be useful for the generalization to multifractals.

Let us come back to the structure functions. Since it is difficult to distinguish between *longitudinal* structure functions and those involving other components, we'll simply denote the structure function of order p by $\langle \delta v_l^p \rangle$. There are two contributions to this quantity:

a factor v_l^p coming from active eddies and an "intermittency factor" $p_l = (l/l_0)^{3-D}$ which gives the fraction of the volume filled by active eddies of scale l. Using $(3.15)_1$, we obtain

$$S_p(l) = \langle \delta v_l^p \rangle \sim v_0^p (l/l_0)^{\zeta_p}, \quad \zeta_p = p/3 + (3-D)(1-p/3).$$

For β-model the exponent ζ_p is a linear-plus-constant function of the order p. For $p = 6$ the discrepancy from the K41 value is equal to the codimension $3 - D$. For $p = 2$ we find that the second order structure function has the exponent $\frac{2}{3} + \frac{1}{3}(3 - D)$; hence, the energy spectrum in the inertial range obeys

$$E(k) \sim k^{-[5/3+(3-D)/3]},$$

which is *steeper* than the Kolmogorov-Obukhov $k^{-5/3}$ spectrum. For $p = 3$, we obtain $\zeta_3 = 1$, as required by Kolmogorov's 4/5-law [36]. The viscous cutoff for the β-model may be obtained by equating the eddy turnover time t_l in $(3.15)_2$ and the viscous diffusion time. This gives the dissipation scale

$$\eta \sim l_0 \mathrm{Re}^{-3/(1+D)} \sim l_0 \mathrm{Re}^{-1/(1+h)}, \tag{3.17}$$

where $\mathrm{Re} = l_0 v_0/\nu$. Thus the dissipation scale η in (3.17) can become smaller than the mean-free-path. When $D < 0$, the spectrum becomes steeper than $k^{-8/3}$. Concerning scaling, there is a single adjustable parameter in the β-model, the dimension D associated with the intermittent cascade.

(a) Bifractal model. It was shown that the β-model is equivalent to the statement that the velocity field has a scaling exponent h on a set J of fractal dimension D, such that h and D are related (3.16). A natural extension is to assume *bifractality*: there are two sets J_1 and J_2, both imbedded in the physical space $\mathbb{R}^3$. Near J_1 the velocity has scaling exponent h_1 and near J_2 it has scaling exponent h_2. Specifically, assume that

$$\frac{\delta v_l(r)}{v_0} \sim \begin{cases} (l/l_0)^{h_1}, & r \in J_1, \quad \dim J_1 = D_1, \\ (l/l_0)^{h_2}, & r \in J_2, \quad \dim J_2 = D_2, \end{cases}$$

where $\delta v_l(r)$ is the velocity increment between the point r and another point a distance l away. The above scaling laws are meant to hold at inertial-range scales. Thus, after the spatial increment, l and the velocity increment $\delta v_l(r)$ divided by l_0 and v_0, are all remaining constants of order one. With this assumption it can be calculated the structure function of an order p (at inertial-range separations). The scaling exponent h_1 gives a contribution $(l/l_0)^{ph_1}$ which must be multiplied by the probability $(l/l_0)^{3-D_1}$ of being within a distance l of the set J_1, and similarly for other exponent. We obtain

$$\frac{\langle \delta v_l^p \rangle}{v_0^p} = \mu_1 (\frac{l}{l_0})^{ph_1} (\frac{l}{l_0})^{3-D_1} + \mu_2 (\frac{l}{l_0})^{ph_2} (\frac{l}{l_0})^{3-D_2}, \tag{3.18}$$

where μ_1 and μ_2 are order unity constants.

Thus, all the structure functions comprise the superposition of two power-laws. In the inertial range, when $l \ll l_0$, the power-law with the smallest exponent will dominate. We obtain

$$\langle \delta v_l^p \rangle \sim l^{\zeta_p}, \quad \zeta_p = \min(ph_1 + 3 - D_1, ph_2 + 3 - D_2).$$

As an illustration of bifractality, take a mixture of K41 turbulence and β-model turbulence: $D_1 = 3$, $h_1 = \frac{1}{3}$, $0 < D_2 < 3D$ and $h_2 = \frac{1}{3} - \frac{1}{3}(3 - D_2)$. Thus, we obtain

$$\zeta_p = \begin{cases} p/3, & 0 \le p \le 3, \\ p/3 + (3 - D_2)(1 - p/3), & p \ge 3. \end{cases}$$

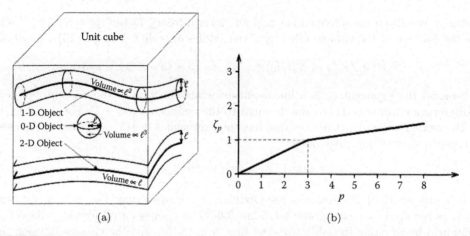

FIGURE 3.1: (a) The probability that a sphere of radius l encounters an object of dimension D behaves as l^{3-D}, as $l \to 0$. (b) Exponent ζ_p for the "bifractal model"; it can be seen slope at $p = 3$.

Observe that the parameters have been chosen to be consistent with the 4/5-law in such a way that $\zeta_3 = 1$. Note that ζ_p, see the graph on Fig. 3.1(b), can be defined for non-integer $p > 0$; in that case it is necessary to take the absolute value of δv_l in the definition of the structure function. It has a kink at $p = 3$. This is known as a "phase transition": for structure functions S_p this model behaves exactly as in the K41 theory for $p \leq 3$ and displays intermittency only for larger values of p.

(b) Multifractal model. Based on our experience with bifractality it is rather clear that a function ζ_p will be obtained if we assume a *continuous infinity* of scaling exponent rather than just two. Actually, there is a more basic reason not to limit ourselves to a finite number of scaling exponents. In the K41 theory, a single value of h was permitted because we imposed *global scale-invariance*. This can be weakened into *local scale-invariance*: all h (within some range) are permitted and, for each h, there is a fractal set with an *h-dependent dimension $D(h)$* near which scaling holds with exponent h.

Specially, we keep hypotheses:

H1. In the limit of infinite Reynolds numbers, all possible symmetries of NSE, usually broken by the mechanism producing the turbulent flow, are restored in a statistical sense at small scales and away from boundaries,

H2. Under the same assumptions as in **H1**, *the turbulent flow has a finite non-vanishing mean rate of dissipation energy ε per unit mass*,

and replace hypothesis

H3. Under the same assumptions as in **H1**, *the turbulent flow is self-similar at small scales, i.e., it possesses a unique scaling exponent h of the K41 theory by $h = 2/3$ for $n = 2$ and $h = 1$ for $h = 3$.*

H4. Under the same assumption as in **H1**, *the turbulent flow is assumed to possess a range of scaling exponents $I = (h_{\min}, h_{\max})$. For each h in this range, there is a set $J_h \in \mathbb{R}^3$ of fractal dimension $D(h)$, such that, as $l \to 0$,*

$$\frac{\delta v_l(\mathbf{r})}{v_0} \sim (l/l_0)^h, \quad \mathbf{r} \in J_h. \tag{3.19}$$

The exponents $h_{\min}$ and $h_{\max}$, and the function $D(h)$ are postulated to be universal, i.e., independent of the mechanism of production of the flow.

From this *multifractal* assumption, proceeding as in the preceding subsection on bifractals, we derive the expression for the structure function of order p:

$$\frac{S_p(l)}{v_0^p} \equiv \frac{\langle \delta v_l^p \rangle}{v_0^p} \sim \int_I (l/l_0)^{ph+3-D(h)} \, d\mu(h).$$

Here, the measure $d\mu(h)$ gives the weight of the different exponents. The factor $(l/l_0)^{ph}$ is the contribution from (3.19) and the factor $(l/l_0)^{3-D(h)}$ is the probability of being within a distance $\sim l$ of the set J_h of dimension $D(h)$. The sum in (3.18) has an integral over the range I of scaling exponents. In the limit $l \to 0$ the power-law with smallest exponent dominates and we obtain by a steepest descent argument $\lim_{l \to 0} \frac{\ln S_l(l)}{\ln l} = \zeta_p$, where

$$\zeta_p = \inf_h [ph + 3 - D(h)]. \tag{3.20}$$

The weights $d\mu(h)$ have indeed disappeared from the asymptotic expressions of the structure functions. More loosely, ignoring logarithmic corrections, etc., (3.20) may be written as $S_p(l)/v_0^p \sim (l/l_0)^{\zeta_p}$ for $l \to 0$. By Kolmogorov's four-fifth law (K45),
the exponent for the 3rd order structure function is $\zeta_3 = \inf_h [3h + 3 - D(h)] = 1$. In the relation (3.20) between the dimensions $D(h)$ and exponents of structure functions ζ_p we recognize *Legendre transformation*. There is an associated geometrical construction; see Fig. 3.2a): the quantity $3 - \zeta_p$ is the maximum, signed vertical distance between the graph of $D(h)$ and the line through the origin with slope p. If $D(h)$ has a non-increasing derivative, i.e., is concave, then for a given value of p the maximum is attained at the unique value $h_*(p)$ such that

$$D'(h_*(p)) = p, \tag{3.21}$$

and ζ_p is given by

$$\zeta_p = ph_*(p) + 3 - D(h_*(p)). \tag{3.22}$$

From there it follows that (3.20) can be written as

$$D(h) = \inf_h [ph + 3 - \zeta_p]. \tag{3.23}$$

Indeed, from (3.21) and (3.22) we obtain

$$\frac{d\zeta_p}{dp} = h_*(p) + [p - D'(h_*(p))]\frac{dh_*(p)}{dp} = h_*(p). \tag{3.24}$$

Remark that (3.22) and (3.24) imply (3.23). Even if $D(h)$ is not concave, its Legendre transform ζ_p defined by (3.20) will be concave; but the inversion formula returns then not $D(h)$ but its *concave hull*, i.e., the lowest concave graph lying above the graph of $D(h)$.

The inversion formula (3.23) can be used to extract the function $D(h)$ from experimental or numerical data about the exponents ζ_p. The corresponding geometrical construction is shown in Fig. 3.2(b). An important consequence of (3.24) is that the scaling exponent h is the slope of the graph ζ_p for the value of p which minimizes $ph + 3 - \zeta_p$. Hence, the range I of possible values of p is the range of slopes of the graph of ζ_p. The finiteness of the Mach number requires ζ_p to be a non-decreasing function of p. Hence, *negative scaling exponents h are ruled out in the multifractal model* for incompressible flow.

3.2.2 Random variables and correlation functions

Generally we may have n random variables $\xi_1, \ldots \xi_n$. As an example we may consider n dice, where ξ_i is the random variable corresponding to the i-th die. Making N experiments

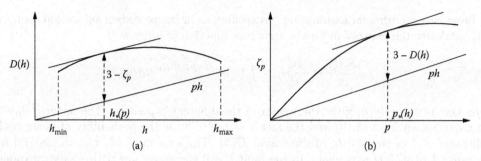

FIGURE 3.2: Geometrical construction of (a) the Legendre transform, and (b) the inverse Legendre transform.

for each random variable ξ_i, we get N numbers $\xi_{i1}, \ldots \xi_{iN}$. As in the case of one random variable we may take averages of an arbitrary function $f(\xi_1, \ldots \xi_n)$ according to

$$\langle f(\xi_1, \ldots \xi_n)\rangle = \lim_{N \to \infty} \frac{1}{N} \sum_{i=1}^{N} f(\xi_{1i}, \ldots \xi_{ni}). \tag{3.25}$$

We introduce the n-dimensional distribution function

$$W_{\xi_1, \ldots \xi_n}(x_1, \ldots x_n) = W_n(x_1, \ldots x_n) = \langle \delta(x_1 - \xi_1) \ldots \delta(x_n - \xi_n)\rangle. \tag{3.26}$$

The index $\xi_1, \ldots \xi_n$ of the distribution function will be omitted to be replaced by number n of variables. With the help of (3.26) the averages (3.25) over a statistical ensemble can be calculated by integration

$$\langle f(\xi_1, \ldots \xi_n)\rangle = \int f(x_1, \ldots x_n) W_n(x_1, \ldots x_n)\, dx.$$

By (3.25), we obtain probability density of the first $i < n$ random variables by integration over the other variables

$$W_i(x_1, \ldots x_i) = \int W_n(x_1, \ldots x_i, x_{i+1} \ldots x_n)\, dx_{i+1} \ldots dx_n. \tag{3.27}$$

In general case the distribution function may be arbitrary probability density function satisfying the normalization condition

$$\int W_n(x_1, \ldots x_n)\, dx = \|W(x)\|_{L^1} = 1.$$

Conditional probability density. If the last $n - 1$ random variables take fixed values $\xi_2 = x_2, \ldots \xi_n = x_n$, then we obtain certain probability density for the first random variable. This probability density is called *conditional probability density* and will be written as $F(x_1|x_2, \ldots x_n)$.

Obviously $W_n(x_1, \ldots x_n)\, dx$ that the random variables ξ_i are in the interval $x_i \leq \xi_i \leq x_i + dx_i$ is the probability $F(x_1|x_2, \ldots, x_n)dx_1$ that the first variable is in the interval $x_1 \leq \xi_1 \leq x_1 + dx_1$ and that the other variables have sharp values $\xi_i = x_i$ $(2 \leq i \leq n)$ times the probability $W_{n-1}(x_2, \ldots x_n)\, dx_2 \ldots dx_n$ that the last $n - 1$ variables are in the interval $x_i \leq \xi_i \leq x_i + dx_i$ $(2 \leq i \leq n)$, i.e., we have

$$W_n(x_1, \ldots x_n) = F(x_1|x_2, \ldots x_n) W_{n-1}(x_2, \ldots x_n).$$

Because W_{n-1} follows from W_n, see (3.27), we may express the conditional probability density by W_n:

$$F(x_1|x_2,\ldots x_n) = W_n(x_1,\ldots x_n)/W_{n-1}(x_2,\ldots x_n) = W_n(x_1,\ldots x_n)/\int W_n(x_1,\ldots x_n)\, dx_1.$$

Especially for two random variables the conditional probability density in terms of W_2, which is called the *joint probability density*, reads

$$F(x_1|x_2) = W_2(x_1, x_2)\, /\int W_2(x_1, x_2)\, dx_1.$$

Time-dependent random variables. We now consider a random variable ξ which depends on the time t, i.e., $\xi = \xi(t)$. Assume that we have an ensemble of systems and that each system leads to a number ξ, which depends on time. Though the outcome for one system cannot precisely predicted, we assume that ensemble averages exist and that these averages can be calculated. For the fixed time $t = t_1$ we may define a probability density by $W_1(x_1, t) = \langle \delta(x_1 - \xi(t_1)) \rangle$. The bracket $\langle\,\rangle$ indicates the ensemble average. The probability to find the random variable $\xi(t_1)$ in the interval $x_1 \leq \xi(t_1) \leq x_1 + dx_1$, $\xi(t_2)$ is in the interval $x_2 \leq \xi(t_2) \leq x_2 + dx_2$ and so on. This probability may be written as $W_n(x_1, t_1; \ldots x_n, t_n)\, dx$, where the probability density W_n is given by

$$W_n(x_1, t_1; \ldots x_n, t_n) = \langle \delta(x_1 - \xi(t_1)) \ldots \delta(x_n - \xi(t_n)) \rangle.$$

Correlation function. In statistical mechanics, the *correlation function* is a measure of the order in a system, as characterized by a mathematical correlation function. Correlation functions describe how microscopic variables, such as spin and density, at different positions are related. More specifically, correlation functions quantify how microscopic variables co-vary with one another on average across space and time. The most common definition of a correlation function is the canonical ensemble average of the scalar product of two random variables, $\mathbf{u}_1$ and $\mathbf{u}_2$ at positions $\mathbf{r}_0$ and $\mathbf{r}_0 + \mathbf{r}$, and times t and $t + \tau$,

$$C(\mathbf{r}, \tau) = \langle \mathbf{u}_1(\mathbf{r}_0, t) \cdot \mathbf{u}_2(\mathbf{r}_0 + \mathbf{r}, t + \tau) \rangle - \langle \mathbf{u}_1(\mathbf{r}_0, t) \rangle \langle \mathbf{u}_2(\mathbf{r}_0 + \mathbf{r}, t + \tau) \rangle.$$

Here the brackets $\langle\,\rangle$, indicate the above-mentioned average. It is a matter of convention whether one subtracts the uncorrelated average product of $\mathbf{u}_1$ and $\mathbf{u}_2$, $\langle \mathbf{u}_1(\mathbf{r}_0, t) \cdot \mathbf{u}_2(\mathbf{r}_0 + \mathbf{r}, t + \tau) \rangle$ from the correlated product $\langle \mathbf{u}_1(\mathbf{r}_0, t) \cdot \mathbf{u}_2(\mathbf{r}_0 + \mathbf{r}, t + \tau) \rangle$ with the convention differing among fields. The most common uses of correlation functions are when $\mathbf{u}_1$ and $\mathbf{u}_2$ describe the same variable, such as a spin-spin correlation function, or a particle position-position correlation function in an elementary liquid cell (often called a radial distribution function or a pair correlation function). Correlation functions between the same random variable are autocorrelation functions. In statistical mechanics, not all correlation functions are autocorrelation functions. For example, in multi-component condensed phases, the pair correlation function between different elements is often of interest. Such mixed-element pair correlation functions are an example of cross-correlation functions, as the random variables $\mathbf{u}_1$ and $\mathbf{u}_2$ represent the average variations in density as a function position for two distinct elements.

Equilibrium equal-time (spatial) correlation functions. Often, one is interested in solely the spatial influence of a given random variable, say the direction of a spin, on its local environment, without considering later times, τ. In this case, we neglect the time evolution of the system, so the above definition is rewritten with $\tau = 0$. This defines the equal-time correlation function $C(\mathbf{r}, 0)$. It is written as

$$C(\mathbf{r}, 0) = \langle \mathbf{u}_1(\mathbf{r}_0, t) \cdot \mathbf{u}_2(\mathbf{r}_0 + \mathbf{r}, t) \rangle - \langle \mathbf{u}_1(\mathbf{r}_0, t) \rangle \langle \mathbf{u}_2(\mathbf{r}_0 + \mathbf{r}, t) \rangle.$$

Often, one omits the reference time τ and reference radius $\mathbf{r}$ by assuming equilibrium (and thus time invariance of the ensemble) and averaging over all sample positions, yielding:

$$C(\mathbf{r}) = \langle \mathbf{u}_1(0) \cdot \mathbf{u}_2(\mathbf{r}) \rangle - \langle \mathbf{u}_1(0) \rangle \langle \mathbf{u}_2(\mathbf{r}) \rangle,$$

where, again, the choice of whether to subtract the uncorrelated variables differs among fields. The radial distribution function is an example of an equal-time correlation function where the uncorrelated reference is generally not subtracted.

Equilibrium equal-position (temporal) correlation functions. One might also be interested in the *temporal* evolution of microscopic variables. In other words, how the value of a microscopic variable at a given position and time $\mathbf{r}_0$ and τ influences the value of the same microscopic variable at a later time $t + \tau$ (and usually at the same position). Such temporal correlations are quantified via equal-position correlation functions $C(0, \tau)$. They are defined analogously to above equal-time correlation functions, but we now neglect spatial dependencies by setting $\mathbf{r} = 0$, yielding:

$$C(0, \tau) = \langle \mathbf{u}_1(\mathbf{r}_0, t) \cdot \mathbf{u}_2(\mathbf{r}_0, t + \tau) \rangle - \langle \mathbf{u}_1(\mathbf{r}_0, t) \rangle \langle \mathbf{u}_2(\mathbf{r}_0, t + \tau) \rangle.$$

Assuming equilibrium (and thus time invariance of the ensemble) and averaging over all sites in the sample gives a simpler expression for the equal-position correlation function as for the equal-time correlation function:

$$C(\tau) = \langle \mathbf{u}_1(0) \cdot \mathbf{u}_2(\tau) \rangle - \langle \mathbf{u}_1(0) \rangle \langle \mathbf{u}_2(\tau) \rangle.$$

The above assumption may seem non-intuitive at first: how can an ensemble which time-invariant has a non-uniform temporal correlation function? Temporal correlations remain relevant to talk about in equilibrium systems because a time-invariant, macroscopic ensemble can still have non-trivial temporal dynamics *microscopically*. One example is in diffusion. A single-phase system at equilibrium has a homogeneous composition macroscopically. If one watches the microscopic movement of each atom, fluctuations in composition are constantly occurring due to the quasi-random walks taken by the individual atoms. Statistical mechanics allows one to make insightful statements about the temporal behavior of such fluctuations of equilibrium systems.

Radial distribution functions. Higher-order correlation functions involve multiple reference points, and are defined through a generalization of the above correlation function by simply taking the expected value of the product of more than two random variables:

$$C_{i_1,\dots i_n}(\mathbf{r}_1, \dots \mathbf{r}_n) = \langle X_{i_1}(\mathbf{r}_1) \dots X_{i_n}(\mathbf{r}_n) \rangle.$$

Such higher order correlation functions are relatively difficult to interpret and measure.

3.2.3   Richardson–Kolmogorov concept of turbulence

Straightforward attempts to assess the solutions of NSE may still be very non-realistic. For example, we could estimate the velocity of water flow in any one of the mighty rivers, which drop hundreds of meters in a course of about a thousand kilometers. The typical river depth L is about 10 meters. Equating the gravity force ag ($g \simeq 10^3 \mathrm{cm/sec}^2$) and the viscous drag $\nu\, d^2 u / dz^2 \sim \nu u / L^2$ we find that u has the order $10^7 \mathrm{cm/sec}$ instead of the observed value of $10^2 \mathrm{cm/sec}$. (The relation $a \sim b$ means that a and b have the same order). Example: $a = 967$ means $a \sim 1000$).

This is, of course, absurd, perhaps to the regret of the white water rafting industry.

The resolution of this discrepancy was suggested by Reynolds,[1] who stressed the importance of a dimensionless ratio of the nonlinear term to the viscous term in (3.2). With a velocity drop of the order of U on a scale L the nonlinear term e is estimated as U^2/L. The viscous term is about $\nu U/L^2$. The ratio of the two, known as the Reynolds number Re, is UL/ν. The magnitude of Re measures how large is the nonlinearity compared to the effect of the viscous dissipation in a particular fluid flow. For the case of Re $\ll 1$, one can neglect the nonlinearity and the solutions of NSE can be found in close form in many instances [62]. In natural circumstances Re is very large. For example, in the rivers discussed above, the value Re $\sim 10^7$. Reynolds understood that for Re $\gg 1$ there is no stable stationary solution for the equations of motion. The solutions are strongly affected by the nonlinearity, and the actual flow pattern is complicated, convoluted and vortical. Such flows are called *turbulent*.

Modern concepts about high Re number turbulence started to evolve with Richardson's insightful contributions [90] which contained the famous "poem" that can be paraphrased as: "*Big whirls have little whirls that feed on their velocity, and little whirls have lesser whirls and so on to viscosity in the molecular sense.*" In this way distances Richardson conveyed an image of the creation of turbulence by large scale forcing, setting up a cascade of energy transfers to smaller and smaller scales by the nonlinearities of fluid motion, until the energy dissipates at small scales by viscosity, turning into heat. This picture led in time to innumerable *cascade models* that tried to capture the statistical physics of turbulence by assuming something about the *cascade process*. Indeed, no one in their right mind is interested in the full solution to of the turbulent velocity field at all points in space-time. The interest is in the statistical properties of the turbulent flow. Moreover, the statistics of the velocity field itself is ding too heavily dependent on the particular boundary conditions of the flow. It was understood [90] that universal in properties may be found in the statistics of *velocity differences* $\delta\mathbf{u}(\mathbf{r}_1, \mathbf{r}_2) \equiv \mathbf{u}(\mathbf{r}_2) - \mathbf{u}(\mathbf{r}_1)$ across a separation $\mathbf{r} = \mathbf{r}_2 - \mathbf{r}_1$. Taking such a difference, we subtract the non-universal large scale motions (known as the *wind*) in atmospheric flows). In experiments it is common to consider one dimensional cuts of the velocity field,

$$\delta\mathbf{u}(R) = \delta\mathbf{u}(\mathbf{r}_1, \mathbf{r}_2) \cdot \mathbf{r}/R,$$

where R is the only available length for the development of dimensional analysis. The interest routes in the probability distribution function of $\delta\mathbf{u}(R)$ and its moments. These moments are known as the *structure functions* defined by $S_n(R) = \langle \delta\mathbf{u}(R)^n \rangle$, where angle-form scratches stands for a suitably defined ensemble average. For *Gaussian statistics* the whole distribution function is governed by the second moment $S_2(R)$, and there is no information to be gained from higher order moments. In contrast, hydrodynamic experiments indicate that turbulent statistics is understood extremely non-Gaussian, and the higher order moments contain important new information about the distribution functions.

Possibly the most ingenious attempt to understand the *statistics of turbulence* is due to Kolmogorov [53] who proposed the idea of universality (turning the study o of small scale turbulence from mechanics to fundamental physics) based on the notion of the *inertial range*. The idea is that for very large values of Re there is a wide separation between the *scale of energy input* L and the typical *viscous dissipation scale* η at which viscous friction become important and dumps the energy into heat. In the stationary situation, when the statistical characteristics of the turbulent flow are time independent, the he rate of energy input at large scales (L) is balanced by ion the rate of energy dissipation at small scales (η), and must be also the same as the flux of energy from larger to smaller scales (denoted $\bar{\varepsilon}$) as it is measured at any scale R in the so called *inertial* interval $\eta \ll R \ll L$. Kolmogorov proposed that the only relevant parameter in the inertial interval is $\bar{\varepsilon}$, and that L and η are irrelevant

---

[1]Reynolds O. On the dynamical theory of incompressible viscous fluids and the determination of the criterion. *Phil. Trans. Roy. Soc.*, A186, 1895, 123–164.

the for the statistical characteristics of motions on the scale of R. This assumption means, that R is the only available length for the development of dimensional analysis. In addition we have the dimensional parameters $\bar{\varepsilon}$ and the mass density of the fluid ρ. From these three parameters we form combinations $\rho^x \bar{\varepsilon}^y R^z$ such that with a proper choice of the exponents x, y, z we form any dimensionality that we want. This leads to detailed predictions about the statistical physics of turbulence. For example, to predict $S_n(R)$ we note that the only combination of $\bar{\varepsilon}$ and R that gives the right dimension for S_n is $(\bar{\varepsilon}R)^{n/3}$. In particular, for $n = 2$ this is the famous Kolmogorov "2/3"-*law*, which in Fourier representation is also known as the "5/3"-*law*. The idea that one extracts universal he properties by focusing on statistical quantities can be applied also to the correlations of gradients of the velocity his field. An important example is the rate $\varepsilon(\mathbf{r}, t)$ at which energy is dissipated into heat due to viscous damping. This rate is roughly $\nu |\nabla \mathbf{u}|^2$. One is interested in the fluctuations of the energy dissipation $\varepsilon(\mathbf{r}, t)$ about their mean $\bar{\varepsilon}$, $\hat{\varepsilon}(\mathbf{r}, t) = \varepsilon(\mathbf{r}, t) - \bar{\varepsilon}$, and how these fluctuations are correlated in space. The answer is given by the often-studied correlation function

$$K_{\varepsilon\varepsilon}(R) = \langle \hat{\varepsilon}(\mathbf{r} + \mathbf{r}, t)\hat{\varepsilon}(\mathbf{r}, t)\rangle.$$

If the fluctuations at different points were uncorrelated, this function would vanish for all $R \neq 0$. Within the Kolmogorov theory one estimates $K_{\varepsilon\varepsilon}(R) \simeq \nu^2 \bar{\varepsilon}^{-4/3} R^{-8/3}$, which means that the correlation decays as a power, like about $R^{-8/3}$.

The major aspect of Kolmogorov predictions, i.e., that the statistical quantities depend on the length scale R as power laws, is corroborated by ear experiments. On the other hand, the predicted exponents seem not to be exactly realized. For example, the experimental correlation $K_{\varepsilon\varepsilon}(R)$ decays according to a power the law, $K_{\varepsilon\varepsilon}(R) \sim R^{-\mu}$ for $\eta \ll R \ll L$, where $\eta = (\nu^3/\langle\varepsilon\rangle)^{1/4}$ is the dissipative scale with $\mu = \nu\rho$ having a numerical value of $0.2 - 0.3$ instead of $8/3$, see [99]. The structure functions also behave as power laws, $S_n(R) \simeq R^{\zeta_n}$, but the numerical values of ζ_n deviate progressively from $n/3$ when n increases, see [4, 9]. Something fundamental seems n to be missing. One might think that the numerical value of this exponent or another is not a fundamental issue. It is important to understand that the Kolmogorov theory exhausts the dimensions of the statistical quantities under the assumption that $\bar{\varepsilon}$ is the only relevant parameter. A deviation in the numerical value of an exponent from the prediction of dimensional analysis requires the appearance of another dimensional parameter. Of course, there exist two dimensional parameters, i.e., L and η, which may turn out to be relevant. Indeed, experiments indicated that for the statistical quantities mentioned above the energy input scale on L is indeed relevant and it appears as a normalization scale for the deviations from Kolmogorov's predictions:

$$S_n(R) \simeq (\bar{\varepsilon}R)^{n/3}(L/R)^{\delta_n},$$

where $\zeta_n = n/3 - \delta_n$ (called a *scaling exponent set*). Such form of scaling which deviates from the predictions of dimensional analysis is referred to as *anomalous scaling*. The realization that the experimental results for the structure functions were consistent with L rather than η as the normalization scale developed over a long time and involved a large number of experiments; recently the accuracy of determination of the exponents increased appreciably as a result of a clever method of data analysis time, see [9]. Similarly a careful he demonstration of the appearance of L in the dissipation correlation was achieved in [99]. A direct analysis of scaling exponents ζ_n and μ for high Reynolds number flow was presented in [87] leading to the same conclusions.

3.2.4 Scaling of the structure functions

Richardson have understood that the universal properties of turbulence may be found in the statistics of *velocity differences* W_R across a separation R in the inertial interval in which the nonuniversal large scale motions, such as the "wind" in atmospheric flows, have been eliminated. The value W_R is dominated by the eddies of scale R. All information about statistics of W_R contains in the moments (mean values of powers) of W_R known as *velocity structure function* $W_R = \mathbf{u}(R)$. The physical meaning of some structure functions are $S_n(R) = \langle |W_R|^n \rangle \sim R^{\zeta_n}$ for inertial interval $\eta \ll R \ll L$. For instance, $S_2(R)$ is the energy of R, and $S_3(R)$ is the flux energy. (*Scaling* $f(x) \sim x^\alpha$ means $\lim\limits_{x \to 0,\ x \to \infty} \frac{f(x)}{x^\alpha} = C -$ const).

In *isotropic homogeneous turbulence*, these structure functions are observed to behave as *scale invariant functions* of separation R with (static) *scaling exponents* ζ_n which may be universal. The statistical behavior of three-dimensional fully developed turbulence at small scales has been intensively investigated in the last five decades. A common way to board this problem is through the velocity structure functions. The n-th order structure function is defined as: $\langle \delta u(R)^n \rangle = \langle u(x + R) - u(x) \rangle$, where u is the component of the velocity along of R. Kolmogorov's original theory [53] predicts that at very large Reynolds numbers the scaling $\langle \delta u(R)^n \rangle = R^{n/3}$ holds for R from the inertial interval $\eta \ll R \ll L$. In 1962, Oboukhov and Kolmogorov [54, 83] proposed an inertial interval range scaling of the n-order structure function which includes the intermittency in the energy dissipation rate: $\langle \delta u(R)^n \rangle \sim \langle \varepsilon_R^{n/3} \rangle R^{n/3}$, where ε_R is the average of the energy dissipation rate ε over and $\varepsilon = 15\nu (du/dx)^2$ for homogeneous and isotropic turbulent power law behavior of $\langle \varepsilon_R^{n/3} \rangle$ in the inertial range. Some works [14] use a new scaling, which extends beyond the inertial range. This is the basis to reproduce any velocity structure function for all scales in terms of $\langle \varepsilon_R^{n/3} \rangle$ and universal function f:

$$\langle \delta u(R)^n \rangle = C_n \langle \varepsilon_R^{n/3} \rangle [R f(R/\eta)]^{n/3}, \tag{3.28}$$

where C_n and f are dimensionless, related to the third order structure function that is

$$f(R/\eta) = \frac{D}{\langle \varepsilon \rangle} \frac{\langle |\delta u(R)|^3 \rangle}{R}.$$

In order to define $f(R/\eta)$ it may be used $\langle |\delta u(R)|^3 \rangle$ rather than $\langle \delta u(R)^3 \rangle$ because the computation of the latter is much more demanding. The constant D should be introduced to normalize this function, that, is $f(R/\eta)$ must go to 1 in the inertial interval. On the other hand, the continuity in velocity implies $\langle |\delta u(R)|^3 \rangle \simeq R^3$ when R lies in the vicinity of the Kolmogorov scale η, then $f(R/\eta)$ becomes proportional to R^2. At first sight the right hand side of the relation (3.28) reproduces the characteristics of a structure function in two regions. The first one is the dissipative range where the power law behavior is completely dominated by the term $(R f(R/\eta))^{n/3}$, the quantity $\langle \varepsilon_R^{n/3} \rangle$ going to a constant value when small scales are considered. The second one is the inertial range where the power law behavior is composed of a dependence on $R^{n/3}$ and a correction related to the energy dissipation rate. At intermediate scales the relation also holds as shown by the experimental data. It must be noticed that relation $\langle \delta u(R)^n \rangle \sim \langle \varepsilon_R^{n/3} \rangle R^{n/3}$ is a particular case of relation (3.28). The anomalous exponent ζ_n associated with the n-th order structure function is calculated as a slope in a $\log - \log$ plot of $\langle \delta u(R)^n \rangle$ versus $\langle |\delta u(R)|^3 \rangle$. Now a self scaling broader than that observed (Fig. 3.3). In all cases the slopes adopt values between 0.995 and 1.005. This implies a linear relation between $\langle \varepsilon_R^{n/3} \rangle [R f(R/\eta)]^{n/3}$ and $\langle u(R)^n \rangle$.

3.2.5 Dissipative and dynamical scaling

Suppose that we want to calculate the average response of a turbulent fluid at some point $\mathbf{r}_0$ to forcing at a point $\mathbf{r}_1$. The field theoretic approach allows us to consider this response

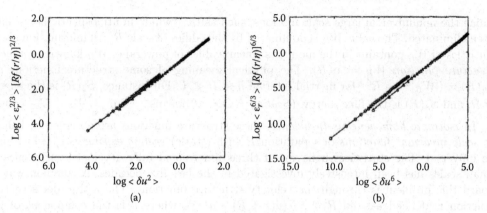

FIGURE 3.3: (a) – $\langle \varepsilon_R^{2/3}\rangle[Rf(R/\eta)]^{2/3}$ versus $\langle \delta u^2\rangle$; (b) – $\langle \varepsilon_R^{6/3}\rangle[Rf(R/\eta)]^{6/3}$ versus $\langle \delta u^6\rangle$ at two Reynolds numbers: Re = 11000 ($\circ$) and Re = 300000 ($\triangle$).

as an infinite sum of all the following processes: firstly there is the direct response at point r_0 due to the forcing at r_1. This response is instantaneous if we assume that the fluid is incompressible (and therefore the speed of sound is infinite). Then there is the process of forcing at r_1 with a response at an intermediate point r_2, which then acts as a forcing for the response at r_0. This intermediate process can take time, and we need to integrate over all the possible positions of point r_2 and all times. This is the second order term in perturbation theory. Then we can force at r_1, the response at r_1 acting as combinations a forcing for r_3 and the response at r_3 forces a response at r_0. We need to integrate over all possible intermediate positions r_2 and r_3 and all the intermediate times. This is the third order term in perturbation theory, and so on. The actual response is the infinite sum of all these contributions. In applying this field theoretical method one encounters three main difficulties:

(A) The theory has no small parameter. The usual procedure is to develop the perturbation theory around the tracts linear part of the equation of motion. In other words, the ties zero order solution of equation (3.2) is obtained by discarding quadratic in the velocity field terms. The expansion parameter is obtained from the ratio of the viscous quadratic to the linear terms; this ratio is of the order of Reynolds number Re which was defined above. Since are interested in Re $\gg$ 1, naive perturbation expansions are badly divergent. In other words, the contribution of the various processes described above increases as $(\mathrm{Re})^n$ on with the number n of intermediate points in space-time.

(B) The theory exhibits two types of nonlinear interactions. Both are hidden in the nonlinear term $\mathbf{u} \cdot \nabla \mathbf{u}$ in (3.2). The larger of the two is known to any person who watched how a small floating object is entrained in eddies of a river and swept along a complicated path with the turbulent flow. In a similar way any fluctuation of small scale is swept along by all the larger eddies. This sweeping couples any given scale of motion to all the larger scales. Unfortunately the largest energy scales contain most of the energy of the flow; these large scale motions are what is experienced as gusts of wind in the atmosphere or the swell in the ocean. In the perturbation theory for $S_n(R)$ one has the consequences of the sweeping effect from all the scale larger than R, with the so main contribution coming from the largest, most intensive gusts on the scale of L. These contributions diverge when $L \to \infty$.

This is sometimes called as "infrared divergences." Such divergences are common in other field theories, with the best known example being quantum electrodynamics. In this theory the divergences are of similar strength in higher order terms in the series, and they can be removed by introducing finite constants to the theory, like the charge and the mass of the electron. Such a theory is sometimes called "not renormalizable". Sweeping is just a kinematic effect that does not lead to energy redistribution between scales. This redistribution of energy results from the second type of interaction, that stems from the shear and torsion effects that are sizable only if they couple fluid motions of comparable scales. The second type of nonlinearity is smaller in size but crucial in consequence, and it may certainly lead to a scale-invariant theory.

(**C**) Nonlocality of interaction in **r** space. One recognizes that the gradient of the pressure is dimensionally the same as $(\mathbf{u} \cdot \nabla \mathbf{u})\mathbf{u}$, and the fluctuations in the pressure are quadratic in the fluctuations of the velocity. This means on that the pressure term is also nonlinear in the velocity. The pressure at any given point is determined he by the velocity field everywhere. Theoretically one sees this effect by taking the divergence of (3.2). This leads to

$$\nabla^2 p = \nabla \cdot [(\mathbf{u} \cdot \nabla \mathbf{u})\mathbf{u}].$$

The inversion of the Laplacian operator involves an integral over all space. Physically this stems from the fact that in the incompressible limit of NSE sound speed is infinite and velocity fluctuations in all distant points are instantaneously coupled.

The first task of a successful theory of turbulence is to overcome the existence of the interwoven nonlinear effects that were explained in difficulty (**B**). This is not achieved by direct applying a formal field theory tools to NSE. It does not matter whether one uses standard field theoretic perturbation theory, path integral formulation, large N-limit or one's formal method of choice. One needs to take care of the particular nature of hydrodynamic turbulence as embodied in difficulty (**B**) first, and then proceed using formal tools.

The basic field under study is the difference of the Eulerian velocity field **u** across a scale $\mathbf{r} \equiv \mathbf{r}' - \mathbf{r}$: $\mathbf{w}(\mathbf{r}, \mathbf{r}', t) = \mathbf{u}(\mathbf{r}', t) - \mathbf{u}(\mathbf{r}, t)$. The fundamental statistical quantities are the simultaneous "fully unused" n-rank tensor correlation function of velocity differences:

$$\mathbf{F}_n(\mathbf{r}_1, \mathbf{r}_1'; \ldots; \mathbf{r}_n, \mathbf{r}_n') = \langle \mathbf{w}(\mathbf{r}_1, \mathbf{r}_1', t) \ldots \mathbf{w}(\mathbf{r}_n, \mathbf{r}_n', t) \rangle, \tag{3.29}$$

where pointed brackets denote the cumulant part of the average over a time-stationary ensemble. In this quantity all coordinates are distinct. The more usual structure function

$$S_n(R) = \langle |\mathbf{w}(\mathbf{r}, \mathbf{r}', t)|^n \rangle, \quad \text{where} \quad \mathbf{r} \equiv \mathbf{r}' - \mathbf{r},$$

is obtained by fusing all the coordinates $\mathbf{r}_i$ into one point **r**, and all the coordinates $\mathbf{r}_i'$ into another point $2\,\mathbf{r}$. Obviously, in using the functions of many variables $\mathbf{F}_n$ instead of the one variable function $S_n(R)$ one is paying a heavy price. On the other hand, this has an enormous advantage: when one develops the theory for $S_n(R)$ on the basis of NSE one encounters the notorious closure problem: knowing $S_n(R)$ requires information about $S_{n+1}(R)$, etc. It is well known that arbitrary closures of this hierarchy of equations are doomed, leading to predictions that are in contradiction with experiments. On the other hand, no one succeeded to solve the hierarchy in its entirety. In contradistinction, the theory for the fully unfused $\mathbf{F}_n$ does not suffer from this problem: it was shown that there exist homogenous equations for $\mathbf{F}_n$ in terms of $\mathbf{F}_n$, without any hierarchic connections to higher or lower order correlation functions. This fact allows one to proceed [66] to derive a variety of exact bridge relations between the scaling exponents of gradient fields and the scaling exponents of the structure functions themselves, and to study the nature of the dissipative scales in turbulence, showing that in fact they are scaling functions with well defined scaling exponents.

In considering the decorrelation times of many "fully unfused" space-*time* correlation functions we need to make a choice of which velocity field we take as our fundamental field. The Eulerian velocity field won't do, simply because its decorrelation time is dominated by the sweeping of small scales by large scale flows. In [67] it was shown that at least from the point of view of the perturbation theory one can get rid of the sweeping effect using the Belinicher-L'vov (BL) *velocity fields*, whose decorrelation time is intrinsic to the scale of consideration. The field $\mathbf{u}(\mathbf{r}_D, t_D | \mathbf{r}, t)$ is defined in [7] using the Eulerian velocity as

$$\mathbf{u}(\mathbf{r}_D, t_D | \mathbf{r}, t) \equiv \mathbf{u}[\mathbf{r} + \boldsymbol{\rho}_L(\mathbf{r}_D, t_D | t), t], \quad \boldsymbol{\rho}_L(\mathbf{r}_D, t_D | t) = \int_{t_D}^{t} \mathbf{u}[\mathbf{r}_D + \boldsymbol{\rho}_L(\mathbf{r}_D, \tau), \tau] \, d\tau.$$

Note that $\mathbf{r}_D$ is precisely the Lagrangian trajectory of a fluid particle that is positioned at $\mathbf{r}_D$ and time $t = t_D$. Under assumption that the variables $\mathbf{u}(\mathbf{r}_D, t_D | \mathbf{r}, t)$ (called BL–*velocities*) to be satisfied Navier–Stokes type equations in the limit of incompressible fluid, and that their simultaneous correlators are identical to the simultaneous correlators of $\mathbf{u}(\mathbf{r}, t)$. Introduce a difference of two (simultaneous) BL-velocities at points $\mathbf{r}$ and $\mathbf{r}'$

$$\mathbf{w}(\mathbf{r}_D, t_D | \mathbf{r}, \mathbf{r}', t) \equiv \mathbf{u}(\mathbf{r}_D, t_D | \mathbf{r}, t) - \mathbf{u}(\mathbf{r}_D, t_D | \mathbf{r}', t).$$

The equation of the motion for $\mathbf{w}$ can be calculated starting from NSE for the Eulerian field,

$$\left[\partial_t + \hat{\mathcal{L}}(\mathbf{r}, \mathbf{r}', t) - \nu (\nabla_r^2 - \nabla_r'^2) \right] \mathbf{w}(\mathbf{r}_D, t_D | \mathbf{r}, \mathbf{r}', t) = 0. \tag{3.30}$$

Let us introduce operator $\hat{\mathcal{L}}(\mathbf{r}_D, t_D | \mathbf{r}, \mathbf{r}', t)$

$$\hat{\mathcal{L}}(\mathbf{r}_D, t_D | \mathbf{r}, \mathbf{r}', t) = \mathbf{P} W(\mathbf{r}_D, t_D | \mathbf{r}, \mathbf{r}_D, t) \cdot \nabla_r + \mathbf{P}' W(\mathbf{r}_D, t_D | \mathbf{r}', \mathbf{r}_D, t) \cdot \nabla_r', \tag{3.31}$$

where $\mathbf{P}$ is the transverse projection operator defined by $\mathbf{P} : \mathbf{u} \to \nabla \times \nabla \times \mathbf{u}$. The application of $\mathbf{P}$ to any given vector field $\mathbf{u}(\mathbf{r})$ is nonlocal, and it has the form

$$[\mathbf{P} \, \mathbf{u}(\mathbf{r})]^\alpha = \int P^{\alpha\beta}(\mathbf{r} - \tilde{\mathbf{r}}) u_\beta(\tilde{\mathbf{r}}) \, d\mathbf{r}. \tag{3.32}$$

The explicit form of the kernel can be found in [8, 67]. In (3.31), $\mathbf{P}$ and $\mathbf{P}'$ are *projection operators* acting on fields that depend on the corresponding coordinates $\mathbf{r}$ and $\mathbf{r}'$. The equation of the motion (3.30) forms the basis of the following discussion of the time correlation functions. To simplify the appearance of the fully unfused, multitime correlation function of BL-velocity differences we choose the economic notation $\mathbf{w}_j \equiv \mathbf{w}(\mathbf{r}_D, t_D | \mathbf{r}_j, \mathbf{r}_j', t)$:

$$\mathcal{F}_n(\mathbf{r}_D, t_D | \mathbf{r}_1, \mathbf{r}_1', t_1; \dots; \mathbf{r}_n, \mathbf{r}_n', t_n) = \langle \mathbf{w}_1 \dots \mathbf{w}_n \rangle. \tag{3.33}$$

We start with the simplest non-simultaneous case with two different times in (3.33). Chose $t_i = t + \tau$ for every $i \le p$ and $t_i = t$ for every $i > p$. Denote the correlation function with this choice of time as $\mathcal{F}_{n,1}^{(p)}$, omitting for brevity the rest of the arguments. Compute the time derivative of $\mathcal{F}_{n,1}^{(p)}$ with respect to τ: $\partial_\tau \mathcal{F}_{n,1}^{(p)}(\tau) = \sum_{j=1}^{p} \langle \mathbf{w}_1 \dots \partial_t \mathbf{w}_j \dots \mathbf{w}_n \rangle$. Using the motions equation (3.30) we find

$$\partial_\tau \mathcal{F}_{n,1}^{(p)}(\tau) + \mathcal{D}_{n,1}^{(p)} = \mathcal{J}_{n,1}^{(p)}, \tag{3.34}$$

$$\mathcal{D}_{n,1}^{(p)} = \sum_{j=1}^{p} \langle \mathbf{w}_1 \dots \hat{\mathcal{L}}_j \mathbf{w}_j \dots \mathbf{w}_n \rangle, \quad \mathcal{J}_{n,1}^{(p)} = \nu \sum_{j=1}^{p} (\nabla_j^2 + \nabla_j'^2) \langle \mathbf{w}_1 \dots \mathbf{w}_j \dots \mathbf{w}_n \rangle,$$

with $\hat{\mathcal{L}}_j \equiv \hat{\mathcal{L}}(\mathbf{r}_D, t_D | \mathbf{r}_j, \mathbf{r}_j', t)$. Recall that $\hat{\mathcal{L}}_j \mathbf{w}_j$ is a nonlocal object that is quadratic in BL-velocity differences, cf. (3.31). Then we reiterate that all the functions depend on $2n$ space coordinates that we do not display for notational economy. To understand the role of the various terms in (3.34) we will make use of the analysis of a similar equation $\mathbf{F}_n = \mathcal{F}_{n,1}^{(p)}(0)$ that was presented in [68]. In that case we considered the simultaneous object

$\mathbf{F}_n = \mathcal{F}_{n,1}^{(p)}(0)$ and computed, as above, its t-time derivative. Obviously, in the stationary ensemble this derivative vanishes. Instead of $\mathcal{D}_{n,1}^{(p)}(\tau)$ and $\mathcal{J}_{n,1}^{(p)}(\tau)$ we got $\mathcal{D}_{n,1}^{(p)}(\tau)$ and $\mathcal{J}_{n,1}^{(p)}(\tau)$ which differ from the present ones only in that the summation go up to n instead of p. The crucial observations of [68] are the following ones: $\lim_{\nu \to 0} \mathcal{J}_{n,1}^{(n)}(\tau) = 0$. For fully unfused simultaneous correlation functions the viscous term in the balance equation disappears in the limit of vanishing viscosity. The integral in $\mathcal{D}_{n,1}^{(n)}(\tau)$ which originates from the projection operator converges in the infra-red and the ultraviolet regimes. This means that every term in the sum over j that contributes to $\mathcal{D}_{n,1}^{(n)}(\tau)$ can be estimated as $S_{n+1}(R)/R$ when all the separation $\mathbf{r}_j \equiv \mathbf{r}_j - \mathbf{r}_j'$ are of the same order R. When we take the full sum up to n there exist internal canceling between all these terms, leading to the homogeneous equation $\mathcal{D}_{n,1} = 0$.

Proposition 28 *The viscous term $\mathcal{D}_{n,1}^{(p)} = 0$ is negligible when $\nu \to 0$; the integrals in this term converge.*

Proof. In the present analysis the first claim is an immediate consequence of the previous result. Since we are taking only partial sums on j, the internal cancelation in $\mathcal{D}_{n,1}^{(p)}(\tau)$, disappears, and it has a finite limit when $\nu \to 0$. On the other hand, $\mathcal{J}_{n,1}^{(p)}(\tau)$ can only increase if we take $\tau = 0$. Thus again, when $\nu \to 0$, we neglect $\mathcal{J}_{n,1}^{(p)}(\tau)$ in the fully unfused situation. Accordingly for $\nu \to 0$ we have

$$\partial_\tau \mathcal{F}_{n,1}^{(p)} + \mathcal{D}_{n,1}^{(p)}(\tau) = 0. \tag{3.35}$$

The proof of convergence of the integrals in $\mathcal{D}_{n,1}^{(p)}(\tau)$ follows from the previous results, since the time correlation functions are bounded from above by the simultaneous ones. Accordingly, when all the coordinates are fully unfused and the separations are all of the order of R,

$$\mathcal{D}_{n,1}^{(p)}(\tau) \sim \mathcal{F}_{n+1,1}^{(p+1)}(\tau)/R. \tag{3.36}$$

Introduce the typical decorrelation time $\tau_{n,1}^{(p)}(R)$ that is associated with the one-time difference quantity $\mathcal{F}_{n,1}^{(p)}(\tau)$ when all the separations are of the order of R:

$$\int_{-\infty}^{0} \mathcal{F}_{n,1}^{(p)}(\tau)\, d\tau \equiv \tau_{n,1}^{(p)} \mathcal{F}_{n,1}^{(p)}(0).$$

Remember that the simultaneous correlation functions of BL-velocity differences (3.33) are identical [67] to the simultaneous correlation functions of Eulerian velocity differences (3.29), i.e., $\mathcal{F}_{n,1}^{(p)}(0) = \mathbf{F}_n$. Integrate (3.35) in the interval $(-\infty, 0)$, use the evaluation (3.36), and derive $R\mathbf{F}_n \sim \tau_{n+1,1}^{(p+1)} \mathbf{F}_{n+1}$. We see that from the point of view of scaling there is no p dependence in this equation: for different values of p only the coefficients can change. We thus estimate

$$\tau_{n,1}(R) \sim RS_{n-1}(R)/S_n(R) \sim R^{z_{n,1}}. \tag{3.37}$$

Here we introduced the dynamical scaling exponent $z_{n,1}$ that characterizes this time and found that

$$z_{n,1} = 1 + \zeta_{n-1} - \zeta_n. \tag{3.38}$$

Consider next the three-time quantity that is obtained from $\mathcal{F}_n$ by choosing $t_i = t + \tau_1$ for $i \leq p$, $t_i = t + \tau_2$ for $p < i \leq p+q$, and $t_i = t$ for $i > p+q$. Denote this quantity as

$\mathcal{F}_{n,2}^{(p,q)}(\tau_1, \tau_2)$, omitting again the rest of the arguments. Define the *decorrelation time* $\tau_{n,2}^{(p,q)}$ of this quantity by

$$\int_{-\infty}^{0} \int_{-\infty}^{0} \mathcal{F}_{n,2}^{(p,q)}(\tau_1, \tau_2)\, \mathrm{d}\tau_1\, \mathrm{d}\tau_2 \equiv \left[\tau_{n,2}^{(p,q)}\right]^2 \mathcal{F}_{n,2}^{(p,q)}(0).$$

From the naive point of view, the decorrelation time $\tau_{n,2}^{(p,q)}$ is of the same order as in (3.37). The calculation leads to a different result. To show this, calculate the double derivative of $\mathcal{F}_{n,2}^{(p,q)}(\tau_1, \tau_2)$ with respect to τ_1 and τ_2. This results in a new balance equation. For the fully unfused situation, and in the limit $\nu \to 0$ we find

$$\partial_{\tau_1 \tau_2}^2 \mathcal{F}_{n,2}^{(p,q)}(\tau_1, \tau_2) + \mathcal{D}_{n,2}^{(p,q)}(\tau_1, \tau_2) = 0, \qquad (3.39)$$

where

$$\mathcal{D}_{n,2}^{(p,q)}(\tau_1, \tau_2) = \sum_{j=1}^{p} \sum_{k=p+1}^{p+q} \left\langle \mathbf{w}_1 \ldots \hat{\mathcal{L}}_j \mathbf{w}_j \ldots \hat{\mathcal{L}}_k \mathbf{w}_k \ldots \mathbf{w}_n \right\rangle.$$

In the RHS of (3.39) we neglected two terms that vanish in the limit $\nu \to 0$. The expression for $\mathcal{D}_{n,2}^{(p,q)}(\tau_1, \tau_2)$ contains two space-integrals that originate from the two projection operators which are hidden in $\hat{\mathcal{L}}_j$ and $\hat{\mathcal{L}}_k$. Using the same ideas when all the separations are of the order of R we estimate with impunity

$$\mathcal{D}_{n,2}^{(p,q)}(\tau_1, \tau_2) \sim \mathcal{F}_{n+2,2}^{(p+1,q+1)}(\tau_1, \tau_2)/R^2.$$

Integrating (3.39) over τ_1 and τ_2 in the interval $(-\infty, 0)$ and using $\mathcal{F}_{n,2}^{(p,q)}(0,0) = \mathbf{F}_n$ we get

$$R\mathbf{F}_n \sim \left[\tau_{n+2,2}^{(p+1,q+1)}\right]^2 \mathbf{F}_{n+2}.$$

As before the scaling exponents are independent of p and q, and we introduce a notation $\tau_{n,2} \sim \tau_{n,2}^{(p,q)}$. Up to p,q-dependent coefficients $[\tau_{n,2}(R)]^2 \sim R^2 S_{n-2}(R)/S_n(R) \sim [R^{z_{n,2}}]^2$. We see that the naive expectation is not realized. The scaling exponent of the present time is different from (3.38): $z_{n,2} = 1 + (\zeta_{n-2} - \zeta_n)/2$. The difference between the two scaling exponents is $z_{n,1} - z_{n,2} = \zeta_{n-1} - (\zeta_n + \zeta_{n-1})/2$. This difference is zero for linear scaling, meaning that in that case the naive expectation that the time scales are identical is correct. On the other hand, for the situation of multiscaling the Holder inequalities require the difference to be positive. Accordingly, for $R \ll L$ we have $\tau_{n,2}(R) \gg \tau_{n,1}(R)$.

Proceed with correlation functions that depend on m time differences. Omitting the upper indices which are irrelevant for the scaling exponents we denote the correlation function as $\mathcal{F}_{n,m}(\tau_1 \ldots \tau_m)$, and establish the exact scaling law for its decorrelation time. The definition of the *decorrelation time* is

$$\int_{-\infty}^{0} \cdots \int_{-\infty}^{0} \mathcal{F}_{n,m}(\tau_1, \ldots, \tau_m)\, \mathrm{d}\tau_1 \ldots \mathrm{d}\tau_m \equiv [\tau_{n,m}]^m \mathcal{F}_{n,m}(0, \ldots, 0). \qquad (3.40)$$

Repeating the steps described above we find the dynamical scaling exponent that characterizes $\tau_{n,m}$ when all the separations are of the order of R, $\tau_{n,m} \sim R^{z_{n,m}}$: $z_{n,m} = 1 + (\zeta_{n-m} - \zeta_n)/m$ $(n - m \le 2)$. One can see, using the Hölder inequalities, that $z_{n,m}$ is a non-increasing function of m for fixed n, and in a multiscaling situation they are decreasing. The meaning is that the larger m is the *longer* is the decorrelation time of the corresponding $m + 1$-time correlation function, $\tau_{n,p}(R) \gg \tau_{n,q}(R)$ for $p < q$.

To gain further understanding of the properties of the time correlation functions, we consider higher order temporal moments of the two-time correlation functions:

$$\int_{-\infty}^{0} \int_{-\infty}^{0} \mathcal{F}_{n,1}(\tau)\, \mathrm{d}\tau\, \mathrm{d}\tau^{k-1} \equiv (\bar{\tau}_k)_{n,1}^{(p)} \cdot \mathcal{F}_{n,1}^{(p)}(0). \qquad (3.41)$$

The intuitive sense of $(\bar{\tau}_k)_{n,1}^{(p)}$ is a *k-order decorrelation moment* of $\mathcal{F}_{n,1}^{(p)}(R,\tau)$, and note that its dimension is $(\text{time})_k$. The first order decorrelation moment is the previously defined decorrelation time $\tau_{n,1}^{(p)}$. To find the scaling exponents of these quantities we start with (3.35), multiply by τ^k, and integrate over in the interval $(-\infty, 0)$. Using the evaluation (3.36) and assuming convergence of the integrals over document we derive

$$-k \int_{-\infty}^{0} \mathcal{F}_{n,1}^{(p)} \, \tau^{k-1} \mathrm{d}\tau \sim \frac{1}{R} \int_{-\infty}^{0} \mathcal{F}_{n+1,1}^{(p+1)} \, \tau^k \mathrm{d}\tau,$$

where we have integrated by parts on the LHS. In deriving this equation we assert that $k+1$ moments exist; this is not known *a priori*. Using the definition (3.41) and for all the separations of the order of R we find the recurrence relation

$$R S_n(R)(\bar{\tau}^k)_{n,1}^{(p)} \sim S_{n+1}(R)(\bar{\tau}^k)_{n+1,1}^{(p+1)}.$$

The solution is $(\bar{\tau}^k)_{n,1}^{(p)} \sim (\tau_{n,k})^k \sim \frac{R^k S_{n-k}(R)}{S_n(R)}$ for $k \leq n-2$. The procedure does not yield information about higher k values. We learn from the analysis of the moments that there is no single typical time which characterizes the dependence of $\mathcal{F}_{n,1}^{(p)}$. There is no simple "dynamical scaling exponent" z that can be used to collapse the time dependence in the form $\mathcal{F}_{n,1}^{(p)} \sim R^{\zeta_n} \int (\tau/R^z) \mathrm{d}\tau$. Even the two-time correlation function is not a scale invariant object. In this respect it is similar to the probability distribution function of the velocity differences across a scale R, for which the spectrum of ζ_n is a reflection of the lack of scale invariance.

The main conclusion is that all the dynamical scaling exponents can be determined from the knowledge of the scaling exponents ζ_n of the standard structure functions $S_n(R)$. All the scaling relations that were obtained above can be remembered using the following simple rule.

To get the dynamical scaling exponent every integral over in the definition of the decorrelation time (3.40) and every factor τ in the definition of the moments (3.41) can be traded for a factor of R/W *within the average* of the correlation function involved. The dynamical exponent is determined by the resulting scaling exponents of the resulting simultaneous correlation function. The deep reason for this simple rule is the non-perturbation locality (convergence) of the integrals appearing in $\mathcal{D}_n$. Because of this locality one can estimate from the equation of motion (3.30)

$$1/\tau \sim \partial/\partial\tau \sim \mathbf{w} \cdot \nabla \sim W(R)/R.$$

Thus, we can use the substitutions $\tau \int \mathrm{d}\tau \sim R/W(R)$ *as long as we use them within the average*, and when all the separations are of the order of R. We propose to refer to this substitution rule as "weak dynamical similarity," where "weak" stands for a reminder that the rules can be used *only* under the averaging procedure, and *only* for scaling purposes. The same property of locality of the interaction integrals was shown [68] to yield another set of bridge relations between scaling exponents of correlation functions of gradient fields and the scaling exponents ζ_n. Those relations can be summarized by another substitution rule that we refer to as "weak dissipative similarity"; It follows from equating the viscous and nonlinear terms in the equations of motion: $\nu \nabla^2 \sim W(R)/R$.

Again "weak" refers to the reminder that we are only allowed to use these substitutions for scaling purposes within the average. Our weak dissipative similarity rule is weaker than the Kolmogorov refined similarity hypothesis which states the *dynamical* relationship $(\nabla W(R))^2 \sim W^3/R$. Both our rules are derived from first principles, while Kolmogorov's hypothesis is a guess.

Finally, we should ask whether the results presented above are particular to the time correlation function of BL-velocity differences, or do they reflect intrinsic scaling properties that are shared by other dynamical presentations like the standard Lagrangian velocity fields. The answer is that the results are general; all that we have used are the property of convergence of the interaction integrals, and the fact that the simultaneous correlation functions of the BL-fields are the same as those of the Eulerian velocities. These properties hold also for Lagrangian velocities, and in fact for any sensible choice of velocity representation in which the sweeping effect is eliminated. Accordingly we state that the dynamical exponent are invariant to the representation and in particular will be the same for many-time correlation functions of Lagrangian velocity differences.

3.2.6   Fusion rules in turbulence systems

Fusion rules in turbulence specify the analytic structure of many-point correlation functions of the turbulent field when a group of coordinates coalesce. We show that the existence of universal flux equilibrium in fully developed turbulent systems combined with a direct cascade induces universal fusion rules. In certain examples these fusion rules suffice to compute the multiscaling exponents exactly, and in other examples they give rise to an infinite number of scaling relations that constrain enormously the structure of the allowed theory.

In a series of papers elements of the analytic theory of Navier–Stokes turbulence [67, 69] and passive-scalar turbulent advection [60] were presented. We will explain that the structure of the essential part of these theories is economically summarized by a set of "fusion rules" that determine the analytic structure of n-point correlation functions when a group of coordinates tend together. We show here that the fusion rules can be deduced from very few general assumptions about the nature of the universal flux equilibrium that exists in fully developed turbulent systems. Of course, the same fusion rules can be also established by direct calculations in specific examples. First, deduce the fusion rules, then we exemplify their utility in determining scaling exponents, and lastly we demonstrate how in one example the fusion rules follow from first principles.

Let $\mathbf{u}(\mathbf{r}, t)$ be a vector or a scalar turbulent field and denote the difference $\mathbf{w}(\mathbf{r}_1|\mathbf{r}_2, t) = \mathbf{u}(\mathbf{r}_2, t) - \mathbf{u}(\mathbf{r}_1, t)$. We consider the statistical properties of the turbulent field in terms of simultaneous many-point correlation functions of differences with respect to one, two or more reference points:

$$S_n(\mathbf{r}_D|\mathbf{r}_1, \ldots, \mathbf{r}_n) = \langle \mathbf{w}(\mathbf{r}_D|\mathbf{r}_1) \ldots \mathbf{w}(\mathbf{r}_D|\mathbf{r}_n) \rangle, \tag{3.42a}$$

$$\begin{aligned} S_{n,m}(\mathbf{r}_D, \mathbf{r}'_D|\mathbf{r}_1, \ldots, \mathbf{r}_n; \mathbf{r}_{n+1}, \ldots, \mathbf{r}_{n+m}) \\ = \langle \mathbf{w}(\mathbf{r}_D|\mathbf{r}_1)\mathbf{w}(\mathbf{r}_D|\mathbf{r}_2) \ldots \mathbf{w}(\mathbf{r}_D|\mathbf{r}_n)\mathbf{w}(\mathbf{r}_D|\mathbf{r}_{n+1}) \ldots \mathbf{w}(\mathbf{r}_D|\mathbf{r}_{n+m}) \rangle, \end{aligned} \tag{3.42b}$$

etc. When $\mathbf{u}$ is a vector the n-point correlation is an n-rank tensor. The class of systems to be discussed are driven on a characteristic scale referred to as the outer scale L. This driving can be achieved by either a time dependent low frequency "stirring force" or by specifying given values of $\mathbf{u}$ at a set of "boundary" points with a characteristic separation L away from our observation points $\mathbf{r}_D, \mathbf{r}'_D, \mathbf{r}_1$, etc. The systems have dissipation (viscosity, diffusivity etc.) and in the dissipation-less limit there exists an integral of motion which we refer to as "energy." We consider systems with a "direct" cascade in which the intake of energy on the scale L is balanced by dissipation on a small scale $\eta \ll L$.

We invoke two Kolmogorov type assumptions [54]:

1. Scale invariance: all the correlation functions are homogeneous functions of their arguments in the core of the inertial interval $\eta \ll |\mathbf{r}_i - \mathbf{r}_D| \ll L$: $S_n(\lambda\mathbf{r}_D|\lambda\mathbf{r}_1, \ldots, \lambda\mathbf{r}_n) = \lambda^{\zeta_n} S_n(\mathbf{r}_D|\mathbf{r}_1, \ldots, \mathbf{r}_n)$, where ζ_n are *scaling exponents*.

2. Universality of the fine scale structure of turbulence; in its strong version this means that the correlation functions of the type (1) have a universal functional dependence on the separation distances when they are all in the interior of the inertial interval (η, L).

Thus, we can fix an arbitrary set of velocity differences on the scale of L, and the correlation functions will depend on their precise choice only via an overall factor determined by the L-scale motions. Mathematically this is expressed as the following property of the conditional average:

$$
\begin{aligned}
&\langle \mathbf{w}(\mathbf{r}_D|\mathbf{r}_1) \ldots \mathbf{w}(\mathbf{r}_D|\mathbf{r}_n) | \mathbf{w}(\mathbf{r}_D|\mathbf{r}_1') \ldots \mathbf{w}(\mathbf{r}_D|\mathbf{r}_N') \rangle \\
&= \tilde{S}_n(\mathbf{r}_D|\mathbf{r}_1, \ldots, \mathbf{r}_n) \Phi_{n,N}(\mathbf{r}_D|\mathbf{r}_1', \ldots, \mathbf{r}_N')
\end{aligned}
$$

for $|\mathbf{r}_i' - \mathbf{r}_D| \sim L$ and $|\mathbf{r}_i - \mathbf{r}_D| \ll L$. The functions $\tilde{S}_n$ coincide with S_n in the inertial interval. They can differ in their crossover to viscous behavior, and their (different) crossover scales may depend on the large scale motions. A weaker version of the universality assumption concerns with the scaling exponent only. In this version the functions S_n and $\tilde{S}_n$ may differ, but their scaling exponents coincide. This weaker version is sufficient for most of our developments below. In both versions ζ_n can be identified with the scaling exponent of the n-order structure function $\langle |\mathbf{w}(\mathbf{r}_D|\mathbf{r})|^n \rangle$. We are interested in multiscaling systems in which ζ_n is a nonlinear function of n. The derivation of these two properties from first principles differs from system to system. Discuss the fusion rules and their consequences in systems for which these assumptions are valid. The first set of fusion rules that we derive concerns S_n when p points $(p < n)$ tend to $\mathbf{r}_D$, (so that the typical separation from $\mathbf{r}_D$ is r) and all the other separations remain much larger, of the order of R, $r \ll R \ll L$. Without loss of generality, choose these p coordinates as $\mathbf{r}_1, \mathbf{r}_2, \ldots, \mathbf{r}_p$. We claim that

$$
S_n(\mathbf{r}_D|\mathbf{r}_1, \ldots, \mathbf{r}_n) = \tilde{S}_n(\mathbf{r}_D|\mathbf{r}_1, \ldots, \mathbf{r}_p) \psi_{n,p}(\mathbf{r}_D|\mathbf{r}_{p+1}, \ldots, \mathbf{r}_n), \tag{3.43}
$$

where $\psi_{n,p}(\mathbf{r}_D|\mathbf{r}_{p+1}, \ldots, \mathbf{r}_n)$ is a homogeneous function with a scaling exponent $\zeta_n - \zeta_p$. The derivation of fusion rule (3.43) follows from Bayes' theorem and assumptions **1** and **2**. Write

$$
\begin{aligned}
S_n(\mathbf{r}_D|\mathbf{r}_1, \ldots, \mathbf{r}_n) = \int d\mathbf{w}(\mathbf{r}_D|\mathbf{r}_{p+1}) \ldots d\mathbf{w}(\mathbf{r}_D|\mathbf{r}_n) \times \mathbf{w}(\mathbf{r}_D|\mathbf{r}_{p+1}) \ldots \\
\mathbf{w}(\mathbf{r}_D|\mathbf{r}_n) \wp(\mathbf{w}(\mathbf{r}_D|\mathbf{r}_{p+1}) \ldots \mathbf{w}(\mathbf{r}_D|\mathbf{r}_n)) \times \langle \mathbf{w}(\mathbf{r}_D|\mathbf{r}_1) \mathbf{w}(\mathbf{r}_D|\mathbf{r}_1) \ldots \\
\mathbf{w}(\mathbf{r}_D|\mathbf{r}_p) \mathbf{w}(\mathbf{r}_D|\mathbf{r}_p) | \mathbf{w}(\mathbf{r}_D|\mathbf{r}_{p+1}) \mathbf{w}(\mathbf{r}_D|\mathbf{r}_{p+1}) \ldots \mathbf{w}(\mathbf{r}_D|\mathbf{r}_{p+n}) \mathbf{w}(\mathbf{r}_D|\mathbf{r}_{p+n}) \rangle,
\end{aligned} \tag{3.44}
$$

where $\wp(\mathbf{w}(\mathbf{r}_D|\mathbf{r}_{p+1}) \ldots \mathbf{w}(\mathbf{r}_D|\mathbf{r}_n))$ is the probability to see $\mathbf{w}(\mathbf{r}_D|\mathbf{r}_{p+1}) \ldots \mathbf{w}(\mathbf{r}_D|\mathbf{r}_n)$. Note the consequence of assumption **2**: the scaling laws of the correlation functions at scale r are the same independent of whether we force the system on the scale $L \gg r$ or on the scale $R \gg r$. The conditional average in (3.44) is proportional to $\tilde{S}_p$, and hence (3.43). This result seems rather obvious at this point, but we will see that it leads to a totally unconventional scaling structure of the theory. For NSE and passive scalar advection these fusion rules for $p = 2$ were derived from first principles [69].

The next set of fusion rules is obtained for the structure function $S_{n,m}$ of (3.42a,b) when two groups of $p \le n$ and $q \le m$ points tend to $\mathbf{r}_D$ and $\mathbf{r}_D'$, respectively. The separation between these groups of point is of the order of R. The derivation of the fusion rules is obvious, with the result

$$
\begin{aligned}
S_{n,m}(\mathbf{r}_D, \mathbf{r}_D'|\mathbf{r}_1, \ldots, \mathbf{r}_n; \mathbf{r}_{n+1}, \ldots, \mathbf{r}_m) = \tilde{S}_p(\mathbf{r}_D|\mathbf{r}_1, \ldots, \mathbf{r}_p) \tilde{S}_q(\mathbf{r}_D'|\mathbf{r}_{n+1}, \ldots, \mathbf{r}_{n+q}) \\
\times \Psi_{n,m,p,q}(\mathbf{r}_D, \mathbf{r}_D'|\mathbf{r}_{p+1}, \ldots, \mathbf{r}_n; \mathbf{r}_{n+q+1}, \ldots, \mathbf{r}_{n+m}).
\end{aligned}
$$
$$\tag{3.45}$$

The scaling exponent of $\Psi_{n,m,p,q}$ is $\zeta_{n+m} - \zeta_p - \zeta_q$. The fusion rules (3.43) and (3.45) are *not* decompositions into products of lower order correlation functions, and the functions Ψ are not correlations of velocity differences across large separations. In fact we will show that Ψ is much larger than the corresponding correlation functions in all situations with multiscaling. Evidently one can derive similar fusion rules for three, four or more groups of "coalescing" points with large separations between the groups. The structure of the resulting correlation function will be a product of the correlation function associated with each group times some function Ψ of big separations which carries the overall exponent.

Next, we discuss fusion rules for correlation functions that include gradient fields. These rules depend on the type of rotational invariant that one can define from the tensors that appear after taking gradients. Consider only the lowest order invariant which is a scalar under rotation. For passive scalars T this is $|\nabla T \cdot \nabla T|^2$ and for a vector $\mathbf{u}$ the quantity $|\nabla \mathbf{u}|^2$ is the square of the strain tensor $g_{ij}g_{ij}$, where $g_{ij} \equiv (\partial u_i/\partial r_j + \partial u_j/\partial r_i)/2$. Consider the quantity

$$J_{2p,n}(\mathbf{r}_D|\mathbf{r}_{2p+1},\ldots,\mathbf{r}_n) = \langle |\nabla\mathbf{u}(\mathbf{r}_D)|^{2p}\mathbf{w}(\mathbf{r}_D|\mathbf{r}_{2p+1})\ldots\mathbf{w}(\mathbf{r}_D|\mathbf{r}_n)\rangle.$$

To evaluate $J_{2p,n}$ we consider a related object in which all the gradients are taken at different points:

$$\tilde{J}_{2p,n} = \nabla_{r_1}^{i_1}\nabla_{r_1'}^{j_1}\nabla_{r_2}^{i_2}\nabla_{r_2'}^{j_2}\ldots\nabla_{r_p}^{i_p}\nabla_{r_p'}^{j_p}\mathcal{C}_{k_1,l_1,\ldots,k_p,j_p}^{i_1,j_1,\ldots,i_p,j_p} \times \langle w^{k_1}(\mathbf{r}_D|\mathbf{r}_1)w^{l_1}(\mathbf{r}_D|\mathbf{r}_1')\ldots$$

$$w^{k_p}(\mathbf{r}_D|\mathbf{r}_p)w^{l_p}(\mathbf{r}_D|\mathbf{r}_p') \times \mathbf{w}(\mathbf{r}_D|\mathbf{r}_{2p+1})\ldots\mathbf{w}(\mathbf{r}_D|\mathbf{r}_n)\rangle, \qquad (3.46)$$

where the contraction tensor $\mathcal{C}$ ensures that the required scalar is obtained. Represent this quantity in a compact form without displaying all the tensor indices as

$$\tilde{J}_{2p,n}(\mathbf{r}_D|\mathbf{r}_1,\mathbf{r}_2,\ldots,\mathbf{r}_n) = \nabla_{r_1}\nabla_{r_1'}\ldots\nabla_{r_p'}\mathcal{C}S_n(\mathbf{r}_D|\mathbf{r}_1,\mathbf{r}_1',\ldots,\mathbf{r}_n).$$

The quantity (3.46) gives us $J_{2p,n}$ when the $2p$ first points coalesce together with $\mathbf{r}_D$, whereas all the rest of the coordinates remain a typical distance R from $\mathbf{r}_D$. When R is in the inertial interval we expect scaling behavior in terms of R,

$$J_{2p,n} \sim R^{-\xi(2p,n)}. \qquad (3.47)$$

Considering the distances between all the coalescing points to be in the inertial range we apply (3.43) and find for $2p$ coalescing points

$$J_{2p,n}(\mathbf{r}_D|\mathbf{r}_1,\ldots,\mathbf{r}_n) = \nabla_{r_1}\ldots\nabla_{r_p}\tilde{S}_{2p}(\mathbf{r}_D|\mathbf{r}_1,\mathbf{r}_1',\ldots,\mathbf{r}_p')\Psi_{n,2p}(\mathbf{r}_D|\mathbf{r}_{2p+1},\ldots,\mathbf{r}_n).$$

We expect that $J_{2p,n}$ is independent of the first $2p$ coordinates when the distances between them are well in the viscous regime. Denote by $\eta(2p,n,R)$ the characteristic viscous length at which $\tilde{S}_{2p}$ crosses smoothly from inertial range behavior to dissipative behavior. Of course this is also the cross over scale of $J_{2p,n}$ with respect to first $2p$ coordinates. This allows us to evaluate the coalescing gradients by taking the $2p$ separations to be $\eta(2p,n,R)$

$$J_{2p,n}(\mathbf{r}_D|\mathbf{r}_1,\ldots,\mathbf{r}_n) \sim \eta(2p,n,R)^{\zeta_{2p}-2p}\Psi_{n,2p}(\mathbf{r}_D|\mathbf{r}_{2p+1},\ldots,\mathbf{r}_n), \qquad (3.48)$$

with $2p$ coalescing points. If there are two groups of coalescing points with gradients, with p points coalescing onto $\mathbf{r}_D$ and q points coalescing on $\mathbf{r}_D'$ respectively, we consider $J_{p,q,n,m}$ (where as before $n+m \geq p+q$ is the total number of points). The rule for p and q coalescing points is

$$J_{p,q,n,m}(\mathbf{r}_D|\mathbf{r}_1,\mathbf{r}_2,\ldots\mathbf{r}_n) \sim \eta(p,n,R)^{\zeta_p-p}\eta(q,n,R)^{\zeta_q-q}\Psi_{n,m,p,q}(\mathbf{r}_D|\mathbf{r}_{p+1},\mathbf{r}_{p+2},\ldots\mathbf{r}_n). \qquad (3.49)$$

The generalization of this fusion rule for three or more groups of coalescing points with gradients is obvious.

This is as much as one can do in general. Now the crucial issue is how $\eta(2p, n, R)$ depends on its arguments. The simplest version of the theory comes about when the dissipative length is independent of R, $\eta(2p, n, R) = \eta(2p, n)$. This is realized for example in passive scalar convection as is shown below. In this case the fusion rules imply various sets of scaling relations. For example the exponents $\xi(2p, n) = \zeta_n - \zeta_{2p}$ of $J_{2p,n}$ are given by

$$\xi(2p, n) = \zeta_n - \zeta_{2p}. \tag{3.50}$$

The fusion rules equivalent to this simple version were proposed in [25] on the basis of formal assumptions used in field theory, namely, existence of an ultraviolet universal Renormalization-Group, fixed point and "asymptotic completeness".

As another example of scaling relations consider the correlation functions

$$K_{LL}^{(2s)}(R) \equiv \langle |\nabla \mathbf{u}(\mathbf{r})|^{2s} |\nabla \mathbf{u}(\mathbf{r} + \mathbf{r})|^{2s} \rangle \sim R^{-\mu(2s)}. \tag{3.51}$$

From (3.49) in the case $n = m = p = q = 2s$ we get a set of scaling relations $\mu(2s) = 2\zeta_{2s} - \zeta_{4s}$. Next, consider a correlation of l gradient fields with the same power, (i.e., $|\nabla \mathbf{u}|^{2s}$) at l different points which are separated by a distance of the order of R. The corresponding exponent $\mu(l, 2s)$ is

$$\mu(l, 2s) = l\zeta_{2s} - \zeta_{2sl}. \tag{3.52}$$

This algebra can be generalized to any correlation of powers of $|\nabla \mathbf{u}|^2$. For example, the scaling exponent $\mu(p_1, \ldots, p_n)$ of a correlation of fields $\langle |\nabla \mathbf{u}(\mathbf{r}_1)|^{p_1} |\nabla \mathbf{u}(\mathbf{r}_2)|^{p_2} \ldots |\nabla \mathbf{u}(\mathbf{r}_n)|^{p_n} \rangle$, in which all the separations is of the order of R is

$$\mu(p_1, p_2, \ldots, p_n) = \sum_{j=1}^{n} \zeta_{p_j} - \zeta_{\bar{p}}, \quad \bar{p} = \sum_{j=1}^{n} p_j.$$

In usual operator algebras [52, 106] every local field is associated with a leading exponent and the correlation function scales with the sum of these exponents. In this case the algebra is different. There is a global exponent $\zeta_{\bar{p}}$ from which one subtracts the sum of individual exponents ζ_{p_j}. In multiscaling situations the global exponent is a nonlinear function of $\bar{p} = \sum p_j$. Accordingly it is not a property of every individual field. Nonscalar invariants of the gradient field tensors are associated with other individual exponents.

The range of applicability of these fusion rules should be understood on the basis of the equations of motion for any given system. As an example we explain here briefly why they are applicable for model [58] of passive scalar convection with a driving velocity field that is δ-correlated in time. It was shown [58] that the cumulant part F_{2n}^{ε} of the 2nd order correlation function $F_{2n} = \langle T(\mathbf{r}_1, t) T(\mathbf{r}_2, t) \ldots T(\mathbf{r}_{2n}, t) \rangle$ satisfies for $n > l$ the exact homogeneous differential equation

$$\left[-D \sum_{\alpha=1}^{2n} \nabla_\alpha^2 + \hat{B}_{2n} \right] F_{2n}^{\varepsilon}(\mathbf{r}_1, \mathbf{r}_1, \ldots, \mathbf{r}_{2n}) = 0, \tag{3.53}$$

where D is the molecular diffusivity and ∇_α^2 is the Laplace operator acting on $\mathbf{r}_\alpha$. The operator $\hat{B}_{2n}$ is the sum of the binary operators $\hat{B}_{\alpha,\beta}$:

$$\hat{B}_{\alpha,\beta} \equiv h_{ij}(\mathbf{r}_\alpha - \mathbf{r}_\beta) \frac{\partial^2}{\partial r_{\alpha,i} \partial r_{\beta,j}}, \quad \hat{B}_{2n} = \sum_{\alpha > \beta=1}^{2n} \hat{B}_{\alpha,\beta}.$$

Here $h_{ij}(\mathbf{r})$ is the eddy diffusivity that behaves like $H R^{\zeta_h}$ with $0 < \zeta_h < 2$ and $\zeta_2 = 2 - \zeta_h$.

In the inertial range of scales we can disregard the Laplace operators in this equation. For deriving the fusion rules (3.43) we consider the p coalescing points with characteristic separation r and denote their coordinates by the index α or α'. The remaining $2n - p$

coordinates have characteristic separations R and are denoted by β or β'. We assemble [26] the $\hat{B}$ operators in three groups: $\hat{B}_p = \sum_{\alpha > \alpha'} \hat{B}_{\alpha\alpha'}$, $\hat{B}_{2n-p} = \sum_{\beta > \beta'} \hat{B}_{\beta\beta'}$ and $\hat{B}_R = \sum_{\alpha\beta} \hat{B}_{\alpha\beta}$. The evaluation of the action of the operators in the first and second groups is H/r^{ζ_2} and H/R^{ζ_2}, respectively. The evaluation of the action of each term in the third group is HR/rR^{ζ_2}. Space homogeneity results in a cancellation of this evaluation in the sum of the terms in this group [69]. The next order surviving evaluation is again H/R^{ζ_2}. We thus combine the second and third group into an effective operator $\tilde{B}$. The equation to consider is $[\hat{B}_p + \tilde{B}]F_{2n} = 0$. When $p = 2$ we find the solution of this equation as the following expansion in powers of the small difference r_{12}: $A_2\{R\} + r_{12}^{\zeta_2}C_2\{R\} + r_{12}^{2\zeta_2}D_2\{R\} + r_{12}^2 E_2\{R\} + \ldots$, where A_2, C_2, D_2, etc. are some functions of the large separations of the order of R When we use this solution to compute S_{2n} the leading contribution $A\{R\}$ drops and we find (3.43) for $p = 2$. For $p > 2$ we need to distinguish between even and odd p because of the special property of passive advection in which $S_{2n+1} = 0$. The next even p is $p = 4$. For this case we find a solution in the form

$$F_4^\varepsilon = A_4\{R\} + C_4\{R\}\left[\sum_{\alpha\alpha'}^4 r_{\alpha\alpha'}^{\zeta_2}\right] + D_4\{R\}\left[(r_{12}r_{13})^{\zeta_2} + (r_{12}r_{14})^{\zeta_2} + (r_{12}r_{23})^{\zeta_2} + \ldots\right]$$
$$+ F_4^\varepsilon(\mathbf{r}_1, \mathbf{r}_2, \mathbf{r}_3, \mathbf{r}_4)\Psi_{2n,4}\{R\} + \ldots,$$

where F_4^ε is a contribution of a new type, as it solves the homogeneous equation (3.53). Computing S_{2n} the first two terms disappear and in a multiscaling situation the leading contribution becomes the last. In fact, this is the general rule for any even order, and is the explicit mechanism for the fusion rules in this particular case. It arises here from the possibility to split the total operator $\hat{B}_{2n}$ into the two groups $\hat{B}_p$ and $\tilde{B}$ such that for p coalescing points $\hat{B}_p$ carries the leading contribution. Since the sum of Laplacians in (3.53) is also dominated by the sum up to p, the crossover scale $\eta(p, 2n, R)$ from inertial range to dissipative behavior is determined in this case by a balance between $-k\sum_{\alpha=1}^p \nabla_\alpha^2 \hat{B}_p$ and $\hat{B}_p$. It therefore cannot depend on n or on R: $\eta(p, 2n, R) = \eta(p)$. The fusion rules (3.51)-(3.52) which were based on the independence of η on R follow. In fact, these results, and in particular the scaling relations (3.50) seem sufficient to determine the exponents ζ_n in their entirety. The necessary steps were detailed in [26] and will not be repeated here. The case of Navier–Stokes turbulence calls for additional considerations. The fusion rules (3.43), (3.45) were found from first principles for $p = 2$ [69] and we believe that similar techniques can be used to establish them for any p. Equations (3.48) and (3.49) follow, but in the NSE case it is possible that the dissipative scale $\eta(p, n, R)$ is R dependent. If we assume that this is not the case the scaling relations obtained above should apply also to the NSE case. To explore another possibility we follow Kolmogorov's refined similarity hypothesis [54] in assuming that the conditional average

$$\nu\langle|\nabla\mathbf{u}|^2|\mathbf{w}(0|\mathbf{r})\rangle \sim w(0|\mathbf{r})^3/R. \tag{3.54}$$

This assumption means $\nu J_{2,n}\{R\} \sim S_{n+1}(R)/R$. Comparing with (3.48) this can be consistent only if

$$[\eta(2, n, R)/\eta(2)]^{2-\zeta_2} \sim (R/L)^{\zeta_n - \zeta_{n-1} + \zeta_3 - \zeta_2},$$

where $\eta(2)$ is the *viscous cutoff for the second order structure function*, $\nu S_2(\eta(2)) \sim \eta(2)^2\bar{\varepsilon}$. The Hölder inequalities guarantee that $\zeta_n - \zeta_{n-1}$ is a decreasing function of n in a multiscaling situation. Accordingly, the effective dissipative scale of $J_{2,n}$ for two coalescing points $\eta(2, n, R)$ is much smaller than the viscous cutoff for $S_2(\eta(2))$. The consequences of the above two assumptions were discussed in [69]. Needles to say, with assumption (3.54) all our scaling relations change. For example consider $\xi(2, n)$ of (3.47). Instead of (3.50) we have $\xi(2, n) = \zeta_{n+1} - 1$. Another example is $K_{LL}^{(2)}$, and we find $\mu(1) = 2 - \zeta_6$. This result is known as "the bridge" in the phenomenological theory of multiscaling turbulence, c.f. [36, 81]. Notwithstanding the different scaling relation, the operator algebra that is induced has "global" and individual scaling exponents as discussed above. The values of these exponents may be changed due to the R dependence of the dissipative cutoff as it appears in different models.

Summary: We proposed fusion rules for multiscaling turbulent statistics that induce an unusual operator algebra. These rules are of two classes. The first does not involve gradients and is universal for all turbulent systems with a direct cascade of "energy," in which there is a universal flux equilibrium. The second class involves gradients and these bring in an explicit dependence on a viscous scale that in general is not universal. We explained why in the case of passive scalar advection this problem may be solved. Accordingly one can derive an infinite set of non trivial scaling relations that allow the expression of the scaling exponents of the correlation functions of gradient fields with the exponents ζ_n of the structure functions. For hydrodynamic turbulence the fusion rules that involve gradients must by supplemented with a NSE based theory for the R dependence of the viscous cutoff.

3.3 Calculation of scaling exponents

The viscous scale η appears via a rather standard mechanism that arise in perturbation theory as logarithmic divergences, but in order to see it one needs to consider the statistics of gradient fields rather than the velocity differences themselves. For example, considering the perturbation series for $K_{\varepsilon\varepsilon}(R)$, which is the correlation function of the rate of energy dissipation $\nu|\mathbf{u}|^2$, leads immediately to the discovery of logarithmic ultraviolet divergences in every order of the perturbation theory. These divergences are controlled by an ultraviolet cutoff scale which is identified as the viscous-dissipation scale η acting here as the renormalization scale. The summation of the infinite series results in a factor $(R/\eta)^{2\Delta}$ with some anomalous exponent Δ which is, generally speaking, of the order of unity. The appearance of such a factor means that the actual correlation of two R-separated dissipation fields is much larger, when R is much larger than η, than the naive prediction of dimensional analysis. The physical explanation of this renormalization is the effect of the multi-step interaction of two R-separated small eddies of scale η with a large eddy of scale R via an infinite set of eddies of intermediate scales. The net result on the scaling exponent is that the exponent μ changes from 8/3 as expected in the Kolmogorov theory to $8/3 - 2\Delta$.

3.3.1 Basic formulas

It is important to calculate the value of the anomalous exponent Δ. In [67] there was found an exact sum that forces a relation between the numerical value of Δ and the numerical value of the exponent ζ_2 of $S_2(R)$, $\Delta = 2 - \xi_2$. Such a relation between different exponents is sometimes called as a *scaling relation* or a *bridge relation*. It is a consequence of the existence of a universal nonequilibrium stationary state that supports an energy flux from large to small scales [67]. The scaling relation for Δ has implications for the theory of structure functions. With this value of Δ the series for the structure functions $S_n(R)$ diverge when the energy-input scale L approaches ∞ as powers of L, like $(L/R)^{\delta_n}$. The *anomalous exponents* δ_n are the derivations of $S_n(R)$ from their Kolmogorov value. We expand on this very delicate and important point. Let us think about low the series representation of $S_n(R)$ in terms of lower order w quantities, and imagine that one succeeded to resume it into an operator equation for $S_n(R)$. Typically such a resumed equation looks like

$$[\hat{I} - \hat{O}]S_n(R) = \text{RHS},$$

where $\hat{O}$ is some integro-differential operator which is not small compared to unity. If we expand this equation in powers of $\hat{O}$ around the RHS we regain the infinite perturbation series that we started with. Now we realize that the equation possesses also homogeneous solutions, solutions of $[\hat{I} - \hat{O}]S_n(R) = 0$, which are inherently non-perturbative since they can no longer be expanded around a RHS. These homogeneous solutions n may be much larger than the inhomogeneous perturbative solutions. Of course, homogeneous solutions must be matched with the boundary conditions at $R = L$, and this is the way that the energy input scale L appears in the theory. This is important when the homogeneous solution diverge in size when $L \to \infty$ as is indeed the case for the problem at hand.

The divergence of the perturbation theory for $S_n(R)$ with $L \to \infty$ forces us to seek a non-perturbative handle on the correlations theory. One finds this in the idea that there exists a global balance between energy input and dissipation, which may be turned into a non-perturbative constraint on each n-th order structure function [67]. Using (3.42b) one derives the set of equations of motion

$$\partial_t S_n(R, t) + D_n(R, t) = \nu J_n(R, t),$$

where D_n and J_n stem from the nonlinear and the viscous terms in (3.42b), respectively. To understand the physical meaning of this equation note that $S_2(R)$ is precisely the mean kinetic energy of motions of size R. The term $D_2(R)$ whose meaning is the rate of energy flux through the scale R is known exactly: $D_2(R) = S_3'(R)$. The term $\nu J_2(R)$ is precisely the rate of energy dissipation due to viscous effects. The higher order equation for $n > 2$ are direct generalizations of this to higher order moments. In the stationary state the time derivative vanishes and one has the balance equation $D_n(R) = \nu J_n(R)$. For $n = 2$ it reflects the balance between energy flux and energy dissipation. The evaluation of $D_n(R)$ for $n > 2$ requires dealing with the difficulty (**C**) of the nonlocality of the beyond interaction, but it does not pose conceptual difficulties. It was shown [67] that $D_n(R)$ is of the order of $S_{n+1}'(R)$. On the other hand, the evaluation of $J_n(R)$ raises a number of very interesting issues whose resolution lies at the heart of the universal scaling properties of turbulence. From the derivation of equations of motion one finds that $J_n(R)$ consists of a correlation of $\nabla^2 u$ with $n - 2$ velocity differences across a scale $R = |\mathbf{r}_1 - \mathbf{r}_2|$: $\langle \nabla^2 u(\mathbf{r}_1)[\delta u(\mathbf{r}_1, \mathbf{r}_2)]^{n-2} \rangle$.

How to evaluate such a quantity in terms of the usual structure functions $S_n(R)$? Recall that a gradient of a field is the difference in the field values at two points divided by the separation when the latter goes to zero. In going to zero one necessarily crosses the dissipative scale. To understand what happens in this process one needs first to introduce many point correlation functions of a product of n velocity differences: $\mathbf{F}_n(\mathbf{r}_0 | \mathbf{r}_1, \dots, \mathbf{r}_n) \equiv \langle \delta u(\mathbf{r}_0, \mathbf{r}_1) \dots \delta u(\mathbf{r}_0, \mathbf{r}_n) \rangle$. There we need to formulate rules for the evaluation of such bridge, correlation functions of velocity differences when some of the coordinates get very close to each other. For example, a gradient $\partial/\partial r_\alpha$ can be formed from the limit $\mathbf{r}_1 \to \mathbf{r}_0$ when we divide by $r_{1,\alpha} \to r_{0,\alpha}$. These so called *fusion rules* for hydrodynamic turbulence were presented in [70]. They show that when p coordinates in F_n are separated by a small distance r, and the remaining $n - p$ coordinates are separated by a large distance R, then the scaling dependence on r is like that of $S_p(r)$, i.e., r^{ζ_p}. This is true until r crosses the dissipative scale. Assuming that below the viscous-dissipation scale η derivatives exist and the fields are smooth, one can estimate gradients at the end of the smooth range by dividing differences across η by η. The question is, what is the appropriate cross-over scale to smooth behavior? Is there just one cross-over scale η, or is there a multiplicity of such scales, depending on the function one is studying? For example, when does the above n-point correlator become differentiable as a function of r when p of its coordinates approach $\mathbf{r}_0$? Is that typical scale the same as the one exhibited by $S_p(r)$ itself, or does it depend on p and n, and on the remaining distances of the remaining $n - p$ coordinates that are still far away from $\mathbf{r}_0$? The detail mechanism for calculation of the scaling exponent is presented below.

We discuss the first order steps of this scheme, identify the mechanism for anomalous scaling in turbulence, and the series of steps available if one wants to improve the numerical values of the computed anomalous exponents. The main ideas to achieve the lowest order closure without breaking the rescaling symmetry of the Euler equation are formulated such that they repeat essentially unchanged in any higher order closure; it seems that no new ideas are necessary. We will review essential ideas and results. First we stress that a dynamical theory of the turbulence statistics calls for a convenient transformation of variables that removes the effects of kinematic sweeping. We use the Belinicher–L'vov (BL) transformation [67] in which the field $\mathbf{u}(\mathbf{r}_0, t_0 | \mathbf{r}, t)$ is defined through the Eulerian velocity $\mathbf{u}(\mathbf{r}, t)$:

$$\mathbf{u}(\mathbf{r}_0, t_0 | \mathbf{r}, t) \equiv \mathbf{u}[\mathbf{r} + \boldsymbol{\rho}(\mathbf{r}_0, t), t], \quad \text{where} \quad \boldsymbol{\rho}(\mathbf{r}_0, t) = \int_{t_0}^{t} \mathbf{u}[\mathbf{r}_0 + \boldsymbol{\rho}(\mathbf{r}_0, s), s] ds.$$

Note that $\boldsymbol{\rho}(\mathbf{r}_0, t)$ is precisely the Lagrangian trajectory of a fluid particle that is positioned at $\mathbf{r}_0$ at time $t = t_0$. The field $\mathbf{u}(\mathbf{r}_0, t_0 | \mathbf{r}, t)$ is simply the Eulerian field in the frame of reference of a single chosen material point $\boldsymbol{\rho}(\mathbf{r}_0, t)$. Next, define the field of velocity differences $W(\mathbf{r}_0, t_0 | \mathbf{r}, \mathbf{r}', t)$:

$$W(\mathbf{r}_0, t_0 | \mathbf{r}, \mathbf{r}', t) \equiv \mathbf{u}(\mathbf{r}_0, t_0 | \mathbf{r}, t) - \mathbf{u}(\mathbf{r}_0, t_0 | \mathbf{r}', t).$$

It was shown that the equation of motion of this field is independent of t_0, and we omit this label altogether. The fundamental statistical quantities in our study are the many-time, many-point, "fully unfused" n-rank-tensor correlation function of the BL-velocity differences $W_j \equiv W_j(\mathbf{r}_0 | \mathbf{r}_j, \mathbf{r}'_j, t_j)$. To simplify the notation we choose the following short hand notation:

$$X_j \equiv \{\mathbf{r}_j, \mathbf{r}'_j, t_j\}, \quad x_j \equiv \{\mathbf{r}_j, t_j\}, \quad W_j \equiv W(X_j), \quad F_n(\mathbf{r}_0 | X_1, \ldots, X_n) = \langle W_1, \ldots, W_n \rangle.$$

By the term "fully unfused" we mean that all the coordinates are distinct and all the separations between them lie in the inertial range. We have

$$X_j \equiv \{\mathbf{r}_j, \mathbf{r}'_j, t_j\}, \quad x_j \equiv \{\mathbf{r}_j, t_j\}, \quad W_j \equiv W(X_j),$$
$$F_n(\mathbf{r}_0 | \mathbf{r}_1, \mathbf{r}'_1, t_1; \mathbf{r}_2, \mathbf{r}'_2, t_2; \ldots; \mathbf{r}_n, \mathbf{r}'_n, t_n) = F_n(\mathbf{r}_0 | X_1, \ldots, X_n) = \langle W_1, \ldots, W_n \rangle$$

For example, the 2nd order correlation function written explicitly is

$$F_2^{\alpha\beta}(\mathbf{r}_0 | \mathbf{r}_1, \mathbf{r}'_1, t_1; \mathbf{r}_2, \mathbf{r}'_2, t_2) = \langle W^\alpha(\mathbf{r}_0 | \mathbf{r}_1, \mathbf{r}'_1, t_1) W^\beta(\mathbf{r}_0 | \mathbf{r}_2, \mathbf{r}'_2, t_2) \rangle.$$

In addition to the n-th order correlation functions the statistical theory calls for the introduction of a similar array of response or Green's functions. The most familiar is the 2nd order Green's function $G^{\alpha\beta}(\mathbf{r}_0 | X_1; x_2)$ defined by the functional derivative

$$G^{\alpha\beta}(\mathbf{r}_0 | X_1; x_2) = \left\langle \frac{\delta W^\alpha(\mathbf{r}_0 | X_1)}{\delta \varphi^\beta(\mathbf{r}_0 | X_2)} \right\rangle. \tag{3.55}$$

In stationary turbulence these quantities depend on $t_1 - t_2$ only, and it may be denoted this time difference as t. The simultaneous correlation function T_n is obtained from F_n when $t_1 = t_2 = \cdots = t_n$. In this limit one can use differences of Eulerian velocities, $\mathbf{w}(\mathbf{r}, \mathbf{r}', t) \equiv \mathbf{u}(\mathbf{r}', t) - \mathbf{u}(\mathbf{r}, t)$ instead of BL-differences, the result is the same. In statistically stationary turbulence the equal time correlation function $T_n(\mathbf{r}_1, \mathbf{r}'_1; \mathbf{r}_2, \mathbf{r}'_2; \ldots; \mathbf{r}_n, \mathbf{r}'_n)$ is time independent, and we denote it as

$$T_n(\mathbf{r}_1, \mathbf{r}'_1; \ldots; \mathbf{r}_n, \mathbf{r}'_n) = \langle \mathbf{w}(\mathbf{r}_1, \mathbf{r}'_1, t) \ldots \mathbf{w}(\mathbf{r}_n, \mathbf{r}'_n, t) \rangle.$$

One expects that when all the separations $R_i \equiv |\mathbf{r}_i - \mathbf{r}'_i|$ are in the inertial range, $\eta \ll R_i \ll L$, the simultaneous correlation function is scale invariant in the sense that

$$T_n(\lambda\mathbf{r}_1, \lambda\mathbf{r}'_1; \ldots; \lambda\mathbf{r}_n, \lambda\mathbf{r}'_n) = \lambda^{\zeta_n} T_n(\mathbf{r}_1, \mathbf{r}'_1; \ldots; \mathbf{r}_n, \mathbf{r}'_n).$$

The goal of our work is: *the calculation of the exponents ζ_n from first principles. This is first aim of the statistical theory of turbulence.* One could assume that also the time correlation functions F_n are homogeneous functions of their arguments, including the time arguments. It should be stressed that this is not the case, and that in the context of turbulence, when the exponents ζ_n are anomalous, dynamical scaling is broken. The time correlation functions are "multiscaling" in the time variables. In [70] it was shown that the scaling properties of the time correlation functions can be parameterized conveniently with the help of one scalar function $L(h)$. To understand this presentation, consider first the simultaneous function $T_n(\mathbf{r}_1, \mathbf{r}'_1, \ldots, \mathbf{r}_n, \mathbf{r}'_n)$. Following the ideas of multifractals [36, 46] the simultaneous function can be represented as

$$T_n(\mathbf{r}_1, \mathbf{r}'_1, \ldots, \mathbf{r}_n, \mathbf{r}'_n) = U^n \int_{h_{\min}}^{h_{\max}} (R_n/L)^{nh+L(h)} \tilde{T}_{n,h}(\boldsymbol{\rho}_1, \boldsymbol{\rho}'_1, \ldots, \boldsymbol{\rho}_n, \boldsymbol{\rho}'_n) d\mu(h), \quad (3.56)$$

where U is a typical velocity scale, and Greek coordinates stand for dimensionless (rescaled) coordinates, i.e., $\boldsymbol{\rho}_j = \mathbf{r}_j/R_n$, $\boldsymbol{\rho}'_j = \mathbf{r}'_j/R_n$. In (3.56) there is introduced the "typical scale of separation" of the set of coordinates

$$R_n^2 = \frac{1}{n} \sum_{j=1}^{n} |\mathbf{r}_j - \mathbf{r}'_j|^2.$$

At this point L is an undetermined renormalization scale. At the end of the calculation we will find that it is the outer scale of turbulence. The function $L(h)$ is defined as $L(h) \equiv 3 - D(h)$. The function $L(h)$ is related to the *scaling exponents* ζ_n via the saddle point requirement

$$\zeta_n = \min_h [nh + L(h)]. \quad (3.57)$$

This identification stems from the fact that the integral in (3.56) is computed in the limit $R_n/L \to 0$ via the steepest descent method. Neglecting logarithmic corrections one finds that $T_n \sim R_n^{\zeta_n}$. The physical intuition behind the representation (3.56) is that there are velocity field configurations that are characterized by different scaling exponents h. For different orders n the main contribution comes from that value of h that determines the position of the saddle point in the integral (3.56). It is convenient to introduce a dimensional quantity $T_{n,h}$ according to

$$T_{n,h}(\mathbf{r}_1, \mathbf{r}'_1, \ldots, \mathbf{r}_n, \mathbf{r}'_n) = U^n(R_n/L)^{nh+L(h)} \tilde{T}_{n,h}(\boldsymbol{\rho}_1, \boldsymbol{\rho}'_1, \ldots, \boldsymbol{\rho}_n, \boldsymbol{\rho}'_n).$$

Dimensional quantities of this type will play an important role in our theory. Especially their rescaling properties will be used to organize a L-covariant theory. This quantity is rescaled like

$$T_{n,h}(\lambda\mathbf{r}_1, \lambda\mathbf{r}'_1, \ldots, \lambda\mathbf{r}_n, \lambda\mathbf{r}'_n) = \lambda^{nh+L(h)} T_{n,h}(\mathbf{r}_1, \mathbf{r}'_1, \ldots, \mathbf{r}_n, \mathbf{r}'_n).$$

Below we'll use a shorthand notation $T_{n,h} \to \lambda^{nh+L(h)} T_{n,h}$ for such rescaling laws.

The intuition behind the representation (3.56) is extended to the time domain. The particular velocity configurations that are characterized by an exponent h also display a typical time scale $t_{R,h}$ written as $t_{R,h} \sim (R/U)(L/R)^h$. Accordingly we chose a temporal multiscaling representation for the time dependent function

$$F_n(\mathbf{r}_0|X_1, \ldots, X_n) = U^n \int_{h_{\min}}^{h_{\max}} (R_n/L)^{nh-L(h)} \tilde{F}_{n,h}(\mathbf{r}_0|\Xi_1, \ldots, \Xi_n) d\mu(h), \quad (3.58)$$

where $\tilde{F}_{n,h}$ depends on the dimensionless (rescaled) space and time coordinates $\Xi_j = (\boldsymbol{\rho}_j, \boldsymbol{\rho}'_j, \tau_j)$, $\tau_j = t_j / t_{R_n,h}$. As before we introduce the related dimensional quantity $F_{n,h}$ according to

$$F_n(\mathbf{r}_0 | X_1, \ldots, X_n) = U^n (R_n/L)^{nh - L(h)} \tilde{F}_{n,h}(\mathbf{r}_0 | \Xi_1, \ldots, \Xi_n).$$

We require that $\tilde{F}_{n,h}(\mathbf{r}_0 | \Xi_1, \ldots, \Xi_n)$ is identical to the function $\tilde{T}_{n,h}(\boldsymbol{\rho}_1, \boldsymbol{\rho}'_1, \ldots, \boldsymbol{\rho}_n, \boldsymbol{\rho}'_n)$ when its rescaled time arguments are all the same. This representation reproduces all the scaling laws that were found for time integrations and differentiations [70]. When the multiscaling representation of turbulent structure functions was introduced for the first time ([36] and references therein) it was a phenomenological idea, that could be understood as the inversion of the data on ζ_n; see (3.57). It will be shown below that in the context of our theory it appears as a result of an exact symmetry of the equations of motion. For our purposes it turns out easier to compute the function $L(h)$ than to compute the exponents ζ_n directly. In [8] it was shown that the n-th order correlation functions satisfy, in the limit of infinite Reynolds number, an exact hierarchy of equation:

$$\partial_{t_1} F_n(\mathbf{r}_0 | X_1, X_2, \ldots, X_n) + \int F_{n+1}\big(\mathbf{r}_0 | \tilde{X}, \tilde{X}, X_2, \ldots, X_n\big) \gamma(\mathbf{r}_1, \mathbf{r}_2, \tilde{\mathbf{r}}) d\tilde{\mathbf{r}} = 0. \qquad (3.59)$$

The *vertex function* $\gamma(\mathbf{r}_1, \mathbf{r}'_1, \tilde{\mathbf{r}}) \equiv \gamma^{\alpha\mu\sigma}(\mathbf{r}_1, \mathbf{r}'_1, \tilde{\mathbf{r}})$ is defined as

$$\gamma^{\alpha\mu\sigma}(\mathbf{r}_1, \mathbf{r}'_1, \tilde{\mathbf{r}}) = \frac{1}{2}\{[P^{\alpha\mu}(\mathbf{r}_1 - \tilde{\mathbf{r}}) - P^{\alpha\mu}(\mathbf{r}'_1 - \tilde{\mathbf{r}})]\partial_{\tilde{r}_\sigma} + [P^{\alpha\sigma}(\mathbf{r}_1 - \tilde{\mathbf{r}}) - P^{\alpha\sigma}(\mathbf{r}'_1 - \tilde{\mathbf{r}})]\partial_{\tilde{r}_\mu}\}.$$

The *projection operator* $P^{\alpha\beta}$ was defined in (3.32). At this point that the equations of motion do not contain a dissipative term since we choose to deal with fully unfused correlation functions. For such quantities the viscous term becomes negligible in the limit of Reynolds number Re $\to \infty$ or the viscosity $\nu \to 0$. This is the main advantage of working with unfused correlation functions; the more commonly used structure functions do not enjoy this property, and in their equations of motion the viscous term remain relevant also in the limit Re $\to \infty$. The price of working with unfused correlation functions is that we have to deal with functions of many variables.

We'll consider systematic method of solving this chain of equations. We begin with the introduction of the central idea that this chain of equations can solved on an "h-slice."

(a) **Correlation functions and the rescaling group.** Examining (3.59) one realized that they are linear in the set of correlation functions F_n with $n = 2, 3, \ldots$ in light if the representation (3.58) of F_n in terms of $F_{n,h}$. Rewrite (3.59) in the form

$$\int_{h_{min}}^{h_{max}} \left\{ \partial_{t_1} F_{n,h}(\mathbf{r}_0 | X_1, \ldots, X_n) + \int \gamma(\mathbf{r}_1, \mathbf{r}_2, \tilde{\mathbf{r}}) F_{n,h}\big(\mathbf{r}_0 | \tilde{X}, \tilde{X}, X_2, \ldots, X_n\big) d\tilde{\mathbf{r}} \right\} d\mu(h) = 0.$$

Since we integrate over a positive measure, the equations are satisfied only if the terms in curly parentheses vanish. In other words, we will seek solutions to the equations

$$\partial_{t_1} F_{n,h}(\mathbf{r}_0 | X_1, \ldots, X_n) + \int \gamma(\mathbf{r}_1, \mathbf{r}_2, \tilde{\mathbf{r}}) F_{n,h}\big(\mathbf{r}_0 | \tilde{X}, \tilde{X}, X_2, \ldots, X_n\big) d\tilde{\mathbf{r}} = 0. \qquad (3.60)$$

We refer to these equations as the equations on an "h-slice." It is important to analyze their properties under rescaling. To this aim recall that the Euler equation is invariant to rescaling according to $\mathbf{r} \to \lambda \mathbf{r}$, $t \to \lambda^{1-h} t$, $\mathbf{u} \to \lambda^h \mathbf{u}$, for any value λ and h. On the basis of this alone one could guess that Equations (3.60) are invariant under the rescaling group

$\mathbf{r}_i \to \lambda \mathbf{r}_i$, $t_i \to \lambda^{1-h} t_i$ and $F_{n,h} \to \lambda^{nh} F_{n,h}$. In fact, (3.60) are invariant to a broader rescaling group, i.e.,

$$\mathbf{r}_i \to \lambda \mathbf{r}_i, \quad t_i \to \lambda^{1-h} t_i, \quad F_{n,h} \to \lambda^{nh+L(h)} F_{n,h} \tag{3.61}$$

with n-th independent function $L(h)$. This extra freedom is an exact result of the structure of (3.60). It is worthwhile to reiterate that this function appeared originally in the phenomenology of turbulence as an ad hoc model of multifractal turbulence. At this point it appears as a nontrivial and *exact* property of the chain of equations of the statistical theory of turbulence. We will show that the preservation of this rescaling symmetry will lead to a theory in which power counting plays no role. As a consequence the information about the scaling exponents ζ_n is obtainable only from the solvability conditions of this equation. In other words, the information is buried in coefficients rather than in power counting. The spatial derivative in the vertex on the RHS brings down the unknown function $L(h)$, and its calculation will be an integral part of the computation of the exponent. It will serve the role of a generalized eigenvalue of the theory. Of course, we cannot consider the hierarchy of (3.60) in its entirety. We need to find ways to close this equation, and this is the main subject of Section 3.2.3. The main idea in choosing an appropriate closure is to preserve the essential rescaling symmetry of the problem. We will argue below that there are many different possible closures, but most of them do not preserve this rescaling symmetry. We will introduce the notion of L-covariance, and demand that the closure does not destroy the h-slice rescaling symmetry (3.61).

(b) **Temporal multiscaling in the Green's functions**. In addition to the n-order correlation functions the statistical theory calls for the introduction of a similar array of nonlinear response or Green's functions $G_{m,n}$. These represent the response of the direct product of m BL-velocity differences to n perturbations. In particular,

$$G_{2,1}(\mathbf{r}_0|X_1, X_2, x_3) = \Big\langle \frac{\delta[W(\mathbf{r}_0|X_1)W(\mathbf{r}_0|X_2)]}{\delta\varphi(\mathbf{r}_0|x_3)} \Big\rangle,$$

$$G_{3,1}(\mathbf{r}_0|X_1, X_2, X_3, x_4) = \Big\langle \frac{\delta[W(\mathbf{r}_0|X_1)W(\mathbf{r}_0|X_2)W(\mathbf{r}_0|X_3)]}{\delta\varphi(\mathbf{r}_0|x_4)} \Big\rangle.$$

The Green's function G of (3.55) is $G_{1,1}$ in this notation. The set of Green's functions $G_{n,1}$ satisfies a hierarchy of equations that in the limit of infinite Reynolds number is written as

$$\partial_{t_1} G_{n,1}^{\alpha\beta\cdots}(\mathbf{r}_0|X_1,\ldots,X_n;x_{n+1}) + \int \gamma^{\alpha\mu\sigma}(\mathbf{r}_1,\mathbf{r}_1',\tilde{\mathbf{r}}) G_{n+1,1}^{\mu\sigma\beta\gamma\cdots}(\mathbf{r}_0|\tilde{X}_1,\tilde{X}_1,X_2,\ldots X_n;x_{n+1}) d\tilde{\mathbf{r}}$$

$$= G_{n,1}^{(0)\alpha\beta\cdots}(\mathbf{r}_0|X_1,\ldots,X_n;\mathbf{r}_{n+1},t_1+0)\delta(t_1 - t_{n+1}). \tag{3.62}$$

The bare Green's functions of $(n,1)$ on the RHS of this equation are the following decomposition:

$$G_{n,1}^{(0)\alpha\beta\cdots\psi\omega}(\mathbf{r}_0|X_1,\ldots X_n;\mathbf{r}_{n+1},t_1+0) \equiv G^{(0)\alpha\omega}(\mathbf{r}_1,\mathbf{r}_1',\mathbf{r}_{n+1}) F_{n-1}^{\beta\gamma\cdots\psi}(\mathbf{r}_0|X_2,\ldots X_{n-1}). \tag{3.63}$$

These functions serve as the initial conditions for (3.62) at times $t_{n+1} = t_1$. The form of these equations is very close to the hierarchy satisfied by the correlation functions, and it is advantageous to use a similar temporal multiscaling representation for the nonlinear Green's functions:

$$G_{n,1}(\mathbf{r}_0|X_1,\ldots,X_n;x_{n+1}) = \int_{h_{\min}}^{h_{\max}} G_{n,1,h}(\mathbf{r}_0|X_1,\ldots,X_n;x_{n+1})\, d\mu(h). \tag{3.64}$$

By the rescaling symmetry of the Euler equation we could guess the rescaling properties

$G_{n,1,h} \to \lambda^{(n-1)h-3}G_{n,1,h}$. As before, the equations support a broader rescaling symmetry group,

$$\mathbf{r} \to \lambda\mathbf{r}, \quad t \to \lambda^{1-h}t, \quad G_{n,1,h} \to \lambda^{(n-1)h-3+L_G(h)}G_{n,1,h}. \qquad (3.65)$$

In fact the choice of the scalar function $L_G(h)$ is constrained by the initial conditions on the Green's functions. From (3.63) we learn that the Green's functions depend on $L_G(h)$ via the appearance of the correlation functions. This means, that $L(h)$ and $L(h)$ are related. A simple calculation leads to the conclusions that $L_G(h) = L(h)$.

We displayed explicitly only the hierarchy of equations satisfied by $G_{n,1}$. Similar hierarchies can be derived for any family of higher order Green's function $G_{n,m}$ with $m = 2, 3, \dots$. After introducing the multiscaling representation, consider the corresponding Green's functions on an "h-slice", $G_{n,m,h}$ and show that the equations they satisfy have the rescaling symmetry of the Euler equation with a $L(h)$ freedom. In all these equations the initial value terms force the scalar function $L(h)$ to be m-independent. Faced with infinite chains of equations, one needs to truncate at a certain order. Simply by truncating one obtains a set of equations which is not closed upon it self. It is then customary to express the highest order quantities in the truncated set of equations in terms of lower order quantities. This turns the set of equation into a nonlinear set. Such an approach is not guaranteed to preserve the essential rescaling symmetries (3.62) and (3.65) of the equations. In this section we develop a systematic method to close the hierarchies of equations for correlation and Green's functions on an "h-slice" in a way that preserves the rescaling symmetry. The lowest order closure involves five equations on an "h-slice." The first two belong to the F_n hierarchy:

$$\partial_{t_1} F_{2,h}(X_1, X_2) + \int \gamma(\mathbf{r}_1, \mathbf{r}_2, \tilde{\mathbf{r}})F_{2,h}(\tilde{X}, \tilde{X}, X_2) \, d\tilde{\mathbf{r}} = 0, \qquad (3.66a)$$

$$\partial_{t_1} F_{3,h}(X_1, X_2, X_3) + \int \gamma(\mathbf{r}_1, \mathbf{r}_2, \tilde{\mathbf{r}})F_{4,h}(\tilde{X}, \tilde{X}, X_2, X_3) \, d\tilde{\mathbf{r}} = 0. \qquad (3.66b)$$

The next pair of equations belongs to the hierarchy of $G_{n,1}$. Using representation (3.62) in (3.64) we derive

$$\partial_{t_1} G_{1,1,h}(X_1; x_2) + \int \gamma(\mathbf{r}_1, \mathbf{r}_2, \tilde{\mathbf{r}})G_{2,1,h}(\tilde{X}, \tilde{X}; x_2) \, d\tilde{\mathbf{r}} = G_h^{(0)}(X_1; x_2)\delta(t_1 - t_2), \quad (3.67a)$$

$$\partial_{t_1} G_{2,1,h}(X_1, X_2; x_3) + \int \gamma(\mathbf{r}_1, \mathbf{r}_2, \tilde{\mathbf{r}})G_{3,1,h}(\tilde{X}, \tilde{X}, X_2; x_3) \, d\tilde{\mathbf{r}} = 0. \quad (3.67b)$$

Here $G_h^{(0)}$ stands for the bare Green's function on an "h-slice." The fifth needed equation is the first equation from the third hierarchy of Green's function $G_{n,2,h}$:

$$\partial_{t_1} G_{1,2,h}(X_1; x_2; x_3) + \int \gamma(\mathbf{r}_1, \mathbf{r}_2, \tilde{\mathbf{r}})G_{2,2,h}(\tilde{X}, \tilde{X}, x_2; x_3) \, d\tilde{\mathbf{r}} = 0. \qquad (3.68)$$

3.4 Bifurcations for the Kuramoto–Sivashinsky equation

The problems of computing the Hausdorff and entropy dimensions of attractors occupy an important place in the study of partial differential equations. Of particular importance there is the entropy dimension, since for it the Whitney easy theorem on one-to-one projection is fulfilled and, in addition, in numerical calculation this dimension is usually calculated. The equation, which will be considered below, first arose in [95], devoted to wave flows of liquid flowing along the vertical plane, and also in [61], where diffusion chaos in the reaction systems was studied; hence the name of the Kuramoto–Sivashinsky equation. Later, this equation was intensively studied, both numerically and analytically. In particular, the existence of an attractor was proved and its dimension was estimated.

Definition 53 The *orthogonal group of degree* n over a field F (written as $O(n,F)$) is the group of $n \times n$ orthogonal matrices with entries from F, with the group operation that of matrix multiplication.

This is a subgroup of the general linear group $GL(n,F)$ given by

$$O(n,F) = \{Q \in GL(n,F): \ Q^TQ = QQ^T = I\},$$

where Q^T is the transpose of Q. The classical orthogonal group over the real numbers is usually just written $O(n)$. More generally the orthogonal group of a nonsingular quadratic form over F is the group of matrices preserving the form. The Cartan–Dieudonne theorem describes the structure of the orthogonal group. Every orthogonal matrix has determinant either 1 or -1.

Definition 54 The orthogonal $n \times n$ matrices with determinant 1 form a normal subgroup of $O(n,F)$ known as the *special orthogonal group* $SO(n,F)$.

If the characteristic of F is 2, then 1=-1, hence $O(n,F)$ and $SO(n,F)$ coincide; otherwise the index of $SO(n,F)$ in $O(n,F)$ is 2. In characteristic 2 and even dimension, many authors define the $SO(n,F)$ differently as the kernel of the Dickson invariant; then it usually has index 2 in $O(n,F)$. Both $O(n,F)$ and $SO(n,F)$ are algebraic groups, because the condition that a matrix be orthogonal, i.e., has its own transpose as inverse, can be expressed as a set of polynomial equations in the entries of the matrix.

Note that $O(n)$ and $SO(n)$ are real compact Lie groups of dimension $n(n-1)/2$; $O(n)$ has two connected components, with $SO(n)$ being the identity component, i.e., the connected component containing the identity matrix. The real orthogonal and real special orthogonal groups have the following geometric interpretations: $O(n)$ is a subgroup of the Euclidean group $E(n)$, the group of isometries of $\mathbb{R}^n$; it contains those which leave the origin fixed. It is the symmetry group of the sphere ($n = 3$) or hypersphere and all objects with spherical symmetry, if the origin is chosen at the center. $SO(n)$ is a subgroup of $E^+(n)$, which consists of *direct* isometries, i.e., isometries preserving orientation; it contains those which leave the origin fixed. It is the rotation group of the sphere and all objects with spherical symmetry, if the origin is chosen at the center. $\{I, -I\}$ is a normal subgroup and even a characteristic subgroup of $O(n)$, and, if n is even, also of $SO(n)$. If n is odd, $O(n)$ is the direct product of $SO(n)$ and $\{I, -I\}$. The cyclic group of k-fold rotations C_k is for every positive integer k a normal subgroup of $O(n)$ and $SO(n)$. Relative to suitable orthogonal bases, the isometries are of the form:

$$\begin{pmatrix} R_1 & 0 & \cdots & \cdots & \cdots & 0 \\ 0 & \ddots & \ddots & \cdots & \cdots & \vdots \\ \vdots & \ddots & R_k & 0 & \cdots & 0 \\ \vdots & \cdots & 0 & \pm1 & \ddots & \vdots \\ \vdots & \cdots & \cdots & \ddots & \ddots & 0 \\ 0 & 0 & \cdots & \cdots & 0 & \pm1 \end{pmatrix}, \quad \text{where } R_1,\ldots,R_k \text{ are } 2 \times 2 \text{ rotation matrices.}$$

Definition 55 The symmetry group of a circle, $O(2)$, is called a *Dihedral group*, $\mathrm{Dih}(S^1)$, where S^1 denotes the multiplicative group of complex numbers of absolute value 1.

$SO(2)$ is isomorphic (as a *Lie group*) to the circle S^1 (circle group). This isomorphism sends the complex number $\exp(i\varphi) = \cos\varphi + i\sin\varphi$ to the orthogonal matrix $\begin{pmatrix} \cos\varphi & -\sin\varphi \\ \sin\varphi & \cos\varphi \end{pmatrix}$. The group $SO(3)$, understood as the set of rotations of 3-dimensional space, is of major importance in the sciences and engineering (see rotation group and the general formula for a 3-by-3 rotation matrix in terms of the axis and the angle). In terms of algebraic topology, for $n > 2$ the fundamental group of $SO(n)$ is cyclic of order 2, and the spinor group $\text{Spin}(n)$ is its universal cover. For $n = 2$, the fundamental group is infinite cyclic and the universal cover corresponds to the real line. The Lie algebra associated to the Lie groups $O(n)$ and $SO(n)$ consists of the skew-symmetric real $n \times n$ matrices, with the Lie bracket given by the commutator. This Lie algebra is denoted by $o(n)$ or $so(n)$.

3.4.1 Symmetry: translations, reflections, and $O(2)$-equivariance

Symmetry is a central and powerful concept in dynamics. It allows one to reduce the complexity, and often the dimension, of a system. Symmetries have profound implications for the robustness of structures in system under perturbation. A given solution, e.g., a homoclinic connection, may be destroyed by the smallest general perturbation: the effect of weak damping on the energy-conserving saddle separatrix loop in the classical pendulum equation. If only perturbations consistent with the symmetries of the system are allowed (reversibility, for the pendulum), the same solution might be structurally stable and hence survive such perturbations.

We'll discuss how the group $O(2)$ acts on a set of Fourier modes. Let a scalar variable $w(x,t)$ be expanded in a Fourier series $w(x,t) = \sum_j z_j(t)e^{2\pi ijx/L}$. For translations and reflections we have

$$T_\alpha w(x) = w(x+\alpha) = \sum_j z_j e^{2\pi ij(x+\alpha)/L} = \sum_j \left(z_j e^{2\pi ij\alpha/L}\right)e^{2\pi ijx/L},$$

$$R_f w(x) = w(-x) = \sum_j z_j e^{2\pi ij(-x)/L} = \sum_j z_{-j} e^{2\pi ijx/L}.$$

Remembering that $w(x,t)$ is generally real, forcing $z_j^* = z_{-j}$, we conclude that translation and reflection induces the following actions of the coefficients of the Fourier modes:

$$T_\alpha : z_j \to e^{2\pi ij\alpha/L} z_j, \quad R_f : z_j \to z_j^*. \tag{3.69}$$

So, T_α is a rotation in the complex plane and R_f is a reflection across the real axis. For z_1, this is precisely the standard action of $O(2)$ on $\mathbb{C}$, the space where the coefficients live. Every member of $O(2)$ can be constructed out of a finite combination of these transformations. The T_α together with R_f generate $O(2)$. For general j, the only difference is that elements are rotated by angle $j\alpha$, but this amounts to a simple rescaling. Thus, $O(2)$ represents the induced action on the phase-space $\mathbb{C}^n$ by the transformations (3.69) for any n. In summary, if we start with translation invariant system and project it onto Fourier basis, the resulting system of ODEs for the evolution of the coefficients will exhibit $O(2)$ symmetry. We'll take a closer look at consequences forced by $O(2)$ symmetry on a system.

Definition 56 Given a group of transformations Γ of a space X (e.g., $X = \mathbb{R}^n$), we say that a function $g : X \to Y$ (where Y is some space, e.g., $X = Y$ or $X = \mathbb{R}$) is Γ-*invariant* if $g(\mathbf{x}) = g(\gamma(\mathbf{x}))$, $\forall \gamma \in \Gamma$. A function $f : X \to X$ is called Γ-*equivariant* if $\gamma(f(\mathbf{x})) = f(\gamma(\mathbf{x}))$ ($\gamma \in \Gamma$). Similarly, a system

$$\dot{\mathbf{x}} = f(\mathbf{x}), \quad \mathbf{x} \in \mathbb{R}^n \tag{3.70}$$

of ODEs is called Γ-*invariant* if the vector field f is equivariant, i.e.,

$$\gamma(\dot{\mathbf{x}}) = \gamma(f(\mathbf{x})) = f(\gamma(\mathbf{x})), \quad \forall \gamma \in \Gamma. \tag{3.71}$$

The key implication of (3.71) is that, if $\hat{\mathbf{x}}(t)$ is a particular solution of (3.70), then so is $\gamma\,\hat{\mathbf{x}}(t)$ for any $\gamma \in \Gamma$: each solution belongs to a whole group orbit of solutions.

In general, equivariant (scalar- or vector-valued) functions are built out of *pure* equivariant elements multiplied by invariant scalar-valued functions. For example, consider the group D_n, the group of rotation and reflection symmetries of an n-gon in the plane. Identifying the plane $\mathbb{R}^2$ with the complex plane $\mathbb{C}$, for notating convenience, the D_n-invariant functions (under standard action on $\mathbb{C}$) are of the form $g(z) = p(u,v)z + q(u,v)(z^*)^{n-1}$, with $u = zz^*$ and $v = z^n + (z^*)^n$. A simple calculation shows that $(z^*)^{n-1}$ and z are equivariant, and that u and v are invariant. Thus the scalar functions $p(u,v)$ and $q(u,v)$ are invariant and the totality, $g(z)$, is equivariant. As already noted, we are primary interested in the implications of $O(2)$ symmetry. In anticipation of the types of equations we'll extract later, we consider a set of $O(2)$-equivariant ODE of the form

$$\dot{z}_1 = f_1(z_1, z_2), \quad \dot{z}_2 = f_2(z_1, z_2), \quad z_1, z_2 \in \mathbb{C},$$

i.e., the most general form of an $O(2)$-equivariant ODE on $\mathbb{C}^2$. In this case the invariant functions turn out depend only on the three combinations $|z_1|^2$, $|z_2|^2$ and $z_1^2 z_2^* + (z_1^*)^2 z_2$, and the equivariant vector fields are (z_1, z_2) and $(z_1^* z_2, z_1^2)$. Hence, up to third order terms, we have the normal form

$$\begin{cases} \dot{z}_1 = \mu_1 z_1 + c_{12} z_1^2 z_2 + (d_{11}|z_1|^2 + d_{12}|z_2|^2)z_1, \\ \dot{z}_2 = \mu_2 z_2 + c_{21} z_1 z_2^2 + (d_{21}|z_1|^2 + d_{22}|z_2|^2)z_2, \end{cases}$$

where all coefficients are purely real. Assuming that $c_{12}, c_{11} \neq 0$ we can further rescale, reversing the direction of time if necessary, to obtain

$$\begin{cases} \dot{z}_1 = z_1^* z_2 + (\mu_1 + e_{11}|z_1|^2 + e_{12}|z_2|^2)z_1, \\ \dot{z}_2 = \pm z_1^2 + (\mu_2 + e_{21}|z_1|^2 + e_{22}|z_2|^2)z_2, \end{cases} \tag{3.72}$$

where $e_{11} = d_{11}/|c_{11}c_{12}|$, $e_{12} = d_{12}/c_{12}^2$, $e_{21} = d_{21}/|c_{11}c_{12}|$ and $e_{22} = d_{22}/c_{12}^2$. Through these rescalings, we reduce the complexity, but there still remain several cases depending on the signs of various coefficients. We will analyze only relevant to the boundary layer model in the heteroclinic cycle range. Then, the quadratic term in the second equation of (3.72) is $-z_1^2$, and the parameters satisfy $e_{ij} < 0 < \mu_j$, and $\mu_1 - \mu_2(e_{12}/e_{22}) - \sqrt{-\mu_2/e_{22}} < 0 < \mu_1 - \mu_2(e_{12}/e_{22}) + \sqrt{-\mu_2/e_{22}}$. Thus we have the following.

Proposition 29 For $e_{ij} < 0 < \mu_j$ and $\mu_1 - \mu_2(e_{12}/e_{22}) - \sqrt{-\mu_2/e_{22}} < 0 < \mu_1 - \mu_2(e_{12}/e_{22}) + \sqrt{-\mu_2/e_{22}}$, the system (3.72) has a circle of equilibrium connected by heteroclinic cycles. If $\mu_1 - \mu_2 e_{12}/e_{22} < 0$ the cycle is attracting.

Proof. First, observe that the z_2-plane, defined by $z_1 = 0$, is invariant. Seeking equilibrium in this plane and remembering that $e_{ij} < 0 < \mu_j$, we find a circle (group orbit) of fixed points defined by $|z_2| = \sqrt{-\mu_2/e_{22}}$. Set $z_j = x_j + iy_j$ and consider (3.72) as a system on $\mathbb{R}^4$ with $(z_1, z_2) = (x_1, y_1, x_2, y_2)$. We obtain

$$\begin{cases} \dot{x}_1 = x_1 x_2 + y_1 y_2 + x_1(\mu_1 + e_{11}r_1^2 + e_{12}r_2^2), \\ \dot{y}_1 = x_1 y_2 - y_1 x_2 + y_1(\mu_1 + e_{11}r_1^2 + e_{12}r_2^2), \\ \dot{x}_2 = -(x_1^2 - y_1^2) + x_2(\mu_2 + e_{21}r_1^2 + e_{22}r_2^2), \\ \dot{y}_2 = -2x_1 y_1 + y_2(\mu_2 + e_{21}r_1^2 + e_{22}r_2^2), \end{cases} \tag{3.73}$$

where $r_j^2 = x_j^2 + y_j^2$. Linearizing about $(0, 0, \sqrt{-\mu_2/e_{22}}, 0)$, we get the diagonal matrix

$$\begin{pmatrix} \lambda_+ & 0 & 0 & 0 \\ 0 & \lambda_- & 0 & 0 \\ 0 & 0 & -2\mu_2 & 0 \\ 0 & 0 & 0 & 0 \end{pmatrix}, \quad \text{where } \lambda_\pm = \mu_1 - (\mu_2 e_{12}/e_{22}) \pm \sqrt{-\mu_2/e_{22}}.$$

Since these equations are equivariant under rotation about the origin, this structure will be preserved around the circle of equilibrium. Due to our assumptions on the parameters, we have two-dimensional stable and one-dimensional unstable manifold emanating from each point of the circle. By (3.73), one watches that the real subspace, Span$\{(1, 0, 0, 0), (0, 0, 1, 0\}$, is invariant. For the moment, we focus on the two points where the circle of fixed points intersects this invariant plane, namely $A = (0, 0, \sqrt{-\mu_2/e_{22}}, 0)$ and $B = (0, 0, -\sqrt{-\mu_2/e_{22}}, 0)$. Looking at the equations as expressed in (3.73) and setting $y_1 = y_2 = 0$, we see that A is a saddle, and B is a sink. Orbits leaving A are trapped in a bounded *disc*, and since there are no equilibrium or limit sets in the half spaces $\{x_1 > 0\}$ and $\{x_1 < 0\}$, they must limit on B. Thus we have a heteroclinic orbit from A to B. By judicious application of the $O(2)$ symmetry given by transformation (3.69), we complete the picture. The application of translation $T_{\pi/2}$ amounts to a rotation by $\pi/2$ in z_1-plane, and by π in z_2-plane. In cartesian coordinates, this gives $T_{\pi/2} : (x_1, y_1, x_2, y_2) \to (y_1, -x_1, -x_2, -y_2)$. Using of $T_{\pi/2}$ permutes the connections from A to B, giving a connection from B to A in the plane defined by $\{y_2 = x_1 = 0\}$. Thus, we establish the presence of a heteroclinic cycle containing A and B. Application of T_α for arbitrary α shows that every diametrically opposed pair of equilibrium points on the circle $|z_2| = \sqrt{-\mu_2/e_{22}}$ is connected in such a heteroclinic cycle. Determining the asymptotic stability of this connection is a subtle process. It turns out that the stability is determined solely by the ratio of λ_-/λ_+: If $-\lambda_-/\lambda_+ > 1$, the heteroclinic cycle is attracting while if $-\lambda_-/\lambda_+ < 1$ it is unstable. $\qquad\square$

The heteroclinic cycles exist for an open set of parameter values. Any perturbations, which respect $O(2)$ symmetry, will leave the equations, up to third order, in the functional form dictated by (3.72). Thus, a small perturbation, measured in a suitable function space, will only give rise to a small change in the coefficients and therefore leave the heteroclinic cycle intact.

A similar heteroclinic cycle occurs in the 5-mode model of [5]. Here the circle of equilibrium lies in the (a_2, a_4) even mode subspace, intersecting the real subspace at the points $(x_1, x_2, x_3, x_4) = (0, \mp\rho_2, 0, -\rho_4, 0)$, and the unstable manifolds of each point are two-dimensional. Within the real subspace, A is a saddle and B is a sink; the unstable manifold of A lies in (x_1, x_3, x_5) subspace and a connection from A to B can be observed numerically, although a rigorous construction involving positively invariant regions is still lacking. Accepting this "first" connection, the cycle follows just as with the simpler four-dimensional case above.

With the above structures in hand, the model's "bursting" and "sweeping" behavior is understandable in terms of motion in the phase plane. Bursting and sweeping correspond to the rapid transitions from the neighborhood of one fixed point to that another along a trajectory shadowing a heteroclinic connection, the momentary activity 1, 3, 5 modes causing break-up and coalescence of the streamwise vortices. In the light of the symmetries of the problem and the above discussion, such connections are natural and generic. The even modes experience a net phase shift after each circuit, reminiscent of the observed spatial displacement between successive burst/sweep events. There is no particular time scale associated with attracting cycles: solutions spend successively longer periods lingering near saddle points as they approach the cycle.

3.4.2 Kuramoto–Sivashinsky equation

The premixed gas flame (a self-sustained wave of an exothermic chemical reaction) is frequently situated in a nonuniform flow field and subjected to large-scale flame stretch [55]. The latter, apart from affecting the burning rate intensity, may also have a significant impact on the flame stability. It has long been observed that wrinkled structures, occurring spontaneously in freely propagating flames, may be suppressed in a stagnation point flow provided its intensity is high enough. To study the dynamics of an intrinsically unstable flame held in a stagnation point flow, two model equations for the flame interface dynamics have recently been proposed. Some models [56] deal with diffusively unstable flames while the other ones [47] deal with hydrodynamically unstable flames. These models may be combined to describe the situation where the flame is subject to diffusive and hydrodynamic instabilities occurring simultaneously. One thus ends up with a unified formulation, which in appropriately chosen units results in the following initial-boundary value problem

$$
\begin{aligned}
&\partial_t u + \partial_{xxxx} u + \partial_{xx} u + \tfrac{1}{2}(\partial_x u)^2 - \gamma I(u) + \alpha \partial_x(xu) = 0, \quad (x,t) \in Q, \\
&\partial_x u = 0, \quad \partial_{xxx} u = 0, \quad x = \pm l, \quad t \in \mathbb{R}^+, \\
&u(x,0) = u_0(x), \quad x \in (-l, l),
\end{aligned}
\tag{3.74}
$$

where $Q = (-l, l) \times \mathbb{R}^+$ and

$$
I(u) = \sum_{n=1}^{\infty} \frac{n}{l} \Big\{ \int_{-l}^{l} u(y,t) \cos(\frac{n\pi}{l} y) dy \Big\} \cos(\frac{n\pi}{l} x).
$$

The differential equation in BVP (3.74) generalizes the *Kuramoto–Sivashinsky (KS) equation*

$$
\partial_t u + \partial_{xxxx} u + \partial_{xx} u + \frac{1}{2}(\partial_x u)^2 = 0.
\tag{3.75}
$$

In these equations, $u(x,t)$ is the profile of the flame interface. The diffusion term $\partial_x^2 u$ represents the diffusive flame instability occurring in premixed gas flames with sufficiently light deficient reactant (e.g., lean hydrogen-air mixtures). The term $\partial_x^4 u$ models the dissipation of small-scale disturbances, and the nonlinear term $\frac{1}{2}(\partial_x u)^2$ controls saturation of the growing large-scale disturbances.

Two terms in the BVP (3.74) have been added to the KS equation (3.75): the nonlocal term $\gamma I(u)$ is caused by hydrodynamic flame instability induced by the thermal expansion of the burnt gas, γ being the thermal expansion intensity ($0 < \gamma < 1$). The term $-\alpha \partial_x(xu)$ describes the stabilizing effect of the stretch induced by the flow, α being the stretch intensity ($\alpha > 0$).

The KS equation shares two general features with the NSE. The term $\partial_{xx} u$ is destabilizing and acts as an energy source, while $\partial_{xxxx} u$ is stabilizing or energy sink term whose effect increases with spatial wavenumber. The Fourier transform $\partial_t \hat{u}(k,t) = k^2(1 - k^2)\hat{u}(k,t)$ of the linear part of (3.75) with respect to x makes this clear, exhibiting a range of unstable wavenumbers $|k| < 1$ with a peak $|k| = 2^{-1/2}$ corresponding to the wavenumber of the most rapidly growing linear mode. It can be observed that the terms $\partial_{xx} u$ and $\partial_{xxxx} u$ in (3.75) provide one-dimensional analogues to the energy production and dissipation terms in the NSE. More significantly, the translation and reflection symmetries of the KS equation with periodic boundary conditions are the same as the spanwise symmetries of the NSE in the boundary layer. Hence the Fourier series representations and the resulting modal interactions are similar in the two problems. In fact, when only a single eigenfunctions family and just $k_1 = 0$ streamwise modes are included in the boundary layer model, the structure of the linear and quadratic terms in the projected equations is identical: only the numerical values of the coefficients differ. Thus, the KS equation is doubly relevant to our purpose.

First, we will treat the period l in (3.74) as a bifurcation parameter and concentrate for the most part on *small* values of l, for which only few spatial modes are active. Observe that, because only even orders of spatial derivatives appear in each term and we have constant coefficients and periodic boundary conditions, the system is equivariant under spatial translations and reflections: $T_\beta : x \to x + \beta$ and $R_f : x \to -x$. The translation invariance implies that the empirical eigenfunctions obtained from appropriate ensemble averages will be Fourier modes. It is reasonable to examine sets of ODE obtained by projection into subspaces spanned by the basis functions $\varphi_k(x) = \exp(\pi i k x/l)$. It will be shown that such choice will also diagonalise the linear part of (3.75), simplifying our calculations. We represent the dependent variable $u = u(x,t)$ as

$$u(x,t) = \sum_k a_k(t)\varphi_k(x). \tag{3.76}$$

Here $a_k(t)$ are complex modal coefficients and reality of the scalar field u implies that $a_{-k}(t) = a_k^*(t)$, where a_k^* denotes the complex conjugate of a_k. The Fourier modes form an orthogonal basis with respect to the L^2 inner product $(f,g) = \int_{-l}^{l} f(x)g^*(x)dx$; in fact,

$$(\varphi_k, \varphi_m) = \int_{-l}^{l} \exp[\pi i(k-m)/l]\,dx = 2l\delta_{km}. \tag{3.77}$$

Now, using the relationships $\partial_x \varphi_k = \frac{\pi i k}{l}\varphi_k$, $\partial_x^2 \varphi_k = -\left(\frac{\pi k}{l}\right)^2 \varphi_k$ and $\partial_x^4 \varphi_k = \left(\frac{\pi k}{l}\right)^4 \varphi_k$, application of the Galerkin procedure to (3.75) yields

$$\int_{-l}^{l} \sum_k \varphi_m [\dot{a}_k\varphi_k - \left(\frac{\pi k}{l}\right)^2 a_k\varphi_k + \left(\frac{\pi k}{l}\right)^4 a_k\varphi_k + \frac{1}{2}\left(\frac{\pi i k}{l}\right)a_k\varphi_k \sum_j \left(\frac{\pi i j}{l}\right)a_j\varphi_j]\,dx = 0. \tag{3.78}$$

Using of (3.77) reduces (3.78) to

$$2l[\dot{a}_k - \left(\frac{\pi m}{l}\right)^2 a_k + \left(\frac{\pi m}{l}\right)^4 a_k - \frac{1}{2}\left(\frac{\pi}{l}\right)^2 \sum_j j(m-j)a_j a_{m-j}] = 0, \tag{3.79}$$

since in the integrals involving the quadratic terms, we have $\int_{-l}^{l} \exp\left[\frac{\pi i(k+j-m)x}{l}\right]dx = 0$, unless $k + j = m$. Finally, recalling time the multiplicative factor $(\pi/l)^2$, rewrite (3.79) in the compact form:

$$\dot{a}_m = m^2\left[1 - (\pi m/l)^2\right]a_m + \frac{1}{2}\sum_j j(m-j)a_j a_{m-j}. \tag{3.80}$$

So far we have been reticent the sum appearing in the representation (3.76) and in the ODE (3.80). In fact, the form (3.80) applies to any truncation order K originally assumed in (3.76) and, in general, (3.80) defines an ODE in complex $(2K+1)$-dimensional space. If we interpret sequences $\{a_k\}$ to belong to the space l^2 of infinite sequences having *finite energy* $\sum_{k=-\infty}^{\infty} |a_k|^2 < \infty$, we may even let $K = \infty$. We can modestly reduce the dimension of (3.80) by noting that the equation for a_0, the spatial average, uncouples and is merely driven by the other modes according to $\dot{a}_0 = -\frac{1}{2}\sum_j j^2|a_j|^2$; a_0 does not enter the equations for the other components. Moreover the reality condition $a_{-k}(t) = a_k^*(t)$ implies that we need only consider the equations for a_m with $m \geq 1$. Thus, for system truncated at order K, we have only to solve K complex or $2K$ real, first order ODE. To give a feeling for the structure of the system, and for the convenience in center manifold calculations, we display the equations obtained by truncation at order $K = 4$:

$$\begin{cases} \dot{a}_1 = \left[1 - (\pi/l)^2\right]a_1 - 2a_1^*a_2 - 6a_2^*a_3 - 12a_3^*a_4, \\ \dot{a}_2 = 4\left[1 - (2\pi/l)^2\right]a_2 + \frac{1}{2}a_1^2 - 3a_1^*a_3 - 8a_2^*a_4, \\ \dot{a}_3 = 9\left[1 - (3\pi/l)^2\right]a_3 - 2a_1a_2 - 4a_1^*a_4, \\ \dot{a}_4 = 16\left[1 - (4\pi/l)^2\right]a_4 + 2a_2^2 + 3a_1a_3. \end{cases} \tag{3.81}$$

Due to symmetries and equivariance of Fourier representation, observe that (3.80) in general and (3.81) in particular, are equivariant under $O(2)$ with representation $T_\beta : a_m \to \exp\left(\frac{\pi i m \beta}{l}\right)a_m$ and $R_f : a_m \to a_m^*$.

(a) **Local bifurcations from** $u = 0$. The stability characteristics of the equilibrium at the origin ($a_m = 0$ for all m) can be determined directly from the linearization $\dot{a}_m = m^2[1 - (\pi m/l)^2]a_m$ ($m \in \mathbb{N}$), in which all modes decouple and each is simply governed by the two real equations

$$\dot{x}_m = m^2[1 - (\pi m/l)^2]x_m, \quad \dot{y}_m = m^2[1 - (\pi m/l)^2]y_m.$$

Every eigenvalue, given by

$$\lambda_m = m^2[1 - (\pi m/l)^2], \tag{3.82}$$

has multiplicity two. Evidently the origin is asymptotically stable if $l < \pi$ and unstable if $l > \pi$. At $l = \pi$ we have a double zero eigenvalue and a bifurcation occurs: a situation repeated at $l = l_n = \pi n$ for all $n \in \mathbb{N}$. These bifurcations to be $O(2)$-equivariant pitchforks, and we'll show that this is indeed the case. At $l = l_n$ the center eigenspace is the complex a_p-plane and so we seek one (complex-) dimensional center manifold given by a set of complex-valued functions of the form

$$a_m = h_m(a_p, a_p^*), \quad m \neq p, -p, \quad h_m = O(|a_p|^2), \tag{3.83}$$

satisfying $h_{-m} = h_m^*$. In adapting the form for the center manifold in the real case to complex coordinates, we must allow dependence on both a_p and a_p^*. The reduced system is then given, as usual, by substitution of (3.83) into p-th component of (3.80) and it will provide an accurate picture of solutions on the center manifold for $l \approx l_p = \pi p$. After this substitution, most of nonlinear terms $a_j a_{m-j}$ in (3.80) will be of order four or higher; only those involving a_p or $a_{-p} = a_p^*$ contain cubic terms. We therefore explicitly separate the latter, corresponding to $j = -p$ and $j = 2p$ (the terms $j = 0$, $j = p$ vanish identically):

$$\dot{a}_p = p^2[1 - (\pi p/l)^2]a_p - 2p^2 a_p^* h_{2p}(a_p, a_p^*) + O(|a_p|^4). \tag{3.84}$$

Only the single function h_{2p} enters at third order. To compute the leading (quadratic) term in h_{2p} we apply the procedure modestly generalized to accommodate complex functions. We augment each complex ODE with its conjugate

$$\dot{a}_m^* = m^2[1 - (\pi m/l)^2]a_m^* + \frac{1}{2}\sum_j j(m - j)a_j^* a_{m-j}^*,$$

and write for the function h_{2p}

$$\left(\frac{\partial h_{2p}}{\partial a_p}, \frac{\partial h_{2p}}{\partial a_p^*}\right)\begin{pmatrix} \dot{a}_p \\ \dot{a}_p^* \end{pmatrix} = (2p)^2[1 - (2\pi p/l)^2]h_{2p} + \frac{1}{2}p^2 a_p^2 + \frac{1}{2}\sum_{j \neq p} j(2p - j)h_j h_{2p-j}. \tag{3.85}$$

At $l = \pi p$ the linear terms in $\dot{a}_p$ and $\dot{a}_p^*$ vanish and the LHS of (3.85) is consequently of $O(|a_p|^3)$. The "Σ"-term in (3.85) is also of order $O(|a_p|^3)$, and so, balancing the quadratic terms, we obtain

$$h_{2p} = \frac{1}{24}a_p^2 + O(|a_p|^3). \tag{3.86}$$

Substitution of this into the reducing system (3.84) yields

$$\dot{a}_p = p^2[1 - (\pi p/l)^2]a_p - \frac{1}{12}p^2|a_p|^2 a_p + O(|a_p|^4),$$

and finally, letting $l = \pi p(1 + \mu)$, with $\mu \ll 1$, we obtain the bifurcation equation

$$\dot{a}_p = p^2 a_p\left(\mu - \frac{1}{12}|a_p|^2\right) + O(|a_p|^4, \mu^2, \ldots). \tag{3.87}$$

In Cartesian coordinates on each (real) modal plane, this is simply

$$(\dot{x}_p, \dot{y}_p) = p^2 \left[\mu - \frac{1}{12}(x_p^2 + y_p^2) \right](x_p, y_p),$$

which, truncated at cubic order, is precisely the normal form of the $O(2)$-equivariant pitch-fork bifurcation, as claimed. Conclude that, as l increases through each bifurcation value $l_p = \pi p$, a circle of equilibrium branches from origin and that, for $l - l_p > 0$ sufficiently small, this circle lies on an invariant real two-dimensional manifold tangent to the a_p-plane and given by (3.86).

The local bifurcation analysis is also valid for the original PDEs. The fact that Fourier representation diagonalizes the self-adjoint linear operator of (3.75) implies that the (un-called) eigenvalues $\lambda_m = (\pi m/l)^2 [1 - (\pi m/l)^2]$ of the linearized ODE coincide with the eigenvalues of the differential operator, which are also each of multiplicity two, with eigen-functions $\cos(\pi mx/l), \sin(\pi mx/l)$. The center manifold theorem (Theorem 14) for evolution equations then guarantees the existence, for $l = l_p$, of two-dimensional invariant manifold tangent to the center eigenspace $\mathrm{Span}\{\cos(\frac{\pi mx}{l}), \sin(\frac{\pi mx}{l})\}$ in the space of $2l$-periodic func-tions, which can be extended to a family of center manifolds for $l \approx l_p$ that carry the circles of equilibrium for $l - l_p > 0$ sufficiently small.

(b) **The second bifurcation point.** The circle of equilibrium created in the first $O(2)$-equivariant pitchfork bifurcation at $l_1 = \pi$ is asymptotically stable for $l - l_1 > 0$ and sufficiently small. This follows from (3.87) with $p = 1$ and the fact that all other eigenvalues of the equation linearized at the origin are strictly negative, see (3.82). Near the second bifurcation point $l_2 = 2\pi$, therefore, we have two-dimensional unstable manifold tangent to the a_1-plane and two-dimensional center manifold tangent to the a_2-plane. Other directions, tangent to the subspaces spanned by Fourier modes φ_m, $m \geq 3$, all lie in the stable manifold. It therefore seems reasonable to seek a reduced system restricted to four-dimensional *center-unstable* manifold if we wish to study the dynamics for parameter values $l \approx l_2$. First, define a new (small) parameter $\mu = (\pi/l_2)^2 - (\pi/l)^2 = \frac{1}{4} - (\pi/l)^2$, so that

$$[1 - (\pi/l)^2] = 3/4 + \mu, \quad 4[1 - (\pi/l)^2] = 16\mu.$$

Next, parameterize family of center manifolds as graphs of the form $a_m = h_m(a_1, a_1^*, a_2, a_2^*, \mu)$ ($\mu = 3, 4, 5, \dots$). Much as in our derivation of the reduced system (3.87) on two-dimensional center manifold, it turns out that we need only explicitly compute the leading terms of the functions h_m for $m = 3$ and 4 obtain accuracy up to $O(|a_p|^4)$. The reduced system takes the form

$$\begin{aligned}
\dot{a}_1 &= (\tfrac{3}{4} + \mu)a_1 - 2a_1^* a_2 - 6a_2^* h_3 - 12h_3^* h_4 - \textstyle\sum_{k \geq 4} k(k+1)h_k^* h_{k+1}, \\
\dot{a}_2 &= 16\mu a_2 - \tfrac{1}{2}a_1^2 - 3a_1^* h_3 - 8a_2^* h_4 - \textstyle\sum_{k \geq 3} k(k+2)h_k^* h_{k+2}.
\end{aligned} \tag{3.88}$$

We claim that $h_m = O(|a_m|^3)$ for $m \geq 5$, implying that the terms in the sums on the extreme right of (3.88) are all of $O(|a_m|^5)$ and so will not appear in the reduced system approximated up to $O(|a_m|^4)$.

To verify this claim, we consider the exponents of (3.88) which is each of the form

$$\begin{aligned}
&\frac{\partial h_m}{\partial a_1}\dot{a}_1 + \frac{\partial h_m}{\partial a_1^*}\dot{a}_1^* + \frac{\partial h_m}{\partial a_2}\dot{a}_1 + \frac{\partial h_m}{\partial a_2^*}\dot{a}_2^* = m^2\left[1 - (\pi m/l)^2\right]h_m \\
&+ \tfrac{1}{2}\textstyle\sum_{j=1}^{m-1} j(m-j)a_j a_{m-j} - \textstyle\sum_{j=1}^{m-1} j(m+j)a_j^* a_{m+j}, \quad m \geq 3,
\end{aligned} \tag{3.89}$$

where we understand that $a_m = h_m(a_1, a_1^*, a_2, a_2^*, \mu)$ for $m \geq 3$ and we have rewritten the nonlinear terms so that only positive indices appear. It is known *a priori* that $h_m = O(|a_m|^2)$ and from (3.88), that $\dot{a}_1 = O(|a_1|)$ and $\dot{a}_2 = O(\mu|a_2|)$. One order is lost in differentiating

h_m. Thus, for $m = 5$, the LHS of (3.84) is $O(|h_5|)$ while the three terms on the RHS are, respectively, $O(|h_5|)$, $O(|a_1h_4|, |a_2h_3|)$, and $O(|a_1h_6|, |a_2h_7|, \ldots, |a_kh_{k+5}|)$. Balancing the lowest order terms find that $h_5 = O(|a_m|^3)$, and similarly

$$h_6 = O(|a_1h_5|, |a_2h_4|, |h_3|^2) + O(|a_1h_7|, |a_2h_8|, \ldots) = O(|a_p|^3),$$
$$h_7 = O(|a_1h_6|, |a_2h_5|, |h_3h_4|) + O(|a_1h_8|, |a_2h_9|, \ldots) = O(|a_p|^4).$$

In general, we have $h_m = O(|a_p|^{[(m+1)/2]})$, where the square brackets denote the integer part. Our claim is justified. To compute h_3 and h_4 up to and including terms of $O(|a_m|^3)$ and $O(\mu|a_m|^3)$, we return to (3.88) with specific components indicated in (3.89)

$$\begin{bmatrix} \frac{\partial h_3}{\partial a_1} & \frac{\partial h_3}{\partial a_1^*} & \frac{\partial h_3}{\partial a_2} & \frac{\partial h_3}{\partial a_2^*} \\ \frac{\partial h_4}{\partial a_1} & \frac{\partial h_4}{\partial a_1^*} & \frac{\partial h_4}{\partial a_2} & \frac{\partial h_4}{\partial a_2^*} \end{bmatrix} \cdot \begin{pmatrix} (\frac{3}{4} + \mu)a_1 - 2a_1^*a_2 + \ldots \\ (\frac{3}{4} + \mu)a_1^* - 2a_1a_2 + \ldots \\ 16\mu a_2 + \frac{1}{2}a_1^2 \ldots \\ 16\mu a_2^* + \frac{1}{2}(a_1^*)^2 \ldots \end{pmatrix}$$

$$= \begin{pmatrix} 9(-\frac{5}{4} + 9\mu)h_3 + 2a_1a_2 - 4a_1^*h_4 + \ldots \\ 16(-3 + 16\mu)h_4 + 2a_2^2 + 3a_1h_3 + \ldots \end{pmatrix}.$$

This system of two simultaneous complex differential equations is solved, up to the other required, by

$$h_3 = \frac{1}{6}(1 + \frac{16}{3}\mu)a_1a_2 - \frac{1}{162}a_1^3 + \frac{1}{72}a_1^*a_2^2 + \ldots$$
$$h_4 = \frac{1}{24}(1 + \frac{14}{3}\mu)a_2^2 + \frac{1}{108}a_1^2a_2 + \ldots \tag{3.90}$$

Observe that these expressions satisfy

$$h_m(e^{i\vartheta}a_1, e^{-i\vartheta}a_1^*, e^{2i\vartheta}a_2, e^{-2i\vartheta}a_2^*, \mu) = e^{im\vartheta}h_m(a_1, a_1^*, a_2, a_2^*, \mu) \quad \text{and} \quad h_m^* = h_{-m},$$

so that our approximation to the center-unstable manifold is also invariant under the action of $O(2)$. Finally, substituting the expression of (3.90) into (3.88), yields the reduced system

$$\dot{a}_1 = (\tfrac{3}{4} + \mu)a_1 - 2a_1^*a_2 - (1 - \tfrac{16}{3}\mu)a_1|a_2|^2 + \tfrac{1}{27}a_1^3a_2^* - \tfrac{1}{6}a_1^*a_2|a_2|^2 + \ldots,$$
$$\dot{a}_2 = 16\mu a_2 + \tfrac{1}{2}a_1^2 - \tfrac{1}{2}(1 + \tfrac{16}{3}\mu)a_2|a_1|^2 - \tfrac{1}{3}(1 + \tfrac{14}{3}\mu)a_2|a_2|^2$$
$$- \tfrac{1}{54}a_1^2|a_1|^2 - \tfrac{1}{24}(a_1^*)^2a_2^2 - \tfrac{2}{27}a_1^2|a_2|^2 + \ldots, \tag{3.91}$$

in which the error of $O(|a_m|^5)$ and $O(\mu^2|a_m|^3)$. Truncating at cubic order we have

$$\begin{cases} \dot{a}_1 = (\tfrac{3}{4} + \mu)a_1 - 2a_1^*a_2 - a_1|a_2|^2, \\ \dot{a}_2 = 16\mu a_2 + \tfrac{1}{2}a_1^2 - a_2(\tfrac{1}{2}|a_1|^2 + \tfrac{1}{3}|a_2|^2), \end{cases} \tag{3.92}$$

which is a special case of the $O(2)$-equivariant normal form. In fact, making use of the change of variables $-2a_2 \mapsto a_2$, we obtain the form

$$\begin{cases} \dot{a}_1 = (\tfrac{3}{4} + \mu)a_1 + a_1^*a_2 - \tfrac{1}{4}a_1|a_2|^2, \\ \dot{a}_2 = 16\mu a_2 - a_1^2 - a_2(\tfrac{1}{2}|a_1|^2 + \tfrac{1}{12}|a_2|^2), \end{cases} \tag{3.93}$$

and we read off the behavior directly from our earlier analysis. As in the simpler bifurcation analysis, the center manifold given by (3.90) provides an algebraic relation expressing the neglected modes a_3, a_4 in terms of the modes a_1, a_2 of the reduced system.

    We are interested in dynamical behavior and bifurcations, which occur near $\mu = 0$ ($l = l_2 = 2\pi$) in neighborhood of $a_1 = a_2 = 0$. For $\mu > 0$ the circle of fixed points $\{a_1 = 0; |a_2| = 8(3\mu)^{1/2}\}$ has eigenvalues with eigenvectors in the a_1-plane given by

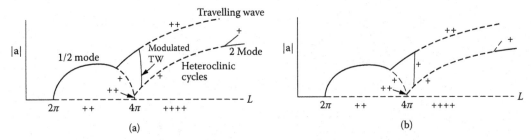

FIGURE 3.4: The bifurcation diagrams for the reduced systems: (a) (3.93), (b) (3.91), '-' includes fourth order terms. Asymptotically stable branches are shown bold, heteroclinic cycles hatched, unstable cycles dotted, '+' indicates the number of positive (unstable) eigenvalues on each branch. $L = 2l$.

$\lambda_\pm = \frac{3}{4} - 47\mu \pm \sqrt{3\mu}$. Thus, these fixed points are saddle with one-dimensional unstable manifolds for all μ satisfying $0.00218\ldots < \mu < 0.11665\ldots$; these extreme μ values correspond to $\lambda_- = 0$ and $\lambda_+ = 0$ resp. Moreover, since $(3/4 + \mu) > 0$ if $\mu > 0$, we also have heteroclinic cycles in this μ range; these cycles are asymptotically stable provided $-\lambda_-/\lambda_+ > 1$, or $\mu > \mu_H = \frac{3}{4\times47} \approx 0.01596\ldots$ (where index H means "heteroclinic"), and at $\mu = \mu_H$ a branch of quasiperiodic modulated traveling waves emerges. This family of invariant tori in turn coalesces with the traveling wave solutions at $16\mu = \frac{7}{25}\left(\frac{3}{4} + \mu\right)$ or $\mu = \mu_{\text{Hopf}} \approx 0.01336$. The inequality $\mu_H > \mu_{\text{Hopf}}$ implies that tori are attracting. Figure 3.4(a) shows a bifurcation diagram for (3.93) which schematically illustrates this behavior and also shows the branch of *mixed mode* equilibrium $\{a_1, a_2 \neq 0\}$ which appear at $l = l_1 = \pi$ and the branch of traveling waves which bifurcates from these mixed modes. The traveling wave bifurcation appears as a Pitchfork bifurcation for the reduced system corresponding to (3.93) and occurs at $\mu \approx -0.05093\ldots$

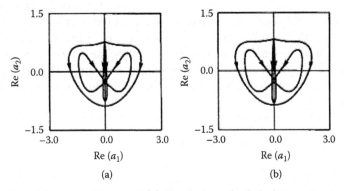

FIGURE 3.5: Numerical simulations of (a) the four-mode Galerkin projection (3.81), and (b) the fourth order reduced system (3.91), both projected onto the Re(a_1)-Re(a_2)-plane. Here $L = 2l \approx 13.03$.

It is perhaps even more striking, that simulations of the fourth order reduced system (3.91) and of the four complex mode truncation (3.81) of the full system agree so well as shown in Fig. 3.5. Truncations include 8, 16 or more modes. The similarity, both qualitative and quantitative, is remarkable. Note coexistence of attracting heteroclinic cycle and traveling waves in both cases.

We conclude the section by showing reconstructions of solution $u(x,t)$ of (3.75) derived from the fourth-order reduced system via $u(x,t) = \sum_{k=-4}^{4} a_k(t)\varphi_k(t)$, $a_{-k}(t) = a_k(t)$ with $a_1(t), a_2(t)$ obtained by integration of (3.81) and $a_3(t)$ and $a_4(t)$ from the center-unstable manifold approximation (3.92).

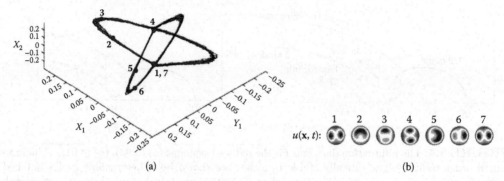

(a)   (b)

FIGURE 3.6: (a) Phase-space depiction of a heteroclinic cycle, and (b) heteroclinic solution $u(\mathbf{x}, t)$ for the Kuramoto–Sivashinsky equation.

Example 81 A generic example of a cellular-pattern-forming dynamical system is described by the KS equation, which can be written in the form

$$\partial_t u = (1 + \nabla^2)^2 u - \eta_2 (\nabla u)^2 - \eta_3 u^3 + \zeta(\mathbf{x}, t),$$

where $u(\mathbf{x}, t)$ represents the perturbation of a planar front (normally assumed to be a flame front) in the direction of propagation, η_1 measures the strength of the perturbation force, η_2 is a parameter associated with growth in the direction normal to the domain (burner) of the front, $\eta_3 u^3$ is a term that has been added to help stabilize its numerical integration, and $\zeta(\mathbf{x}, t)$ represents Gaussian white noise, which models thermal fluctuations, dimensionless in space and time. The KS equation describes the perturbations of a uniform wave front by thermo-diffusive instabilities. It has been studied in different contexts, including the existence of heteroclinic connections [41] have also conducted numerical explorations of the effects of noise on the KS equation in various regions of parameter space. The figure below shows a phase-space diagram of a low-dimensional system of ODEs derived from the KS equation in a region of parameter space where a heteroclinic cycle exists near a 1 : 2 mode interaction. Phase-space depiction of a heteroclinic cycle found in the KS equation is shown in Fig. 3.6(a). The parameters are: $R = 4.285$ is the radius domain of integration, $D = 0.008$ the noise density, and $(\eta_1, \eta_2, \eta_3) = (0.32, 1.0, 0.17)$. In physical space, the 1:2 heteroclinic cycle represents repetitive excursions between a one-cell pattern and a two-cell pattern; see Fig. 3.6(b), with the same parameters.

3.5 Strange attractors and turbulence

It is not clear how to reconstruct the *strange attractors* from a *power spectrum* (with continuous parts); even worse: how can one see whether a given power spectrum (with continuous parts) might be generated by strange attractor? In this section we present procedures to decide whether one may attribute certain experimental data, as in the onset of turbulence, to the presence of strange attractors. These procedures consist of algorithms, to be applied to the experimental data itself and not to the power spectrum; in fact, one doubt whether the power spectrum contains the relevant information.[2]

---

[2]This section is based on the theory of F. Takens [102].

3.5.1 The Taylor–Couette experiment

Consider fluid in the region D between two cylinders. We study its motion when the outer cylinder, the top and bottom are at rest, while the inner cylinder has an angular velocity Ω. Fix a point p in the interior of D. For a number of values of Ω, one component of the fluid velocity at p is measured as a function of time. The idea [93] is the following: for each value of Ω the set of all "possible states" in Hilbert space $\mathcal{H}_\Omega$ consisting of (divergence free) vector fields on D satisfying the appropriate boundary conditions (these vector fields represent velocity distributions of the fluid). For each Ω there is an evolution semi-flow $\{\varphi_t^\Omega : \mathcal{H}_\Omega \to \mathcal{H}_\Omega\}_{t \in \mathbb{R}_+}$, $\mathbb{R}_+ = \{t \in \mathbb{R} : t \geq 0\}$ such that if $X \in \mathcal{H}_\Omega$ represents the state at the time $t = 0$ when $\varphi_{t_0}^\Omega(X)$ represents the state at time t_0. Assume that for all values of Ω there is an attractor Λ_Ω to which (almost) all evolution curves $\varphi_t^\Omega(X)$ tend as $t \to \infty$. The states Λ_Ω and $\varphi_t^\Omega|\Lambda_\Omega$ then describe the asymptotic behavior of all evolution curves $\varphi_t^\Omega(X)$. Roughly the main assumptions in [93] could be rephrased as: $\varphi_t^\Omega|\Lambda_\Omega$ behaves just as an attractor in a finite dimensional dynamic system.

More detail, the assumption was that for all values of Ω under consideration there is a smooth finite dimensional manifold $M_\Omega \in \mathcal{H}_\Omega$, smoothly depending on Ω, such that

(i) M_Ω is invariant in the sense that $\varphi_t^\Omega(X) \in M_\Omega$ for $X \in M_\Omega$;

(ii) M_Ω is attractive, that is evolution curves $\varphi_t^\Omega(X)$, starting outside of M_Ω tend to M_Ω for $t \to \infty$;

(iii) the flow, induced in M_Ω by φ_t^Ω, is smooth, depends smoothly on Ω and has an attractor Λ_Ω.

As a justification we used genericity assumptions: if Z_Ω denotes the vector field on M_Ω which is the infinitesimal generator $\varphi_t^\Omega|M_\Omega$, we assume (M_Ω, Z_Ω) to be a generic one-parameter family of vector fields. If the system under consideration has symmetry, like the case of the Couette flow, then a same type of symmetry must hold for M_Ω, φ_t^Ω and hence for Z_Ω. In this case, genericity should be understood within the class of vector fields with this symmetry.

In the Landau–Lifshitz picture (see [63]), one assumes that the limiting motion (or attractor) is quasi-periodic, i.e., of the form $\varphi_t^\Omega(X) = f^\Omega(X, a_1 e^{2\pi i \omega_1 t}, a_2 e^{2\pi i \omega_2 t}, \ldots)$, where ω_i and a_i depend on Ω and where for each Ω only a finite number of a_i is nonzero. One can imagine that, as more and more a_i become nonzero, the motion becomes more and more turbulent. In this last description we have a smooth finite dimensional manifold as attractor, namely an n-torus, but such attractors do not occur for generic parameter values of one-parameter families of vector fields. For generic one-parameter families of vector fields there may be a set of parameter values with positive measure for which quasi-periodic motion occurs [93]. This n-torus has topological entropy zero and its dimension is an integer. On the other hand, *strange attractors* have in general positive entropy and dimension from "experimental data." Thus, we have to add the observable function from the state space to the real giving the experimental output when composed with φ_t^Ω.

In the present example of the Taylor–Couette experiment, this function $y_\Omega : \mathcal{H}_\Omega \to \mathbb{R}$ assigns to each $X \in \mathcal{H}_\Omega$ the measured component of $X(p)$. Concerning the asymptotic behavior, we only have to deal with $y_\Omega|M_\Omega$ (or with $y_\Omega|\Lambda_\Omega$). Since M_Ω depends smoothly on Ω, all M_Ω are diffeomorphic and so we may drop on the Ω. Summarizing, we have a manifold M with a smooth one-parameter family of vector fields Z_Ω and a smooth one-parameter family of functions y_Ω. For a number of values of Ω the function $y_\Omega(\varphi_t^\Omega(x))$ is known by measurement (for some x in or near M which may depend on Ω); φ_t^Ω denotes the flow on M generated by Z_Ω. The point is to obtain information about the attractor(s) of Z_Ω from these measurements, i.e., from the functions $t \mapsto y_\Omega(\varphi_t^\Omega(x))$. For this we will make genericity assumptions on $(M, Z_\Omega, y_\Omega, x)$.

We will show that under suitable genericity assumptions on $(M, Z_\Omega, y_\Omega, x)$ the positive limit set $L^+(x)$ of x is determined by the function $y_\Omega(\varphi_t^\Omega(x))$. The *main The-orem* (see Section 3.5.4) describes the algorithm which, when applied to a sequence $\{a_i = y_\Omega(\varphi_{\alpha,i}^\Omega(x))\}_{i=1,\overline{N}}$, $\overline{N}$ is sufficiently big, yields an approximation for the dimension of $L^+(x)$, respectively, for topological entropy $\varphi_\alpha^\Omega|L^+(x)$. This leads to possibility of testing and comparing the hypothesis in [63, 93].

3.5.2   Dynamical systems with one observable

Let M be a compact manifold. A dynamic system on M is a diffeomorphism $\varphi : M \to M$ (discrete time) or a vector field X on M (continuous time). In both cases the time evolution corresponding with an initial position $x_0 \in M$ is denoted by $\varphi_t(x_0)$: in the case of discrete time $t \in \mathbb{N}$ and $\varphi_i = (\varphi)^i$, and in the case of continuum time $t \in \mathbb{R}$ and $t \mapsto \varphi_t(x_0)$ is the X integral curve through x_0. An observable is a smooth function $y : M \to \mathbb{R}$. The first problem is obtain information about original dynamic system (and manifold) for some dynamical system from the time evolution φ_t and known functions $t \mapsto y(\varphi_t(x))$, $x \in M$. The next three theorems deal with this problem.

Theorem 47 *Let M be a compact manifold of dimension m. For pairs (φ, y), $\varphi : M \to M$ a smooth diffeomorphism and $y : M \to \mathbb{R}$ a smooth function, it is a generic property that the map $\Phi_{(\varphi,y)} : M \to \mathbb{R}^{2m+1}$ defined by $\Phi_{(\varphi,y)}(x) = (y(x), y(\varphi(x)), \dots, y(\varphi^{2m}(x)))$ is an embedding; by smooth we mean at least C^2.*

Proof. Assume that if x is a point with period k of φ, $k \le 2m + 1$, all eigenvalues of $(d\varphi^k)_x$ are different and different from 1. Also assume that no two different fixed points of φ are in the same level of y. For $\Phi_{(\varphi,y)}$ to be an immersion near a fixed point x, the co-vectors $(dy)_x, d(y\varphi)_x, \dots, d(y\varphi^{2m})_x$ must span T_x^*M. This is the case for generic y if $d\varphi$ satisfies the above condition at each fixed point.

Similarly, $\Phi_{(\varphi,y)}$ is generically an immersion and even an embedding when restricted to the periodic points with period $\tau \le 2m + 1$. Thus, assume that for generic $(\bar\varphi, \bar y)$ we have that $\Phi_{(\bar\varphi,\bar y)}$, restricted to a compact neighborhood V of the set of points with period $\tau \le 2m + 1$, is an embedding; $\Phi_{(\varphi,y)}|V$ is an embedding for some neighborhood U of $(\bar\varphi, \bar y)$, whenever $(\varphi, y) \in$ U.

We will show that for some $(\varphi, y) \in$ U, arbitrarily near $(\bar\varphi, \bar y)$, $\Phi_{(\varphi,y)}$ is an embedding. Indeed, for any point $x \in M$, which is not a point of period $\tau \le 2m + 1$ for $\bar\varphi$, the co-vectors $(d\bar y)_x, d(\bar y\varphi)_x, d(\bar y\varphi^2)_x, \dots, d(\bar y\varphi^{2m})_x \in T^*(M)$ can be perturbed independently by perturbing $\bar y$. Hence arbitrarily near $\bar y$ there is $\bar{\bar y}$ such that $(\bar\varphi, \bar{\bar y}) \in$ U and such that $\Phi_{(\bar\varphi,\bar{\bar y})}$ is an immersion. Then there is a positive ε such that whenever $0 < \rho(x, x') \le \varepsilon$ (ρ is some fixed metric on M), $\Phi_{(\bar\varphi,\bar{\bar y})}(x) \ne \Phi_{(\bar\varphi,\bar{\bar y})}(x')$. There is a neighborhood U$' \subset$ U of $(\bar\varphi, \bar{\bar y})$ such that for any $(\varphi, y) \in$ U$'$, $\Phi_{(\varphi,y)}$ is an immersion and $\Phi_{(\varphi,y)}(x) \ne \Phi_{(\varphi,y)}(x')$, whenever $x \ne x'$ and $\rho(x, x') \le \varepsilon$. Assume that each component of V has diameter smaller than ε.

Finally, we have to show that in U$'$ there is a pair (φ, y) with injective $\Phi_{(\varphi,y)}$. For this we need a finite collection $\{U_i\}_{i=1,N}$ of open subsets of M, covering the closure of $M \backslash \{\bigcap_{j=0}^{2m} \varphi^j(V)\}$, and such that:

(i) for each $i = 1, \dots, N$ and $k = 0, 1, \dots, 2m$, diameter $(\bar\varphi^k(U_i)) < \varepsilon$;

(ii) for each $i, j = 1, \dots, N$ and $k = 0, 1, \dots, 2m$, $\bar\varphi^k(U_i) \cap U_j \ne \emptyset$ and $\bar\varphi^l(U_i) \cap U_j \ne \emptyset$ imply $k = l$;

(iii) for $\bar\varphi^j(x) \in M \backslash (\bigcup_i U_i)$, $j = 0, \dots, 2m$, $x' \notin V$ and $\rho(x, x') > \varepsilon$,
no two points of the sequence $x, \bar\varphi(x), \dots, \bar\varphi^{2m}(x), x', \bar\varphi(x'), \dots, \bar\varphi^{2m}(x')$ belong to the same U_i.

Note that (ii) implies (but is not implied by)

(ii)' no two points of the sequence $x, \bar\varphi(x), \dots, \bar\varphi^{2m}(x)$ belong to the same U_i.

We take a corresponding partition $\{\lambda_i\}$ of unity, i.e., λ_i is a nonnegative function with support $\bar{U}_i$ and $\sum_{i=1}^{N} \lambda_i(x) = 1$ for all $x \in \overline{M \backslash V}$. Consider the map $\psi : M \times M \times \mathbb{R}^N \to \mathbb{R}^{2m+1} \times \mathbb{R}^{2m+1}$ which is defined by $\psi(x, x', \varepsilon_1, \dots, \varepsilon_N) = \big(\Phi_{(\bar{\varphi}, \bar{\bar{y}}_\varepsilon)}(x), \Phi_{(\bar{\varphi}, \bar{\bar{y}}_\varepsilon)}(x')\big)$, where ε stands for $(\varepsilon_1, \dots, \varepsilon_N)$ and $\bar{\bar{y}}_\varepsilon = \bar{\bar{y}} + \sum_{i=1}^{N} \varepsilon_i \lambda_i$. Define $W \subset M \times M$ as

$$W = \{(x, x') \in M \times M | \rho(x, x') \geq \varepsilon \text{ and not both } x \text{ and } x' \text{ are in int}(V)\};$$

ψ, restricted to a small neighborhood of $W \times \{0\}$ in $(M \times M) \times \mathbb{R}^N$, is transverse with respect to the diagonal of $\mathbb{R}^{2m+1} \times \mathbb{R}^{2m+1}$. This transversality follows from all the conditions imposed on the covering $\{U_i\}_{i=1,\dots,N}$. From this transversality we conclude that there are arbitrarily small $\bar{\varepsilon} \in \mathbb{R}^N$ such that $\psi(W \times \{\bar{\varepsilon}\}) \cap \Delta = \emptyset$. If also for such an $\bar{\varepsilon}$, $(\bar{\varphi}, \bar{\bar{y}}_{\bar{\varepsilon}}) \in U'$ then $\Phi_{(\bar{\varphi}, \bar{\bar{y}}_{\bar{\varepsilon}})}$ is injective and hence an embedding.

Thus, $\Phi_{(\varphi, y)}$ is an embedding for a dense set of pairs (φ, y). Since the set of embeddings is open in the set of all mappings, there is an open and dense set of pairs (φ, y) for which $\Phi_{(\varphi, y)}$ is an embedding. $\qquad\square$

This theorem is also valid for noncompact M if we assume our observables to be proper functions.

Theorem 48 *Let M be a compact manifold of dimension m. For pairs (X, y), X – a smooth (i.e., C^2) vector field and y a smooth function on M, it is a generic property that $\Phi_{X,y} : M \to \mathbb{R}^{2m+1}$, defined by $\Phi_{X,y} = (y(x), y(\varphi_1(x)), \dots, y(\varphi_{2m}(x)))$ is an embedding, where φ_t is the flow of X.*

Proof. This is similar to the proof of Theorem 47. We impose the following generic properties on X: if $X(x) = 0$ then all eigenvalues of $(d\varphi_1)_x : T_x(M) \to T_x(M)$ are different and differ from 1; no periodic integral curve of X has integer period $\tau \leq 2m+1$. In this case, φ_1 satisfies the same conditions as $\bar{\varphi}$ in the previous proof. The rest of the proof carries over immediately. $\qquad\square$

Theorem 49 *Let M be a compact manifold of dimension m. For pairs (X, y), X is a smooth vector field and y is a smooth function on M, it is a generic property that the map $\tilde{\Phi}_{X,y} : M \to \mathbb{R}^{2m+1}$ given by*

$$\tilde{\Phi}_{X,y}(x) = \big[y(x), \frac{d}{dt}(y(\varphi_t(x)))\big|_{t=0}, \dots, \frac{d^{2m}}{dt^{2m}}(y(\varphi_t(x)))\big|_{t=0}\big]$$

is an embedding. Here φ_t denotes the flow of X; this time, smooth means at least C^{2m+1}.

Proof. Also this proof is quite analogous to that of Theorem 47. First we may, and do, assume that a generic vector field X has the property that whenever $X(x) = 0$, all eigenvalues of $(dX)_x$ are different and different from zero. $\text{Sing}(X)$ denotes the set of points where X is zero; this set is finite.

As in the proof of Theorem 47, for such a vector field X the set of functions $y : M \to \mathbb{R}$ such that $\tilde{\Phi}_{X,y}$ is an immersion and, when restricted to a neighborhood of $\text{Sing}(X)$, an embedding; it is residual.

Finally, to obtain an embedding for $(X, \bar{y})$, $\bar{y}$ near y, we do not need an open covering in the present case. One can construct directly a map y_v, $v \in V$, where V is some finite vector space which is the analogue of y_ε with the following properties:

(i) $y_0 = y$;

(ii) for $x \in \text{Sing}(X)$, the 1-jet of y_v is independent of v;

(iii) for $x, x' \in \text{Sing}(X)$, $x \neq x'$ the map $j_x^{2m} \times j_x'^{2m} : V \to J_x^{2m}(M) \times J_x'^{2m}(M)$ has a surjective derivative for all (x, x') in $v = 0$; $J_x^{2m}(M)$ is the vector space of $2m$-jets, of functions on M in x; $j_x^{2m}(v)$ is the $2m$-jet on y_v in x.

Using y_v one defines a map $\tilde{\Phi} : M \times M \times V \to \mathbb{R}^{2m+1} \times \mathbb{R}^{2m+1}$ as before. The rest of the proof of Theorem 47 carries over to the present situation. $\square$

Theorems 47–49 show how a dynamical system with time evolution φ_t and observable y is determined generally by the set of all functions $t \to y(\varphi_t(x))$. In practice the following situation may occur for dynamical system with continuous time, but the value of observable y is only determined for discrete set $\{0, \alpha, 2\alpha, \ldots\}$ of values of t; $\alpha > 0$. This happens, for example, in the experiments of the onset of turbulence. Also instead of all sequences of the form $\{y(\varphi_{i \cdot \alpha}(x))\}_{i=0,\infty}$, $x \in M$, we only know such a sequence for one, or few values of x (depending on the number of experiments) and these sequences are not known entirely but only for $i = 1, \ldots \bar{N}$, for some finite big $\bar{N} \gg 1$. In this light we should know whether, under generic assumptions, the topology and dynamics in the positive limit set $L^+(x) = \{x' \in M \mid \exists t_i \to \infty \text{ with } \varphi_{t_i}(x) \to x'\}$ of x is determined by sequence $\{y(\varphi_{i \cdot \alpha}(x))\}_{i=0,\infty}$. This question is treated in the next theorem and its corollary; in Sections 3.5.3 and 3.5.4 we come back to the point that these sequences are only known up to some finite $\bar{N}$.

Theorem 50 *Let M be a compact manifold, X a vector field on M with flow φ_t and p a point in M. Then there is a residual subset $C_{X,p}$ of positive real numbers such that for $\alpha \in C_{X,p}$, the positive limit sets of p for the flow φ_t of X and for the diffeomorphism φ_α are the same. In other words, for $\alpha \in C_{X,p}$ we have that each point $q \in M$ which is the limit of sequence $\{\varphi_{t_i}(p)\}$, $t_i \in \mathbb{R}$, $t_i \to \infty$, is the limit of a sequence $\{\varphi_{n_i \cdot \alpha}(p)\}$, $n_i \in \mathbb{N}$, $n_i \to \infty$.*

Proof. Take $q \in L^+(p)$. For a small positive ε real number define $C_{\varepsilon,q} = \{\alpha > 0 \mid \forall n \in \mathbb{N} \text{ such that } \rho(\varphi_{n \cdot \alpha}(p), q) < \varepsilon\}$, ρ is some fixed metric on M. Clearly $C_{\varepsilon,q}$ is open; it is also dense. To prove this last statement we observe that for any $\bar{\alpha} > 0$ and $\bar{\varepsilon} > 0$, there is a point of $C_{\varepsilon,q}$ in $(\bar{\alpha}, \bar{\alpha} + \bar{\varepsilon})$ if and only if there is a $t \in (n \cdot \bar{\alpha}, n \cdot (\bar{\alpha} + \bar{\varepsilon}))$ with $\rho(\varphi_t(p), q) < \varepsilon$ for some integer n. The existence of such t follows from the fact that for big n the intervals $(n \cdot \bar{\alpha}, n \cdot (\bar{\alpha} + \bar{\varepsilon}))$ overlap in the sense that for big n, $n \cdot (\bar{\alpha} + \bar{\varepsilon}) > (n+1) \cdot \bar{\alpha}$ and the fact that there are arbitrary big values of t with $\rho(\varphi_t(p), q) < \varepsilon$. Since $C_{\varepsilon,q}$ is open and dense we can take for $C_{X,p} \subset \mathbb{R}_+$ the following residual set: $C_{X,p} = \bigcap_{i,j=1}^{\infty} C_{1/i, q_j}$ where $\{q_j\}$ is countable dense sequence in $L^+(p)$. $\square$

Corollary 12 *Let M be a compact manifold of dimension m. Consider quadruples, consisting of a vector field X, a function y, a point p, and a positive real number α. For generic such (X, y, p, α) (more precisely: for generic (X, y) and α satisfying generic conditions depending on X and p), the positive limit set $L^+(p)$ is diffeomorphic with the set of limit points of the following sequence in $\mathbb{R}^{2m+1}$: $\{(y(\varphi_{k \cdot \alpha}(p)), y(\varphi_{(k+1) \cdot \alpha}(p)), \ldots, y(\varphi_{(k+2m) \cdot \alpha}(p)))\}_{k=0,\infty}$.*

The meaning of *diffeomorphic* should be clear here: it means that there is a smooth embedding of M into $\mathbb{R}^{2m+1}$ mapping $L^+(p)$ bijectively to this set of limit points.

For further reference, remark that the metric properties of $\overline{\{\varphi_{i \cdot \alpha}(p)\}}_{i=0,\infty} \subset M$, with $\{\varphi_{i \cdot \alpha}(p)\}$ as a sequence of distinguished points are the same as $\overline{\{b_i\}}_{i=1,\infty} \subset \mathbb{R}^{2m+1}$ with $\{b_i\}$ as a sequence of distinguished points: $b_i = (y(\varphi_{i \cdot \alpha}(p)), \ldots, y(\varphi_{(1+2m) \cdot \alpha}(p))) \in \mathbb{R}^{2m+1}$. These metric properties are analogous to the distances in M and the corresponding distances in $\mathbb{R}^{2m+1}$ have a quotient which is uniformly bounded and bounded away from zero.

3.5.3   Limit capacity and dimension

There are several ways to define the notion of dimension for compact metric spaces. The definition which we use here gives so-called *limit capacity*. Since this limit capacity is not well known we treat here some of its properties.

Definition 57 Let (S, ρ) be a compact metric space. For $\varepsilon > 0$ we define the following:

(a) $s(S, \varepsilon)$ is a maximal cardinality of a subset of S such that no two points have distance less than ε; such set is called *maximal ε-separated set*;

(b) $r(S, \varepsilon)$ is the maximal cardinality of a subset of S such that S is the union of all ε-neighborhoods of its points; such a set is also called *minimal ε-spanning set*.

Note that $r(S, \varepsilon) \leq s(S, \varepsilon) \leq r(S, \varepsilon/2)$. The first inequality follows from the fact that maximal ε-separated set is ε-spanning one. The second inequality follows from the fact that in an $\frac{1}{2}\varepsilon$-neighborhood of any point (of minimal ε-spanning set) there can be at most one point of ε-separated set.

Definition 58 Next we define the limiting capacity $D(S)$ of S as

$$D(S) = \lim_{\varepsilon \to 0} \inf \frac{\ln(r(S, \varepsilon))}{-\ln \varepsilon} = \lim_{\varepsilon \to 0} \inf \frac{\ln(s(S, \varepsilon))}{-\ln \varepsilon}.$$

The fact that the last two expressions are equal follows from the inequalities defined above. The notion of capacity, of rather ε-capacity, was originally used for $s(\varepsilon)$. This limit capacity is strongly related to the Hausdorff dimension, (see Section 1.5), which is clear from the following equivalent definition.

Let U be a finite covering $\{U_i\}_{i \in I}$ of S. Then for $a > 0$ $D_{a,\mathrm{U}} = \sum_{i \in I} (\mathrm{diam}(U_i))^a$. Next we define $D_{a,\varepsilon}$ as the infimum of $D_{a,\mathrm{U}}$ where U runs over all finite covers of S each of whose elements has diameter ε. Notice that $D_{a,\varepsilon} \in [r(S, \varepsilon t) \cdot \varepsilon^a, r(S, \varepsilon/2) \cdot \varepsilon^a]$. One can see that there is a unique number, which is in fact the limit capacity $D(S)$, such that for $a > D(S)$, respectively $a < D(S)$, $\lim_{\varepsilon \to 0} \inf D_{a,\varepsilon}$ is zero, respectively infinite. This definition of limit capacity goes over in the dimension of Hausdorff dimension if we replace "each of whose elements has diameter ε" by "each of whose elements has diameter $\leq \varepsilon$."

For later reference we indicate the third definition of limit capacity.

Definition 59 Let $\{b_i\}_{i \in \mathbb{N}}$ be some countable dense sequence in S. For $\varepsilon > 0$ define the subset $J_\varepsilon \subset \mathbb{N} \cup \{0\}$ by

$$\text{for } i > 0 \text{ and } i \in J_\varepsilon : \ \rho(b_i, b_j) \geq \varepsilon \text{ for all } 0 \leq j \leq i \text{ and } j \in J_\varepsilon.$$

Here C_ε denotes cardinality of J_ε.

From these definitions it follows that $r(S, \varepsilon') \leq C_\varepsilon \leq s(S, \varepsilon)$ whenever $0 < \varepsilon < \varepsilon'$. Hence we may also define $D(S)$ by $D(S) = \lim_{\varepsilon \to 0} \inf \frac{\ln C_\varepsilon}{-\ln \varepsilon}$. From Section 1.5 we know that the Hausdorff dimension is equal or greater than the topological dimension and from the above consideration follows that the limit capacity is equal or greater than Hausdorff dimension. Both Hausdorff dimension and the limit capacity depend on the metric and not only on the topology. If ρ and ρ' are metrics on S such that $C\rho(x, y) \geq \rho'(x, y) \geq C^{-1}\rho(x, y)$ for some constant C and any $x, y \in S$, then the limit capacity and Hausdorff dimension are the same for the metrics ρ and ρ'. In this case the metrics ρ and ρ' are called *metrically equivalent*. Finally, if S is a compact manifold with a metric ρ, which is metrically equivalent with a metric induced by Riemannian structure, then the limit capacity equals the topological dimension.

Hausdorff dimension and the limit capacity can occur for positive limit sets of smooth vector fields on compact manifolds; if the answer is no then for all our purposes Hausdorff dimension and the limit capacity are the same. Contrary to the topological dimension, Hausdorff dimension and the limit capacity need not be integers. If we take, for example, for $\tilde{S}$ a Cantor set in $\mathbb{R}$, define as $\tilde{S} = \bigcap_{i=0}^{\infty} S_i$, where $S_0 = [0, 1]$, $S_{i+1} \subset S_i$, S_i has 2^i

intervals of the length α^i, $\alpha < \frac{1}{2}$, and S_{i+1} is obtained from S_i by removing in the middle of each segment of S_i a segment of the length $\alpha^i \cdot (1 - 2\alpha)$. We take as countable dense subset S the union of the left endpoints of the intervals of S_i for all i. If we compute C_ε for $\varepsilon = \alpha^i$, we find $C_{\alpha^i} = 2^i$. From this one may deduce that $D(S) = -\ln 2/\ln \alpha$. In determining the limit capacity of a closed subset of a compact manifold, note that there is only one metric equivalence class on the manifold which contains a metric induced by Riemannian structure. Limit capacity is always assumed to be defined with respect to a metric in this class.

3.5.4   Dimension and entropy

Let M be a compact manifold with a smooth vector field X, a smooth function y : $M \to \mathbb{R}$. Assume that a point $p \in M$ is a part of its own positive limit set $L^+(p)$; and that for some fixed $\alpha > 0$, the sequence $\{\varphi_{i \cdot \alpha}(p)\}_{i=0,\infty}$ is dense in $L^+(p)$ and that (φ_α, y) is generic in the sense of Theorem 47; φ_t denotes flow of X. Only nongeneric assumption we made on (M, X, y, p, α) is $p \in L^+(p)$. This assumption can in some sense be justified: if the orbit $\varphi_t(q)$ goes to an "attractor for $t \to \infty$", then if we replace q by $\varphi_T(q) = \tilde{q}$, $T \gg 1$, it is almost true that $\tilde{q} \in L^+(\tilde{q})$. So the assumption $p \in L^+(p)$ is a way to include in the description the fact that we can only start measuring after the experiment is already going for quite a time with fixed Ω.

In this situation we have the sequence $\{a_i = y(\varphi_{i \cdot \alpha}(p))\}_{i=0,\infty}$, which represents the experimental output (so for the moment we assume the experiment has been carried out for an infinite amount of time). From this sequence we obtain subsets $J_{n,\varepsilon} \subset \mathbb{N} \cup \{0\}$ by the inductive definition (see Section 3.5.3),

> for $i > 0$: $i \in J_{n,\varepsilon}$ if and only if
> $\max([a_i - a_j], [a_{i+1} - a_{j+1}], \dots, [a_{i+n} - a_{j+n}]) \geq \varepsilon$ for all $0 \leq j \leq i$ with $j \in J_{n,\varepsilon}$.

Here $C_{n,\varepsilon}$ denotes the cardinality of $J_{n,\varepsilon}$.

Theorem 51 (Main theorem) *The limit capacity of $L^+(p)$ equals*

$$D(L^+(p)) = \lim_{n \to \infty} \left(\liminf_{\varepsilon \to 0} \left(\frac{\ln C_{n,\varepsilon}}{-\ln \varepsilon} \right) \right),$$

where $\lim$ for $n \to \infty$ reaches the limit value for every $n \geq 2 \dim M$. The topological entropy H of the flow $\varphi_\alpha | L^+(p)$ equals $H(L^+(p)) = \lim_{\varepsilon \to 0} \left(\lim_{n \to \infty} \sup \left(\frac{1}{n} \ln C_{n,\varepsilon} \right) \right)$, where $\lim$ for $\varepsilon \to 0$ often (e.g., if $L^+(p)$ is an expansive basic set, [102]) reaches the limit value for every $0 < \varepsilon < \varepsilon_0$ for some ε_0.

Proof. Take $N \geq 2 \dim M$. The map Φ : $M \to \mathbb{R}^{N+1}$, defined by $q \mapsto (y(q), y(\varphi_\alpha(q)), \dots, y(\varphi_{N\alpha}(q)))$, is an embedding. We use the metric

$$\rho((x_0, \dots, x_N), (x_0', \dots, x_N')) = \max_i |x_i - x_i'|$$

on $\Phi(M)$, which is equivalent in metric sense to any metric on $\Phi(M)$ derived from Riemannian metric. So we may use ρ to compute $D(L^+(p)) = D(\Phi(L^+(p)))$. The first statement in the theorem follows by applying Section 3.5.3 to the sequence $\{\varphi_{i \cdot \alpha}(p)\}_{i=1,\infty}$ in $L^+(p)$. $\square$

Next we shall determine the topological entropy of $\varphi_\alpha | L^+(p)$. For this we have to find the cardinality of minimal ε-spanning set of orbits of the length n. A minimal ε-spanning set of orbits of the length n of φ_α is a finite set $\{q_i\}_{i \in I}$ in $L^+(p)$ such that for every $q \in L^+(p)$ there is some $i_0 \in I$ such that $\rho(\varphi_{i \cdot \alpha}(q), \varphi_{i \cdot \alpha}(q_{i_0})) < \varepsilon$ for $\forall i \in [0, n]$; among all subsets of $L^+(p)$ satisfying (i), $\{q_i\}_{i \in I}$ has minimal cardinality. Let $r(n, \varepsilon)$ be this cardinality. There is also a maximal cardinality of ε-separated orbits of the length n, denoted by $s(n, \varepsilon)$. The entropy can be defined as

$$H(L^+(p)) = \lim_{\varepsilon \to 0} \left(\lim_{n \to \infty} \sup \left(\frac{1}{n} r(n, \varepsilon) \right) \right) = \lim_{\varepsilon \to 0} \left(\lim_{n \to \infty} \sup \left(\frac{1}{n} \ln(s(n, \varepsilon)) \right) \right).$$

Using the metric ρ, defined above, replace $s(n, \varepsilon)$ or $r(n, \varepsilon)$ by $C_{n+N, \varepsilon}$ (Section 3.5.3). From this we obtain

$$H(L^+(p)) = \lim_{\varepsilon \to 0} \left(\lim_{n \to \infty} \sup \left(\frac{1}{n} \ln(C_{n+N, \varepsilon}) \right) \right) = \lim_{\varepsilon \to 0} \left(\lim_{n \to \infty} \sup \left(\frac{1}{n} \ln(C_{n, \varepsilon}) \right) \right).$$

Remark 20 If we denote $(\ln C_{n, \varepsilon})/(n - \ln \varepsilon)$ by $Z(n, - \ln \varepsilon)$ and regard both n and $- \ln \varepsilon$ as continuous variables one can see from few examples (Anosov automorphisms on the torus and horseshoes) that often $\lim_{\substack{\alpha, \beta \to \infty \\ \alpha / \beta \to \gamma}} Z(\alpha, \beta)$ exists for all $\gamma > 0$ forming one-parameter family of "topological invariants" connecting entropy with limit capacity.

Observation. Application of this main theorem to the output Taylor–Couette flow, described in the beginning of this section, gives some complications due to the fact that $\{a_i\}_{i=1, \bar{N}}$ is finite in this case. For such finite sequence one should proceed as follows: for n, ε, m with $n + m \le \bar{N}$ define subsets $J_{n, \varepsilon, m} \subset \mathbb{N} \cup \{0\}$ as follows:

for $i > 0$; $i \in J_{n, \varepsilon, m} \Leftrightarrow \max_{0 \le k \le n} |a_{i+k} - a_{j+k}| \ge \varepsilon$ for $i \le m$ and for all $j < i$, $j \in J_{n, \varepsilon, m}$.

Here $C_{n, \varepsilon, m}$ denotes cardinality of $J_{n, \varepsilon, m}$. For $\bar{N} \to \infty$, one has $\lim_{m \to \infty} C_{n, \varepsilon, m} = C_{n, \varepsilon}$, i.e., $C_{n, \varepsilon, m}$ is nondecreasing in m. Hence it seems reasonable to take $C_{n, \varepsilon, \bar{N}-n}$ as an approximation of $C_{n, \varepsilon}$ provided the difference between $C_{n, \varepsilon, \bar{N}-n}$ and say, $C_{n, \varepsilon, [(\bar{N}-n)/2]}$ is sufficiently small, say, of the order of $1 - 2$ percent. In this way there is possibility to calculate $C_{n, \varepsilon}$ in a certain region of (n, ε)-plane; also one should consider these values for $C_{n, \varepsilon}$ only reliable if ε is well above the expected errors in the measurement. From these numerical values for $C_{n, \varepsilon}$ one should decide, on the basis of the main theorem what the values of $D(L^+(p))$ and $H(L^+(p))$ are or whether the limits defining these values "do not exist numerically."

If, in the calculation of $D(L^+(p))$, the $\lim_{n \to \infty}$ would have the tendency of going to infinity, this would imply that representing the evolution on a finite dimensional manifold is a mistake. On the other hand this limit would go to a noninteger, this would be evidence in favor of a strange attractor. Namely (see Section 3.5.3), for Cantor set C we may have $D(C)$ a noninteger, and strange attractors have in general a Cantor set like structure [102]. If the experimental data do not clearly indicate the limits in the calculation of $D(L^+(p))$ and $H(L^+(p))$ to exist and to be finite, then both [63] and [93] pictures are to be rejected as explanation of the experimental data.

3.6 Global attractor for Navier–Stokes equation

For an infinite dimensional dynamical system in a box of side L and spatial dimension d, the problem of resolving the typical length scale l of natural features in the flow is normally related to the (integer) number of degrees of freedom $\mathcal{N}$ in the system by

$$\mathcal{N} \sim (L/l)^d. \tag{3.94}$$

It desirable to associate the number of degrees of freedom with the average number of data required to configure all possible motions starting from all initial states. In the descriptive way it has been introduced in (3.94), $\mathcal{N}$ has not been properly defined. To do this we need firstly to show that a global attractor $\mathcal{A}$ exists for the system in question. Secondly it is necessary to devise a way of associating $\mathcal{N}$ with a typical length scale (or scales) of solutions on this attractor. In fact, $\mathcal{N}$ is often loosely identified with the Lyapunov dimension $d_L(\mathcal{A})$ of the global attractor. It will be seen in this chapter that certain estimates for the two-dimensional NSE coincide with some found in the chapter.

In three-dimensional case, because there is yet no proof that strong solutions exist, it must be assumed that one or another time average; such as $\langle \|D\mathbf{u}\|_\infty^{1/2} \rangle$, where $\mathbf{u} = (u_j)$ is the *velocity vector*, is finite and can be manipulated as if strong solutions exist and make sense. The predictions about minimum length scales are made under this assumption. Unless one can elevate weak to strong solutions, we have no a priori estimates; nevertheless, it is useful to do this anyway to see what it is that needs to be proved.

We will discuss the so called *ladder inequality for the NSE on* $\Omega = [0, L]^d$ *with periodic boundary conditions and zero mean*, and discuss its consequences in both two- and three-dimensional cases. This will enable us to relate the evolution of a seminorm of solutions of the NSE, containing a given number of derivatives, to one containing a lower number of derivatives. The proof of the ladder inequality uses seminorms, which contain higher order derivatives of the velocity field. Denote the L^2-norm of the velocity vector by $\|\mathbf{u}\|_2$, and the sum of the L^2-norms of gradients of the components of $\mathbf{u}$ by $\|\nabla\mathbf{u}\|_2$. The notion of the general derivative (called the *D-operator*) of the velocity field of order $N \geq 1$ will be used. In terms of the D-operator, N derivatives on u in d dimensions are defined as

$$|D^N\mathbf{u}|^2 = \sum\nolimits_{|n|=N} \Big| \frac{\partial^N \mathbf{u}}{\partial x_1^{n_1} \dots \partial x_d^{n_d}} \Big|^2,$$

where $n = (n_1, n_2, \dots, n_d)$, $n_i \in \mathbb{N}$ and $|n| = \sum_i n_i$. For example, in two-dimension we have $|D^N\mathbf{u}|^2 = \sum_{i=1}^2 (u_{i,xx}^2 + 2\,u_{i,xy}^2 + u_{i,yy}^2)$. Since the Laplacian Δ and the vector gradient ∇ appear in the NSE, and the rot operator appears in the vorticity ω, we will discuss how D, Δ, ∇ and rot are related. Define the squares of seminorms H_N by

$$H_N = \sum\nolimits_{i=1}^d \int_\Omega |D^N u_i|^2 \, \mathrm{d}x = \sum\nolimits_{i=1}^d \int_\Omega \sum\nolimits_{|n|=N} \Big| \frac{\partial^N u_i}{\partial x_1^{n_1} \dots \partial x_d^{n_d}} \Big|^2 \mathrm{d}x. \tag{3.95}$$

Write $\mathrm{rot}\,\mathbf{u} = \varepsilon_{ijk} \frac{\partial u_k}{\partial x_j}$, where ε_{ijk} (the totally antisymmetric Levi–Civita symbol) is 1 if (i, j, k) is an even permutation of (1,2,3), is -1 if it is an odd permutation, and 0 if any index is repeated. Notice that $\varepsilon_{ijk}\varepsilon_{ilm} = \delta_{jl}\delta_{km} - \delta_{jm}\delta_{kl}$. To show how seminorms of D, ∇ and rot are related on periodic domains for divergence-free fields u, write

$$\int_\Omega |\mathrm{rot}\,\mathbf{u}|^2 \, \mathrm{d}x = \int_\Omega \varepsilon_{ijk}\varepsilon_{ilm} u_{j,k} u_{l,m} \, \mathrm{d}x = \int_\Omega \Big[\Big(\frac{\partial u_j}{\partial x_k} \Big)^2 - \frac{\partial u_j}{\partial x_k} \cdot \frac{\partial u_k}{\partial x_j} \Big] \mathrm{d}x.$$

Integrating by parts twice on the last term produces $-u_{jj}u_{kk}$, which is zero because div $\mathbf{u} = 0$, we obtain

$$\|\mathrm{rot}\,\mathbf{u}\|_2^2 = \|\nabla\mathbf{u}\|_2^2 = \|D\mathbf{u}\|_2^2 = H_1.$$

Thus, D and ∇ are equivalent, but what about D^2 and Δ? The equality $\mathrm{rot} \circ \mathrm{rot}\,\mathbf{u} = -\Delta\mathbf{u} + \nabla\nabla \cdot \mathbf{u}$ allows us to write $\|\mathrm{rot} \circ \mathrm{rot}\,\mathbf{u}\|_2^2 = \|\Delta\mathbf{u}\|_2^2$. By generalizing these calculations, we obtain

$$H_N = \|D^N\mathbf{u}\|_2^2 = \|\mathrm{rot}^N\mathbf{u}\|_2^2 \quad (N > 2).$$

From the above one may conclude that *while the operators* D, rot *and* ∇ *are not equivalent operators pointwise, nevertheless they are effectively equivalent under the* L^2-norm. This equivalence is not true in L^p ($p \neq 2$). For instance, when $p = \infty$, look at $\|D\mathbf{u}\|_\infty$, and the

L^∞ norm of the vorticity $\|\omega\|_\infty$. ($\|D\mathbf{u}\|_\infty$ is interchangeable with $\|\nabla\mathbf{u}\|_\infty$). These objects have the same dimension of frequency, but

$$\|\omega\|_\infty = \|\text{rot}\,\mathbf{u}\|_\infty \leq \|D\mathbf{u}\|_\infty. \qquad (3.96)$$

The reason is that $\|D\mathbf{u}\|_\infty$, is the supnorm of every derivative of every component of $\mathbf{u}$ while $\|\omega\|_\infty$ has derivatives of various components of $\mathbf{u}$ missing because of the way the rot operation is defined. The RHS of (3.96) can be infinite, while the left-hand side is finite. In the following sections we will treat the NSE and its solutions as though everything is defined classically. Really, the operations performed below should be applied to the Galerkin truncations and their associated equations of motion. None of the actual calculations depends on the order of the truncation, though, so we omit reference to it. The produced estimates are uniform in the order of the Galerkin approximation, thus hold for any limit of them, e.g., weak solutions.

3.6.1 The ladder inequality

The energy $2E(t) \equiv H_0 = \|\mathbf{u}\|_2^2$ is bounded from above in every dimension. But this tells us little about the velocity field $\mathbf{u}(x,t)$ because functions bounded in L^2 can still display spatial singularities. For better control of the derivatives of the velocity and vorticity fields, consider behavior of H_N for arbitrary t. Although the time integral of H_1 is bounded from above (Leray's inequality) [22]:

$$\frac{1}{2}\|\mathbf{u}(\cdot,t)\|_2^2 + \nu\int_0^t \|\nabla\mathbf{u}(\cdot,\tau)\|_2^2\,\mathrm{d}\tau \leq \frac{1}{2}\|\mathbf{u}(0,t)\|_2^2 + \int_0^t \Big(\int_\Omega \mathbf{u}(x,\tau)\cdot f(x)\,\mathrm{d}x\Big)\mathrm{d}\tau, \qquad (3.97)$$

this does not tell us much about H_1, as temporal singularities in the associated seminorm could occur.

The control of the L^2- and the time average of the H_1 norms is needed to understanding of whether regularity can be achieved in two- and three-dimensional cases. It is also necessary not only to show how H_N evolves in time but also how the body forcing affects the flow. The desired "Navier–Stokes ladder inequality" relates H_N to H_{N-s} for general N and $1 < s < N$. The main tools used in the proof of this inequality are the Divergence Theorem and the Gagliardo–Nirenberg calculus inequalities (see Section 3.9).

Define the set of seminorms associated with the forcing by

$$\Phi_N = \int_\Omega |D^N f|^2\,\mathrm{d}x, \qquad (3.98)$$

where div $f = 0$ without loss of generality. Assume a forcing which is independent of time, and in which the forcing function $f(x)$ is has a cutoff in its spectrum with a smallest length scale

$$\lambda_f^{-2} = \sup{}_N(\Phi_{N+1}/\Phi_N), \qquad (3.99)$$

i.e. the cutoff k_f in the spectrum is given by $k_f = 2\pi/\lambda_f$. To include the forcing, let introduce a natural "time" $\tau = L^2\nu^{-1}$ for the system based on the length of the box L and the viscosity ν. The forcing function f is dimensionally an acceleration, so $\mathbf{u}_f = \tau f$ is dimensionally a velocity vector. Define

$$F_N = H_N + \tau^2\Phi_N = H_N + \int_\Omega |D^N\mathbf{u}_f|^2\,\mathrm{d}x. \qquad (3.100)$$

This F_N not only contains as N derivatives on the three components of velocity so the three components of the forcing. Define an inverse squared length scale as the following combination:

$$\lambda_0^{-2} = \lambda_f^{-2} + L^{-2}. \tag{3.101}$$

The following theorem is the main theorem of the section:

Theorem 52 *In $d = 2, 3$ dimensions for periodic boundary conditions and $1 < s < N$, the following inequalities hold:*

$$-\nu F_{N+1} - (c_N \|D\mathbf{u}\|_\infty + \nu \lambda_0^{-2}) F_N \leq \frac{1}{2} \dot{F}_N \leq -\nu \frac{F_N^{1+1/s}}{F_{N-s}^{1/s}} + (c_N \|D\mathbf{u}\|_\infty + \nu \lambda_0^{-2}) F_N. \tag{3.102}$$

Proof. We shall use the following vector identity:

$$\text{div}(A \times B) = B \cdot \text{rot}\, A - A \cdot \text{rot}\, B. \tag{3.103}$$

Differentiate the H_N with respect to time

$$\frac{1}{2} \dot{H}_N = \int_\Omega (\text{rot}^N \mathbf{u}) \cdot [\text{rot}^N (\nu \Delta \mathbf{u} - (\mathbf{u} \cdot \nabla \mathbf{u}) - \nabla p + f)] \, dx \tag{3.104}$$

and then proceed with the proof in nine steps.

STEP 1 (The Laplacian term). By the Divergence Theorem, integration of (3.103) yields

$$\int_\Omega (B \cdot \text{rot}\, A) \, dx = \int_\Omega (A \cdot \text{rot}\, B) \, dx. \tag{3.105}$$

Integration by parts without the sign change for (3.105) enables us to get the Laplacian term in (3.104),

$$\text{Laplasian term} = \nu \int_\Omega (\text{rot}^N \mathbf{u}) \cdot (\text{rot}^N \Delta \mathbf{u}) \, dx. \tag{3.106}$$

Since $\text{rot} \circ \text{rot}\, u = -\Delta u$, (3.105) can be used to move one of the rot operators across in (3.106):

$$\text{Laplasian term} = -\nu \int_\Omega (\text{rot}^N \mathbf{u}) \cdot (\text{rot}^{N+2} \mathbf{u}) \, dx = -\nu H_{N+1}.$$

STEP 2 (The nonlinear term). Using the D notation, this term becomes

$$\text{Non-linear term} = \left| \int_\Omega (D^N \mathbf{u}) \cdot (D^N [-(\mathbf{u} \cdot \nabla)\mathbf{u}]) \, dx \right|. \tag{3.107}$$

Performing a Leibnitz expansion on the term with N derivatives and separating the first term, the i-th component of nonlinear term (NLT) is

$$|NLT_i| \leq -\int_\Omega \phi_i (\mathbf{u} \cdot \nabla) \phi_i \, dx + \left| \sum_{i=1}^N \sum_j C_l^N \int_\Omega (D^N u_i)(D^l u_j)(D^{N+1-l} u_i) \, dx \right|, \tag{3.108}$$

where $\phi_i = D^N u_i$. The first term on the RHS of (3.108) can be shown to be identically zero:

$$\int_\Omega \phi_i (\mathbf{u} \cdot \nabla) \phi_i \, dx = \int_\Omega \left[\text{div}\, (\frac{1}{2} \phi_i^2 \mathbf{u}) - \frac{1}{2} \phi_i^2 \, \text{div}\, \mathbf{u} \right] dx.$$

The last term is zero because $\text{div}\, u = 0$ and the first term is zero because of the Divergence Theorem and the periodic boundary conditions. After two applications of Hölder's inequality, with $p^{-1} + q^{-1} = \frac{1}{2}$, inequality (3.108) therefore becomes

$$|NLT| \leq 2^N H_N^{1/2} \sum_{i,j} \sum_{l=1}^N \|D^l u_j\|_p \cdot \|D^{N+1-l} u_i\|_q. \tag{3.109}$$

The constant 2^N is the sum of the binomial coefficients. Since both $p, q > 2$, it is necessary to use calculus inequalities on the L^p and L^q norms in (3.109). Define $A_j \equiv Du_j$, and then use the calculus inequalities [22] for spatial dimensions $d = 1, 2, 3$,

$$\|f\|_\infty^2 \leq c_1 \|\nabla f\|_2 \cdot \|f\|_2, \quad \|f\|_\infty^2 \leq c_2 \|\nabla f\|_2^2 [1 + \log(L\|\Delta f\|_2/\|\nabla f\|_2)],$$
$$\|f\|_\infty^2 \leq c_3 \|\Delta f\|_2 \cdot \|\nabla f\|_2$$

to obtain

$$\|D^{l-1} A_j\|_p \leq c \|D^{N-1} A_j\|_2^a \cdot \|A_j\|_\infty^{1-a}. \tag{3.110}$$

In (3.110), the exponent a is calculated through the formula (3.265) in Section 3.9, which balances the dimensional weighting across the norms. From this $\frac{1}{p} = \frac{l-1}{d} + a(\frac{1}{2} - \frac{N-1}{d})$, where the value of a is restricted to be in the range

$$\frac{l-1}{N-1} \leq a < 1. \tag{3.111}$$

The value of a is deliberately chosen to be at the minimum of its range in inequality (3.111), in which case, for $l < N$, a is independent of the dimension d and takes the value $a = \frac{l-1}{N-1}$, $ap = 2$. The case $l = N$ does not require the use of (3.110). The same procedure on the L^q norm gives

$$\|D^{l-1} A_j\|_q \leq c \|D^{N-1} A_j\|_2^b \cdot \|A_j\|_\infty^{1-b}$$

with $b = \frac{N-l}{N-1}$ ($l \geq 1$) and $bq = 2$. Using the fact that $a + b = 1 \Rightarrow p^{-1} + q^{-1} = \frac{1}{2}$, the combination of these results in (3.109) means that we have proved the following lemma which will be useful later on:

Lemma 7 *An upper estimate for the integral $I_{ij} = \left| \int_\Omega (D^N u_i)(D^l u_j)(D^{N+1-l} u_i) \, dx \right|$ is given by*

$$I_{ij} \leq c_N \big(H_N^{(i)}\big)^{1/2+1/q} \big(H_N^{(j)}\big)^{1/p} \|Du_j\|_\infty^{1-a} \cdot \|Du_i\|_\infty^{1-b},$$

where $H_N^{(i)} = \int_\Omega |D^N u_i|^2 \, dx$.

Continuing with Step 2 of the proof of Theorem, note that because $H_N^{(i)} \leq H_N$, $\|Du_j\|_\infty \leq \|Du\|_\infty$ and $a + b = 1$, we end up with a simple upper bound for (3.109)

$$|\text{NLT}| \leq c_N H_N \|Du\|_\infty. \tag{3.112}$$

The constants c_N are dimensionless and that the RHS of (3.112) has the same dimension as that of $\dot{H}_N$. This is the end of Step 2.

STEP 3. The pressure term: it vanishes identically by the action of the rot operation in H_N.

STEP 4. The forcing term: this is the last term in (3.104) and, after an application of the Cauchy–Schwarz inequality, becomes

$$\left| \int_\Omega (\text{rot}^N \mathbf{u}) \cdot (\text{rot}^N f) \, dx \right| \leq H_N^{1/2} \Phi_N^{1/2}.$$

The Schwarz inequality has been used.

STEP 5. The combination of Steps 1-4. Put all these results together

$$\frac{1}{2} \dot{H}_N \leq -\nu H_{N+1} + c_N H_N \|Du\|_\infty + H_N^{1/2} \Phi_N^{1/2}. \tag{3.113}$$

STEP 6. An upper bound on $-H_{N+1}$: This can be achieved through the following

Lemma 8 *For $1 \le s \le N$ and $r \ge 1$, H_N defined in (3.95) satisfy the inequality*

$$H_N \le H_{N-s}^{\frac{r}{r+s}} \cdot H_{N+r}^{\frac{s}{r+s}}.$$

Proof. This proceeds by induction and is performed in four steps A to D

Step A. Taking the definition $M_N = \int_\Omega |D^N \mathbf{u}|^2 \, dx$, and integrating by parts and the Cauchy–Schwarz inequality, we obtain

$$M_N \le \Big(\int_\Omega |D^{N-1} \mathbf{u}|^2 \, dx \Big)^{1/2} \Big(\int_\Omega |D^{N+1} \mathbf{u}|^2 \, dx \Big)^{1/2}.$$

Hence,

$$M_N \le M_{N+1}^{1/2} \cdot M_{N-1}^{1/2}. \tag{3.114}$$

This is also true separately for each component.

Step B. We show that for $s \ge 1$,

$$M_s \le M_{s+1}^{\frac{s}{s+1}} \cdot M_0^{\frac{1}{s+1}}. \tag{3.115}$$

By (3.114), (3.115) holds for $s = 1$. Assume (3.115) for $s > 1$. Then

$$M_{s+1} \le M_{s+2}^{1/2} \cdot M_s^{1/2} \le M_{s+2}^{1/2} \cdot M_{s+1}^{\frac{s}{2(s+1)}} \cdot M_0^{\frac{1}{2(s+1)}},$$

so $M_{s+1} \le M_{s+2}^{\frac{s+1}{s+2}} \cdot M_0^{\frac{1}{s+2}}$. Hence (3.115) is true for any s by induction.

Step C. We show that for any s, r

$$M_s \le M_{s+r}^{\frac{s}{s+r}} \cdot M_0^{\frac{r}{s+r}}. \tag{3.116}$$

By (3.115), (3.116) holds for $r = 1$. Assume (3.116) holds for $r > 1$. Then

$$M_s \le M_{s+r}^{\frac{s}{s+r}} \cdot M_0^{\frac{r}{s+r}} \le M_{s+r+1}^{\frac{s}{s+r+1}} \cdot M_0^{\frac{r+1}{s+r+1}},$$

where we have used (3.115). Hence (3.116) is true for for all s, r by induction.

Step D. We know (3.116) holds with $M_s = \|D^s \mathbf{u}\|_2^2$. Now suppose that $v_i = D^{N-s} u_i$, then $M_s = H_N$ and $M_0 = H_{N-s}$. Equation (3.116) becomes $H_N \le H_{N+r}^{\frac{s}{s+r}} \cdot H_{N-s}^{\frac{r}{s+r}}$. Clearly, the choice $r = 1$ in the above lemma gives the required term in the theorem. We have demonstrated that for $1 \le s \le N$ and in dimension d, the H_N satisfy

$$\frac{1}{2} \dot{H}_N \le -\nu \frac{H_N^{1+1/s}}{H_{N-s}^{1/s}} + c_N H_N \|D\mathbf{u}\|_\infty + H_N^{1/2} \Phi_N^{1/2}.$$

This is a form of ladder inequality for the H_N. It is inconvenient to leave it in this form because of the way the forcing term is expressed here. Later on, we will want to divide through by H_N and the square root in the forcing term will cause complications. For this reason we set about dealing with the forcing term.

STEP 7. Returning to the theorem proof at the stage of (3.113), add and subtract a $\nu \tau^2 \Phi_{N+1}$ term considering (3.100) to obtain

$$\frac{1}{2} \dot{F}_N \le -\nu F_{N+1} + c_N F_N \|D\mathbf{u}\|_\infty + \nu \tau^2 \Phi_{N+1} + F_N^{1/2} \Phi_N^{1/2},$$

which can be written as

$$\frac{1}{2} \dot{F}_N \le -\nu F_{N+1} + (c_N \|D\mathbf{u}\|_\infty + \nu F_N^{-1} \tau^2 \Phi_{N+1} + \tau^{-1}) F_N. \tag{3.117}$$

Since from (3.100) inequality $F_N \geq \tau^2 \Phi_N$ can appear, (3.99), (3.101), and (3.117) can be written as

$$\frac{1}{2}\dot{F}_N \leq -\nu F_{N+1} + (c_N \|D\mathbf{u}\|_\infty + \nu \lambda_0^{-2})F_N \qquad (3.118)$$

with λ_0 defined as in (3.101).

STEP 8. The result of Lemma 8 which we established for $s \geq 1$, $r = 1$ is also true for every "component" of $H_N = \sum_{i=1} H_N^{(i)}$ where the suffix i refers to the i-th component u_i; of the velocity field:

$$(H_N^{(i)})^{1+s} \leq (H_{N+1}^{(i)})^s \cdot H_{N-s}^{(i)}.$$

We can therefore take the $(s+1)$-th root and sum over i from 1 to 6 which, in effect, includes all six "components" of F_N: three from H_N and three from the forcing. Therefore

$$F_N \leq \sum_{i=1}^6 \left(H_{N+1}^{(i)}\right)^{\frac{s}{1+s}} \cdot \left(H_{N-s}^{(i)}\right)^{\frac{1}{1+s}}.$$

Using the Hölder–Schwarz inequality, we have proved.

Lemma 9 *For $1 \leq s \leq N$, the F_N defined in (3.100) satisfy the inequality $F_N \leq F_{N-s}^{\frac{1}{1+s}} \cdot F_{N+1}^{\frac{s}{1+s}}$. This gives the first term in RHS of (3.102).*

STEP 9. Proof of the LHS of (3.102): Step 1, which concerns the Laplacian term, contains no inequalities. In consequence, we bound the time derivative of H_N below as well as above

$$\frac{1}{2}\dot{H}_N \geq -\nu H_{N+1} - c_N H_N \|D\mathbf{u}\|_\infty - H_N^{1/2} \Phi_N^{1/2}.$$

Finally, to obtain the last step in the proof of LHS (3.102), and noting that $H_{N+1} \leq F_{N+1}$, it is not difficult to see that

$$\frac{1}{2}\dot{F}_N \geq -\nu F_{N+1} - c_N F_N \|D\mathbf{u}\|_\infty - \tau^{-1} F_N,$$

where $\tau = L^2 \nu^{-1}$. Because $\tau^{-1} \leq \nu \lambda_0^{-2}$, the result follows.

Definition of a length scale. Computing upper bounds on norms of the Navier–Stokes velocity field $\mathbf{u}(x,t)$ is a vacuous exercise if pursued without recourse to physical interpretation. The ultimate aim of these calculations is to produce a lower bound on the smallest length scale in a forced flow. The term "smallest length scale" is not uniquely defined and so requires some thought. Ideally, we would like to resolve the smallest "eddy" in the flow starting from smooth initial conditions. To see how Theorem 52 naturally defines a set of length scales, we show that the F_N ladder suggests length scales associated with moments of the power spectrum. A scale in turbulence theory may be defined in the following way. From the energy in the flow $2E(t) \equiv H_0 = \int_\Omega |\mathbf{u}(x,t)|^2 \, dx$ we can use *Parseval's theorem*

$$E(t) = \frac{1}{2}\int |\hat{\mathbf{u}}(\mathbf{k},t)| \, dk \equiv \int E(\mathbf{k},t) \, dk,$$

where $\hat{\mathbf{u}}(\mathbf{k},t)$ is the Fourier image of $\mathbf{u}(x,t)$, to define the instantaneous energy spectrum $E(\mathbf{k},t)$. This defines the distribution of energy among scales and allows us to consider the normalized "probability distribution" $P(\mathbf{k},t) = \frac{E(\mathbf{k},t)}{E(t)}$. Thus, time-dependent moments of this distribution can be written as

$$[k^{2N}]_{s.a.} = \frac{\int_\Omega |\mathbf{k}|^{2N} |\mathbf{u}(x,t)|^2 \, dx}{\int_\Omega |\mathbf{u}(x,t)|^2 \, dx} := \frac{H_N}{H_0},$$

where the *s.a.* on $[k^{2N}]_{s.a.}$ means "spatial average." The relevant time-dependent length scale associated with this quantity is

$$(\text{length})^{-1} \sim \{[k^{2N}]_{s.a.}\}^{1/2N}.$$

Not only can the forcing be included in the definitions to make the H_N into F_N, but the ratio of moments $[k^{2N}]_{s.a.}/[k^{2r}]_{s.a.} = \frac{F_N}{F_R}$ also naturally defines a length scale $(\text{length})^{-1} \sim (F_N/F_r)^{1/2(N-r)}$. Clearly, we can define a set of *wave-numbers* defined by the ratio of moments above.

Definition 60 *Wave-numbers* are defined by the ratio

$$\kappa_{N,r}(t) = (F_N/F_r)^{1/2(N-r)}, \quad r = N - s \geq 0. \tag{3.119}$$

It is noteworthy that if one divides the ladder inequality (in Theorem 52) through by F_N, then the square of our time dependent *wave-numbers* $\kappa_{N,r}$ appear in the Laplacian term. It is convenient and natural, therefore, to time average the square of these objects to get an associated set of inverse squared lengths.

Definition 61 *Inverse square lengths* are defined as

$$l_{N,r}^{-2} \equiv \langle \kappa_{N,r}^2 \rangle = \langle (F_N/F_r)^{1/(N-r)} \rangle,$$

where the time average $\langle g(t) \rangle$ is $\langle g(t) \rangle = \lim_{t \to \infty} \sup_g \frac{1}{t} \int_0^t g(\tau) d\tau$.

Let us compare our definition of a length scale with the way time averages of moments are taken to define a length scale in turbulence theory. Consider the time averaged energy spectrum $\langle E(\mathbf{k}) \rangle$. The relevant length scales are defined in terms of the average distribution of energy

$$\langle P \rangle(\mathbf{k}) = \frac{\langle E(\mathbf{k}', \cdot) \rangle}{\int_\Omega \langle E(\mathbf{k}', \cdot) \rangle dk'}.$$

This, of course, is not the same as time averaged instantaneous distribution of energy

$$\langle P(\mathbf{k}, \cdot) \rangle = \frac{\langle E(\mathbf{k}', \cdot) \rangle}{\int_\Omega \langle E(\mathbf{k}', \cdot) \rangle dk'}.$$

A conventional definition of a length scale is given via $\langle P \rangle(\mathbf{k})$, and so a *ratio of time averages* of the F_N and F_r would occur. For the scale given in Definition 61 we have a *time average of a ratio*. Moments computed by taking the time average last are going to be more sensitive to rare fluctuations driving energy deep down to shorter scales than those computed from the time averaged energy spectrum. This is because, all other things being equal, rare events contribute little to the time averaged energy spectrum. Moreover, when a significant fraction of energy is at high wave-numbers, the (relative) energy dissipation is necessarily higher, so these events will be characterized by lower than normal total energy. They will then not even count that much toward the time average of the energy spectrum. Dividing by the total energy before averaging therefore amplifies the role of the low energy but high wave-number configurations in the distribution, skewing the distribution toward high wave-numbers. We therefore assert that the natural scale defined by computing averages on the $\kappa_{N,r}^2$ are likely to be more sensitive to intermittent fluctuations than a length scale determined by the time averaged energy spectrum alone. Using the operation of the time average, it is still possible that we may not have resolved all length scales in the fluid, a problem to which we will return in later sections.

The dynamical wave-numbers $\kappa_{N,r}$. We discuss how such quantities as the time average of ratios can be obtained from the ladder. Dividing through RHS of (3.102) by F_N and time averaging according to Definition 61 we find

$$\frac{1}{2\nu}\Big\langle \frac{d}{dt}[\log F_N]\Big\rangle + \langle \kappa_{N,r}^2 \rangle \le c_N \nu^{-1}\langle \|D\mathbf{u}\|_\infty \rangle + \lambda_0^{-2}.$$

The integral in the average $\langle \cdot \rangle$ integrates the time derivative perfectly leaving the LHS as $\log(F_N(t)/F_N(0))$. For each N, provided $\int_\Omega |D^N\mathbf{u}_f|^2\,dx > 1$, the function F_N is bounded from below by a fixed positive constant dependent only on the forcing and so the $\log F_N$ term is bounded from below. These terms are therefore bounded by a quantity that vanishes in the large time limit because of the $1/t$ in Definition 61. We have proved

Theorem 53 *For each value of N, if $\|D^N\mathbf{u}_f\|_2^2 > 1$, then*

$$\langle \kappa_{N,r}^2 \rangle \le c_N \nu^{-1}\langle \|D\mathbf{u}\|_\infty \rangle + \lambda_0^{-2}. \tag{3.120}$$

Remark 21 In its most important part, the $\|D\mathbf{u}\|_\infty$ term, the RHS of (3.120) is as if independent of the domain length L.

Since we have a ladder inequality for the time evolution of F_N and F_r, it is worth seeing how the $\kappa_{N,r}$ evolve with respect to time:

$$2(N-r)\dot\kappa_{N,r}\kappa_{N,r}^{2(N-r)-1} = \frac{\dot F_N}{F_r} - \frac{F_N}{F_r}\cdot\frac{\dot F_r}{F_r}.$$

We apply Theorem 53 for an upper bound on $\dot F_N$ and LHS in (3.102) for a lower bound on $\dot F_r$. In terms of $\kappa_{N,r}$, we get

$$\frac{1}{2}(N-r)\dot\kappa_{N,r} \le -\frac{1}{2}\nu\kappa_{N,r}(\kappa_{N,r}^2 - \kappa_{r+1,r}^2) + (c_N\|D\mathbf{u}\|_\infty + \nu\lambda_0^{-2})\kappa_{N,r}. \tag{3.121}$$

The consequences of (3.121) in the $d = 2, 3$ cases are discussed in Section 3.6.2. The inequality in (3.121) also gives us a clue about the nature of three-dimensional Euler singularities, a topic of Section 3.6.2.

Turning back to Definition 60 for $\kappa_{N,r}$ and noting that it has a dimension of inverse length, remark that it is possible to show that it has some meaning in terms of the calculus inequalities of Section 3.9, just used in the proof of the ladder inequality. If we take the r-th derivative of $\mathbf{u}$, namely $D^r\mathbf{u}$, and ask what the calculus inequality looks like which interpolates between $\|D^r\mathbf{u}\|_\infty$ and $\|D^r\mathbf{u}\|_2$, we ought to obtain a quantity which has dimension (inverse length)$^{d/2}$ because of compensation for the volume integral in the L^2-norm,

$$\|D^r u_i\|_\infty \le c\,\|D^N u_i\|_2^a \cdot \|D^r u_i\|_2^{1-a} \tag{3.122}$$

with $a = \frac{d}{2(N-r)}$ and $N > r + \frac{d}{2}$. To form $\kappa_{N,r}$, notice that

$$\|D^N u_i\|_2^2 \le F_N, \quad \|D^r u_i\|_2^2 \le F_r.$$

Dividing through (3.122) by $F_r^{a/2}$ we find that

$$\|D^r u_i\|_\infty \le c\kappa_{N,r}^{d/2}\|D^r u_i\|_2.$$

Here $\kappa_{N,r}^{d/2}$ is the object of dimension of (inverse length)$^{d/2}$. In other words, $\kappa_{N,r}$ mediates between L^∞- and L^2- norms of $D^r u_i$. The $\kappa_{N,r}$ are ordered in increasing values of N for fixed r and increasing values of r for fixed N.

TABLE 3.1: A summary of results so far; see exercises for the second F_N ladder.

N	Definition F_N	$F_N = \|D^N\mathbf{u}\|_2^2 + \|D^N\mathbf{u}_f\|_2^2$
1	First F_N ladder	$\frac{1}{2}\dot{F}_N \leq -\nu(F_N^{1+1/s}/F_{N-s}^{1/s}) + (c_N\|D\mathbf{u}\|_\infty + \nu\lambda_0^{-2})F_N$
2	Second F_N ladder	$\frac{1}{2}\dot{F}_N \leq -\frac{1}{2}\nu(F_N^{1+1/s}/F_{N-s}^{1/s}) + (c_N\nu^{-1}\|D\mathbf{u}\|_\infty + \nu\lambda_0^{-2})F_N$
3	Definition $\kappa_{N,r}$	$\kappa_{N,r} = (F_N/F_r)^{1/2(N-r)}, \quad r < N$
4	Time average $\kappa_{N,r}^2$	$\langle\kappa_{N,r}^2\rangle \leq c_N\nu^{-1}\langle\|D\mathbf{u}\|_\infty\rangle + \lambda_0^{-2}$
5	Ladder for $\kappa_{N,r}$	$\frac{N-r}{2}\dot{\kappa}_{N,r} \leq -\frac{1}{2}\nu\kappa_{N,r}(\kappa_{N,r}^2 - \kappa_{r+1,r}^2) + (c_N\|D\mathbf{u}\|_\infty + \nu\lambda_0^{-2})\kappa_{N,r}$

3.6.2 Estimates

The results of the previous sections are summarized in Table 3.1 Although two versions of the ladder given in Table 3.1 are the most general results for arbitrary N, for the three lowest rungs ($N = 0$, 1, and 2), there are more sensitive estimates for F_0, F_1, and F_2. Good bounds on these are important as the ladder is expressed as a type of a recurrence relation. It has already been shown that the energy and the enstrophy are important in both two- and three-dimensional flows, although for very different reasons.

Estimates for F_0. The only level where it is not necessary to use F_N-notation is at $N = 0$. By the vector identity,

$$\nabla(A \cdot B) = A \cdot \nabla B + B \cdot \nabla A + A \times \operatorname{rot} B + B \times \operatorname{rot} A,$$

the NSE in (3.3) can be rewritten as

$$\partial_t\mathbf{u} + \omega \times \mathbf{u} = \nu\Delta\mathbf{u} - \nabla\left(\frac{1}{\rho}p + \frac{1}{2}\mathbf{u}^2\right) + f. \tag{3.123}$$

Taking the scalar product of $\mathbf{u}$ with (3.123) and integrating over the domain gives the evolution of the energy, where the nonlinear term vanishes,

$$\frac{d}{dt}\left(\frac{1}{2}\|\mathbf{u}\|_2^2\right) = -\nu\|\nabla\mathbf{u}\|_2^2 + \int_\Omega \mathbf{u} \cdot f\,dx.$$

The L^2-norm of the gradient of a function and the L^2-norm of the function itself are related by Poincare's inequality.

Theorem 54 (Poincare's inequality) *Suppose Ω is a set for which the negative Laplacian $-\Delta$ along with boundary conditions is a strictly positive self-adjoint linear operator with a discrete spectrum and smallest eigenvalues $\lambda_1 > 0$. Suppose that $f(x)$ and its gradient $\nabla f(x)$ are square integrable on a set Ω, that $f(x)$ satisfies the boundary conditions, and that the boundary conditions allow for the integration by parts $\int_\Omega f(x)[-\Delta f(x)]\,dx = \int_\Omega |\nabla f(x)|^2\,dx$. Then,*

$$\|f\|_2^2 \leq \frac{1}{\lambda_1}\|\nabla f\|_2^2. \tag{3.124}$$

Although this proof is phrased in terms of a scalar function $f(x)$, the same argument holds for the Laplacian restricted to divergence-free vector fields. An appeal to Poincare's inequality (3.124), which is used on the Laplacian term, and use of the Cauchy–Schwarz inequality on the forcing term, give

$$\frac{d}{dt}\left(\frac{1}{2}\|\mathbf{u}\|_2^2\right) \leq -\nu k_1^2 \|\nabla \mathbf{u}\|_2^2 + \|\mathbf{u}\|_2 \|f\|_2, \tag{3.125}$$

where $k_1 = 2\pi/L$. One may express $\|\mathbf{u}\|_2^2$ in terms of forcing using the dimensionless Grashof numbers

$$\mathrm{Gr} = \frac{L^{3-d/2}}{\rho\nu^2}\|f\|_2 = \begin{cases} \frac{L^2\|f\|_2}{\nu^2}, & d = 2, \\ \frac{L^{3/2}\|f\|_2}{\nu^2}, & d = 3. \end{cases}$$

Using inequality (3.125), lim–sup estimates for $\|u\|_2^2$ are given

$$\varlimsup_{t\to\infty} \|u\|_2^2 \leq \begin{cases} c\nu^2\,\mathrm{Gr}^2, & 2d-\text{case} \\ cL\nu^2\,\mathrm{Gr}^2, & 3d-\text{case}, \end{cases} \tag{3.126}$$

where the dimensionless constants are denoted by c. The energy estimates given in (3.126) above, while useful, are only part of what we need in order to find an estimate for F_0. Because all the F_N contain forcing terms, estimates for these must also be included. From (3.100), we recall that $F_0 = \|u\|_2^2 + \nu^{-2}L^4\|f\|_2^2$. It is desirable to express the extra term $\nu^{-2}L^4\|f\|_2^2$ in terms of the Grashof numbers defined above. From (3.98) we recall that our choice of forcing function is such that it possesses a smallest scale, see (3.99), called λ_f, which appears in the definition of the standard length scale λ_0, see (3.101), $\lambda_0^{-2} = L^{-2} + \lambda_f^{-2}$. When the forcing is included, (3.126) becomes

$$\varlimsup_{t\to\infty} F_0 \leq \begin{cases} c\nu^2\,\mathrm{Gr}^2, & 2d-\text{case} \\ cL\nu^2\,\mathrm{Gr}^2, & 3d-\text{case}. \end{cases} \tag{3.127}$$

These results are listed in Tables 3.2 and 3.3, the first one for the two-dimensional case and the second one for the three-dimensional case.

Estimates for $\langle F_1 \rangle$ and $\langle \kappa_{1,0}^2 \rangle$. The ladder for $N = 0$ is

$$\frac{1}{2}\dot{F}_0 \leq \nu F_1 + \nu\lambda_0^{-2}F_0. \tag{3.128}$$

Time averaging (3.128) and using (3.126) therefore produces

$$\langle F_1 \rangle \leq \begin{cases} c\lambda_0^{-2}\nu^2\,\mathrm{Gr}^2, & 2d-\text{case}, \\ cL\lambda_0^{-2}\nu^2\,\mathrm{Gr}^2, & 3d-\text{case}. \end{cases} \tag{3.129}$$

The three-dimensional estimate above, in particular, is the time averaged version of Leray's inequality, see (3.97). Our estimate for F_1 in (3.129) includes λ_0^{-2} and thereby takes account of spectral information of the forcing. To obtain a Leray's type inequality in terms of $\kappa_{N,r}^2$ we see that dividing (3.128) by F_0 and time averaging gives $\langle \kappa_{1,0}^2 \rangle \leq \lambda_0^{-2}$, which is true for $d = 2, 3$. These results are listed in Tables 3.2 and 3.3.

Estimates for $\varlimsup_{t\to\infty} F_1$, $\langle F_2 \rangle$ and $\langle \kappa_{2,1}^2 \rangle$. Now we turn to the evolution of the enstrophy. The evolution of the enstrophy is given by

$$\frac{1}{2}\frac{d}{dt}\int_\Omega |\omega|^2\,dx = -\nu\int_\Omega |\nabla\omega|^2\,dx - \int_\Omega \omega\cdot(\mathbf{u}\cdot\nabla\omega)\,dx + \int_\Omega \omega\cdot(\omega\cdot\nabla\mathbf{u})\,dx + \int_\Omega (\omega\cdot\mathrm{rot}\,f)\,dx.$$

By periodicity of the boundary conditions and the assumption $\mathrm{div}\,\mathbf{u} = 0$, the integral

$\int_\Omega \omega \cdot (\mathbf{u} \cdot \nabla \omega) \, dx$ vanishes for $d = 2, 3$. In the two-dimensional case, $\omega \cdot \nabla \mathbf{u} = 0$ because ω is perpendicular to two-dimensional flow in the $x - y$ plane. This enables us to find an absorbing ball for the two-dimensional enstrophy. In terms of F_1, we find

$$\frac{1}{2} \dot{F}_1 \le -\nu F_2 + \nu \lambda_0^{-2} F_1. \tag{3.130}$$

By Lemma 9, $F_1^2 \le F_0 F_2$ and so, using (3.127), we get

$$\varlimsup_{t \to \infty} F_1 \le c \lambda_0^{-2} \nu^2 \, \text{Gr}^2. \tag{3.131}$$

Considering (3.130) again, we may time average the equation to obtain $\langle F_2 \rangle$ or divide by F_1 and then time average to find $\langle \kappa_{2,1}^2 \rangle$. In both cases the results appear in Table 3.2. The estimate for F_1 in (3.131) (showing up also in $\langle F_2 \rangle$ and $\langle \kappa_{2,1}^2 \rangle$ both being a priori bounded above) is one of the fundamental differences between two- and three-dimensional flows. It will be shown in Section 3.6.3 that the bound (3.131) will allow us to control the whole ladder in the two-dimensional case. To date, no such control has been achieved for the three-dimensional case.

TABLE 3.2: A summary of two-dimensional Navier–Stokes estimates.

N	Definition	Estimation
1	Two-dimensional Grashof number	$\text{Gr} = L^2 \nu^{-2} \|f\|_2$
2	Absorbing ball for F_0	$\varlimsup_{t \to \infty} F_0 \le c \nu^2 \, \text{Gr}^2$
3	Absorbing ball for F_1	$\varlimsup_{t \to \infty} F_1 \le c \lambda_0^{-2} \nu^2 \, \text{Gr}^2$
4	Time average of F_1	$\langle F_1 \rangle \le c \lambda_0^{-2} \nu^2 \, \text{Gr}^2$
5	Time average of F_2	$\langle F_2 \rangle \le c \lambda_0^{-4} \nu^2 \, \text{Gr}^2$
6	Time average of $\kappa_{1,0}^2$	$\langle \kappa_{1,0}^2 \rangle \le \lambda_0^{-2}$
7	Time average of $\kappa_{2,1}^2$	$\langle \kappa_{2,1}^2 \rangle \le \lambda_0^{-2}$

For a general three-dimensional flow, the vortex stretching term $\omega \cdot \nabla \mathbf{u} = 0$ is nonzero. This means that we need to estimate the integral $\int_\Omega \omega \cdot (\omega \cdot \nabla \mathbf{u}) \, dx$. We can bound it in the following way: $\left| \int_\Omega \omega \cdot (\omega \cdot \nabla \mathbf{u}) dx \right| \le \|\omega\|_\infty \|\omega\|_2^2$. Expressing the enstrophy in terms of F_1, we find, in the F_N notation,

$$\frac{1}{2} \dot{F}_1 \le -\nu F_2 + (c\|\omega\|_\infty + \nu \lambda_0^{-2}) F_1. \tag{3.132}$$

A division of (3.132) by F_1 and a time averaging operation gives

$$\langle \kappa_{2,1}^2 \rangle \le c \nu^{-1} \langle \|\omega\|_\infty \rangle + \lambda_0^{-2}. \tag{3.133}$$

The extra $\|\omega\|_\infty$ term in (3.133), in comparison to the two-dimensional case in Table 3.2 makes a very great difference. Indeed, with the methods used here, it is mathematically the great difference between the two cases. This extra term is the one that has so far stood in the way of finding an acceptable regularity proof for the three-dimensional NSE. The results we have achieved so far in three-dimensional are given in Table 3.3.

3.6.3 Length scales in the two-dimensional case

The ladder inequality (in Theorem 52), and summarized in Table 3.1, can be used to prove that there is an absorbing ball for each of the F_N. The main piece of extra information

TABLE 3.3: A summary of three-dimensional Navier–Stokes estimates

N	Definition	Estimation
1	Three-dimensional Grashof number	$\mathrm{Gr} = L^{3/2}\nu^{-2}\|f\|_2$
2	Absorbing ball for F_0	$\lim\limits_{t\to\infty} F_0 \leq cL\nu^2\,\mathrm{Gr}^2$
3	Time average of F_1	$\langle F_1 \rangle \leq cL\lambda_0^{-2}\nu^2\,\mathrm{Gr}^2$ (Leray)
4	Time average of $\kappa_{1,0}^2$	$\langle \kappa_{1,0}^2 \rangle \leq \lambda_0^{-2}$
5	Time average of $\kappa_{2,1}^2$	$\langle \kappa_{2,1}^2 \rangle \leq c\nu^{-1}\langle \|\omega\|_\infty \rangle + \lambda_0^{-2}$

in the two-dimensional case is that we have an absorbing ball for F_1 (see Table 3.2). The task, therefore, is to control the $\|D\mathbf{u}\|_\infty$ term. We use a two-dimensional calculus inequality $\|D\mathbf{u}\|_\infty \leq c\|D^N\mathbf{u}\|_2^a \cdot \|D\mathbf{u}\|_2^{1-a}$ (see Section 3.9), where $a = (N-1)^{-1}$ and $N \geq 3$, to obtain $\|D\mathbf{u}\|_\infty \leq cF_N^{a/2} \cdot F_1^{(1-a)/2}$. For $s = N-1$ in the ladder, we get

$$\frac{1}{2}\dot{F}_N \leq -\nu\big(F_N^{1+1/(N-1)}/F_1^{1/(N-1)}\big) + \big(c_N F_N^{a/2} \cdot F_1^{(1-a)/2} + \nu\lambda_0^{-2}\big)F_N. \tag{3.134}$$

By (3.134), the exponent of the negative term is stronger than that of the nonlinear term enabling us to find an absorbing ball. The result of this is

$$\overline{\lim\limits_{t\to\infty}} F_N \leq c_N\nu^{-2(N-1)}\big(\overline{\lim\limits_{t\to\infty}} F_1\big)^N + \lambda_0^{-2(N-1)}\overline{\lim\limits_{t\to\infty}} F_1. \tag{3.135}$$

From Table 3.2, we have an estimate for $\overline{\lim}_{t\to\infty}F_1$ in terms of the two-dimensional Grashof number Gr, which, when substituted into (3.135), gives

$$\overline{\lim\limits_{t\to\infty}} F_N \leq c\lambda_0^{-2N}\nu^2(\mathrm{Gr}^{2N} + \mathrm{Gr}^2). \tag{3.136}$$

In two-dimensional case the limit of the Galerkin approximations $\|\omega^N\|_2^2$ gives finite enstrophy [40]. The weak solutions can therefore be turned into strong solutions, with all the consequences for uniqueness discussed there. Consequently the bound on F_1 implies the bound on F_N given in (3.136) for every $N \geq 1$, so no singularities in any derivative can develop from smooth initial data in a two-dimensional flow. Since two-dimensional flows are not particularly physical and their inability to amplify vorticity, due to the absence of the vortex stretching term, makes them a poor substitute for three-dimensional flows. It is the question of control over F_1, around which the difference between the two- and three-dimensional cases revolves. These results allow us to consider the idea of a global attractor and length scales of solutions.

Global attractor. The global attractor $\mathcal{A}$ for a finite dimensional dynamical system needs to be extended when dealing with infinite dimensional systems. If $\mathcal{B}$ is a compact, connected set, absorbing all trajectories (the absorbing ball for F_1) and $S(t)$ is the nonlinear semigroup flow such that $\mathbf{u}(t) = S(t)\mathbf{u}(0)$, then the *global attractor* is $\mathcal{A} = \bigcap_{t>0} S(t)\mathcal{B}$ with the following properties:

- $S(t)\mathcal{A} = \mathcal{A}$ both forward and backward in time,
- For the ω-limit set of any bounded set $J \subset \mathcal{A}$ then $\omega(J) \subset \mathcal{A}$,
- $\mathcal{A}$ is compact and $\mathrm{dist}_{t\to\infty}\{u(t, \mathcal{A})\} = 0$.

By (3.136), solutions for the two-dimensional NSE on $\mathcal{A}$ are smooth. Estimate for $d_L(\mathcal{A})$ is given in Section 3.6.5.

Length scales in the two-dimensional case. Now we turn our attention to defining and estimating $\mathcal{N}$. First we use the objects

$$\kappa_{N,r}(t) = (F_N/F_r)^{1/2(N-r)} \tag{3.137}$$

introduced in (3.119). These represent time dependent mean wave-numbers for the system as they all have a dimension of $(\text{length})^{-1}$. They are ordered such that $\kappa_{N,r} \leq \kappa_{N+1,r}$ and $\kappa_{N,r} \leq \kappa_{N,r+1}$ for $r < N$. Since the smallest $\kappa_{1,0}$ is bounded from below, thus all of them are bounded away from zero. Next, define the dimensionless quantities

$$\mathcal{N}_{N,r} = \lambda_0^2 \kappa_{N,r}^2, \tag{3.138}$$

where $\lambda_0^{-2} = L^{-2} + \lambda_f^{-2}$ and $\lambda_f \leq L$. Assume λ_0 to be the smallest length scale on the domain. The set of time averages

$$\langle \mathcal{N}_{N,r} \rangle \sim (\lambda_0/l_{N,r})^2$$

plays the role of $\mathcal{N}$ in (3.94) for $d = 2$, although it actually has a doubly infinite number of members, λ_0 replaces L and $l_{N,r}^{-2} \sim \langle \kappa_{N,r}^2 \rangle$. To use this, take the ladder from (3.118) in the form

$$\frac{1}{2}\dot{F}_N \leq -\nu F_{N+1} + (c_{N,r}\langle \|D\mathbf{u}\|_\infty \rangle + \nu \lambda_0^{-2})F_N. \tag{3.139}$$

We need to convert this differential inequality in F_N and F_{N+1} into one in $\kappa_{N,r}$ and $\kappa_{N+1,r}$ in a slightly different way from that given in (3.121). Taking $s = N - r$ and differentiating (3.137) with respect to t, we find

$$2s\dot{\kappa}_{N,r}\kappa_{N,r}^{2s-1} = (\dot{F}_N/F_r) - (\dot{F}_r/F_r)\kappa_{N,r}^{2s}$$

and (3.139) can be used for F_N, but for F_r a lower bound is necessary. This is available from (3.102)

$$\frac{1}{2}\dot{F}_r \leq -F_{r+1} - (c_{N,r}\langle \|D\mathbf{u}\|_\infty \rangle + \nu \lambda_0^{-2})F_r.$$

By the above,

$$(N - r)\dot{\kappa}_{N,r} \leq -\nu \kappa_{N+1,r}^3 + \nu \kappa_{r+1,r}^2 \kappa_{N,r} + 2(c_{N,r}\langle \|D\mathbf{u}\|_\infty \rangle + \nu \lambda_0^{-2})\kappa_{N,r}. \tag{3.140}$$

In order to estimate the term $\|D\mathbf{u}\|_\infty$ we first prove the following lemma for a general two-dimensional scalar function $A(x)$ and then follow it with a corollary:

Lemma 10 *In two-dimension, the $\|A\|_\infty$ norm of a scalar function is bounded as*

$$\|A\|_\infty \leq c\|DA\|_2 \left[1 + \log\left(L\|D^2A\|_2/\|DA\|_2\right)\right]^{1/2}. \tag{3.141}$$

Thus

$$\|D\mathbf{u}\|_\infty \leq cF_2^{1/2}[1 + \log(\lambda_0\kappa_{3,2})]^{1/2}. \tag{3.142}$$

Proof. Consider the Fourier transform $A(k)$ of $A(x)$ and split the spectrum into low modes $|k| \leq k$ and high modes $|k| \geq k$ (the division point k will be determined later),

$$\begin{aligned}
|A(x)| &= \left|\sum_k \exp(ik \cdot x)\hat{A}(k)\right| \leq \sum_{|k|\leq k} |\hat{A}(k)| + \sum_{|k|\geq k} |\hat{A}(k)| \\
&= \sum_{|k|\leq k} |k|^{-1}|k||\hat{A}(k)| + \sum_{|k|\geq k} |k|^{-2}|k|^2|\hat{A}(k)| \\
&\leq \left(\sum_{|k|\leq k} |k|^{-2}\right)^{1/2}\left(\sum_{|k|\leq k} |k|^2|\hat{A}(k)|^2\right)^{1/2} + \left(\sum_{|k|\geq k} |k|^{-4}\right)^{1/2}\left(\sum_{|k|\geq k} |k|^4|\hat{A}(k)|^2\right)^{1/2},
\end{aligned} \tag{3.143}$$

where we used the Cauchy–Schwarz inequality in the last step. For the leading factor we get

$$\sum\nolimits_{|k| \leq k} |k|^{-2} \leq cL^2 \int_{2\pi/L}^{\kappa} dk/k \leq cL^2 \log Lk, \tag{3.144}$$

where $2\pi/L$ is the smallest k in the box. The leading factor in the second term in (3.143) obeys

$$\sum\nolimits_{|k| \geq k} |k|^{-4} \leq cL^2 \int_{k}^{\infty} dk/k^3 \leq cL^2 k^{-2}.$$

The second factors in two terms in (3.143) are bounded by seminorms $\|DA\|_2$ and $\|D^2A\|_2$:

$$\sum\nolimits_{|k| \leq k} |k|^2 |\hat{A}(k)|^2 \leq cL^{-2}\|DA\|_2^2, \quad \sum\nolimits_{|k| \geq k} |k|^4 |\hat{A}(k)|^2 \leq cL^{-2}\|D^2A\|_2^2. \tag{3.145}$$

Inserting (3.144)–(3.145) into (3.143) gives

$$\|A\|_\infty \leq c\|DA\|_2 (\log Lk)^{1/2} + \frac{1}{k}\|D^2A\|_2. \tag{3.146}$$

Since $k \geq 2\pi/L$ is arbitrary, we may choose $k = \|D^2A\|_2/\|DA\|_2$ yielding (3.141).

Choose a set of scalar functions A_i such that $A_i = Du_i$; for every component of $\mathbf{u}$. Then we return to (3.146) and, noting that $\|D^2u_i\|_2^2 \leq F_2$ and $\|D^3u_i\|_2^2 \leq F_3$, rewrite it as

$$\|D\mathbf{u}\|_\infty \leq c(\log \lambda_0 \kappa)^{1/2} F_2^{1/2} + F_3^{1/2}/k,$$

where $2\pi/\lambda_0$ is chosen instead of $2\pi/L$ as the lowest mode. Then we take k as $k^2 = F_3/F_2$ so k^2 can be identified as $\kappa_{3,2}^3 = F_3/F_2$. This leaves us with the final result (3.142). $\qquad\square$

Now we choose $r = 1$ in (3.140),

$$(N-1)\dot{\kappa}_{N,1} \leq -\nu\kappa_{N+1,1}^3 + \nu\kappa_{2,1}^2\kappa_{N,1} + 2\nu\lambda_0^{-2}\kappa_{N,1} + c_3 F_2^{1/2}[1 + \log(\lambda_0 \kappa_{3,2})]^{1/2}\kappa_{N,1}. \tag{3.147}$$

Since F_N is bounded from below, we find from (3.136) pointwise in time estimates for $\kappa_{N,r}$,

$$\varlimsup_{t \to \infty} \kappa_{N,r} \leq \lambda_0^{-1} \mathrm{Gr}^{(N-1)/(N-r)}. \tag{3.148}$$

Using these and the $\mathcal{N}_{N,r}$ notation of (3.138), a time average of (3.147) gives

$$\langle \mathcal{N}_{N+1,1}^{3/2} \rangle \leq c\nu^{-1}\lambda_0^2 \langle F_2^{1/2} \mathcal{N}_{N,1}^{1/2} \rangle (1 + \log \mathrm{Gr})^{1/2} + 2\langle \mathcal{N}_{N,1}^{1/2} \rangle + c\,\mathrm{Gr}\langle \mathcal{N}_{2,1} \rangle.$$

Invoking the Cauchy–Schwarz inequality and using the estimate for $\langle F_2 \rangle$ from Table 3.2, we obtain a recursion relation for $N \geq 2$

$$\langle \mathcal{N}_{N+1,1} \rangle^{3/2} \leq c\,\mathrm{Gr}(1 + \log \mathrm{Gr})^{1/2} \langle \mathcal{N}_{N,1} \rangle^{1/2} + 2\langle \mathcal{N}_{N,1} \rangle^{1/2} + c\,\mathrm{Gr}\langle \mathcal{N}_{2,1} \rangle. \tag{3.149}$$

Because $\langle \kappa_{2,1}^2 \rangle \leq \lambda_0^2$, we have $\langle \mathcal{N}_{2,1} \rangle \leq 1$ as our bottom rung. For $N = 2$, absorbing the last two terms in (3.149) into the constant (because $\mathrm{Gr} > 1$) we obtain

$$\langle \mathcal{N}_{3,1} \rangle = \lambda_0^2 \langle \kappa_{3,1}^2 \rangle \leq c\,\mathrm{Gr}^{2/3}(1 + \log \mathrm{Gr})^{1/3}. \tag{3.150}$$

This estimate for $\langle \mathcal{N}_{3,1} \rangle$ coincides with that for the Lyapunov dimension of the attractor derived in Section 3.6.2. Using the recursion relation (3.149) we find the first few $\langle \mathcal{N}_{N,1} \rangle$:

$$\begin{aligned}
\langle \mathcal{N}_{2,1} \rangle &= \lambda_0^2 \langle \kappa_{2,1}^2 \rangle \leq 1, \\
\langle \mathcal{N}_{3,1} \rangle &= \lambda_0^2 \langle \kappa_{3,1}^2 \rangle \leq c_8\,\mathrm{Gr}^{2/3}(1 + \log \mathrm{Gr})^{1/3}, \\
\langle \mathcal{N}_{4,1} \rangle &= \lambda_0^2 \langle \kappa_{4,1}^2 \rangle \leq c_9\,\mathrm{Gr}^{8/9}(1 + \log \mathrm{Gr})^{4/9}.
\end{aligned} \tag{3.151}$$

The result for $N \gg 1$ is $\langle \mathcal{N}_{N,1} \rangle = \lambda_0^2 \langle \kappa_{N,1}^2 \rangle \leq c_{10} \, \mathrm{Gr}(1 + \log \mathrm{Gr})^{1/2}$. For values of $r \geq 2$, direct estimation through (3.120) together with the logarithmic estimate (3.142) gives

$$\langle \mathcal{N}_{N,r} \rangle = \lambda_0^2 \langle \kappa_{N,r}^2 \rangle \leq c_{N,r} \, \mathrm{Gr}(1 + \log \mathrm{Gr})^{1/2}.$$

It is also possible to improve upon the pointwise in time estimate given in (3.148). This can be performed by turning (3.140) into a differential inequality for $\kappa_{N,r}$ using $-\kappa_{N+1,r} \leq -\kappa_{N,r}$. In terms of the $\mathcal{N}_{N,r}$, we then obtain

$$\overline{\lim_{t \to \infty}} \, \mathcal{N}_{N,r} = \overline{\lim_{t \to \infty}} \{ \lambda_0^2 \kappa_{N,r}^2 \} \leq \tilde{c}_{N,r} \, \mathrm{Gr}^2.$$

The result in the first inequality of (3.151) follows from the fact that $\langle \mathcal{N}_{2,1} \rangle$ is capturing a combination of the box length L and the smallest forcing scale λ_f, which both are built into λ_0 whereas the estimate for $\langle \mathcal{N}_{3,1} \rangle$ is that which coincides with the estimate for the attractor dimension.

3.6.4 Three-dimensional regularity

In previous sections we discussed how the vortex stretching term $(\omega \cdot \nabla)\mathbf{u}$ makes a difference between two- and three-dimensional cases. This term can stretch, tangle and twist vortex lines producing significant dynamics down to very small scales. The three-dimensional NSE are notorious for the absence of any proof which shows them to be regular beyond a finite time. We shall consider two cases and then show what can be proved.

Problems with three-dimensional regularity. Certain time averaged quantities are bounded a priori. Apart from the L^2-norm of the velocity $\mathbf{u}(x,t)$, which is bounded above for all t, the first of these is the time integral of F_1, which is Leray's inequality. In terms of time average this is

$$\langle F_1 \rangle \leq \nu^2 L \lambda_0^{-2} \, \mathrm{Gr}^2, \tag{3.152}$$

where Gr is the three-dimensional Grashof number, see Table 3.3. In addition to (3.152), it is possible to show that for weak solutions the infinite set $\langle \kappa_{N,1} \rangle$ is bounded for all N and, in addition, $\langle \|\mathbf{u}\|_\infty \rangle$ is also bounded. Leray's inequality in (3.152) is a time average and not a pointwise bound so F_1 can still exhibit singularities in time. In aim to find an absorbing ball for F_1, we write down a differential inequality for F_1 by going direct to the equation for the vorticity,

$$\frac{1}{2}\dot{F}_1 \leq -\nu F_2 + \int_\Omega \omega \cdot (\omega \cdot \nabla)\mathbf{u}\,dx + \nu \lambda_0^{-2} F_1.$$

In the two-dimensional case, the integral term would be absent but here, we integrate by parts once, pull out $\|\mathbf{u}\|_\infty$, break it up using the Cauchy–Schwarz inequality and obtain

$$\frac{1}{2}\dot{F}_1 \leq -\nu F_2 + \|\mathbf{u}\|_\infty F_1^{1/2} F_2^{1/2} + \nu \lambda_0^{-2} F_1. \tag{3.153}$$

Using the calculus inequality, see Section 3.9, (3.264),

$$\|\mathbf{u}\|_\infty \leq c F_2^{1/4} F_1^{1/4} \tag{3.154}$$

and then a Hölder inequality in (3.153), we obtain

$$\frac{1}{2}\dot{F}_1 \leq -\frac{1}{4}\nu F_2 + c\nu^{-3} F_1^3 + \nu \lambda_0^{-2} F_1. \tag{3.155}$$

Using $F_1^2 \leq F_0 F_2$, we finally get

$$\frac{1}{2}\dot{F}_1 \leq -\frac{1}{4}\nu \left(F_1^2/F_0\right) + c\nu^{-3} F_1^3 + \nu \lambda_0^{-2} F_1. \tag{3.156}$$

The cube in the nonlinear term is too large for the square in the viscosity term thereby preventing us finding an absorbing ball for arbitrarily large initial data for fixed ν. The two exceptions are when either initial data on F_1 is small or ν is large enough so that the negative term dominates. Remark that the $\nu^{-3}F_1^3$ term in (3.156) is sharp in the following sense. The dimension of every term in the inequality is L^3T^{-3} and if we were to ask whether it is possible to perform inequalities in an alternative way to produce an exponent of F_1 lower than 3, this term should still have the same dimension. In order to estimate in terms of the combination $\nu^\alpha F_0^\beta F_1^\gamma$ one may show that the lowest value of y that can be achieved with $\beta \geq 0$ is $\gamma = 3$ with $\alpha = -3$, $\beta = 0$. Now look at the problem a different way and examine the $\|D\mathbf{u}\|_\infty$ term in the F_N ladder. A calculus inequality from Section 3.9 yields

$$\|D\mathbf{u}\|_\infty \leq cF_N^{a/2}\|\mathbf{u}\|_2^{1-a}, \tag{3.157}$$

where $a = \frac{5}{2}N^{-1}$ with $N > \frac{5}{2}$. The ladder becomes

$$\frac{1}{2}\dot{F}_1 \leq -\nu\big(F_N^{1+1/s}/F_{N-s}^{1/s}\big) + cF_N^{1+a/2}\|\mathbf{u}\|_2^{1-a} + \nu\lambda_0^{-2}F_N,$$

from which a necessary condition for an absorbing ball is

$$1/s > a/2, \tag{3.158}$$

but where the F_{N-s} term needs to be controlled. We wish to go down the ladder as far as $s = N$ because F_0 is bounded above for all t whereas no proof exists that F_1 is bounded in the three-dimensional case, as it is in the two-dimensional case. An absorbing ball cannot be achieved this way because (3.158) means that we need $s < \frac{4}{5}N$. This cannot be fulfilled by taking $s = N$, and we can get down to $s = N - 1$ without violating the condition for an absorbing ball. Consequently, F_1 is the bottom rung of the ladder, as in the two-dimensional case, and not F_0. Since we have no control over this bottom rung in three-dimension, there is no means of controlling the other rungs of the ladder either. Herein lies the root of the problem. There are several different approaches but they all give the same result. Since it is impossible to prove regularity by these methods, one may try the weakest assumption that yields regularity. The idea is to relax the requirement that we go down to $\|\mathbf{u}\|_2^2$ in the inequality in (3.157) and go down to L^q (for some q to be calculated) and not L^2. Hence, instead of (3.157), we write

$$\|D\mathbf{u}\|_\infty \leq cF_N^{a/2}\|\mathbf{u}\|_q^{1-a},$$

where $a = \frac{2(q+3)}{(2N-3)q+6}$ with $N \geq 3$. The requirement in (3.158) that s must satisfy $as < 2$ to get an absorbing ball means the restriction $q > 3\frac{N-2}{N-3}$. Since N is large, $q > 3$ must be satisfied. In conclusion, if we assume that $\|\mathbf{u}\|_q$ ($q > 3$) is bounded for all t, then all F_N are bounded for all t and no singularities can occur. No independent proof exists that shows that $\|\mathbf{u}\|_{3+\varepsilon}$, is bounded. Thus, there is barrier between what we have proved is bounded $\|\mathbf{u}\|_2$ and what we need to prove is bounded, $\|\mathbf{u}\|_{3+\varepsilon}$.

Bound on $\langle\kappa_{N,1}\rangle$ in the three-dimensional case. Unlike the two-dimensional case of the previous section, no regularity proof about the objects $\kappa_{N,r}$ in three-dimension is known, so a finite time singularity in any of the $\kappa_{N,r}$ cannot be ruled out. We will show that there are bounds on the infinite set of time averaged quantities $\langle\kappa_{N,1}\rangle$.

First, using (3.140) with $r = 1$ and $-\kappa_{N+1,1}^3 \leq -\kappa_{N,1}^3$, we get

$$(N-1)\dot{\kappa}_{N,1} \leq -\nu\kappa_{N,1}^3 + \nu\kappa_{2,1}^2\kappa_{N,1} + 2(c_{N,1}\|D\mathbf{u}\|_\infty + \nu\lambda_0^{-2})\kappa_{N,1}. \tag{3.159}$$

Recall the definition $\lambda_K = (\nu^3/\varepsilon)^{1/4}$ of the *Kolmogorov dissipation length*, where the energy dissipation rate, an a priori bounded quantity for weak solutions, is given by $\varepsilon = \nu\langle F_1\rangle L^{-3}$: sometimes we use either $\langle H_1\rangle$ or $\langle F_1\rangle$, the only difference being the constant forcing terms.

Lemma 11 *In three-dimension, weak solutions of the NSE satisfy*

$$\langle \kappa_{2,1} \rangle \leq cL\lambda_0^{-2} \operatorname{Gr}^2, \tag{3.160}$$

or, alternatively, $L\langle \kappa_{2,1} \rangle \leq c(L/\lambda_K)^4$.

Proof. We already have an inequality for F_1 in the shape of (3.155), namely,

$$\frac{1}{2}\dot{F}_1 \leq -\frac{1}{4}\nu F_2 + c\nu^{-3}F_1^3 + \nu\lambda_0^{-2}F_1.$$

Divide by F_1^2 and time average to get

$$\langle F_2/F_1^2 \rangle \leq c\nu^{-4}\langle F_1 \rangle \leq cL\lambda_0^{-2}\nu^{-2}\operatorname{Gr}^2. \tag{3.161}$$

We have ignored a correction term proportional to $\langle F_1^{-1} \rangle$ which is small and bounded because F_1 is bounded from below. Rewriting (3.161) in terms of the Kolmogorov length λ_K, we find

$$\langle F_2/F_1^2 \rangle \leq c\nu^{-4}\langle F_1 \rangle \leq c\nu^{-2}L^3/\lambda_K^4.$$

Then $\langle \kappa_{2,1} \rangle$ can be written as

$$\langle \kappa_{2,1} \rangle = \langle (F_2/F_1)^{1/2} \rangle = \langle (F_2^{1/2}/F_1) \cdot F_1^{1/2} \rangle \leq \langle F_2/F_1^2 \rangle^{1/2} \cdot \langle F_1 \rangle^{1/2} \leq cL\lambda_0^{-2}\operatorname{Gr}^2,$$

which is (3.160). The alternative upper bound in terms of λ_K follows. $\quad\square$

This result enables us to prove the following:

Theorem 55 *For $N \geq 2$, weak solutions of the three-dimensional NSE satisfy* $\langle \kappa_{N,1} \rangle \leq c_N L\lambda_0^{-2} \operatorname{Gr}^2$, *or, alternatively,*

$$L\langle \kappa_{N,1} \rangle \leq c_N(L/\lambda_K)^4. \tag{3.162}$$

Proof. To handle the $\|D\mathbf{u}\|_\infty$, term we use a calculus inequality from Section 3.9,

$$\|D\mathbf{u}\|_\infty \leq c\|D^N\mathbf{u}\|_2^a \cdot \|D\mathbf{u}\|_2^{1-a} \leq cF_N^{a/2}F_1^{(1-a)/2} = c\kappa_{N,1}^{3/2}F_1^{1/2} \tag{3.163}$$

for $N \geq 3$ where $a = \frac{3}{2(N-1)}$. Divide (3.159) through by $\kappa_{N,1}^2$ and use the facts that $\kappa_{N,1}$ are bounded from below and $\kappa_{2,1} \leq \kappa_{N,1}$ for $N \geq 2$ to get

$$\langle \kappa_{N,1} \rangle \leq c\nu^{-1}\langle \kappa_{N,1}^{1/2}F_1^{1/2} \rangle + \langle \kappa_{2,1} \rangle + 2\lambda_0^{-2}\kappa_{N,1}^{-1}. \tag{3.164}$$

Splitting up the first term on the RHS of (3.164) and using Hölder's inequality we find

$$\langle \kappa_{N,1} \rangle \leq c\nu^{-2}\langle F_1 \rangle + 2\langle \kappa_{2,1} \rangle, \quad N \geq 3,$$

where we ignored the last term in (3.164), which is bounded and small. Use of Lemma 11 allows us to control $\langle \kappa_{2,1} \rangle$ which gives the advertised result. The upper bound (3.162) follows as in Lemma 11. $\quad\square$

Bounds on $\langle \|\mathbf{u}\|_\infty \rangle$ **and** $\langle \|D\mathbf{u}\|_\infty^{1/2} \rangle$. There are two more quantities (due to Theorem 55) that can be bounded from above. The first one follows from the identity

$$\langle F_N^{1/(2N-1)} \rangle = \langle \kappa_{N,r}^{2(N-r)/(2N-1)}F_r^{1/(2N-1)} \rangle.$$

By Hölder inequality, this becomes

$$\langle F_N^{1/(2N-1)} \rangle \leq \langle \kappa_{N,r} \rangle^{2(N-r)/(2N-1)} \langle F_r^{1/(2N-1)} \rangle^{(2r-1)/(2N-1)}.$$

Starting with $r = 1$ and using the fact that both $\langle F_1 \rangle$ and $\langle \kappa_{N,1} \rangle$ are bounded above, we get

$$\langle F_N^{1/(2N-1)} \rangle \leq c_N \nu^{2/(2N-1)} L \lambda_0^{-2} \mathrm{Gr}^2 \quad \text{or} \quad \langle F_N^{1/(2N-1)} \rangle \leq c_N \nu^{2/(2N-1)} L^3 \lambda_K^{-4}.$$

Secondly, an inequality for $\|\mathbf{u}\|_\infty$ comes by using a higher order version of (3.154), expressed in terms of $\kappa_{N,1}$ and $\|\mathbf{u}\|_\infty \leq c \kappa_{N,1}^{1/2} F_1^{1/2}$. Using the Cauchy–Schwarz inequality, we find

$$\langle \|\mathbf{u}\|_\infty \rangle \leq c \nu L \lambda_0^{-2} \mathrm{Gr}^2.$$

The method here not only makes plain the inter-relation between the triad of objects $\langle \kappa_{N,1} \rangle$, $\langle F_N^{1/(N-1)} \rangle$ and $\langle \|\mathbf{u}\|_\infty \rangle$, it also shows that controlling $\langle \kappa_{N,1} \rangle$ is the keystone to all these results. Proceeding with these methods and using (3.163), we estimate $\langle \|D\mathbf{u}\|_\infty^{1/2} \rangle$ as

$$\langle \|D\mathbf{u}\|_\infty^{1/2} \rangle \leq c \langle \kappa_{N,1}^{3/4} F_1^{1/4} \rangle \leq c \langle \kappa_{N,1}^{3/4} \rangle \langle F_1 \rangle^{1/4}.$$

From the above, (3.162) and definition of λ_K we get $\langle \|D\mathbf{u}\|_\infty^{1/2} \rangle \leq c \nu^{1/2} L^3 \lambda_K^{-4}$. This will be used in Section 3.6.5 when estimating the attractor dimension in the three-dimensional case.

The Kolmogorov length and intermittency. Because no proof exists which shows that a three-dimensional flow remains regular beyond a finite time, any predictions about the flow and its natural length scales must be based on an assumption that one of the norms involved remains regular for time averages to make sense. We are interested in the possibility that even if the flow remains regular, the occurrence of large, rare, intermittent fluctuations in the vorticity field away from averages, may act as a source of small scale structures in the flow. In this case, a flow may remain close to its spatial average for large periods of time but such bursts in the vorticity field might cause excursions of energy to short scales. The deeper the excursions, the rarer these events must be to avoid violating the bounds on space and time averages. This phenomenon, called *intermittency*, is associated with energy dissipation being concentrated on a set of dimension smaller than the background space. Because of the relatively long quiescent periods between these events, one might think how to compute a length scale associated with the flow during these periods, and then a second smaller scale associated with a burst, if or when it occurs. In this situation, we should like to discuss the role of the Kolmogorov length λ_K. In particular, whether λ_K is the smallest scale, as is commonly assumed, or whether length scales smaller than this could possibly occur. The estimate from Theorem 53,

$$\langle \kappa_{N,r}^2 \rangle \leq c_N \nu^{-1} \langle \|D\mathbf{u}\|_\infty \rangle + \lambda_0^{-2}$$

is distinguished by the fact that it is formally an intensive quantity, i.e., it not explicitly dependent on the system size. We could define a *Kolmogorov length* from this by defining an *energy dissipation rate* $\varepsilon_\infty = \nu \langle \|D\mathbf{u}\|_\infty \rangle^2$. Dropping the suffices on the scales, this gives

$$l_{N,r}^{-2} \leq c \lambda_{K,\infty}^{-2} + \lambda_0^{-2} \quad \text{with} \quad \lambda_{K,\infty} = (\nu^3 / \varepsilon_\infty)^{1/4}.$$

Although $\lambda_{K,\infty}$ is a length scale, it is not the quantity which is understood to be the Kolmogorov length. A difficult problem is to make comparison between $\lambda_{K,\infty}$ and λ_K. Using the inequalities in Section 3.9 to bound above the L^∞-norm of a function by the L^2-norm of its N-th derivative (say) introduces the system volume into the problem, thereby sacrificing the intuitive notion of an intensive length scale determined only by local properties of the flow. Let us write down how many fixed (or bounded) scales exist against which we want to compare $\nu^{-1} \langle \|D\mathbf{u}\|_\infty \rangle$.

- $\lambda_0^{-2} = L^{-2} + \lambda_f^{-2}$; the box length L and the smallest forcing scale λ_f.
- The *conventional Kolmogorov length* $\lambda_K = (\nu^3/\varepsilon)^{1/4}$.
- The *alternative Kolmogorov length*

$$\Lambda_K = (\nu^3/\bar{\varepsilon})^{1/4} \tag{3.165}$$

defined with an energy dissipation rate $\bar{\varepsilon} = \nu L^{-3} \sup_t F_1$. While we have no absolute certainty that $\bar{\varepsilon}$ is bounded, we assume here that it is. We return to this idea in Section 3.6.5. There are two alternative approximations that can be pursued. The first is to we make the assumption

$$\langle \|D\mathbf{u}\|_\infty \rangle \approx L^{-3/2} \langle \|D\mathbf{u}\|_2 \rangle,$$

which provides $l_{N,r}^{-2} \leq c\lambda_{K,\infty}^{-2} + \lambda_0^{-2} \approx c\lambda_K^{-2} + \lambda_0^{-2}$. This assumption means that the dissipation in the flow is close to its spatial average in the mean square sense most of the time, and that any large deviations do not contribute significantly to the time average. Strong intermittency would certainly violate the assumption (3.165) in a fundamental way. In addition, making this approximation in (3.159) gives $\langle \kappa_{N,1}^3 \rangle \leq c_N \lambda_K^{-3}$, $N \geq 2$, thereby making the estimate for the number of degrees of freedom defined in (3.94) to be $\mathcal{N} \leq c_N (L/\lambda_K)^3$. This uniformly makes λ_K the average small scale when fluctuations are ignored. The second way of pursuing this idea in order to take into account fluctuations away from $\|D\mathbf{u}\|_2$ is not to approximate $\|D\mathbf{u}\|_\infty$ but to estimate it in terms of $\|D\mathbf{u}\|_2$. Using the calculus inequalities from (3.163), we obtain

$$4\langle \kappa_{N,1}^2 \rangle \leq c\nu^{-1} \langle \kappa_{N,1}^{3/2} \|D\mathbf{u}\|_2 \rangle + \lambda_0^{-2}.$$

Using a Hölder inequality, we get

$$\langle \kappa_{N,1}^2 \rangle \leq c\nu^{-4} \langle F_1^2 \rangle + 4\lambda_0^{-2}.$$

Now the $\nu^{-4}\langle F_1^2 \rangle$ term has no known bound but if we use the Kolmogorov length Λ_K defined in (3.165) we can pull one F_1 outside the time average to get

$$l/L \geq c(\lambda_K/L)^2 (\Lambda_K/L)^2, \tag{3.166}$$

where $l^{-2} \sim \langle \kappa_{N,1}^2 \rangle$. This estimate for l is certainly not intensive because of its explicit L dependence. Because $\lambda_K^{-1} \leq \Lambda_K^{-1}$, (3.166) comes out as $l/L \geq c(\Lambda_K/L)^4$.

Singularities and the Euler equations. An important issue in the theory of inviscid fluid dynamics is whether the three-dimensional Euler equations

$$\mathbf{u}_t + (\mathbf{u} \cdot \nabla)\mathbf{u} = -\nabla p$$

with div $\mathbf{u} = 0$, can generate a finite time singularity from smooth initial data. Certain results concerning the three-dimensional Euler equations can be demonstrated using the methods developed here. Consider the results for the evolution of $\kappa_{N,r}$ for the NSE from Section 3.6.2, expressed in both row 5 of Table 3.1 and also in (3.159), but with $\nu = 0$ and $f = 0$

$$(N - r)\dot{\kappa}_{N,r} \leq c_{N,d} \|D\mathbf{u}\|_\infty \kappa_{N,r}. \tag{3.167}$$

This is valid for the Euler equations with the forcing terms removed from the F_N. A simple integration with respect to time gives

$$\kappa_{N,r}(t) \leq \kappa_{N,r}(0) \exp\left[c_N \int_0^t \|D\mathbf{u}\|_\infty(\tau)d\tau\right]. \tag{3.168}$$

In addition, because $\kappa_{N,r} \le \kappa_{N+1,r}$ (see the exercises), (3.167) is also valid in the case of the three-dimensional NSE, with an additional $(\nu\lambda_0^{-2})\kappa_{N,r}$ term. An equality (3.168) shows that $\int_0^t \|D\mathbf{u}\|_\infty(\tau)d\tau$ controls any singularities that might form. For the Euler equations a more natural question to ask is whether $\int_0^t \|\omega\|_\infty(\tau)d\tau$ controls possible singularity formation. We state a theorem which was first proved in [6, 75], valid for the periodic domain $[0, L]^3$:

Theorem 56 *We have*

$$\|D\mathbf{u}\|_\infty \le c\|\omega\|_\infty[1 + \log^+(L\kappa_{N,r})] + L^{-3/2}\|\omega\|_2$$

for $N \ge 3$ and $0 \le r < N$. The "+" sign on the logarithm is defined by $\log^+ a = \log a$ for $a \ge 1$ and $\log^+ a = 0$ otherwise.

Proof. Consider solutions for the velocity field $\mathbf{u}(x, t)$ through the Biot–Savard law which inverts $\omega = \mathrm{rot}\,\mathbf{u}$:

$$\mathbf{u}(x) = \int_\Omega K(x - \xi) \times \omega(\xi)\,d\xi,$$

where K is the gradient of the kernel of the inverse of $-\Delta$ on mean zero periodic functions on $[0, L]^3$. Introduce a cut-off function $\xi_\rho(x)$, satisfying $\xi_\rho(x) = 1$ for $|x| < \rho$, $\xi_\rho(x) = 0$ for $|x| > 2\rho$, and $|D\xi_\rho(x)| \le c/\rho$. Here $\rho\,(0 < \rho \le L/4)$ is a radius chosen suitably small later. Introduce a factor $\xi_\rho(x - \xi) + [1 - \xi_\rho(x - \xi)]$ under the integral sign and split $D\mathbf{u}$ into two terms, the first being

$$D\mathbf{u}^{(1)} = \int_\Omega \xi_\rho(x - \xi)K(x - \xi) \times D\omega(\xi)\,d\xi.$$

The coincidence point singularity in K is bounded according to $|K(x - \xi)| \le c|x - \xi|^{-2}$, thus, by Hölder's inequality

$$\|D\mathbf{u}^{(1)}\|_\infty \le \|K\|_p \|D\omega\|_q \le c\rho^{1-3/q}\|D\omega\|_q$$

with the restriction $q > 3$ where $1/p + 1/q = 1$. Hence $p < 3/2$. We will use a calculus inequality

$$\|D\omega\|_q \le c\|D^{N-1}\omega\|_2^a \|\omega\|_2^{1-a}$$

with $N > 3$, $a = \frac{3}{N-1}\left(\frac{5}{6} - \frac{1}{q}\right)$, and so $\|D\mathbf{u}^{(1)}\|_\infty^2 \le c\rho^{2(1-3/q)}\kappa_{N,1}^{2(N-1)a}F_1$. After certain manipulation, we are left with the second term

$$D\mathbf{u}^{(2)} = \int_\Omega D\{[1 - \xi_\rho(x - \xi)]K(x - \xi)\} \times \omega(\xi)\,d\xi,$$

where the integral is over $\rho \le |x - \xi| \le L$. For $D\mathbf{u}^{(2)}$, we estimate the two terms in the gradient separately and use $|DK| \le |x - \xi|^{-3}$ to obtain

$$\|D\mathbf{u}^{(2)}\|_\infty \le c\left(\int_\rho^L r^{-3}r^2 dr + \int_\rho^{2\rho} r^{-2}\rho^{-1}r^2 dr\right)\|\omega\|_\infty,$$

which finally gives

$$\|D\mathbf{u}\|_\infty \le c\rho^{1-3/q}\kappa_{N,1}^{5/2-3/q}F_1^{1/2} + c\|\omega\|_\infty[1 + \log(L/\rho)]. \tag{3.169}$$

Now we choose $\rho^{-1} = c\kappa_{N,1}^{(5q-6)/2(q-3)}L^{3q/2(q-3)}$, which in (3.169) gives the result of the theorem. $\qquad\square$

The main point here is that $\kappa_{N,r}$ ($N \geq 3$) are natural mediators between $\|Du\|_\infty$ and $\|\omega\|_\infty$. Another consequence of Theorem 56 is the explicit estimate or $\kappa_{N,r}$ as functions of time in terms of the time integral of $\|\omega\|_\infty$. An integration of (3.167) using Theorem 56 gives

$$L\kappa_{N,r}(t) \leq c \exp\left(\int_0^t g(\tau) \exp[I(t) - I(\tau)] d\tau\right),$$

where $I(t) = \int_0^t \|\omega\|_\infty(\tau) \, d\tau$ and $g(t) = L^{-3/2}\|\omega\|_2$. Thus, no singularity can develop at t^* without the quantity $\int_0^{t^*} \|\omega\|_\infty(\tau) \, d\tau \to \infty$ at t^* also. This result guarantees that a solution of the three-dimensional Euler equations cannot develop a singularity through the following processes:

- The development of kinks or curvature singularities in vortex lines.
- The development of "vorticity shocks" where ω becomes discontinuous but $\|\omega\|_\infty$ remains bounded.
- The rate of strain matrix $S_{ij} = \frac{1}{2}(u_{i,j} + u_{j,i})$ becomes singular while the vorticity remains bounded.

Thus, if any of these types of behavior appears in a numerical solution, then it must be an artifact of the numerical scheme. Furthermore, it shows that $I(t)$ *alone* controls any singularities that might form. It is not sufficient to monitor $\|\omega\|_2$ as this may still remain bounded while $\|\omega\|_\infty \to \infty$. If a numerical integration scheme indicates that a singularity of the type $\|\omega\|_\infty \sim (t^* - t)^{-\gamma}$ forms at t^* then the BKM theorem says that $\int_0^{t^*} \|\omega\|_\infty(\tau) \, d\tau$ must also become singular at t^*. Hence, we must have $\gamma \geq 1$ for the singularity to be genuine and not an artefact of the numerics.

3.6.5 The attractor dimension

This section is devoted to estimation of dimension of the global attractor $\mathcal{A}$ for the NSE. The approach is an extension of that developed in the section for ODEs, where it was shown that if N-dimensional volume elements in the system phase space contract to zero, then the attractor dimension $d_L(\mathcal{A})$ is bounded by N. For PDEs the technical chore remains the same; namely, to derive estimates on the spectrum of the linearized (around solutions on the attractor) evolution operator, and to perform this operation in some function space instead of a finite dimensional phase space. As we saw for the Lorentz equations, this requires some knowledge of the location of the attractor, i.e., a priori estimates on the solutions. Due to Section 3.6.4, a global attractor $\mathcal{A}$ exists in this case, and we have good control of solutions on the attractor. Moreover, the result for periodic boundary conditions is quite sharp, within logarithms of both the conventional heuristic estimate for the number of degrees of freedom in a two-dimensional turbulent flow and rigorous lower bounds. To achieve any estimate of the attractor dimension it is necessary to assume that H_1 remains bounded for all t. Then we produce estimates for the attractor dimension in terms of a Kolmogorov length Λ_K based on this quantity instead of using the more conventional definition of λ_K which uses $\langle H_1 \rangle$. This estimate turns out to being close to sharp.

The two-dimensional attractor dimension estimate. We begin with the two-dimensional forced, incompressible NSE (3.3). The vorticity is a scalar $\omega = \hat{k} \cdot \nabla \times \mathbf{u} = \partial_x u_2 - \partial_y u_1$, and its evolution equation is

$$\partial_t \omega + (\mathbf{u} \cdot \nabla) \omega = \nu \Delta \omega + \operatorname{rot} f. \tag{3.170}$$

Let (3.170) be the defining equation, with periodic boundary conditions on the domain $\Omega \equiv [0, L]^2$. The velocity vector field is expressed as $\mathbf{u} = (-\partial_y(\Delta^{-1}\omega), \partial_x(\Delta^{-1}\omega))$ where we may assume that the spatially averaged vorticity vanishes so that the Laplacian may be inverted. Upper bounds on the attractor dimension for the system are determined by considering the time evolution of volume elements in the system's configuration space which, in this case, is $L^2(\Omega)$. If all the N-dimensional volumes contract to zero volume as $t \to \infty$, then the attractor cannot contain any N-dimensional subsets and hence $d_L \leq N$. The goal is to determine the smallest possible N with this property, as it constitutes an upper bound on the dimension of the global attractor in both the fractal and Hausdorff sense.

Restrict ourselves to infinitesimal volume elements, whose evolutions are controlled by the NSE linearized about an arbitrary solution on the attractor. For (3.170), the linearized equation for the difference $\delta\omega$ between two neighboring solutions is

$$\partial_t(\delta\omega) = -A(t)\delta\omega = -\mathbf{u} \cdot \nabla\delta\omega - \delta\mathbf{u} \cdot \nabla\omega + \nu\Delta\delta\omega \qquad (3.171)$$

where the variation in the velocity is $\delta\mathbf{u} = (-\partial_y(\Delta^{-1}\delta\omega), \partial_x(\Delta^{-1}\delta\omega))$. In $L^2(\Omega)$, these $\delta\omega_i$ form an N-dimensional volume element or parallelepiped of volume $V_N(t) = |\delta\omega_1(t) \wedge \ldots \wedge \delta\omega_N(t)|$ each edge of which develops according to (3.171). Volume V_N itself evolves according to

$$V_N(t) = V_N(0) \exp\left(-\int_0^t \text{Tr}[A(t')P_N(t')]\, dt'\right). \qquad (3.172)$$

Here, $P_N(t)$ is the orthogonal projection onto the *finite* dimensional linear subspace $P_N L^2(\Omega)$. In order for N to be an upper bound on the attractor dimension, the volume elements V_N about any solution $\omega(t)$ on the attractor must vanish as $t \to \infty$. Rewrite (3.172),

$$V_N(t) = V_N(0) \exp\left[-t\left(\frac{1}{t}\int_0^t \text{Tr}[A(t')P_N(t')]dt'\right)\right] \xrightarrow[t\to\infty]{} V_N(0) \exp(-t\langle\text{Tr}[AP_N]\rangle), \quad (3.173)$$

where $\langle\cdot\rangle$ denotes the largest possible time average. Thus, to determine an upper bound on the attractor dimension we look for the smallest N which for all solutions $\omega(t)$ satisfies

$$-\langle\text{Tr}[AP_N]\rangle < 0. \qquad (3.174)$$

Let $\{\varphi_1(t), \ldots, \varphi_N(t)\}$ be orthonormal functions spanning $P_N L^2(\Omega)$ and let the associated vector fields $\{v_1(t), \ldots, v_N(t)\}$ be $v_n = (-\partial_y(\Delta^{-1}\varphi_n), \partial_x(\Delta^{-1}\varphi_n))$. Estimate the trace as follows:

$$\begin{aligned}
\text{Tr}[A(t)P_N(t)] &= \sum_{n=1}^{N} \int_\Omega \phi_n(t)A(t)\phi_n(t)\, dx = \nu \sum_{n=1}^{N} \int_\Omega |\nabla\phi_n|^2\, dx \\
&+ \sum_{n=1}^{N} \int_\Omega \phi_n(u \cdot \nabla\phi_n + v_n \cdot \nabla\omega)\, dx = \nu\text{Tr}[-\Delta P_N] + \sum_{n=1}^{N} \int_\Omega \phi_n v_n \cdot \nabla\omega\, dx,
\end{aligned} \qquad (3.175)$$

where we used the fact that $\mathbf{u}$ is divergence free to eliminate one of the terms. Because the spectrum of the Laplacian is known explicitly, the real work consists of finding good sharp upper bounds on the last sum in (3.175). Using the Schwarz inequality, we have

$$\left|\sum_{n=1}^{N}\int_\Omega \phi_n v_n \cdot \nabla\omega\, dx\right| \leq \int_\Omega \left(\sum_{n=1}^{N}\phi_n^2\right)^{1/2}\left(\sum_{n=1}^{N}|v_n|^2\right)^{1/2}|\nabla\omega|\, dx.$$

Using a Hölder inequality we pull out the sum of squares of the v_n in the L^∞-norm to get

$$\left|\sum_{n=1}^{N}\int_\Omega \phi_n v_n \cdot \nabla\omega\, dx\right| \leq \left\|\sum_{n=1}^{N}|v_n|^2\right\|_\infty^{1/2}\int_\Omega \left(\sum_{n=1}^{N}\phi_n^2\right)^{1/2}|\nabla\omega|\, dx.$$

The Cauchy–Schwarz inequality is used to separate the two factors inside the integral above:

$$\left|\sum_{n=1}^{N}\int_\Omega \phi_n v_n \cdot \nabla\omega\, dx\right| \leq \left\|\sum_{n=1}^{N}|v_n|^2\right\|_\infty^{1/2}\int_\Omega \left(\sum_{n=1}^{N}\phi_n^2\, dx\right)^{1/2}\|\nabla\omega\|_2. \qquad (3.176)$$

To estimate the first factor above we use Constantin's theorem which provides L^∞ estimates on collections of functions whose gradients are orthonormal. We use it in the form where it is applicable to the sum of the squares of the v_n's.

Theorem 57 *If gradients of functions v_n $(1 \leq n \leq N)$ are orthonormal, that is $\int_\Omega (\nabla v_{n\beta})(\nabla v_{m\beta}) \, \mathrm{d}x = \delta_{mn}$, then*

$$\| \sum_{n=1}^{N} |v_n|^2 \|_\infty \leq c\{1 + \log(L^2 \mathrm{Tr}[-\Delta P_N])\}, \tag{3.177}$$

where $\mathrm{Tr}[-\Delta P_N] = \sum_{n=1}^{N} \int_\Omega |\nabla \phi_n|^2 \, \mathrm{d}x$ and the constant c is independent of N.

Proof. Consider vector valued functions $A_\alpha(x) = \sum_{n=1}^{N} \xi_n v_{n\alpha}$ (which are arbitrary linear combinations of the $v_n(x)$ defined above) with

$$v_{n\alpha} = -\varepsilon_{\alpha\beta} \partial_\beta \Delta^{-1} \phi_n. \tag{3.178}$$

Since $\{\phi_n\}$ are orthonormal, and $\{v_n\}$ have orthonormal gradients:

$$\int_\Omega (\nabla v_{n\beta}) \cdot (\nabla v_{m\beta}) \, \mathrm{d}x = \delta_{nm}.$$

Then simply $\|A\|_2^2 = \sum_{n=1}^{N} \xi_n^2 = |\xi|^2$. Now we apply Lemma 10 to A:

$$| \sum_{n=1}^{N} \xi_n v_{n\alpha} | \leq c|\xi| \left[1 + \log \left(L|\xi|^{-1} \| \sum_{n=1}^{N} \xi_n \nabla v_n \|_2 \right) \right]^{1/2}$$

$$\leq c|\xi| \left[1 + \log \left(L^2 \sum_{n=1}^{N} \|\Delta v_n\|_2^2 \right) \right]^{1/2},$$

where we have used the Schwarz inequality in the last step. Rewrite this as

$$|\xi|^{-1} | \sum_{n=1}^{N} \xi_n v_n | \leq c \left[1 + \log \left(L^2 \sum_{n=1}^{N} \|\Delta v_n\|_2^2 \right) \right]^{1/2}$$

and because this is true for every N-vector ξ, we have

$$\| \sum_{n=1}^{N} |v_n|^2 \|_\infty \leq c \left[1 + \log \left(L^2 \sum_{n=1}^{N} \|\Delta v_n\|_2^2 \right) \right]. \tag{3.179}$$

Using (3.178), we note that the sum of $\|\Delta v_n\|_2^2$ inside the log-term above is simply estimated,

$$\sum_{n=1}^{N} \|\Delta u\|_2^2 = \sum_{n=1}^{N} \int_\Omega (\varepsilon_{\beta\gamma} \partial_\gamma \phi_n)(\varepsilon_{\beta\delta} \partial_\delta \phi_n) \, \mathrm{d}x = \sum_{n=1}^{N} \int_\Omega |\nabla \phi_v|^2 \, \mathrm{d}x = \mathrm{Tr}[-\Delta P_N]. \tag{3.180}$$

Inserting (3.180) above into (3.179) finishes the proof. □

Continuing with estimation of (3.176), its middle factor is evaluated by recalling that ϕ_n are orthonormal functions so that

$$\sum_{n=1}^{N} \int_\Omega \phi_n^2 \, \mathrm{d}x = N. \tag{3.181}$$

The first term in (3.175), the trace of the Laplacian in an N-dimensional subspace, is estimated as

$$\mathrm{Tr}[-\Delta P_N] \geq cL^{-2} N^{(2+d)/d}. \tag{3.182}$$

Hence we rewrite (3.181) as

$$\int_\Omega \sum_{n=1}^{N} \phi_n^2 \, \mathrm{d}x \leq c(\mathrm{Tr}[-\Delta P_n])^{1/2} L. \tag{3.183}$$

Putting together (3.176), (3.177) and (3.183), we arrive at

$$\left|\sum_{n=1}^{N}\int_{\Omega}\phi_n v_n\cdot\nabla\omega\,dx\right|\le c\|\nabla\omega\|_2[g(L^2\text{Tr}[-\Delta P_N])]^{1/2},$$

where $g(\zeta)=\sqrt{\zeta}(1+\log\zeta)$. Taking the time average we have

$$\left\langle\left|\sum_{n=1}^{N}\int_{\Omega}\phi_n v_n\cdot\nabla\omega\,dx\right|\right\rangle\le c\left\langle\|\nabla\omega\|_2[g(L^2\text{Tr}[-\Delta P_N])]^{1/2}\right\rangle.$$

Using the Cauchy–Schwarz inequality on the time average on the above gives

$$\left\langle\left|\sum_{n=1}^{N}\int_{\Omega}\phi_n v_n\cdot\nabla\omega\,dx\right|\right\rangle\le c\langle\|\nabla\omega\|_2^2\rangle^{1/2}\langle g(L^2\text{Tr}[-\Delta P_N])\rangle^{1/2}.$$

Now, the function $g(\zeta)$ is concave for $\zeta>1/e$. For large values of N which are appropriate for turbulent flows, $L^2\text{Tr}[-\Delta P_N]\gg 1$, we invoke Jensen's inequality $\langle g(\zeta)\rangle\le g(\langle\zeta\rangle)$ for g concave to find

$$\left\langle\left|\sum_{n=1}^{N}\int_{\Omega}\phi_n v_n\cdot\nabla\omega\,dx\right|\right\rangle\le c\langle\|\nabla\omega\|_2^2\rangle[g(\langle L^2\text{Tr}[-\Delta P_N]\rangle)]^{1/2}.$$

Therefore, the time averaged trace in (3.173) and (3.174), controlling the exponential growth or contraction of volume elements, is estimated by

$$\langle\text{Tr}[A\,P_N]\rangle\ge\frac{\nu}{L^2}\langle L^2\text{Tr}[-\Delta P_N]\rangle-c\langle\|\nabla\omega\|_2^2\rangle^{1/2}[g(\langle L^2\text{Tr}[-\Delta P_N]\rangle)]^{1/2}.\tag{3.184}$$

Next we use the two-dimensional Grashof number $\text{Gr}=L^2\nu^{-2}\|f\|_2$, see Table 3.2 in Section 3.6.2. Multiplying the vorticity version of the NSE (3.170) by ω, integrating over the spatial variables and taking the time average, we have

$$\nu\langle\|\nabla\omega\|_2^2\rangle=\left\langle\int_{\Omega}\omega\hat{k}\cdot\text{rot}\,f\,dx\right\rangle.$$

Integration by parts and application of Cauchy's inequality yields

$$\nu\langle\|\nabla\omega\|_2^2\rangle\le\langle\|\nabla\omega\|_2\rangle\|f\|_2\le\langle\|\nabla\omega\|_2^2\rangle^{1/2}\|f\|_2,$$

so that

$$\langle\|\nabla\omega\|_2^2\rangle^{1/2}\le\nu^{-1}\|f\|_2=\nu\,\text{Gr}/L^2.$$

Define $\tilde{N}$ by $\tilde{N}=L^2\langle\text{Tr}[-\Delta P_N]\rangle$. Then the trace formula in (3.184) is

$$\langle\text{Tr}[AP_N]\rangle\ge\frac{\nu}{L^2}\left[\tilde{N}^2-c\,\text{Gr}\,\tilde{N}^{1/2}(1+\log\tilde{N})^{1/2}\right].\tag{3.185}$$

As N increases, then so does $\tilde{N}$, and for a given Gr it eventually forces the right-hand side of (3.185) to become positive so that all volume elements of dimension higher than N contract to zero. If the logarithmic term was absent in (3.185), then this crossover point would occur when $\tilde{N}\sim\text{Gr}^{2/3}$. The logarithm introduces corrections to this. The precise answer, including the logarithmic correction, allows us to conclude that all N-dimensional volume elements contract to zero when $N\ge c\,\text{Gr}^{2/3}(1+\log\text{Gr})^{1/3}$. Indeed, we must show that

$$\tilde{N}\ge c\,\text{Gr}^{2/3}(1+\log\text{Gr})^{1/3}\ \Rightarrow\ \tilde{N}^2-c'\,\text{Gr}\,\tilde{N}^{1/2}(1+\log\tilde{N})^{1/2}\ge 0.$$

We prove this via transposition by showing that

$$\tilde{N}^2 - c'\,\mathrm{Gr}\tilde{N}^{1/2}(1+\log\tilde{N})^{1/2} < 0 \;\Rightarrow\; \tilde{N} < c\,\mathrm{Gr}^{2/3}(1+\log\mathrm{Gr})^{1/3}. \tag{3.186}$$

Toward this end, assume that

$$\tilde{N}^2 - c'\,\mathrm{Gr}\,\tilde{N}^{1/2}(1+\log\tilde{N})^{1/2} < 0. \tag{3.187}$$

Then $\frac{\tilde{N}^3}{1+\log\tilde{N}} < c''\,\mathrm{Gr}^2$ and

$$3\log\tilde{N} - \log\left(1+\log\tilde{N}\right) < 2\log\mathrm{Gr} + \log c''. \tag{3.188}$$

But it is easy to show that for the relevant regime where $\tilde{N} > 1$, $\log\left(1+\log\tilde{N}\right) \le \log\tilde{N}$, so (3.188) implies

$$1+\log\tilde{N} < \tilde{c}(1+\log\mathrm{Gr}). \tag{3.189}$$

Now, the premise (3.187) is $\tilde{N}^{3/2} < c'\,\mathrm{Gr}\left(1+\log\tilde{N}\right)^{1/2}$, and inserting (3.189) we get $\tilde{N}^{3/2} < c'\sqrt{\tilde{c}}\,\mathrm{Gr}(1+\log\mathrm{Gr})^{1/2}$. This is equivalent to the right-hand side of (3.186) and establishes the result. Thus, the attractor dimension is bounded by

$$d_L < c\,\mathrm{Gr}^{2/3}(1+\log\mathrm{Gr})^{1/3}. \tag{3.190}$$

Finally, to express d_L in terms of the *Kraichnan length* λ_{Kr} [22] instead of Gr, we return to (3.184) and express $\langle\|\nabla\omega\|_2^2\rangle$ in terms of the average *enstrophy dissipation rate* χ given by $\chi = \nu L^{-2}\langle\|\nabla\omega\|_2^2\rangle$. From (3.184), we have $\lambda_{Kr} = (\nu^3/\chi)^{1/6}$, so we finally obtain

$$d_L \le c(L/\lambda_{Kr})^2[1+\log(L/\lambda_{Kr})]^{1/3}. \tag{3.191}$$

Apart from the logarithmic correction, the estimate in (3.191) fits with the scaling theory of two-dimensional turbulence, that is the number of degrees of freedom in a two-dimensional domain is $(L/\lambda_{Kr})^2$. Note that (3.190) is the same as (3.150) for the first of the nontrivial number of degrees of freedom $\mathcal{N}_{3,1}$. This result shows that the rigorous methods of attractor dimension analysis, using the full properties of the two-dimensional NSE without uncontrolled approximations, coincide with the predictions arising from heuristic considerations.

The three-dimensional attractor dimension estimate. By Section 3.6.4, the lack of a regularity proof for the three-dimensional NSE precludes us from asserting the existence of a global attractor. This does not prevent us using the Kolmogorov length λ_K, though, as it is a properly bounded quantity, being dependent on $\langle F_1\rangle$. Recall that the energy dissipation rate and the Kolmogorov length are respectively defined by $\varepsilon = \nu L^{-3}\langle H_1\rangle$ and $\lambda_K^{-1} = (\varepsilon/\nu^3)^{1/4}$. Because of the lack of a solid proof of the existence of a global attractor, the device used in Section 3.6.4 to circumvent this problem, see (3.165), was to introduce $\sup_t H_1$ instead of the time average $\langle H_1\rangle$ (or $\langle F_1\rangle$) in order to define an alternative energy dissipation rate and an alternative Kolmogorov length according to

$$\bar{\varepsilon} = \nu L^{-3}\sup_t H_1, \qquad \Lambda_K^{-1} = (\bar{\varepsilon}/\nu^3)^{1/4}. \tag{3.192}$$

Let $\sup_t H_1$ be bounded and proceed to estimate the attractor dimension based on Λ_K. Turning to the trace formula, no advantage is gained in three-dimension by working with $\omega \in L^2$. Hence we consider the velocity field orbit $\mathbf{u}(t) \in L^2$. The linearized evolution of $\delta\mathbf{u}$ about $\mathbf{u}$ is

$$\delta\mathbf{u}_t + \mathbf{u}\cdot\nabla\delta\mathbf{u} + \delta\mathbf{u}\cdot\nabla\mathbf{u} = \nu\Delta\delta\mathbf{u} - \nabla\delta p,$$

which can formally be written $\delta\mathbf{u}_t = -A\delta\mathbf{u}$. As in (3.174), we wish to find such N that turns the sign of $\langle \mathrm{Tr}[AP_N] \rangle$ from negative to positive. This value of N, (say, $\tilde{N}$) bounds above d_L. The orthonormal basis $\{\phi_1, \phi_2, \ldots, \phi_N\}$ in L^2 consists of divergence-free vector functions. The trace formula we need to explicitly estimate is

$$\mathrm{Tr}[AP_N] = -\sum_{n=1}^{N} \int_{\Omega} \phi_n \cdot \{\nu \Delta \phi_n - \mathbf{u} \cdot \nabla \phi_n - \phi_n \cdot \nabla \mathbf{u} - \nabla \tilde{p}(\phi_n)t\} \, dx.$$

Since div $\varphi_n = 0$, the pressure term integrates away, as does the second term, to give

$$\mathrm{Tr}[AP_N] \geq \nu \sum_{n=1}^{N} \int_{\Omega} |\nabla \varphi_n|^2 \, dx - \sum_{n=1}^{N} \int_{\Omega} |\nabla \mathbf{u}| \, |\nabla \varphi_n|^2 \, dx. \tag{3.193}$$

The φ_n satisfy the Lieb–Thirring inequalities for orthonormal functions (see Section 3.9),

$$\int_{\Omega} \left| \sum_{n=1}^{N} |\nabla \phi_n|^2 \right|^{5/3} \, dx \leq c \sum_{n=1}^{N} \int_{\Omega} |\nabla \phi_n|^2 \, dx, \tag{3.194}$$

where c is independent of N. Moreover, by (3.182), $\mathrm{Tr}[(-\Delta)P_N]$ is bounded from below. We can exploit the Lieb–Thirring inequality (3.194) and estimate the last term in (3.193) as

$$\sum_{n=1}^{N} \int_{\Omega} |D\mathbf{u}| \, |\varphi_n|^2 \, dx \leq \left(\int_{\Omega} |D\mathbf{u}|^{5/2} \, dx \right)^{2/5} \left[\int_{\Omega} \left(\sum_{n=1}^{N} |\varphi_n|^2 \right)^{5/3} \, dx \right]^{3/5}.$$

Hence, using (3.194) and time averaging $\langle \cdot \rangle$, we find that (3.193) can be written as

$$\langle \mathrm{Tr}[AP_N] \rangle \geq \frac{2}{5} \nu \langle \mathrm{Tr}[(-\Delta)P_N] \rangle - c \nu^{-3/2} \langle \int_{\Omega} |D\mathbf{u}|^{5/2} \, dx \rangle. \tag{3.195}$$

The problem lies in $\langle \|D\mathbf{u}\|_{5/2}^{5/2} \rangle$ term for which we have no a priori estimate, so we rewrite (3.195) as

$$\mathrm{Tr}[AP_N] \geq \frac{2}{5} \nu \langle \mathrm{Tr}[(-\Delta)P_N] \rangle - c \nu^{-3/2} \langle \|D\mathbf{u}\|_{\infty}^{1/2} \rangle \left(\sup_t H_1 \right). \tag{3.196}$$

We have an estimate for $\langle \|D\mathbf{u}\|_{\infty}^{1/2} \rangle$ from Section 3.6.4:

$$\langle \|D\mathbf{u}\|_{\infty} \rangle \leq c \langle \kappa_{N,1}^{3/4} F_1^{1/4} \rangle \leq c \langle \kappa_{N,1} \rangle^{3/4} \langle F_1 \rangle^{1/4},$$

so (3.196) becomes

$$\mathrm{Tr}[AP_N] \geq c_1 \nu L^{-2} N^{5/3} - c_2 \nu L^2 (L/\lambda_K)^4 \Lambda_K^{-4},$$

where we have used the definition of Λ_K given in (3.192). Hence, the trace becomes positive when $N \geq \tilde{N}$ and so

$$d_L(\mathcal{A}) \leq \tilde{N} = c(L/\Lambda_K)^{12/5} (L/\lambda_K)^{12/5}.$$

If d_L is also associated with the number of degrees of freedom $\mathcal{N}$ of the system discussed at the beginning of the present chapter, then this can be written in terms of the natural small scale of system l

$$L/l \leq c(L/\Lambda_K)^{8/5},$$

where we have exploited the fact that $\lambda_K^{-1} \leq \Lambda_K^{-1}$. The exponent 8/5 is significantly closer to conventional heuristic estimate of unity and it is certainly much better than the exponent 4 found in (3.166).

3.7 Hierarchical shell models

In this section, we consider models based on the idea of applying a functional basis of special type that most closely corresponds to the structure of turbulent fields. The idea of such basis is to use self-similar functions of progressively decreasing scale. The basis was called *hierarchical* and based on it were built numerous models [21, 34, 42, 107] also called hierarchical. In the literature of the 1980, the word "wavelet" appeared, and by the beginning of the 1990s the term "wavelet analysis" turned into a self-sustaining, well-developed area of mathematical physics. The ideas underlying the theory of wavelets coincide with the ideas of the hierarchical representation of turbulent fields, and in terms of this novel science, hierarchical models are those constructed using the wavelet-representation of the described fields.

Shell models as well as Navier–Stokes turbulence pose an infinite hierarchy of dynamical equations for the n-order correlation functions. This hierarchy is *linear* in the correlation functions, and in the limit of infinite Reynolds number is also homogeneous. It was recently discovered [21, 72, 73, 84] that this hierarchy obeys a rescaling symmetry which stems from the rescaling symmetry of the Euler equation. This rescaling symmetry foliates the space of solutions into slices of different scaling exponents h of the velocity; these are referred to as h-slices. On each h-slice one finds "normal scaling" with the given value of h. The full solution is a linear combination of all the solutions on the h-slices with nonuniversal weights which are determined by the forcing on the integral scale of turbulence. Different orders of the correlation functions are dominated by different h-slices, and accordingly the full solution has anomalous scaling. The anomalous exponents are expected to be universal.

3.7.1 Gledzer–Ohkitani–Yamada shell model

The simplest shell model [21], is governed by the following infinite set of coupled nonlinear ODE, indexed by $n = 0, 1, 2, \ldots$ (called as the shell index):

$$\left(\frac{d}{dt} + \nu k_n^2\right)u_n = \alpha(k_{n-1}u_{n-1}^2 - k_{n+1}u_n u_{n+1}) + \beta(k_n u_{n-1}u_n - k_{n+1}u_{n+1}^2) + f_n. \quad (3.197)$$

Here, the dynamical variables $u_n(t)$ are real numbers; u_{-1} which appears in the equation for $n = 0$ is taken equal to zero, the forcing item f_n is prescribed, time-independent and restricted usually to a single shell: $f_n = f\delta_{n,n_0}$ where δ_{n,n_0} is Kronecker delta function. The wavenumbers k_n are given by

$$k_n = k_0 2^n,$$

with $k_0 > 0$ is a reference wavenumber. The ratio of a factor 2 between the wavenumber of successive shells is an arbitrary choice, easily modified. The parameter $\nu > 0$ is a viscosity, and the control parameters α and β are real numbers, one of which may vanish. Remark that this model has a number of properties in common with the NSE. The nonlinear term is quadratic and has dimension [velocity]^2/[length]; it converges the energy $\frac{1}{2}\sum_n u_n^2$. It is invariant under time translations and under a discrete form of scaling transformations: when $f = \nu = 0$ and the "boundary condition" for u_{-1} is ignored, the equations are invariant under $n \to n+1$, $u_n \to (1/2)^h u_n$, $t \to (1/2)^{1-h}t$, where h is an arbitrary scaling exponent. Finally, the shell model has exact *static* (time-independent) K41 solutions: at those scales where the force and the viscosity are negligible, thus the nonlinear term vanishes when $u_n = Ck_n^{-1/3}$ for arbitrary C. Shell models originated in [74] and the Russian school [21, 42], etc. Of particular interest is the Gledzer–Ohkitani–Yamada (GOY) model [42, 84] which has one complex mode per shell and next-nearest shell interactions and may be viewed as a complex version of a model introduced in [42]. The governing equations are

$$\left(\frac{d}{dt} + \nu k_n^2\right)u_n = i(a_n u_{n+1}u_{n+2} + b_n u_{n-1}u_{n+1} + c_n u_{n-1}u_{n-2})^* + f_n, \qquad (3.198)$$

where the star denotes complex conjugation, $a_n = k_n$, $b_n = -k_{n-2}$, $c_n = -k_{n-3}$, $k_n = k_0 2^n$, and the force is applied to the fourth shell $f_n = f\delta_{4,n}$. There was presented in [84] qualitative and quantitative numerical evidence that the GOY model is chaotic. Then, there have been calculated the structure functions, defined for shell models as $S_p(n) = \langle |u_n|^p \rangle$, and found

(i) at inertial-range scales, $S_p(n) \sim k_n^{-\zeta_p}$;

(ii) the exponents ζ_p nontrivially depend on the order p, suggesting *multifractality*.

This multifractal behavior is indeed supported by further simulations, involving up to 250×10^6 time steps. Figures 3.7(a)(b) show the structure functions $S_p(n)$ of order $2-10$ and their exponents ζ_p. Calculations with parameters $f = (1+i) \times 5 \cdot 10^{-3}$, $k_0 = 2^{-4}$, $\nu = 10^{-7}$ use 22 shells. Intermittency in shell models is the subject of much current work and is quite poorly understood yet. All shell models possess *pulse solutions* of the form

$$u_n(t) = k_n^{-h}g_h[k_n^{-h}(t - t_*)], \qquad (3.199)$$

where t_* and the scaling exponent $h < 1$ are arbitrary and where the functions g_h are obtained by substitution into the dynamical equations ignoring forcing and dissipation terms. When the function g_h is localized, i.e., decreases rapidly to zero at small and large arguments, the temporal evolution of $u_n(t)$ has the form of pulse of width $O(k_n^{h-1})$ centered around the time t_*. If well-separated pulses are generated by an instability at random times t_*, it follows that the structure function S_p of order p scales with an exponent $\zeta_p = 1 + (p-1)h$. Indeed, there are p factors k_n^{-h} (the prefactor in (3.199)) and a probability $\sim k_n^{1-h}$ of being within a pulse in the shell number n. To obtain a finite nonvanishing energy flux, one must take $h = 0$; then $\zeta_p = 1$, for all p, an extreme form of intermittency.

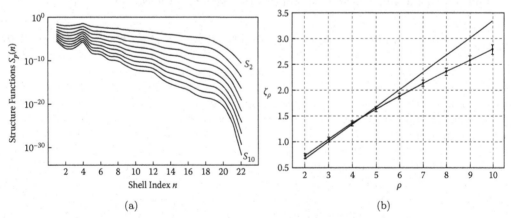

(a) (b)

FIGURE 3.7: (a) Structure functions of order $2 - 10$ (top to bottom) for the GOY model. The shell index n is $\log_2$ wavenumber. The data are averaged over 250×10^6 time steps. (b) Exponent ζ_p for the structure function of order p. The dashed line corresponds to K41 ($\zeta_p = p/3$). The error bars are from a least squares fit of the structure functions shown in Fig. 3.7(a) to a power-law in the interval $6 \leq n \leq 18$.

One aspect of the both shell models (3.197), (3.198), which was investigated systematically, is the stability of K41 solutions and the bifurcations away from K41 solutions. Since K41 solutions emerge only in the limit $\nu \to 0$, the control parameter must be other than the viscosity, for example, the ratio β/α in the Desnyansky-Novikov model (3.197), or a similar parameter which can be introduced into the GOY model (3.198). In addition to the technical problems in the linear stability study, there is a conceptual difficulty: an intermittent solution cannot be "close" to K41 and thus amenable to small perturbation techniques.

3.7.2 $(\mathbf{N}, \varepsilon)$-sabra shell models

In trying to evaluate the scaling exponents appearing in this way from first principles, in [72, 73] it was proposed to truncate the hierarchy of equations, preserving the fundamental rescaling symmetry that gives rise to anomalous scaling. Truncation is problematic; in turbulence there is no natural small parameter, and therefore any closure of an infinite hierarchy is uncontrolled. It is therefore worthwhile to introduce a one-parameter family of models, characterized by a parameter $\varepsilon \in [0, 1]$, which shows normal scaling when $\varepsilon = 0$ and recovers the anomalous scaling of the original model when $\varepsilon = 1$.

Following an idea in [59], we'll construct such a family, and show that the transition from normal to anomalous behavior occurs at some finite value of $\varepsilon > 0$. We will use $\varepsilon > 0$ as a small parameter to regularize the closure procedure; we will show that our closed equations are valid to $O(\varepsilon^4)$, whereas the neglected terms are of $O(\varepsilon^6)$. We can improve the closure scheme systematically by including terms of $O(\varepsilon^6)$, neglecting terms of $O(\varepsilon^8)$, etc.

A way to achieve this small parameter is to couple N copies of the same turbulent system, be it a shell model or the NSE, and choose the coupling to have both a deterministic and a random contributions, with relative amplitudes ε and $\sqrt{1-\varepsilon^2}$. For $\varepsilon = 1$ we loose the coupling between the copies, and recover the initial anomalous problem for any value of N. For $\varepsilon = 0$ we will show that that in the limit $N \to \infty$ we get normal scaling. Thus, for some value of $\varepsilon = 0$ and for large enough N we expect to see the birth of anomalous scaling, hopefully in a perturbative fashion. The existence of this transition is the main discovery of this paper, and we study it analytically using the ε-controlled closure procedure, and by direct simulations of the N-copied model.

The Sabra model. The starting point is the *Sabra shell model* due to [71]

$$\frac{du_n(t)}{dt} = i(ak_{n+1}u_{n+1}^*u_{n+2} + bk_nu_{n-1}^*u_{n+1} - ck_{n-1}u_{n-2}u_{n-1}) - \nu k_n^2 u_n + f_n(t). \quad (3.200)$$

Here u_n refers to the amplitude associated with "wave vector" k_n, where the spacing in this reduced model is determined by $k_n \equiv k_0\lambda^n$ – the spacing parameter, star $(^*)$ is complex conjugation, ν – the "viscosity", $f_n(t)$ – a random Gaussian force which is operating on the lowest shells. The parameters a, b and c are restricted by $a+b+c = 0$, which guarantees the conservation of the energy $E = \sum_n |u_n(t)|^2$, in the inviscid, unforced limit. The equations (3.200) are invariant under the phase transformation $u_n \to u_n \exp(i\theta_n)$, where the phases θ_n are restricted by the set of equations $\theta_{n-1} + \theta_n = \theta_{n+1}$. Choosing θ_1 and θ_2 arbitrarily, θ_n is determined for all $n \geq 2$. Evidently, the physical results of the model must be independent of the choice of the phases θ_1 and θ_2. In particular, the only nonzero correlation functions are those which are independent of the phases θ_n. The nonvanishing second and third order correlators are $S_2(k_n) = \langle |u_n|^2 \rangle$ and $S_3(k_n) = \text{Im}\langle u_{n-1}u_nu_{n+1}^* \rangle$. Note that (3.200) respects additional symmetry $u_n \to -u_n^*$. One of the consequences of this symmetry is that $\text{Re}\langle u_{n-1}u_nu_{n+1}^* \rangle = 0$, explaining why we only need to consider the imaginary part for $S_3(k_n)$. The symmetries for this model were selected explicitly in [71] to give rise to a small number of nonzero correlation functions, with the aim of simplifying the calculations.

$(\mathbf{N}, \varepsilon)$-**generalization of the Sabra model.** The standard available procedures to generalize dynamical systems to N copies involve *real* variables. In shell models in general, and in the Sabra model in particular, the amplitudes u_n are complex. Therefore, we rewrite

(3.200) following [9] in terms of the real and imaginary parts $u'_n \equiv \mathrm{Re}(u_n)$, $u''_n \equiv \mathrm{Im}(u_n)$. Doing this we guarantee, after the generalization to N copies, that the $N \to 1$ limit coincides with the original model. The equations are

$$\frac{du'_n}{dt} = \left[\gamma_{a,n+1}\left(-u'_{n+1}u''_{n+2} + u''_{n+1}u'_{n+2}\right) + \gamma_{b,n}\left(-u'_{n-1}u''_{n+1} + u''_{n-1}u'_{n+1}\right) \right.$$
$$\left. +\gamma_{b,n}\left(-u'_{n-1}u''_{n+1} + u''_{n-1}u'_{n+1}\right)\right] - \nu k_n^2 u'_n + f'_n(t),$$

$$\frac{du''_n}{dt} = \left[\gamma_{a,n+1}\left(-u'_{n+1}u''_{n+2} + u''_{n+1}u''_{n+2}\right) + \gamma_{b,n}\left(-u'_{n-1}u''_{n+1} + u''_{n-1}u''_{n+1}\right) \right.$$
$$\left. +\gamma_{b,n}\left(-u'_{n-1}u'_{n+1} + u''_{n-1}u''_{n+1}\right)\right] - \nu k_n^2 u''_n + f''_n(t),$$

where $\gamma_{a,n} \equiv ak_n$, $\gamma_{b,n} \equiv bk_n$ and $\gamma_{c,n} \equiv ck_n$. Denoting $u_{n,-1} \equiv u'_n$, $u_{n,+1} \equiv u''_n$, these equations can be written more compactly in the matrix form

$$\frac{du_{n,\sigma}}{dt} = [\mathbf{A}^\sigma_{\sigma'\sigma''}\left(\gamma_{a,n+1}u_{n+1,\sigma'}u_{n+2,\sigma''} + \gamma_{b,n}u_{n-1,\sigma'}u_{n+1,\sigma''}\right)$$
$$+\mathbf{C}^\sigma_{\sigma'\sigma''}\gamma_{c,n-1}u_{n-2,\sigma'}u_{n-1,\sigma''}] - \nu k_n^2 u_{n,\sigma} + f_{n,\sigma}, \tag{3.201}$$

where $\mathbf{A}^\sigma_{\sigma'\sigma''}$ and $\mathbf{C}^\sigma_{\sigma'\sigma''}$ with $\sigma, \sigma', \sigma'' = \pm 1$, so that

$$\mathbf{A}^{+1} = \begin{pmatrix} 1 & 0 \\ 0 & 1 \end{pmatrix}, \quad \mathbf{A}^{-1} = \begin{pmatrix} 0 & -1 \\ 1 & 0 \end{pmatrix}, \quad \mathbf{C}^{+1} = \begin{pmatrix} -1 & 0 \\ 0 & 1 \end{pmatrix}, \quad \mathbf{C}^{-1} = \begin{pmatrix} 0 & 1 \\ 1 & 0 \end{pmatrix}.$$

The first subscript σ' denotes line ($+1$ corresponds to the upper line), second subscript σ'' denotes column ($+1$ corresponds to the left one). Clearly, $\mathbf{A}^\sigma_{\sigma'\sigma''} = \mathbf{A}^\sigma_{\sigma''\sigma'}$, $\mathbf{A}^\sigma_{\sigma'\sigma''} = \mathbf{C}^{\sigma'}_{\sigma''\sigma}$.

Consider N copies of (3.201), indexed by i, j, or l, and these indices take on values $-J, ..., +J$, $2J + 1 = N$, N odd. The i-th copy is denoted by $u^{[i]}_{n,\sigma}$. Let $D^{[ijl]}$ be coupling between copies, which will be chosen later. Thus (3.201) for a collection of copies reads

$$\frac{du^{[i]}_{n,\sigma}}{dt} = \sum_{jl} D^{[ijl]}[\mathbf{A}^\sigma_{\sigma'\sigma''}\left(\gamma_{a,n+1}u^{[j]}_{n+1,\sigma'}u^{[l]}_{n+2,\sigma''} + \gamma_{b,n}u^{[j]}_{n-1,\sigma'}u^{[l]}_{n+1,\sigma''}\right)$$
$$+\mathbf{C}^\sigma_{\sigma'\sigma''}\gamma_{c,n-1}u^{[j]}_{n-2,\sigma'}u^{[l]}_{n-1,\sigma''}] - \nu k_n^2 u^{[i]}_{n,\sigma} + f^{[i]}_{n,\sigma}. \tag{3.202}$$

The index l is defined modulo N, and introduce a Fourier transform in the *copy space*, $u^\alpha_{n,\sigma} = \frac{1}{\sqrt{N}}\sum_{l=-J}^{J} u^{[l]}_{n,\sigma}\exp(\frac{2i\pi\alpha l}{N})$. The index α is also defined modulo $N = 2J + 1$. It is convenient to consider a value α within *the first Brillouin zone*, i.e., from $-J$ to J. We'll refer to it as the α-momentum. Since $u^{[i]}_{n,\sigma}$ is real, $u^{-\alpha}_{n,\sigma} = (u^\alpha_{n,\sigma})^* \equiv u^{\alpha*}_{n,\sigma}$. In α-*Fourier space*, (3.202) may be represented as

$$\frac{du^\alpha_{n,\sigma}}{dt} = \sum_{\beta\gamma}\Phi^{\alpha\beta\gamma}(\Delta_{\alpha,\beta+\gamma} + \Delta_{\alpha+N,\beta+\gamma} + \Delta_{\alpha,\beta+\gamma+N})[\mathbf{A}^\sigma_{\sigma'\sigma''}(\gamma_{a,n+1}u^\beta_{n+1,\sigma'}u^\gamma_{n+2,\sigma''}$$
$$+\gamma_{b,n}u^\beta_{n-1,\sigma'}u^\gamma_{n+1,\sigma''}) + \mathbf{C}^\sigma_{\sigma'\sigma''}\gamma_{c,n-1}u^\beta_{n-2,\sigma'}u^\gamma_{n-1,\sigma''}] - \nu k_n^2 u^\alpha_{n,\sigma} + f^\alpha_{n,\sigma}, \tag{3.203}$$

where $\Delta_{\alpha\beta}$ is the Kronecker delta function, $\Delta_{\alpha\alpha} = 1$ and $\Delta_{\alpha\beta} = 0$ for $\alpha \neq \beta$. We use Greek indices for denoting component in α-Fourier space, and Latin indices for copies in the copy space. As a consequence of the discrete translation symmetry of the copy index $[i]$ equations (3.203) conserve α-momentum modulo N at the nonlinear vertex, as one can see explicitly in the above equation. The coupling amplitudes $\Phi_{\alpha\beta\gamma}$ in these equations are the Fourier transforms of the coupling amplitudes $D^{[ijl]}$. Our choice of these amplitudes will be presented below. Now we choose the coupling amplitudes according to

$$\Phi^{\alpha\beta\gamma} = \frac{1}{\sqrt{N}}(\varepsilon + \sqrt{1 - \varepsilon^2}\Psi^{\alpha\beta\gamma}), \tag{3.204}$$

where $\Psi^{\alpha\beta\gamma}$ are quenched random phases chosen with uniform distribution of the phase, independently distributed with zero average. The meaning of quenched randomness in the context of a direct numerical simulation is that we run (3.203) with a given random choice of Φ, obtain results averaged over the randomness of the forcing f, and then rerun with a fresh random choice of Φ. Only at the end we average the statistical objects over the runs. Clearly, when $N \to \infty$, the statistical functions are self averaging, and the last average is indeed. The couplings satisfy the following symmetries:

$$\Psi^{\alpha\beta\gamma} = \Psi^{\alpha\gamma\beta}, \quad (\Psi^{\alpha\beta\gamma})^* = \Psi^{-\alpha,-\beta,-\gamma}, \quad \Psi^{\alpha\beta\gamma} = \Psi^{-\gamma,\beta,-\alpha}.$$

The first of these conditions stems from the identity of copies, leading to the invariance of equations of motion to an interchange of copies in the nonlinear term. The second one is the reality conditions, the third imposes energy conservation in the inviscid, unforced limit. The requirement $\Psi^{\alpha\beta\gamma} = 1$ (if $\alpha\beta\gamma = 0$) guaranties that for $N = 1$ we recapture the original model at any ε. For $\varepsilon = 0$ we have so-called *Random Coupling Model* proposed in the context of the Navier–Stokes statistics in [57]. For $\varepsilon = 1$ the coupling coefficients in the α-Fourier space (3.204) are index-independent. This corresponds to uncoupling equation (3.202) in the copy space, because $D^{[ijl]} = \delta_{ij}\,\delta_{il}$. Thus for $\varepsilon = 1$ we recover the original Sabra model with anomalous scaling [71]. It was shown that with the choice of the couplings (3.204) the model interpolates continuously between *Random Coupling Model* for $\varepsilon = 0$ whose scaling is normal K41 (at $N \to \infty$) and the Sabra model with anomalous scaling for $\varepsilon = 1$. A model of this type was proposed in the context of the Navier–Stokes statistics in [57].

Global attractors and their dimensions. The first mathematical concept, which we use to establish the finite dimensional long-term behavior of the viscous Sabra shell model is the global attractor. The global attractor, A, is the maximal bounded invariant subset of the space H. It encompasses all of the possible permanent regimes of the dynamics of the shell model. It is also a compact subset of the space H, which attracts all the trajectories of the system. Establishment of finite Hausdorff and fractal dimensionality of the global attractor implies the possible parametrization of the permanent regimes of the dynamics in terms of a finite number of parameters. For the definition and further discussion of the concept of the global attractor, see [29, 103]. In analogy with Kolmogorov's mean rate of dissipation of energy in turbulent flow we define $\varepsilon = \nu\langle\|u\|^2\rangle$, the mean rate of dissipation of energy in the shell model system. The symbol $\langle\cdot\rangle$ represents the ensemble average or a long-time average. Since we don't know whether such long-time averages converge for trajectories we will replace the above definition of ε by

$$\varepsilon = \nu \sup_{u^{in} \in A} \limsup_{t \to \infty} \frac{1}{t} \int_0^t \|u(s)\|^2\, ds.$$

Define also the viscous dissipation length scale l_d. Due to Kolmogorov's theory, it should only depend on the viscosity ν and the mean rate of energy dissipation ε. Hence, pure dimensional arguments lead to the definition $l_d = (\nu^3/\varepsilon)^{1/4}$, which represents the largest spatial scale at which the viscosity term begins to dominate over the nonlinear inertial term of the shell model equation. In analogy with the conventional theory of turbulence this is also the smallest scale that one needs to resolve in order to get the full resolution for turbulent flow associated with the shell model system. We will obtain an estimate of the fractal dimension of the global attractor for the system

$$du/dt + \nu Au + B(u, u) = f, \quad u(0) = u^{in} \tag{3.205}$$

in terms of another nondimensional quantity – the generalized Grashoff number. Here

$$Au = (k_1 u_1, k_2 u_2, \ldots), \quad \forall\, u = (u_1, u_2, \ldots),$$

$$B(u,v) = -i \sum_{n=1}^{\infty} (ak_{n+1} v_{n+2} u_{n+1}^* + bk_n v_{n+1} u_{n-1}^* + ak_{n-1} u_{n-1} v_{n-2} + bk_{n-1} v_{n-1} u_{n-2})\, \phi_n,$$

where $\{\phi_j\}_{j=1}^{\infty}$ is the canonical basis in H, and $u_0 = u_{-1} = v_0 = v_{-1} = 0$. Then, our definition of bilinear operator B, with the energy conservation condition $a + b + c = 0$ imply

$$B(u,v) = -i \sum_{n=1}^{\infty} (ak_{n+1} u_{n+2} u_{n+1}^* + bk_n u_{n+1} u_{n-1}^* - ck_{n-1} u_{n-1} u_{n-2})\phi_n.$$

Suppose the forcing term satisfies $f \in L^{\infty}([0,T], H)$ and denote $\|f\|_{L^{\infty}([0,\infty], H)} = |f|_{\infty}$. Then we define the generalized Grashoff number for our system to be

$$G = |f|_{\infty}/(\nu^2 k_1^3). \tag{3.206}$$

The generalized Grashoff number was firstly introduced in the context of the study of the finite dimensionality of long-term behavior of turbulent flow in [30]. To check that it is indeed nondimensional we note that $\|f\|_{L^{\infty}([0,\infty], H)}$ has the dimension of L/T^2, where L is a length scale and T is a time scale. The value k_1 has the dimension of $1/L$ and the kinematic viscosity v has the dimension of L^2/T. In order to obtain an estimate of the generalized Grashoff number, which will be used in the proof of the main result of this section, apply inequality [16]

$$\nu \int_0^t \|u^m(s)\|^2 ds \leq |u^m(0)|^2 + (k_1^2 \nu)^{-1} |f|_{\infty}^2\, t$$

to get

$$\limsup_{t \to \infty} \frac{1}{t} \int_0^{\infty} \|u(s)\|^2 ds \leq |f|_{\infty}^2/(\nu^2 k_1^2) = \nu^2 k_1^4 G^2. \tag{3.207}$$

Theorem 58 *The Hausdorff and fractal dimensions of the global attractor of the system of equations (3.205), $d_H(A)$ and $d_F(A)$, respectively, satisfy*

$$d_H(A) \leq d_F(A) \leq \log_\lambda(L/l_d) + \frac{1}{2}\log_\lambda(C_1(\lambda^2 - 1)).$$

In terms of the Grashoff number G the upper bound takes the form

$$d_H(A) \leq d_F(A) \leq \log_\lambda G^{1/2} + \frac{1}{2}\log_\lambda(C_1(\lambda^2 - 1)). \tag{3.208}$$

Proof. We follow [17, 18] and linearize the shell model system of equations (3.205) about the trajectory $u(t)$ in the global attractor. The solution $u(t)$ is differentiable with respect to the initial data [16], hence the resulting linear equation has the view

$$du/dt + \nu Au + B(u,u) = f.$$

To simplify the notation, set $\Lambda(t) = -\nu A - B_0(t)$, where $B_0(t)U = B(u(t), U(t)) + B(U(t), u(t))$ is a linear operator. Let $U_j(t)$ $(j = 1, \ldots, N)$ be solutions of the above system satisfying $U_j(0) = U_j^{in}$. Assume that $U_1^{in}, \ldots, U_N^{in}$ are linearly independent in H, and consider $A_N(t)$ – an H-orthogonal projection onto the span$\{U_j(t)\}_{j=1}^N$. Let $\{\varphi_y(t)\}_{j=1}^N$ be orthonormal basis of span$\{U_j(t)\}_{j=1}^N$. Then $\varphi_y \in D(A)$ since span$\{U_j(t)\}_{j=1}^N \subset D(A)$. By definition of $\Lambda(t)$ and the inequalities [16],

$$\|B(u,v)\|_H \leq C_1|u|\,\|v\| \ (u \in H, v \in V), \quad \|B(u,v)\|_H \leq C_2\|u\|\,|v| \ (u \in V, v \in H),$$
$$\|B(u,v)\|_{V'} \leq C_1|u|\,\|v\| \ (u,v \in H), \quad \|B(u,v)\|_V \leq C_3|u|\,\|Av\| \ (u \in H, v \in D(A)),$$
$$\mathrm{Re}(B(u,v),v) = 0, \quad u \in H, v \in V,$$

with

$$C_1 = |a|(\lambda^{-1} + \lambda) + |b|(\lambda^{-1} + 1), \quad C_2 = 2|a| + 2|b|\lambda, \quad C_3 = |a|(\lambda^{-3} + \lambda^3) + |b|(\lambda^{-2} + \lambda),$$

and for bounded bilinear operators $B : H \times V \to H$, $B : H \times H \to V$, $B : H \times H \to V'$ and $B : H \times D(A) \to V$, we get

$$\mathrm{Re}(\mathrm{Tr}[\Lambda(t) \circ Q_N(t)]) = \mathrm{Re}\sum_{j=1}^N (\Lambda(t)\phi_j, \phi_j) = \sum_{j=1}^N [-\nu\|\phi_j\|^2 - \mathrm{Re}(B(\phi_j, u(t)), \phi_j)]$$
$$\leq -\nu\sum_{j=1}^N (k_j^2 + |(B(\phi_j, u(t)), \phi_j)|) \leq -\nu k_0^2\lambda^2(\lambda^{2N} - 1)/(\lambda^2 - 1) + C_1 N\|u(t)\|, \quad (3.209)$$

where the symbol "$\circ$" means binary operation. Due to [15], if N is large enough, such that

$$\limsup_{t\to\infty} \frac{1}{t}\int_0^t \mathrm{Re}(\mathrm{Tr}[\Lambda(t) \circ A_N(s)])ds < 0,$$

then the fractal dimension of the global attractor is bounded by N. We need to estimate N in terms of the energy dissipation rate. Using (3.209), the definition of ε, we find that it is sufficient to require N to be large enough such that it satisfies

$$\nu k_0^2\lambda^2(\lambda^{2N} - 1)/(\lambda^2 - 1) > C_1 N(\limsup_{t\to\infty}\int_0^t \|u(t)\|^2 dt)^{1/2} = C_1 N(\varepsilon/\nu)^{1/2}.$$

Finally, $\varepsilon\nu^{-3} = l_d^{-4}$ implies

$$\lambda^{2N} > (\lambda^{2N} - 1)/N > C_1(\lambda^2 - 1)/(k_1 l_d)^2 = C_1(\lambda^2 - 1)(L/l_d)^2,$$

which proves (3.208). Applying the estimate (3.207) to the inequality (3.209) we get the bound (3.208) in terms of the generalized Grashoff number. $\qquad\square$

Determining modes. Consider two solutions of the shell model equations u, v corresponding to the forces $f, g \in L^2([0, \infty], H)$

$$du/dt + \nu Au + B(u, u) = f, \quad dv/dt + \nu Av + B(v, v) = g. \quad (3.210)$$

We give a more general definition of a notion of the determining modes, than the one that is introduced previously in literature [31].

Definition 62 Define a set of determining modes as a finite set of indices $\mathcal{M} \subset \mathbb{N}$ such that whenever the forces f, g satisfy

$$|f(t) + g(t)| \to 0 \quad as \quad t \to \infty, \quad (3.211)$$

and $\sum_n |u_n(t) - v_n(t)|^2 \to 0$ *as* $t \to \infty$, it follows that $|u(t) - v(t)| \to 0$ as $t \to \infty$. The number of determining modes N of the equation is the size of the smallest such a set $\mathcal{M}$.

Recall the following generalization of the Gronwall's lemma which was proved in [50, 51].

Lemma 12 *Let $\alpha = \alpha(t)$ and $\beta = \beta(t)$ be locally integrate real-valued functions on $[0,\infty)$ that satisfy the following conditions for some $T > 0$:*

$$\lim_{t\to\infty} \frac{1}{T} \int_t^{t+T} \alpha(\tau)d\tau > 0, \ \lim_{t\to\infty} \frac{1}{T} \int_t^{t+T} \alpha^-(\tau)d\tau < \infty, \ \lim_{t\to\infty} \frac{1}{T} \int_t^{t+T} \beta^+(\tau)d\tau = 0, \ (3.212)$$

where $\alpha^+(t) = \max\{-\alpha(t), 0\}$ and $\beta^+(t) = \max\{-\beta(t), 0\}$. Let $\zeta - \zeta(t)$ be an absolutely continuous nonnegative function on $[0,t)$ obeying the inequality $d\zeta/dt + \alpha\zeta \leq \beta$ almost everywhere on $[0,\infty)$. Then $\zeta(t) \to 0$ as $t \to \infty$.

A weaker version of the above statement (sufficient for our purposes) was given in [30].

Theorem 59 *Let G be a Grashoff number defined in (3.206).*

(i) Then the first N modes are determining modes for the shell model equation provided

$$N > \frac{1}{2} \log_\lambda(C_1 G). \tag{3.213}$$

(ii) Let u, v be two solutions of (3.210)(a,b), correspondingly. Let the forces f, g satisfy (3.211) and integer N be defined as in (3.213). If $\lim_{t\to\infty} |u_N(t) - v_N(t)| = 0$ and $\lim_{t\to\infty} |u_{N-1}(t) - v_{N-1}(t)| = 0$ then $\lim_{t\to\infty} |Q_N u(t) - Q_N v(t)| = 0$.

Proof. Denote $w = u - v$, which satisfies the equation

$$dw/dt + \nu Aw + B(u,w) + B(v,w) = f - g. \tag{3.214}$$

Fix an integer $m > 1$ and define $P_m : H \to H$ to be an orthogonal projection onto the first m coordinates, namely, onto the subspace of H spanned by m vectors $\{\phi_i\}_{i=1}^m$ of the standard basis. Also denote $A_m = I - P_m$. The second part of the theorem implies the first statement. Thus, it is enough to prove that if

$$\lim_{t\to\infty} |w_N(t)| = 0, \quad \lim_{t\to\infty} |w_{N-1}(t)| = 0,$$

then $\lim_{t\to\infty} |A_m w| = 0$ for every $m > \frac{1}{2} \log_\lambda(C_1 G)$. Multiplying (3.214) by $A_m w$ in the scalar product of H we get

$$\frac{1}{2}\frac{d}{dt}|Q_m w|^2 + \nu\|Q_m w\|^2 = \mathrm{Re}(f - g, Q_m w) - \mathrm{Re}(B(u,w), Q_m w) - \mathrm{Re}(B(w,v), Q_m w). \tag{3.215}$$

For (3.215) we use the inequality $k_{m+1}^2 |Q_m w| \leq \|Q_m w\|$, for the RHS observe that

$$\mathrm{Re}(B(u,w), Q_m w) = \mathrm{Im} \sum_{n=m+1}^\infty (ak_{n+1}w_{n+2}u_{n+1}^*w_n^* + bk_n w_{n+1}u_{n-1}^*w_n^* + ak_{n-1}w_n^*u_{n-1}w_{n-2}$$

$$+ bk_{n-1}w_n^*u_{n-2}w_{n-1}) \leq |ak_m w_{m+1}^* u_m w_{m-1} + ak_{m+1}w_{m+2}^* u_{m+1}w_m + bk_m w_{m+1}^* u_{m-1}w_m|.$$

Moreover,

$$\mathrm{Re}(B(u,v), Q_m w) = \mathrm{Im}\sum_{n=m+1}^\infty (ak_{n+1}v_{n+2}w_{n+1}^*w_n^* + bk_n v_{n+1}w_n^*w_{n-1}^*$$

$$+ ak_{n-1}w_n^*w_{n-1}v_{n-2} + bk_{n-1}v_{n-1}w_n^*w_{n-2}) \leq |ak_m w_{m+1}^* w_m v_{m-1}$$

$$+ ak_{m+1}w_{m+2}^* w_{m+1}v_m + bk_m w_{m+1}^* w_{m-1}v_m| + C_1\|v\| \cdot |Q_m w|^2.$$

Denoting $\zeta = |Q_m w|^2$, rewrite (3.215) as $\frac{1}{2} d\zeta/dt + a\zeta \leq \beta$, where $\alpha = \nu k_{m+1}^2 - C_1\|v(t)\|$ and

$$\beta(t) = (f - g, Q_m w) + |(ak_m w_{m+1}^* u_m w_{m-1} + ak_{m+1}w_{m+2}^* u_{m+1}w_m + bk_m w_{m+1}^* u_{m-1}w_m)|$$

$$+ |ak_m w_{m+1}^* v_{m-1}w_m + ak_{m+1}w_{m+2}^* v_m w_{m+1} + bk_m w_{m+1}^* v_m w_{m-1}|.$$

To see that $\beta(t)$ satisfies condition (3.212c) of Lemma 12 it is enough to show that $\beta(t) \to 0$ as $t \to \infty$. This is true, if we assume that the forces f, g satisfy (3.211) and $|w_m(t)| \to 0$, $|w_{m-1}(t)| \to 0$ as $t \to \infty$. Moreover, it is easy to see that $\alpha(t)$ satisfies (3.212b). Hence, in order to apply Lemma 12 we need to show that $\alpha(t)$ also satisfies (3.212a), namely,

$$\lim_{t\to\infty} \sup \frac{1}{T} \int_t^{t+T} \alpha(\tau)\mathrm{d}\tau \geq \nu k_{m+1}^2 - C_1 \big(\lim_{t\to\infty} \sup \frac{1}{T} \int_t^{t+T} \|v(\tau)\|\mathrm{d}\tau \big)^{1/2} > 0.$$

To estimate the last quantity, we use the bound (3.207) to conclude that if m satisfies $\lambda^{2m} > C_1 G$ then $|Q_m w| \to 0$ as $t \to \infty$. $\square$

Remark 22 The existence of the determining modes for the NSE both in two and three dimensions is known (see, e.g., [19, 29, 51] and reference therein). There exists a gap between the upper bounds for the lowest number of determining modes and the dimension of the global attractor for the two-dimensional NSE both for the no-slip and periodic boundary conditions. Our upper bounds for the dimension of the global attractor and for the number of determining modes for the Sabra shell model equation coincides. Recently it was shown in [48] that a similar result is also true for the damped-driven NSE and the Stommel–Charney barotropic model of ocean circulation.

Existence of inertial manifolds. In this section we prove the existence of a finite-dimensional Inertial Manifold (IM). The concept of inertial manifold for nonlinear evolution equations was first introduced in [32]. An inertial manifold is a finite dimensional Lipschitz, globally invariant manifold which attracts all bounded sets in the phase space at an exponential rate and, consequently, contains the global attractor. In fact, one can show that the IM is smoother, in particular C^1 (see, e.g., [20, 85, 94]). The smoothness and invariance under the reduced dynamics of the IM implies that a finite system of ordinary differential equations is equivalent to the original infinite system. This is the ultimate and best notion of system reduction that one could hope for. In other words, IM is an exact rule for parametrization the large modes (infinite many of them) in terms of the low ones (finitely many of them).

In this section we use [33, Theorem 3.1*] to show the existence of inertial manifolds for (3.205) and to estimate its dimension. We formulate [33, Theorem 3.1] in the following way.

Theorem 60 *Let the nonlinear term of the equation*

$$du/dt + \nu Au + R(u) = 0, \tag{3.216}$$

$R(\cdot)$ *be a differentiable map from* $D(A)$ *into* $D(A^{1-\beta})$ *satisfying*

$$|R'(u)v| \leq C_4|Au||A^\beta v|, \quad |A^{1-\beta}R'(u)v| \leq C_5|Au||Av|, \tag{3.217}$$

for some $0 \leq \beta < 1$ *for all* $u, v \in D(A)$ *and appropriate constants* $C_4, C_5 > 0$, *which depend on physical parameters* a, b, c, ν *and* f. *Let* $\tilde{\beta} > 0$ *and* $l \in [0, 1)$ *be two fixed numbers. Assume that there exists* N, *large enough, such that the eigenvalues of* A *satisfy*

$$(k_{N+1} - k_N)/(k_{N+1}^\beta - k_N^\beta) \geq \max\{2K_2(1+l)/l, \tilde{\beta}/(1-\beta)K_1\}, \tag{3.218}$$

where $K_1 = |A^{1-\beta}f| + 4C_5\rho^2$, $K_2 = 8|A^{1-\beta}f|/\rho + 26C_5\rho$, *and* ρ *is the radius of the absorbing ball in* $D(A)$, *whose existence is provided by* [16, Proposition 14]. *Then* (3.216) *possesses an inertial manifold of dimension* N *which is the graph of a function* $\Phi : P_N H \to A_N D(A)$ *with* $\tilde{b} = \sup_{p \in P_N H} |A\Phi(p)|$ *and*

$$l = \lim_{p_1, p_2 \in P_N, \, p_1 \neq p_2} |A(\Phi(p_1) - \Phi(p_2))|/|A(p_1 - p_2)|$$

is the Lipschitz constant of Φ.

In order to apply the above Theorem, we need to show that the Sabra model equation satisfies the properties of the abstract framework [33]. Rewrite our system (3.205) in the form

$$du/dt + \nu Au + R(u) = 0,$$

where the nonlinear term $R(u) = B(u, u) - f$. First, we need to show that $R(u)$ of (3.216) satisfies (3.217) for $\beta = 0$. Indeed, and based on inequality [16],

$$\|B(u, v)\| \leq C_3 |u| \, |Av|, \quad u \in H, \ v \in F(A),$$

the term $R(u)$ is a differentiable map from $D(A)$ to $D(A^{1/2}) = V$, where $R'(u)w = B(u, w) + B(w, u)$ $(u, w \in D(A))$. Moreover, the estimates (3.217) are satisfied due to [16, Proposition 1], namely,

$$|R'(u)v| = |B(u, w) + B(w, u)| \leq (C_1 + C_2)\|u\| \, |v| \leq (C_1 + C_2)k_1^{-1}|Au| \, |v|,$$

where the last inequality results from the fact that $k_1\|u\| \leq |Au|$ and the constant

$$C_4 = (C_1 + C_2)/k_1 = (|a|(\lambda + \lambda^{-1}) + |\beta|(1 + \lambda^{-1}) + 2(|a| + |b|\lambda))/k_1.$$

In addition, one can prove as in [16, Proposition 1] that

$$|AR'(u)v| \leq |AB(u, v)| + |AB(v, u)| \leq C_5'\|u\| \, |Av| \leq C_5'|Au| \, |Av|,$$

where again the last inequality results from the fact that $k_1\|u\| \leq |Au|$, and the constant $C_5' = C_5/k_1$. Finally, because of the form of the wave-numbers $k_n = k_0\lambda^n$, the spectral gap condition (3.218) is satisfied for $\beta = 0$. Therefore, we apply Theorem 60 to (3.205), and conclude that the Sabra shell model possesses an inertial manifold.

Dimension of inertial manifold. One may calculate the dimension of the inertial manifold of the Sabra shell model equation for the specific choice of parameters $\beta = 0$, $l = 1/2, \tilde{b} = \rho$, where ρ is the radius of the absorbing ball in the norm of the space $D(A)$, see [16, Proposition 14]. In that case, the conditions (3.218) take the form $k_{N+1} - k_N \geq \max\{4K_2, 2\rho/K_1\}$, and the following holds.

Corollary 13 *The equation (3.205) possesses an inertial manifold of dimension*

$$N \geq \max\{\log_2[4K_2/(\lambda - 1)], \log_\lambda[2\rho/(\lambda - 1)K_1]\}.$$

3.7.3 Navier–Stokes equations in the common wavelets representation

In 1984 there was introduced a new technique [45], the wavelet transform, which allowed the decomposition of a signal into contributions from both space and scales. A modern wavelet analysis is intruding into very different spheres of scientific activities, including the study of turbulent flows. The wavelet series give an optimum the investigator's attention in the study of the coherent structures in fully developed turbulence [28, 78]. The concept of a hierarchy of vortices is widely used in the theory of turbulence. Together with scale theory a model of turbulence as a vortex system of different series with random amplitudes allows us to obtain various dependencies for the turbulence statistic. In this way there was obtained, by Kolmogorov [53], the spectral law (K41) for homogeneous and isotropic turbulence.

A traditional way of representing fluid motion is a system of differential equations for the functions of spatial coordinates and time. The spatial-temporal representation of hydrodynamics equations does not conform to the hierarchy of model vortices. If the velocity field is approximated by the Kolmogorov–Shannon series and maximum values of spatial and time frequencies tend to infinity the velocity field in the spatial-temporal representation becomes the sum of the infinitely small vortices with infinitely small life-time. In order to build in the idea of hierarchy of vortices the velocity field may be represented in terms of Fourier integrals. This representation is not satisfactory. Each Fourier mode in the decomposition of the velocity vector potential corresponds to a coherent vortex system filling all the space. But under the turbulent motion the strong nonlinear interaction of the spatial-temporal modes results in the effect that periodic solutions, which correspond to coherent vortices are not typical structural elements of the motion. These models named hierarchical.

Turbulence is an essential three-dimensional (3D) phenomenon – the turbulent flows are practically always three-dimensional and the cascade energy from the large scales to the small ones favors the isotropisation of flow. The hierarchical bases were created specially for describing turbulent motion and, unlike the wavelet theory. It is well known that the laws of two-dimensional (2D) turbulence evolution are qualitatively different from the laws, which determine the development of ordinary three-dimensional turbulence. In inviscid limit, the two-dimensional leads to the appearance of the total enstrophy Ω (the mean square of vorticity) is conserved at the same time as the total energy E, while the three-dimensional leads to the appearance of the total helicity H as the second integral of motion (Section 3.1).

The cascade equations, written for the quantities A_i, each define the velocity oscillations in some interval of wave numbers and describe the principal characteristics of energy redistribution processes between different scales. The cascade equations minimize the dimensionality of systems, which describe the turbulent flows in wide range of wave numbers, and have a form

$$\frac{dA_i}{dt} = \sum\nolimits_{jk} T_{ijk} A_j A_k + K_i A_i + F_i, \qquad (3.219)$$

where F_i characterize the energy sources in corresponding interval of spectrum. Cascade models like (3.219) were constructed by a number of authors, for example, in [42], [84], etc. The models differ by the way of wave number space separation and definition of variables A_i, by limitation of considered triad interactions and by other details, but all the models preserve the fundamental properties of initial NSE – satisfy in nondissipative (inviscid) limit the conservation laws and describe adequately the nonlinear interactions between the selected three modes. The total shell model holds a procedure to construct cascade equations which use the hierarchical bases and the conservation laws and which takes into consideration the interactions between all the excited scales of motion.

Hierarchical base. The hierarchical model of turbulence is based on the natural assumption that the turbulence is an ensemble of vortices of progressively diminishing scales. The hierarchical base for two-dimensional turbulence describes the ensemble of the vortices, in which any vortex of the given size consists of four vortices of half size and so on. The ensemble of vortices of the same size forms a *level*. The functions of the hierarchical base are constructed in such a way that Fourier-images of vortices of single level occupy only single octave in the wave-number space and regions of localization of different levels in the Fourier space do not overlap. The Fourier-images space is to be divided by infinite number of shells. This gives the name: shell-models. For three-dimensional turbulence the building procedure is similar.

For visualization of the building procedure of the hierarchical base return to the two-dimensional case. The wave-number space may be divided, for example, at ring zones, that

$$\pi 2^N < |\mathbf{k}| < \pi 2^{N+1}, \qquad (3.220)$$

see Fig. 3.8(a). The 2D-velocity field $\mathbf{u}(\mathbf{r}, t)$ is filtrated as $\mathbf{u}(\mathbf{r}, t) = \int \mathbf{u}(\mathbf{r}', t) \cdot g_N(\mathbf{r} - \mathbf{r}') d\mathbf{r}$, where the Fourier transform of g_N is equal to unity inside the zone (3.220) and equal to zero outside it. Then the velocity $\mathbf{u}(\mathbf{r}, t)$ and vorticity $\boldsymbol{\omega}(\mathbf{r}, t)$ fields can be written as

$\mathbf{u}(\mathbf{r},t) = \sum_N \mathbf{u}_N(\mathbf{r},t)$ and $\boldsymbol{\omega}(\mathbf{r},t) = \sum_N \boldsymbol{\omega}_N(\mathbf{r},t)$, where $\boldsymbol{\omega} = \nabla \times \mathbf{u}$. The kinetic energy and enstrophy can be represented as sums

$$E = \sum_N E_N, \quad E_N = \frac{1}{2}\int |\mathbf{u}_N|\, d\mathbf{r}, \quad \Omega = \sum_N \Omega_N, \quad \Omega_N = \frac{1}{2}\int |\boldsymbol{\omega}_N|\, d\mathbf{r},$$

because $\int \mathbf{u}_N \mathbf{u}_M\, d\mathbf{r} = \delta_{NM}$ and $\int \boldsymbol{\omega}_N \boldsymbol{\omega}_M\, d\mathbf{r} = \delta_{NM}$. The functions $\mathbf{u}_N$, $\boldsymbol{\omega}_M$ can be decomposed in a sum of functions each of them describes the velocity or vorticity oscillations of concrete scale in concrete region space

$$\mathbf{u}(\mathbf{r},t) = \sum_{N,n} A_{N,n}(t)\mathbf{u}_{N,n}(\mathbf{r} - \mathbf{r}_{N,n}), \quad \boldsymbol{\omega}(\mathbf{r},t) = \sum_{N,n} B_{N,n}(t)\boldsymbol{\omega}_{N,n}(\mathbf{r} - \mathbf{r}_{N,n}),$$

where $\mathbf{r}_{N,n}$ is the radius-vector of the vortex center. The Fourier form of the N-th level function for the vorticity is in zone (3.220)

$$\Psi_{N,n} = \begin{cases} \frac{2^{1-N}}{\sqrt{3}} \exp(-i\mathbf{k}\cdot\mathbf{r}_{N,n}), & \pi 2^N < |\mathbf{k}| < \pi 2^{N+1}, \\ 0, & |\mathbf{k}| < \pi 2^N, \; \pi 2^{N+1} < |\mathbf{k}|. \end{cases}$$

In the physical space the base functions for the velocity and vorticity are

$$\mathbf{u}_{N,n}(\mathbf{r}) = \frac{\mathbf{r}\times\mathbf{e}}{\sqrt{3\pi}r}\cdot\frac{2J_0(2s) - J_0(s)}{s}, \quad \boldsymbol{\omega}_{N,n}(\mathbf{r}) = \sqrt{\pi/3}\,2^N\frac{2J_1(2s) - J_1(s)}{s},$$

where $s = 2^N\pi$, and J_0, J_1 are Bessel functions. These base is orthogonal only by the index N, but not by the index n. It is the result of choice of the zone (3.220), which is to be continued by the wish to obtain the axisymmetrical base functions.

The full orthogonal base can be obtained by the same procedure for the $3D$-case. In this case the vector potential of the velocity may be represented in the form of a Fourier integral

$$\mathbf{a}(\mathbf{r},t) = \int \mathbf{A}(\mathbf{k},t)\exp(2\pi i\mathbf{k}\cdot\mathbf{r})d\mathbf{k},$$

where the function $\mathbf{A}(\mathbf{k},t)$ must be represented inside each layer $N: 2^N < |\mathbf{k}| < 2^{N+1}$ ($N \in \mathbb{Z}$) in the form of an expansion in terms of the discrete totality of the basis functions $(2\pi k)^{-1}\mathbf{e}_\nu \exp(2\pi i\, \mathbf{r}_{N,n}\cdot\mathbf{k})$, where $\mathbf{e}_\nu$, is the unit vector in the direction of the coordinate axes, and $\mathbf{r}_{N,n}$ are random points, uniformly distributed in the volume of the region U filled with a fluid with an average density ρ_N. The value ρ_N should be chosen such that the product of the volume of the Nth spherical layer by the volume ρ_N^{-1} of the basis function in $\mathbf{r}$-space is equal to unity. We obtain $\rho_N = (7/9)\pi 2^{3N}$. The calculation of the Fourier transform $\mathbf{v}^F(\mathbf{k}) = 2\pi i(\mathbf{k}\times\mathbf{A})$ of the velocity $\mathbf{u}(\mathbf{r},t)$ gives the normalized basis functions

$$\mathbf{u}_{N,n,\nu}^F(\mathbf{k}) = \begin{cases} \frac{3i}{\sqrt{7\pi}}2^{-\frac{3}{2}N}\frac{\mathbf{k}\times\mathbf{e}_\nu}{k}e^{-2\pi i\mathbf{k}\cdot\mathbf{r}_{N,n}}, & 2^N < |\mathbf{k}| < 2^{N+1}, \\ 0, & |\mathbf{k}| < 2^N, \; 2^{N+1} < |\mathbf{k}|. \end{cases}$$

Thus, the divergence-free vector wavelets $\mathbf{u}_{N,\nu,n}(\mathbf{r},t)$ may be defined as [107]

$$\mathbf{u}_{N,\nu,n}(\mathbf{r}) = -\frac{9}{14}\rho_N^{1/2}\mathbf{e}_\nu\times\nabla_s\left(\frac{\cos(s) - \cos(2s)}{s^2}\right)$$

with $\mathbf{s} = \pi 2^N(\mathbf{r} - \mathbf{r}_{N,n})$ and $s = |\mathbf{s}|$. Here, each wavelet can be considered as an axisymmetric vortex with its axis along unit vector $\mathbf{e}_\nu$ (Fig. 3.8(b)). The functions of a given basis, as a whole, correspond to a hierarchy of eddies of different scale. In fact, after substitution

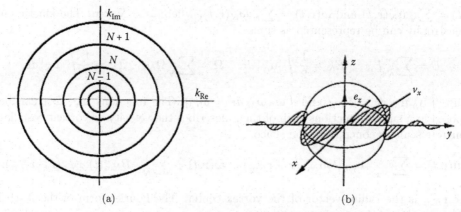

FIGURE 3.8: (a) The wave-number space is divided at ring zones. (b) The axisymmetric vector wavelet aligned with the z-axis. The velocity profile of this wavelet in the x-direction is shown.

$\mathbf{u}_{N,\nu,n}$ into the NSE, multiplying the resulting relation by conjugated wavelets $\mathbf{u}^*_{N\nu n}$ and integration the result over the entire fluid-filled volume we obtain

$$\sum_{m,\mu} \hat{P}_{Nn\nu,M\mu m} \dot{A}_{M\mu m}(t) + \sum_{M\mu m} \sum_{L\lambda l} \hat{T}_{N\nu n,M\mu m,L\lambda l} A_{Mm\mu}(t) A_{L\lambda l}(t)$$
$$= \nu_0 \sum_{\mu m} \hat{\Lambda}_{N\nu m,N\mu m} A_{N\mu m}(t),$$

where ν_0 is the fluid viscosity and

$$\hat{T}_{N\nu n,M\mu m,L\lambda l} = \int \mathbf{u}^*_{N\nu n}(\mathbf{r})(\mathbf{u}_{M\mu m}(\mathbf{r}) \cdot \nabla)\mathbf{u}_{L\lambda l}(\mathbf{r})d\mathbf{r},$$
$$\hat{\Lambda}_{N\nu m,N\mu m} = \int \mathbf{u}^*_{N\nu n}(\mathbf{r}) \cdot \nabla^2 \mathbf{u}_{N\mu m}(\mathbf{r})d\mathbf{r},$$
$$\hat{P}_{Nn\nu,M\mu m} = \int \mathbf{u}^*_{N\nu n}(\mathbf{r}) \cdot \mathbf{u}_{M\mu m}(\mathbf{r})d\mathbf{r}.$$

Notice that the term of the pressure gradient vanishes due to the basis functions to be satisfied the equation of continuity, i.e., $\mathbf{u}_{N,\nu,n}$ tend to zero as $N \to \infty$. The basis functions should be orthogonal with respect to the shell index N follow the Plancherel theorem.

Using representation $A_{N\nu n}(t) = a_{N\nu n}(t)A_N(t)$ the kinetic energy may be expressed as

$$E = \frac{1}{2}\sum_N A_N^2(t) \sum_{nm,\nu\mu} a_{N\mu m}(t)a_{N\nu n}(t) = \frac{1}{2}\sum_N \alpha_N A_N^2(t),$$

where $\alpha_N = \sum_{nm,\nu\mu} a_{N\nu n}(t)a_{N\mu m}(t)$. These transform help us to obtain the hierarchical system

$$\alpha_N \frac{dA_N}{dt} = -\sum_{M,L} \hat{\Phi}_{NML} A_M(t)A_L(t) + \nu_0 \hat{K}_N A_N(t) + \tilde{f}_N, \qquad (3.221)$$

where

$$\hat{\Phi}_{NML} = \sum_{nml,\nu\mu\lambda} a_{N\nu n}a_{M\mu m}a_{L\lambda l}\hat{T}_{N\nu n,M\mu m,L\lambda l}, \quad \hat{K}_N = \sum_{nm,\nu\mu} a_{N\nu n}a_{N\mu m}\hat{\Lambda}_{N\nu n,N\mu m}.$$

Solutions of NSE must satisfy the main conservation laws: the energy and the enstrophy (for the two-dimensional turbulence), or the helicity (for the three-dimensional turbulence).

Proposition 30 *The conservation kinetic energy law has a form*

$$\hat{Q}_{NML} + \hat{Q}_{MLN} + \hat{Q}_{LNM} = 0; \qquad (3.222a)$$

the conservation laws for enstrophy (two-dimensional case) and helicity (three-dimensional case) have a form

$$2^{(4-D)N}\hat{Q}_{NML} + 2^{(4-D)M}\hat{Q}_{MLN} + 2^{(4-D)L}\hat{Q}_{LNM} = 0, \qquad (3.222b)$$

where $D = 2,3$ is the dimension of the turbulence problem, and

$$\hat{Q}_{NML} = \hat{\Phi}_{NML} + \hat{\Phi}_{NLM}, \quad \forall N, M, L.$$

Proof. Obviously that, for the inviscid limit, the system (3.221) may be rewritten as

$$\alpha_N \frac{dA_N}{dt} = -\sum_{p,q>0} \left\{ \left[\hat{\Phi}_{N(N+p)(N+p+q)} + \hat{\Phi}_{N(N+p+q)(N+p)} \right] A_{N+p} A_{N+p+q} \right.$$
$$+ \left[\hat{\Phi}_{N(N+p)(N-q)} + \hat{\Phi}_{N(N-q)(N+p)} \right] A_{N+p} A_{N-q}$$
$$+ \left[\hat{\Phi}_{N(N-p)(N-p-q)} + \hat{\Phi}_{N(N-p-q)(N-p)} \right] A_{N-p} A_{N-p-q} \right\}$$
$$= -\sum_{p,q>0} \left\{ \hat{Q}_{N(N+p)(N+p+q)} A_{N+p} A_{N+p+q} + \hat{Q}_{N(N+p)(N-q)} A_{N+p} A_{N-q} \right.$$
$$+ \hat{Q}_{N(N-p)(N-p-q)} A_{N-p} A_{N-p-q} \right\}.$$

Accordingly, the energy conservation law may be written as

$$-\frac{dE}{dt} = -\sum_N \alpha_N A_N \frac{dA_N}{dt}$$
$$= \sum_{N,p,q} \left\{ \hat{Q}_{N(N+p)(N+p+q)} A_N A_{N+p} A_{N+p+q} + \hat{Q}_{N(N+p)(N-q)} A_N A_{N+p} A_{N-q} \right.$$
$$+ \hat{Q}_{N(N-p)(N-p-q)} A_N A_{N-p} A_{N-p-q} \right\} = 0.$$

Since, due to invariance of the tensor $\hat{Q}_{NML}$ under transforms of all indices $N \mapsto N + k$, $M \mapsto M + k$, $N \mapsto N + k$, we obtain

$$\sum_{N,p,q} \hat{Q}_{N(N+p)(N-q)} A_N A_{N-p} A_{N-q} = \sum_{N,p,q} \hat{Q}_{(N+q)(N+p+q)q} A_{N+q} A_{N+p+q} A_N,$$
$$\sum_{N,p,q} \hat{Q}_{N(N-p)(N-p-q)} A_N A_{N-p} A_{N-p-q} = \sum_{N,p,q} \hat{Q}_{(N+p+q)(N+q)N} A_{N+p+q} A_{N+q} A_N.$$

Thus, the energy conservation law is $\hat{Q}_{NML} + \hat{Q}_{MLN} + \hat{Q}_{LNM} = 0$.

Consider the set of functions $\varphi_{N\nu n}(\mathbf{r}) = \nabla \times \mathbf{u}_{N\nu n}(\mathbf{r})$. Following Plancherel theorem the functions $\varphi_{N\nu n}(\mathbf{r})$ are orthogonal with respect to the shell indices N, since their Fourier images are orthogonal due to shells construction. As a result, it is easy to obtain the conservation law for enstrophy $\Omega \equiv \langle \frac{1}{2}|\nabla \times \mathbf{u}|^2 \rangle$ and for helicity $H \equiv \langle \frac{1}{2}\mathbf{u} \cdot (\nabla \times \mathbf{u}) \rangle$:

$$\dot{\Omega} = \sum_N \sum_{nm,\nu\mu} \left\langle a_{N\nu n} a_{N\mu m} \int \varphi_{N\nu n}^*(\mathbf{r}) \cdot \varphi_{N\mu m}(\mathbf{r}) d\mathbf{r} \right\rangle 2^{2N} A_N \frac{dA_N}{dt} = 0,$$
$$\dot{H} = \sum_N \sum_{nm,\nu\mu} \left\langle a_{N\nu n} a_{N\mu m} \int \mathbf{u}_{N\nu n}^*(\mathbf{r}) \cdot \varphi_{N\mu m}(\mathbf{r}) d\mathbf{r} \right\rangle 2^N A_N \frac{dA_N}{dt} = 0.$$

Under use the similar manipulations as in the first part of the proof, we obtain

$$2^{2N}\hat{Q}_{NML} + 2^{2M}\hat{Q}_{MLN} + 2^{2L}\hat{Q}_{LNM} = 0, \quad 2^N\hat{Q}_{NML} + 2^M\hat{Q}_{MLN} + 2^L\hat{Q}_{LNM} = 0. \qquad \square$$

Proposition 31 *Solutions of the system (3.222a,b) present us the common relation*

$$\frac{\hat{Q}_{NML}}{\hat{Q}_{LNM}} = \frac{1 - 2^{(4-D)(L-M)}}{2^{(4-D)(N-M)} - 1}, \quad (D = 2,3). \qquad (3.223)$$

Proof. Exclude Q_{LMN} from (3.222a,b). $\qquad\qquad\qquad\qquad\qquad\qquad\qquad\square$

Formula (3.223) shows us that all diagonal elements of a tensor $\hat{Q}_{NML}$ are zero. So, if we neglect nondiagonal elements that is, e.g., in [34, 107], we get only the trivial solution $A_N = \text{const}$. Thus, we take into consideration, namely, nondiagonal elements, but not diagonal. Also note that the phase-space is conserved. In fact, $V = \int dV = \prod_k \int dA_k$, then

$$\partial_t V \equiv \partial_t \Big(\int \prod_k dA_k \Big) = \sum_k \int \Big(\frac{\partial \dot{A}_k}{\partial A_k} \Big) \prod_k dA_k = \sum_k \int \Big(\frac{\partial \dot{A}_k}{\partial A_k} \Big) dV = 0.$$

Finally we'll obtain the analytical solutions to (3.223). Assume in what follows $a = 2^{4-D}$.

Proposition 32 *The solution of (3.223) has the form*

$$Q_{NML} = a^{\frac{(M-N)+(M-L)}{3}}(1 - a^{L-M}) \sum_{k \geq 1} c_{2k-1}\{(1 - a^{L-M})(1 - a^{M-N})(1 - a^{N-L})\}^{2k-1}. \tag{3.224}$$

Proof. Since the turbulent flow satisfy to the symmetries (Section 3.1), and the tensor elements $\hat{Q}_{NML}$ are similar to each other with respect to the tensor diagonal, we obtain

$$\frac{\hat{Q}_{N,N-i,N-j}}{\hat{Q}_{N-j,N,N-i}} = \frac{1 - a^{i-j}}{a^i - 1}. \tag{3.225}$$

This is equivalent to the discrete dynamical equations to be similar for all levels, i.e., the elements $Q_{N,N-i,N-j}$ do not depend on three indices but only from their differences, in particular, from the difference between the left and the middle ones:

$$\hat{Q}_{N \xrightarrow{(-)} N-i \xrightarrow{(-)} N-j} = F(i - j, i), \quad \hat{Q}_{N-j \xrightarrow{(-)} N \xrightarrow{(-)} N-i} = F(-j, -i).$$

The subscript transitions mean: the first and the second variables of the function F should be calculated as the differences between left and middle, and between right and middle subscripts, of the tensor $\hat{Q}$, correspondingly, i.e., the transitions $N \xrightarrow{(-)} N-i \xleftarrow{(-)} N-j$ mean that $N - (N - i) \to i$, $(N - j) - (N - i) \to i - j$, as a result we receive $F(i, i - j)$. Simultaneously, the transitions $N - j \xrightarrow{(-)} N \xleftarrow{(-)} N-i$ mean that $(N - j) - N \to -j$, $(N - i) - N \to -i$, and as a result we receive $F(-j, -i)$. Hence, the tensor relation (3.225) is equivalent to the functional relation

$$F(i, i - j) = F(-j, -i)\frac{1 - a^{i-j}}{a^i - 1}. \tag{3.226}$$

By Remark 23, its solution, among symmetric functions, may be found in the form

$$F(i, j) = a^{-(2i-j)/3}(1 - a^{i-j})\Upsilon(f), \quad f \equiv f(i, j) = (1 - a^{i-j})(1 - a^{-i})(1 - a^j),$$

where $\Upsilon(f) = -\Upsilon(-f)$ is an odd function. By inverse transformations $i = N - M$, $j = N - L$ we get

$$\hat{Q}_{NML} = a^{(2M-N-L)/3}(1 - a^{L-M})\Upsilon(\zeta_{NML}),$$
$$\zeta_{NML} = (1 - a^{L-M})(1 - a^{M-N})(1 - a^{N-L}). \tag{3.227}$$

For $\Upsilon \in C^\infty(-\infty, \infty)$ the using its Taylor series competes the proof. $\qquad\qquad\square$

Lemma 13 *The functional equation of the form*

$$\frac{F(\alpha_1 x + \beta_1 y, \alpha_2 x + \beta_2 y)}{F(\gamma_1 x + \delta_1 y, \gamma_2 x + \delta_2 y)} = \frac{a^{\xi_1 x + \eta_1 y} \pm a^{\xi_2 x + \eta_2 y}}{a^{\xi_3 x + \eta_3 y} \pm a^{\xi_4 x + \eta_4 y}}, \tag{3.228}$$

where $\alpha_1\beta_2 \neq \alpha_2\beta_1$, $\gamma_1\delta_2 \neq \gamma_2\delta_1$, $(\alpha_1 - \gamma_1)(\beta_2 - \delta_2) \neq (\beta_1 - \delta_1)(\alpha_2 - \gamma_2)$, *has a particular solution*

$$F(x, y) = \left[a^{\frac{\beta_2(\xi_1 - \xi_2) - \alpha_2(\eta_1 - \eta_2)}{\alpha_1\beta_2 - \alpha_2\beta_1} x + \frac{\alpha_1(\eta_1 - \eta_2) - \beta_1(\xi_1 - \xi_2)}{\alpha_1\beta_2 - \alpha_2\beta_1} y} \pm 1 \right]$$
$$\times a^{\frac{(\beta_2 - \delta_2)(\xi_2 - \xi_4) - (\alpha_2 - \gamma_2)(\eta_2 - \eta_4)}{(\alpha_1 - \gamma_1)(\beta_2 - \delta_2) - (\beta_1 - \delta_1)(\alpha_2 - \gamma_2)} x + \frac{(\alpha_1 - \gamma_1)(\eta_2 - \eta_4) - (\beta_1 - \delta_1)(\xi_2 - \xi_4)}{(\alpha_1 - \gamma_1)(\beta_2 - \delta_2) - (\beta_1 - \delta_1)(\alpha_2 - \gamma_2)} y}. \tag{3.229}$$

Remark 23 Particularly, in the case of (3.226), when

$$\alpha_1 = 0, \ \beta_1 = -1, \quad \alpha_2 = -1, \ \beta_2 = 0, \quad \gamma_1 = 1, \ \delta_1 = 0, \quad \gamma_2 = 1, \ \delta_2 = -1,$$
$$\xi_1 = 1, \ \eta_1 = 0, \quad \xi_2 = 0, \ \eta_2 = 0, \quad \xi_3 = 0, \ \xi_4 = 1, \quad \eta_3 = 0, \ \eta_4 = -1,$$

the solution has a form $F(x, y) = a^{-\frac{i+j}{3}}(1 - a^j)$.

Proof of Lemma 13. Find the solution in the form $F(x, y) = (a^{hx + gy} \pm 1)C(x, y)$. After its substituting into (3.228) and simple manipulations we obtain

$$\begin{cases} h\alpha_1 + g\alpha_2 = \xi_1 - \xi_2 \\ h\beta_1 + g\beta_2 = \eta_1 - \eta_2 \end{cases}; \quad \begin{cases} h\gamma_1 + g\gamma_2 = \xi_3 - \xi_4 \\ h\delta_1 + g\delta_2 = \eta_3 - \eta_4 \end{cases}.$$

The first of the systems gives

$$h = \frac{\beta_2(\xi_1 - \xi_2) - \alpha_2(\eta_1 - \eta_2)}{\alpha_1\beta_2 - \alpha_2\beta_1}, \quad g = \frac{\alpha_1(\eta_1 - \eta_2) - \beta_1(\xi_1 - \xi_2)}{\alpha_1\beta_2 - \alpha_2\beta_1}, \tag{3.230}$$

while the second system leads to

$$h = \frac{\delta_2(\xi_3 - \xi_4) - \gamma_2(\eta_3 - \eta_4)}{\gamma_1\delta_2 - \gamma_2\delta_1}, \quad g = \frac{\gamma_1(\eta_3 - \eta_4) - \delta_1(\xi_3 - \xi_4)}{\gamma_1\delta_2 - \gamma_2\delta_1}. \tag{3.231}$$

It is naturally to require the following:

$$\frac{\beta_2(\xi_1 - \xi_2) - \alpha_2(\eta_1 - \eta_2)}{\alpha_1\beta_2 - \alpha_2\beta_1} = \frac{\delta_2(\xi_3 - \xi_4) - \gamma_2(\eta_3 - \eta_4)}{\gamma_1\delta_2 - \gamma_2\delta_1},$$
$$\frac{\alpha_1(\eta_1 - \eta_2) - \beta_1(\eta_1 - \eta_2)}{\alpha_1\beta_2 - \alpha_2\beta_1} = \frac{\gamma_1(\eta_3 - \eta_4) - \delta_1(\xi_3 - \xi_4)}{\gamma_1\delta_2 - \gamma_2\delta_1}.$$

These equations are compatible only under conditions $\alpha_1\beta_2 \neq \alpha_2\beta_1$ and $\gamma_1\delta_2 \neq \gamma_2\delta_1$. Hence, one may choose for h, g any of the representations (3.230) or (3.231). Similarly we can find the form of the function $C(x, y)$: $C(x, y) = a^{g_0 x + h_0 y}$. Its substitution into (3.222a,b) gives

$$g_0 = \frac{(\beta_2 - \delta_2)(\xi_2 - \xi_4) - (\alpha_2 - \gamma_2)(\eta_2 - \eta_4)}{(\alpha_1 - \gamma_1)(\beta_2 - \delta_2) - (\beta_1 - \delta_1)(\alpha_2 - \gamma_2)}, \quad h_0 = \frac{(\alpha_1 - \gamma_1)(\eta_2 - \eta_4) - (\beta_1 - \delta_1)(\xi_2 - \xi_4)}{(\alpha_1 - \gamma_1)(\beta_2 - \delta_2) - (\beta_1 - \delta_1)(\alpha_2 - \gamma_2)}.$$

The unique of solution needs follow conditions $(\alpha_1 - \gamma_1)(\beta_2 - \delta_2) \neq (\beta_1 - \delta_1)(\alpha_2 - \gamma_2)$. Hence, the solution has a form

$$F(x, y) = (a^{hx + gy} \pm 1)a^{h_0 x + g_0 y}. \tag{3.232}$$

The final expression (3.229) follows from (3.232) under using

$$\frac{\beta_2(\xi_1 - \xi_2) - \alpha_2(\eta_1 - \eta_2)}{\alpha_1(\eta_1 - \eta_2) - \beta_1(\xi_1 - \xi_2)} = \frac{\delta_2(\xi_3 - \xi_4) - \gamma_2(\eta_3 - \eta_4)}{\gamma_1(\eta_3 - \eta_4) - \delta_1(\xi_3 - \xi_4)},$$

which may be obtained after dividing the first equal (3.226) by the second one. □

Remark 24 For example, the tensor elements [107]

$$Q_{NML} \sim \begin{cases} 2^{5M/2}, & M = L,\ N < L, \\ 2^{5L/2}, & N = M,\ L < N, \\ 2^{N+3M/2}, & N = L,\ M < N \end{cases}$$

cannot satisfy to (3.223), (3.224), (3.227) and so they don't satisfy the conservation laws.

3.8   Entropy principle maximum

Variation methods play a cardinal role in formulation of the fundamental physics laws and the derivation of the motion equations [44, 86]. These principles aren't more fundamental than the equations themselves, and both may serve as a basic postulate for the construction of the theory. The Hamilton motion equations and the Hamilton principle, for example, can be derived from one another.

The aim of the work is to explore the idea of relating the various motion equations to a single concept, the maximum entropy law [38]. This principle survived practically unchanged the upheaval of mechanics and modern physics in 20th century. The common notion concerning the split between statistic and dynamics relates to the entropy absence in the basic motion equations, which do not posses a mechanism for entropy increase [3, 101, 104]. Any of the probability distribution function is a constant of the motion [88], and the entropy increase occurs as a result of an additional mechanism, such as coarse graining [64, 101, 104]. In the present work the motion equations to be related at the fundamental level to the entropy law, which replaces the established variation principles as the basic postulate.

The information theoretic approach to the entropy principle maximum[3] has the advantage of not being related to a particular experiment, but rather is a statement on the assignment of probability to the various possible outcomes [2, 12, 49]. The basic postulate of the information theory can be formulated as follows: After calculating the probabilities of the available choices subject to the pertinent constraints of the system under investigation, select the one with the largest probability. The constraints should include all the known and none of the unknown information. A common application of this principle to physics is the calculation of probability distributions and partition functions for a large number of freedom degrees, using as constraints the conservation laws derived from the relevant motion equations [2, 49].

An interesting variant of this approach is to use the motion equation itself rather than the associated conserved quantities to derive Boltzmann-like expressions for nonequilibrium distributions [89]. The last decades saw the migration of this method from statistical mechanics of the large systems down to molecular collisions [2] and nuclear physics [98]. In these cases it is common to assume additional constraints that do not derive from the basic equations, but rather ones that are approximately maintained under the specific conditions of the experiment. Agreement between the experimental results and the prediction of the maximum entropy distribution using these constraints can inform us something about the mechanism of the process [79].

---

[3]Szilard L., On the Decrease of Entropy in a Thermodynamic System by the Intervention of Intelligent Beings. Translated by A. Rapoport and M. Knoller from: Uber dis Entropieverminderung in einem thermodynamischen System bei Eingriffen intelligenter Wesen. *Z. Physik*, 53, 1929, 840–856; Reprinted: *Behavioral Science*, 9, No. 4, Oct. 19.

In the present work we try to answer the following question: "Do the basic equations to be used and their associated conservation laws constitute known information?" After all the Hamilton and Schrödinger equations and their equivalent formulations involve postulates whose justification is based on the condition that they properly predict all the known experimental data in the domains where each equation applies [44, 97]. The variation problem will be posed the assumption that a motion equation exists, except its form to be unknown, and subject to the process of entropy maximum. The possibility that no such equation exists is also variable but it describes a trivial case where nothing changes, and thus of no interest here.

3.8.1 Entropy and probability

In statistical physics and information theory, a maximum entropy probability distribution is a probability distribution whose entropy is at least as great as that of all other members of a specified class of distributions. Due to the principle of maximum entropy, if nothing is known about a distribution except that it belongs to a certain class, then the distribution with the largest entropy should be chosen as the default. The motivation is twofold: first, maximizing entropy minimizes the amount of prior information built into the distribution; second, many physical systems tend to move towards maximal entropy configurations over time.

Definition 63 Let X be a *discrete random variable* with distribution given by $P(X = x_k) = p_k$ for $k \in \mathbb{N}$ then the *entropy* of X is defined as $H(X) = -\sum_{k \geq 1} p_k \log p_k$. Let X be a *continuous random variable* with probability density $p(x)$, then the entropy of X is defined as

$$H(x) = \int_{-\infty}^{\infty} p(x) \log p(x)\, dx,$$

where $p(x) \log p(x)$ is assumed to be zero when $p(x) = 0$. In connection with maximum entropy distribution, this form of definition is often the only one given, or at least it is taken as the standard form.

Examples of maximum entropy distributions.

a. The *normal distribution* $F(x, \mu, \sigma^2) = \frac{1}{\sqrt{2\pi\sigma^2}} \exp\left[-\frac{(x-\mu)^2}{2\sigma^2}\right]$ has maximum entropy among all real-valued distributions with specified *mean* μ and *standard deviation* σ.

b. The *uniform distribution* on the interval $[a, b]$ is the maximum entropy distribution among all continuous distributions which are supported in $[a, b]$ (which means that probability density is zero outside of this interval). More generally, if we have a given subdivision $a = a_0 < a_1 < \ldots < a_k = b$ of the interval $[a, b]$ and probabilities $p_1, \ldots, p_k$ which add up to one, then consider the class of all continuous distributions such that $P(a_{j-1} \leq X < a_j) = p_j$ for $j = 1, \ldots, k$. The density of the maximum entropy distribution for this class is constant on each of the intervals $[a_{j-1}, a_j)$; it looks somewhat like a histogram.

c. The *exponential distribution* (with parameter λ of the distribution)

$$f(x, \lambda) = \begin{cases} 1 - \exp(-\lambda x), & x \geq 0 \\ 0, & x < 0 \end{cases}$$

is the maximum entropy among all continuous distributions supported in $[0, \infty)$ with a mean $1/\lambda$. This occurs when gravity acts on a gas that is kept at constant pressure and temperature: if X describes the height of a molecule, then the variable X is exponentially distributed (which also means that the density of the gas depends on height proportional to the exponential distribution). The reason: X is positive and its mean, which corresponds to the average potential energy, is fixed. Over time, the system will attain its maximum entropy configuration, according to the second law of thermodynamics.

d. *Discrete distribution with given mean.* Among all the discrete distributions supported on the set $\{x_1, \ldots, x_n\}$ with mean μ, the maximum entropy distribution has the shape:

$$P(X = x_k) = Cr^{x_k} \quad \text{for} \quad k = 1, \ldots, n,$$

where the positive constants C and r can be determined by the requirements that the sum of all the probabilities must be 1 and the expected value must be μ. For example, if a large number N of dice is thrown, and you are told that the sum of all the shown numbers is S. Based on this information alone, what would be a reasonable assumption for the number of dice showing $1, 2, \ldots, 6$? This is an instance of the situation considered above, with $\{x_1, \ldots, x_n\} = \{1, \ldots, 6\}$ and $\mu = S/N$. Finally, among all the discrete distributions supported on the infinite set $\{x_1, x_2, \ldots\}$ with mean μ, the maximum entropy distribution has the shape: $P(X = x_k) = Cr^{x_k}$ for $k = 1, 2, \ldots$, where the constants C and r are determined by the requirements that the sum of all the probabilities must be 1 and the expected value must be μ. For example, if $x_k = k$, this gives $C = \frac{1}{1+\mu}$ and $r = \frac{\mu}{1+\mu}$.

Boltzmann Theorem. Suppose S is a closed subset of the real numbers R and for given n there are n measurable functions $f_1, \ldots, f_n$ and n numbers $a_1, \ldots, a_n$. Consider the class C of all continuous random variables which are supported on S (i.e., whose density function is zero outside of S) and which satisfy the n expected value conditions $E(f_j(X)) = a_j$ for $j = 1, \ldots, n$ If there is a member in C whose density function is positive everywhere in S, and if there exists a maximal entropy distribution for C, then its probability density $p(x)$ has the following shape: $p(x) = A \exp\left(\sum_{j=1}^{n} \lambda_j f_j(x)\right)$ for all $x \in S$, where the constants A and λ_j have to be determined so that the integral of $p(x)$ over S is 1 and the above conditions for the expected values are satisfied. Conversely, if constants A and λ_j like this can be found, then $p(x)$ is indeed the density of the (unique) maximum entropy distribution for our class C. This theorem is proved with the *calculus of variations* and *Lagrange multipliers*. The proof of this theorem can be found in most classical textbooks on statistical mechanics.

Discrete version. Suppose $S = \{x_1, x_2, \ldots\}$ is a (finite or infinite) discrete subset of the real values and for given n there are n functions $f_1, \ldots, f_n$ and n numbers $a_1, \ldots, a_n$. Consider the class C of all discrete random variables X which are supported on S and which satisfy the n conditions $E(f_j(X)) = a_j$ for $j = 1, \ldots, n$ If there exists a member of C which assigns positive probability to all members of S, and if there exists a maximum entropy distribution for C, then this distribution has the following shape:

$$p(X = x_k) = A \exp\left(\sum_{j=1}^{n} \lambda_j f_j(x)\right) \quad \text{for} \quad k \in \mathbb{N},$$

where the constants A and λ_j have to be determined so that the sum of the probabilities is 1 and the above conditions for the expected values are satisfied. Conversely, if constants A and λ_j like this can be found, then the above distribution is indeed the maximum entropy distribution for our class C. This version of the theorem can be proved with the tools of ordinary calculus and *Lagrange multipliers*.

3.8.2   Derivation of the motion equations

The maximum entropy problem may be stated as follows: maximize the integral

$$J(t_0, t_1) = -\int_{t_0}^{t_1} dt \int_{-\infty}^{\infty} dq \int_{-\infty}^{\infty} F(q, p, t) \ln F(q, p, t)\, dp \tag{3.233}$$

subject to the normalization constraint

$$\int_{-\infty}^{\infty} dq \int_{-\infty}^{\infty} F(q,p,t)dp = 1,$$ (3.234)

and the motion equation be written symbolically as

$$\Omega[\{F(\mathbf{x})\}, \{A_j(\mathbf{x})\}] = 0, \quad j = 1, \dots, N$$ (3.235)

with unknown functional Ω of the probability distribution $F(\mathbf{x})$ and N additional variables $A_j(\mathbf{x})$ governing its evolution. The notation $\{F\}$ is shorthand for F and all derivatives, and the vector $\mathbf{x} = (q, p, t)$ includes coordinate, momentum and time. The solution requires the extremum of the Lagrangian

$$\mathcal{L} = -F(\mathbf{x}) \ln F(\mathbf{x}) + \mu(t)F(\mathbf{x}) + \lambda(\mathbf{x})\Omega[\{F(\mathbf{x}), A_j(\mathbf{x})\}],$$ (3.236)

where μ and λ are the Lagrange coefficients. The basic variation calculus indicates that the parameter μ associated with an integral like (3.234) depends only on t, while λ depends on $\mathbf{x}$ when (3.235) is a differential or algebraic equation. The distribution F must be of entropy maximum at each instant t if (3.233) is stationary independently of endpoints t_0 and t_1.

Typical problems in the information theory involve constraints such as (3.234) or other known information [2, 49, 104]. Here we have a different approach. Assume that nothing except normalization to be known, and the existence of an unknown motion equation. Other constraints will be introduced involve limitations of mathematical nature on the motion equation. The role of these limitations is to reduce the complexity of the solution, expressing prudence regarding our ability of finding a full general solution, at least in the beginning. The constraints are mathematical simplifications, and aren't related to the outcomes or experiments of any kind. The first restriction imposed on the constraint Ω be the simplest form of the motion equation, namely a first order linear differential equation

$$\Omega[\{F(\mathbf{x})\}, \{A_j(\mathbf{x})\}] \equiv \partial_t F + A_0(\mathbf{x})F(\mathbf{x}) + A_1(\mathbf{x})\,\partial_q F + A_2(\mathbf{x})\,\partial_p F = 0.$$ (3.237)

The Euler–Lagrange (EL) equations may be obtained by differentiating Lagrangian $\mathcal{L}$ of (3.236) with respect to F and A_j:

$$\mu(t) - 1 - \ln F(\mathbf{x}) + \lambda(\mathbf{x})A_0(\mathbf{x}) - \partial_q[\lambda(\mathbf{x})A_1(\mathbf{x})] - \partial_p[\lambda(\mathbf{x})A_2(\mathbf{x})] = 0, \quad (3.238a)$$
$$\lambda(\mathbf{x}) = 0, \quad \lambda(\mathbf{x})\,\partial_q F = 0, \quad \lambda(\mathbf{x})\,\partial_p F = 0.$$ (3.238b)

Equations (3.238b) hold only under $\lambda = 0$ leading to the trivial "no equation" case discussed above. The only way for our problem to have a significant solution is for A_j to depend on each order, namely there can be at most two independent variables. In present article we limit ourselves by the case where all A_j be derived from a single control variable $H(q, p)$, which depends on p and q only. The existence of a second independent control variable and the extension to t dependent H will not be considered in present study. In general case, the variables $A_j(\mathbf{x})$ may be described as

$$A_j(\mathbf{x}) = \sum_{k=1}^{2} \sum_{n=0}^{\infty} \alpha_{j,k,n} \frac{\partial^n H}{\partial x_k^n}, \quad j = 1, \dots, N,$$

with constant coefficients $\alpha_{j,k,\beta}$. Then we should replace

$$\sum_{k=1}^{2} \sum_{n=0}^{\infty} \alpha_{0kn} \lambda F \frac{\partial^n H}{\partial x_k^n} + \sum_{j,k=1}^{2} \sum_{n=0}^{\infty} \alpha_{jkn} \lambda \frac{\partial^n H}{\partial x_k^n} \frac{\partial F}{\partial x_j}$$

instead of $\lambda(A_0 F + A_1 \partial_q F + A_2 \partial_p F)$. Since

$$\lambda F \frac{\partial^n H}{\partial x_k^n} = \sum_{s=0}^{n-1} (-1)^s C_n^s \frac{\partial^{n-s}}{\partial x_k^{n-s}} \left[H \frac{\partial^s (\lambda F)}{\partial x_k^s} \right] + (-1)^n H \frac{\partial^n (\lambda F)}{\partial x_k^n},$$

$$\lambda \frac{\partial^n H}{\partial x_k^n} \frac{\partial F}{\partial x_j} = \sum_{s=0}^{n-1} (-1)^s C_n^s \frac{\partial^{n-s}}{\partial x_k^{n-s}} \left[H \frac{\partial^s}{\partial x_k^s} \left(\lambda \frac{\partial F}{\partial x_j} \right) \right] + (-1)^n H \frac{\partial^n}{\partial x_k^n} \left(\lambda \frac{\partial F}{\partial x_j} \right)$$

then (3.238b) may be replaced by a single EL equation for the variation H:

$$\sum_{j,k=1}^{2} \sum_{n=0}^{\infty} \alpha_{j,k,n} (-1)^n \frac{\partial^n}{\partial x_k^n} \left(\lambda(\mathbf{x}) \frac{\partial F}{\partial x_j} \right) + \sum_{k=1}^{2} \sum_{n=0}^{\infty} \alpha_{0,k,n} (-1)^n \frac{\partial^n}{\partial x_k^n} (\lambda(\mathbf{x}) F(\mathbf{x})) = 0. \quad (3.239)$$

If (3.238a) and (3.239) be satisfied for arbitrary F and H, each of the individual terms in the multiple sums must vanish. Hence

$$\sum_{k=1}^{2} \left[\sum_{n=0}^{\infty} \alpha_{0kn} (-1)^n \frac{\partial^n}{\partial x_k^n} (\lambda F) + \sum_{n=0}^{\infty} \alpha_{1kn} (-1)^n \frac{\partial^n}{\partial x_k^n} \left(\lambda \frac{\partial F}{\partial p} \right) + \sum_{n=0}^{\infty} \alpha_{2kn} (-1)^n \frac{\partial^n}{\partial x_k^n} \left(\lambda \frac{\partial F}{\partial q} \right) \right] = 0,$$

$$\mu - 1 - \ln F + \sum_{k=1}^{2} \left[\sum_{n=0}^{\infty} \alpha_{0kn} \lambda F \frac{\partial^n H}{\partial x_k^n} + \sum_{n=0}^{\infty} \alpha_{1kn} \frac{\partial^n H}{\partial x_k^n} \frac{\partial F}{\partial p} + \sum_{n=0}^{\infty} \alpha_{2kn} \frac{\partial^n H}{\partial x_k^n} \frac{\partial F}{\partial q} \right] = 0.$$

The only surviving coefficients are $\alpha_{1,2,1} = -\alpha_{2,1,1}$ (other α_{jkn} are zero) reducing (3.239) to

$$\frac{\partial \lambda}{\partial p} \partial_q F - \frac{\partial \lambda}{\partial q} \partial_p F = 0. \quad (3.240)$$

Substituting these results into (3.237) and (3.238a) we obtain

$$\frac{\partial \lambda}{\partial t} + \frac{\partial H}{\partial p} \frac{\partial \lambda}{\partial q} - \frac{\partial H}{\partial q} \frac{\partial \lambda}{\partial p} = \mu - 1 - \ln F, \quad (3.241a)$$

$$\partial_t F + \frac{\partial H}{\partial p} \partial_q F - \frac{\partial H}{\partial q} \partial_p F = 0. \quad (3.241b)$$

Equation (3.241b) is the classical Liouville equation [44, 88, 89]. To complete its proof we have to show that there the solution λ to (3.241a) compatible with relation (3.239) can be found. Equation (3.241a) is inhomogeneous version of (3.241b) with the solution

$$\lambda = \int_0^1 \mu(\tau) d\tau - [1 - \ln F]t + \lambda_h(\mathbf{x}), \quad (3.242)$$

where the first two terms contain a particular solution and $\lambda_h(\mathbf{x})$ is the general solution of the homogeneous equation [79, 98]. In the proof of (3.242) there it was used well-known fact that any function F is a solution of the Liouville equation [88]. Also $\lambda_h(\mathbf{x})$ can be chosen as an arbitrary combination of functions F. All terms in (3.242) satisfy condition (3.240), completing the proof of (3.241b) as entropy maximum for the motion equation. In other words, the most probable (in the information theoretic sense) first order linear differential equation is none other than the classical Liouville equation.

The Liouville equation (3.241b) derived as a minimum information, or most probable equation without resorting to any experimental evidence or conservation laws (other than normalization). The conservation laws of classical mechanics, as well as the Hamilton or Newton laws result from the Liouville equation [89], that has been shown originate in the entropy principle maximum. Extrapolating this statement to the macroscopic world means that the first thermodynamic law has its roots in the second law, which emerges here as the single fundamental law that governs the evolution of classical system.

Other corollaries of (3.241b) are the basic symmetries of Galilei invariance, time reversal, space inversion, etc. The information theoretic standpoint is that the most probable law is the same, independent of phase space translations or other symmetry operations. This argument can be extended to other invariance involved with conservation laws, such as enstrophy, or helicity in turbulent motion. Thus it is permissible to exploit known symmetry properties as constraints in future more difficult problems to reduce them to manageable proportions. In principle it should be possible to obtain the same result without imposing symmetry, just as we watched above. In variation calculus there constraints are labeled as "noninformational" [2]. They don't change the entropy or the solutions, and occur, for example, when the constraint in question depends on the others. No harm other than possible extra work is done, and in our case the symmetry constraint actually reduces work.

An important property of the Poisson bracket is its invariance to canonical transformations [44], which play a fundamental role in classical mechanics. The extension to three dimensions and many bodies should be especially simple in terms of the Poisson brackets. Application to statistical mechanics may be done by the method of [97], where was formulated a variation problem similar to the present, except that the unknown motion equation (3.237) was replaced by the Liouville equation from the beginning. In [97] there were also derived Boltzmann-like expressions for the nonequilibrium distribution and were postulated useful formulas for thermodynamic quantities.

The new concept introduced in this work gets rid of the postulates that form the basis for the Liouville equation and quantum mechanics. Starting with the minimum information principle, the motion equation emerges as the most probable one, subject to the specified constraints. Classical mechanics is the first order approximation to the general linear equation. Within the above restriction the motion equation does not offer a mechanism for entropy increase. In particular, allowing arbitrary linear equations leads to an infinite order linear differential equation, with the Wigner–Moyal equation for the Wigner distribution function being a special case [80, 105]. This equation leads in a natural way to the Schrödinger equation.

3.8.3 The hierarchical dynamical system

Consider spontaneous process when the random function $\xi(t)$ is $\mathbf{A} = (A_1, \ldots, A_N)$ at the time t. Define $F(\mathbf{A}) = \text{prob}\{\xi(t) = \mathbf{A}\}$ as a probability of this event; thus we obtain the distribution function defined on trajectories $\mathbf{A}(t) = (A_1(t), \ldots, A_N(t), \ldots)$. The fully developed turbulence represents stationary process satisfying to ergodic condition. For the entropy $S(t) = -F(\mathbf{A}(t)) \ln F(\mathbf{A}(t))$ we may postulate following problem:

$$J = \lim_{T \to \infty} \left\{ -\frac{1}{T} \int_{t_0}^{t_0+T} F(A_1(t), \ldots, A_N(t)) \ln F(A_1(t), \ldots, A_N(t)) dt \right\} \to \max \quad (3.243)$$

under additional condition

$$\lim_{T \to \infty} \int_{t_0}^{t_0+T} F(A_1(t), \ldots, A_N(t)) dt = 1. \quad (3.244)$$

The distribution function $F(\mathbf{A})$ may accept nonzero values only on the trajectories $A_k(t)$ corresponded to the dynamic system received from the hierarchical system [39] by symmetrization of tensor T_{kml}

$$\dot{A}_k = -\frac{1}{2} \sum_{m,l} Q_{kml} A_m(t) A_l(t) + \nu_0 K_k A_k(t) + f_k, \quad Q_{kml} = T_{kml} + T_{klm}, \quad Q_{kml} = Q_{kml}. \quad (3.245)$$

One may neglect the external term f_N. Differentiation of (3.244) gives the Liouville equation

$$\partial_t F + \sum_{k=1}^{N} \dot{A}_k \cdot \partial_{A_k} F = 0. \qquad (3.246)$$

It is true only for closed subsystems, and so the last equation has a place for not so long intervals of time when our subsystem may be considered as a closed. Further it will be shown that in general case we should use Fokker–Planck equation [91], which has additional diffusion term. In general case we should use Fokker–Planck equation, which has additional diffusion term. The Liouville equation (3.246) may be considered as a partial case of Fokker-Planck equation, and so the mathematically correct conclusion of the equation (see Section 3.8.4) leads also to strictly correct statement for (3.246). By (3.246), the distribution function is a constant along the phase trajectories. Thus, the phase volume, which must be remained for the closed systems, has not be constant on long time intervals. Its evolution can be described as follows

$$\frac{dV}{dt} = \sum_{k=1}^{N} \int \frac{dA_k}{dt} \prod_{s \neq k} dA_s = V \sum_{k=1}^{N} \frac{\partial \dot{A}_k}{\partial A_k} = \text{div } \mathbf{f}(\mathbf{A}), \quad N \in \mathbb{Z} \cup \infty,$$

where

$$f_k(A_1, \ldots, A_N, \ldots) = -\frac{1}{2} \sum_{m,l \neq k} Q_{kml} A_m(t) A_l(t) + \nu_0 K_k A_k(t).$$

Thus, div $\mathbf{f}(\mathbf{A}) = \nu_0 \sum_{k=1}^{N} K_k \neq 0$. The sum of K_k represented the dissipative term responsible for the mechanics energy being transformed into the heat, $E_{\nu_0} = \nu_0 \sum_{k=1}^{N} K_k A_k^2$ is negative. Hence, we obtain an important consequence div $\mathbf{f}(\mathbf{A}) \equiv -\nu_0 \sum_{k=1}^{N} K_k = -h < 0$, and so $V(t) = V_0 e^{-ht} \to 0$ as $t \to \infty$, i.e., the phase volume is pulled together in a point. So the solutions of the problem (3.243)–(3.245) should be found on the invariant manifold of the dimension less than the dimension of our phase space $V(t)$. Finally, pose the problem:

$$J = \lim_{T \to \infty} \left\{ -\frac{1}{T} \int_{t_0}^{t_0+T} F(A_1(t), \ldots, A_N(t)) \ln F(A_1(t), \ldots, A_N(t)) dt \right\} \to \max,$$

$$\lim_{T \to \infty} \int_{t_0}^{t_0+T} F(A_1(t), \ldots, A_N(t)) dt = 1, \qquad (3.247)$$

$$\frac{\partial F}{\partial t} + \sum_{k=1}^{N} \dot{A}_k \frac{\partial F}{\partial A_k} = 0, \quad \dot{A}_k = -\frac{1}{2} \sum_{m,l \neq k} Q_{kml} A_m(t) A_l(t) + \nu_0 K_k A_k(t).$$

We build Lagrangian

$$\mathcal{L}(t; F; \mathbf{A}; \dot{\mathbf{A}}) = -F(\mathbf{A}) \ln F(\mathbf{A}) + \mu(t) F(\mathbf{A}) + \lambda_0(\mathbf{A}; t) \left(\frac{\partial F}{\partial t} + \sum_{k=1}^{N} \dot{A}_k \frac{\partial F_k}{\partial A_k} \right)$$

$$+ \sum_{k=1}^{N} \lambda_k(\mathbf{A}; t) \left[\dot{A}_k + \frac{1}{2} \sum_{m,l \neq k}^{N} Q_{kml} A_m A_l - \nu_0 K_k A_k \right].$$

In view of

$$\frac{\partial \mathcal{L}}{\partial A_k} = -\nu_0 \lambda_k K_k + \frac{1}{2} \sum_{s \neq k} \lambda_s \sum_{m \neq s,k} Q_{smk} A_m; \quad \frac{\partial \mathcal{L}}{\partial \dot{A}_k} = \lambda_0 \frac{\partial F}{\partial A_k} + \lambda_k;$$

$$\frac{d}{dt} \left(\frac{\partial \mathcal{L}}{\partial \dot{A}_k} \right) = \frac{\partial \lambda_0}{\partial t} \frac{\partial F}{\partial A_k} + \frac{\partial \lambda_k}{\partial t},$$

the Euler–Lagrange equations are

$$\frac{\partial \mathcal{L}}{\partial F} = \mu(t) - 1 - \ln F = 0; \quad \frac{\partial \lambda_k}{\partial t} + \nu_0 \lambda_k K_k - \frac{1}{2} \sum_{s \neq k} \lambda_s \sum_{m \neq s,k} Q_{smk} A_m + \frac{\partial F}{\partial A_k} \frac{\partial \lambda_0}{\partial t} = 0.$$

3.8.4 Fokker–Planck equation

Let $P(\mathbf{A}, t)$ be a probability of the event $\{\xi \in [\mathbf{A}, \mathbf{A} + d\mathbf{A}], t\}$, let $W(\mathbf{A}_2; \mathbf{A}_1)$ be the probability of the transition $\mathbf{A}_1 \to \mathbf{A}_2$ per unit time, i.e., $W(\mathbf{A}_2; \mathbf{A}_1) = \text{prob}\{\mathbf{A}_1 \to \mathbf{A}_2, \Delta t \to 0\}$. Then

$$\partial_t P(\mathbf{A}, t) = \int [W(\mathbf{A}; \mathbf{A}')P(\mathbf{A}', t) - W(\mathbf{A}'; \mathbf{A})P(\mathbf{A}, t)] \, d^N \mathbf{A}'$$

is the balance equation for the probabilities to be found on the trajectory $\mathbf{A}$. Consider $W(\mathbf{A}; \mathbf{A}')$ as a function of the bounce $\mathbf{r} = \mathbf{A} - \mathbf{A}'$ and of the initial point $\mathbf{A}'$, i.e., $W(\mathbf{A}; \mathbf{A}') = W(\mathbf{A}', \mathbf{r})$. Hence

$$\partial_t P(\mathbf{A}, t) = \int [W(\mathbf{A} - \mathbf{r}, \mathbf{r})P(\mathbf{A} - \mathbf{r}, t) - W(\mathbf{A}, -\mathbf{r})P(\mathbf{A}, t)] \, d^N \mathbf{r}$$
$$= \int W(\mathbf{A} - \mathbf{r}, \mathbf{r})P(\mathbf{A} - \mathbf{r}, t)dr - P(\mathbf{A}, t) \int W(\mathbf{A}, -\mathbf{r})d^N \mathbf{r}. \tag{3.248}$$

The first our assumption is $W(\mathbf{A}', \mathbf{r})$ be a smooth function of the initial condition $\mathbf{A}'$ but has a sharp peak on $\mathbf{r}$, i.e., there may be only low bounces. The second: the solution $P(\mathbf{A}, t)$ is a continuous function of $\mathbf{A}$. Then it can be possible to make the shift $\mathbf{A} \to \mathbf{A} - \mathbf{r}$ in the first integral under using Taylor's series

$$W(\mathbf{A} - \mathbf{r}, \mathbf{r})P(\mathbf{A} - \mathbf{r}, t) = W(\mathbf{A}, \mathbf{r})P(\mathbf{A}, t) - \sum_{k=1}^{N} r_k \frac{\partial}{\partial A_k}[W(\mathbf{A}, \mathbf{r})P(\mathbf{A}, t)]$$
$$+ \frac{1}{2} \sum_{k,m=1}^{N} r_k r_m \frac{\partial^2}{\partial A_k \partial A_m}[W(\mathbf{A}, \mathbf{r})P(\mathbf{A}, t)]. \tag{3.249}$$

After substitution (3.249) into (3.248) we obtain

$$\partial_t P(\mathbf{A}, t) = \int W(\mathbf{A}, \mathbf{r})P(\mathbf{A}, t)d^N \mathbf{r} - \sum_{k=1}^{N} \int r_k \frac{\partial}{\partial A_k}[W(\mathbf{A}, \mathbf{r})P(\mathbf{A}, t)]d^N \mathbf{r}$$
$$+ \frac{1}{2} \sum_{k,m=1}^{N} \int r_k r_m \frac{\partial^2}{\partial A_k \partial A_m}[W(\mathbf{A}, \mathbf{r})P(\mathbf{A}, t)]d^N \mathbf{r} - P(\mathbf{A}, t) \int W(\mathbf{A}, -\mathbf{r})d^N \mathbf{r}.$$

Consider the sum of the first and the third terms $\int P(\mathbf{A}, t)[W(\mathbf{A}, \mathbf{r}) - W(\mathbf{A}, -\mathbf{r})]d^N \mathbf{r}$. Since $W(\mathbf{A}, \mathbf{r})$ has a sharp peak on $\mathbf{r}$, i.e.,

$$W(\mathbf{A}, \mathbf{r}) = \begin{cases} W(\mathbf{A}, \mathbf{0}), & |\mathbf{r}| < \delta \\ 0, & |\mathbf{r}| > \delta, \end{cases}$$

then

$$\int P(\mathbf{A}, t)[W(\mathbf{A}, \mathbf{r}) - W(\mathbf{A}, -\mathbf{r})]d^N \mathbf{r} = P(\mathbf{A}, t) \int_{-\delta}^{\delta}[W(\mathbf{A}, \mathbf{r}) - W(\mathbf{A}, -\mathbf{r})]d^N \mathbf{r}$$
$$\leq 2\delta P(\mathbf{A}, t)W(\mathbf{A}, \mathbf{0}) \to 0$$

as $\delta \to 0$, and thus,

$$\partial_t P(\mathbf{A}, t) = -\sum_{k=1}^{N} \frac{\partial}{\partial A_k}\left[P(\mathbf{A}, t)\int r_k W(\mathbf{A}, \mathbf{r})d^N \mathbf{r}\right]$$
$$+ \frac{1}{2} \sum_{k,m=1}^{N} \frac{\partial^2}{\partial A_k \partial A_m}\left\{P(\mathbf{A}, t)\int r_k r_m W(\mathbf{A}, \mathbf{r})d^N \mathbf{r}\right\}.$$

Note that $\int r_k W(\mathbf{A}, \mathbf{r})d^N \mathbf{r}$ is a k-th component of the first order moment for the transition $\mathbf{A} - \mathbf{A}' \to \mathbf{A}$. In our case we have dynamic system with deterministic variables, and so the first moment is equal to the same variable itself. Thus $\int r_k W(\mathbf{A}, \mathbf{r}) d^N \mathbf{r} \equiv \dot{A}_k$. The second moments $\int r_k r_m W(\mathbf{A}, \mathbf{r}) d^N \mathbf{r}$ may be considered as the additional "dynamic noise," or as a diffusion coefficient $D_{km}(\mathbf{A}, t)$ in the phase space; in our case $D_{km}(\mathbf{A}, t)$ is equal $\frac{1}{2}(\dot{A}_k A_m + A_k \dot{A}_m)$. Hence

$$\partial_t P(\mathbf{A}, t) = -\sum_{k=1}^{N} \frac{\partial}{\partial A_k}\left[P(\mathbf{A}, t)\dot{A}_k\right] + \frac{1}{2}\sum_{k,m=1}^{N} \frac{\partial^2}{\partial A_k \partial A_m}\{P(\mathbf{A}, t)D_{km}(\mathbf{A}, t)\}$$
$$= -\sum_{k=1}^{N} \dot{A}_k \frac{\partial}{\partial A_k}P(\mathbf{A}, t) + \frac{1}{2}\sum_{k,m=1}^{N} \frac{\partial^2}{\partial A_k \partial A_m}\{P(\mathbf{A}, t)D_{km}(\mathbf{A}, t)\} \tag{3.250}$$

with $\dot{A}_k = -\frac{1}{2}\sum_{m,l \neq k} Q_{kml}A_m(t)A_l(t) + \nu_0 K_k A_k(t)$. Thereby, in general case (for open systems), for the problem (3.247) we should consider Fokker-Planck equation (3.250) instead of Liouville equation (3.246). Consider a problem

$$\partial_t P(\mathbf{A}, t) = -\sum_{k=1}^{N} \dot{A}_k \frac{\partial}{\partial A_k}P(\mathbf{A}, t) + \frac{1}{2}\sum_{k,m=1}^{N} \frac{\partial^2}{\partial A_k \partial A_m}\{P(\mathbf{A}, t)D_{km}(\mathbf{A}, t)\},$$
$$\int_V P(\mathbf{A}, t)d^N \mathbf{A} = 1 \tag{3.251}$$

with the boundary conditions

$$\lim_{\|\mathbf{A}\| \to \infty} \|\mathbf{A}\|^n \cdot P(\mathbf{A}, t) = 0, \quad n \in \mathbb{N}; \quad P(\mathbf{A}, t)\big|_{\partial V = \{(A_1, \ldots, A_N): \|\mathbf{A}\| \to \infty\}} = 0 \tag{3.252}$$

due to $\|\mathbf{A}\| = \sqrt{\sum_k |A_k|^2} < \infty$. There ∂V is the boundary of the phase space $V(t)$.

To solve (3.251)–(3.252) we use Jeynes's the maximum entropy principle [49], where the entropy with N constraints on the differential equations and norm conditions are written as

$$S(t) = \int_{V(t)} \left\{-P(\mathbf{A}, t)\ln P(\mathbf{A}, t) - \lambda_0 P(\mathbf{A}, t) - \sum_{n=1}^{M} \lambda_n\left[\sum_{j=1}^{N} \alpha_{nj} A_j(t)P(\mathbf{A})\right]\right.$$
$$\left. -\beta \sum_{j=1}^{N} A_j^2(t)P(\mathbf{A})\right\}d^N \mathbf{A}, \tag{3.253}$$

where λ_i and β are the Lagrange multipliers, and $\mathbf{A}(t) = (A_1(t), \ldots, A_N(t))$ is represented N-dimensional vector-function. In order to estimate the probability density $P(\mathbf{A}, t)$, we choose $\hat{P}(\mathbf{A}, t)$ such that it satisfies these constraints and maximize the entropy function (3.253). The variation of S is

$$\delta S = \int_{V(t)} \left[-1 - \ln P(\mathbf{A}, t) - \lambda_0 - \sum_j \Gamma_j A_j(t) - \beta \sum_j A_j^2(t)\right]\delta P d^N \mathbf{A} = 0,$$

where $\Gamma_j = \sum_{k=1}^{N} \alpha_{kj}\lambda_k$. Then we obtain the probability density in the form

$$\hat{P}(\mathbf{A}) = \frac{1}{Z}\exp\left[-\sum_j (\Gamma_j A_j(t) + \beta A_j^2(t))\right], \tag{3.254}$$

where the partition function Z is defined as $Z = \exp(1 + \lambda_0)$ or $Z = \prod_j (\pi/\beta)^{1/2}\exp\left(-\frac{\Gamma_j^2}{4\beta}\right)$. Using (3.254) we may calculate the moments $\langle \mathbf{A} \rangle$ and $\langle \mathbf{A}^2 \rangle$,

$$\langle A_j \rangle = \int A_j(t)\hat{P}(\mathbf{A}, t)d\mathbf{A} = \frac{1}{Z}\int \cdots \int A_j \exp[-(\Gamma_j A_j + \beta A_j^2)]\,dA_1 \ldots dA_N = -\frac{\Gamma_j}{2\beta},$$

$$\langle A_j^2 \rangle = \int A_j^2(t)\hat{P}(\mathbf{A}, t)d\mathbf{A} = \frac{1}{Z}\int \cdots \int A_j^2 \exp[-(\Gamma_j A_j + \beta A_j^2)]\,dA_1 \ldots dA_N = \frac{1}{2\beta} - \frac{\Gamma_j^2}{4\beta^2}.$$

To complete the defining equations for Lagrange multipliers we use the definition of variance

$$\sigma^2 = \frac{1}{N}\sum_{j=1}^{N} \left|\langle A_j^2 \rangle - \langle A_j \rangle^2\right| = \frac{1}{2\beta N}, \quad \text{or} \quad \frac{1}{2\beta} = \sigma^2 N.$$

Hence

$$Z = (2\pi\sigma^2 N)^{N/2} \exp\left\{\frac{1}{2\sigma^2 N}\sum_j \langle A_j^2\rangle - \frac{N}{2}\right\}$$
$$= \left[2\pi\sum_j |\langle A_j^2\rangle - \langle A_j\rangle^2|\right]^{N/2} \exp\left[\frac{1}{2}\frac{\sum_j \langle A_j^2\rangle - N}{\sum_j |\langle A_j^2\rangle - \langle A_j\rangle^2|}\right]$$

and, accordingly,

$$\left((2\pi\sigma N)^{N/2}\exp\left(\frac{1}{2\sigma^2 N}\sum_{j=1}^{N}\langle A_j^2\rangle - \frac{N}{2}\right)\right)^{-1}\exp\left\{-\frac{1}{\sigma^2 N}\sum_j\left[-A_j(t)\langle A_j\rangle + \frac{1}{2}A_j^2(t)\right]\right\}$$
$$= \left[2\pi\sum_j |\langle A_j^2\rangle - \langle A_j\rangle^2|\right]^{-N/2}\exp\left\{-\frac{1}{2}\frac{\sum_j[\langle A_j^2\rangle - \langle A_j\rangle^2]^2 - N}{\sum_j |\langle A_j^2\rangle - \langle A_j\rangle^2|}\right\}.$$
$$(3.255)$$

So we will solve the problem (3.251)–(3.252) by use multiplication both sides of (3.251) by A_j^n and integrating over phase volume V with further summing on j. It yields the closed forms for the moments in terms of $D_{ij}(\mathbf{A},t)$ and $A_k(t)$. The LHS gives

$$\int_V A_j^n \frac{\partial P(\mathbf{A},t)}{\partial t}d^N\mathbf{A} = \frac{d}{dt}\int_V A_j^n P(\mathbf{A},t)d^N\mathbf{A} - n\int_V A_j^{n-1}\dot{A}_j P(\mathbf{A},t)d^N\mathbf{A}$$
$$= \frac{d}{dt}\langle A_j^n(t)\rangle - n\langle A_j^{n-1}(t)\dot{A}_j\rangle.$$
$$(3.256)$$

The first term of the RHS yields

$$-\sum_{k=1}^{N}\dot{A}_k\int_V A_j^n\frac{\partial}{\partial A_k}P(\mathbf{A},t)d^N\mathbf{A} = -\sum_{k=1}^{N}\dot{A}_k\int_V A_j^n dA_1\cdots dA_{k-1}dP(\mathbf{A},t)dA_{k+1}\ldots dA_N$$

$$= -\sum_{k=1}^{N}\dot{A}_k\left\{\int_V A_j^n P(\mathbf{A},t)\Big|_{A_k=-\infty}^{A_k=\infty}dA_1\cdots dA_{k-1}dA_{k+1}\cdots dA_N - n\int_V A_j^{n-1}P(\mathbf{A},t)d^N\mathbf{A}\right\}$$

$$= n\sum_{k=1}^{N}\dot{A}_k\langle A_j^{n-1}\rangle,$$
$$(3.257)$$

because of the first term is equal zero follow (3.252). The second term on the right-hand side, similarly, gives

$$\frac{1}{2}\sum_{k,m=1}^{N}\int_V A_j^n\frac{\partial^2}{\partial A_k\partial A_m}\{P(\mathbf{A},t)D_{km}(\mathbf{A},t)\}d^N\mathbf{A}$$
$$= \frac{1}{2}\sum_{k,m=1}^{N}n\langle A_j^{n-1}(t)\frac{\partial}{\partial A_k}D_{km}(\mathbf{A},t) + (n-1)_j^{n-2}(t)D_{km}(\mathbf{A},t)\rangle.$$
$$(3.258)$$

Here $\langle A_j^n(t)\rangle = \int_V A_j^n P(\mathbf{A},t)dA_1\ldots dA_N$. After substituting (3.256), (3.257) and (3.258) into (3.251) we obtain

$$\frac{d}{dt}\langle A_j^n(t)\rangle - n\langle A_j^{n-1}(t)\dot{A}_j\rangle = n\langle A_j^{n-1}\rangle\sum_{k=1}^{N}\dot{A}_k$$
$$+\frac{1}{2}\sum_{k,j=1}^{N}n\langle A_j^{n-1}(t)\frac{\partial}{\partial A_k}D_{kj}(\mathbf{A},t) + (n-1)A_j^{n-2}(t)D_{kj}(\mathbf{A},t)\rangle.$$
$$(3.259)$$

For the partial case $n = 1, 2$, which we will consider below, Equation (3.259) is simplified

$$\frac{d}{dt}\langle A_j(t)\rangle - \langle\dot{A}_j\rangle = \sum_{k=1}^{N}\dot{A}_k + \frac{1}{2}\sum_{k,m=1}^{N}\langle\frac{\partial}{\partial A_k}D_{km}(\mathbf{A},t)\rangle, \tag{3.260a}$$

$$\frac{d}{dt}\langle A_j^2(t)\rangle - 2\langle A_j(t)\dot{A}_j\rangle = 2\langle A_j\rangle\sum_{k=1}^{N}\dot{A}_k + \sum_{k,m=1}^{N}\langle A_j(t)\frac{\partial}{\partial A_k}D_{km}(\mathbf{A},t) + D_{km}(\mathbf{A},t)\rangle. \tag{3.260b}$$

Thus, solution of (3.251)–(3.252) may be written in the form (3.255) (compare with [23, 24]). First, consider the simple case without diffusion between trajectories $A_j(t)$. Then

$$\frac{d}{dt}\sum\nolimits_{j=1}^{N}[\langle A_j(t)\rangle - \langle \dot{A}_j\rangle] = N\sum\nolimits_{k=1}^{N}\dot{A}_k,$$

$$\frac{d}{dt}\sum\nolimits_{j=1}^{N}\langle A_j^2(t)\rangle = 2\sum\nolimits_{j=1}^{N}\left[\langle A_j(t)\dot{A}_j\rangle + \langle A_j\rangle\sum\nolimits_{k=1}^{N}\dot{A}_k\right].$$

Note that

$$\int\langle A_j(t)(t)\dot{A}_j\rangle dt = \int dt \int A_j(t)\dot{A}_j P(\mathbf{A},t)d^N\mathbf{A}$$
$$= \int dA_j \int A_j(t)P(\mathbf{A},t)d^N\mathbf{A} = \int\langle A_j\rangle dA_j = A_j\langle A_j\rangle \int\langle A_j(t)\rangle \sum_k \dot{A}_k dt$$
$$= \sum_k \int dA_k \int A_j(t)P(\mathbf{A},t)d^N\mathbf{A} = \sum_k \int dA_k = \sum_k \int\langle A_j\rangle dA_k = \langle A_j\rangle \sum_k A_k,$$

so it can be obtained

$$\langle A_j(t)\rangle = \int dt \int \dot{A}_j P(\mathbf{A},t)d^N\mathbf{A} + \sum_k A_k = \int d^N\mathbf{A}\int P(\mathbf{A},t)dA_j + \sum_k A_k = A_j + \sum_k A_k,$$
$$\langle A_j^2(t)\rangle = 2\left[\int d^N\mathbf{A}\int A_j P(\mathbf{A},t)dA_j + \sum_{k=1}^{N}\int d^N\mathbf{A}\int A_j P(\mathbf{A},t)dA_k\right]$$
$$= 2A_j^2 + 2A_j\sum_{k\neq j}A_k = 2A_j\sum_k A_k,$$
$$|\langle A_j^2(t)\rangle - \langle A_j(t)\rangle^2| = (\sum_k A_k)^2 + A_j^2.$$

$$(3.261)$$

After substitution (3.261) into (3.255) we finally obtain

$$P(A_1,\ldots,A_N) = [2\pi\sum_j |\langle A_j^2(t)\rangle - \langle A_j(t)\rangle^2|]^{-N/2}\exp\left\{-\frac{1}{2}\frac{\sum_j[\langle A_j^2(t)\rangle - \langle A_j(t)\rangle^2]^2 - N}{\sum_j|\langle A_j^2(t)\rangle - \langle A_j(t)\rangle^2|}\right\}$$
$$= \left\{2\pi[N(\sum_j A_j^2)^2 + \sum_j A_j^2]\right\}^{-N/2}\exp\left\{\frac{1}{2}\frac{\sum_j[(\sum_k A_k)^2 + A_j^2(t)]^2 - N}{N(\sum_j A_j)^2 + \sum_j A_j^2}\right\}.$$

Thus, the problem (3.247) may be formulated as

$$J = \lim_{T\to\infty}\left\{-\frac{1}{T}\int_{t_0}^{t_0+T}F(A_1(t),\ldots,A_N(t))\ln F(A_1(t),\ldots,A_N(t))\,dt\right\} \to \max$$

$$F(A_1,\ldots,A_N) = \left\{2\pi[N(\sum_j A_j^2)^2 + \sum_j A_j^2]\right\}^{-N/2}\exp\left\{\frac{1}{2}\frac{\sum_j[(\sum_k A_k)^2 + A_j^2]^2 - N}{N(\sum_j A_j)^2 + \sum_j A_j^2}\right\},$$

$$\dot{A}_k = -\frac{1}{2}\sum_{m,l\neq k}Q_{kml}A_m(t)A_l(t) + \nu_0 K_k A_k(t).$$

Now we consider the general case with the diffusion between trajectories A_k; we should also take into account the second terms in the RHSs of (3.260a). The integration of the second term in the first equation gives zero. Really

$$\frac{1}{2}\sum_{k,m=1}^{N}\int\langle\frac{\partial}{\partial A_k}D_{km}(\mathbf{A},t)\rangle dt = \frac{1}{2}\sum_{k,m=1}^{N}\int dt\int P(\mathbf{A},t)\frac{\partial}{\partial A_k}D_{km}(\mathbf{A},t)d^N\mathbf{A}$$
$$= \frac{1}{2}\sum_{k,m=1}^{N}\int dt\int P(\mathbf{A},t)dA_1\cdots dA_{k-1}dD_{km}(\mathbf{A},t)dA_{k+1}\cdots dA_N$$
$$= \frac{1}{2}\sum_{k,m=1}^{N}\int\int\frac{\partial}{\partial A_k}D_{km}(\mathbf{A},t)dA_k dt\int P(\mathbf{A},t)dA_1\cdots dA_{k-1}dA_{k+1}\cdots dA_N$$
$$= \frac{1}{2}\sum_{k,m=1}^{N}\int D_{km}(\mathbf{A},t)dt\int P(\mathbf{A},t)dA_1\cdots dA_{k-1}dA_{k+1}\cdots dA_N$$
$$= \frac{1}{2}\sum_{k,m=1}^{N}\int D_{km}(\mathbf{A},t)dt\int\frac{\partial P(\mathbf{A},t)}{\partial A_k}d^N\mathbf{A}$$
$$= \frac{1}{2}\sum_{k,m=1}^{N}\int D_{km}(\mathbf{A},t)P(\mathbf{A},t)|_{A_k\in\partial V}\,dt = \int dA_1\cdots dA_{k-1}dA_{k+1}\cdots dA_N = 0,$$

because of

$$\sum_{k,m=1}^{N}\left|\int D_{km}(\mathbf{A},t)P(\mathbf{A},t)\big|_{A_k\in\partial V}dt\int dA_1\cdots dA_{k-1}dA_{k+1}\cdots dA_N\right|$$

$$\leq \frac{1}{2}\sum_{k,m=1}^{N}\int\left|D_{km}(\mathbf{A},t)\right|P(\mathbf{A},t)\big|_{A_k\in\partial V}V(t)dt = 0$$

and $P(\mathbf{A}, t)|_{\partial V} = 0$. The integration of the second term of (3.260b) yields

$$\sum_{k,m=1}^{N} \int \langle \mathbf{A}(t) \tfrac{\partial}{\partial A_k} D_{km}(\mathbf{A}, t) + D_{km}(\mathbf{A}, t) \rangle dt$$

$$= \sum_{k,m=1}^{N} \int dt \Big[\int \mathbf{A}(t) P(\mathbf{A}, t) \tfrac{\partial}{\partial A_k} D_{km}(\mathbf{A}, t) d^N \mathbf{A} + \int D_{km}(\mathbf{A}, t) P(\mathbf{A}, t) d^N \mathbf{A} \Big]$$

$$= \sum_{k,m=1}^{N} \int dt \Big[\int \tfrac{\partial}{\partial A_k} D_{km}(\mathbf{A}, t) dA_k \int \mathbf{A}(t) P(\mathbf{A}, t) dA_1 \dots dA_{k-1} dA_{k+1} \dots dA_N$$

$$+ \int D_{km}(\mathbf{A}, t) P(\mathbf{A}, t) d^N \mathbf{A} \Big]$$

$$= \sum_{k,j=1}^{N} \int dt \int \tfrac{\partial}{\partial A_k} D_{kj}(\mathbf{A}, t) dA_k \int \mathbf{A}(t) P(\mathbf{A}, t) dA_1 \dots dA_{k-1} dA_{k+1} \cdots dA_N$$

$$+ \sum_{k,m=1}^{N} \int dt \int D_{km}(\mathbf{A}, t) P(\mathbf{A}, t) d^N \mathbf{A} = \sum_{k,m=1}^{N} \int D_{km}(\mathbf{A}, t)|_{A_k \in \partial V} dt \int \tfrac{\partial \mathbf{A}}{\partial A_k} P(\mathbf{A}, t) d^N \mathbf{A}$$

$$+ \sum_{k,m=1}^{N} \int dt \int D_{km}(\mathbf{A}, t) P(\mathbf{A}, t) d^N \mathbf{A} = \sum_{k,m=1}^{N} \int D_{km}(\mathbf{A}, t)|_{A_k \in \partial V} dt$$

$$+ \sum_{k,m=1}^{N} \int dt \int D_{km}(\mathbf{A}, t) P(\mathbf{A}, t) d^N \mathbf{A}$$

$$= \sum_{k,m=1}^{N} \int [D_{km}(\mathbf{A}, t)|_{A_k \in \partial V} + \langle D_{km}(\mathbf{A}, t) \rangle] dt = \sum_{k,m=1}^{N} (\tfrac{1}{2} A_k A_m + \int \langle D_{km}(\mathbf{A}, t) \rangle dt)$$

$$= \tfrac{1}{2} \sum_{k,m=1}^{N} [3 A_k A_m + (A_k + A_m) \textstyle\sum_k A_k],$$

because of

$$\int \langle D_{km}(\mathbf{A}, t) \rangle dt = \tfrac{1}{2} \int dt \int (\dot{A}_k A_m + A_k \dot{A}_m) P(\mathbf{A}, t) d\mathbf{A} = \tfrac{1}{2} \big(\int \dot{A}_k \langle A_m \rangle dt + \int \dot{A}_m \langle A_k \rangle dt \big)$$
$$= \tfrac{1}{2} \big(\int \langle A_m \rangle dA_k + \int \langle A_k \rangle dA_m \big) = \tfrac{1}{2} (A_k \langle A_m \rangle + A_m \langle A_k \rangle) = A_k A_m + \tfrac{1}{2} (A_k + A_m) \textstyle\sum_k A_k.$$

Thus, in the general case, (3.261) to be transformed

$$\langle A_j(t) \rangle = A_j + \textstyle\sum_k A_k$$
$$\langle A_j^2(t) \rangle = 2 A_j \textstyle\sum_k A_k + \tfrac{3}{2} \sum_{k,m} A_k A_m + N(\sum_k A_k)^2 \tag{3.262}$$
$$|\langle A_j^2(t) \rangle - \langle A_j(t) \rangle^2| = (N+1)(\textstyle\sum_k A_k)^2 + A_j^2 + \tfrac{3}{2} \sum_{k,m} A_k A_m.$$

After summation (3.262) on j and substitution the result in (3.255) we obtain

$$P(A_1, \dots, A_N) = \Big\{ 2\pi N \big[(N+1) \big(\textstyle\sum_j A_j \big)^2 + \tfrac{1}{N} \sum_j A_j^2 + \tfrac{3}{2} \sum_{k,m} A_k A_m \big] \Big\}^{-N/2}$$
$$\times \exp \Big\{ - \tfrac{1}{2} \frac{\tfrac{1}{N} \sum_j [(N+1)(\sum_j A_j)^2 + A_j^2 + \tfrac{3}{2} \sum_{k,m} A_k A_m]^2 - 1}{(N+1)(\sum_j A_j)^2 + \tfrac{1}{N} \sum_j A_j^2 + \tfrac{3}{2} \sum_{k,m} A_k A_m} \Big\}. \tag{3.263}$$

Thus we obtain the problem, where we should find the optimal tensor Q_{nml}, which makes the functional J to be maximal,

$$J = \lim_{T \to \infty} \Big\{ - \tfrac{1}{T} \int_{t_0}^{t_0+T} F(A_1(t), \dots, A_N(t)) \ln F(A_1(t), \dots, A_N(t)) dt \Big\} \to \max$$

$$F(A_1, \dots, A_N) = \Big\{ 2\pi N \big[(N+1)(\textstyle\sum_j A_j)^2 + \tfrac{1}{N} \sum_j A_j^2 + \tfrac{3}{2} \sum_{k,m} A_k A_m \big] \Big\}^{-N/2}$$
$$\times \exp \Big\{ - \tfrac{1}{2} \frac{\tfrac{1}{N} \sum_j [(N+1)(\sum_j A_j)^2 + A_j^2 + \tfrac{3}{2} \sum_{k,m} A_k A_m]^2 - 1}{(N+1)(\sum_j A_j)^2 + \tfrac{1}{N} \sum_j A_j^2 + \tfrac{3}{2} \sum_{k,m} A_k A_m} \Big\}$$
$$\dot{A}_k = -\tfrac{1}{2} \sum_{m,l \neq k} Q_{kml} A_m(t) A_l(t) + \nu_0 K_k A_k(t).$$

The solution (3.263) to the problem (3.251)–(3.252) has been obtained under choice its form $\hat{P}(\mathbf{A}, t)$ to maximize the entropy function (3.253). Further it will be shown that there it can be found another solutions. Moreover we can define the whole class of the solutions, i.e., the problem (3.251)–(3.252) accepts nonunique solution.

3.9 Appendix: inequalities

1. A version of Hölder's inequality. If a and b are real numbers, and p and q are positive real numbers satisfying $\frac{1}{p} + \frac{1}{q} = 1$, then $ab \leq \frac{a^p}{p} + \frac{b^q}{q}$.

Proof. We prove it for p and q rational. Let m and n be integers with $n > m$. Let $p = n/m$ and $q = n/(n - m)$. Start with the Arithmetic-mean-Geometric-mean inequality $\prod_{n=1}^{N} a_n \leq \frac{1}{N} \sum_{n=1}^{N} a_n^N$ for n real positive numbers, choosing the first m equal to $a^{1/m}$ and the remaining $n - m$ equal to $b^{1/(n-m)}$. Then $ab \leq \frac{ma^{n/m} + (n-m)b^{n/(n-m)}}{n} = \frac{a^p}{p} + \frac{b^q}{q}$.

2. Cauchy's inequality. Let x and y be N-dimensional vectors. Then $x \cdot y \leq |x| \cdot |y|$, where the inner product is $x \cdot y = \sum_{n=1}^{N} x_n y_n$, and the norm $|x|^2 = \sum_{n=1}^{N} x_n^2$.

Proof. Start with Arithmetic-mean-Geometric-mean inequality $ab \leq \frac{1}{2}(a^2 + b^2)$ for $a_n = x_n/|x|$ and $b_n = y_n/|y|$, summing over n:

$$\frac{x \cdot y}{|x|\,|y|} = \sum_{n=1}^{N} \frac{x_n}{|x|}\frac{y_n}{|y|} \leq \frac{1}{2} \sum_{n=1}^{N} \left(\frac{x_n^2}{|x|} + \frac{y_n^2}{|y|}\right) = 1.$$

3. Cauchy–Schwarz inequality. Let f and g be square integrable functions on Ω. Then

$$\int_{\Omega} fg\,dV \leq \left(\int_{\Omega} |f|^2 dV\right)^{1/2} \left(\int_{\Omega} |g|^2 dV\right)^{1/2} = \|f\|_2 \cdot \|g\|_2,$$

where $\|f\|_2 = \left(\int_{\Omega} |f|^2 dV\right)^{1/2}$ is the L^2 norm.

4. Hölder's inequality. Let f and g be functions on a domain Ω such that if $|f|^p$ and $|g|^q$ are integrable for some positive real p and q satisfying $1/p + 1/q = 1$. Then

$$\int_{\Omega} fg\,dV \leq \left(\int_{\Omega} |f|^p dV\right)^{1/p} \left(\int_{\Omega} |g|^q dV\right)^{1/q} = \|f\|_p \cdot \|g\|_q,$$

where the L^p and L^q norms are introduced. This also holds for $p = 1$ and $q = \infty$ where $\|f\|_\infty = \text{ess} - \sup_{x \in \Omega} |f(x)|$ which, for continuous functions, is just the sup-norm.

5. Minkowski's inequality. Let f and g be functions on a domain Ω such that If $|f|^p$ and $|g|^p$ are integrable for some positive real $p \geq 1$. Then $\|f + g\|_p \leq \|f\|_p + \|g\|_p$.

6. Some calculus inequalities. Let f be a smooth, square integrable, mean zero periodic function $\Omega = [0, L]^d$. For spatial dimensions $d = 1$, 2, and 3, there exist finite absolute constants c_d (i.e., no dependence on L or the choice of f) such that

$$\|f\|_\infty^2 \leq \begin{cases} c_1 \|\nabla f\|_2 \cdot \|f\|_2, & d = 1, \\ c_2 \|\nabla f\|_2^2 [1 + \log(L\|\Delta f\|_2/\|\nabla f\|_2)], & d = 2, \\ c_3 \|\Delta f\|_2 \cdot \|\nabla f\|_2, & d = 3. \end{cases} \tag{3.264}$$

Proof. The upper two inequalities are proved in Section 3.6.3. The third inequality mimics those proofs, and is left as an exercise for the reader. Similar results can be proved in higher dimensions, involving norms of successively higher derivatives.

7. Gagliardo–Nirenberg inequality. Let f be a smooth, square integrable, mean zero function on $\Omega = [0, L]^d$, then for $1 \leq q, r < \infty$ and integers $0 \leq j < m$, we have

$$\|D^j f\|_p \leq c \|D^m f\|_r^a \cdot \|f\|_q^{1-a}, \quad \frac{1}{p} = \frac{j}{d} + a\left(\frac{1}{r} - \frac{m}{d}\right) + \frac{1-a}{q}, \tag{3.265}$$

where a is restricted to j/m, the constant c depends only on n, m, j, q, r, a, with the next exception cases:

(a) If $j = 0$, $rm < n$, $q = \infty$ then we should make the additional assumption that either f tends to zero at infinity or $f \in L_{\tilde{q}}$ for some finite $\tilde{q} > 0$.

(b) If $1 < r < \infty$ and $m - j - d/r$ is a nonnegative integer then (3.265) holds only for $a \in [j/m, 1)$.

Proof. We will not give the detailed proof, it can be found in [1, 11, 82], we will indicate the main steps. First, we remark that:

– The value p is determined simply by dimensional analysis.

– For $a = 1$ the fact that $f \in L_q$ does not enter in the estimate (3.265), and the estimate is equivalent to the results of Sobolev (note that we permit r to be unity).

– j/m is the smallest possible value for a may be seen by taking $f = \sin \lambda x_1 \zeta(x)$, $\zeta \in C_0^\infty$: For large λ we have, $\|f\|_q = O(1)$, $\|D^j f\|_p = O(\lambda^j)$, $\|D^m f\|_r = O(\lambda^m)$.

– It will be clear from proof that the result holds also for f defined in a product domain $x_s \in (-\infty, \infty)$, $x_t \in (0, \infty)$: $s = 1, \ldots, k$, $t = k + 1, \ldots, n$ and hence for any domain that can be mapped in a one-to-one way onto such domain by sufficiently *nice* mapping.

– Similar estimates hold for L_p norms of $D^j f$ on linear subspaces of lower dimensions.

– This theorem, in its generality is useful in treating nonlinear problems. We mention that the form (3.265) for $a = j/m$, $q = \infty$ it follows that the set $\{f\}$ consists of bounded functions which have derivatives of order m belonging to L_r forms of Banach Algebra.

The proof of the theorem is elementary and contains in particular an elementary proof for the Sobolev case $a = 1$. In order to prove (3.265) for any given j, one has only to examine the extreme values of a, j/m and the unity. For in general case there is a simple.

8. Lieb–Thirring inequalities for orthogonal functions. Let $\{\varphi_1, \ldots, \varphi_N\}$ the set of orthonormal functions in d spatial dimensions. These must obviously satisfy $\int_\Omega \sum_{n=1}^N |\varphi_n|^2 \, dx = N$. Lieb and Thirring have shown that they obey the inequality

$$\int_\Omega \left(\sum_{n=1}^N |\varphi_n|^2\right)^{(d+2)/d} dx \leq c \sum_{n=1}^N \int_\Omega |\nabla \varphi_n|^2 \, dx,$$

where the constant c is independent of N. This is also true for vector functions φ_i.

3.10 Exercises

1. For a general vector field u defined on the periodic domain Ω in two or three dimensions, show that if $S = \operatorname{div} u$ and $\omega = \operatorname{rot} u$, then

(a) $\int_\Omega |\nabla u|^2 \, dx = \int_\Omega [|\nabla \times u|^2 + (\operatorname{div} u)^2] \, dx$, (b) $\int_\Omega |\Delta u|^2 \, dx = \int_\Omega (|\operatorname{rot} \omega|^2 + (\nabla S)^2) \, dx$.

Solution: (a) Prove the first equality. The RHS integral is

$$\int_\Omega [|\nabla \times u|^2 + (\operatorname{div} u)^2] \, dx = \int_\Omega \left\{ \left[\left(\frac{\partial u_x}{\partial x}\right)^2 + \left(\frac{\partial u_x}{\partial y}\right)^2 + \left(\frac{\partial u_x}{\partial z}\right)^2\right] + \left[\left(\frac{\partial u_y}{\partial x}\right)^2 + \left(\frac{\partial u_y}{\partial y}\right)^2 + \left(\frac{\partial u_y}{\partial z}\right)^2\right] \right.$$
$$\left. + \left[\left(\frac{\partial u_z}{\partial x}\right)^2 + \left(\frac{\partial u_z}{\partial y}\right)^2 + \left(\frac{\partial u_z}{\partial z}\right)^2\right] - 2\left[\frac{\partial u_x}{\partial y}\frac{\partial u_y}{\partial x} + \frac{\partial u_x}{\partial z}\frac{\partial u_z}{\partial x} + \frac{\partial u_y}{\partial z}\frac{\partial u_z}{\partial y}\right] \right\} dx.$$

The LHS integral gives the same under the definition

$$|\nabla u|^2 = \left[\left(\frac{\partial u_x}{\partial x}\right)^2 + \left(\frac{\partial u_x}{\partial y}\right)^2 + \left(\frac{\partial u_x}{\partial z}\right)^2\right] + \left[\left(\frac{\partial u_y}{\partial x}\right)^2 + \left(\frac{\partial u_y}{\partial y}\right)^2 + \left(\frac{\partial u_y}{\partial z}\right)^2\right] + \left[\left(\frac{\partial u_z}{\partial x}\right)^2 + \left(\frac{\partial u_z}{\partial y}\right)^2\right.$$
$$\left. + \left(\frac{\partial u_z}{\partial z}\right)^2\right] - 2\left[\frac{\partial u_x}{\partial y}\frac{\partial u_y}{\partial x} + \frac{\partial u_x}{\partial z}\frac{\partial u_z}{\partial x} + \frac{\partial u_y}{\partial z}\frac{\partial u_z}{\partial y}\right].$$

(b) The second equality can be proven simultaneously.

2. It is possible to formally "solve" the F_N ladder by integration. By defining $Y_N = F_N^{-s}$, show that the ladder in LHS of (3.102) in Theorem 52 can be rewritten as $\dot{Y}_N + A(t)Y_N \geq \frac{2}{s}Y_{N-s}$, where $A(t) = c\|D\mathbf{u}\|_\infty + \nu\lambda_0^{-2}$. Show that the ladder inequality is equivalent to

$$Y_N I(t) \geq \frac{2}{s} \int_0^t Y_{N-s}(\zeta)I(\zeta)d\zeta, \quad \text{where} \quad I(t) = \exp\int_0^t A(\tau)\,d\tau.$$

3. There is an alternative version of the H_N and F_N ladders which depend on $\|\mathbf{u}\|_\infty$ instead of $\|D\mathbf{u}\|_\infty$. To obtain this, first show that an alternative to either (3.107) or (3.112) can be found by using the vector identity (3.115) to obtain the NSE in the form

$$\mathbf{u}_t + \omega \times \mathbf{u} = \nu\Delta\mathbf{u} - \nabla\left(p/\rho + \frac{1}{2}u^2\right) + f.$$

Use this and Lemma 8 to prove $|NLT| \leq c_N H_N^{1/2} H_{N+1}^{1/2}\|\mathbf{u}\|_\infty$. Show that the alternative full ladder is

$$\frac{1}{2}\dot{F}_N \leq -\frac{1}{2}\nu(F_N^{1+1/s}/F_{N-s}^{1/s}) + (c_N\nu^{-1}\|\mathbf{u}\|_\infty + \nu\lambda_0^{-2})F_N.$$

4. (The ladder in full norm form) Let $M_N = \sum_{j=0}^{N-s} L^{-2j}F_{N-j}$. This definition has coefficients which are powers of L which keep each term the same dimension. In full, one may write $M_N = F_N + L^{-2}F_{N-1} + \ldots + L^{-2(N-s)}F_s$. For $1 \leq s \leq N$, use Lemma 9 to show

$$\frac{1}{2}\dot{M}_N \leq -\nu\left(M_N^{1+1/s}/M_{N-s}^{1/s}\right) + (c_N\|D\mathbf{u}\|_\infty + \nu\lambda_0^{-2})M_N.$$

Hence, high norms depend on low norms: $M_{N-s} = F_{N-s} + L^{-2}F_{N-s-1} + \ldots + L^{-2(N-s)}F_0$.

5. Show that for $r < N_1 < N_2$, $\kappa_{N_1,r} \leq \kappa_{N_2,r}$. Show that $r_1 < r_2 < N$, $\kappa_{N,r_1} \leq \kappa_{N,r_2}$.

6. Use the inequality $\|D\mathbf{u}\|_\infty \leq cF_N^{a/2}\|\mathbf{u}\|_q^{1-a}$ for $q > 1$ in the ladder inequality for the three-dimensional NSE to show that an absorbing ball for F_N can be found only if $q > 3$.

7. For the two-dimensional NSE, use the ladder for $\kappa_{N,r}$. Given in Table 3.1 and Lemma 10 to verify (3.148), namely that $\overline{\lim}_{t\to\infty}\kappa_{N,r} \leq \lambda_0^{-1}\operatorname{Gr}^{(N-1)/(N-r)}$ by considering λ_0 as the smallest scale on the domain. Show also that an estimate which is pointwise in t exists in the form $\kappa_{N,2}(t) \leq c\kappa_{N,2}(0)\exp(t\lambda_0^{-2}\nu\operatorname{Gr})$.

8. Verify the estimate $\langle\mathcal{N}_{N,1}\rangle = \lambda_0^2\langle\kappa_{N,1}^2\rangle \leq c_{10}\operatorname{Gr}(1 + \log\operatorname{Gr})^{1/2}$ given in Section 3.6.3 for the second case, namely (using Lemma 10),

$$\langle\mathcal{N}_{N,r}\rangle = \lambda_0^2\langle\kappa_{N,r}^2\rangle \leq c_{N,r}\operatorname{Gr}(1 + \log\operatorname{Gr})^{1/2}.$$

9. For the three-dimensional Euler equations show that $\frac{d}{dt}\int_\Omega \mathbf{u}\cdot\omega\,dx = 0$ when $\omega\cdot n = 0$ on $\partial\Omega$. The quantity $\int_\Omega u\cdot\omega\,dx$ is called the helicity, see Chapter 3.

10. Use (3.155) to show that in three-dimension $\langle\kappa_{2,1}^2\rangle \leq c\nu^{-4}\langle F_1^2\rangle + \lambda_0^{-2}$.

11. Show that $\langle\|\mathbf{u}\|_\infty^2\rangle^{2(N-1)} \leq \langle F_N\rangle(L\lambda_0^{-2}\nu^2\operatorname{Gr}^2)^{2N-3}$. Show also that the result in (3.156) also implicitly gives an inverse length scale λ_T^{-1} such that $(\lambda_T/L)^{-1} \leq c(L/\lambda_0)^2\operatorname{Gr}^2$.

12. By considering the eigenvalues of the Stokes' operator $-\Delta$, show that

$$\operatorname{Tr}[-\Delta P_N] \geq cL^{-2}N^{(2+d)/d}. \tag{3.266}$$

13. Use the inequality of Lieb and Thirring given in Section 3.9 to prove (3.266).

14. Consider the PDE $\mathbf{u}_t = R\mathbf{u} + (1 + i\nu)\Delta\mathbf{u} - (1 + i\mu)\mathbf{u}\cdot|\mathbf{u}|^2$ on the two-dimensional periodic domain $\Omega \equiv [0,1]^2$, known as the complex Ginzburg–Landau (CGL) equation. The parameters R, μ and ν are real and lie in the range $R > 1$, $|\mu| < \infty$ and $|\nu| < \infty$. Show that
    (i) when $|\mu| \leq \sqrt{3}$, then $d_L \leq R$;
    (ii) when $|\mu| > \sqrt{3}$ then $d_L \leq cR$, where c is independent of ν but not of μ.

Bibliography

[1] Adams R.A. *Sobolev Spaces*. Pure and Applied Math., 140. Elsevier/Academic Press, 2003, 305 pp.

[2] Alhassid Y. and Levine R.D. Connection between the maximal entropy and the scattering theoretic analyses of collision processes. *Phys. Rev. A*, 18, 1978, 89–116.

[3] Andrew K. Entropy. *Amer. J. Phys.*, 52, 1984, 492–496.

[4] Anselmet F et al. High-order velocity structure functions in turbulent shear flows. *J. Fluid Mech.*, 140, 1984, 63–89.

[5] Aubry N. et al. The dynamics of coherent structures in the wall region of a turbulent boundary layer. *J. Fluid Mech.*, 192, 1988, 115–173.

[6] Beale J.T., Kato T. and Majda A. Remarks on the breakdown of smooth solutions for the 3-D Euler equations. *Commun. Math. Phys.*, V. 94, 1984, 61–66.

[7] Belinicher V.I. and L'vov V.S. A scale-invariant theory of developed hydrodynamic turbulence. *Sov. Phys. JETP*, 66, 1987, 303–313.

[8] Belinicher V.I., L'vov V.S., and Procaccia I. A new approach to computing the scaling exponents in fluid turbulence from first principles. *Physica A*, 254, 1998, 215–230.

[9] Benzi R. et al. Extended self-similarity in turbulent flows. *Phys. Rev. E*, 48, 1993, R29–R32.

[10] Birkhoff G.D. Dynamical Systems. *AMS Colloquium*, Vol. 9, 1927.

[11] Brezis H. and Gallouet T. Nonlinear Schrödinger evolution equations. *Nonlinear Analysis, Theory, Methods and Applications*, 4(4): 677–681, 1980.

[12] Brillouin L. Science and Information Theory. Academic, New York, 1962, 351 pp.

[13] Cartwright M. and Littlewood J. On nonlinear differential equations of the second order, the equation $y''+k(1-y^2)y'+y = b\lambda k \cos(\lambda t+a)$, k large. *J. of London Math. Society*, 20, 1945, 180–189.

[14] Chavaria G.R. Anomalous scaling of velocity structure functions in turbulence: a new approach. *J. Phys. II France*, 4, 1994, 1083–1088.

[15] Chepyzhov V.V. and Ilyin A.A. On the fractal dimension of invariant sets; applications to Navier–Stokes equations. *Disc. Cont. Syn. Systems*, 10 (1 and 2), 2004, 117–135.

[16] Constantin P. et al. Analytic study of shell models of turbulence, *Phys. D: Nonlinear Phenomena*, 219, Issue 2, 2006, 120–141.

[17] Constantin P. and Foias C. *Navier–Stokes Equations*, University of Chicago Press, 1988.

[18] Constantin P. and Foias C. Global Lyapunov exponents, Kaplan-Yorke formulas and the dimensional of the attractors for 2D Navier–Stokes equations, *Comm. Pure Appl. Math.*, 38, 1985, 1–27.

[19] Constantin P. et al. Determining modes and fractal dimension of turbulent flows, *J. Fluid Mech.*, 150, 1985, 427–440.

[20] Demengel F. and Ghidaglia J.M. Some remarks on the smoothness of inertial manifolds, *Nonlinear Analysis–TMA*, 16 (1991), 79–87.

[21] Desnyansky V.N. and Novikov E.A. The evolution of turbulence spectra to the similarity regime. *Izv. Akad. Nauk SSSR, Fiz. Atm. Okeana*, 10, 1974, 127–136.

[22] Doering Ch.R. and Gibbon J.D. *Applied Analysis of the Navier–Stokes Equations.* Cambridge Univ. Press, 2004.

[23] El-Wakil S.A. et al. Solution of Fokker-Planck equation by means of maximum entropy. *JQSRT*, 69, 2001, 41–48.

[24] El-Wakil S.A. et al. On the maximum-entropy method for kinetic equation of radiation particle and gas. *Phys. Scr.*, 51, 1995, 129–136.

[25] Eyink G. Lagrangian field-theory, multifractals, and universal scaling in turbulence. *Phys. Lett A.* 172, 1993, 335–360.

[26] Fairhall A. et al. Anomalous scaling in a model of passive scalar advection: exact results. *Phys. Rev. E*, 53, 1996, 3518–3535.

[27] Falconer K. *Fractal Geometry*, 3rd Ed., Wiley, 2014, 368 pp.

[28] Farge M. Ondelettes continues: application a la turbulence. *J. Ann. Soc. Math. De France*, 1990, 17–62.

[29] Foias C. et al. *Navier–Stokes Equations and Turbulence*, Cambridge University Press, 2001.

[30] Foias C. et al. Asymptotic analysis of the Navier–Stokes equations, *Phys. D: Nonlinear Phenomena*, 9, 1983, 157–188.

[31] Foias C. and Prodi G. Sur le comportement global des solutions non stationnaires des équations de Navier–Stokes en dimension two. *Rend. Sem. Mat. Univ. Padova*, 39, 1967, 1–34.

[32] Foias C., Sell G.R. and Temam R. Inertial manifolds for nonlinear evolutionary equations. *J. Diff. Equations*, 73, 1988, 309–353.

[33] Foias C., Sell G.R. and Titi E. Exponential tracking and approximation of inertial manifolds for dissipative nonlinear equations. *J. Dynamics and Differential Equations*, 1 (2), 1989, 199–244.

[34] Fric P.G. Hierarchical model of two-dimensional turbulence. *J. Magnetic Hydrodynamic*, 1, 1983, 60–66.

[35] Frisch U. and Orszag S.A. Turbulence: challenges for theory and experiment. *Phys. Today*, 43, 1990, 24–32.

[36] Frisch U. *Turbulence. The Legacy of A.N. Kolmogorov.* Cambridge Univ. Press, 1996, 296 pp.

[37] Frisch U. Fully Developed Turbulence and Intermittency, 71–88, In *Turbulence and Predictability in Geophysical Fluid Dynamics and Climate Dynamics*, eds. M. Ghil, R. Benzi and G. Parisi, Proc. Int. School of Physic E. Fermi, North Holland, Amsterdam, 1985.

[38] Gaissinski I. and Rovenski V. The entropy principle maximum for hierarchical dynamic systems and the turbulence problem. *SITA Journal*, 10, No. 4, 2008, 147–158.

[39] Gaissinski I. and Rovenski V. The test for adequacy of shell models for turbulence problem. *Int. J. Pure and Appl. Math.*, 36, N. 4, 2007, 407–418.

[40] Gaissinski I. and Rovenski V. *Non-linear Models in Mechanics*. Lambert Acad. Publ., Germany, 2010, 664 pp.

[41] Gassner, S., Blomgren, P., and Palacios A. Noise-induced intermittency in pattern-forming systems. *Int. J. Bifurcation and Chaos*, 17, Issue 8, 2007, 2765–2779.

[42] Gledzer E.B. System of hydrodynamic type admitting two quadratic integral of motion. *Sov. Phys. Dokl.*, 18, 1973, 216–217.

[43] Goldstein S. Fluid mechanics in the first half of this century. *Ann. Rev. Fluid Mech.*, 1, 1969, 1–29.

[44] Goldstein H. *Classical Mechanics*. 2nd edition. Addison Wesley, 1980, 672 pp.

[45] Grossmann A. and Morlet J. Decomposition of Hardy functions into square integrable wavelets of constant shape. *SIAM, J. Math. Annual.*, 15, 1984, 1417–1423.

[46] Halsey T. C. et al. Fractal measures and their singularities: The characterization of strange sets. *Phys. Rev. A*, 33, 1986, 1141–1151.

[47] Hilhorst D. et al. Global attractor and inertial sets for a nonlocal Kuramoto–Sivashinsky equation. *Report Math. Institute MI 06-99*, Leiden University, 1999.

[48] Ilyin A.A. and Titi E.S. Sharp estimates for the number of degrees of freedom for the damped-driven 2D Navier–Stokes. *J. of Nonlinear Science*, 16 (3), 2006, 233–253.

[49] Jeynes E.T. Information theory and statistical mechanics I, II. *Phys. Rev.*, 106, 1955, 620–630; *Phys. Rev.*, 108, 1957, 171–190.

[50] Jones D.A. and Titi E.S. Determination of solutions of the Navier–Stokes equations by finite volume elements. *Phys. D*, 60 (1992), 165–174.

[51] Jones D.A. and Titi E.S. Upper bounds on the number of determining modes, nodes, and volume elements for the Navier–Stokes equations. *Indiana Univ. Math. J.*, 42, 1993, 875–887.

[52] Kadanoff L.P. Operator algebra and determination of critical indices. *Phys. Rev. Lett.*, 23, 1969, 1430–1438.

[53] Kolmogorov A.N. The local structure of turbulence in incompressible viscous fluid for very large Reynolds number. *Dokl. Akad. Nauk USSR*, 30, 1941, 9–13.

[54] Kolmogorov A.N. A refinement of previous hypothesis concerning the local structure of turbulence in a viscous incompressible fluid at high Reynolds number. *J. of Fluid Mech.*, 13, 1962, 82–85.

[55] Kotsarts Y. et al. On the stability of stretched flames. *Combust. Theory & Model.* 2, 1997, 153–156.

[56] Kotsarts Y., Brailovsky I. and Sivashinsky G. On hydrodynamic instability of stretched flames. *Combustion Science and Technology*, 110, 1997, 524–529.

[57] Kraichnan R.H. Dynamics of nonlinear stochastic systems. *J. Math. Phys.*, 2, 1961, 124–148.

[58] Kraichnan R.H. Small-scale structure of a scalar field convected by turbulence. *Phys. Fluids*, 11, 1968, 245–253.

[59] Kraichnan R.H. Convergents of turbulence functions. *J. Fluid Mech.*, 41, 1970, 189–217.

[60] Kraichnan R.H. Anomalous scaling of a randomly advected passive scalar. *Phys. Rev. Lett.*, 72, 1994, 1016–1019.

[61] Kuramoto Y. Diffusion-induced chaos in reactions systems. *Supp. Progr. Theor. Phys.*, V. 64, 1978, 346–367.

[62] Lamb H. *Hydrodynamics*. Reprint of 5th edition. Cambridge Univ. Press, Cambridge, 1993, 738 pp.

[63] Landau L. and Lifshitz E. *Fluid Mechanics, Course of Theor. Phys.*, v. 6. Pergamon Press, 1959.

[64] Landsberg P.T. *Thermodynamics and Statistical Mechanics*. Dover Publ. Inc., 1990, 461 pp.

[65] Libchaber A.J. and Maurer J. Une expirience de Rayleigh–Btnard de gkometrie rtduite: multiplication, accrochage et demultiplication de frtquences. *J. Phys. Lett.*, 40, 1979, L419.

[66] L'vov V.S., Podivilov E. and Procaccia I. Scaling behavior in turbulence is doubly anomalous. *Phys. Rev. Lett.*, 76(21), 1996, 3963–3966.

[67] L'vov V. S. and Procaccia I. Exact resummation in the theory of hydrodynamic turbulence. Parts I, II, and III. *Phys. Rev. E*, 52, 1995, 3840–3857; 52, 1995, 3858–3875; 53, 1996, 3468–3490.

[68] L'vov V. S. and Procaccia I. Towards a nonperturbation theory of hydrodynamic turbulence: fusion rules, exact bridge relations and anomalous viscous scaling functions. *Phys. Rev. E*, 54(6), 1996, 6268–6284.

[69] L'vov V.S. and Procaccia I. Fusion rules in turbulent systems with flux equiliblium. *Phys. Rev. Lett.*, 76, 1996, 2898–2901.

[70] L'vov V.S., Podivilov E. and Procaccia I. Temporal multiscaling in hydrodynamic turbulence. *Phys. Rev. E*, 55, 1997, 7030–7035.

[71] L'vov V.S. et al. Improved shell model of turbulence. *Phys. Rev. E*, 58, 1998, 1811–1822.

[72] L'vov V.S. and Procaccia I. Computing the scaling exponents in fluid turbulence from first principles: the formal setup. *Physica A*, 257, 1998, 165–196.

[73] L'vov V.S. et al. Anomalous scaling from controlled closure in a shell model of turbulence. *Phys. Fluid.*, 12, 2000, 803–821.

[74] Lorentz E.N. Low order models representing realizations of turbulence. *J. Fluid Mech.* 55, 1972, 545–563.

[75] Majda A. Vorticity, turbulence and acoustics in fluid flow. *SIAM Review*, V. 33, 349–388, 1991.

[76] Mandelbrot B.B. Intermittent turbulence in self-similar cascades: divergence of high moments and dimension of the carrier. *J. of Fluid Mech.*, 62, 1974, 331–358.

[77] Mandelbrot B.B. *The Fractal Geometry of Nature.* San Francisco, Freeman, 1982, 460 pp.

[78] Meneveau C. Analysis of turbulence in the orthonormal wavelet representation. *CTR Manuscript*, 120, Standford Univ., 1990.

[79] Morse P.M. and Feshbach H. *Methods of Theoretical Physics, 2 Vols.*, McGraw-Hill, New York, 1953.

[80] Moyal J.E. Quantum mechanics as a statistical theory. *Proc. Camb. Phil. Soc.*, 45, 1949, v. 45, 99–124.

[81] Nelkin M. Universality and scaling in fully developed turbulence. *Adv. in Physics*, 43, 1994, 143–181.

[82] Nirenberg L. On elliptic partial diff. equations. *Annali delta Scuola Norm. Sup.*, 13, 1959, 115–162.

[83] Oboukhov A.M. Some specific features of atmospheric turbulence. *J. Fluid Mech.*, 13, 1962, 77–81.

[84] Ohkitani K. and Yamada M. Temporal intermittency in the energy cascade process and local Lyapunov analysis in fully-developed model turbulence. *Progr. Theor. Phys.*, 89, 1989, 329–341.

[85] Ou Y.-R. and Sritharan S.S. Analysis of regularized Navier–Stokes equations. I, II, *Quart. Appl. Math.*, 49, 1991, 651-685, 687–728.

[86] Polyanin A.D. et al. *Handbook of First Order PDEs.* Taylor & Francis, 2002, 500 pp.

[87] Praskovsky A. and Oncley S. Measurements of the Kolmogorov constant and intermittency exponent at very high Reynolds numbers. *Phys. Fluids*, 6(9), 1994, 2886–2888.

[88] Prigogine I. *From Being to Becoming.* Freeman, San Francisco, 1980, 272 pp.

[89] Prigogine I. *Non-Equilibrium Statistical Mechanics.* Wiley, New York, 1962, 319 pp.

[90] Richardson L.F. *Weather Prediction by Numerical Process.* Cambridge Univ. Press, 2007, 236 pp.

[91] Risken H. *The Fokker-Planck equation: Methods of solution and applications*, Springer Series in Synergetics, 1996, 413 pp.

[92] Ruelle D. *Chance and Chaos.* Princeton University Press, 1991, 195 pp.

[93] Ruelle D. and Takens F. On the nature of turbulence. *Comm. Math. Phys.*, 20, 1971, 167–192; 23, 1971, 343–344.

[94] Sell G.R. and You Y. Inertial manifolds: the non-self-adjoint case, *J. Diff. Eq.*, 96, 1992, 203–255.

[95] Sivashinsky G. and Michelson D.M. On irregular wavy flow of a liquid down a vertical plane. *Prog. Theor. Phys.*, V. 63, 1980, 2112–2114.

[96] Smale S. Morse inequalities for a dynamical system. *Bulletin of the AMS*, 66, 1960, 43–49.

[97] Sobouti I. A Lagrangian formulation for classical systems. *Physica A*, 168, 1990, 1021–1034.

[98] Sokolnikoff I.S. and Redheffer R.M. *Mathematics of Physics and Modern Engineering.* McGraw-Hill, New York, 1966, 752 pp.

[99] Sreenivasan K.R. and Kailasnath P. An update on the intermittency exponent in turbulence. *Phys. Fluids A*, 5, 1993, 512–514.

[100] Sulem P.-L. and Frisch U. Bounds on energy flux for finite energy turbulence. *J. Fluid. Mech.*, 72, 1975, 417–423.

[101] Tapp M.C. The Boltzmann function as an embedded property of Liouville's equation. *J. Phys. A: Math. Gen.*, 23, 1990, L427–L432.

[102] Takens F. Detecting strange attractors in turbulence. Dynamical Systems and Turbulence, Springer-Verlag, 1981, 366–381.

[103] Temam R. *Infinite-Dimensional Dynamical Systems in Mechanics and Physics*, Springer-Verlag, 1988.

[104] Wehrl A. General properties of entropy. *Rev. Mod. Phys.*, 110, 1978, 221–260.

[105] Wigner E.P. On quantum correction for thermodynamic equilibrium. *Phys. Rev.*, 40, 1932, 749–759.

[106] Wilson K.G. Non-Lagrangian models of current algebra. *Phys. Rev.*, 179, 1969, 1499–1508.

[107] Zimin V. and Hussain F. Wavelet based model for small-scale turbulence. *Phys. Fluids Letters*, 7, 1995, 2925–2927.

Chapter 4

Modeling of Flow over Blunted Bodies

Modeling of a flow over blunted bodies plays an important role in investigations of supersonic (with Mach number higher than 1) and hypersonic (with Mach number higher than 4–5) flows, while there is not enough such publications in scientific literature. Therefore, we try to fill this gap by including a separate chapter in this book.

The precise Mach number at which a craft can be said to be fully hypersonic is elusive, especially since physical changes in the airflow (molecular dissociation, ionization) occur at quite different speeds. Generally, a combination of effects becomes important "as a whole" around Mach number equal to five. The hypersonic regime is often associated with a speed of ramjet propulsion that does not produce net thrust. This is a nebulous definition in itself, as there are known restrictions to allow them (ramjets, sometimes referred to as a flying stovepipe) to operate in the hypersonic regime (scramjets – supersonic combusting ramjets). While the definition of a hypersonic flow can be quite vague and is generally debatable (especially due to the lack of discontinuity between supersonic and hypersonic flows), such flow may be characterized by certain physical phenomena that can no longer be analytically discounted as in supersonic flow. These phenomena include:

• *Small shock stand-off distance*: As Mach numbers increase, the density behind the shock also increases, which corresponds to a decrease in volume behind the shock wave due to conservation of mass. Consequently, the distance between the shock and the body generating it reduces at high Mach numbers.

• *Viscous interaction*: A portion of the large kinetic energy associated with flow at high Mach numbers transforms into internal energy in the fluid due to viscous effects. The increase in internal energy is realized as an increase in temperature. Since the pressure gradient normal to the flow within a boundary layer is approximately zero for low to moderate hypersonic Mach numbers, the increase of temperature through the boundary layer coincides with a decrease in density. Thus, the boundary layer over the body grows and can often merge with the thin shock layer.

• *High temperature flow*: High temperatures are a consequence of manifestation of viscous dissipation due to nonequilibrium chemical flow properties such as dissociation and ionization of molecules resulting in convection and radiation.

• *Effects*: The hypersonic flow regime is characterized by a number of effects which are not found in typical aircraft operating at low subsonic Mach numbers. The effects depend strongly on the speed and type of vehicle under investigation.

Prediction of heat accruing to the surface of vehicles (re-)entering planetary atmosphere is important for heat-shield design. Turbulent flow induces much higher heating in comparison with laminar flow. Therefore, prediction of the location of laminar-turbulent transition is a key factor to define the geometry sizes and materials used for the thermal protection system (TPS). The fundamental physical processes related to the laminar turbulent transition in high-speed boundary layers are not well understood yet.

Early experiments on laminar-turbulent transition were undertaken by Osbourne Reynolds at the University of Manchester, and further experiments have examined different circumstances under which transition can occur over many orders of magnitude of

Reynolds numbers (at least 1,000 to 100,000). The location of transition is called the *transition point*. Examples of transition include Tollmien–Schlichting waves and boundary layer transition. Transition can also be strongly affected by other factors such as upstream disturbances, surface roughness and cross-flow. Therefore transition can occur over a wide range of Reynolds numbers. For small free-stream disturbances and negligible surface roughness, the laminar-turbulent transition leads to amplification of unstable modes in the boundary layer. For essentially two-dimensional supersonic and hypersonic flows, the initial phase of transition is associated with excitation and amplification of the first and/or second modes. The first mode is an extension to high speeds of the Tollmien-Schlichting (TS) waves, which represent viscous instability at low Mach numbers. The inviscid nature of the first mode begins to dominate when the Mach number increases, since compressible boundary layer profiles contain a generalized inflection point. The second mode results from an inviscid instability driven by a region of supersonic mean flow relative to the disturbance phase velocity. This instability belongs to the family of trapped acoustic modes propagating in a waveguide between the wall and the sonic line.

Unlike incompressible flows, high-speed boundary layers can possess more than one instability mode. These modes usually correspond to separate frequency bands in a stability diagram. The first mode is similar to the incompressible instability with the difference that often first-mode oblique waves are more amplified than two-dimensional disturbances. At low temperature gas usually behaves as calorically perfect gas. In high-speed flows the temperature inside the boundary layer might be very high and, as a result, the gas properties such as specific heat become a function of temperature (thermally perfect gas). At even higher temperatures, chemical reaction processes occur and change the composition of the gas. If the flow velocity is small in comparison with chemical reaction rates, the gas will be in a state of chemical equilibrium. For higher speeds or smaller reaction rates, nonequilibrium effects may occur.

One of the most important applications of boundary layer theory is the calculation of the friction drag of bodies in a flow, e.g., the drag of a flat plate at zero incidence; the friction drag of a ship, an airfoil, the body of an airplane, or a turbine blade. One particular property of the boundary layer is that, under certain conditions, a reverse flow can occur directly at the wall. A *separation* of the boundary layer from the body and the formation of large or small eddies at the back of the body, can then occur. This results in a great change in the pressure distribution at the back of the body, leading to the *form or pressure drag* of the body. This can also be calculated using boundary layer theory. Boundary layer theory answers the important question of what shape a body must have in order to avoid this detrimental separation. It is not only in flow past a body, where separation can occur, but also in flow through a duct. In this way, boundary layer theory can be used to describe the flow through blade cascades in compressors and turbines, as well as through diffusers and nozzles. The processes involved in maximum lift of an airfoil, where separation is also important, can only be understood using boundary layer theory. The boundary layer is also important for heat transfer between a body and the fluid around it.

Recognizing this importance, the purpose of the present chapter is to introduce the reader to the basic fundamentals of hypersonic flow, including an emphasis on high-temperature multi-component gas dynamics which, as we will see is an important aspect of high-speed flows in general. Wherever pertinent, we will also discuss modern computational fluid dynamics applications in hypersonic flow and high temperature flow.

4.1 Onsager's theory

Our system of equations becomes closed when we add equations defining the following values: $\mathbf{J}_q$, $\mathbf{J}_k$, $\mathbf{J}_k$, κ_α, τ^{ij}. These equations are called *kinetic equations*. One of the main goals of thermodynamics or irreversible processes is kinetic equations. To establish them we will use Onsager's theory [28]. To clarity the situation we will consider one simple particular example of an irreversible process: heat conduction. In fact there are two main questions:

A. What is the source of this process?

B. How this process can be described from macroscopic point of view? In other words, how the process intensity can be estimated?

The answers are rather clear:

A. The heat conductivity source is the gradient of temperature in the different point of medium.

B. The process intensity is estimated by the value of heat flux.

The basic problem of thermodynamics of this process is the establishment of link between heat flux and temperature gradient. For our particular case this link is defined by Fourier's law.

4.1.1 General concept of a multi-component gas mixture

What do we mean by the *viscous hypersonic flow*? The simplest and rather obvious answer is that during our consideration we have to take into account two things [2, 32]:

- Viscous effects or, in other words, molecular transfer effects.
- Gas flow with very high Mach numbers M or high velocity flow.

The last point has two aspects: mathematical and physical-chemical one. The mathematical aspect is connected with the fact that some parameters of a gas flow have an asymptotic limit for high Mach numbers. In other words, we say about some semi-similar behavior of the flow with respect to high Mach number. One asks additional small questions: What does it mean *high*? $M = 3$, or $M = 100$? The answer depends on some other parameters, but in average if we say *hypersonic flow regime* it means that $M \geq 6 - 8$. The physical-chemical aspect of the second point is based on the fact that for a high velocity gas flow we have to take into account additional physical-chemical processes, which are neglected for supersonic gas flow with relative small Mach numbers. To explain why the situation is changed for high Mach number flow we will consider a very simple example. Firstly we should estimate the gas temperature at the stagnation point for high Mach numbers. It is known from gasdynamics that for inviscid gas a full enthalpy is conserved along the stream line 3rd conservation law),

$$\frac{1}{2} V_\infty^2 + c_p T_\infty = c_p T_0, \tag{4.1}$$

where V is the flow velocity, T – gas mixture temperature, c_p – heat capacity at the constant pressure, subscripts "0" and "∞" correspond to stagnation and continuity points, accordingly. Taking into account the general thermodynamic law $P_\infty = R\rho_\infty T_\infty$, with $R = c_p - c_V$ – universal gas constant and $\gamma = c_p/c_V$ – heat ratio, where c_V – heat capacity at the constant volume, one can write

$$c_p T_\infty = \frac{c_p}{R} \frac{P_\infty}{\rho_\infty} = \frac{c_p}{c_p - c_V} \frac{P_\infty}{\rho_\infty} = \frac{\gamma}{\gamma - 1} \frac{P_\infty}{\rho_\infty}, \quad c_p T_0 = c_p T_\infty \frac{T_0}{T_\infty} = \frac{\gamma}{\gamma - 1} \frac{P_\infty}{\rho_\infty} \frac{T_0}{T_\infty}.$$

Rewriting (4.1) as $\frac{V_\infty^2}{2} + \frac{\gamma}{\gamma-1}\frac{P_\infty}{\rho_\infty} = \frac{\gamma}{\gamma-1}\frac{P_\infty}{\rho_\infty}\frac{T_0}{T_\infty}$, we estimate the ratio T_0/T_∞:

$$T_0/T_\infty = 1 + \frac{1}{2}(\gamma-1)M_\infty^2.$$

For example, for ideal gas air flow with $\gamma = 1.4$ one will have the following good estimation: $T_0/T_\infty = 1 + 0.2M_\infty^2$. Thus, for $M = 10.0$ we have $T_0/T_\infty = 21$, for $M = 20.0$, $T_0/T_\infty = 81$ and, if $M = 25.0$ then $T_0/T_\infty = 126$. Backward, for $T_\infty = 300K$ we have, respectively (see table below):

TABLE 4.1: Dependency of T_0 on M.

T_0	M
6300 K	10.0
24300 K	20.0
37800 K	25.0

For high Mach number the gas temperature at the stagnation region is very high and for precise description of the gas flow we have to account real gas effects, namely: multi-component nature of the gas, chemical reactions as well as diffusion effects. So, the *viscous hypersonic flow* means that we have to account molecular transfer and real gas effects as well as asymptotic properties of the initial governing equations for high values M.

Before justifying the governing equations we will introduce the concept of multi-component gas mixture. It is known that the concept of a continuous media is basic in gas dynamics. From microscopic point of view (10^{-10} m) gas is a lot of particles: atoms, molecular etc., and from this point of view gas is a discontinuous media. Assume that from macroscopic point of view (10^0 m) for our purposes we can consider a gas as a continuous media. It means we assume that the gas exists at each point of our domain. Considering the continuous media approach we say that multi-component gas mixture consisting of N components is the continuous media of N medias and *all* of these media *are in the same domain*, which is occupied by our gas mixture. To describe this type of media we have to use some microscopic parameters which will be characterized each media.

Introduce some definitions: ρ_i – the density of i-th component, and v_i – an averaged velocity of i-th component. For example, write $\rho_i = \lim_{V\to 0}(\Delta m_i/\Delta V)$, where Δm_i is the mass of i-th component in an elementary volume ΔV. Introduce a density ρ and velocity $\mathbf{v}$ for gas mixture as whole as

$$\Delta m = \sum_{i=1}^{N} m_i, \quad \rho = \lim_{\Delta V\to 0}\frac{\Delta m}{\Delta V} = \lim_{\Delta V\to 0}\sum_{i=1}^{N}\frac{\Delta m_i}{\Delta V} = \sum_{i=1}^{N}\rho_i,$$

$$\rho\mathbf{v} = \sum_{i=1}^{N}\rho_i\mathbf{v}_i. \qquad (4.2)$$

The velocity in (4.2) is the mass average velocity, that is the velocity of the mass center of N continuous medias corresponding to N components. Let $n_i[cm^{-3}]$ and $n[cm]^{-3}$ be the number of particles of kind i and the number of all particles in unite volume, respectively. Then $n = \sum_{i=1}^{N} n_i$. The following nondimensional variables are often used to characterize the species in mixture: $c_i = \rho_i/\rho$ – the mass concentration of i-th component, $x_i = n_i/n$ – the molar concentration of i-th component. If $m_i[g/mol]$ is a molecular mass and N_A is Avogadro number, we write $\rho_i = \frac{n_i m_i}{N_A}$ and $\rho = \frac{1}{N_A}\sum_{i=1}^{N} n_i m_i$. On the other hand, $\rho = \frac{n}{N_A}m$, where m is the molecular mass of the mixture. Using these formulae, rewrite the relations between c_i and x_i:

$$c_i = \frac{\rho_i}{\rho} = \frac{n_i m_i N_A}{nmN_A} = x_i\frac{m_i}{m}. \qquad (4.3)$$

From the definitions of c_i and x_i it follows

$$\sum_{i=1}^{N} c_i = 1, \quad \sum_{i=1}^{N} x_i = 1. \tag{4.4}$$

Using (4.3) and (4.4), we write the relations

$$m = \sum_{i=1}^{N} x_i m_i, \quad \frac{1}{m} = \sum_{i=1}^{N} \frac{c_i}{m_i}$$

for m_i via x_i and c_i. Introduce a vector of diffusion flux $\mathbf{J}_i$ (instead of velocity component $\mathbf{v}_i$), which characterizes the velocity of i−th component with respect to the gas mixture as a whole $\mathbf{J}_i = \rho_i(\mathbf{v}_i - \mathbf{v})$. Using definition for ρv, write the equation for the diffusion flux conservation law $\sum_{i=1}^{N} \mathbf{J}_i = 0$. We will additionally assume a gas mixture to be a mixture of perfect gases. Thus, each component must satisfy the following classical relation: $P_i = \frac{\rho_i}{m_i} R_0 T_i$, where P_i is the pressure of i-th component, and R_0 is the universal gas constant. In the framework of Dalton's law the gas mixture pressure P is

$$P = \sum_{i=1}^{N} P_i = R_0 \sum_{i=1}^{N} (\rho_i/m_i)T_i.$$

Assuming the temperature of each component of the mixture to be the same for infinitesimal small particle, rewrite the important state equation as

$$P = \sum_{i=1}^{N} P_i = R_0 \sum_{i=1}^{N} \frac{\rho_i}{m_i} T_i = \rho R_0 T \sum_{i=1}^{N} \frac{c_i}{m_i} = \frac{\rho R_0 T}{m}.$$

Then we have

$$\frac{P_i}{P} = \frac{(\rho_i/m_i)R_0 T}{(\rho/m)R_0 T} = \frac{\rho_i}{\rho} \frac{m}{m_i} = c_i \frac{m}{m_i} = x_i.$$

4.1.2 Thermodynamic potentials, forces and flows

The basic thermodynamic potential is internal energy. In a fluid system, the energy density u depends on the *matter density* r and the *entropy density* s as $du = Tds + mdr$, where T is the temperature and m is a combination of pressure and chemical potential. We write

$$ds = (1/T)du - (m/T)dr.$$

The extensive quantities u and r are conserved and their flows satisfy continuity equations:

$$\frac{\partial u}{\partial t} + \text{div } \mathbf{J}_u = 0, \quad \frac{\partial r}{\partial t} + \text{div } \mathbf{J}_r = 0,$$

see Section 1.8, where the divergence of the flux densities $\mathbf{J}$ is used. The gradients of the conjugate variables (thermodynamics) of u and r, are thermodynamic forces $1/T$ and $-m/T$, and they cause flows of the corresponding extensive variables. In the absence of matter flows we have, $\mathbf{J}_u = k\nabla(1/T)$ and, in the absence of heat flows, $\mathbf{J}_r = k'\nabla(m/T)$.

The reciprocity relations. When there are both heat and matter flows, there are "cross-terms" in the relationship between flows and forces (the proportionality coefficients are denoted by L):

$$\mathbf{J}_u = L_{uu}\nabla(1/T) - L_{ur}\nabla(m/T), \quad \mathbf{J}_r = L_{ru}\nabla(1/T) - L_{rr}\nabla(m/T).$$

The Onsager reciprocity relations state the equality of the cross-coefficients L_{ur} and L_{ru},

[28]. Proportionality follows from simple dimensional analysis (i.e., both coefficients are measured in the same units of temperature times mass density).

Abstract formulation. Let E_i be the extensive variables on which entropy s depends. In the following analysis, these symbols will refer to densities of these thermodynamic quantities. Then, $ds = \sum_{i=1}^{n} I_i dE_i$, where $I_i = \partial s/\partial E_i$ defines the intensive quantity I_i conjugate to the extensive quantity E_i. Gradients of intensive quantities are thermodynamic forces are $\mathbf{F}_i = -\nabla I_i$, and they cause fluxes J_i of the extensive quantities satisfying continuity equations $\frac{\partial E_i}{\partial t} + \operatorname{div} \mathbf{J}_i = 0$. The fluxes are proportional to the thermodynamic forces: $\mathbf{J}_i = \sum_{j=1}^{n} L_{ij} \mathbf{F}_j$ by a matrix $\{L_{ij}\}$. Thus,

$$\frac{\partial E_i}{\partial t} = \operatorname{div} \sum_{j=1}^{n} L_{ij} \nabla I_j.$$

Introducing a susceptibility matrix $\sigma_{ij} = \frac{\partial E_i}{\partial I_j}$, we get

$$\sum_{j=1}^{n} \sigma_{ij} \frac{\partial I_j}{\partial t} = \operatorname{div} \sum_{j=1}^{n} L_{ij} \nabla I_j.$$

4.1.3 Closing relations

For more complicated case with several irreversible processes we will use concepts of *generalized thermodynamic forces* X_i and *general thermodynamic fluxes* I_i. The tensor's type of these values can be different for various processes: scalar, vector or second order tensor.

Generally, each force can be considered as a source for each flux and therefore the following relation can be written: $I_j = I_j(X_1, \ldots, X_n)$. At equilibrium state all forces are zero: $X_i = 0$, and all fluxes are also zero: $I_j = 0$. We will consider small displacement from local equilibrium point and write

$$I_j = I_j(0, \ldots, 0) + \sum_{k=1}^{n} \frac{\partial I_j}{\partial X_k} dX_k = \sum_{k=1}^{n} L_{jk} X_k. \tag{4.5}$$

Onsager's theory assumes the following relation between generalized thermodynamic forces and generalized thermodynamic fluxes:

$$\sigma = \sum_{k=1}^{n} X_k I_k, \quad L_{jk} = L_{kj}. \tag{4.6}$$

Using this theory and (1.322) for dissipation σ, we obtain the relation between X_i and I_i:

$$\begin{aligned}
-T^{-2}\nabla T &= \omega = \omega^i \mathbf{e}^i \Leftrightarrow \mathbf{J}_q, \\
\mathbf{F}_k/T - \nabla\left(\mu_k/T\right) &= \omega_k = \omega_k^i \mathbf{e}^i \Leftrightarrow \mathbf{J}_k, \\
-\sum_{k=1}^{N} \left(\mu_k M_k \nu_{k\alpha}/T\right) &= \gamma_\alpha \Leftrightarrow \kappa_\alpha, \\
e^{ij} &\Leftrightarrow \tau^{ij}/T.
\end{aligned} \tag{4.7}$$

Note that σ is scalar and therefore its arguments can be only scalars P, T, c_k, γ_α and invariant combinations built on the basis of vectors and tensors ω^i, ω_i^k, e_{ij}, g_{ij}. Using (4.5)–(4.6), write the formula for dissipation σ in view of quadratic form of generalized thermodynamic forces X_i

$$\sigma = \sum_{j,k=1}^{n} L_{jk} X_j X_k,$$

where the matrix elements L_{jk} should consist of scalars and tensors components. In general case this quadratic form may be written as follows:

$$\begin{aligned}
\sigma &= a_0 T^2 g_{ij} \omega^i \omega^j + 2a^k g_{ij} \omega^i \omega_k^j + a^{kl} g_{ij} \omega_k^i \omega_l^j + b^{sm} T \gamma_s \gamma_m \\
&+ 2b^s \gamma_s g^{ij} e_{ij} + (\lambda/T)(g^{ij} e_{ij})^2 + 2(\mu/T) g^{ij} g^{mn} e_{ij} e_{mn}.
\end{aligned} \tag{4.8}$$

Hence, from (4.8) for full description of dissipative function σ the following coefficients a_0, a^k, $a^{kl} = a^{lk}$, $b^{sm} = b^{ms}$, b^s, λ, μ have to be known. In accordance with Onsager's theory, $I_j = \frac{1}{2}\frac{\partial \sigma}{\partial X_j}$. Considering links (4.7) between fluxes and forces we obtain:

$$(J_q)_i = \frac{1}{2}\frac{\partial \sigma}{\partial \omega^i} = a_0 T^2 \omega_i + \sum_{k=1}^{N} a^k (\omega_k)_i \quad \text{for heat fluxes } \mathbf{J}_q,$$

or in vector form

$$\mathbf{J}_q = -a_0 \nabla T + \sum_{k=1}^{N} a^k \left[\mathbf{F}_k/T - \nabla\left(\mu_k/T \right) \right], \tag{4.9}$$

$$\mathbf{J}_k \Rightarrow (J_k)_i = \frac{1}{2}\frac{\partial \sigma}{\partial \omega_k^i} = a^k \omega_i + \sum_{l=1}^{N} a^{kl}(\omega_l)_i \quad \text{for diffusion fluxes } \mathbf{J}_k,$$

or in vector form

$$\mathbf{J}_q = -(a_k/T^2)\nabla T + \sum_{k=1}^{N} a^{kl} \left[\mathbf{F}_l/T - \nabla\left(\mu_l/T \right) \right],$$

$$\kappa_\alpha = \frac{1}{2}\frac{\partial \sigma}{\partial \gamma_\alpha} = b^\alpha \text{div } \mathbf{v} - \sum_{s=1}^{r}\sum_{k=1}^{N} b^{s\alpha} \mu_k M_k \nu_{ks} \quad \text{for values } \kappa_\alpha,$$

$$\tau^{ij} = \frac{1}{2}T\frac{\partial \sigma}{\partial e_{ij}} = -g^{ij}\sum_{s=1}^{r}\sum_{k=1}^{N} b^s \mu_k M_k \nu_{ks} + \lambda g^{ij}\text{div } \mathbf{v} + 2\mu e^{ij} \quad \text{for } \tau^{ij}. \tag{4.10}$$

In fact the links (4.9)–(4.10) are the closing relations for our governing equations for multi-component chemical nonequilibrium gas mixture. The transport coefficients in these relations are determined either from experimental data or from molecular theory of gas flow as functions of gas mixture temperature, pressure, gas-species molecular mass and some additional parameters.

4.2 Governing equations for hypersonic viscous gas flow

4.2.1 Conditions on the surface of discontinuity

Consider the situation when there exist discontinuous surfaces in our flow domain.

Definition 64 Surfaces at which unknown functions are continuous while only several derivatives with respect to space coordinates and time are discontinuous are called *weak discontinuous surfaces*; when unknown functions themselves are discontinuous these surfaces are called *strong discontinuous surfaces*.

Consider the moving surface S satisfied to equation $f(x^1, x^2, x^3, t) = 0$. The vector $\mathbf{D}$ of the displacement velocity in space of each point of the surface S can be determined under the following formula:

$$\mathbf{D} = -\frac{\partial f/\partial t}{|\nabla f|}\mathbf{n}.$$

In relation to a certain part of an isolated discontinuity surface S consider the closed surface Σ as the boundary of a volume obtained by the way shown in Fig. 4.1.

For each point of S draw normal and mark off along it the segments of the length $h/2$ on the both sides of S, where h is a very small constant length. All points bounded by the surface Σ form the corresponding fixed volume V. Together with the fixed volume V there

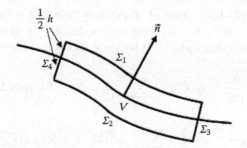

FIGURE 4.1: The element of liquid volume.

should be considered the moving volume V^* which coincides with V for given time t. As it follows from integral formulae (considered in Section 1.7.4) for any integrable function $A(x^1, x^2, x^3, t)$ it can be written expression below,

$$\frac{d}{dt}\int_{V^*} A d\tau = \frac{d}{dt}\int_V A d\tau + \int_\Sigma A \cdot (\mathbf{v}_n - \mathbf{D}_n)\, d\sigma. \qquad (4.11)$$

For fixed volume V we have

$$I = \frac{d}{dt}\int_V A d\tau = h\frac{d}{dt}\int_S A^* d\tau,$$

where A^* is the mean value of A over corresponding segment of length h orthogonal to the surface S. It is obviously that if A is finite and continuous function together with its derivatives on the both sides of the surface S, the value I tends to zero as $h \to 0$. If we consider infinitesimal element of S then

$$\lim_{h\to 0,\ \Delta\sigma\to 0} \frac{1}{\Delta\sigma}\frac{d}{dt}\int_V A d\tau = 0. \qquad (4.12)$$

Denote the parameters on the both sides of S by subscripts "1" and "2," accordingly, and assume the direction of normal to the surface S be corresponded to transition from Side 2 to Side 1. Considering (4.12) one obtains from (4.11) the following kinematic relation:

$$\lim_{h\to 0,\ \Delta\sigma\to 0} \frac{1}{\Delta\sigma}\frac{d}{dt}\int_{V^*} A d\tau = A_1(v_{n1} - D) - A_2(v_{n2} - D).$$

Applying this expression to integral forms of governing equations helps to transform them as
1. Continuity equation (1.298) for the species:

$$\frac{d}{dt}\int_{V^*} \rho_i d\tau - \int_{V^*} \dot{w}_i d\tau = 0. \qquad (4.13)$$

2. Continuity equation (1.301) for gas mixture as whole:

$$\frac{d}{dt}\int_{V^*} \rho d\tau = 0.$$

3. Momentum equation (1.305):

$$\frac{d}{dt}\int_{V^*} \rho \mathbf{v} d\tau - \int_{V^*} \rho \mathbf{F} d\tau - \int_\Sigma \mathbf{P}_n\, d\sigma = 0.$$

4. Energy equation (1.320) for the total energy per unit mass:

$$\frac{d}{dt}\int_{V^*} \rho(U + \tfrac{1}{2}v^2)d\tau - \int_{V^*} \rho\mathbf{F}\cdot\mathbf{v}d\tau - \sum_{k=1}^N \int_{V^*} \mathbf{F}_k\cdot\mathbf{J}_k d\tau - \int_\Sigma \mathbf{P}_n\cdot\mathbf{v}\, d\sigma + \int_\Sigma \mathbf{J}_q\cdot\mathbf{n}\, d\sigma = 0.$$

$$(4.14)$$

Accept some properties of behavior of the volume integrals in formulae (4.13)–(4.14) and suppose that $\lim\limits_{h \to 0} \int_{V_*} \dot{w}_i d\tau = \int_S \dot{r}_i \, d\sigma$ and

$$\lim_{h \to 0} \int_{V_*} \rho \mathbf{F} d\tau = \int_S \mathbf{R} \, d\sigma, \quad \lim_{h \to 0} \int_{V_*} (\rho \mathbf{F} \cdot \mathbf{v} + \sum_{k=1}^{N} \mathbf{F}_k \cdot \mathbf{J}_k) d\tau = \int_S W \, d\sigma,$$

where $\dot{r}_i$, $\mathbf{R}$ and W are the surface density distributions, over S, of the mass of i-th species, forces and energy fluxes which are external to media. One can write the conditions on the surface of discontinuity for multicomponent nonequilibrium gas mixture:

– From continuity equations for species,

$$(\rho_i)_1 (v_{in} - D)_1 + \dot{r}_i = (\rho_i)_2 (v_{in} - D)_2, \tag{4.15}$$

or, introducing the vector of diffusion flux,

$$(\rho c_i)_1 (v_{in} - D)_1 + (J_{in})_1 + \dot{r}_i = (\rho c_i)_2 (v_{in} - D)_2 + (J_{in})_2, \tag{4.16}$$

– From continuity equation for gas mixture as whole,

$$\rho_1 (v_{n1} - D) = \rho_2 (v_{n2} - D),$$

– From momentum equation,

$$\mathbf{R} + \mathbf{P}_{n1} - \rho_1 \mathbf{v}_1 (v_{n1} - D) = \mathbf{P}_{n2} - \rho_2 \mathbf{v}_2 (v_{n2} - D),$$

– From equation for the total energy,

$$W + \mathbf{P}_{n1} \cdot \mathbf{v}_1 - \rho_1 (v_{n1} - D) \cdot (U + \tfrac{1}{2} v^2)_1 - (\mathbf{J}_{qn})_1$$
$$= \mathbf{P}_{n2} \cdot \mathbf{v}_2 - \rho_2 (v_{n2} - D) \cdot (U + \tfrac{1}{2} v^2)_2 - (\mathbf{J}_{qn})_2. \tag{4.17}$$

Under conditions $v_{n1} - D = 0$, $v_{n2} - D = 0$ the gas mixture particles do not pass from one side of discontinuity to another one and we have $v_{n1} = v_{n2}$.

Definition 65 This type of discontinuity is called *tangential discontinuity*.

The relations (4.15)–(4.17) at the discontinuity surface S are true in any coordinate system and for all points of surface discontinuity. Note that these conditions take into account molecular transfer effect near the surface of discontinuity. But if this effect is neglected the classical gasdynamics conditions at the discontinuity surface are true:

$$(\rho c_i)_1 (v_{in} - D)_1 = (\rho c_i)_2 (v_{in} - D)_2,$$
$$p_1 \mathbf{n} - \rho_1 \mathbf{v}_1 (v_{n1} - D) = p_2 \mathbf{n} - \rho_2 \mathbf{v}_2 (v_{n2} - D),$$
$$p_1 v_{n1} - \rho_1 (v_{n1} - D) \cdot (U + \tfrac{1}{2} v^2)_1 = p_2 v_{n2} - \rho_2 (v_{n2} - D) \cdot (U + \tfrac{1}{2} v^2)_2.$$

In the conclusion, we use the conditions above to justify the boundary conditions (1.332) for diffusion equations at the catalytic body surface (for nonpermeable surface) and (1.333) (for permeable surface). This surface can be considered as particular case of discontinuity surface and these boundary conditions can be obtained from (4.16) with following values of the parameters of discontinuity surface:

– For permeable surface, $v_{n1} = D = v_{n2} = 0$,
– For nonpermeable surface, $D = 0$, $(\rho v_n)_1 = (\rho v_n)_2 = (\rho v_n)_w$.

4.2.2   Governing parameters

Our main goal is to analyze the influence of the governing parameters for hypersonic viscous gas flow over blunted bodies with permeable surface on gasdynamic regime of the flow. For simplicity, this analysis will be done for steady three-dimensional viscous hypersonic flow of homogeneous gas <u>under zero mass forces</u>. As a basic methodology the asymptotic methods are used. In this case three-dimensional NSE have the form, see Section 1.8:

$$\nabla_i(\rho v^i) = 0, \quad \rho v^j \nabla_j v^i = -g^{ij}\nabla_j P + \nabla_j \tau^{ij},$$
$$\rho v^j \nabla_j h = v^j \nabla_j P - \nabla_j J_q^j + \Phi, \quad P = \rho RT, \tag{4.18}$$

where (Fig. 4.2)

$$\tau^{ij} = g^{ij}(\zeta - \tfrac{2}{3}\mu)\nabla_k v^k + 2\mu e^{ij},$$
$$e^{ij} = \tfrac{1}{2}(g^{ik}\nabla_k v^j + g^{jk}\nabla_k v^i),$$
$$J_q^j = -\lambda g^{kj}\nabla_k T, \quad h = c_p T, \quad \Phi = 2\mu e^{ij}e_{ij} + (\zeta - \tfrac{2}{3}\mu)(\nabla_k v^k)^2.$$

The boundary conditions at the body surface for (4.18) are

$$\rho \mathbf{v}_w \cdot \mathbf{n}_w = G, \quad \mathbf{v}_w - v_{nw}\mathbf{n}_w = \mathbf{R}, \quad T = T_w.$$

The boundary conditions at infinity are the classical ones:

$$\mathbf{v}|_{\mathbf{r}\to\infty} = \mathbf{V}_\infty, \quad P|_{\mathbf{r}\to\infty} = P_\infty, \quad T|_{\mathbf{r}\to\infty} = T_\infty.$$

Further we assume to have a deal with perfect gas with constant c_p and c_V, Prandtl number $\mathrm{Pr} = \lambda/(\mu c_p) = \mathrm{const}$, and $\mu = \mu(T) = T^\omega$. For given boundary conditions and given flow geometry the dimensionless governing parameters of the problem are:

$$M_\infty = \frac{V_\infty}{\sqrt{\gamma P_\infty/\rho_\infty}}, \quad Re_\infty = \frac{\rho_\infty V_\infty L}{\mu_\infty}, \quad g = \frac{G^*}{\rho_\infty V_\infty},$$
$$\theta = \frac{T_w^*}{T_\infty(\gamma - 1)M_\infty^2}, \quad \gamma = \frac{c_p}{c_V}, \quad \sigma = \sum_{k=1}^n X_k I_k, \quad \omega = T^{-2}\nabla T. \tag{4.19}$$

The parameters at the body surface are marked by subscription (*). The parameters (4.19) may be divided on three groups.

– Parameters determined by flow conditions at infinity, i.e., Mach and Reynolds numbers M_∞, Re_∞;

– Parameters determined by physical-chemical properties of the gas mixture, i.e., γ, σ, ω.

For example, for nondissociated air $\gamma \approx 1.4$, for dissociated air $\gamma \approx 1.2$, and for ionized air $\gamma \approx 1.05$. Note that instead of parameters (4.19) it is more convenient to use their combinations

$$\varepsilon = \frac{\gamma - 1}{\gamma + 1}, \quad \delta = \frac{1}{(\gamma - 1)M_\infty^2}, \quad e = \frac{1}{Re_\infty\delta^\omega}, \quad \Lambda = \varepsilon g^2\theta, \quad \kappa = \max\{\varepsilon, \sqrt{\Lambda}\}.$$

To understand physical meaning of these parameters, remember from gasdynamics that for supersonic and hypersonic regimes the flow over blunted bodies is characterized by strong shock wave which is not attached to the body surface [9]. Near the stagnation point the free stream velocity vector is very close to normal direction with respect to the shock wave surface [31]. It means that for estimations of the flow parameters at this region of the shock layer it can be used the classical Rankine–Hugoniot conditions at the normal shock wave for inviscid gas. If the subscripts ∞ and s are refer to conditions upstream and downstream of the normal shock wave (Fig. 4.1), respectively, for the limit case of very high Mach numbers there can be written the following relations:

$$\frac{\rho_\infty}{\rho_s} = \frac{\gamma - 1}{\gamma + 1}, \quad \frac{V_\infty}{v_s} = \frac{\gamma + 1}{\gamma - 1}, \quad \frac{T_\infty}{T_s} = \frac{2}{(\gamma - 1)M_\infty^2}, \quad \frac{P_\infty}{P_s} = \frac{2}{(\gamma + 1)M_\infty^2}, \quad \frac{\mu_\infty}{\mu_s} = \left(\frac{T_\infty}{T_s}\right)^\omega.$$
$$(4.20)$$

Considering the relations (4.20) one can speak about the following physical meaning of the governing parameters of the problem:

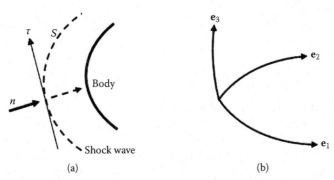

FIGURE 4.2: (a) Structure of the shock wave near a body surface. (b) Basis vectors.

– Parameter ε is equal to the ratio of gas densities before and after normal shock wave for very high values of Mach numbers,

$$\varepsilon = \frac{\gamma - 1}{\gamma + 1} = \frac{\rho_\infty}{\rho_s},$$

– Parameter δ is equal to the half of the ratio of gas temperatures before and after normal shock wave for very high values of Mach numbers,

$$\delta = \frac{1}{(\gamma - 1)M_\infty^2} = \frac{1}{2}(T_\infty/T_s),$$

– Parameter $1/e$ has the same order as Reynolds number Re_s which is calculated together with parameters before and after normal shock wave for very high values of Mach numbers,

$$Re_s = \frac{\rho_s v_s L}{\mu_s} = \frac{\rho_\infty v_\infty L}{\mu_\infty} \frac{\mu_\infty}{\mu_s} \approx Re_\infty \delta^\omega = \frac{1}{e},$$

– Parameter θ is equal to the half of the ratio of gas temperatures at the body surface and gas temperature after normal shock wave for very high values of Mach numbers,

$$\theta = \frac{T_w^*}{(\gamma - 1)M_\infty^2 T_\infty} = \frac{1}{2}(T_w^*/T_s),$$

– Parameter Λ has the same order as a ratio of mass impulse through permeable body surface and of mass impulse at infinity,

$$\Lambda = \frac{1}{2}\frac{\rho_\infty}{\rho_s}\frac{(\rho_w^* v_w^*)^2}{(\rho_\infty V_\infty)^2}\frac{T_w^*}{T_s} \approx \frac{\rho_w^*}{\rho_\infty}\frac{(v_w^*)^2}{(V_\infty)^2}\frac{\rho_w^* T_w^*}{\rho_s T_s} \approx \frac{\rho_w^*}{\rho_\infty}\frac{(v_w^*)^2}{(V_\infty)^2}\frac{P_w^*}{P_s} \approx \frac{\rho_w^*}{\rho_\infty}\frac{(v_w^*)^2}{(V_\infty)^2}.$$

Suppose the following assumptions for the parameters above:

$$\frac{1}{Re_\infty} \to 0, \quad \frac{1}{M_\infty} \to 0, \quad \varepsilon \to 0, \quad \delta \to 0, \quad e \to 0, \quad \Lambda \to 0,$$
$$g < 1, \quad \theta \leq \frac{1}{2}, \quad 0 < \omega \leq 1.0, \quad v_w/\sqrt{\gamma R T_w} < 1.0, \quad \frac{1}{2} < \sigma < \frac{3}{2}.$$

In order words, we assume that

    – Gas flow is hypersonic with high Reynolds number at infinity as well as downstream to the chock wave at the stagnation point,

    – Density and temperature at infinity are small by comparison with the corresponding parameters at the stagnation point,

    – Gas injection (gas suction) is subsonic with moderate mass flux through permeable surface,

    – Mass impulse through permeable surface is small by comparison with corresponding parameters at infinity,

    – Temperature at the body surface does not exceed the temperature after normal shock wave.

4.2.3   Transformation of initial equations

For asymptotic analysis of the problem is more convenient to use curvilinear coordinate system $\{x^i\}$. Let coordinate x^3 be directed along outward normal to the body surface and therefore the surfaces $x^3 = $ const are parallel to the body surface. Coordinates x^1 and x^2 should be selected on the surface with origin at the stagnation point (Fig. 4.3). Then we assume additionally that coordinate system (x^1, x^2) is nondegenerate coordinate system. From differential geometry it is known that for coordinate system of this type the metric of

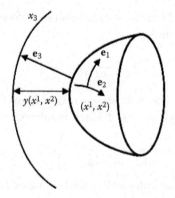

FIGURE 4.3: Coordinates (x_1, x_2) on the surface with the origin at the stagnation point.

the space can be written as follows, $ds^2 = g_{ij}dx^i dx^j$, where

$$g_{\alpha\beta} = g_{\beta\alpha} = a_{\alpha\beta} - 2b_{\alpha\beta}x^3 + a^{\gamma\lambda}b_{\gamma\alpha}b_{\lambda\beta}(x^3)^2, \quad g_{3\alpha} = g_{\alpha3}, \quad g_{33} = 1,$$

and indices $\alpha, \beta, \gamma, \lambda = 1, 2$ are associated with $2d$ space $\mathbb{R}^2$, indices $i, j, k = 1, 2, 3$ are associated with $3d$ space $\mathbb{R}^3$, values $a_{\alpha\beta}, b_{\alpha\beta}$ – covariant components of the first and the second basic quadratic form of the body surface which are known functions of the body surface coordinates x^1 and x^2. Introduce also another curvilinear coordinate system (ξ^1, ξ^2, ζ):

$$\xi^1 = x^1, \quad \xi^2 = x^2, \quad \zeta = x^3 - y(x^1, x^2), \tag{4.21}$$

where $y(x^1, x^2)$ is known function which can be considered as the distance between the surface $x^3 = y(x^1, x^2)$ and the body surface. For asymptotic analysis it is necessary to use dimensionless form of the governing equations. To do it we introduce the following dimensionless variables:

$$\rho_0 = \rho/\rho_\infty, \quad P_0 = P/P_\infty, \quad T_0 = T/T_\infty, \quad \mu_0 = \mu/\mu_\infty, \quad u_0^i = (v^i/V_\infty)\sqrt{g_{(ii)}}, \quad ds_0 = ds/L,$$

where u_0^i are the physical components of the velocity vector. Using these designations the initial equations (4.18) may be written in coordinate system (4.21) in the following dimensionless form (the subscript "0" is omitted), [8]:

$$\frac{\partial}{\partial \xi^\alpha}\left(\rho u^\alpha \sqrt{g/g_{(\alpha\alpha)}}\right) + \frac{\partial}{\partial \zeta}(\rho V \sqrt{g}) = 0, \quad g = \det g_{\alpha\beta},$$

$$\rho D u^\alpha + \rho A_{jk}^\alpha u^j u^k = -\frac{2\varepsilon\delta g^{\alpha\beta}}{1+\varepsilon}\sqrt{g_{(\alpha\alpha)}}\left(\frac{\partial P}{\partial \xi^\beta} - \frac{\partial P}{\partial \zeta}\frac{\partial y}{\partial \xi^\beta}\right)$$
$$+\frac{1}{Re_\infty}\left[\frac{\partial}{\partial \zeta}\left(\mu\frac{\partial u^\alpha}{\partial \zeta}\right)\right] - \frac{1}{3}\sqrt{g_{(\alpha\alpha)}}g^{\alpha\beta}\frac{\partial}{\partial \zeta}\left(\mu\frac{\partial u^3}{\partial \zeta}\right)\frac{\partial y}{\partial \xi^\beta} + \cdots$$

$$\rho D u^3 + \rho A_{jk}^3 u^j u^k = -\frac{2\varepsilon\delta}{1+\varepsilon}\frac{\partial P}{\partial \zeta} + \frac{1}{Re_\infty}\left[\frac{4}{3}\frac{\partial}{\partial \zeta}\left(\mu\frac{\partial u^3}{\partial \zeta}\right) + \cdots\right],$$

$$\rho D T = \frac{2\varepsilon}{1+\varepsilon}DP + \frac{1}{Re_\infty}\frac{\partial}{\partial \zeta}\left(\frac{\mu}{\sigma}\frac{\partial T}{\partial \zeta}\right) + \frac{\mu}{\delta Re_\infty}\left[\frac{g_{\alpha\beta}}{\sqrt{g_{(\alpha\alpha)}g_{(\beta\beta)}}}\frac{\partial u^\alpha}{\partial \zeta}\frac{\partial u^\beta}{\partial \zeta} + \frac{4}{3}\left(\frac{\partial u^3}{\partial \zeta}\right)^2\right] + \cdots$$

$$P = \rho T, \quad \mu = T^\omega, \quad D = \frac{u^\alpha}{\sqrt{g_{(\alpha\alpha)}}}\frac{\partial}{\partial \xi^\alpha} + V\frac{\partial}{\partial \zeta}, \quad V = u^3 - \frac{u^\alpha}{\sqrt{g_{(\alpha\alpha)}}}\frac{\partial}{\partial \xi^\alpha},$$

$$A_{jk}^i = \frac{1}{2}\left(g_{(jj)}g_{(kk)}\right)^{-1/2}\left(2\Gamma_{jk}^i - \delta_j^i\frac{\partial \sqrt{g_{(jj)}}}{\partial x^k} - \delta_k^i\frac{\partial \sqrt{g_{(kk)}}}{\partial x^j}\right). \tag{4.22}$$

Here by dots are designated omitted terms which have small orders (in asymptotic sense and which will not be used during our analysis. Also, summation is not carried out over the indices in parentheses.

For example, we present the transformation of the continuity equation, see Section 1.7.2,

$$\nabla_i(\rho v^i) \equiv \frac{\partial}{\partial x^i}(\rho v^i) + \rho v^k \sum_{i=1}^3 \Gamma_{ik}^i.$$

Using Γ_{jk}^j, see Section 1.7.3, we get $\sum_{i=1}^3 \Gamma_{ik}^i = \frac{1}{2g}\frac{\partial g}{\partial x^k} = \frac{1}{\sqrt{g}}\frac{\partial \sqrt{g}}{\partial x^k}$, and therefore

$$\sqrt{g}\,\nabla_i(\rho v^i) \equiv \sqrt{g}\frac{\partial}{\partial x^i}(\rho u^i/\sqrt{g_{(ii)}}) + \rho(u^i/\sqrt{g_{(ii)}})\frac{\partial \sqrt{g}}{\partial x^i} = \frac{\partial}{\partial x^i}(\rho u^i\sqrt{g/g_{(ii)}}).$$

To transfer this equation to coordinates (ξ^1, ξ^2, ζ) we should take into account

$$\frac{\partial}{\partial x^1} = \frac{\partial}{\partial \xi^1} - \frac{\partial y}{\partial \xi^1}\frac{\partial}{\partial \zeta}, \quad \frac{\partial}{\partial x^2} = \frac{\partial}{\partial \xi^2} - \frac{\partial y}{\partial \xi^2}\frac{\partial}{\partial \zeta}, \quad \frac{\partial}{\partial x^3} = \frac{\partial}{\partial \zeta}.$$

As the result we obtain the first equation in the system (4.22).

Note that (4.22) have mixed nature. On one hand, these equations are written in the coordinate system (ξ^1, ξ^2, ζ), but on the other hand, the known functions (u^1, u^2, u^3) in this system are physical components of the velocity vector in the coordinate system (x^1, x^2, x^3). Of course, if $y(x^1, x^2) \equiv 0$ these coordinate systems coincide but for opposite case it is not true. What is more, together with components u^i the function V (which is the physical component in ζ-direction of the velocity vector velocity in the coordinate system (ξ^1, ξ^2, ζ)) is also used. We will use these properties of the system of initial equations in the further analysis. The relations for coefficients A_{ij}^k are

$$A_{\alpha\alpha}^\alpha = \frac{g_{12}}{g}\left[\frac{\partial \sqrt{g_{(\alpha\alpha)}}}{\partial x^\beta} + \frac{g_{12}}{g_{(\alpha\alpha)}}\frac{\partial \sqrt{g_{(\alpha\alpha)}}}{\partial x^\alpha} - \frac{1}{\sqrt{g_{(\alpha\alpha)}}}\frac{\partial g_{12}}{\partial x^\alpha}\right], \quad g = g_{11}g_{22} - g_{12}^2,$$

$$A_{12}^\alpha = A_{21}^\alpha = \frac{1}{2g}\left[\sqrt{g_{11}g_{22}}\left(1 + \frac{g_{12}^2}{g_{11}g_{22}}\right)\frac{\partial \sqrt{g_{(\alpha\alpha)}}}{\partial x^\beta} - 2g_{12}\frac{\partial \sqrt{g_{(\beta\beta)}}}{\partial x^\alpha}\right],$$

$$A_{\alpha\alpha}^\beta = \frac{\sqrt{g_{(\beta\beta)}}}{g}\left[\frac{\partial g_{12}}{\partial x^\alpha} - \sqrt{g_{(\alpha\alpha)}}\frac{\partial \sqrt{g_{(\alpha\alpha)}}}{\partial x^\beta} - \frac{g_{12}}{\sqrt{g_{(\alpha\alpha)}}}\frac{\partial \sqrt{g_{(\alpha\alpha)}}}{\partial x^\alpha}\right],$$

$$A_{3\alpha}^\beta = A_{\alpha3}^\beta = \frac{1}{2}\sqrt{\frac{g_{(\beta\beta)}}{g_{(\alpha\alpha)}}}\left[g_{(\alpha\alpha)}\frac{\partial g_{12}}{\partial x^3} - g_{12}\frac{\partial g_{(\alpha\alpha)}}{\partial x^3}\right], \tag{4.23}$$

$$A_{3\alpha}^\alpha = A_{\alpha3}^\alpha = \frac{1}{2g}\left[\frac{1}{2}\left(g_{(\beta\beta)} + \frac{g_{12}^2}{g_{(\alpha\alpha)}}\right)\frac{\partial g_{(\alpha\alpha)}}{\partial x^3} - g_{12}\frac{\partial g_{12}}{\partial x^3}\right],$$

$$A_{\alpha\beta}^3 = -\frac{1}{2\sqrt{g_{(\alpha\alpha)}g_{(\beta\beta)}}}\frac{\partial g_{\alpha\beta}}{\partial x^3}, \quad A_{3\alpha}^3 = A_{\alpha3}^3 = A_{33}^\alpha = A_{33}^3 = 0.$$

It is assumed in all formulae (4.23) that $\alpha \neq \beta$. For orthogonal coordinates (with $g_{12} = 0$) the relations (4.23) are simplified:

$$A_{\alpha\alpha}^\alpha = 0, \quad 2A_{12}^\alpha = 2A_{21}^\alpha = -A_{\alpha\alpha}^\beta = \frac{1}{\sqrt{g_{11}g_{22}}} \frac{\partial \sqrt{g_{(\alpha\alpha)}}}{\partial x^\beta}, \quad A_{3\alpha}^\beta = A_{\alpha3}^\beta = A_{\alpha\beta}^3 = 0.$$

The boundary conditions are the following:

$$x^3 \to \infty : \mathbf{u} \to \mathbf{V}_\infty, \quad P = T = 1;$$
$$x^3 = 0 : u^\alpha = u_w^\alpha, \quad pu^3 = G, \quad T = T_w.$$

4.3 Flow regimes for hypersonic viscous gas flow

In accordance with relations between governing parameters of the problem the different gasdynamic regimes are realized for hypersonic viscous gas flow over *blunted bodies with permeable surface*. We will consider six partial regimes [9, 25, 29]. The boundary layers/sublayers corresponding to these regimes were clearly observed in experiments [18, 21, 24].

4.3.1 Viscous shock layer

Swirling hypersonic flow in various formulations have been investigated in a number of papers (most of them were reviewed in [18]). An inviscid formulation of the problem was considered in [36], then a study was carried out in the framework of boundary layer model in an approximation of the parabolized Navier–Stokes equations a solution of the equations of a viscous shock layer in the neighborhood of a stagnation point was obtained. At the same time, the body surface was assumed impermeable, while the solutions are restricted to the case of weak injection. We consider an axisymmetric swirling flow of homogeneous compressible gas in a hypersonic viscous shock layer near permeable surface.

This regime is realized if the following conditions are satisfied:

$$Re_\infty \delta^\omega \varepsilon = O(1) \quad (0 \leq g^2 \leq \varepsilon^{1/2}\theta^{\omega - 1/2}).$$

In the present section we will use the following dimensionless parameters (see Section 4.2.2):

$$\delta = \frac{1}{(\gamma - 1)M_\infty^2}, \quad \theta = \frac{T_w^*}{(\gamma - 1)M_\infty^2 T_\infty} = \frac{1}{2}(T_w^*/T_s), \quad g = \frac{\rho v_n}{\rho_\infty V_\infty}, \quad \Lambda = \varepsilon g^2 \theta.$$

For this regime the flow can be divided into four sublayers. Three sublayers are associated with the shock wave structure, with the following orders of thickness (Fig. 4.4):

- Outer layer of the shock wave – $O(1/Re_\infty)$,
- Middle layer of the shock wave – $O(1/(Re_\infty \delta^\omega))$,
- Inner layer of the shock wave – $O(\varepsilon/(Re_\infty \delta^\omega))$.

The last sublayer between the shock wave and the body surface is called a *viscous shock layer*. The order of thickness of viscous shock layer is equal to $O(\varepsilon)$.

4.3.2 Vortex intersection, nonstrong and strong injection

Vortex intersection and nonstrong injection (suction). Knowledge on transitional allows at elevated Mach numbers is very limited due to the immense difficulty in conducting experiments – be it wind-tunnel or free-flight. Therefore, direct numerical simulation

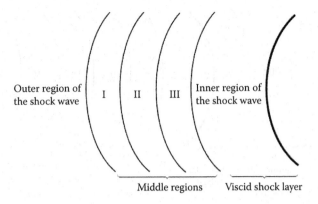

FIGURE 4.4: Viscous shock layer structure.

provides a very powerful tool to gain significant insight into these high-temperature flows. These high-temperature hypersonic flows become chemically reacting creating additional challenges for the modeling of chemical reactions and the thermodynamic properties of such flows. The simulation of laminar-turbulent transition in boundary-layer flows for entry scenarios can deliver estimates of flight relevant physical properties such as drag and heat transfer important for the flight path design and the design of the heat shield of an entry vehicle, respectively. Disturbances should be introduced at the wall through a disturbance strip. Blowing and *nonstrong injection* (suction) is applied simultaneously to ensure that zero net mass is introduced at any one time step.

This regime is realized if the following conditions are satisfied:

$$Re_\infty \delta^\omega \varepsilon^{5/2} = O(1) \quad (0 \le g^2 \le \varepsilon^2 \theta^{\omega - 1/2}).$$

For this regime the flow can be divided into five sublayers, Fig. 4.5(a). Three of them are associated with the structure of the shock wave and have the same orders of thickness as inviscid shock layer regime. The fourth layer called inviscid shock layer is adjacent to the shock wave. The order of thickness of viscous shock layer is $O(\varepsilon)$. The last fifth sublayer, called *inviscid boundary layer*, is arranged near the body surface; its order of thickness is $O(\varepsilon^{3/2})$.

Vortex intersection and strong injection. This regime is realized if the following holds:

$$Re_\infty \delta^\omega \varepsilon^{5/2} = O(1) \quad ((\varepsilon\theta)^{-1} \gg g^2 \gg \varepsilon^2 \theta^{\omega - 1/2}).$$

For this regime the flow can be divided into six sublayers (Fig. 4.5(b)). Three of them are associated with the structure of the shock wave and have the same orders of thickness as inviscid shock layer regime. The fourth layer, called *inviscid shock layer*, is adjacent to the shock wave. The order of thickness of inviscid shock layer – $O(\varepsilon)$. The sixth sublayer called *injection gas layer* is arranged near the body surface; its order of thickness – $O(\Lambda^{1/2})$. The fifth sublayer, viscous mixing layer, is arranged between viscous shock and injection gas layers; its order of thickness – $O(\varepsilon^{3/2})$.

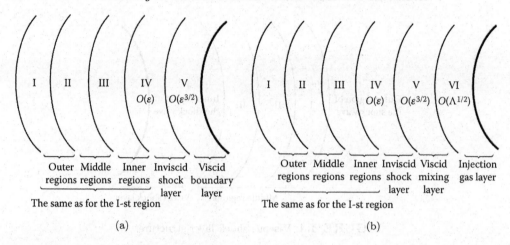

Outer Middle Inner Inviscid Viscid
regions regions regions shock boundary
layer layer

The same as for the I-st region

Outer Middle Inner Inviscid Viscid Injection
regions regions regions shock mixing gas layer
layer layer

The same as for the I-st region

(a)   (b)

FIGURE 4.5: Vortex intersection and (a) nonstrong injection structure, (b) strong injection structure.

4.3.3   Boundary layer, nonstrong and strong injection

Boundary layer and nonstrong injection (suction). This regime is realized if the following conditions are satisfied:

$$Re_\infty \delta^\omega \varepsilon^n = O(1) \quad (n > 5/2, \ 0 \le g^2 \le \frac{\theta^{\omega-1/2}}{\varepsilon^{1/2} Re_\infty \delta^\omega}).$$

For this regime the flow can be also divided into six sublayers (Fig. 4.6). Three of them are associated with the structure of the shock wave and have the same orders of thickness as inviscid shock layer regime. The fourth layer called inviscid shock layer is adjacent to the shock wave. The order of thickness of inviscid shock layer – $O(\varepsilon)$. Then inviscid boundary layer is arranged; its order of thickness is $O(\varepsilon^{3/2})$. The last sixth sublayer, called *viscous boundary layer*, is arranged near the body surface; its order of thickness is $O(\varepsilon^{1/4}/(Re_\infty \delta^\omega)^{1/2})$.

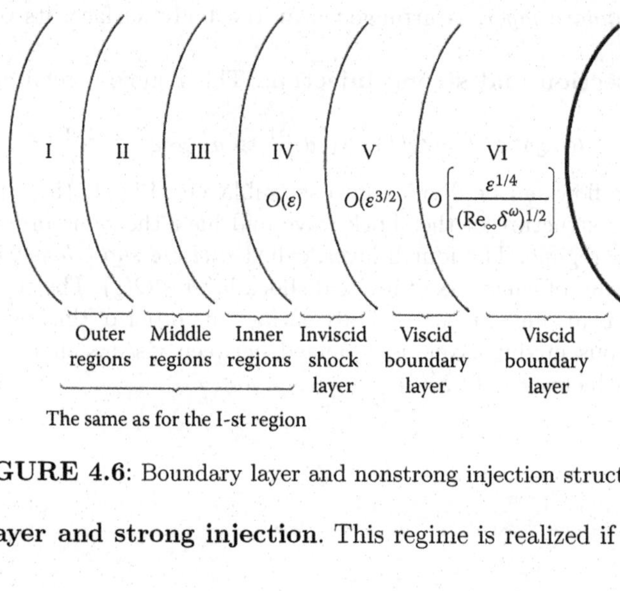

Outer Middle Inner Inviscid Viscid Viscid
regions regions regions shock boundary boundary
layer layer layer

The same as for the I-st region

FIGURE 4.6: Boundary layer and nonstrong injection structure.

Boundary layer and strong injection. This regime is realized if the following conditions are satisfied:

$$Re_\infty \delta^\omega \varepsilon^n = O(1) \quad (n > 5/2), \quad (\varepsilon\theta)^{-1} \gg g^2 \gg \frac{\theta^{\omega-1/2}}{\varepsilon^{1/2} Re_\infty \delta^\omega}.$$

For this regime the flow can be also divided into seven sublayers (Fig. 4.7). Three of them are associated with the structure of the shock wave and have the same orders of thickness as inviscid shock layer regime. The fourth layer, called *inviscid shock layer*, is adjacent to the shock wave; its order of thickness is $O(\varepsilon)$. Then inviscid boundary layer is arranged; its order of thickness – $O(\varepsilon^{3/2})$. The seventh sublayer, called *injection layer*, is arranged near the body surface; its order of thickness – $O(\Lambda^{1/2})$. The sixth sublayer, called *viscous mixed layer*, is arranged between viscous boundary and injection gas layers; its order of thickness – $O(\varepsilon^{1/4}/(Re_\infty \delta^\omega)^{1/2})$.

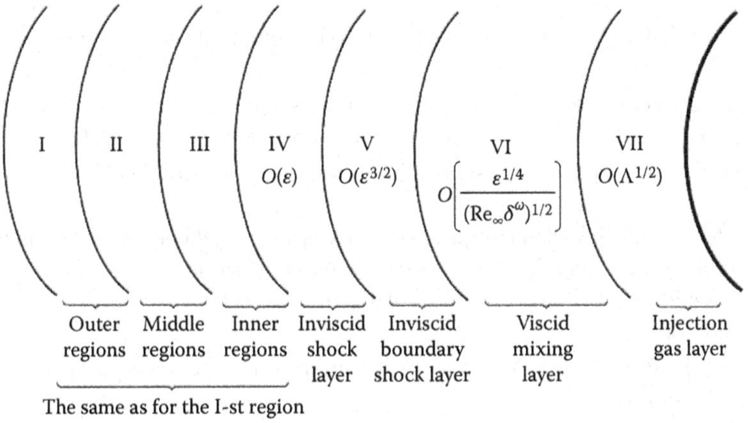

FIGURE 4.7: Boundary layer and strong injection structure

4.3.4   Boundary layer and strong struction

This regime is realized if the following conditions are satisfied:

$$Re_\infty \delta^\omega \varepsilon^n = O(1) \quad (n > 5/2), \quad \varepsilon^2 \geq g^2 \gg \frac{1}{\varepsilon^{1/2} Re_\infty \delta^\omega}.$$

For this regime the flow can be also divided into six sublayers (Fig. 4.8). Three of them are associated with the structure of the shock wave and have the same orders of thickness as inviscid shock layer regime. The fourth layer, called *inviscid shock layer*, is adjacent to the shock wave; its order of thickness – $O(\varepsilon)$. Then inviscid boundary layer is arranged; its order of thickness – $O(\varepsilon^{3/2})$. The last, sixth sublayer, *viscous boundary layer*, is arranged to the body surface; its order of thickness – $O(1/(Re_\infty \delta^\omega)^{1/2})$.

We give some remarks concerning this classification:

1. Transfer from viscous shock layer regime to viscous boundary layer regime via intermediate vortex intersection regime can be obtained under increasing Reynolds number Re_∞. It means that for the body moving along re-entry trajectory the flow regime is changed from viscous shock layer (at altitude 90 – 100 km) to boundary layer regime (at altitude 50 – 55 km).

2. The order of thickness for the shock wave sublayers, for viscous shock layer, for injection gas layer and for inviscid boundary layer are the same for all considered regimes.

3. For all regimes with injection we assume the parameter $\Lambda \to 0$. Because the influence of molecular transfer effects on the flow depending on local Reynolds number, the difference

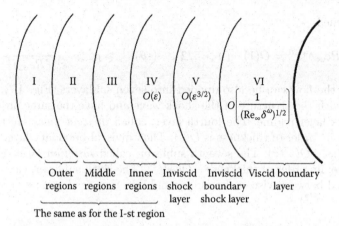

FIGURE 4.8: Boundary layer and strong suction structure.

between strong injection and nonstrong injection is completely determined by Reynolds number calculated for the conditions at the body surface $Re_w = \rho_w^* v_w^* L / \mu_w^*$.

For high values $Re_w \gg 1$ we get strong injection; for opposite case – nonstrong injection $Re_w = O(1)$.

Of course, for strong injection regime it is necessary to check the existence of the governing parameters ranges which have to satisfy the conditions $\Lambda \ll 1$, $Re_w \gg 1$. It will be done during consideration of the regimes with strong injection (regimes 3 and 5).

The main goal is:

• To present the governing equations for the main terms of asymptotic expansion of the solution for each sublayer;

• To obtain the boundary conditions for the listed above equations which have to satisfy matching conditions with the solution for the neighboring domain;

• To find, if it is possible, and to study the analytic solutions for associated boundary value problems.

4.4   Shock wave structure

The general question is: what does it mean – shock wave. This concept was introduced during theoretical investigations of Euler equations [2]. It was obtained that the qualitative character of solutions to Euler equations for gas flow over blunted bodies is completely determined by Mach number at infinity M_∞.

For the steady subsonic flow ($M_\infty < 1$) over blunted body the steady Euler equations are of elliptic type and the smooth solutions for associated boundary value problem (with boundary conditions at infinity and at the body surface) exists. For supersonic flow ($M_\infty > 1$) the steady Euler equations are hyperbolic and the smooth solution for the same boundary value problem does not exist. Detailed analysis demonstrates that for supersonic flow over blunted bodies the solution to associated boundary value problem can be obtained in more wide class of functions with one discontinuity. In fact, the surface of this discontinuity is the shock wave. So, in the framework of gasdynamic based on Euler equations the shock wave is discontinuity surface of the solution to associated boundary value problem. On the other hand, from physical point of view the existence of the discontinuity in the solution is the defect of the theory because in real life the discontinuity in gas flow cannot exist. It means that in real life the shock wave has a very thin layer with very high values of derivatives of flow parameters (pressure, velocity, density, temperature) with respect to normal direction. To understand how we can eliminate this drawback, recall that Euler equations are the limit form of NSE when $Re_\infty \to \infty$. The principal difference between these two systems of equations is that Euler equations are the first order PDE while NSE are the second order ones. It means that it is necessary to consider supersonic flow over blunted bodies in the framework of NSE if we want to find a solution without discontinuity.

Thus, the general analysis demonstrates that for asymptotic solution to our initial problem near shock wave we have to introduce new inner variables in this region. To do it we should take into account the follows principles:

1. The length scale L_n with respect to coordinate ζ (normal to shock wave surface direction) is very small by comparison with the length scale L_α with respect to coordinate ξ^α.

2. The order of the flow parameters in this region can be estimated from Rankine-Hugoniot conditions at the shock wave.

3. We must preserve in the asymptotic equations the second derivatives with respect to normal to shock wave surface direction which accounts molecular transfer effects across shock wave. To analyze the asymptotic solutions we will use the main results in [17].

4.4.1 Outer sublayer

As it was noted above the shock wave has a three-layer structure. It is rather nature to ask: why it is necessary to account three sublayers to obtain the asymptotic solution? The brief answer is following. As it was explained in the introduction the main goal asymptotic analysis (see Section 1.2) is to obtain the smooth transfer between upstream and downstream flow parameters at the discontinuity surface (shock wave). Our goal is not only *transfer* but also *smooth transfer*. That is main reason why it is necessary to consider three asymptotic sublayers. The main layer is the middle sublayer, which enable us to transfer from flow parameters with order at the upstream conditions to the parameters with order at the downstream conditions. The analysis demonstrates that this asymptotic solution for the middle sublayer has not direct smooth matching with upstream solution as well as with downstream one. To obtain these smooth matching we have to introduce two thin additional layers at the external boundaries of the middle layer of the shock wave. The first one – outer sublayer – permits one to find smooth matching with upstream flow parameters at the discontinuity surface. The second one – inner sublayer – enables to transfer smoothly from solution for the middle sublayer to the downstream flow parameters (after shock wave). Using previous reasoning we introduce the following new coordinates for the outer sublayer of the shock wave:

$$\xi^\alpha = \xi_0^\alpha, \quad \zeta = (Re_\infty)^{-1}\xi_0, \quad y = ky_0, \tag{4.24}$$

where subscript "0" is associated with flow parameters at outer sublayer. As for dependent variables we will seek the solution in the form [34]

$$u^\alpha = u_s^\alpha + \delta^{1/\sigma} + \ldots, \quad P = P_0 + \ldots, \quad T = T_0 + \ldots, \quad \rho = 1 + \delta^{\frac{3}{4\sigma}}\rho_0, \quad u^3 = u_s^3 + \delta^{\frac{3}{4\sigma}}v_0, \tag{4.25}$$

where index s corresponds to the values at infinity. Note that functions u_s^α, u_s^3 are introduced for the sake of convenience because in this case we have very simple zero boundary conditions for u_0^α, v_0, ρ at infinity. Another important question is: where we know from that the difference $u^\alpha - u_s^\alpha$ has order $\delta^{1/\sigma}$, or the difference $u^3 - u_s^3$ has order $\delta^{\frac{3}{4\sigma}}$? As it can be seen later these orders arise to match the solution for the outer part of the shock wave with the solution for the middle sublayer.

Substituting expansions (4.24)–(4.25) into initial equations (4.22), we obtain

$$\frac{\partial}{\partial \zeta_0}[\rho V \sqrt{g}] = 0, \quad \rho V = (1 + \delta^{\frac{3}{4\sigma}}\rho_0)(u_s^3 + \delta^{\frac{3}{4\sigma}}v_0),$$

$$\rho V \frac{\partial u_0^\alpha}{\partial \zeta_0} = \frac{\partial}{\partial \zeta_0}\left(\mu \frac{\partial u_0^\alpha}{\partial \zeta_0}\right) - \frac{1}{3}\kappa a^{\alpha\beta}\frac{\sqrt{a_{(\alpha\alpha)}}}{\delta^{\frac{1}{4\sigma}}}\frac{\partial}{\partial \zeta_0}\left(\mu \frac{\partial v_0}{\partial \zeta_0}\right)\frac{\partial y}{\partial \xi_0^\beta}, \qquad (4.26)$$

$$\rho V \frac{\partial v_0}{\partial \zeta_0} = \frac{4}{3}\frac{\partial}{\partial \zeta_0}\left(\mu \frac{\partial v_0}{\partial \zeta_0}\right), \quad \rho V \frac{\partial T_0}{\partial \zeta_0} = \frac{\partial}{\partial \zeta_0}\left(\frac{\mu}{\sigma}\frac{\partial T_0}{\partial \zeta_0}\right).$$

Equations (4.26) have first integrals. Considering boundary conditions at infinity after transformation from variables ζ_0 to v_0 these integrals as independent variables may be written as follows:

$$P_0 = T_0, \quad \rho_0 = -\frac{v_0}{u_s^3}, \quad \frac{\partial \zeta_0}{\partial v_0} = \frac{4}{3}\frac{T_0^\omega}{v_0 u_s^3},$$

$$\frac{\partial u_0^\alpha}{\partial v_0} = \frac{4}{3}\frac{u_0^\alpha}{v_0} + \frac{1}{3}\kappa a^{\alpha\beta}\frac{\sqrt{a_{(\alpha\alpha)}}}{\delta^{1/(4\sigma)}}\frac{\partial y}{\partial \xi_0^\beta}, \quad \frac{\partial}{\partial v_0}(T_0 - 1) = \frac{4}{3}\sigma\frac{T_0 - 1}{v_0}. \qquad (4.27)$$

Equations (4.27) can be integrated, the result is

$$P_0 = T_0 = 1 + T_{0s}u_0^{4\sigma/3}, \quad \zeta_0(\xi^1, \xi^2, v_0) = -\frac{4}{3u_s^3}\int\limits_{v_0}^{1}(T_0^\omega(s)/s)\,ds,$$

$$u_0^\alpha = u_{0s}^\alpha v_0^{4/3} - \kappa a^{\alpha\beta}\frac{\sqrt{a_{(\alpha\alpha)}}}{\delta^{\frac{1}{4\sigma}}}\frac{\partial y}{\partial \xi_0^\beta}, \quad \rho_0 = -\frac{v_0}{u_s^3}. \qquad (4.28)$$

So, we obtain analytic solution (4.28) described the flow-field in outer part of the shock wave. This solution depends on T_{0s} and u_{0s}^α which should be determined based on matching with solution for the middle part of the shock wave.

4.4.2   Middle sublayer

In the middle sublayer we have to modify our assumptions concerning orders of the variables. First of all it should be taken into account that the seeking solutions to NSE is the smooth transfer from parameters corresponding upstream and downstream conditions at discontinuity surface in the framework of solutions to Euler equations. The analysis of these (classical Rankine–Hugoniot) conditions at the shock wave gives useful estimations in the middle sublayer of the shock wave [35]: $T = O(1/\delta)$, $\rho = O(1)$, $P = O(1/\delta)$. In this connection the thickness order of the middle sublayer depending on local Reynolds number is also changed, i.e., to change the order of inner sublayer thickness, i.e., the order $O(1/(\delta^\omega Re_\infty))$ of middle sublayer thickness has to replace $O(1/Re_\infty)$ (for the outer sublayer). That is why it will be used the following variables for asymptotic expansion in the middle part of the shock wave:

$$\xi^\alpha = \xi_m^\alpha, \quad \zeta = \frac{\zeta_m}{Re_\infty \delta^\omega}, \quad y = ky_m \qquad (4.29)$$

where subscript m is associated with flow parameters at the middle sublayer of the shock wave.

As for independent variables it is convenient to seek the solutions in the form:

$$u^\alpha = u_m^\alpha + \ldots, \quad u^3 = v_m + \ldots, \quad \rho = \rho_m + \ldots, \quad P = \frac{1}{\delta}P_m + \ldots, \quad T = \frac{1}{\delta}T_m + \ldots, \quad \mu = \mu_m + \ldots \qquad (4.30)$$

Substituting expansions (4.29) and (4.30) into initial equations (4.22) gives

$$\frac{\partial}{\partial \zeta_m}[\rho_m v_m \sqrt{g}] = 0, \quad \rho_m v_m \frac{\partial u_m^\alpha}{\partial \zeta_m} = \frac{\partial}{\partial \zeta_m}\left(\mu_m \frac{\partial u_m^\alpha}{\partial \zeta_m}\right),$$

$$\rho_m v_m \frac{\partial T_m}{\partial \zeta_m} = \frac{\partial}{\partial \zeta_m}\left(\frac{\mu_m}{\sigma}\frac{\partial T_m}{\partial \zeta_m}\right) + \mu_m\left[B_{\alpha\beta}\frac{\partial u_m^\alpha}{\partial \zeta_m}\frac{\partial u_m^\alpha}{\partial \zeta_m} + \frac{4}{3}\left(\frac{\partial v_m}{\partial \zeta_m}\right)^2\right], \qquad (4.31)$$

$$\rho_m v_m \frac{\partial v_m}{\partial \zeta_m} = \frac{4}{3}\frac{\partial}{\partial \zeta_m}\left(\mu_m \frac{\partial v_m}{\partial \zeta_m}\right), \quad B_{\alpha\beta} = \frac{a_{\alpha\beta}}{\sqrt{a_{(\alpha\alpha)}a_{(\beta\beta)}}}.$$

Equations (4.31) have first integrals. Considering the boundary conditions at infinity after transform from variable ζ_m to $V_m = v_m - u_s^3$ these integrals as independent variables may be written as

$$P_m = \rho_m T_m, \quad \rho_m v_m = \text{const} = u_s^3, \quad \frac{\partial \zeta_m}{\partial V_m} = \frac{4}{3} \frac{T_m^\omega}{V_m u_s^3},$$

$$\frac{\partial^2 T_m}{\partial V_m^2} = \frac{4\sigma - 3}{3 V_m} \frac{\partial T_m}{\partial V_m} - \sigma \left(\frac{4}{3} + B_{\alpha\beta} \frac{\partial V_m^\alpha}{\partial V_m} \frac{\partial V_m^\beta}{\partial V_m} \right), \tag{4.32}$$

$$\frac{\partial V_m^\alpha}{\partial V_m} = \frac{4}{3} \frac{\partial V_m^\beta}{\partial V_m}, \quad V_m^\alpha = u_m^\alpha - u_s^\alpha, \quad V_m = v_m - u_s^3.$$

Equations (4.32) can be also integrated. One can show that temperature equation can be presented in the form

$$\frac{dz}{dy} = \alpha \frac{z}{y} + \beta + \gamma y^{2/3}, \quad z \equiv \frac{dT_m}{d\zeta_m}, \quad y \equiv \zeta_m.$$

Try to find the solution of the form $z = C(y)y^\alpha$ one obtains $\frac{dC}{dy} = \frac{\beta}{y^\alpha} + \frac{\gamma}{y^{2/3-\alpha}}$. The solution for the middle sublayer has a view

$$P_m = \rho_m T_m, \quad T_m = T_{ms} V_m^{4\sigma/3} - \frac{\sigma V_m^2}{3-2\sigma} - \frac{\sigma V_m^{8/3}}{4-2\sigma} B_{\alpha\beta} u_{ms}^\alpha u_{ms}^\beta,$$

$$\rho_m = \frac{u_s^3}{v_m}, \quad V_m^\alpha = u_{ms}^\alpha V_m^{4/3}, \quad \zeta_m(\xi^1, \xi^2, v_m) = \frac{4}{3u_s^3} \int\limits_0^{v_m} \frac{T_m^\omega dv_m}{v_m - u_s^3}. \tag{4.33}$$

Now it is necessary to match the solution (4.33) with solution (4.28) at outer part of the shock wave. To do it, take into account the following relations:

$$u^3 - u_s^3 = v_m - u_s^3 = V_m, \quad u^\alpha - u_s^\alpha = V_m^\alpha, \quad T = \frac{1}{\delta} T_m.$$

Because the inner sublayer is the last one in the shock wave structure, at this part we have to modify our expansions for u^3, ρ and P. Clear that if we want to save second order derivatives in the governing equations, we also have to change the order of inner sublayer thickness, i.e., to replace the order $O(\varepsilon/(\delta^\omega Re_\infty))$ of inner sublayer thickness instead of $O(1/(\delta^\omega Re_\infty))$ for middle sublayer. Thus, the independent variables at the inner part of shock wave has a view

$$\xi^\alpha = \xi_i^\alpha, \quad \zeta = \frac{\varepsilon}{Re_\infty \delta^\omega} \zeta_i, \quad y = k y_i, \tag{4.34}$$

where i is associated with flow parameters at the inner sublayer. As for dependent variables, we will find in the following form:

$$u^\alpha = u_{is}^\alpha(\xi^1, \xi^2) + k u_i + \dots, \quad u^3 = \varepsilon V_i + \frac{k u_{is}^\alpha}{\sqrt{a_{(\alpha\alpha)}}} \frac{\partial y_i}{\partial \xi_i^\alpha} + \dots, \quad \rho = \frac{1}{\varepsilon} + \dots,$$

$$P = \frac{1}{\varepsilon\delta} P_i + \dots, \quad T = \frac{1}{\delta} [T_{is}(\xi^1, \xi^2) + k T_i] + \dots, \quad \mu = \frac{1}{\delta^\omega} \mu_i + \dots \tag{4.35}$$

Substituting expansions (4.34) and (4.35) into initial equations (4.22), one obtains the equations for the main expansion terms

$$P_i = T_{is}\rho_i, \quad \frac{\partial}{\partial \zeta_i}(\rho_i V_i) = 0, \quad \frac{4}{3} \frac{\partial}{\partial \zeta_i}\left(\mu_i \frac{\partial V_i}{\partial \zeta_i}\right) - 2\frac{\partial P_i}{\partial \zeta_i} = 0,$$

$$\frac{\partial}{\partial \zeta_i}\left(\mu_i \frac{\partial u_i^\alpha}{\partial \zeta_i}\right) = 0, \quad \frac{\partial}{\partial \zeta_i}\left(\frac{\mu_i}{\sigma} \frac{\partial T_i}{\partial \zeta_i}\right) = 0. \tag{4.36}$$

These equations are very simple and have the following first integrals:

$$\rho_i V_i = \text{const} = u_s^3, \quad \mu_i \frac{\partial V_i}{\partial \zeta_i} - 2P_i = C_3, \quad \mu_i \frac{\partial u_i^\alpha}{\partial \zeta_i} = C_\alpha, \quad \frac{\mu_i}{\sigma} \frac{\partial T_i}{\partial \zeta_i} = C_4.$$

Here, the coefficients C_1, C_2, C_3 and C_4 are the constants of integration. They should be determined using matching with the solution for the middle sublayer. Demonstrate this procedure only with one variable – the temperature. Using the definitions of variables at the middle sublayer (4.29)–(4.30) it can be written

$$\frac{\partial T}{\partial \zeta} = \frac{1}{\delta} \frac{\partial T_m}{\partial \zeta_m} \frac{\partial \zeta_m}{\partial \zeta} = \frac{1}{\delta} Re_\infty \delta^\omega \frac{\partial T_m}{\partial \zeta_m}.$$

On the other hand, the same value $\partial T / \partial \zeta$ may be presented via variables at the inner sublayer as

$$\frac{\partial T}{\partial \zeta} = \frac{1}{\delta} \frac{\partial}{\partial \zeta_i} \big[T_{is}(\xi^1, \xi^2) + k T_i \big] \frac{\partial \zeta_i}{\partial \zeta} = \frac{1}{\delta} \frac{k}{\varepsilon} Re_\infty \delta^\omega \frac{\partial T_i}{\partial \zeta_i} \sim \frac{1}{\delta} Re_\infty \delta^\omega \frac{\partial T_i}{\partial \zeta_i}.$$

As the result one can write simultaneously the matching conditions for main terms of asymptotic expansions

$$\lim_{\zeta_m \to -\infty} \Big[\mu_m \frac{\partial u_m^\alpha}{\partial \zeta_m}, \frac{\mu_m}{\sigma} \frac{\partial T_m}{\partial \zeta_m}, \mu_m \frac{\partial v_m}{\partial \zeta_m} \Big] = \lim_{\zeta_i \to \infty} \Big[\mu_i \frac{\partial u_i^\alpha}{\partial \zeta_i}, \frac{\mu_i}{\sigma} \frac{\partial T_i}{\partial \zeta_i}, \mu_i \frac{\partial v_i}{\partial \zeta_i} \Big],$$

$$\lim_{\zeta_m \to -\infty} V_m = -u_s^3, \qquad \lim_{\zeta_m \to -\infty} u_m^\alpha = u_{is}^\alpha, \qquad \lim_{\zeta_m \to -\infty} T_m = T_{is}. \tag{4.37}$$

Note that there are two types of the matching conditions, the first conditions in (4.37) are for derivatives, and the second ones in (4.37) enable to determine the constants C_1, C_2, C_3, C_4: via solution for the outer boundary of the middle sublayer, i.e.,

$$C_\alpha = \lim_{\zeta_m \to -\infty} \mu_m \frac{\partial u_m^\alpha}{\partial \zeta_m}, \quad C_4 = \lim_{\zeta_m \to -\infty} \frac{\mu_m}{\sigma} \frac{\partial T_m}{\partial \zeta_m}, \quad C_3 = \lim_{\zeta_m \to -\infty} \Big(\frac{4}{3} \mu_m \frac{\partial v_m}{\partial \zeta_m} - \varepsilon P_m \Big).$$

Using the government equations (4.31) for the flow at the middle sublayer, one obtains the relations for these limit values

$$C_\alpha = \lim_{\zeta_m \to -\infty} \mu_m \frac{\partial u_m^\alpha}{\partial \zeta_m} = - \lim_{\zeta_m \to -\infty} [u_s^3 (u_s^3 - u_m^\alpha)] = -u_s^3 (u_s^3 - u_{im}^\alpha),$$

$$C_3 = \lim_{\zeta_m \to -\infty} \Big(\frac{4}{3} \mu_m \frac{\partial v_m}{\partial \zeta_m} - \varepsilon P_m \Big) = - \lim_{\zeta_m \to -\infty} [u_s^3 (u_s^3 - v_m)] = -(u_s^3)^2,$$

$$C_4 = \lim_{\zeta_m \to -\infty} \frac{\mu_m}{\sigma} \frac{\partial T_m}{\partial \zeta_m} = \lim_{\zeta_m \to -\infty} \big\{ u_s^3 \big[\tfrac{1}{2} (u_s^3)^2 - T_m + \tfrac{1}{2} B_{\alpha\beta} (u_m^\alpha - u_s^\alpha)(u_m^\beta - u_s^\beta) \big] \big\}$$

$$= u_s^3 \big[\tfrac{1}{2} (u_s^3)^2 - T_{is} + \tfrac{1}{2} B_{\alpha\beta} (u_{is}^\alpha - u_s^\alpha)(u_{is}^\beta - u_s^\beta) \big].$$

By these means the final relations for the first integrals of equations (4.36) may be written as follows:

$$\rho_i V_i = u_s^3, \quad P_i = \rho_i T_{is}, \quad \frac{4}{3} T_{is}^\omega \frac{\partial V_i}{\partial \zeta_i} = 2 u_s^3 \frac{T_{is}}{V_i} - (u_s^3)^2, \quad T_{is}^\omega \frac{\partial u_i^\alpha}{\partial \zeta_i} = -u_s^3 (u^\alpha - u_{is}^\alpha),$$

$$\frac{T_{is}^\omega}{\sigma} \frac{\partial T_i}{\partial \zeta_i} = -u_s^3 \big[\tfrac{1}{2} (u_s^3)^2 - T_{is} + \tfrac{1}{2} B_{\alpha\beta} (u_{is}^\alpha - u_s^\alpha)(u_{is}^\beta - u_s^\beta) \big]. \tag{4.38}$$

Note that (4.38) are rather simple and they can be also integrated. As for matching conditions for (4.37) these relations can be used to determine the links between u_{is}^α, T_{is} and u_{ms}^α, T_{ms},

$$u_{ms}^\alpha = -\frac{u_s^\alpha - u_{is}^\alpha}{(u_s^3)^{4/3}}, \quad -u_s^3 \Big[T_{is} + \frac{\sigma}{3 - 2\sigma} (u_s^3) + \frac{\sigma}{4 - 2\sigma} B_{\alpha\beta} u_{ms}^\alpha u_{ms}^\beta (u_s^3)^{8/3} \Big].$$

We obtain three-sublayer asymptotic solution for $3d$ NSE across shock the wave. This solution depends on three free parameters u_{is}^α, T_{is} which should be determined using matching procedure with the solution for viscous shock layer $u_{is}^\alpha = u_L^\alpha(\xi^1, \xi^2, y_L)$ and $T_{is} = T_L(\xi^1, \xi^2, y_L)$, where subscript L is associated with flow parameters at viscous shock layer region and the value y_L is the thickness of viscous shock layer (Fig. 4.9(a)). On the other hand,

$$u^3 - u_s^3 = \delta^{\frac{3}{4\sigma}} v_0, \quad u^\alpha - u_s^\alpha = \delta^{1/\sigma} u_0^\alpha, \quad T = T_0,$$

and therefore $V_m = \delta^{\frac{3}{4\sigma}} v_0$, $V_m^\alpha = \delta^{1/\sigma} u_0^\alpha$, $T_m = T_0$. Using solutions (4.33) one obtains

$$V_m^\alpha = u_{ms}^\alpha V_m^{4/3} = u_{ms}^\alpha (\delta^{\frac{3}{4\sigma}} v_0)^{4/3} = u_{ms}^\alpha \delta^{1/\sigma} v_0^{4/3},$$
$$T_m \sim T_{ms} V_m^{4\sigma/3} = T_{ms} (\delta^{\frac{3}{4\sigma}} v_0)^{4\sigma/3} = T_{ms} \delta v_0^{4\sigma/3}.$$

On the other hand, from solution (4.28) for the outer part of the shock wave (Fig. 4.9(b))

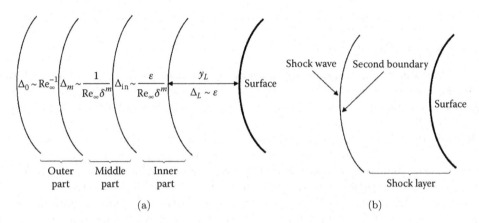

FIGURE 4.9: (a) Middle sublayer structure. (b) Shock layer.

we have

$$\delta^{1/\sigma} u_0^\alpha = u_{0s}^\alpha \delta^{1/\sigma} v_0^{4/3}, \quad \delta T_0 \sim T_{0s} \delta v_0^{4\sigma/3}$$

and therefore the following matching conditions can be written as follows $u_{0s}^\alpha = u_{ms}^\alpha$ and $T_{0s} = T_{ms}$. As a result, the solution for the middle sublayer is known. This solution depends on the functions u_{ms}^α, T_{ms} which should be determined based on matching with the solution for the inner part of the shock wave.

4.4.3 Inner sublayer

In the inner sublayer, disturbances are effected by viscosity and the Mach number remains small enough for compressibility to be neglected. Exact mathematical analysis for small perturbations as applied in hydrodynamical stability problems has been made in this case of flow without separation and the logarithmic decrement of the upstream influence under zero-heat transfer condition is estimated which confirms experimental results. The wall pressure distribution shows [13] that the pressure first jumps at the impact point to an intermediate value and then progressively reaches the constant level corresponding to shock reflection in a Mach 4 uniform flow. This behavior, which is observed in high Mach number flows, can thus be interpreted by inviscid arguments. At lower Mach number, below 2.5, an overshoot is observed in the wall pressure distributions, which cannot be explained simply by rotational effects. In these circumstances, the influence of the subsonic layer close to the wall and also the viscous inner layer can no longer be neglected and a purely inviscid analysis captures only a part of the solution.

For the inner sublayer our assumptions concerning the variables order have to be modified at this part of the shock wave. Considering the solution for the middle sublayer it is possible to write the estimations for the flow parameters

$$u^\alpha \sim O(1), \quad u^3 \sim O(1), \quad \rho \sim O(1), \quad T \sim O(1/\delta), \quad P \sim O(1/\delta).$$

On the other hand, based on the classical Rankine–Hugoniot conditions at the shock wave there can be written the estimations for the flow parameters after shock wave [35]:

$$u^\alpha \sim O(1), \quad u^3 \sim O(\varepsilon), \quad \rho \sim O(1/\varepsilon), \quad T \sim O(1/\delta), \quad P \sim O(1/(\varepsilon\delta)).$$

4.5 Viscous shock layer

The three-dimensional problems of hypersonic rarefied gas flow over blunt bodies are important in investigating the heat transfer and aerodynamics of space vehicles in the upper atmospheres of the Earth and other planets. They are characterized by large Kn (Knudsen) or small Re (Reynolds) numbers.

With decrease in Re continuum flow models, such as the NS equations (see Section 1.8), the viscous shock layer, and the boundary layer, become inapplicable to solving these problems, even using boundary conditions with account for slippage and a surface temperature jump, since above a certain Re value (which decreases if the slippage effects are allowed) these models yield values of the heat transfer and skin friction coefficients which are higher than in free-molecular flow and unboundedly increase as Re approaches zero. The solution of the three-dimensional problems of hypersonic flow over bodies within the framework of kinetic equations (Boltzmann equation or its simplified models) is still difficult. Therefore, for the regime transitional from continuum to free-molecular flow, such problems are mainly solved using the direct simulation Monte Carlo method. Recently, hybrid methods that mate the solution obtained using the kinetic equations or the Direct Simulation Monte Carlo method with the solution of the continuum equations have started to be developed. A review of various methods of investigating hypersonic rarefied gas flows over bodies is given in [11].

Let us consider flow at viscous shock layer. As it was noted above this regime is realized if the following conditions are satisfied [2]:

$$K = Re_\infty \delta^\omega \varepsilon = O(1), \quad 0 \le g^2 \le \varepsilon^{1/2}\theta^{-1/2}. \tag{4.39}$$

This layer is arranged between the shock wave and the body surface. Thus, to estimate the order of the variables at this layer it is necessary to use Rankine–Hugoniot conditions at the shock wave. These conditions give the following estimations [35]:

$$u^\alpha \sim O(1), \quad u^3 \sim O(\varepsilon), \quad \rho \sim O(1/\varepsilon), \quad T \sim O(1/\delta), \quad P \sim O(1/\varepsilon\delta).$$

The coordinate system must be connected with body surface and therefore $y = 0$. On the other hand, with the aim to save all terms in the continuity equation (and by this way to satisfy mass conservation law at the considered layer) one obtains the estimation $\Delta_L \sim O(\varepsilon)$ for the layer thickness order.

4.5.1 Main equations

Now it can be written the relations for independent variables at viscous shock layer,

$$\xi^\alpha = \xi_L^\alpha, \quad \zeta = \varepsilon\zeta_L, \quad y = 0, \tag{4.40}$$

where subscript L is associated with flow parameters at viscous shock layer.

As for independent variables we will seek the solutions in the form:

$$u^\alpha = u_L^\alpha + \ldots, \quad u^3 = \varepsilon v_L + \ldots, \quad \rho = \tfrac{1}{\varepsilon}\rho_L + \ldots, \quad P = \tfrac{1}{\varepsilon\delta}P_L + \ldots,$$
$$T = \tfrac{1}{\delta}T_L + \ldots, \quad \mu = \tfrac{1}{\delta^\omega}\mu_L + \ldots, \quad g_{\alpha\beta} = a_{\alpha\beta} + \ldots. \tag{4.41}$$

Substituting expansions (4.40) and (4.41) into initial equations (4.22) equations for the main terms of expansion (index L is omitted) to be obtained [34],

$$\frac{\partial}{\partial\xi^\alpha}(\rho u^\alpha\sqrt{a/a_{(\alpha\alpha)}}) + \frac{\partial}{\partial\zeta}(\rho v\sqrt{a}) = 0,$$
$$\rho D^* u^\alpha + \rho A^\alpha_{\beta\gamma}u^\beta u^\gamma = -2a^{\alpha\beta}\frac{\varepsilon}{1+\varepsilon}\sqrt{a_{(\alpha\alpha)}}\frac{\partial P}{\partial\xi^\beta} + \frac{\partial}{\partial\zeta}\left(\frac{\mu}{K}\frac{\partial u^\alpha}{\partial\zeta}\right),$$
$$2\frac{\partial P}{\partial\zeta} = -\rho A^3_{\beta\gamma}u^\beta u^\gamma, \quad \rho D^* T = 2\frac{\varepsilon}{1+\varepsilon}\frac{u^\alpha}{\sqrt{a_{(\alpha\alpha)}}}\frac{\partial P}{\partial\xi^\alpha} + \frac{\partial}{\partial\zeta}\left(\frac{\mu}{K\sigma}\frac{\partial T}{\partial\zeta}\right) + \frac{\mu}{K}\left[B_{\alpha\beta}\frac{\partial u^\alpha}{\partial\zeta}\frac{\partial u^\beta}{\partial\zeta}\right], \tag{4.42}$$
$$D^* = \frac{u^\alpha}{\sqrt{a_{(\alpha\alpha)}}}\frac{\partial}{\partial\xi^\alpha} + v\frac{\partial}{\partial\zeta}, \quad P = \rho T, \quad \mu = T^\omega.$$

Definition 66 Equations (4.42) are called *thin viscous shock layer equations* or *hypersonic viscous shock layer equations*.

For $2d$ flow these equations were justified in [22]. Note that the terms with longitudinal pressure gradients of order $O(\varepsilon)$ are saved in these equations. Why was it done? The detailed explanations to this important question will be done later. The brief answer is given below. The asymptotic order of these terms depends on Reynolds number. For low and moderate Reynolds numbers $Re_\infty\delta^\omega \sim O(1/\varepsilon)$ the terms with longitudinal pressure gradients have order $O(\varepsilon)$. For high Reynolds numbers the order of these terms near the body surface is increased up to $O(1)$ and they play key role at the boundary layer. Therefore, there is the possibility to use also these composite equations for regimes with higher Reynolds numbers if these terms are saved at the thin viscous shock layer.

Another important problem is the boundary conditions for thin viscous shock layer equations (4.42). The body surface boundary conditions are rather clear:

$$\zeta = 0, \quad u^\alpha = u_w^\alpha, \quad (\rho v)_w = G, \quad T = T_w.$$

The situation with the boundary conditions at the external boundary of viscous shock layer is more complex. First of all note that these conditions are matching with asymptotic solutions across shock wave $u_{is}^\alpha = u_L^\alpha(\xi^1,\xi^2,y_L)$, $T_{is} = T_L(\xi^1,\xi^2,y_L)$, where subscript L is associated with flow parameters at viscous shock layer and the value y_L is the layer's thickness.

Using these conditions the solutions for viscous shock layer cannot be separated from the solutions across shock wave and the problem has to be solved jointly. If one is not interesting of the detailed information about the shock wave structure he can considerably simplify matters and to separate the solution to these two problems one from another.

4.5.2 Generalized Rankine–Hugoniot conditions at the shock wave

To do it, remember that the orders of shock wave sublayers thickness are as follows $\Delta_0 \sim \frac{1}{Re_\infty}$, $\Delta_m \sim \frac{1}{Re_\infty\delta^\omega}$, $\Delta_i \sim \frac{\varepsilon}{Re_\infty\delta^\omega}$. We conclude that from asymptotic point of view the shock wave thickness is determined by the middle sublayer thickness Δ_m because $\Delta_0/\Delta_m \ll 1$, $\Delta_i/\Delta_m \ll 1$. It is also known the order of the main parameters inside the shock wave

$$T \sim P \sim O(1/\delta), \quad u^\alpha \sim v \sim \rho \sim O(1), \quad \zeta \sim O(1 / Re_\infty\delta^\omega).$$

The equations described the flow-field across the shock wave (in the coordinate system associated with the shock wave surface) are

$$\frac{\partial}{\partial \zeta}(\rho V \sqrt{g}) = 0, \quad \rho V \frac{\partial u^\alpha}{\partial \zeta} = \frac{\partial}{\partial \zeta}\left(\mu \frac{\partial u^\alpha}{\partial \zeta}\right), \quad \rho V \frac{\partial v}{\partial \zeta} + 2\varepsilon \frac{\partial P}{\partial \zeta} = \frac{4}{3}\frac{\partial}{\partial \zeta}\left(\mu \frac{\partial v}{\partial \zeta}\right),$$
$$\rho V \frac{\partial T}{\partial \zeta} = 2\varepsilon V \frac{\partial P}{\partial \zeta} + \frac{\partial}{\partial \zeta}\left(\frac{\mu}{\sigma}\frac{\partial T}{\partial \zeta}\right) + \mu\left[B_{\alpha\beta}\frac{\partial u^\alpha}{\partial \zeta}\frac{\partial u^\beta}{\partial \zeta} + \frac{4}{3}\left(\frac{\partial v}{\partial \zeta}\right)^2\right]. \tag{4.43}$$

The explanation why terms $2\varepsilon \partial P/\partial \zeta$ and $2\varepsilon V \partial P/\partial \zeta$ are saved in asymptotic equations (4.43) will be presented later. The system (4.43) can be integrated. Using boundary conditions at infinity $\partial/\partial \zeta = 0$ as $\zeta \to +\infty$ the integrals of the first three equations (4.43) are

$$\rho V = u_s^3, \quad u_s^3(u_\infty^\alpha - u^\alpha) = -\mu \frac{\partial u^\alpha}{\partial \zeta}, \quad u_s^3(u_\infty^3 - v) + 2\varepsilon(P_\infty - P) = -\frac{4}{3}\mu \frac{\partial v}{\partial \zeta}. \tag{4.44}$$

To integrate the last equation we have to use the following relations:

$$\mu \frac{\partial u^\alpha}{\partial \zeta}\frac{\partial u^\beta}{\partial \zeta} = \frac{1}{2}\frac{\partial}{\partial \zeta}\left(\mu u^\alpha \frac{\partial u^\beta}{\partial \zeta} + \mu u^\beta \frac{\partial u^\alpha}{\partial \zeta}\right) - \frac{1}{2}u^\alpha \frac{\partial}{\partial \zeta}\left(\mu \frac{\partial u^\beta}{\partial \zeta}\right) - \frac{1}{2}u^\beta \frac{\partial}{\partial \zeta}\left(\mu \frac{\partial u^\alpha}{\partial \zeta}\right)$$
$$= \frac{1}{2}\frac{\partial}{\partial \zeta}\left(\mu u^\alpha \frac{\partial u^\beta}{\partial \zeta} + \mu u^\beta \frac{\partial u^\alpha}{\partial \zeta}\right) - \frac{1}{2}u_s^3 \frac{\partial}{\partial \zeta}(u^\alpha u^\beta),$$
$$\frac{4}{3}\mu\left(\frac{\partial v}{\partial \zeta}\right)^2 = \frac{4}{3}\frac{\partial}{\partial \zeta}\left(\mu v \frac{\partial v}{\partial \zeta}\right) - v \frac{\partial}{\partial \zeta}\left(\frac{4}{3}\mu \frac{\partial v}{\partial \zeta}\right) = \frac{\partial}{\partial \zeta}\left(\frac{4}{3}\mu v \frac{\partial v}{\partial \zeta}\right) - \frac{1}{2}u_s^3 \frac{\partial}{\partial \zeta}(v^2) - 2\varepsilon v \frac{\partial P}{\partial \zeta}.$$

The following formula is obtained:

$$u_s^3\left[T_\infty - T + \frac{1}{2}(u_s^3 - v)^2 + \frac{1}{2}B_{\alpha\beta}(u_s^\alpha - u^\alpha)(u_s^\beta - u^\beta)\right] = -\frac{\mu}{\sigma}\frac{\partial T}{\partial \zeta}. \tag{4.45}$$

The conditions (4.44)–(4.45) may be rewritten for the inner boundary of the shock wave. Taking into account the values of variables

$$T_\infty \sim P_\infty \sim O(\delta), \quad v \sim y_L \sim O(\varepsilon), \quad \varepsilon P \sim O(1)$$

at this region give the answer to the question why it is necessary to save the term $2\varepsilon \partial P/\partial \zeta$ and at the inner boundary $P \sim O(1/\varepsilon)$ of the shock wave in asymptotic equations (4.43) and therefore why this term should be preserved. As the result one can obtain the relations at the shock wave in coordinate system associated with body surface,

$$\zeta_L = y_L : \quad \frac{\mu_L}{K}\frac{\partial u_L^\alpha}{\partial \zeta_L} = -u_s^3(u_s^\alpha - u_L^\alpha), \quad v_L = 2\frac{T_L}{u_s^3} + \left(u_L^\alpha/\sqrt{a_{(\alpha\alpha)}}\right)\frac{\partial y_L^\alpha}{\partial \xi_L^\alpha},$$
$$P_L = \frac{1}{2}(u_s^3)^2, \quad \rho_L = \frac{P_L}{T_L}, \quad \frac{\mu_L}{K\sigma}\frac{\partial T_L}{\partial \zeta_L} = -u_s^3\left(\frac{1}{2}(u_s^3)^2 + \frac{1}{2}B_{\alpha\beta}(u_s^\alpha - u_L^\beta) - T_L\right). \tag{4.46}$$

Definition 67 Conditions (4.46) are called *hypersonic approximation of generalized Rankine–Hugoniot conditions at the shock wave*. In fact, these conditions can be considered as external boundary conditions for thin viscous shock later equations.

Thus, there firstly appears the possibility to solve thin viscous shock layer equations without any connection with the solution across the shock wave.

We verify: is the boundary value problem for thin viscous shock layer equations correct or not. To do it there should be compared the total number of the boundary conditions and the total order of the system equations. Obviously that for six unknown variables u^α, v, P, T, ρ there are five differential equations and one algebraic one. The total number of the boundary conditions is equal nine – four at the body surface and five at the shock wave. The total order of the system is equal to eight. There appears an additional boundary condition which should be used to determine unknown distance y_L between the shock wave and body surface.

4.6 Models for shock sublayers

4.6.1 Inviscid shock and boundary layers

Inviscid shock layer. Consider flow at the layer which is adjacent to the shock wave but for the case when the following conditions are realized:

$$Re_\infty \delta^\omega \varepsilon^n = O(1), \quad n \geq 5/2. \tag{4.47}$$

It is necessary to note that if condition (4.47) is satisfied the parameter K introduced in previous section (see (4.39)) tends to unity. This layer is called *inviscid shock layer*.

For asymptotic solution of this layer, first of all, there should be determined the orders of flow variables. To do this we will take into account that this layer is arranged near the shock wave. It means that to estimate the variables order at the layer as the first approximation it can be used Rankine–Hugoniot conditions at the shock wave; they give following estimations: $u^\alpha \sim O(1)$, $u^3 \sim O(\varepsilon)$, $\rho \sim O(1/\varepsilon)$, $T \sim O(1/\delta)$, $P \sim O(1/\varepsilon\delta)$. Another important question connected with the choice of coordinate system. Assume that for strong injection regime $y = y_c(\xi^1, \xi^2) = \Lambda^{1/2} y_I$, where here and hereafter subscript I is associated with flow parameters at viscous shock layer, and function $y_c(\xi^1, \xi^2)$ describes the contact surface which is generated in flow with strong injection.

For other regimes it is assumed $y = y_I = 0$. So, for inviscid shock layer our coordinate system will be connected either with the contact surface, – for strong injection regime, or with the body surface – for other regimes. Based on these estimations it is convenient to seek asymptotic solutions for viscous shock layer in the form:

$$u^\alpha = u_I^\alpha + \dots, \quad V = \varepsilon V_I + \dots, \quad u^3 = \varepsilon V_I + \Lambda^{1/2}\big(u_I^\alpha / \sqrt{a_{(\alpha\alpha)}}\big)\frac{\partial y_I}{\partial \xi_I^\alpha} + \dots,$$
$$P = \frac{1}{\varepsilon\delta}P_I + \dots, \quad \rho = \frac{1}{\varepsilon}\rho_I + \dots, \quad \mu = \frac{1}{\delta^\omega}\mu_I + \dots, \quad g_{\alpha\beta} = a_{\alpha\beta} + \dots \tag{4.48}$$

On the other hand, nature desire to save all terms of continuity equation, and by this way to satisfy mass conservation law inside considered layer, leads to the estimation $\Delta_I \sim O(\varepsilon)$ for the layers thickness order. It means that for independent variables at inviscid shock layer we have to use the relations

$$\xi^\alpha = \xi_I^\alpha, \quad \zeta = \varepsilon\zeta_I. \tag{4.49}$$

Substituting expansions (4.48) and (4.49) into initial equations (4.22) helps to obtain the relations for the main terms of the expansion (index I is omitted),

$$L_0^* \equiv \frac{\partial}{\partial \xi^\alpha}\big(\rho u^\alpha \sqrt{a/a_{(\alpha\alpha)}}\big) + \frac{\partial}{\partial \zeta}(\rho V \sqrt{a}) = 0, \quad L_\alpha^* \equiv \rho D^* u^\alpha + \rho A_{\alpha\beta}^\alpha u^\beta u^\gamma = 0,$$
$$L_3^* \equiv 2\frac{\partial P}{\partial \zeta} = -A_{\alpha\beta}^3 u^\alpha u^\beta, \quad L_4^* \equiv \rho D^* T = 0, \quad P = \rho T, \quad D^* = \big(u^\alpha / \sqrt{a_{(\alpha\alpha)}}\big)\frac{\partial}{\partial \xi^\alpha} + V\frac{\partial}{\partial \zeta}. \tag{4.50}$$

Definition 68 Equations (4.50) are called *thin inviscid shock layer equations or hypersonic approximation of inviscid shock layer equations.*

For 2d flows these equations were suggested and justified in [12, 16].

Consider the situation with the boundary conditions for thin inviscid shock layer (4.50). Because these equations are or the first order and the flow is directed from the shock wave to the body surface it is necessary to use the boundary conditions at the external boundary of inviscid shock layer which is adjacent to the shock wave. To obtain these conditions one can use the limit of hypersonic approximation of generalized Rankine–Hugoniot conditions

(4.47) at the shock wave for $K \to \infty$ [35]. The boundary conditions to viscous shock layer equations at the shock wave have a view:

$$\zeta = \zeta_s, \quad u_I^\alpha = u_s^\alpha, \quad P_I = T_I = \frac{1}{2}(u_s^3)^2, \quad V_I - (u_s^\alpha / \sqrt{a_{(\alpha\alpha)}}) \frac{\partial \zeta}{\xi_i^\alpha} = u_s^3. \tag{4.51}$$

Definition 69 Conditions (4.51) are called hypersonic approximation of Rankine–Hugoniot conditions at the shock wave.

As for the body surface boundary condition or the contact surface boundary condition – for strong injection regime, this condition results from the fact that the body surface (contact surface) is the stream surface (Fig. 4.10) and therefore normal component of the velocity vector with respect to this surface should be equal zero: $\zeta = 0 : V_I = 0$. Note that is the main question about the boundary value problem for thin inviscid shock layer equations to be correct or not. To do it, compare the total number of the boundary conditions and the total order of the system of equations. Obviously that for six unknown variables u^α, V, P, T, ρ there was obtained five PDE and one algebraic equation. The total number of boundary conditions is equal to six, one at the body surface and five at the shock wave. The total order of the system is equal to five. There appears an additional boundary condition which should be used to determine unknown distance ζ_s between the shock wave and body surface. For future consideration it is useful to introduce the function: $W^\alpha(\zeta^1, \zeta^2) = \lim_{\zeta^I \to 0} \frac{\partial u_I^\alpha}{\partial \zeta_I}$.

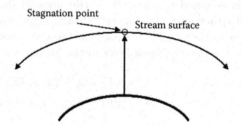

Stagnation point

Stream surface

FIGURE 4.10: Stream surface.

Inviscid boundary layer. Let the following conditions are realized:

$$Re_\infty \delta^\omega \varepsilon^n = O(1), \quad n > 5/2. \tag{4.52}$$

Inviscid boundary layer is adjacent to inviscid shock layer. It is nature to understand why inviscid boundary layer generates and why is the reason of this generation. Comparison conditions (4.52) for inviscid boundary layer with the conditions (4.47) for inviscid shock layer shows they be essentially the same. What does it mean? It means that the asymptotic solution to inviscid shock layer equations has sufficient defect for small values of the coordinate ζ_I; to remedy the situation we must introduce new sublayer named *inviscid boundary layer* in this region.

To understand this defect we will consider the special case of $2d$ plane flow at inviscid shock layer. The case of $3d$ flow at inviscid shock layer is considered by [20]. The mass conservation and the momentum equations to the tangent direction with respect to the body surface (contact surface) have the form:

$$\frac{\partial}{\partial \xi}(\rho u) + \frac{\partial}{\partial \zeta}(\rho V) = 0, \quad \rho\left(u \frac{\partial u}{\partial \xi} + V \frac{\partial u}{\partial \zeta}\right) = 0, \quad \rho\left(u \frac{\partial T}{\partial \xi} + V \frac{\partial T}{\partial \zeta}\right) = 0.$$

Determine the stream function ψ as $\rho u = \frac{\partial \psi}{\partial \zeta}$ and $\rho V = -\frac{\partial \psi}{\partial \xi}$, and introduce new independent variables (x, y), $x = \xi$, $y = \psi(\xi, \zeta)$. Thus, in new variables, the continuity equation is identity and the momentum equation has a form $\frac{\partial u}{\partial x} = 0$ and $\frac{\partial T}{\partial x} = 0$. It means that the tangent component of the velocity vector u and the temperature T are constant along stream line and are equal to their values at the boundary condition at the shock wave. In particular case, for the stagnation line, the result is

$$u_0 = u_{s0} = 0, \quad T_0 = T_{s0} = (u_{s0}^3)^2/2 = 1/2.$$

But the stagnation stream line coincides with the stream line along the body surface (contact surface) and therefore near the body surface (contact surface) the tangent component of the velocity vector u tends to zero. Compare this estimation for u near the surface $\zeta = 0$ with our assumption (4.48) for inviscid shock layer $\sim O(1)$ one understands the character of the defect which the asymptotic solution for inviscid shock layer has near the surface $\zeta = 0$.

It is time to clarify the situation and to understand the nature of the defect of the asymptotic solution for inviscid shock layer. To improve this solution we must introduce new asymptotic sublayer and to modify the orders of the main variables at this layer.

First of all it is necessary to modify our assumption to the order of tangent components of the velocity vector u^α. To do it note that the nearest term (from asymptotic point of view), neglected in the equations for u^α, is the term with longitudinal and circumferential gradients of pressure, which has order $O(\varepsilon)$. Therefore to correct the defect of solutions we have to remain this term in the government equations for new sublayer. It is easy to watch that it can be done if this zone $u^\alpha \sim O(\sqrt{\varepsilon})$.

On the other hand, if we want to save all terms in continuity equation and by this way to satisfy mass conservation law at the considered layer, we obtain the estimations $\Delta_{IB} \sim O(\varepsilon^\gamma)$ for order of the sublayer thickness Δ_{IB} and $V \sim O(\varepsilon^{\gamma+1/2})$ for the order of normal component V of the velocity vector $\mathbf{u}$. To determine the value γ we should use the matching condition between the solution for new sublayer and the solution for inviscid shock layer for tangent components u^α of the velocity vector. First, note that the matching procedure cannot be provided for the value u^α directly because this function has a defect at $\zeta_I \to 0$ and the main goal of this new sublayer is to correct u^α near the surface $\zeta_I = 0$. Thus we have to use the matching condition for the first derivatives of this function, which has a form

$$\lim_{\zeta_I \to 0} \frac{\partial u^\alpha}{\partial \zeta} = \lim_{\zeta_{IB} \to \infty} \frac{\partial u^\alpha}{\partial \zeta}.$$

We estimate the order of $\partial u^\alpha/\partial \zeta$. On one hand, for inviscid shock layer variables, we have $\partial u^\alpha/\partial \zeta = O(1/\varepsilon)$. On the other hand, using our previous estimations for inviscid boundary layer variables, it can be obtained $\partial u^\alpha/\partial \zeta = O(\varepsilon^{1/2-\gamma})$. The matching conditions are satisfied if only $\varepsilon^{1/2-\gamma} = 1/\varepsilon \Rightarrow \gamma = \frac{3}{2}$. Considering above listed reasoning there appears possibility to introduce the variables (see below) for inviscid boundary layer. Based on the above listed reasoning, the asymptotic solution for inviscid boundary layer should be sought in the form:

$$u^\alpha = \varepsilon^{\frac{1}{2}} u_{IB}^\alpha + \dots, \quad V = \varepsilon^2 V_{IB} + \dots, \quad \rho = \frac{1}{\varepsilon} \rho_{IB} + \dots, \quad P = \frac{1}{\varepsilon\delta} P_{IB} + \dots, \quad g_{\alpha\beta} = a_{\alpha\beta} + \dots,$$
$$T = \frac{1}{\delta} T_{IB} + \dots, \quad u^3 = \varepsilon^2 V_{IB} + \Lambda^{1/2} \varepsilon^{1/2} \left(u_{IB}^\alpha / \sqrt{a_{(\alpha\alpha)}}\right) \frac{\partial y_{IB}}{\partial \xi_{IB}^\alpha} + \dots, \quad \mu = \frac{1}{\delta^\omega} \mu_{IB} + \dots$$
$$(4.53)$$

For independent variables, for inviscid boundary layer, we have to use the following relations:

$$\xi^\alpha = \xi_{IB}^\alpha, \quad \zeta = \varepsilon^{3/2} \zeta_{IB}. \tag{4.54}$$

In addition, assume that for strong injection regime $y = y_c(\xi^1, \xi^2) = \Lambda^{1/2} y_{IB}$ and $y = y_I = 0$

for other regimes. Substitution of expansions (4.53) and (4.54) into (4.22) gives the following equations for the main terms of expansion (index IB is omitted):

$$
\begin{aligned}
&L_0^* \equiv \frac{\partial}{\partial \xi^\alpha}\left(\rho u^\alpha \sqrt{a/a_{(\alpha\alpha)}}\right) + \frac{\partial}{\partial \zeta}\left(\rho V \sqrt{a}\right) = 0, \\
&L_\alpha^* \equiv \rho D^* u^\alpha + \rho A_{\beta\gamma}^\alpha u^\beta u^\gamma = -2a^{\alpha\beta}\sqrt{a_{(\alpha\alpha)}}\frac{\partial P}{\partial \xi^\beta}, \\
&L_3^* \equiv 2\frac{\partial P}{\partial \zeta} = 0, \quad L_4^* \equiv \rho D^* T = 0, \quad P = \rho T, \quad D^* = \left(u^\alpha/\sqrt{a_{(\alpha\alpha)}}\right)\frac{\partial}{\partial \xi^\alpha} + V\frac{\partial}{\partial \zeta}.
\end{aligned}
\qquad (4.55)
$$

Now it is interesting to understand why namely conditions (4.52) are used as the conditions for inviscid boundary layer. To do it we should estimate the order of the main viscous term of the momentum equation for u^α in inviscid boundary layer variables. It is easy to write the estimations

$$
\frac{1}{Re_\infty}\frac{\partial}{\partial \zeta}\left(\mu \frac{\partial u^\alpha}{\partial \zeta}\right) \sim \frac{\varepsilon^{1/2}}{Re_\infty \delta^w \varepsilon^3} \sim \frac{1}{Re_\infty \delta^w \varepsilon^{5/2}}.
$$

For $n > 5/2$ this effects are small in comparison with convective terms and so they may be neglected. That is the reason why this layer is called *inviscid boundary layer*.

Consider the situation with the boundary condition for viscous boundary layer equations (4.55). The boundary condition for $\zeta_{IB} \to +\infty$ must be obtained from matching conditions with the solution for viscous shock layer at $\zeta_I \to 0$. If we remember that the surface $\zeta_I = 0$ coincides with stagnation streamline surface, we obtain the conditions for pressure and temperature,

$$
\lim_{\zeta_{IB} \to +\infty} T_{IB} = \lim_{\zeta_I \to 0} T_I = \frac{1}{2}, \quad \lim_{\zeta_{IB} \to +\infty} P_{IB} = P_I(\xi^1, \xi^2, 0).
$$

The matching procedure for tangent values of the velocity component cannot be used due to $\lim_{\zeta_I \to 0} u_I^\alpha = 0$ follow to results of previous section. How this difficulty can be resolved? In our case it is necessary to use higher matching orders, in particular, for the first derivatives of matching functions. Therefore we must use the following matching conditions:

$$
\lim_{\zeta_{IB} \to +\infty} \frac{\partial u^\alpha}{\partial \zeta} = \frac{\varepsilon^{1/2}}{\varepsilon^{3/2}}, \quad \lim_{\zeta_{IB} \to +\infty} \frac{\partial u_{IB}^\alpha}{\partial \zeta_{IB}} = \lim_{\zeta_I \to 0} \frac{\partial u^\alpha}{\partial \zeta} = \frac{1}{\varepsilon}\lim_{\zeta_I \to 0} \frac{\partial u_I^\alpha}{\partial \zeta_I}.
$$

One obtains

$$
\lim_{\zeta_{IB} \to +\infty} \frac{\partial u_{IB}^\alpha}{\partial \zeta_{IB}} = \lim_{\zeta_I \to 0} \frac{\partial u_I^\alpha}{\partial \zeta_I} = W^\alpha(\xi^1, \xi^2, 0).
$$

The boundary condition at the surface $\zeta_{IB} = 0$ depends on flow regimes. For regimes of the flow in the boundary layer region with injection and nonstrong suction this condition is $\zeta_{IB} : V_{IB} = 0$. This condition results from the fact, that the body surface (contact surface) is the stream surface and therefore the normal component of the velocity vector should be equal to zero. For the boundary layer and strong suction regimes we must use conditions

$$
\zeta_{IB} : \rho_{IB} V_{IB} = -\frac{1}{\varepsilon}G(\xi^1, \xi^2, 0).
$$

Check if the boundary value problem for inviscid boundary layer equations is correct or not. It is easy to watch that for six unknown variables u^α, V, P, T, ρ there are five PDEs and one algebraic equation. The total number of boundary conditions is equal to five, one at the body surface and four at infinity. The total order of the system is equal to five and therefore the boundary value problem for inviscid boundary layer equations is correct. Introduce the following notation: $u_{IB}^\alpha(\xi^1, \xi^2, 0) = u_+^\alpha(\xi^1, \xi^2)$.

4.6.2 Injection gas layer

Definition 70 Consider flow at the layer near the body surface for strong injection conditions

$$Re_\infty \delta^\omega \varepsilon^n = O(1) \quad (n \geq 5/2), \quad (\varepsilon\theta)^{-1} \gg g^2 \gg \frac{\theta^{\omega-1/2}}{Re_\infty \delta^\omega \varepsilon^{1/2}}, \tag{4.56}$$

Fig. 4.11(a). This layer is called *injection gas layer*.

First of all we it is necessary to understand why these conditions are called strong injection. The influence of the viscous effect near the body surface depends on local Reynolds number calculated for the body surface boundary conditions $Re_w = \rho_w^* v_w^* L/\mu_w^*$. If $Re_w \sim O(1)$ the influence of molecular transfer effects is important while if $Re_w \gg 1$ this influence may be neglected and one can say about strong injection regime. We demonstrate that if relations (4.56) are satisfied the condition $Re_w \gg 1$ is also realized. By definition of Re_w and the first relation (4.56) there can be obtained the estimation below,

$$Re_\infty = \frac{\rho_w^* v_w^*}{\rho_\infty V_\infty} \frac{\mu_\infty}{\mu_w^*} \frac{\rho_\infty V_\infty L}{\mu_\infty} \sim g Re_\infty \delta^\omega \sim g\varepsilon^{-n}.$$

On the other hand, based on the combination of the first relation and of the right inequality of the second one in (4.56) it can be written

$$g \gg \varepsilon^{n/2-1/4} \quad \Rightarrow \quad Re_w \gg \varepsilon^{-n/2-1/4} \gg 1.$$

As for the left inequality of the second relation (4.56) it is easy to watch that this condition can be presented in the form $\varepsilon\theta g^2 \equiv \Lambda \ll 1$ and therefore we consider the case when the ratio of the mass impulse through permeable surface to the mass impulse at infinity be a very small value.

The next step is to introduce new independent and dependent variables in the region of injection gas layer. To do it note that the flow at this layer is completely determined by the body surface boundary conditions due to $Re_w \gg 1$. Therefore, we must estimate the main flow parameters orders at the body surface if we want to find the estimations for the same parameters inside injection gas layer. Remember definitions and estimations to be done during introduction to the governing parameters of our problem,

$$\Lambda = \varepsilon\theta g^2 \sim \frac{\rho_w^*(v_w^*)^2}{\rho_\infty V_\infty^2}, \quad g = \frac{\rho_w^* v_w^*}{\rho_\infty V_\infty}, \quad \frac{T_w}{T_\infty} \sim \frac{\theta}{\delta}, \quad \frac{\mu_w}{\mu_\infty} \sim \frac{\theta^\omega}{\delta^\omega}.$$

It can be obtained the estimation for v_∞: $v_w^*/V_\infty \sim O(\varepsilon\theta g)$. Because for large local Reynolds numbers the viscous effects are negligible the flow structure is completely determined by the balance between convective and the pressure gradient terms. It means that these terms should be preserved in asymptotic equations and therefore we have, for pressure gradient terms,

$$\frac{P}{P_\infty}\varepsilon\delta \sim O(1) \quad \Rightarrow \quad \frac{P}{P_\infty} \sim \frac{1}{\varepsilon\delta}.$$

Considering estimations for P and T, and using the state equations we obtain for ρ: $\rho/\rho_\infty \sim (\varepsilon\delta)^{-1}$. By the same way for convective terms it can be obtained

$$\frac{\rho}{\rho_\infty}\frac{u^\alpha}{V_\infty}\frac{u^\alpha}{V_\infty} \sim O(1) \quad \Rightarrow \quad \frac{u^\alpha}{V_\infty} \sim \sqrt{\varepsilon\theta}.$$

To estimate the thickness Δ_B of injection gas layer, use rather nature assumption that all terms in the convective operator have the same order, i.e.,

$$(u^\alpha/V_\infty)^2 \sim \frac{v}{V_\infty}\frac{\partial}{\partial\zeta}(u^\alpha/V_\infty) \quad \Rightarrow \quad \varepsilon\theta \sim \varepsilon\theta\frac{g\sqrt{\varepsilon\theta}}{\Delta_B} \quad \Rightarrow \quad \Delta_B \sim g\sqrt{\varepsilon\theta} = \Lambda.$$

So, there is known the sufficient information which gives possibility to introduce inner variables for injection gas layer. For independent variables the relations to be used:

$$\xi^\alpha = \xi_B^\alpha, \quad y = 0, \quad \zeta = \sqrt{\Lambda}\zeta_B. \tag{4.57}$$

As for dependent variables it will be sought asymptotic solutions for this layer of the similar form as for inviscid boundary layer

$$u^\alpha = \sqrt{\varepsilon\theta}u_B^\alpha + \dots, \quad u^3 = g\varepsilon\theta v_B + \dots, \quad T = \tfrac{\theta}{\delta}T_B + \dots,$$
$$\mu = \tfrac{\theta^\omega}{\delta^\omega}\mu_B + \dots, \quad \rho = \tfrac{1}{\varepsilon\theta}\rho_B + \dots, \quad P = \tfrac{1}{\varepsilon\theta}P_B + \dots \tag{4.58}$$

Substituting expansions (4.57)–(4.58) into initial system we obtain the equations [10] for the main terms (index B is omitted)

$$L_0^* \equiv \tfrac{\partial}{\partial\xi^\alpha}\left(\rho u^\alpha \sqrt{a/a_{(\alpha\alpha)}}\right) + \tfrac{\partial}{\partial\zeta}(\rho V\sqrt{a}) = 0,$$
$$L_\alpha^* \equiv \rho D^* u^\alpha + \rho A_{\beta\gamma}^\alpha u^\beta u^\gamma = -2a^{\alpha\beta}\sqrt{a_{(\alpha\alpha)}}\tfrac{\partial P}{\partial\xi^\beta},$$
$$L_3^* \equiv 2\tfrac{\partial P}{\partial\zeta} = 0, \quad L_4^* \equiv \rho D^* T = 0, \quad P = \rho T, \quad D^* = \left(u^\alpha/\sqrt{a_{(\alpha\alpha)}}\right)\tfrac{\partial}{\partial\xi^\alpha} + V\tfrac{\partial}{\partial\zeta}. \tag{4.59}$$

To understand why relations (4.56) are used as the strong injection conditions it is necessary to check viscous terms orders for injection gas layer variables. It is easy to obtain the following estimations:

$$\frac{1}{Re_\infty}\frac{\partial}{\partial\zeta}\left(\mu\frac{\partial u^\alpha}{\partial\zeta}\right) \sim \frac{\theta^\omega}{Re_\infty\delta^\omega}\sqrt{\varepsilon\theta}/(\sqrt{\Lambda})^2 \sim \frac{\theta^\omega\sqrt{\varepsilon\theta}}{Re_\infty\delta^\omega\varepsilon\theta g^2} \sim \frac{\theta^{\omega-1/2}}{Re_\infty\delta^\omega g^2\sqrt{\varepsilon}}.$$

The last estimation enables one to understand condition (4.56): if this condition be satisfied the influence of the molecular transfer effects on the flow may be neglected as the first approximation. As it is easy to see, injection gas layer equations (4.59) are the first order equations. In accordance with this property these equations are solved with the body surface and the contact surface boundary conditions. The body surface boundary conditions are

$$u_B^\alpha = u_w^\alpha(\xi^1, \xi^2), \quad \rho_B v_B = G(\xi^1, \xi^2), \quad T_B = T_w.$$

The boundary conditions at the contact surface are connected with matching condition with the solution for mixing gas layer

$$\zeta_B = \zeta_c(\xi^1, \xi^2), \quad P_B = P_I(\xi^1, \xi^2, 0), \quad v_n = v_B + (u_B^\alpha/\sqrt{a_{(\alpha\alpha)}})\frac{\partial\zeta_c}{\partial\xi^\alpha} = 0$$

where $\zeta_c(\xi^1, \xi^2)$ describes the contact surface which is not known in advance before the solution to the associated boundary value problem not to be solved for injection gas layer.

We check: if the boundary value problem for injection gas layer equations is correct.

To do it we have to compare the total number of the boundary conditions and the total order of the system equations. For six unknown variables u^α, V, P, T, ρ there are five PDEs and one algebraic equation. The total number of boundary conditions is equal six, four at the body surface and two at the contact surface. The total order of the system is equal five and therefore there is an additional boundary condition. Nevertheless the boundary value problem for injection gas layer is correct because this additional condition is used to determine the position of the contact surface ζ_c to be unknown in advance.

For future consideration, introduce the following definition: $\sqrt{\theta}u_B^\alpha(\xi^1, \xi^2, 0) = u_-^\alpha(\xi^1, \xi^2)$.

4.6.3 Viscous boundary and mixing gas layers

Consider the flow at the layer near the body surface for nonstrong injection or nonstrong suction, or the layer near the contact surface for strong injection under the condition,

$$Re_\infty \delta^\omega \varepsilon^n = O(1) \quad (n \geq 5/2). \tag{4.60}$$

Definition 71 The layer near the body surface is called *viscous boundary layer*, and the layer near the contact surface is named viscous *mixing layer* (Fig. 4.11(b)).

Why is one interested in these asymptotic sublayers under relation (4.60) to be satisfied? First, answer this question for the nonstrong injection (or nonstrong suction) case. As it is known from previous analysis for moderate Reynolds number the flow-field between the shock wave and body surface is described by equations for thin viscous shock layer. Note that the solution to these equations satisfies to the initial boundary conditions at the shock wave and at the body surface.

For high Reynolds numbers when condition (4.60) is realized the flow structure is more complicated. The layer that is adjacent to shock wave is inviscid shock layer. On contrary to viscous shock layer, the solution to the associated boundary value problem, for this layer, satisfies to the boundary conditions at the shock wave region only. Why is it so? The answer is simple – because inviscid shock layer equations are of the first order. As a result one obtains discontinuity between the solution to inviscid shock layer equations at $\zeta_I = 0$ and the body surface conditions. That is why it is necessary to introduce additional asymptotic sublayer and by this way to eliminate the discontinuity above and to find the smooth transfer from the solution at viscous shock layer and the body surface conditions.

The situation with additional asymptotic sublayer near the contact surface for strong injection regime is similar: it is necessary to introduce this sublayer named viscous mixing layer to eliminate discontinuity between the solutions to inviscid shock layer equations and to injection gas layer equations at the contact surface. This explanation helps to understand that viscous terms along normal direction to the body surface (contact surface) should be saved for asymptotic equations, for this layer.

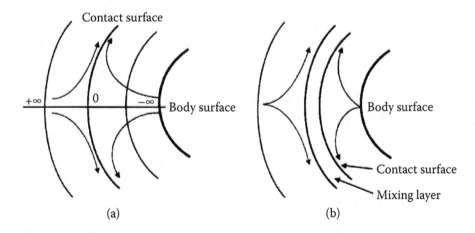

FIGURE 4.11: (a) Strong injection regime. (b) Viscous mixing layer.

The next step is the estimation of the flow parameters for this layer. First of all, note that $\xi^\alpha \sim O(1)$. Because viscous boundary layer is adjacent to viscous shock layer it is nature to use the estimations for ρ and T (obtained for viscous shock layer) for viscous boundary layer,

$$\rho \sim O(1/\varepsilon), \quad T \sim O(1/\delta) \Rightarrow \mu \sim O(1/\delta^\omega).$$

Based on the state equation one obtains for pressure $P \sim O(1/(\varepsilon\delta))$.

Another simple estimation is connected with y value. For nonstrong injection or for nonstrong suction our coordinate system is associated with body surface and therefore $y = 0$. For strong injection using the previous results for injection gas layer it can be written $y \sim O(\sqrt{\Lambda})$. The initial physical conditions should be used so that the flow-field at this asymptotic layer be determined from the balance between molecular transfer effects in normal direction, convective and the pressure gradient terms. It means that all these terms are of order $O(1)$ for this layer. One obtains the following estimations:

– For the tangent component u^α of the velocity vector,

$$\rho(u^\alpha)^2 \sim O(1) \Rightarrow (u^\alpha)^2/\varepsilon \sim O(1) \Rightarrow u^\alpha \sim O(\sqrt{\varepsilon}),$$

– For the layers thickness Δ_{VB},

$$\frac{1}{Re_\infty}\frac{\partial}{\partial\zeta}\left(\mu\frac{\partial u^\alpha}{\partial\zeta}\right) \sim O(1) \Rightarrow \frac{\sqrt{\varepsilon}}{Re_\infty\delta^\omega(\Delta_{VB})^2} \sim O(1) \Rightarrow \Delta_{VB} \sim \frac{\varepsilon^{1/4}}{\sqrt{Re_\infty\delta^\omega}},$$

– For normal component V_{VB} of the velocity vector,

$$\rho V_{VB}\frac{\partial u^\alpha}{\partial\zeta} \sim O(1) \Rightarrow V_{VB}\frac{\sqrt{\varepsilon}}{\varepsilon\Delta_{VB}} \sim O(1) \Rightarrow V_{VB} \sim \frac{\varepsilon^{3/4}}{\sqrt{Re_\infty\delta^\omega}}.$$

Thus, it is enough information to introduce inner variables for viscous boundary layer. For asymptotic analysis there will be used the following independent variables:

$$\zeta^\alpha = \zeta^\alpha_{VB}, \ \zeta = \frac{\varepsilon^{1/4}}{\sqrt{Re_\infty\delta^\omega}}\zeta_{VB}, \ y = y_c = \begin{cases} \Lambda^{1/2}y_{VB}, & \text{for strong injection} \\ 0, & \text{for other cases.} \end{cases}$$

As for dependent variables, it is convenient to sought asymptotic solutions in the form of the expansions below [3],

$$u^\alpha = \sqrt{\varepsilon}u^\alpha_{VB} + \ldots, \quad V = \frac{\varepsilon^{1/4}}{\sqrt{Re_\infty\delta^\omega}}V_{VB} + \ldots, \quad T = \frac{1}{\delta}T_{VB} + \ldots, \quad \mu = \frac{1}{\delta^\omega}\mu_{VB} + \ldots,$$
$$\rho = \frac{1}{\varepsilon}\rho_{VB} + \ldots, \quad P = \frac{1}{\varepsilon\delta}P_{VB} + \ldots, \quad u^3 = \frac{\varepsilon^{3/4}}{\sqrt{Re_\infty\delta^\omega}}V_{VB} + \Lambda^{1/2}\varepsilon^{1/2}\frac{u^\alpha_{VB}}{\sqrt{a_{(\alpha\alpha)}}}\frac{\partial y_{VB}}{\partial\xi^\alpha}.$$
$$(4.61)$$

Substituting expansions (4.61) into the system of initial equations one obtains the equations [27] for viscous boundary and for viscous mixing layers (index VB is omitted),

$$L_0^* \equiv \frac{\partial}{\partial\xi^\alpha}\left(\rho u^\alpha\sqrt{a/a_{(\alpha\alpha)}}\right) + \frac{\partial}{\partial\zeta}(\rho V\sqrt{a}) = 0,$$
$$L_\alpha^* \equiv \rho D^*u^\alpha + \rho A^\alpha_{\beta\gamma}u^\beta u^\gamma = -2a^{\alpha\beta}\sqrt{a_{(\alpha\alpha)}}\frac{\partial P}{\partial\xi^\beta} + \frac{\partial}{\partial\zeta}\left(\mu\frac{\partial u^\alpha}{\partial\zeta}\right),$$
$$L_4^* \equiv \rho D^*T = \frac{\partial}{\partial\zeta}\left(\frac{\mu}{\sigma}\frac{\partial T}{\partial\zeta}\right) + \varepsilon\mu B_{\alpha\beta}\frac{\partial u^\alpha}{\partial\zeta}\frac{\partial u^\beta}{\partial\zeta} + 2\left(u^\alpha/\sqrt{a_{(\alpha\alpha)}}\right)\varepsilon\frac{u^\alpha}{\sqrt{a_{(\alpha\alpha)}}}\frac{\partial P}{\partial\xi^\alpha},$$
$$L_3^* \equiv 2\frac{\partial P}{\partial\zeta} = 0, \quad P = \rho T, \quad D^* = \left(u^\alpha/\sqrt{a_{(\alpha\alpha)}}\right)\frac{\partial}{\partial\xi^\alpha} + V\frac{\partial}{\partial\zeta}.$$
$$(4.62)$$

Note that together with the terms of order $O(1)$ the terms of order $O(\varepsilon)$ are remained. The reason is because it is easy to extend the applicability range of these equations up to the moderate supersonic flow.

To finish the problem statement for viscous boundary layer region it is necessary to indicate the boundary conditions to the system (4.62). These conditions depend on the flow regimes that are completely determined by parameter n from condition (4.60).

If we return to our classification of the flow regime it is nature to call the regimes with $n = 5/2$ – *vortex intersection regime*. The nature questions are: what is the reason of this title and why this regime exists only for $n = 5/2$. To clarify the situation, consider the structure of hypersonic flow over blunted bodies with nonpermeable surface in very high Reynolds number limit case. For this case the flow between the shock wave, to be considered as discontinuity surface, and the body surface can be divided into three sublayers.

The first sublayer, nearest to the shock wave and named *thin inviscid shock layer*, has the thickness order $\Delta_I \sim O(\varepsilon)$. But the solution for this sublayer has a defect for small values of normal coordinate. To correct this defect new asymptotic sublayer, called *inviscid boundary layer*, has to be of the thickness order $\Delta_{IB} \sim O(\varepsilon^{3/2})$. Finally, the last sublayer named viscous *boundary layer* introduced to eliminate discontinuity between the solution for inviscid boundary layer and the body surface boundary conditions. The thickness order of viscous boundary layer is

$$\Delta_{VB} \sim O\left(\frac{\varepsilon^{1/4}}{\sqrt{Re_\infty \delta^\omega}}\right).$$

Compare the values Δ_{VB}, Δ_{IB} and Δ_I. For all regimes – $\Delta_{IB} \ll \Delta_I$. Therefore there are two possibilities $\Delta_{VB} \sim \Delta_{IB}$ or $\Delta_{VB} \ll \Delta_{IB}$. Using these relations there can be found the conditions below to realize these two regimes,

$$\Delta_{VB} \sim \Delta_{IB} : \quad \frac{\varepsilon^{1/4}}{\sqrt{Re_\infty \delta^\omega}} = \varepsilon^{3/2} \Rightarrow Re_\infty \delta^\omega \varepsilon^{5/2} \sim O(1).$$

The first condition $\Delta_{VB} \sim \Delta_{IB}$ means that the both inviscid and viscous boundary layers are merged into one sublayer, where viscous and inviscid effects are interacted. That's why this regime is called *vortex intersection regime*. To realize the second relation $\Delta_{VB} \ll \Delta_{IB}$, the following condition to be obtained:

$$\Delta_{VB} \ll \Delta_{IB} : \quad \frac{\varepsilon^{1/4}}{\sqrt{Re_\infty \delta^\omega}} \ll \varepsilon^{3/2} \Rightarrow Re_\infty \delta^\omega \varepsilon^n \sim O(1) \quad \left(n > \frac{5}{2}\right).$$

Consider the outer boundary conditions for boundary layer (mixing layer). For vortex intersection regimes with $n = \frac{5}{2}$ we have

$$\zeta_{VB} \to +\infty : \quad \frac{\partial u_{VB}^\alpha}{\partial \zeta_{VB}} = W^\alpha(\zeta^1, \zeta^2), \quad T_{VB} = \frac{1}{2}, \quad P_{VB} = P_I(\zeta^1, \zeta^2, 0).$$

For boundary layer regimes with $n > \frac{5}{2}$ we have

$$\zeta_{VB} \to +\infty : \quad u_{VB}^\alpha = u_+^\alpha(\zeta^1, \zeta^2), \quad T_{VB} = \frac{1}{2}, \quad P_{VB} = P_I(\zeta^1, \zeta^2, 0).$$

The boundary conditions for inner bound of viscous boundary layer (viscous mixing layer) are also depended on the flow regimes. For nonstrong injection or for nonstrong suction this bound coincides with the body surface and these conditions may be written in the form:

$$\zeta_{VB} = 0 : \quad u_{VB}^\alpha = u_w^\alpha(\xi^1, \xi^2), \quad \rho_{VB} u_{VB}^3 = \varepsilon^{1/4}\sqrt{Re_\infty \delta^\omega} G(\xi^1, \xi^2), \quad T_{VB} = T_w(\xi^1, \xi^2).$$

For strong injection regime it is necessary to match the solution for viscous mixing layer with the solution for injection gas layer in the neighborhood of the contact surface. To do it we have to use the conditions

$$\zeta_{VB} \to -\infty : \quad u_{VB}^\alpha = u_-^\alpha(\zeta^1, \zeta^2), \quad T_{VB} = (T_B)_w(0, 0).$$

In addition it is necessary to use the condition that at the contact surface normal component of the velocity vector is equal zero,

$$\zeta_{VB} = 0: \quad V_{VB} \equiv u_{VB}^3 - \frac{\Lambda^{1/2}\sqrt{Re_\infty \delta^\omega}}{\varepsilon^{1/4}}(u_{VB}^\alpha/\sqrt{a_{(\alpha\alpha)}})\frac{\partial y_{VB}}{\partial \xi^\alpha} = 0.$$

We check the correctness of the boundary value problem for (4.62).

To do it we have to compare the total number of the boundary conditions and the total order of the system equations. It is very simple to see that for six unknown variables u^α, V, P, T, ρ there are five PDEs and one algebraic equation.

The total number of boundary conditions for viscous boundary layer is eight: four at the outer bound of the boundary layer and four at the body surface. The total order of boundary conditions for viscous mixing layer is also eight: four at the outer bound of viscous mixing layer, three at the inner bound of mixing layer and one at the contact surface. The total order of the system (4.62) is eight; therefore, the boundary value problem for viscous boundary layer as well as for viscous mixing layer is correct.

Note that we must solve two-point boundary value problem for boundary layer whereas three-point boundary value problem for viscous mixing layer.

4.6.4 Viscous boundary layer with strong suction

Consider the flow at the layer near the body surface for strong under the following condition:

$$Re_\infty \delta^\omega \varepsilon^n \geq O(1) \quad (n \geq 5/2), \quad \varepsilon^2 \geq g \gg \left(Re_\infty \delta^\omega \varepsilon^{1/2}\right)^{-1}. \tag{4.63}$$

The trivial question is: why does one have to take into account asymptotic sublayers in relation (4.63) to be satisfied? First, note that for this case the above formulation is not correct. The reason is in contrary to previous cases when one *would be forced to introduce* new asymptotic sublayers, for present case there appears possibility to introduce new layer and by this way to simplify the solution for viscous boundary layer (Fig. 4.12). This simplification should be connected with additional small parameter for the *strong suction* case [39].

Estimate the orders of the flow parameters at this layer. Clearly, $\xi^\alpha \sim O(1)$. Because this layer is adjacent to inviscid boundary layer it is nature to use the estimations for ρ, u^α and T at inviscid boundary layer for viscous boundary layer:

$$\rho \sim O(1/\varepsilon), \quad T \sim O(1/\delta), \quad u^\alpha \sim O(\varepsilon^{1/2}) \Rightarrow \mu \sim O(1/\delta^\omega).$$

Based on the state equation it can be obtained for the pressure $P \sim O(1/\varepsilon\delta)$. Another simple estimation is connected with value y. Obviously that for strong suction our coordinate system is associated with the body surface and there for $y = 0$. But these estimations are coincided with our previous ones for viscous boundary layer with nonstrong suction. For next steps we should clarify the difference between strong and nonstrong suctions. This difference should be connected with the difference between normal components u_w^3 of the velocity vector $\mathbf{u}_w$ at the body surface. In this connection it is nature to suggest the preliminary criterion: for strong suction the value u_w^3 is large whereas for nonstrong suction the save value is not too large. How one can understand what is large and what is not too large? To do it, assume that for strong injection regime the order of normal component u^3 of the velocity vector at this layer is determined by the boundary condition at the body surface. In this case the estimation for u^3 can be written as

$$u^3 \sim u_w^3 \sim \frac{O(\rho_w u_w^3)}{O(\rho_w)} \sim \frac{O(g)}{O(1/\varepsilon)} \sim O(g\varepsilon).$$

For the last estimation of this layers thickness Δ_{SU}, use rather clear physical reasoning. For viscous boundary layer with nonstrong suction the flow-field is determined by balance between four terms which are of the same orders:

- Molecular transfer in normal direction,
- Convective transfer in normal direction,
- Convective transfer in tangent direction,
- Pressure gradient term.

If the layer's thickness decreased with increasing of the suction velocity and the ratio between these terms is modified. In contrary to the term of convective transfer in tangent direction and to the pressure gradient term which are not changed the other two terms are increased. As for asymptotic limit of strong suction, the flow-field at the considered layer should be determined from balance between only the first two terms: molecular and convective transfer in normal direction. This assumption enables one to write the relation:

$$\frac{1}{Re_\infty}\frac{\partial}{\partial\zeta}\left(\mu\frac{\partial}{\partial\zeta}\right) \sim \rho u^3 \frac{\partial}{\partial\zeta} \Rightarrow \frac{1}{Re_\infty\delta^\omega(\Delta_{SU})^2} = \frac{\varepsilon g}{\varepsilon\Delta_{SU}} \Rightarrow \Delta_{SU} \sim \frac{1}{gRe_\infty\delta^\omega}.$$

There is enough information to introduce the inner variables at viscous boundary layer region for strong suction regime. For asymptotic analysis of this layer there will be used the following independent variables:

$$\xi^\alpha = \xi_{SU}^\alpha, \quad \zeta = \frac{1}{gRe_\infty\delta^\omega}\zeta_{SU}, \quad y = 0. \tag{4.64}$$

As for dependent variables, it is convenient to seek the asymptotic solution in the form of following expansions:

$$u^\alpha = \sqrt{\varepsilon}u_{SU}^\alpha + \dots, \quad u^3 = \varepsilon gv_{SU} + \dots, \quad T = \tfrac{1}{\delta}T_{SU} + \dots,$$
$$\mu = \tfrac{1}{\delta^\omega}\mu_{SU} + \dots, \quad \rho = \rho_{SU} + \dots, \quad P = \tfrac{1}{\varepsilon\delta}P_{SU} + \dots \tag{4.65}$$

Substituting expansions (4.64) and (4.65) into the system of initial equations we obtain the equations for viscous boundary layer for strong suction regime (index SU omitted),

$$\frac{\partial}{\partial\zeta}(\rho v\sqrt{a}) = 0, \quad \rho v\frac{\partial u^\alpha}{\partial\zeta} = \frac{\partial}{\partial\zeta}\left(\mu\frac{\partial u^\alpha}{\partial\zeta}\right), \quad \frac{\partial P}{\partial\zeta} = 0,$$
$$P = \rho T, \quad \rho v\frac{\partial T}{\partial\zeta} = \frac{\partial}{\partial\zeta}\left(\frac{\mu}{\sigma}\frac{\partial T}{\partial\zeta}\right) + \varepsilon\mu B_{\alpha\beta}\frac{\partial u^\alpha}{\partial\zeta}\frac{\partial u^\beta}{\partial\zeta}. \tag{4.66}$$

These equations have the second order and therefore they should be solved together with the boundary conditions at $\zeta_{SU} = 0$ and at infinity, $\zeta_{SU} \to \infty$.

The conditions at infinity are connected under matching conditions with the solution for inviscid boundary layer and so they have the following simple form:

$$\zeta_{SU} \to \infty : u_{SU}^\alpha = C^\alpha + u_{IB}^\alpha(\xi^1,\xi^2,0), \quad T_{SU} = C = T_{IB}(\xi^1,\xi^2,0) = \frac{1}{2}, \quad P_{SU} = P_I(\xi^1,\xi^2,0).$$

At the body surface the boundary conditions are

$$\zeta_{SU} = 0 : u_{SU}^\alpha = B^\alpha = (u_{SU}^\alpha)_w(\xi^1,\xi^2), \quad g\rho_{SU}v_{SU} = G(\xi^1,\xi^2), \quad T_{SU}(T_{SU})_w(\xi^1,\xi^2). \tag{4.67}$$

Two-point boundary value problem (4.66)–(4.67) accepts analytic solution. To obtain it first of all note that the first equation is of the first order and has simple solution (see below) accounted the body surface boundary condition, $\rho v = A(\xi^1,\xi^2) = G/g$. Another first order equation is also integrated (with the boundary condition at infinity), $P = P(\xi^1,\xi^2) = P_I(\xi^1,\xi^2,0)$. To solve the last two equations we use new independent variable

$$\psi = \int_0^\zeta \frac{A}{\mu}\,d\zeta. \tag{4.68}$$

Considering (4.68) helps to rewrite the equations for u^α in the form:

$$A\frac{\partial u^\alpha}{\partial \zeta} = A\frac{A}{\mu}\frac{\partial u^\alpha}{\partial \psi} = \frac{A}{\mu}\frac{\partial}{\partial \psi}\left(\frac{A}{\mu}\mu\frac{\partial u^\alpha}{\partial \psi}\right) \Rightarrow \frac{\partial u^\alpha}{\partial \psi} = \frac{\partial^2 u^\alpha}{\partial \psi^2}. \tag{4.69}$$

Equation (4.69) has the general solution $u^\alpha = Q_1^\alpha + Q_2^\alpha \exp\psi$. Considering the boundary conditions and the fact that $\lim\limits_{\zeta \to +\infty} \psi \to -\infty$ (due to $A < 0$) one obtains the relation connected the values Q_1^α and Q_2^α

$$Q_1^\alpha = C^\alpha, \quad Q_1^\alpha = B^\alpha - C^\alpha.$$

After transformation to new variables the last equation may be rewritten as follows:

$$\frac{\partial T}{\partial \psi} = \frac{1}{\sigma}\frac{\partial^2 T}{\partial \psi^2} + R\exp(2\psi), \quad R \equiv \varepsilon B_{\alpha\beta}C^\alpha C^\beta.$$

The general solution for the temperature can be sought in the form

$$T = Q_3\exp(\sigma\psi) + Q_4\exp(2\psi) + Q_5.$$

Substituting the expression above into initial equation gives the relation

$$Q_4 = \frac{1}{2}[\sigma/(\sigma - 2)].$$

Finally, using the boundary conditions the values Q_3 and Q_5 can be found as follows

$$Q_3 = T_w - \frac{1}{2}[1 + \sigma R/(\sigma - 2)], \quad Q_5 = C = \frac{1}{2}.$$

The solution to the temperature equation for viscous boundary layer with strong suction is

$$T = T_w + \left(\frac{1}{2} - T_w\right)[1 - \exp(\sigma\psi)] + \frac{1}{2}\left(\frac{\sigma}{\sigma - 2}\right)R[\exp(2\psi) - \exp(\sigma\psi)].$$

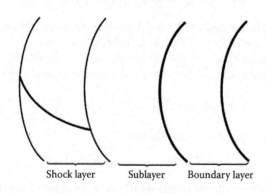

Shock layer   Sublayer   Boundary layer

FIGURE 4.12: The structure of viscous boundary layer with strong suction.

4.6.5 Flow at viscous boundary layer

The main goal in this section is to analyze the flow structure at some sublayers considered above. Start from $3d$ viscous laminar boundary layer [1, 24]. First of all it should be written for this layer the initial governing equations similar that were obtained previously, see (4.22),

$$
\begin{aligned}
&\frac{\partial}{\partial \xi^\alpha}(\rho u^\alpha \sqrt{a/a_{(\alpha\alpha)}}) + \frac{\partial}{\partial \zeta}(\rho V \sqrt{a}) = 0,\\
&\rho D^* u^\alpha + \rho A^\alpha_{\beta\gamma} u^\beta u^\gamma = -\frac{\gamma+1}{\gamma} a^{\alpha\beta} \sqrt{a_{(\alpha\alpha)}} \frac{\partial P}{\partial \xi^\beta} + \frac{\partial}{\partial \zeta}\left(\mu \frac{\partial u^\alpha}{\partial \zeta}\right),\\
&\rho D^* T = \frac{\partial}{\partial \zeta}\left(\frac{\mu}{\sigma} \frac{\partial T}{\partial \zeta}\right) + \frac{\gamma-1}{\gamma+1} \mu B_{\alpha\beta} \frac{\partial u^\alpha}{\partial \zeta} \frac{\partial u^\beta}{\partial \zeta} + \frac{\gamma-1}{\gamma}\left(u^\alpha/\sqrt{a_{(\alpha\alpha)}}\right) \frac{\partial P}{\partial \xi^\alpha},\\
&P(\xi^1, \xi^2) = P_e(\xi^1, \xi^2) = \rho T, \quad \mu = T^\omega, \quad D^* = \left(u^\alpha/\sqrt{a_{(\alpha\alpha)}}\right)\frac{\partial}{\partial \xi^\alpha} + u^3 \frac{\partial}{\partial \zeta}.
\end{aligned}
\tag{4.70}
$$

Equations (4.70) can be solved with boundary conditions at infinity and at the boundary surface

$$
\begin{aligned}
&\zeta \to \infty: \ u^\alpha = u^\alpha_e(\xi^1, \xi^2), \quad T = T_e(\xi^1, \xi^2), \quad P = P_e(\xi^1, \xi^2),\\
&\zeta = 0: \ \rho u^3 = G(\xi^1, \xi^2), \quad T = T_w(\xi^1, \xi^2),\\
&\zeta = 0: \ u^\alpha = u^\alpha_w(\xi^1, \xi^2), \quad \lim_{\xi^1 \to 0, \ \xi^2 \to 0}(u^\alpha_w/u^\alpha_e) = O(1).
\end{aligned}
$$

Here are the subscriptions e, w, o associated with the parameters at the outer bound of the boundary layer, body surface and at the stagnation point, respectively. The pressure P, see (4.70), depends only on the body surface coordinates ξ^1, ξ^2, and that it is known function.

The second important remark to this statement which is connected with boundary conditions at infinity is how we should determine the values $P_e(\xi^1, \xi^2)$, $u^\alpha_e(\xi^1, \xi^2)$, and $T_e(\xi^1, \xi^2)$. From general point of view these values should be known from the solution to Euler equations over the same body with the condition $u^3 = V_w$ at the body surface that is the stream surface for these equations. It is not too simple to obtain this solution. In this connection the following more simple approach is applied. First, as an initial data there is only used the pressure distribution $P_e(\xi^1, \xi^2)$ along the body surface. The reason is that this value is more conservative by comparison with other ones. In addition to determine this distribution a number of approximate theories can be used. As a role they have rather accuracy, and the choice of the particular theory depends on the class of the body shape and on the range of Mach number at infinity. For example, for hypersonic flow over blunted bodies to determine the pressure distribution along the body surface one can use the following expression which is called Newton formula:

$$
P_e/P_\infty = \cos^2 \alpha,
$$

where α is the angle between external normal to the body surface and the direction of the velocity vector to infinity. With known pressure distribution one can obtain the velocity $u^\alpha_e(\xi^1, \xi^2)$ and temperature $T_e(\xi^1, \xi^2)$ distributions based on the solution to the following rather simple system of differential and algebraic equations:

$$
\begin{aligned}
&\rho_e\left\{\left(u^\alpha_e/\sqrt{a_{(\alpha\alpha)}}\right)\frac{\partial u^\alpha_e}{\partial \xi^\alpha} + A^\alpha_{\beta\gamma} u^\beta_e u^\gamma_e\right\} = -\frac{\gamma+1}{\gamma} a^{\alpha\beta}\sqrt{a_{(\alpha\alpha)}}\frac{\partial P}{\partial \xi^\beta},\\
&\rho_e\left(u^\alpha_e/\sqrt{a_{(\alpha\alpha)}}\right)\frac{\partial T_e}{\partial \xi^\alpha} = \left(u^\alpha_e/\sqrt{a_{(\alpha\alpha)}}\right)\frac{\partial P_e}{\partial \xi^\alpha}.
\end{aligned}
\tag{4.71}
$$

In fact, (4.71) are momentum Euler equations in α direction and energy Euler equation for the temperature which are written along the stream surface. Introduction new coordinate l along the streamline $dl = \frac{u^1_e}{\sqrt{a_{11}}}d\xi^1 + \frac{u^2_e}{\sqrt{a_{22}}}d\xi^2$ helps to transfer the energy equation to the form

$$
\rho_e \frac{dT_e}{dl} = \frac{\gamma-1}{\gamma}\frac{dP_e}{dl} \ \Rightarrow \ \rho_e \frac{dT_e}{T_e} = \frac{\gamma-1}{\gamma}\frac{dP_e}{P_e}.
$$

One obtains integrals which are true along the body surface

$$
\frac{T_e}{T_{e0}} = \left(\frac{P_e}{P_{e0}}\right)^{\frac{\gamma-1}{\gamma}}, \quad \frac{\rho_e}{\rho_{e0}} = \left(\frac{P_e}{P_{e0}}\right)^{\frac{1}{\gamma}},
$$

where the index $e0$ is associated with the parameters at the stagnation point of the outer bound of the boundary layer. Consider system (4.70) under differential equation theory point of view. The main questions are:

   – What type have these equations?

   – Where are the singular points (lines, surfaces) for these equations?

Answers to these questions are very important for general study of the flow structure at the boundary layer as well as for numerical simulation and for analytic analysis of solutions [9].

The second question, because all streamlines at the boundary layer begin with the stagnation point (the maximum point of the pressure distribution – along the body surface), the stagnation point is the singular point of system (4.70). To understand the singularity type one should study the flow in the neighborhood of the stagnation point. To do it we have to determine the coordinate system (ξ^1, ξ^2) at the body surface. It is known from differential geometry that a smooth surface in the stagnation point neighborhood may be approximated by the following way:

$$p^3 = \frac{1}{2}[(p^1)^2 + k(p^2)^2].$$

Here (p^1, p^2, p^3) is the Cartesian coordinate system, axis p^3 is directed along external normal to the body surface at the stagnation point, axes p^1, p^2 are directed along the main direction at the stagnation point of the body surface, $k = R_1/R_2 \leq 1.0$ is the ratio of the main curvature radii R_1 and R_2 at the stagnation point.

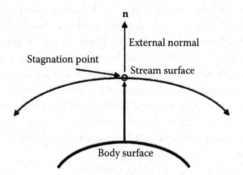

FIGURE 4.13: Geometry of the stream surface and of the stagnation point.

Introduce the parametrization of the body surface as follows (Fig. 4.14),

$$p^1 = \xi^1, \quad p^2 = \xi^2, \quad p^3 = \frac{1}{2}[(\xi^1)^2 + (\xi^2)^2]$$

and to obtain the relations for components $a_{\alpha\beta}$: $a_{11} = 1 + (\xi^1)^2$, $a_{22} = 1 + k^2(\xi^2)^2$ and $a_{12} = a_{21} = k\xi^1\xi^2$. Note that in above introduced coordinate system the stagnation point has the coordinates $\xi^1 = 0$ and $\xi^2 = 0$. Considering these links there can be written the estimations for nonzero values $A^{\alpha}_{\beta\gamma}$ at the stagnation point neighborhood,

$$A^1_{11} \sim \xi^1(\xi^2)^2, \quad A^2_{22} \sim \xi^2(\xi^1)^2, \quad A^1_{22} \sim \xi^1, \quad A^2_{11} \sim \xi^2.$$

On the other hand, because the stagnation point is the maximal point of the pressure distribution it can be used the estimation for P_e in the stagnation point neighborhood,

$$P_e = P_{e0} - \frac{1}{2}[P_1(\xi^1)^2 + P_2(\xi^2)^2] \quad \Rightarrow \quad \frac{\partial P_e}{\partial \xi^\alpha} \sim \xi^\alpha.$$

It is nature to assume that the character of singularity of the velocity vectors components u^α is the same as at the outer bound of the boundary layer. Using previous estimations and based on (4.71), it can be written as

$$\frac{\partial u_e^1}{\partial \xi^1} = \sqrt{\frac{\gamma+1}{\gamma}}P_1, \quad \frac{\partial u_e^2}{\partial \xi^2} = \sqrt{\frac{\gamma+1}{\gamma}}P_2,$$

and therefore $u^1 \sim u_e^1 \sim \xi^1$, $u^2 \sim u_e^2 \sim \xi^2$. Now we have enough information to introduce new variables and by this way to resolve the singularity problem for the boundary layer equations at the stagnation point. Using independent variables

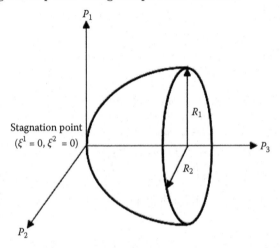

FIGURE 4.14: Parametrization of the coordinate system connected with stagnation point.

$$x = \xi^1, \quad y = \xi^2, \quad z = \Lambda(\xi^1, \xi^2) \int_0^\zeta \rho d\zeta,$$

there can be written the relations for differential operators, as follows below,

$$\frac{\partial}{\partial \xi^1} = \frac{\partial}{\partial x} + \frac{\partial z}{\partial \xi^1}\frac{\partial}{\partial z}, \quad \frac{\partial}{\partial \xi^2} = \frac{\partial}{\partial x} + \frac{\partial z}{\partial \xi^2}\frac{\partial}{\partial z}, \quad \frac{\partial}{\partial \zeta} = \Lambda\rho\frac{\partial}{\partial z}.$$

Before to introduce new variables, define the streamlines functions f_α and using them to satisfy to continuity equation

$$\rho u^\alpha \sqrt{a/a_{(\alpha\alpha)}} = \frac{\partial f_\alpha}{\partial \zeta}, \quad \rho u^3 \sqrt{a} = -\left(\frac{\partial f_1}{\partial \xi^1} + \frac{\partial f_2}{\partial \xi^2}\right). \tag{4.72}$$

In new variables the first two relations have a view:

$$\rho u^1 \sqrt{a/a_{11}} = \Lambda\rho\frac{\partial f_1}{\partial z} \Rightarrow u \equiv \frac{u^1}{u_*^1} = \frac{\partial}{\partial z}\left(\frac{\Lambda}{u_*^1}f_1\sqrt{a_{11}/a}\right) = \frac{\partial\varphi_1}{\partial z},$$
$$\rho u^2 \sqrt{a/a_{22}} = \Lambda\rho\frac{\partial f_2}{\partial z} \Rightarrow w \equiv \frac{u^2}{u_*^2} = \frac{\partial}{\partial z}\left(\frac{\Lambda}{u_*^2}f_2\sqrt{a_{22}/a}\right) = \frac{\partial\varphi_2}{\partial z},$$

where the functions $u_*^\alpha(\xi^1, \xi^2)$ are given below, and φ_α are the modified stream functions

$$\varphi_1 \equiv \frac{\Lambda}{u_*^1}f_1\sqrt{\frac{a_{11}}{a}}, \quad \varphi_2 \equiv \frac{\Lambda}{u_*^2}f_2\sqrt{\frac{a_{22}}{a}}.$$

Using these links the third relation in (4.72) may be rewritten as

$$\rho u^3 \sqrt{a} = -\left[\frac{\partial}{\partial x}(\delta_1\varphi_1) + \frac{\partial}{\partial y}(\delta_2\varphi_2) + \delta_1\frac{\partial\varphi_1}{\partial z}\frac{\partial z}{\partial \xi^1} + \delta_2\frac{\partial\varphi_2}{\partial z}\frac{\partial z}{\partial \xi^2}\right],$$

where the coefficients δ_α are:

$$\delta_1 = (u_*^1/\Lambda)\sqrt{a_{11}/a}, \quad \delta_2 = (u_*^2/\Lambda)\sqrt{a_{22}/a}.$$

It can be obtained formula for operator D^*

$$D^* \equiv \frac{u^1}{\sqrt{a_{11}}}\frac{\partial}{\partial \xi^1} + \frac{u^2}{\sqrt{a_{22}}}\frac{\partial}{\partial \xi^2} + u^3\frac{\partial}{\partial \zeta} = \frac{uu_*^1}{\sqrt{a_{11}}}\frac{\partial}{\partial x} + \frac{wu_*^2}{\sqrt{a_{22}}}\frac{\partial}{\partial y} - \frac{\Lambda}{\sqrt{a}}\left[\frac{\partial}{\partial x}(\delta_1\varphi_1) + \frac{\partial}{\partial y}(\delta_2\varphi_2)\right]\frac{\partial}{\partial z}.$$

Define new dimensionless variables which are called Dorodnitzyn–Lis variables:

$$u \equiv \frac{u^1}{u_*^1} = \frac{\partial\varphi_1}{\partial z}, \quad w \equiv \frac{u^2}{u_*^2} = \frac{\partial\varphi_2}{\partial z}, \quad \theta \equiv \frac{T}{T_e}, \quad l = \frac{\mu\rho}{\mu_e\rho_e}. \tag{4.73}$$

Using our previous relations, write the final form for the boundary layer equations in variables (4.73)

$$\begin{aligned}
\frac{\partial}{\partial z}\left(l\frac{\partial u}{\partial z}\right) &= Du + \beta_1 u^2 + \beta_2 w^2 + \beta_3 uw - \gamma_1\theta, \\
\frac{\partial}{\partial z}\left(l\frac{\partial w}{\partial z}\right) &= Dw + \beta_4 w^2 + \beta_5 u^2 + \beta_6 uw - \gamma_2\theta, \\
\frac{\partial}{\partial z}\left(l\frac{\partial w}{\partial z}\right) &= Dw + \beta_4 w^2 + \beta_5 u^2 + \beta_6 uw - \gamma_2\theta, \\
\frac{\partial}{\partial z}\left(\frac{l}{\sigma}\frac{\partial\theta}{\partial z}\right) &= D\theta - l\left[\alpha_5\left(\frac{\partial u}{\partial z}\right)^2 + \alpha_6\left(\frac{\partial w}{\partial z}\right)^2 + \alpha_7\frac{\partial u}{\partial z}\frac{\partial w}{\partial z}\right], \\
D &= \alpha_1\left(u\frac{\partial}{\partial x} - \frac{\partial\varphi_1}{\partial x}\frac{\partial}{\partial z}\right) + \alpha_2\left(w\frac{\partial}{\partial y} - \frac{\partial\varphi_2}{\partial y}\frac{\partial}{\partial z}\right) - (\alpha_3\varphi_1 + \alpha_4\varphi_2)\frac{\partial}{\partial z}.
\end{aligned} \tag{4.74}$$

The coefficients α_k, β_k and γ_k, in (4.74) depend only on the body surface coordinates (x, y) and are given by

$$\begin{aligned}
&\alpha_1 = \frac{u_*^1}{A\sqrt{a_{11}}}, \quad \alpha_2 = \frac{u_*^2}{A\sqrt{a_{22}}}, \quad \alpha_3 = \frac{\Lambda}{A\sqrt{a}}\frac{\partial\delta_1}{\partial x}, \quad \alpha_4 = \frac{\Lambda}{A\sqrt{a}}\frac{\partial\delta_2}{\partial y}, \\
&\alpha_5 = \frac{(u_*^1)^2}{T_e}\frac{\gamma-1}{\gamma+1}, \quad \alpha_6 = \frac{(u_*^2)^2}{T_e}\frac{\gamma-1}{\gamma+1}, \quad \alpha_7 = 2B_{12}\frac{u_*^1 u_*^2}{T_e}\frac{\gamma-1}{\gamma+1}, \\
&\beta_1 = \frac{1}{A}\left[A_{11}^1 u_*^1 + \frac{1}{\sqrt{a_{11}}}\frac{\partial u_*^1}{\partial x}\right], \quad \beta_2 = \frac{A_{22}^1}{A}\frac{(u_*^2)^2}{u_*^1}, \quad \beta_3 = \frac{u_*^1}{A}\left[2A_{12}^1 + \frac{1}{\sqrt{a_{22}}}\frac{\partial\ln u_*^1}{\partial y}\right], \\
&\beta_4 = \frac{1}{A}\left[A_{22}^2 u_*^2 + \frac{1}{\sqrt{a_{22}}}\frac{\partial u_*^2}{\partial y}\right], \quad \beta_5 = \frac{A_{11}^1}{A}\frac{(u_*^1)^2}{u_*^2}, \quad \beta_6 = \frac{u_*^1}{A}\left[2A_{12}^2 + \frac{1}{\sqrt{a_{11}}}\frac{\partial\ln u_*^2}{\partial x}\right], \\
&\gamma_1 = \frac{1}{Au_*^1}\left[\frac{u_e^1}{\sqrt{a_{11}}}\frac{\partial u_e^1}{\partial x} + \frac{u_e^2}{\sqrt{a_{22}}}\frac{\partial u_e^1}{\partial y} + A_{11}^1(u_e^1)^2 + A_{22}^1(u_e^2)^2 + 2A_{12}^1 u_e^1 u_e^2\right], \\
&\gamma_2 = \frac{1}{Au_*^2}\left[\frac{u_e^1}{\sqrt{a_{11}}}\frac{\partial u_e^2}{\partial x} + \frac{u_e^2}{\sqrt{a_{22}}}\frac{\partial u_e^2}{\partial y} + A_{11}^2(u_e^1)^2 + A_{22}^2(u_e^2)^2 + 2A_{12}^2 u_e^1 u_e^2\right], \quad A = \Lambda\mu_e\rho_e.
\end{aligned} \tag{4.75}$$

The boundary conditions are

$$\begin{aligned}
z \to \infty: \quad &u = u_e^1/u_*^1, \quad w = u_e^2/u_*^2, \quad \theta = 1; \\
z \to 0: \quad &u = u_w^1/u_*^1, \quad w = u_w^2/u_*^2, \quad \theta = \theta_w = T_w/T_e, \\
&\frac{\partial}{\partial x}(\delta_1\varphi_1) + \frac{\partial}{\partial y}(\delta_2\varphi_2) = -G\sqrt{a}.
\end{aligned}$$

To finish the statement of the problem it we have to determine free functions u_*^1, u_*^2 and Λ. It can be done using the following principles:

    1. The coefficients (4.75) involved in (4.74) have no singularities at the stagnation point.

    2. These coefficients should be as simple as possible.

To satisfy to the first condition we have to use our previous estimations concerning the parameters orders at the stagnation point. It was found that this condition is satisfied if $u_*^1 \sim x$, $u_*^2 \sim y$. More precise choice of the listed above free functions depends on the flow geometry. As an example, if there are two symmetry planes in the flow, it is convenient to use the following relations for these functions: $u_*^1 = u_e^1$, $u_*^2 = u_e^2$, $\Lambda = (u_e^1/x)\sqrt{a_{11}/a}$. In this case the coefficients (4.75) are simplified,

$$\delta_1 = x, \quad \delta_2 = u_e^2(x/u_e^1)\sqrt{a_{11}/a_{22}}, \quad \gamma_1 = \beta_1 + \beta_2 + \beta_3, \quad \gamma_2 = \beta_4 + \beta_5 + \beta_6.$$

Consider the flow in a neighborhood of the stagnation point. Because the outer bound of the boundary layer is the stream surface (due to solution to Euler equations) there may be written the relation between flow parameters along stream line for viscous gas flow:

$$\frac{1}{2}(V_e^2/c_p T_{e0}) = 1 - T_e/T_{e0} = 1 - (T_e/T_{e0})^{\frac{\gamma-1}{\gamma}}.$$

To calculate the pressure distribution along the body surface for hypersonic gas flow one can use Newton formula which for the flow-field inside the neighborhood of the stagnation point has a view:

$$\frac{P_e}{P_\infty} = \cos^2 \alpha = (\mathbf{n} \cdot \mathbf{V}_\infty)^2 = \frac{1}{1 + x^2 + k^2 y^2} \simeq 1 - (x^2 + k^2 y^2) - 2k^2 x^2 y^2.$$

On the other hand,

$$V_e^2 = \frac{\gamma - 1}{\gamma + 1}\left[(u_e^1)^2 + (u_e^2)^2 + 2\frac{a_{12}}{\sqrt{a_{11}a_{22}}} u_e^1 u_e^2\right].$$

This yields the formulae for u_e^1 and u_e^2 at the stagnation point neighborhood: $u_e^1 = Cx$, $u_e^2 = Cky$ and $C = \sqrt{2c_p T_{e0}}\sqrt{(\gamma + 1)/\gamma}$. Based on the above one can obtain the expressions for nonzero coefficients of the boundary layer equations, $\alpha_3 = \beta_1 = C/A$, $\alpha_4 = \beta_4 = kC/A$ If the function Λ is determined from $\frac{C}{A} = \frac{C}{\mu_e \rho_e \Lambda^2} = 1$ one obtains the simple form of the boundary layer equations for the neighborhood of stagnation point of blunted bodies

$$\begin{aligned}
&\frac{\partial}{\partial z}\left(l\frac{\partial u}{\partial z}\right) + (\varphi_1 + k\varphi_2)\frac{\partial u}{\partial z} = u^2 - \theta, \\
&\frac{\partial}{\partial z}\left(l\frac{\partial w}{\partial z}\right) + (\varphi_1 + k\varphi_2)\frac{\partial w}{\partial z} = k(w^2 - \theta), \\
&\frac{\partial}{\partial z}\left(l\frac{\partial \theta}{\partial z}\right) + (\varphi_1 + k\varphi_2)\frac{\partial \theta}{\partial z} = 0, \\
&z \to \infty: \ u = w = \theta = 1, \\
&z = 0: \ u = u_w, \quad w = w_w, \quad \theta = \theta_w, \quad \varphi_1 + k\varphi_2 = -G.
\end{aligned}$$

Important conclusions:

1. The solution to the boundary layer equations at the stagnation point is reduced to two-point boundary value problem for the second order ODE system. This system can be obtained independently from the solution to the boundary layer equations at the other domains. It means that the boundary layer equations are of parabolic type and their solution at the stagnation point is the initial condition for the solution at the lateral body surface.

2. The influence of the body shape on the solution to the boundary layer equations at the stagnation point is completely determined by parameter k. The case $k = 1$ corresponds to the stagnation point at the sphere while the case $k = 0$ corresponds stagnation point at the cylinder.

4.7   Flow at injection gas layer

An interest to the flows at *intense gas injection* through the surface of a flight vehicle into the flow is due to the fact, that the use of this technique can reduce heat fluxes and friction and produce a required pressure distribution. Moreover, the intense injection regime can sometimes arise in a "natural" way upon ablation of thermal protection. Here, we should study the flows near thin bodies, whose thickness is not greater in the order than that of the region occupied by injected gas, and bodies with small longitudinal curvatures of the surface (wedge, flat plate, etc.). In these cases, the pressure distribution is determined by the distribution of the displacement thickness of the *injected gas layer*, which, in turn, depends on pressure distribution. The effects that take place at gas injection into a supersonic and hypersonic flows should be studied using NS equations (see Section 1.8) and the asymptotic expansions matching principle (see Section 1.6) and transition from the flow pattern described by the classical layer theory to the intense injection regime, in which a region of an inviscid (in the first approximation) flow of the injected gas is formed near the body surface, should be considered.

The governing equations [15] for the flow at injection gas layer region are:

$$\frac{\partial}{\partial \xi^\alpha}(\rho u^\alpha \sqrt{a/a_{(\alpha\alpha)}}) + \frac{\partial}{\partial \zeta}(\rho u^3 \sqrt{a}) = 0, \quad \rho(D^* u^\alpha + A^\alpha_{\alpha\beta} u^\beta u^\gamma) = -\frac{\gamma+1}{\gamma} a^{\alpha\beta} \sqrt{a_{(\alpha\alpha)}} \frac{\partial P}{\partial \xi^\beta},$$
$$\rho D^* T = \frac{\gamma-1}{\gamma}(u^\alpha/\sqrt{a_{(\alpha\alpha)}})\frac{\partial P}{\partial \xi^\alpha}, \quad P(\xi^1,\xi^2) = P_c(\xi^1,\xi^2) = \rho T,$$
$$D^* = (u^\alpha/\sqrt{a_{(\alpha\alpha)}})\frac{\partial}{\partial \xi^\alpha} + u^3 \frac{\partial}{\partial \zeta},$$

$$(4.76)$$

see Section 4.6.1. Equations (4.76) are solved under the boundary conditions at the body and at the contact surfaces:

$$\zeta = 0: \ \rho u^3 = G(\xi^1,\xi^2), \quad T = T_w(\xi^1,\xi^2), \quad u^\alpha = u^\alpha_w(\xi^1,\xi^2), \quad \lim_{\xi^1 \to 0, \ \xi^2 \to 0} \frac{u^\alpha_w}{\xi^\alpha} = O(1);$$
$$\zeta = \zeta_c: \ u^3 = 0, \quad P = P_c(\xi^1,\xi^2).$$

$$(4.77)$$

Here, the subscripts c, w associated with the parameters at the contact and the body surfaces, respectively; $P_c(\xi^1,\xi^2)$ - known function. Obviously, (4.76) are the first order ones equations. Because the most part of the boundary conditions lie at the body surface one can obtain explicitly the formulae for the heat flux and for the skin friction at the body surface $\tau_\alpha = (\mu \frac{\partial u^\alpha}{\partial \zeta})_w$ and $q = (\lambda \frac{\partial T}{\partial \zeta})_w$. To determine τ_α and q it is necessary to know the derivatives $\partial u^\alpha/\partial \zeta$ and $\partial T/\partial \zeta$ at the body surface. Because the governing equations (4.76) are true in all domains including the body surface the following explicit relations for the derivatives above can be written as [7]

$$\frac{\partial u^\alpha}{\partial \zeta} = -\frac{1}{G}\left[\frac{\gamma+1}{\gamma} a^{\alpha\beta} \sqrt{a_{(\alpha\alpha)}} \frac{\partial P}{\partial \xi^\beta} + \rho_w D^0 u^\alpha_w + \rho_w A^\alpha_{\beta\delta} u^\beta_w u^\delta_w\right],$$
$$\frac{\partial T}{\partial \zeta} = \frac{1}{G}\left[\frac{\gamma-1}{\gamma} D^0 P - \rho_w D^0 P_w\right], \quad D^0 \equiv (u^\alpha_w/\sqrt{a_{(\alpha\alpha)}})\frac{\partial}{\partial \xi^\alpha}.$$

Generally, the solution for the velocity and the temperature profiles across injection gas layer is sought in the form of series in ζ. For a number of special cases this solution can be obtained in quadratures.

4.7.1   General stagnation point

First, consider the solution at $3d$ stagnation point. To do it, use the previous estimations concerning orders of the velocity tangent components u^1 and u^2 in the neighborhood of the stagnation point of the blunted bodies and introduce new dimensionless variables $\bar{u}^1 = u^1/\xi^1$, $\bar{u}^2 = u^2/\xi^2$. By this way we resolve the singularity of the initial injection gas layer equations at the stagnation point. We obtain the following system:

$$\bar{u}^1 + \bar{u}^2 + \frac{du^3}{d\zeta} = 0, \quad (\bar{u}^1)^2 + u^3 \frac{d\bar{u}^1}{d\zeta} = -\frac{\gamma+1}{\gamma}\frac{1}{\rho_w}\frac{\partial}{\partial \xi^1}\left(\frac{\partial P_c}{\partial \xi^1}\right),$$
$$(\bar{u}^2)^2 + u^3 \frac{d\bar{u}^2}{d\zeta} = -\frac{\gamma+1}{\gamma}\frac{1}{\rho_w}\frac{\partial}{\partial \xi^2}\left(\frac{\partial P_c}{\partial \xi^2}\right), \quad \frac{dT}{d\zeta} = 0.$$

$$(4.78)$$

Consider the boundary condition $u^3 = 0$ together with the last two equations from (4.78) which are for the contact surface. We obtain relations

$$(\bar{u}^1)^2 = -\frac{\gamma+1}{\gamma}\frac{1}{\rho_w}\frac{\partial}{\partial \xi^1}\left(\frac{\partial P_c}{\partial \xi^1}\right), \quad (\bar{u}^2)^2 = -\frac{\gamma+1}{\gamma}\frac{1}{\rho_w}\frac{\partial}{\partial \xi^2}\left(\frac{\partial P_c}{\partial \xi^2}\right).$$

It immediately follows important conclusion: the solution to the boundary layer problem (4.78), (4.77) in real domain exists if the following conditions are realized: $\frac{\partial}{\partial \xi^1}\left(\frac{\partial P_c}{\partial \xi^1}\right) < 0$ and $\frac{\partial}{\partial \xi^2}\left(\frac{\partial P_c}{\partial \xi^2}\right) < 0$. In order words, this solution exists if the longitudinal pressure gradients, at the contact surface, which are determined from the solution to inviscid shock layer problem, are negative in all directions. It means that the condition of the existence of this solution is the following: the stagnation point is the point of local maximum in the pressure distribution along the contact surface. Later it will be considered only this case. For convenience to integration, introduce new variables $u = \bar{u}^1/\sqrt{a_1}$, $w = \bar{u}^2/\sqrt{a_2}$ and $v = u^3$ and $z = \sqrt{a_1}\zeta$, where

$$a_1 = -\frac{\gamma+1}{\gamma}\frac{1}{\rho_w}\frac{\partial}{\partial \xi^1}\left(\frac{\partial P_c}{\partial \xi^1}\right), \qquad a_2 = -\frac{\gamma+1}{\gamma}\frac{1}{\rho_w}\frac{\partial}{\partial \xi^2}\left(\frac{\partial P_c}{\partial \xi^2}\right).$$

We obtain the following system:

$$\begin{aligned} u + kw + \tfrac{dv}{dz}, \quad T &= \text{const} = T_w, \quad \rho = \text{const} = \rho_w, \\ v\tfrac{du}{dz} &= 1 - u^2, \quad v\tfrac{dw}{dz} = k(1-w^2), \quad k = \text{const} = \sqrt{a_2/a_1}. \end{aligned} \tag{4.79}$$

For a hypersonic flow the value k is equal to the ratio of the main radii R_1 and R_2 of the body surface at the stagnation point.

At first we obtain the general solution to the system (4.79). From the last two equations it follows,

$$\begin{aligned} \frac{du}{dw} &= k^{-1}\frac{1-u^2}{1-w^2} \Rightarrow \frac{kdu}{u^2-1} = \frac{dw}{w^2-1} \Rightarrow \tfrac{1}{2}k\left(\frac{du}{u-1} - \frac{du}{u+1}\right) = \tfrac{1}{2}k\left(\frac{dw}{w-1} - \frac{dw}{w+1}\right) \\ &\Rightarrow |(w-1)/(w+1)| = C_1|(u-1)/(u+1)|^k, \end{aligned}$$

where C_1 is the constant of integration; it is determined from the boundary condition at the body surface:

$$C_1 = |(w_w - 1)/(w_w + 1)| \cdot |(u_w + 1)/(u_w - 1)|^k.$$

To obtain the second integral, substitute the link between dv and dz from the first equation to the second one,

$$dz = -\frac{dv}{u+kw} \Rightarrow v\frac{du}{dv} = \frac{u^2-1}{u+kw} \Rightarrow \frac{dv}{v} = \frac{(u+kw)du}{u^2-1} = \frac{udu}{u^2-1} + k\frac{wdu}{u^2-1}.$$

Because of the link between du and dw, $\frac{kdu}{u^2-1} = \frac{dw}{u^2-1}$, we have

$$dz = \frac{udu}{u^2-1} + \frac{wdu}{w^2-1} = \frac{1}{2}[d\ln(u^2-1) + d\ln(w^2-1)].$$

This has solution $v = C_2\sqrt{|(u^2-1)(w^2-1)|}$, where $C_2 = v_w/\sqrt{|(u_w^2-1)(w_w^2-1)|}$.

Finally, write the last integral $z = -\int_{v_w}^{v}\frac{dv}{u+kw}$. So, we obtain the solution to the Cauchy problem for injection gas layer equations (4.76). At the same time we have to remember that our initial problem is the boundary value problem: the solution must satisfy to the body surface boundary conditions together with the contact surface condition $v = 0$.

Our goal is to recognize the solution to the boundary value problem between the solutions to the Cauchy problem. To solve this problem, turn to the integral curve pattern of the system (4.79) in the phase space (u, w, v). The solution to the boundary value problem is a part of the integral curve between point $P_w(u_w, w_w, v_w)$ and point $P_c(u_c, w_c, v_c)$ which is intersection point of this integral curve with the plane $v = 0$. It means that the existence of solutions to the boundary value problem completely determined by the existence of this point of intersection. To clarify the situation, remember from differential equations theory, that the topology of integral curves is determined by singular points of the system of these differential equations. In particular, it is known that all integral curves either are closed curves or they are started from one singular point and are finished at another one. To show this, consider infinity as one of singular points of our system. Our system (4.79) has four singular points: stable node $F_1(1, 1, 0)$, unstable node $F_2(-1, -1, 0)$ and two saddle points $F_3(-1, 1, 0)$ and $F_4(1, -1, 0)$; see Fig. 4.15. The projection of the integral curves on the plane $v = 0$ is also given in Fig. 4.15.

All integral curves can be divided on two classes: the first class includes the integral curves which arrive at singular point F_1. The second one includes the integral curves which depart from infinity (or arrive to infinity singular point) without intersection with the plane $v = 0$. Because at the point F_1 the condition $v = 0$ is satisfied we conclude that the solution to the boundary value problem is the part of the curve which belongs to the first class of integral curves. On the other hand, the class of integral curves is completely determined by the starting point P_w or, in other words, by the boundary conditions at the body surface for u and w. It is easy to see that the integral curve is among the first class only if the following condition is satisfied:

$$u_w > -1, \quad w_w > -1. \tag{4.80}$$

Thus, we may conclude that the condition (4.80) can be considered as the condition of existence of two-point boundary value problem for injection gas layer equations at the stagnation point neighborhood.

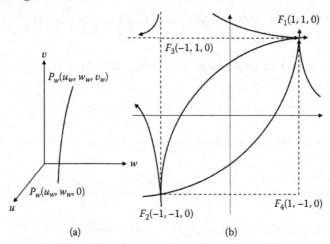

FIGURE 4.15: Integral curves: (a) space curve, (b) projection on the plane $v = 0$.

4.7.2 Wings at incidence at slipping angles

For this flow geometry, choose ξ^2-axis along normal to the wings section. Assuming the body surface boundary conditions be independent from ξ^2 one can conclude that the solution to the boundary value problem (4.76)–(4.77) is also independent from this coordinate and in this connection the terms with derivatives on ξ^2 are zero in (4.76). Introducing designations $u = u^1$, $w = u^2$, $v = u^3$, $x = \xi^1$, $y = \zeta$ the system (4.76) may be rewritten in a more simple view

$$\frac{\partial}{\partial x}(\rho u) + \frac{\partial}{\partial y}(\rho v) = 0, \quad \rho\left(u\frac{\partial u}{\partial x} + v\frac{\partial v}{\partial y}\right) = -\frac{\gamma+1}{\gamma}\frac{\partial P_c}{\partial x},$$
$$\rho\left(u\frac{\partial w}{\partial x} + v\frac{\partial w}{\partial y}\right) = 0, \quad \rho\left(u\frac{\partial T}{\partial x} + v\frac{\partial T}{\partial y}\right) = \frac{\gamma-1}{\gamma}\frac{\partial P_c}{\partial x}.$$

The best way to integrate these equations is to introduce the stream function ψ,

$$\rho u = \frac{\partial \psi}{\partial y}, \quad \rho v = -\frac{\partial \psi}{\partial x},$$

and than to transfer from variables (x, y) to new ones (x, ψ):

$$\frac{\partial}{\partial x} = \frac{\partial}{\partial x} - \rho v \frac{\partial}{\partial \psi}, \quad \frac{\partial}{\partial y} = \rho u \frac{\partial}{\partial \psi}.$$

We get

$$\frac{1}{2} \rho \frac{\partial}{\partial x}(u^2) = -\frac{\gamma + 1}{\gamma} \frac{\partial P_c}{\partial x}, \quad \frac{\partial w}{\partial x} = 0, \quad \rho \frac{\partial T}{\partial x} = \frac{\gamma - 1}{\gamma} \frac{\partial P_c}{\partial x}.$$

These equations have the following solution:

$$u^2(x,t) - u_w^2(t) = 2T_w(t)\frac{\gamma-1}{\gamma+1}\{1 - [P_c(x)/P_c(t)]^\kappa\}, \quad w(x,t) = w_w(t),$$
$$\frac{T(x,t)}{T_w(t)} = [P_c(x)/P_c(t)]^k, \quad \frac{\rho(x,t)}{\rho_w(t)} = [P_c(x)/P_c(t)]^{1/\gamma},$$
$$y = \int_t^x \frac{\rho_w(s)v_w(t)}{\rho(x,s)u(x,s)}ds \quad (0 \le t \le x), \quad \kappa(t) = \frac{\gamma(t)-1}{\gamma(t)},$$

where t is the coordinate x of the stream line leaved the body surface. Note that there is a singularity in this solution at the stagnation point. To resolve this problem, consider the initial equations with the following variables:

$$u_0 = \frac{u}{a_1 x}, \quad y_0 = a_1 y, \quad a_1 = \left(-\frac{\gamma+1}{\gamma}\frac{1}{\rho_w}\frac{d^2 P_c}{dx_2}\right)^{1/2}.$$

Then we obtain (subscript 0 is omitted)

$$\rho = \text{const} = \rho_w, \quad T = \text{const} = T_w, \quad u + \frac{dv}{dy} = 0, \quad u^2 + v\frac{du}{dy} = 1, \quad w = \text{const} = w_w.$$

There are two singular points at the phase space (u, v): stable node $F_1(1, 0)$ and unstable node $F_2(-1, 0)$. All integral curves either arrive to the stable node F_1 (for $u_w > -1$) or depart to infinity for opposite case. The last equations can be integrated,

$$u^2 = 1 + C_1 v^2, \quad y = y(v) = \int_v^{v_w} \frac{ds}{\sqrt{1+C_1 s^2}}, \quad C_1 = \frac{u_w^2 - 1}{v_w^2} \quad (0 \le v \le v_w).$$

Because the contact surface corresponds to the condition $v = 0$ the flow parameters at this surface will be $u_c = 1$, $w_c = w_w$ and $y_c = \int_0^{v_w} \frac{dv}{\sqrt{1+C_1 v^2}}$. Obviously that the solution with limit value of the layer thickness exists only if $u_w > -1$.

4.7.3 Flow over swirling axisymmetric bodies

For this flow geometry, choose ξ^2-axis along the circumferential direction. Assuming the body surface boundary conditions be independent from ξ^2 it can be concluded that the solution to the boundary value problem (4.76)–(4.77) is also independent from this coordinate and in this connection the terms with derivatives on ξ^2 are zero in (4.76). Introducing designations $u = u^1$, $w = u^2$, $v = u^3$, $x = \xi^1$, $y = \zeta$, equations (4.76) may be simplified to the form

$$\frac{\partial}{\partial x}(\rho u r) + \frac{\partial}{\partial y}(\rho v r) = 0, \quad \rho\left(u\frac{\partial w}{\partial x} + v\frac{\partial w}{\partial y} - \frac{w^2}{r}\frac{dr}{dx}\right) = -\frac{\gamma+1}{\gamma}\frac{\partial P_c}{\partial x},$$
$$\rho\left(u\frac{\partial w}{\partial x} + v\frac{\partial w}{\partial y} + \frac{uw}{r}\frac{dr}{dx}\right) = 0, \quad \rho\left(u\frac{\partial T}{\partial x} + v\frac{\partial T}{\partial y}\right) = \frac{\gamma-1}{\gamma}u\frac{\partial P_c}{\partial x}.$$

Introducing the stream function ψ and using transformation from variables (x, y) to new ones (x, ψ) we obtain the system

$$\frac{1}{2}\rho\frac{\partial}{\partial x}(u^2 + v^2) = -\frac{\gamma+1}{\gamma}\frac{\partial P_c}{\partial x}, \quad \frac{\partial}{\partial x}(wr) = 0, \quad \rho\frac{\partial T}{\partial x} = \frac{\gamma-1}{\gamma}\frac{\partial P_c}{\partial x}.$$

This system has the solution

$$u^2(x,t) + w^2(x,t) = u_w^2(t) + w_w^2(t) + 2T_w(t)\tfrac{\gamma-1}{\gamma+1}\{1 - [P_c(x)/P_c(t)]^\kappa\},$$
$$w(x,t) = w_w(t)\tfrac{r(t)}{r(x)}, \quad \tfrac{T(x,t)}{T_w(t)} = [P_c(x)/P_c(t)]^k, \quad \tfrac{\rho(x,t)}{\rho_w(t)} = [P_c(x)/P_c(t)]^{1/\gamma},$$
$$y = \int\limits_t^x \tfrac{\rho_w(t)w_w(t)r(t)}{\rho(x,t)u(x,t)r(x)}\, dt \quad (0 \le t \le x), \quad \kappa(t) = \tfrac{\gamma(t)-1}{\gamma(t)},$$

where t is the coordinate x of the stream line leaved the body surface. To resolve this problem, use the following variables:

$$u_0 = \frac{u}{a_1 x}, \quad w_0 = \frac{w}{a_1 x}, \quad y_0 = a_1 y, \quad a_1 = \Big(-\frac{\gamma+1}{\gamma}\frac{1}{\rho_w}\frac{d^2 P_c}{dx_2}\Big)^{1/2}.$$

Then we obtain (subscript 0 omitted):

$$\rho = \text{const} = \rho_w, \ T = \text{const} = T_w, \ 2u + \frac{dv}{dy} = 0, \ u^2 + v\frac{du}{dy} - w^2 = 1, \ 2uw + v\frac{dw}{dy} = 0.$$

This system can be integrated

$$u^2 = 1 - C_1 v^2 + C_2 v, \ w = C_1 v, \quad y = y(v) = \frac{1}{2}\int_v^{v_w}\frac{ds}{u(s)}, \ C_1 = \frac{w_w}{v_w^2}, \ C_2 = \frac{1}{v_w}(u_w^2 + w_w^2 - 1).$$

There are two singular points of this system at the phase space (u, v): stable node $F_1(1, 0)$ and unstable node $F_2(-1, 0)$. Note that the behavior of the integral curves completely determined by the velocity swirling w_w at the body surface. If $w_w = 0$ integral curves are the parts of parabola and independent from the value u_w either depart from F_2 and arrive at F_1 (for $|u_w| < 1$) or arrive at F_1 from infinity (for $|u_w| \ge 1$) or depart from F_2 and go to infinity (for $|u_w| \le 1$). Note that the last class of integral curves which are departed to infinity correspond to the case when the contact surface boundary condition $v = 0$ cannot be satisfied and the solution to the boundary value problem does not exist.

The situation is changed completely for nonzero swirling body. In this case all integral curves depart from F_2 and arrive at F_1 because they are the parts of ellipses with the center $O(0, C_2/2C_1^2)$ and with the axes $\sqrt{1 + \tfrac{1}{2}(C_2/C_1)}$, $\sqrt{1 + \tfrac{1}{2}(C_2/C_1)}/C_1$. It means that all integral curves satisfy to the conditions at the contact surface $v = 0$ for all boundary conditions at the body surface and therefore the solution to the boundary value problem exists for all values u_w.

Finally, we conclude that the domain of existence of solutions to injection gas layer equations over axisymmetric blunted bodies, at the stagnation point depends on swirling of the body [31]. For nonzero swirling the solution exists for arbitrary values u_w. In contrary, for nonswirling bodies this solution exists only if the condition $u_w > -1$ is realized.

4.8 Nonuniform flow at inviscid shock layer

At the last section we will consider viscous gas flow for the case of nonuniform upstream. What does it mean nonuniform flow – it means that the flow parameters at infinity assumed be not constant. The main goal is now to study the influence of this nonuniformity on the flow structure. At the beginning, remember again the statement of the problem. The composite governing equations described the flow-field at inviscid shock and boundary layers are

$$\frac{\partial}{\partial \xi^\alpha}(\rho u^\alpha \sqrt{a/a_{(\alpha\alpha)}}) + \frac{\partial}{\partial \zeta}(\rho V \sqrt{a}) = 0, \quad \rho(D^* u^\alpha + A^\alpha_{\alpha\beta} u^\beta u^\gamma) = -\frac{\gamma+1}{\gamma} a^{\alpha\beta} \sqrt{a_{(\alpha\alpha)}} \frac{\partial P}{\partial \xi^\beta},$$
$$\frac{\gamma+1}{\gamma} \frac{\partial P}{\partial \zeta} = -\rho A^3_{\beta\gamma} u^\beta u^\gamma, \quad \rho D^* T = \frac{\gamma-1}{\gamma}(u^\alpha/\sqrt{a_{(\alpha\alpha)}}) \frac{\partial P}{\partial \xi^\alpha}, \quad P = \rho T,$$
$$D^* = (u^\alpha/\sqrt{a_{(\alpha\alpha)}}) \frac{\partial}{\partial \xi^\alpha} + V \frac{\partial}{\partial \zeta}.$$

$$(4.81)$$

These equations have to be solved with the boundary conditions at the shock wave and at the body (contact) surface. At the shock wave the boundary conditions have a view [30]:

$$\zeta = \zeta_s: \ u^\alpha = u^\alpha_\infty, \ P = \rho_\infty (u^3_\infty)^2 \frac{\gamma}{\gamma+1}, \ \rho\left[V - (u^\alpha_\infty/\sqrt{a_{(\alpha\alpha)}}) \frac{\partial \zeta_s}{\partial \xi^\alpha}\right] = \rho_\infty u^3_\infty, \ T = \frac{1}{2}(u^3_\infty)^2.$$

$$(4.82)$$

Definition 72 Conditions (4.82) are called *hypersonic approximation of the Rankine-Hugoniot conditions at the shock wave.*

As for boundary condition at the body surface (or at the contact surface – for strong injection regime) this condition results from the fact that the body surface (contact surface) is the stream surface and therefore normal component of the velocity vector with respect to this surface should be equal to zero, $\zeta = 0: \ V = 0$. To calculate the components $u^1_\infty, u^2_\infty, u^3_\infty$ of the upstream velocity vector $\mathbf{V}_\infty$ at the shock wave the shape of the shock wave should be known due to $u^\alpha_\infty = (\mathbf{V}_\infty, \tau_\alpha)$, $u^3_\infty = (\mathbf{V}_\infty, \mathbf{n}_{s.w})$. As it is known from previous results, the thickness of the shock layer is negligible small in comparison with the linear scale of the problem. It means that the shape of the shock wave is differed from the body shape on small value and with rather well accuracy it may be assumed these shapes to be coincided. In this case the vectors $\tau_1, \tau_2, \mathbf{n}_{s.w}$ coincide with the basis vectors of the coordinate system associated with the body surface. There are two particular cases of upstream nonuniformity: nonuniform flow of the far-wake type and the upstream swirling flow which are interesting to be considered there.

4.8.1 Nonuniform flow of the far-wake type

The flow of this type is axisymmetric at infinity. Designating the distance from the symmetry axis via r, the flow parameters at infinity can be presented in the form

$$\mathbf{V}_1(r) = [1 - a \exp(-br^2)] \cdot \mathbf{l}, \ P_1 = \text{const}, \ \rho_1 = \frac{B}{1 + [1 - V^2_1(1-a)^{-2}]}, \ B = 1 + [1 - (1-a)^{-2}],$$

where the unit vector $\mathbf{l}$ is directed along the symmetry axis, a and c are the defect of the velocity the density at the wake axis, $b^{-1/2}$ is the radius of wake. For far-wake flow $a \ll 1$.

We are interested in the flow behavior in neighborhood of the $3d$ stagnation point of the blunted body. Take into account that in this case $\mathbf{l} = -\mathbf{n}$, where the unit vector $\mathbf{n}$ is directed along external normal to the body surface at the stagnation point. On the other hand, in the neighborhood of the stagnation point, the arbitrary smooth surface may be approximated by paraboloid, and the basis vectors $(\mathbf{e}_1, \mathbf{e}_2, \mathbf{n})$ of our curvilinear coordinate system (ξ^1, ξ^2, ζ) can be written as

$$\mathbf{e}_1 = \frac{\mathbf{p}_1 + \xi^1 \mathbf{p}_3}{\sqrt{1 + (\xi^1)^2}}, \quad \mathbf{e}_2 = \frac{\mathbf{p}_2 + k\xi^2 \mathbf{p}_3}{\sqrt{1 + k^2(\xi^2)^2}}, \quad \mathbf{e}_3 = \frac{\mathbf{p}_1 + k\xi^2 \mathbf{p}_2 - \mathbf{p}_3}{\sqrt{1 + (\xi^1)^2 + k^2(\xi^2)^2}},$$

where $\mathbf{p}_i$ are the basis unit vectors of the Cartesian coordinate system (p^1, p^2, p^3) and the axis $\mathbf{p}_3$ coincides with the symmetric axis. The explicit formulae for the flow parameters at the shock wave are

$$u_\infty^1 = \frac{V_1 \xi^1}{(1-a)\sqrt{1+(\xi^1)^2}}, \quad u_\infty^2 = \frac{V_1 k \xi^2}{(1-a)\sqrt{1+k^2(\xi^2)^2}}, \quad u_\infty^3 = -\frac{V_1}{(1-a)\sqrt{1+(\xi^1)^2+k^2(\xi^2)^2}},$$
$$\rho_\infty = \rho_1/B.$$

It is easy to see that at the stagnation point,

$$u_\infty^1 \sim \xi^1, \quad u_\infty^2 \sim k\xi^2, \quad u_\infty^3 = -1, \quad \rho_\infty = 1.$$

To analyze the flow in the neighborhood of the stagnation point it is necessary to resolve the singularity at this point. To do it, introduce new parameters

$$u = \frac{u^1}{\xi^1}, \quad w = \frac{u^2}{\xi^2}, \quad P_1 = \frac{1}{\xi^1}\frac{\partial P}{\partial \xi^1}, \quad P_2 = \frac{1}{k\xi^2}\frac{\partial P}{\partial \xi^2}.$$

The initial equations (4.81) will be transformed to

$$u + kw + \frac{\partial V}{\partial \zeta} = 0, \quad \frac{\partial T}{\partial \zeta} = \frac{\partial P}{\partial \zeta} = \frac{\partial \rho}{\partial \zeta} = 0,$$
$$u^2 + V\frac{\partial u}{\partial \zeta} = -\frac{\gamma-1}{\gamma}\frac{P_1}{\rho}, \quad kw^2 + V\frac{\partial w}{\partial \zeta} = -\frac{\gamma-1}{\gamma}\frac{P_2}{\rho}. \tag{4.83}$$

Obviously, the system (4.83) is not closed. To close it one has to include in accounting an additional equation for longitudinal pressure gradient P_1 and P_2. To obtain these equations, apply the operator of differentiation on tangent coordinates ξ^α to normal projection of momentum equations (the third equation in (4.81)) and then to change the order of differentiation:

$$\frac{\partial}{\partial \xi^\alpha}\left(\frac{\partial P}{\partial \zeta}\right) = \frac{\partial}{\partial \zeta}\left(\frac{\partial P}{\partial \xi^\alpha}\right) = -\frac{\gamma}{\gamma+1}\frac{\partial}{\partial \xi^\alpha}(\rho A_{\delta\beta}^3 u^\delta u^\beta).$$

On the other hand, $A_{\delta\beta}^3 = b_{\delta\beta}/\sqrt{a_{(\delta\delta)}a_{(\beta\beta)}}$. At the stagnation point

$$b_{11} = -1, \quad b_{12} = 0, \quad b_{22} = -k \Rightarrow A_{11}^3 = -1, \quad A_{12}^3 = 0, \quad A_{22}^3 = -k.$$

Using these relations the equations for longitudinal pressure gradients will be written as

$$\frac{\partial P_1}{\partial \zeta} = 2\frac{\gamma}{\gamma+1}\rho u^2 = 4\frac{\gamma^2}{(\gamma+1)^2}u^2, \quad \frac{\partial P_2}{\partial \zeta} = 2\frac{\gamma}{\gamma+1}\rho k^2 w^2 = 4\frac{\gamma^2}{(\gamma+1)^2}k^2 u^2. \tag{4.84}$$

The boundary conditions for system (4.83)–(4.84) are

$$\zeta = \zeta_s : u = w = 1, \quad T = \tfrac{1}{2}, \quad P = \frac{\gamma}{\gamma+1}, \quad \rho = \frac{2\gamma}{\gamma+1}, \quad V = -\frac{\gamma}{\gamma+1},$$
$$P_{1s} = -\frac{2\gamma}{\gamma+1}[1-\Lambda], \quad P_{2s} = -\frac{2\gamma}{\gamma+1}[k-\frac{\Lambda}{k}], \quad \Lambda = 2\frac{ab(1+c)}{1-a}; \tag{4.85}$$
$$\zeta = 0 : V = 0.$$

Preliminary conclusions:

1. The influence of the upstream nonuniformity is completely determined only by parameter Λ.

2. The governing equations for inviscid shock layer coincide with those for injection gas layer except the equations for longitudinal pressure gradients P_1 and P_2. For injection gas layer these values are constant across the layer as for inviscid shock layer, the values P_1 and P_2 depend on normal coordinate.

3. Because these equations are true at the body surface (contact surface) with $V = 0$ and $\zeta = 0$, one can write the relations $u^2 = -\frac{\gamma-1}{\gamma}\frac{P_1(0)}{\rho}$ and $kw^2 = -\frac{\gamma-1}{\gamma}\frac{P_2(0)}{\rho}$.

In fact, it means that the solution to the boundary value problem (4.83)–(4.85) in real space exists only if the values $P_1(0)$ and $P_2(0)$ are negative or equal to zero at the body surface (contact surface).

In general, an analytic solution to the boundary value problem (4.83)–(4.85) is not found. Hence, in order to explore the qualitative effect of free-stream nonuniformity on the flow, a numerical solution has been obtained. It was demonstrated that under certain conditions (see below) it is possible to use the above presented two-sublayer structure of this layer: inviscid shock layer with thickness $O(\varepsilon)$ and inviscid boundary layer with thickness $O(\varepsilon^{3/2})$. This scheme of flow enables one to suggest the following two step approximate approach for solution to the problem:

Step 1. Inviscid shock layer equations without terms of longitudinal pressure gradients are solved. These equations are

$$u + kw + \frac{\partial V}{\partial \zeta} = 0, \quad \frac{\partial T}{\partial \zeta} = \frac{\partial P}{\partial \zeta} = \frac{\partial \rho}{\partial \zeta} = 0, \quad u^2 + V\frac{\partial u}{\partial \zeta} = 0, \quad kw^2 + V\frac{\partial w}{\partial \zeta} = 0 \quad (4.86)$$

with the boundary conditions

$$\zeta = \zeta_s: \quad u = w = 1, \quad T = \tfrac{1}{2}, \quad P = \tfrac{\gamma}{\gamma+1}, \quad \rho = 2\tfrac{\gamma}{\gamma+1}, \quad V = -\tfrac{1}{2}\tfrac{\gamma+1}{\gamma};$$
$$\zeta = 0: \quad V = 0.$$

From the second and the fourth equations of the system (4.86) it can be obtained

$$V\,du = -u^2 d\zeta, \quad V\,dw = -kw^2\,d\zeta \;\Rightarrow\; k\frac{du}{u^2} = \frac{dw}{w^2} \;\Rightarrow\; \frac{k}{u} = \frac{1}{w} + C.$$

Considering the boundary conditions at the shock wave one obtains $w = \frac{u}{u(1-k)+k}$ and $u = \frac{kw}{1-w(1-k)}$. Another integral can be found by combination of the first and the third equations from (4.86),

$$u^2 + V\frac{\partial u}{\partial \zeta} = 0 \;\Rightarrow\; u^2 + V\frac{\partial u}{\partial V}(u + kw) = 0 \;\Rightarrow\; \frac{dV}{V} = \frac{du}{u} + kw\frac{du}{u^2}$$
$$\Rightarrow\; \frac{dV}{V} = \frac{du}{u} + \frac{dw}{w} \;\Rightarrow\; V = C_1 uw \;\Rightarrow\; V = -\tfrac{1}{2}\tfrac{1+\gamma}{\gamma}uv.$$

Step 2. Based on this solution, Equations (4.84) for variables P_1 and P_2 can be integrated,

$$P_1(0) = P_{1s} + \left(2\frac{\gamma}{\gamma+1}\right)^2 \int_{\zeta_s}^{0} u^2 d\zeta = P_{1s} + \left(2\frac{\gamma}{\gamma+1}\right)^2 \int_{0}^{1} V\,du$$
$$= P_{1s} + 2\frac{\gamma}{\gamma+1}\int_{0}^{1}\frac{u^2}{u(1-k)+k}du.$$

This integral can be calculated using new variables $t = u(1-k)+k$, $u = \frac{t-k}{1-k}$, $dt = (1-k)du$,

$$\int_{0}^{1}\frac{u^2}{u(1-k)+k}du = \frac{1}{(1-k)^3}\int_{0}^{1}(t - 2k + k/t^2)\,dt.$$

We obtain

$$P_1(0) = -2\frac{\gamma}{\gamma+1}\Big[1 - 2\frac{ab(1+c)}{1-a} + \frac{k^2}{(1-k)^3}\big(\tfrac{3}{2} + \tfrac{1}{2}k^{-2} - 2k^{-1} - \ln k\big)\Big].$$

By the same way it can be found the value $P_2(0)$,

$$P_2(0) = -2\frac{\gamma}{\gamma+1}\Big[k - 2\frac{ab(1+c)}{k(1-a)} - \frac{k^2}{(1-k)^3}\big(\tfrac{3}{2} - 2k + \tfrac{1}{2}k^2 + \ln k\big)\Big].$$

Note that the relations for $P_1(0)$ and $P_2(0)$ have singularities in two important particular cases: the stagnation point near cylinder ($k = 0$) and the stagnation point near sphere ($k = 1$). Resolving these singularities one has

$$P_1(0; k = 1) = P_2(0; k = 1) = -2\frac{\gamma}{\gamma+1}\left[\frac{4}{3} - 2\frac{ab(1+c)}{1-a}\right],$$
$$P_1(0; k = 0) = -2\frac{\gamma}{\gamma+1}\left[\frac{3}{2} - 2\frac{ab(1+c)}{1-a}\right], \quad P_2(0; k = 0) = 0.$$

So, it is possible to obtain the approximate solution to the boundary value problem under assuming the longitudinal pressure gradients to be constants, equal to $P_\alpha(0)$, across the shock layer. This initial problem is sufficiently simplified: it can be rewritten in the form:

$$u + kw + \frac{\partial V}{\partial \zeta} = 0, \quad u^2 + V\frac{\partial u}{\partial \zeta} = \Pi_1, \quad w^2 + V\frac{\partial w}{\partial \zeta} = \Pi_2, \tag{4.87}$$

where

$$\Pi_1 = \frac{\gamma-1}{\gamma}\left[1 - 2\frac{ab(1+c)}{1-a} + \frac{k^2}{(1-k)^3}\left(\frac{3}{2} + \frac{1}{2}k^{-2} - 2k^{-1} - \ln k\right)\right],$$
$$\Pi_2 = \frac{\gamma-1}{\gamma}\left[1 - 2\frac{ab(1+c)}{1-a} - \frac{k^2}{(1-k)^3}\left(\frac{3}{2} - 2k + \frac{1}{2}k^2 + \ln k\right)\right].$$

This system should be solved with the boundary conditions at the shock wave

$$\zeta = \zeta_s: \quad u = w = 1, \quad V = -\frac{1}{2}\frac{\gamma+1}{\gamma}, \tag{4.88a}$$

and at the body surface (contact surface)

$$\zeta = 0: \quad V = 0. \tag{4.88b}$$

Obviously, the problem (4.87)–(4.88a,b) coincides with two-point boundary value problem (described in previous section) for injection gas layer. Using our designations and considering the boundary conditions this solution may be rewritten as

$$\frac{\sqrt{\Pi_1}+u}{\sqrt{\Pi_1}+1}\frac{\sqrt{\Pi_1}-1}{\sqrt{\Pi_1}-u} = \left[\frac{\sqrt{\Pi_2}+w\sqrt{k}}{\sqrt{\Pi_2}+\sqrt{k}}\frac{\sqrt{\Pi_2}-\sqrt{k}}{\sqrt{\Pi_2}-w\sqrt{k}}\right]^\alpha, \quad \alpha = \sqrt{\frac{\Pi_1}{k\Pi_2}},$$
$$V = -\frac{\gamma+1}{\gamma}\sqrt{\frac{\sqrt{\Pi_1}-u^2}{\sqrt{\Pi_1}-1}\frac{\sqrt{\Pi_2}-kw^2}{\sqrt{\Pi_2}-k}}, \quad \zeta = -\int_0^V \frac{dV}{u+kw}.$$

The numerical simulation [19] shows the asymptotic solution gives good accuracy if the intensity of wake is relatively low. As the flow upstream nonuniformity becomes more intensive the accuracy of solutions deteriorates because of the variation of the longitudinal pressure gradients across the layer strongly influence on the flow structure in viscous shock layer.

Moreover, the calculations demonstrated that there are critical values of the parameters a, b and c (determining degree of the flow nonuniformity) at which the flow regime changes abruptly and the bifurcation of solution arises: there is a sharp discontinuous increase in the layer's thickness and a reverse flow appears at the center of the layer. To clarify the bifurcation nature of solutions, turn to the integral curve pattern of initial system (4.83)–(4.84) in the phase space (u, w, V). The solution to the associated two-pint boundary value problem corresponds to segment of the integral curve leaving the point $R_s(1, 1, -(\gamma+1)/2\gamma)$ up to its intersection with the plane $V = 0$ (Fig. 4.16(a)). An analysis shows that as integral curve approaches the plane $V = 0$, its behavior is completely determined by values $\pi_1^0 = \frac{\gamma-1}{\gamma\rho}\lim_{V\to 0} P_1$ and $\pi_2^0 = \frac{\gamma-1}{\gamma\rho}\lim_{V\to 0} P_2$. It means that the motion of a fluid particle depends only on the values of local longitudinal pressure gradients π_α^0 and it is rather clear from physical point of view. On the other hand, π_α^0 should depend on the intensity of the wake. Simple estimations demonstrate that for uniform flow $\pi_\alpha^0 < 0$; the value π_α^0 increases together with nonuniformity, moreover it satisfies to condition $\pi_1^0 \leq \pi_2^0$ so that there are critical values a, b and c for which $\pi_2^0 = 0$.

From analysis it is evident that if $\pi_1^0 \leq 0$ and $\pi_2^0 \leq 0$ then the integral curve intersects the plane $V = 0$ and the singular point $R_w(\sqrt{-\pi_1^0}, \sqrt{-\pi_2^0/k}, 0)$, which is a stable node and a solution to the boundary value problem exists. Note that as shown by comparison with numerical calculations, for these conditions the asymptotic solution describes flow-field quite accuracy. If at least one of these two quantities, as a maximum one $-\pi_2^0$, is positive. So, there is intersection and the behavior of the integral curve is more complex: after reaching a maximum $V_m < 0$ this curve enters the region $w < 0$ and the reverse flow is developed in the layer (Fig. 4.16(a)). Nevertheless, because as it follows from (4.84) the pressure gradients are monotonically decreased along the integral curve with decreasing of coordinate ζ (Fig. 4.16(b)). In this connection starting from a certain layer thickness $\zeta = \zeta_0$ the quantity π_2^0 becomes negative and the integral curve reapproaches and intersects the plane $V = 0$. Numerical simulations showed that the critical values of nonuniformity parameters, as which

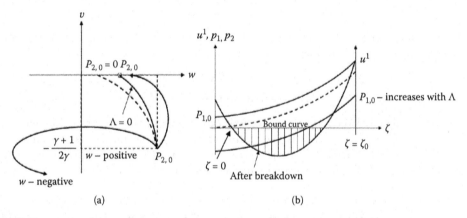

(a)   (b)

FIGURE 4.16: (a) Nonuniform flow of the far-wake type; phase space. (b) Nonuniform flow of the far-wake type; behavior of integral curves.

the change of flow regime takes place, can be quite accuracy determined by the condition $\pi_2^0 = 0$ which can be expressed in the form:

$$\Lambda_* = 2\frac{ab(1+c)}{1-a} = k^2\Big[1 + \frac{1}{2}\frac{k(k-3)}{(1-k)^2} - \frac{k \ln k}{(1-k)^3}\Big].$$

The analysis of this condition leads to a number of conclusions. First, although nonuniformity of the flow in far wake region depends on parameters a, b and c, the separation of flow at the stagnation point is determined by a single parameter $\lambda = 2ab/(1+c)/(1-a)$. On this basis one can obtain the following criterion for separation-less of viscous flow at the stagnation point of the blunted body moving along wake of another body,

$$\Lambda \leq \Lambda_* = k^2\Big[1 + \frac{1}{2}\frac{k(k-3)}{(1-k)^2} - \frac{k \ln k}{(1-k)^3}\Big].$$

Secondary, for the same upstream nonuniformity the answer to the question of whether the flow separation will take a place on the frontal surface will depend on geometry of the body characterized by the ratio k of the body surface's main radii at the stagnation point. In particular, for $k = 1$ (axisymmetric body), the value Λ_* reaches a maximum ($\Lambda_* = \frac{4}{3}$) and $\Lambda_* = 0$, for $k = 0$ (plane body). Thus, for a plane body the separation will take place even at low wake intensity.

This conclusion can be explained as follows. By the boundary condition at the shock wave (4.82), in the free-stream there is positive total pressure gradient in all longitudinal directions. At the same time, the flow intensity is $\sim O(k)$ in the direction defined by the cross-section of the body with the maximum radius of the body surface's curvature at the stagnation point. In this connection, for $k \to 0$ and for fixed parameters a, b and c, there is k_0 such that the positive free-stream pressure gradient, in this direction, becomes to dominate and reverse flow arises in the layer.

4.8.2   Upstream swirling flow over axisymmetric bodies

For this flow geometry, choose ξ^2-axis along the circumferential direction. Assuming that boundary conditions at the body surface are independent from ξ^2, we conclude that the solution to (4.81)–(4.82) does not depend on this coordinate and in this connection the terms with derivatives on ξ^2, in (4.81) are zero. Introduce designations

$$u = u^1, \quad w = u^2, \quad v = V, \quad x = \xi^1, \quad y = \zeta$$

and let take into account the fact that for above-mentioned coordinate system the components $a_{\alpha\beta}$ and coefficients $A^i_{\alpha\beta}$ are simplified:

$$a_{11} = 1, \quad a_{22} = a = r_w^2, \quad a_{12} = 0, \quad 2A^2_{12} = -A^1_{12} = (1/r_w) \cdot (dr_w/dx),$$
$$A^3_{11} = -\kappa, \quad A^3_{11} = -(1/r_w)\sin\alpha, \quad \sin\alpha = [1 + (dz/dr)^2]^{-1/2}(dz/dr),$$

where α is the angle between axisymmetry axis direction and the vector of external normal to the body surface, $r_w = r_w(x)$ is the radius of the body surface, κ is the longitudinal curvature of the body surface, $z = z(r)$ is the equation of the body surface in cartesian coordinate system (r, z). The governing equations are simplified as follows:

$$\frac{\partial}{\partial x}(\rho u r_w) + \frac{\partial}{\partial y}(\rho v r_w) = 0, \quad \rho\left(u\frac{\partial u}{\partial x} + v\frac{\partial u}{\partial y} - \frac{w^2}{r_w}\frac{dr_w}{dx}\right) = -\frac{\gamma-1}{\gamma}\frac{\partial P}{\partial x},$$
$$\rho\left(ku^2 + \frac{w}{r_w}\sin\alpha\right) = \frac{\gamma+1}{\gamma}\frac{\partial P}{\partial y}, \quad u\frac{\partial w}{\partial x} + v\frac{\partial w}{\partial y} + \frac{wuw}{r_w}\frac{dr_w}{dx} = 0,$$
$$\rho\left(u\frac{\partial T}{\partial x} + v\frac{\partial T}{\partial y}\right) = \frac{\gamma-1}{\gamma}u\frac{\partial P}{\partial x}. \tag{4.89}$$

Assume that at the flow upstream the following conditions are realized:

$$\text{rot}\,\mathbf{V}_\infty = (\Omega^*, 0, 0), \quad \rho_\infty = \text{const}, \quad P_\infty = P^0_\infty + \frac{1}{2}\rho_\infty w^2_\infty, \quad w_\infty = \Omega^* r,$$

where r is the distance from the symmetry axis, P^0_∞ is the pressure at the symmetry axis $r = 0$. Let $\Omega = L\Omega^*/V_\infty \le O(1)$. Then, the boundary conditions in new variables are

$$y = y_s: \; u = u_\infty, \quad w = w_\infty, \quad T = \tfrac{1}{2}v^2_\infty, \quad P = \tfrac{\gamma}{\gamma+1}(v^2_\infty + \tfrac{1}{2}w^2_\infty),$$
$$\rho = \left(v - u\frac{dy_s}{dx}\right) = v_\infty, \quad u_\infty = \frac{1}{\sqrt{1+(dz/dr)^2}}\frac{dz}{dr}; \tag{4.90a}$$
$$y = 0: \; v = 0. \tag{4.90b}$$

Consider the solution to two-point boundary value problem (4.89)–(4.90a,b) in a neighborhood of the stagnation point. To resolve the singularity, introduce the following new variables:

$$U = \frac{u}{x}, \quad W = \frac{w}{x}, \quad V = v, \quad P_1 = \frac{1}{2}\frac{\gamma^2-1}{\gamma^2}\frac{\partial^2 P}{\partial x^2}.$$

Then we obtain ODE system that describe the flowfield at inviscid shock layer over axisymmetric blunted bodies in the neighborhood of the stagnation point,

$$\frac{dV}{dy} + 2U = 0, \quad U^2 + V\frac{dU}{dy} - W^2 = -P_1, \quad 2UW + V\frac{dW}{dy} = 0,$$
$$\frac{dP_1}{dy} = 2\frac{\gamma-1}{\gamma+1}(U^2 + W^2), \quad \frac{dT}{dy} = \frac{d\rho}{dy} = 0 \tag{4.91}$$

with the boundary conditions

$$y = y_s: \quad U = 1, \quad W = \Omega, \quad V = -\tfrac{1}{2}\tfrac{\gamma+1}{\gamma}, \quad T = \tfrac{1}{2}, \quad \rho = 2\tfrac{\gamma}{\gamma+1}, \quad P_1 = \tfrac{1}{2}\tfrac{\gamma-1}{\gamma}(\Omega^2 - 2);$$
$$y = 0: \quad V = 0.$$

It is easy to find the first integral. Considering the boundary conditions at the shock wave, the third equation from (4.91) is integrated,

$$2UW + V\frac{dW}{dy} = 2U\left(W - V\frac{dW}{dV}\right) = 0 \quad \Rightarrow \quad \frac{dW}{dW} = \frac{dV}{V} \quad \Rightarrow \quad W = CV = -2\Omega\frac{\gamma}{\gamma+1}V.$$

We obtain the two-point boundary value problem to determine U, V, P_1,

$$\frac{dV}{dy} = -2U, \quad V\frac{dU}{dy} = 4\Omega^2\frac{\gamma^2}{(\gamma^2+1)}V^2 - P_1 - U^2,$$
$$\frac{dP_1}{dy} = 2\frac{\gamma-1}{\gamma+1}\left[U^2 + \Omega^2\frac{\gamma^2}{(\gamma^2+1)}V^2\right] \tag{4.92}$$

with the boundary conditions

$$y = y_s: \quad U = 1, \quad V = -\tfrac{1}{2}\tfrac{\gamma+1}{\gamma}, \quad P_1 = \tfrac{1}{2}\tfrac{\gamma-1}{\gamma}(\Omega^2 - 2),$$
$$y = 0: \quad V = 0.$$

Solution of (4.92) depends on the parameter Ω. As a numerical solution to this boundary value problem there is a critical value of Ω at which the flow regime is changed abruptly and the bifurcation of solutions arises: there is a sharp discontinuous in the layer's thickness and a reverse flow appears in the layer's center. To clarify the bifurcation nature, turn to the integral curve pattern of initial system (4.92) in the phase space (U, V). Solution to associated two-point boundary value problem corresponds to a segment of the integral curve leaving the point $(1, -(\gamma+1)/2\gamma)$ up to its intersection with the plane $V = 0$, Fig. 4.16(b).

By the above, as integral curve approaches the plane $V = 0$ its behavior is completely determined by the value $\pi_1^0 = \lim_{V \to 0} P_1$. Thus, the motion of a fluid particle depends only on the value of local longitudinal pressure gradients π_1^0. On the other hand, π_1^0 should depend only on the intensity of the upstream of flow swirling. Rather simple estimations demonstrate that for uniform flow ($\Omega = 0$) the value π_1^0 is negative. As Ω increases, the value π_1^0 increases, so that there is a critical value Ω_* for which $\pi_1^0 = 0$. From analysis it is evident that if $\pi_1^0 < 0$ then the integral curve intersects the plane $V = 0$ at singular point $(1, -(\gamma+1)/2\gamma)$ which is a stable node and solution to the boundary value problem exists. If π_1^0 is positive, there is no intersection and the behavior of the integral curve is more complex: this curve enters the region $U < 0$ after reaching a maximum $V_m < 0$ and reverse flow develops in the layer. Nevertheless, the pressure gradients monotonically decrease along the integral curve with coordinate y as it follows from (4.92). Starting from a certain layer of thickness y_0 the quantity π_1^0 becomes negative and the integral curve re-approaches and intersects the plane $V = 0$. Numerical calculations show that the critical value Ω_* (which changing in flow regime takes a place) is equal $\Omega_* = 2.18$ for $\gamma = 1.22$ [8, 14].

4.9 Exercises

1. Find the frictional force on an oscillating plane cover by a layer of fluid of thickness h, the upper surface being free.

Solution. The boundary condition at the solid plane is $v = u$ for $x = 0$, and that at the free surface the viscous stress tensor is $\sigma_{xy} = \eta\,\partial_x v = 0$ for $x = h$. Then the velocity $v = u \cos k(h - x)/\cos kh$, where $k = (1 + i)/\delta$, $\delta = \sqrt{2v/\omega}$, $i\omega = vk^2$. The frictional force is $P_x = \eta\partial_x v|_{x=0} = \eta ku \tan kh$.

2. A plane disk of radius R executes rotary oscillations of small amplitude about its axis, the angle of rotation being $\theta = \theta_0 \cos \omega t$, where $\theta_0 \ll 1$. Find the moment of the frictional forces acting on the disk.

Solution. For oscillations of small amplitude the term $(v \cdot \nabla)v$ in the motion equation is always small compared with $\partial_t v$, whatever the frequency ω. If $R \gg \delta = \sqrt{2v/\omega}$, $i\omega = vk^2$, the disk may be regarded as infinite in determining the velocity distribution. Take cylindrical coordinates with z-axis along the axis of rotation, and seek a solution such that $v_r = v_z = 0$, $v_\phi = v = r\Omega(z, t)$. For angular fluid velocity we obtain the equation $\partial_t \Omega = v\partial_{zz}^2 \Omega$. The solution of this equation which is $-\omega\theta_0 \sin \omega t$ for $z = 0$ and zero for $z \to \infty$ is $\Omega = -\omega\theta_0 \exp(-z/\delta) \sin(\omega t - z/\delta)$. The moment of the frictional forces on both sides of the disk is

$$M = 2 \int_0^R r \cdot 2\pi r \eta\,\partial_z v|_{z=0}\, dr = \omega\theta_0 \pi \sqrt{\omega\rho\eta}\, R^4 \cos(\omega t - \pi/4).$$

3. Determine the flow between two parallel planes when there is a pressure gradient which varies harmonically with time.

Solution. Take xz-plane half-way between two planes, with x-axis parallel to the pressure gradient, which has the form $-(1/\rho)\partial_x\rho = a \exp(-i\omega t)$. The velocity is everywhere in x-direction, and is determined by equation

$$\partial_t v = a \exp(-i\omega t) + v\partial_{yy}^2 v.$$

The solution to this equation which satisfies conditions $v = 0$ for $y = \pm h/2$ is

$$\bar{v} = i(a/\omega) \exp(-i\omega t)(1 - \cos ky/\cos(kh/2)).$$

The mean velocity over a cross-section is

$$\bar{v} = i(a/\omega) \exp(-i\omega t)(1 - \tan(kh/2)/(kh/2)).$$

For $h/\delta \ll 1$ this becomes $\bar{v} \approx (a/12) \exp(-i\omega t)(h^2/v)$, while for $h/\delta \gg 1$ we have $\bar{v} \approx i(a/\omega) \exp(-i\omega t)$. In accordance with the fact that in this case the velocity must be almost constant over the cross-section, varying only in a narrow surface layer.

4. Find the drag on a sphere of radius which executes translatory oscillations in a fluid.

Solution. Write the velocity of the sphere in the form $\mathbf{u} = \mathbf{u_0} \exp(-i\omega t)$. Then, we seek the fluid velocity in the form $\mathbf{v} = \mathbf{u_0} \exp(-i\omega t) \operatorname{curl} \operatorname{curl} f\mathbf{u_0}$, where f is a function of $\mathbf{r}$ only (the origin is taken at the instantaneous position of the center of the sphere). Substituting $\mathbf{v}$ into the motion equation

$$\partial_t \operatorname{curl} \mathbf{v} = v \cdot \Delta \operatorname{curl} \mathbf{v},$$

using equality $\Delta \operatorname{curl} v = 0$ and transformations

$$\operatorname{curl} \mathbf{v} = \mathbf{u_0} \exp(-i\omega t) \operatorname{curl} \operatorname{curl} \operatorname{curl}(f\mathbf{u_0}) = -\mathbf{u_0} \exp(-i\omega t)\Delta \operatorname{curl} f\mathbf{u_0},$$
$$\Delta \operatorname{curl} \mathbf{v} = -\mathbf{u_0} \exp(-i\omega t)\Delta^2 \operatorname{curl} f\mathbf{u_0},$$

we obtain $\Delta^2(\nabla f \cdot \mathbf{u_0}) + (i\omega/v)\Delta(\nabla f \cdot \mathbf{u_0}) = 0$. Due to $\mathbf{u_0}$, we get $\Delta^2 f + i(\omega v)\Delta f = 0$. Hence $\Delta f = \operatorname{const} \exp(ikr)/r$. The solution being chosen which decreases exponentially with r. Integrating, we have $\partial_r f = [a \exp(ikr)(r + i/k) + b]/r^2$, the function f is not needed, since only the derivatives f' and f'' appear in the velocity expression. The constants a and b are determined from the condition $\mathbf{v} = \mathbf{u}$ for $r = R$, and are found to be $a = \frac{3}{2} i(R/k) \exp(-ikR)$ and $b = -\frac{3}{2} R^3[\frac{1}{3} + i(kR)^{-1} - (kR)^{-2}]$. Remark that, at large distances $(R >> \delta)$, $a \to \infty$ and $b \to -R^3/2$. The drag is calculated from formula [23]: $F = \int(-p\cos\theta + \sigma'_{rr}\cos\theta - \sigma'_{r\theta}\sin\theta)\, df$, in which the integrating is over the surface on the sphere. The result is

$$F = 6\pi\eta R(1 + R/\delta)u + 3\pi R^2\sqrt{2\eta\rho/\omega}\,[1 + (2/9)R/\delta]\,\partial_t u.$$

For $\omega = 0$ this becomes Stokes' formula while for large frequencies we get $F = \frac{2}{3}\pi\rho R^3 \partial_t u + 3\pi R^2\sqrt{2\eta\rho/\omega}\,u$. The first term in this expression corresponds to the inertial force in potential flow past a sphere, while the second term gives the limit of the dissipative force.

5. Deduce the equation for the flow due to a sphere embedded in a pure straining motion.

Solution. Fluid of viscosity μ and density ρ occupies the space outside a sphere of radius R, and far from the sphere the fluid is in pure straining motion specified by the rate-of-strain tensor ε_{ij}, where $\varepsilon_{ii} = 0$. The velocity and pressure in the fluid may be written as

$$u_i = u'_i + \varepsilon_{ij}c^j, \quad p = p' + P, \tag{4.93}$$

where P is the pressure in the pure straining motion represented by e_{ij} in the absence of the sphere; u'_i and p' represent the changes due to the presence of the sphere and

$$u'_i \to 0, \quad p' \to 0, \tag{4.94}$$

as $r = |x| \to \infty$. Let the center of the sphere be at the origin, thus by symmetry there is no tendency for the sphere to translate and the surface of the sphere is given by $r = R$; hence

$$n \cdot u_{r=R} = 0. \tag{4.95}$$

There are further conditions at the surface of the sphere which depend on the nature of the particle. We may include the various kinds of particle within the scope of the analysis by supposing the sphere to contain incompressible fluid viscosity $\bar{\mu}$ (the case of a rigid particle corresponding to $\bar{\mu}/\mu \to \infty$). The velocity must be continuous across the interface, and so also is the tangential component of stress if we suppose that the interface has no mechanical properties other than a uniform surface tension. Thus

$$u_i|_{r=R} = u'_i + \varepsilon_{ij}x^j = \bar{u}_i, \quad \varepsilon_{klj}n^l n^j(\sigma_{ij} - \bar{\sigma}_{ij})|_{r=R} = 0, \tag{4.96}$$

where the over-bar indicates a quantity referring to the motion within the sphere and $\mathbf{n}$ is the normal to the interface. Also $\bar{p}$ and $\bar{u}_i$ are finite at $r = 0$. The velocities $\mathbf{u}$ and $\bar{\mathbf{u}}$ satisfy Navier–Stokes equation (with different values for the viscosity), but for small spherical particle we may evidently use an approximate form of these equations:

$$\nabla((p - p_0)/\mu) = \nabla^2\mathbf{u} = -\nabla \times \omega, \quad \text{div}\,\mathbf{u} = 0,$$

where p_0 is a constant pressure at the sufficient distance from the particle. For the flow outside the sphere, Navier–Stokes equation becomes, after substitution from (4.93),

$$\rho[\partial_t u'_i + (u'_j + \varepsilon^j_k x^k)\partial_j u'_i] = -\partial_i p' + \mu\nabla^2 u'_i. \tag{4.97}$$

Now the variation of the unperturbed velocity over the region occupied by the particle is of magnitude $|\varepsilon_{ij}|\,R$, and it is evident that the disturbance velocity $\mathbf{u}'$ also has this magnitude in the region not far from the particle. Hence, when the radius R satisfies the condition

$$|\varepsilon_{ij}|\,R^2/\nu << 1 \quad (\nu = \mu/\rho) \tag{4.98}$$

(and provided the imposed rate of strain is not changing rapidly), the flow near the particle is governed by the approximate equation

$$\nabla p' = \mu \nabla^2 \bar{\mathbf{u}}. \tag{4.99}$$

Under the same conditions the velocity $\bar{u}$ and the pressure $\bar{p}$ inside the sphere also satisfy the equation of motion with neglect of inertia forces:

$$\nabla \bar{p} = \bar{\mu} \nabla^2 \bar{\mathbf{u}}.$$

Finally, mass conservation gives two equations

$$\operatorname{div} \mathbf{u}' = \operatorname{div} \mathbf{u} = 0, \quad \operatorname{div} \bar{\mathbf{u}} = 0. \tag{4.100}$$

Equations (4.99)–(4.100) with boundary conditions (4.94)–(4.96) governing the disturbance motion are linear and homogeneous in $\mathbf{u}'$, p', $\mathbf{u}$, $\bar{p}$ and ε_{ij}. No vector occurs in the description of the interface, and the pressures (which are harmonic functions) and velocities are seemed to be of the form

$$p' = C\mu\varepsilon_{ij}x^i x^j r^{-5}, \quad \bar{p} - \bar{p}_0 = C\bar{\nu}\varepsilon_{ij}x^i x^j,$$
$$u'_i = \varepsilon_{ij}x^j M + \varepsilon_{jk}x_i x^j x^k Q, \quad \bar{u}_i = \varepsilon_{ij}x^j \bar{M} + \varepsilon_{jk}x_i x^j x^k \bar{Q},$$

where $M, Q, \bar{M}$ and $\bar{Q}$ are functions of r, and $C, \bar{C}, \bar{p}_0$ are constants. The forms of the functions that satisfy the governing equations and the conditions far from the particle and at $r = 0$ are

$$M = Dr^{-5}, \quad Q = \frac{1}{2}Cr^{-5} - \frac{5}{2}Dr^{-7}, \quad \bar{M} = \bar{D} + \frac{5}{21}\bar{C}r^2, \quad \bar{Q} = -\frac{2}{21}\bar{C},$$

and the conditions at the interface $r = R$ are then satisfied provided

$$[C/(2\mu + 5\bar{\mu})]R^{-3} = (D/\mu)R^{-5} = -\frac{2}{2l}(\bar{C}/\mu)R^2 = \frac{2}{3}\bar{D}/\mu = -1/(\mu + \bar{\mu}).$$

Observe that $u'_i = \frac{1}{2}C\varepsilon_{jk}x_i x^j x^k r^{-5} + O(r^{-4})$ at large distances from the particle. Showing that the disturbance velocity is one order smaller than in the case of flow due to a sphere in translational motion, as might have been expected from dipole nature of the condition on u' ([6]) at the surface of the sphere (see (4.95) with (4.93)) in the present case.

The above solution has been obtained by neglecting the terms representing the inertia force in equation of motion (4.97), and self-consistency of solutions may be examined under using it to evaluate the order of magnitude of the neglecting terms. We find in this way that the ratio of the magnitude of the neglected inertia force to the retained viscous force is of order $|\varepsilon_{ij}|r^2/\nu$. This ratio is small compared with unity in the neighborhood of the sphere when the condition (4.98) is satisfied, as had been supposed, but the ratio is not small in the outer region of the flow field, where r/R is of order $(|\varepsilon_{ij}|R^2/\nu|)^{-1/2}$. Thus (4.99) is not valid approximation to the complete equation of motion (4.97) in the order of field, although again we are reassured by observation that all terms in the complete equation of motion are small (after being made nondimensional with parameters R and $|\varepsilon_{ij}|$) in this same region. An improved and completely self-consistent approximation to the velocity distribution could presumably be obtained (for a steady imposed straining motion) from

$$\rho(\varepsilon_k^j x^k \partial_j u'_i + \varepsilon_i^j u'_j) = -\partial_i p' + \mu \nabla^2 u'_i,$$

which is still linear in u', but we will take it for granted that this improved approximation would not differ significantly from the above solution in the neighborhood of the particle.

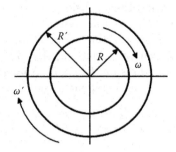

FIGURE 4.17: Viscous fluid between two coaxial cylinders.

6. Motion of viscous fluid between two coaxial cylinders tube rotated with different angular velocities is described (Fig. 4.17).

Solution. The problem can be solved in polar coordinate system so as, $V_r \equiv 0$, $V_z \equiv 0$, $V_\phi = V_\phi(r, \phi)$ with the boundary conditions to the problem $V_{\phi|r=R} = R\omega$, $V_{\phi|r=R'} = R'\omega'$. The NS equation for axisymmetric flow in the polar coordinates reads as

$$\partial_{rr}^2 V_\phi + r^{-1}\partial_r V_\phi - r^{-1}V_\phi = 0, \quad r^{-2}V_\phi^2 = -\partial_r \rho/\rho.$$

Solution of the differential equation is $V_\phi = C_1 r^{-1} + C_2 r$. Determine the constants C_1, C_2 from the boundary conditions, the solution is obtained,

$$V_\phi = \frac{1}{(R')^2 - R^2}\left[(\omega R^2 - \omega(R')^2)r + (\omega - \omega')(R'R)^2/r\right],$$

$$\frac{p}{\rho} = \frac{1}{((R')^2 - R^2)^2}\left[\frac{1}{2}(\omega'R^2 - \omega(R')^2)^2 r^2 + 2(\omega - \omega')(R'R)^2\right.$$

$$\left. \times(\omega'R^2 - \omega(R')^2)\ln(r/R) - \frac{1}{2}(\omega - \omega')^2(R'R)^4/r^2\right] + \text{const}.$$

The stress tensor between the two ring layers is

$$\tau_{r\phi} = -2\mu(\omega - \omega')(R'R)^2/((R')^2 - R^2)r^2,$$

where μ is the fluid viscosity. Summary moment relative rotation axis for the ring with radius r is

$$L = \int_0^{2\pi} \tau_{r\phi} r^2 \, d\phi = -4\pi\mu\frac{(\omega - \omega')(R'R)^2}{(R')^2 - R^2}.$$

7.[1] Find exact solutions to fluid motion equations with the structure of three-dimensional nonstationary motions of viscous incompressible fluid is described by the NSE and continuity equation,

$$\partial_t V_n + V_1\partial_x V_n + V_2\partial_y V_n + V_3\partial_z V_n = -\nabla_n p/\rho + \nu(\partial_{x^2}^2 V_n + \partial_{y^2}^2 V_n + \partial_{z^2}^2 V_n),$$
$$\partial_x V_1 + \partial_y V_2 + \partial_z V_3 = 0, \quad n = 1, 2, 3. \tag{4.101}$$

Solution. It has the form

$$V_n = f_n(z, t) + q_n(z, t), \quad n = 1, 2, \quad V_3 = F(z, t). \tag{4.102}$$

---

[1]This exercise is based on the results in [33].

Substitution (4.102) into (4.101) and the exclusion of the pressure result in linear polynomials of the form $A_m x + B_m y + C_m = 0$ in coordinates x and y, where the coefficients A_m, B_m and C_m depend only on the variables z and t and are expressed in terms of the functions f_n, g_n, F and their derivatives. Equating A_m, B_m and C_m to zero, one obtains an over-determined system of equations for f_n, g_n, F. Analysis of this system provides the following representation for the velocity components and pressure [26, 4],

$$V_1 = x(-\tfrac{1}{2}\partial_z F + w) + yv, \quad V_2 = xu - y(-\tfrac{1}{2}\partial_z F + w), \quad V_3 = F,$$
$$P/p = p_0 - \tfrac{1}{2}\alpha x^2 - \tfrac{1}{2}\beta y^2 - \gamma xy - \tfrac{1}{2}F^2 + \nu\partial_z F - \int \partial_t F\, dz, \tag{4.103}$$

where p_0, α, β and γ are arbitrary functions of time t specifying the pressure distributions and F, u, v and w are unknown functions of the coordinate z and time t. Substitute the velocity components and the pressure (4.103) into hydrodynamic equations (4.94) nonlinear system of four equations is obtained,

$$\partial^2_{tz} F + F\partial^2_{z^2} F - \tfrac{1}{2}(\partial_z F)^2 = \nu\partial^3_{z^3} F + 2(uv + w^2) - \alpha - \beta,$$
$$\partial_t u + F\partial_z u - u\partial_z F = \nu\partial^2_{z^2} u + \gamma, \quad \partial_t v + F\partial_z v - v\partial_z F = \nu\partial^2_{z^2} v + \gamma, \tag{4.104}$$
$$\partial_t w + F\partial_z w - w\partial_z F = \nu\partial^2_{z^2} w + \tfrac{1}{2}(\alpha - \beta).$$

By separating one isolated equation for the longitudinal velocity $V_3 = F$ from the system (4.104) and determining a new auxiliary function G, let set for the velocity components,

$$u = \frac{1}{2}\sin\phi(q - \partial_z F) + a^2 G, \quad v = \frac{1}{2}\sin\phi(q - \partial_z F) - b^2 G, \quad w = \frac{1}{2}\sin\phi(q - \partial_z F) + abG,$$

$$\alpha = \frac{1}{4}q^2 - \frac{1}{2}p(1 - \cos\phi) + \frac{1}{2}\partial_t q\cos\phi, \quad \beta = \frac{1}{4}q^2 - \frac{1}{2}p(1 + \cos\phi) + \frac{1}{2}\partial_t q\cos\phi,$$

$$\gamma = \frac{1}{2}p\sin\phi + \frac{1}{2}\partial_t q\sin\phi,$$

where $p = p(t)$ and $q = q(t)$ are arbitrary functions, a and b are arbitrary constants, $G = G(z, t)$ is the desired function, and ϕ is the constant determined from the transcendental equation $(a^2 - b^2)\sin\phi + 2ab\cos\phi$. The system (4.104) reduces to two equations

$$\partial^2_{tz} F + F\partial^2_{z^2} F - (\partial_z F)^2 = \nu\partial^3_{z^3} F + q\partial_z F + p, \quad \partial_t G + F\partial_z G - G\partial_z F = \nu\partial^2_{z^2} G. \tag{4.105}$$

Nonlinear equation (4.105a) can be considered independently and (4.102) is linear with respect to the desired function G and has a trivial partial solution $G = 0$. Equation (4.105a) has the following general property, if $\tilde{F}(z, t)$ is its certain solution, the function $F = \tilde{F}(z + \psi(t), t) - \psi'_t$, where $\psi(t)$ is an arbitrary function, is also a solution to (4.105a). In addition, the function $F = -\tilde{F}(-z, t)$ is also a solution [4, 26]. Several new exact solutions of (4.105a), which are expressed in terms of elementary functions, are as follows.

(i) A fractionally rational solution with the generalized separation of variables:

$$F = -a'_t(t) + b(t)[z + a(t)] - 6\nu/(z + a(t)), \quad q = -4b, \quad p = b'_t + 3b^2,$$

where $a(t)$ and $b(t)$ are arbitrary functions.

(ii) Solutions with the generalized separation of variables in the exponential form in z:

$$F = -a'_t(t)\exp(-\sigma z) + b(t), \quad q = a'_t - \sigma b - \sigma^2\nu, \quad p = 0,$$

where $a(t)$ and $b(t)$ are arbitrary functions and σ is an arbitrary constant. If $a(t)$ and $b(t)$ are periodic functions, the solution is periodic in time.

(iii) Spatially periodic solutions in the form of a product of functions of different arguments:

$$F = -a'_t(t)\sin(\sigma z + B), \quad a(t) = C\exp\left(-\sigma^2\nu t + \int q(t)\, dt\right), \quad p = -\sigma^2 a^2(t),$$

TABLE 4.2: Regions of the stability and instability of solutions.

Parameter variation region	Longitudinal velocity component F	Stability/instability
$p < 0,\ q \in \mathbb{R}$; or $q < 0,\ p \in \mathbb{R}$; or $p = q = 0$	F is any solution except for $F \equiv$ const	Solutions of (4.105b) are unstable
$p \geq 0,\ q \geq 0$ (are satisfied simultaneously and $|p| + |q| \neq 0$)	F is any solution except for $F \equiv$ const	Solutions of (4.105b) are conditionally stable
$p = 0,\ q > 0$	$F \equiv$ const	Solutions of (4.105a) are unstable and of (4.105b) are stable
$p = 0,\ q \leq 0$	$F \equiv$ const	Solutions of (4.105,b) are stable

where $q = q(t)$ is an arbitrary function and B, C and σ are arbitrary constants. The substitution $q(t) = \nu\sigma^2 + \phi'_t(t)$ with $\phi(t)$ a periodic function provides periodic solution in both arguments z and t.

(iv) A solution with the functional separation of variables

$$F = -6\nu\sigma \tanh(\sigma(z + a(t))) - a'_t(t), \quad p = 0, \quad q = 8\nu\sigma^2,$$

where $a(t)$ is an arbitrary function and σ is an arbitrary constant (this solution is bounded if the derivative is bounded). The solutions of (4.105b) can be represented in terms of solutions of (4.105a). Let $F = F(z,t)$ be a solution of (4.105a). Then, (4.105b) has the solution

$$G = A'_t + qA + (A\partial_z F) \cdot (B\partial^2_{zz} F), \tag{4.106}$$

where the functions $A = A(t)$ and $B = B(t)$ satisfy the ODEs

$$A''_{tt} + qA'_t + (p + q'_t)A = 0, \quad B'_t + Cq = 0. \tag{4.107}$$

This statement is proven by excluding G from (4.105b) and (4.106) with the subsequent comparison of the resulting expression with both (4.105b) and the equation appearing after the differentiation of (4.105a) with respect to z. The general solution of (4.107)(a) is $B = C\exp(-\int q\,dt)$, where C is an arbitrary constant. The stability/instability of solutions is analyzed using (4.106)–(4.107), which correspond to the solutions of (4.105,b). Note that in many cases, it is not necessary to know the explicit form of the function F. Regions of the stability and instability of solutions to the system (4.105,b) are represented in Table 4.2. General mathematical and physical interpretation of solutions is given in [5, 33].

8. Describe full-developed incompressible pulsative viscous fluid motion in cylindrical tube under the action of the pressure drop changed harmonically with time.

Solution. The equation for one dimension steady stage axisymmetric pulsatile flow in cylindrical coordinates has the form:

$$\partial_t V_z - \nu(\partial^2_{rr} V_z + r^{-1}\partial_r V_z) = -\partial_z \rho/\rho, \quad -\partial_z \rho/\rho = f(t). \tag{4.108}$$

We find the solution of the equation with the boundary conditions $V_z = 0,\ r = R$ and the initial conditions $V_z = V_0(r),\ t = 0$. The force is determined by the periodic function $f(t) = \rho A\cos\omega t$, where A is amplitude and ω is frequency of the periodic function. Using $V_z(r,t) = \phi(r,t) + (A/\omega)\sin\omega t$ and introducing the new variable $\tau = \nu t$ instead of t, the previous equation may be rewritten in the form

$$\partial_\tau \phi = \partial^2_{rr}\phi + r^{-1}\partial_r\phi \tag{4.109}$$

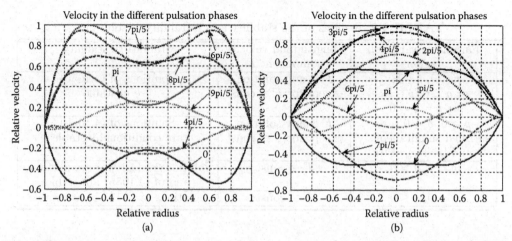

FIGURE 4.18: Velocity profiles for various phases parameter (a) $R\sqrt{\omega/\nu} = 4.14$, (b) $R\sqrt{\omega/\nu} = 3.12$.

with transformed boundary condition $\phi(R,t) = -(A/\omega)\sin(\tau/\nu)\omega$. Particular solution of the equation will be found in the form of exponential function as $\phi = \Phi(r)\exp(-i\lambda\tau)$, where λ is an undetermined real number. Substituting ϕ in (4.109), we obtain second order ordinary linear equation for $\Phi(r)$:

$$\partial^2_{rr}\Phi + t^{-1}\partial_r\Phi + i\lambda\Phi = 0.$$

A finite particular solution at $r = 0$ is

$$\Phi(r) = J_0(r\sqrt{i\lambda}) = ber(r\sqrt{\lambda}) - bei(r\sqrt{\lambda}).$$

Here, $ber(x) = \operatorname{Re} J_0(x\sqrt{i}$ and $bei(x) = -\operatorname{Im} J_0(x\sqrt{i})$ are real and imaginary parts of zero order Bessel function. Using a complex constant $B - iC$ into general solution of (4.109) gives us the function ϕ:

$$
\begin{aligned}
\phi(r,t) &= \operatorname{Re}[(B - iC)(ber(r\sqrt{\lambda}) - bei(r\sqrt{\lambda}))(\cos\lambda\tau - i\sin\lambda\tau)] \\
&= B[ber(r\sqrt{\lambda})\cos\lambda\tau + bei(r\sqrt{\lambda})\sin\lambda\tau] - C[bei(r\sqrt{\lambda})\cos\lambda\tau + ber(r\sqrt{\lambda})\sin\lambda\tau].
\end{aligned}
$$

Using the boundary condition and $\lambda = \omega/\nu$, the constant B and C may be obtained from

$$B\,ber(R\sqrt{\omega/\nu}) - C\,bei(R\sqrt{\omega/\nu}) = 0, \quad B\,ber(R\sqrt{\omega/\nu}) + C\,bei(R\sqrt{\omega/\nu}) = A/\omega.$$

Evaluating these constants we find

$$B = \frac{A\,bei(R\sqrt{\omega/\nu})}{\omega\,ber^2(R\sqrt{\omega/\nu}) + bei^2(R\sqrt{\omega/\nu})}, \quad C = \frac{A\,ber(R\sqrt{\omega/\nu})}{\omega\,ber^2(R\sqrt{\omega/\nu}) + bei^2(R\sqrt{\omega/\nu})}.$$

Solution to (4.108) includes zero order Bessel function and the velocity profiles for time $t = \tau/\nu$ can be found according the formulae

$$
\begin{aligned}
V_x(r,t) &= \frac{A}{\omega}\left\{\left[1 - \frac{bei(R\sqrt{\omega/\nu})\,bei(R\sqrt{\omega/\nu}) - ber(R\sqrt{\omega/\nu})\,ber(R\sqrt{\omega/\nu})}{ber^2(R\sqrt{\omega/\nu}) + bei^2(R\sqrt{\omega/\nu})}\right]\sin(\omega t)\right. \\
&\quad + \left.\frac{bei(R\sqrt{\omega/\nu})\,ber(r\sqrt{\omega/\nu}) - ber(R\sqrt{\omega/\nu})\,bei(r\sqrt{\omega/\nu})}{ber^2(R\sqrt{\omega/\nu}) + bei^2(R\sqrt{\omega/\nu})}\cos(\omega t)\right\}.
\end{aligned}
$$

As for example the velocity profiles for the different phases and the values of the parameter $R\sqrt{\omega/\nu} = 4.1442$ and 3.1081 are shown in Fig. 4.18. From these figures one can conclude that the reverse flow occurs in the tube and the fluid layers close to the wall of the tube lead to the layers at the tube's center line. A similar solution was presented in [38].

Bibliography

[1] Abu-Ghannam B. J. and Shaw R. Natural transition of boundary layers – the effects of turbulence, pressure gradient, and flow history. *J. of Mech. Eng. Science*, 22, N. 5, 1980, 213–228.

[2] Anderson J. D. Jr. *Fundamentals of Aerodynamics*. McGraw-Hill, 2001.

[3] Antonov A., Belov Yu. Solution of the Cauchy's problem for the nonlinear equation describing gas flow with strong blowing. *Int. Appl. Mech.*, 8, No. 10, 1972, 1118–1122.

[4] Aristov S.N. and Polyanin A.D. (2009), New classes of exact solutions and some transformations of the Navier–Stokes equations. *Dokl. Akad. Nauk*, 2009, 427:35 [*Dokl. Phys.*, 54:316 (2009)].

[5] Aristov S.N. and Polyanin, A.D. New classes of exact solutions and some transformations of the Navier–Stokes equations, *Russian J. of Math. Physics*, 17, 2010, No. 1, 2010, 1–18.

[6] Batchelor G.K. An *Introduction to Fluid Dynamics*, Cambrige Univ. Press, 2000, 615 pp.

[7] Borodin A. I. and Peigin V. S. Influence of body shape on the spatial boundary layer characteristics on a permeable surface. *J. Eng. Phys. and Thermophys.*, 53, No. 3, 1986, 994–1000.

[8] Borodin A. I. and Peigin S. V. Three-dimensional gas flow past blunted bodies in the framework of the parabolic viscous shock layer theory. *Math. Models and Comp. Exper.*, 5, No. 1, 1993, 16–25.

[9] Borodin A., Peigin S. and Timchenko S. Numerical analysis of supersonic viscous gas flow past axisymmetric nonpointed bodies. – *Fluid Dynamics*, 34, No. 1, 1999, 116–121.

[10] Borodin A. I. Three-dimensional boundary layer on a permeable and partially permeable surface of a blunt body. *Fluid Dynamics Res.*, 40, No. 6, 2002, 850–855.

[11] Brykina I.G., Rogov B.V., Tirskiy G.A. Continuum Models of Rarefied Gas Flows in Problems of Hypersonic Aerodynamics. *Applied Math. and Mech.*, 70(6), 2006, 990–1016.

[12] Chernyi G. Gas flow around a body at hypersonic speeds. *Dokl. Akad. Nauk SSSR*, 107, 1956, 657–660.

[13] Délery J. and Dussauge J.-P. Some physical aspects of shock wave/boundary layer interactions. *Shock Waves*, 19, 2009, 453–468.

[14] Eremeitsev I. G., Pilyugin N. N. and Yunitskii S. A. Supersonic nonuniform viscous gas flow past a blunted body with supply of gas from the surface. *Fluid Dynamics*, Vol. 23, No. 4, 1988, 585–591.

[15] Franklin M. Orr. Jr. *Theory of Gas Injection Processes*. Tie-Line Technology, 2007.

[16] Freeman N. On the theory of hypersonic flow past bluff bodies. *J. Fluid Mech.*, Vol. 1, No. 4, 1956, 366–387.

[17] Gaissinski I., Kelis O., Rovenski V. *Hydrodynamic Instability Analysis.* VDM, 2009, 218 pp.

[18] Gaissinski I. and Kartvelishvili L., *Viscous Hypersonic Flow.* Verlag VDM, 2010, 220 pp.

[19] Gershbein E. A. and Peigin S. V. Hypersonic viscous shock layer in a swirling gas flow on a permeable. *Fluid Dynamics*, 21, No. 6, 1986, 866–875.

[20] James P. Three-dimensional inviscid waves in buoyant boundary layer flows. *Fluid Dynamics Res.*, 28, 2001.

[21] Kalimuthu R., Rathakrishnan E. and Mehta R.C. *Experimental Investigation of Hypersonic Flow Over Spiked Blunt Body.* Lambert Academic Publisher, 2012, 180 pp.

[22] Kao H. K. The blunt-body problem in hypersonic flow at the stream-line of a blunted body. *AIAA J.*, 2, No. 11, 1964, 1892–1906.

[23] Landau L.D. and Lifshitz E.M. *Fluid Mechanics: Course of Theoretical Physics, Vol. 6*, Butterworth-Heinemann, 7th ed., 1987.

[24] Leballeur J. C. Viscid-inviscid coupling calculations for two and 3D flows. Von Karman Inst. for Fluid Dynamics, *Computational Fluid Dynamics*, 1982, 89 p.

[25] Magomedov K. M. Hypersonic viscous gas flow past blunted bodies. *Fluid Dynamics*, 5, No. 2, 1970, 213–222.

[26] Meleshko S.V. A particular class of partially invariant solutions of the Navier–Stokes equations. *Nonlinear Dynamics*, 2004, 47–68.

[27] Mikhalev A. N. Modeling trans- and supersonic viscous flow over a cone on a ballistic range. *Technica Physics Lett.*, 33, No. 9, 2007, 758–760.

[28] Onsager L. Reciprocal relations in irreversible processes I. *Phys. Rev.* 37, 1931, 405–426.

[29] Pavlov B. M. Calculation of supersonic viscous flow near stagnation line of a blunted body. *Fluid Dynamics*, No. 2, 1967, 91–95.

[30] Peigin S. V. and Filonenko B. F. Unsteady three-dimensional laminar boundary layer on blunted bodies with strong blowing. *J. Appl. Mech and Techn. Phys.*, 28, No. 6, 1987, 849–855.

[31] Peigin S. V. Investigation of swirling flow of a viscous gas near the stagnation line of a blunted body. *J. Appl. Mech and Techn. Phys.*, 29, No. 4, 1988, 649–654.

[32] Perry R., Green D. (Eds). *Perry's Chemical Engineers' Handbook.* McGraw-Hill, 1997.

[33] Polyanin A. D. On the nonlinear instability of solutions of hydrodynamic type systems. *JETP Letters*, 2009, 90, N 3, 2009, 217–221.

[34] Shcherback V. G. Boundary conditions on a shock wave in a supersonic flow. *J. Appl. Mech. and Techn. Phys.*, 30 (1), 1989, 45–52.

[35] Shelkovich V. M. The Rankine-Hugoniot conditions and balance laws for δ-shocks. *J. of Mathematica Sciences*, 151, No. 1, 2008, 2781–2792.

[36] Smith R. Hypersonic swirling flow past blunt bodies. *Aeronautical Quarterly*, 24, 1973, 241–251.

[37] Van Driest F.R. Investigation of laminar boundary layer in compressible fluids using the Crocco method, *NACA TN 2597*, Jan. 1952.

[38] Womersley J.R. Method for the calculation of velocity, rate of flow and viscous drag in arteries when the pressure gradient is known. *J. of Psysiology*, 127, 1955, 553–563.

[39] Yasuhara M. On the hypersonic viscous flow past a flat plate with suction or injection. *J. Phys. Soc. Jap.*, 12, No. 2, 1957, 177–182.

Index

Index